Handbook of The Economics of Innovation
Volume 1

HANDBOOKS IN ECONOMICS

1

Series Editors

KENNETH J. ARROW
MICHAEL D. INTRILIGATOR

ELSEVIER

AMSTERDAM • BOSTON • HEIDELBERG • LONDON
NEW YORK • OXFORD • PARIS • SAN DIEGO
SAN FRANCISCO • SINGAPORE • SYDNEY • TOKYO

N·H

HANDBOOK OF THE ECONOMICS OF INNOVATION

VOLUME 1

Edited by

BRONWYN H. HALL
University of California, Berkeley, California, USA
University of Maastricht, Maastricht, The Netherlands

and

NATHAN ROSENBERG
Stanford University, Standford, California, USA

ELSEVIER

AMSTERDAM • BOSTON • HEIDELBERG • LONDON
NEW YORK • OXFORD • PARIS • SAN DIEGO
SAN FRANCISCO • SINGAPORE • SYDNEY • TOKYO

N·H

North-Holland is an imprint of Elsevier
Radarweg 29, PO Box 211, 1000 AE Amsterdam, The Netherlands
The Boulevard, Langford Lane, Kidlington, Oxford, OX5 1GB

First edition 2010
Reprinted 2011

Library of Congress Cataloging-in-Publication Data
A catalog record for this book is available from the Library of Congress

British Library Cataloguing in Publication Data
A catalogue record for this book is available from the British Library

ISBN: 978-0-444-51995-5 (Volume 1)
ISBN: 978-0-444-53609-9 (Volume 2)

ISSN: 0169-7218 (Handbooks in Economics series)
ISSN: 1573-4471 (Handbook of Development Economics series)

For information on all North-Holland publications visit our website at www.elsevierdirect.com

Printed and bound by CPI Group (UK) Ltd, Croydon, CR0 4YY
Transferred to Digital Print 2011

INTRODUCTION TO THE SERIES

The aim of the *Handbooks in Economics* series is to produce Handbooks for various branches of economics, each of which is a definitive source, reference, and teaching supplement for use by professional researchers and advanced graduate students. Each Handbook provides self-contained surveys of the current state of a branch of economics in the form of chapters prepared by leading specialists on various aspects of this branch of economics. These surveys summarize not only received results but also newer developments, from recent journal articles and discussion papers. Some original material is also included, but the main goal is to provide comprehensive and accessible surveys. The Handbooks are intended to provide not only useful reference volumes for professional collections but also possible supplementary readings for advanced courses for graduate students in economics.

KENNETH J. ARROW *AND* MICHAEL D. INTRILIGATOR

CONTENTS OF THE HANDBOOK

VOLUME 1

VOLUME 2

PART I

INTRODUCTION AND OVERVIEW

Chapter 1

INTRODUCTION TO THE HANDBOOK

BRONWYN H. HALL*,‡ *AND* NATHAN ROSENBERG†

**University of California, Berkeley*

California, USA

†*Stanford University, Stanford*

California, USA

‡*University of Maastricht*

Maastricht, The Netherlands

Although innovation and the production of new goods and services have almost always been a part of economic activity, economic research on innovation has been to some extent scattered among a number of quite disparate economic fields, including macroeconomics (growth accounting), industrial organization (the strategies and interactions of innovative firms), public finance (policies for encouraging private sector innovation), and economic development (innovations systems and technology transfer). However, as Verspagen and Werker (2003) have recently shown using survey data, a large and fairly tightly clustered network of economists working on innovation and technical change has developed, a network that includes both those working within the "evolutionary" paradigm and those using more traditional methods of analysis. By now, this community of scholars has generated a large body of work on the topic, some of which is multidisciplinary. Thus, it seemed to the editors to be an appropriate time to provide a comprehensive overview of the field, bringing together chapters by scholars working in a number of subfields of economics and closely related disciplines in order to provide a coherent picture of the entire landscape of the economics of innovation. In undertaking the production of this handbook, we had several goals beyond the desire to provide a good overview of an increasingly important research area. We hoped to encourage the economics profession to view the economics of innovation as a distinct area of applied economics, and also to encourage researchers working in one of the many subfields in this area to become aware of work by researchers studying similar topics, but who operate in different research domains and perhaps use different methodologies.

When our handbook project was initiated it bore the title *The Economics of Technical Change*. However, as the volume approached publication, it became apparent that the research done in this area had in fact broadened to include new economic dimensions of great significance that did not fit comfortably under the rubric of "technical change." Thus, although this term continues to appear abundantly in these pages, the editors have decided to use the broader term "economics of innovation" to describe the subject matter within. The term "innovation" includes technical change, and also includes many dimensions of economic change that do not fall easily into the category of technical change. The older term conjures up hardware and long assembly lines, but not the software of the digital

Handbooks in Economics, Volume 01

DOI: 10.1016/S0169-7218(10)01001-4

world of computers, the Internet, social networking, nor the reorganization of work that has followed innovation in these areas. But software can also be used in much broader senses to refer to anything that is not hardware. This usage can encompass research carried out in universities and industrial and government labs, or the new ideas that may emerge from the human brain (which some would refer to as "wetware"), but which Romer (1990), for example, has labeled simply as ideas. In so doing, Romer's usage has shaped much of the language of economists over the last couple of decades. To some extent, the evolution of usage from technical change to innovation parallels the rise in the importance of nonmanufacturing sectors in developed economies, and also the importance of productivity and welfare-enhancing change that is not the product of organized Research and Development (R&D).

Innovation economists owe a great debt to Joseph Schumpeter, who can be said to be the father of the field, and whose work contains much verbal theorizing on the topic that is still influential today. In the preface to the Japanese edition of his 1937 book *The Theory of Economic Development*, Schumpeter sketches out what is probably the most precise and succinct statement of his own intellectual agenda that he ever committed to print. That agenda focuses not only upon the understanding of how the economic system generates economic change but also upon how that change occurs as the working out of purely endogenous forces:

> *"If my Japanese readers asked me before opening the book what it is that I was aiming at when I wrote it, more than a quarter of a century ago, I would answer that I was trying to construct a theoretic model of the process of economic change in time, or perhaps more clearly, to answer the question how the economic system generates the force which incessantly transforms it ... I felt very strongly that ... there was a source of energy within the economic system which would of itself disrupt any equilibrium that might be attained. If this is so, then there must be a purely economic theory of economic change which does not merely rely on external factors propelling the economic system from one equilibrium to another. It is such a theory that I have tried to build."*[1]

It should be noted that these words were published in 1937, when Schumpeter was, as we know, already at work on *Capitalism, Socialism, and Democracy*. In fact, *Capitalism, Socialism, and Democracy* is the fulfillment of precisely the intellectual agenda that Schumpeter articulated in the passage to his Japanese readers that was just quoted.

Of course, an account of how and why economic change took place was precisely something that could not be provided within the "rigorously static" framework of neoclassical equilibrium analysis, as Schumpeter referred to it. Schumpeter also observed that it was Walras' view that economic theory was only capable of examining a "stationary process," that is, "a process which actually does not change of its own initiative, but merely produces constant rates of real income as it flows along in time." As Schumpeter interprets Walras:

> *"He would have said (and, as a matter of fact, he did say it to me the only time I had the opportunity to converse with him) that of course economic life is essentially passive and merely adapts itself to the natural and social influences which may be acting on it, so that the theory of a stationary process constitutes really the whole of theoretical economics and that as economic theorists we cannot say much about the factors that account for historical change, but must simply register them."*[2]

[1] Schumpeter (1937), p. 158.
[2] Schumpeter (1937), pp. 2–3.

The critical point here is that Schumpeter directly rejects the view of Walras that economic theory must be confined to the study of stationary processes, and that it cannot go farther than demonstrating how departures from equilibrium, such as might be generated by a growth in population or in savings, merely set into motion forces that restore the system to an equilibrium path. In proposing to develop a theory showing how a stationary process can be disturbed by internal as well as external forces, Schumpeter is suggesting that the essence of capitalism lies not in equilibrating forces but in the inevitable tendency of that system to depart from equilibrium—in a word, to disequilibrate. Equilibrium analysis fails to capture the essence of capitalist reality. Lest there be any doubt about Schumpeter's position on this critical matter, we cite his own forceful formulation: "Whereas a stationary feudal economy would still be a feudal economy, and a stationary socialist economy would still be a socialist economy, stationary capitalism is a contradiction in terms."[3]

As we look over the collection of chapters in this volume, it is clear that this basic understanding of the importance of internally generated economic change for the progress of the economy and the weaknesses of static economic analysis in the face of this phenomenon occupies much of the research in innovation economics. A number of themes that are common to at least several of the chapters touch on this and related ideas.

The first and perhaps the most important theme is the essential dynamism of the innovative process—knowledge, inventions, and innovations created today build on those created in the past, and the benefits of an innovation are often not felt until it undergoes a dynamic, cumulative learning and diffusion process. An understanding of this phenomenon underlies almost all of the chapters, and is perhaps most obvious in those by Thompson on learning by doing, Bresnahan on general purpose technologies, Teece on the innovative firm, and Stoneman and Battista on diffusion. The fact that the central process in which we are interested has dynamic and hysteresis-like properties means that static economic modeling will be of limited value for analysis; this awareness is reflected in many of the papers and a few of them put forth alternative modeling approaches.

Three of the chapters, those by Dosi and Nelson, Teece, and Soete et al., explicitly take as their starting point the limitations of neoclassical theory in analyzing innovation at the industry, firm, or country level. In addition, the chapters by Soete et al. and Steinmueller argue that Arrow and Nelson's market failure rationale for science and technology policy, although valid, is an incomplete guide to policy because it overemphasizes the importance of assigning property rights to innovators and ignores the systemic nature of the needed policies. For example, subsidies for R&D will fail to have the desired result if it takes time to produce trained scientists and engineers, or if the education system is simply not capable of producing them. It is probably safe to say that the topic of innovation systems and institutions is in its infancy empirically; see Röller and Mohnen (2005) for a study of complementarities in European innovation policies. Although numerous studies in the management of innovation literature have been informed by the "new" institutional economics, empirical study at the economy-wide level has lagged behind, probably because of the formidable modeling and data obstacles.

A second major theme of this volume is the importance of the needs of innovation policy in driving the research agenda of the economics of innovation. We can see this reflected in the chapters by Foray and Lissoni on university research and public–private interaction, Rockett on intellectual property rights,

[3] Schumpeter (1951), p. 174. On these matters, see Rosenberg (2010).

Hall, Mairesse, and Mohnen on the measurement of returns to R&D, Hall and Lerner on the financing of innovation, Popp, Newell, and Jaffe on the environment, and Pardey, Alston, and Ruttan on innovation in agriculture. The extensive study of these particular topics has to a great extent been driven by the questions raised in the implementation of various policies toward science and technology, questions that have often been accompanied by more tangible resources to encourage the analysis. In addition to the chapters mentioned, there are several chapters in the final section of the handbook that are directly addressed to policy topics. Steinmueller and Soete, Verspagen and ter Weel address the broad topics of technology policy in general and the systems of innovation approach to its analysis, whereas Mowery discusses one of the most important sources of spillovers from government R&D: the defense sector.

The close relationship between the economics of innovation and policy questions has two related causes. First, as reviewed by Hulten in the chapter on growth accounting, the economic growth literature of the past 50 or so years has identified technical change as a major contributor to productivity growth (Abramovitz, 1956; Solow, 1957). Second, the invention and innovation that are the source of technical change also create knowledge that can spill over to entities that were not responsible for the original creation, and this transfer occurs without a priced transaction taking place. As Arrow (1962) and Nelson (1959) pointed out long ago, this fact immediately suggests a need for policy to encourage the appropriate level of investment in these activities. Because such knowledge transfers can be diffuse and do not necessarily take place in a well-defined market, policy attention also needs to be directed to spillovers across sectors and across national boundaries; attempts to measures these spillovers are prominent in the chapter by Hall, Mairesse, and Mohnen. The importance of cross-national spillovers for technology transfer and development, where these spillovers are mediated via trade and foreign direct investment, also appears in the chapters by Keller and Fagerberg, Srholec, and Verspagen.

A third theme with prominence in several chapters is the importance of the digital revolution that has led to major innovations in information and computing technology (ICT) that have impacted all sectors in the economy. Broadly speaking, the semiconductor and attendant innovations have all the characteristics of a General Purpose Technology, as described by Bresnahan in his chapter. The specific evolution of the computing and Internet sector during the past 50 years is dealt with in the chapter by Greenstein. In general, these technologies are highly cumulative and interactive, requiring a great deal of interoperability between components made by different firms, which has increased the importance of standards, collaboration among firms, and network effects in adoption. This in turn has led to a renewed interest in markets for technology (Arora and Gambardella), user and firm collaboration and networks (von Hippel; Powell and Gianella), and the functioning of the patent system (Rockett). In the case of patents, the complexity of ICT products has meant that the patent system operates very differently for firms in that sector than for those in the traditional patenting sectors such as chemicals and pharmaceuticals; this point is discussed in the chapter by Scherer and touched on elsewhere in the handbook.

One of the consequences of the digital revolution has been the successful entry of innovative new firms that have grown rapidly and are now among the largest in the world. For example, in the United States almost 40% of the top 200 R&D-performing firms in 2005/2006 were founded after 1980, while 32% of the top 200 R&D-performing firms in 1980 had exited by 2005 (Hall and Mairesse, 2009). This is certainly suggestive of the Schumpeterian view that "how capitalism administers existing structures" is essentially irrelevant, since "the relevant problem is how it creates and destroys them."[4] As Schumpeter goes on to say in the same passage:

[4] Schumpeter (1976), p. 84.

> *"The first thing to go is the traditional conception of the* modus operandi *of competition. Economists are at long last emerging from the state in which price competition was all they saw. As soon as quality competition and sales effort are admitted into the sacred precincts of theory, the price variable is ousted from its dominant position. However, it is still competition within a rigid pattern of invariant conditions, methods of production and forms of industrial organization in particular . . . that practically monopolizes attention. But in capitalist reality as distinguished from its textbook picture, it is not that kind of competition which counts but the competition from the new commodity, the new technology, the new source of supply, the new type of organization which commands a decisive cost or quality advantage and which strikes not at the margins of the profits and the outputs of the existing firms but at their foundations and their very lives."*[5]

The transformation of the industrial landscape of innovating firms during the past quarter century certainly confirms the view expressed in this passage. The picture thus painted of the nature of competition is very much to the forefront in the chapters by Teece and Dosi and Nelson, while empirical study of the competitive forces that he emphasized is to be seen in the chapters by Cohen, who considers the firm size–innovation relationship, and Greenstein, who studies the computing industry. The chapter by Hall and Lerner, which reviews the literature on internal finance for innovation, addresses yet another Schumpeterian topic, the importance of past profits in financing future innovation.

A quite different area of development in the economics of innovation is that of the data sources necessary for its study. The analysis of innovation and innovative activity requires data other than conventional economic data: in addition to the usual economic quantities, data on types of innovation, inventions, technologies, arrangements among firms and between firms, and research institutions such as universities are needed. Several authors are concerned with the development of new data sources containing such noneconomic data, possibly merged with the usual currency-denominated economic data such as GDP and R&D spending. The pioneers in this area of development were Mansfield (1968) and Pavitt and coworkers at SPRU (Pavitt, 1984; Townsend et al., 1981) for innovation survey data, and Schmookler (1966) and Griliches (1990) for patent data. The chapters by Nagaoka et al. on patent data and Mairesse and Mohnen on innovation survey data review these sources of data and their uses, but their value is also apparent elsewhere in the handbook, for example in the chapter by Powell and Gianella on collective invention, that by Cohen on empirical studies of innovation, and that by Arora and Gambardella on markets for technology.

Finally, we would like to draw the reader's attention to the fact that some of the papers in this volume are not authored by economists but by those in related fields such as management and sociology. This is not an accident, but reflects the nature of the field, for reasons that again go back to Schumpeter's critique of the static neoclassical framework, a framework that was dominant in economics during some of the time that this field developed. And, of course, we took this into account when selecting chapters for the volume.

The structure of the handbook to some extent follows the "linear model" of innovation, which remains a useful way of thinking about the subject, in spite of the fact that many have pointed out the feedback loops that exist in the system (e.g., see Rosenberg, 1982).[6] The first section of the book

[5] Op. cit.

[6] Kline and Rosenberg (1986) critique the linear model, whereas Balconi et al. (2009) offer a nuanced defense of its value in analysis.

provides an overview, with papers on the economic history of innovation, the evolutionary approach to its analysis, and an overview of empirical work on innovation in firms. The next long section centers on the inventive process and its incentives, looking at the role of science and research organizations, the reward systems, networks, collaboration, and user invention, and including a couple of industry case studies on the information technology and pharmaceutical sectors. This is followed by sections on commercialization and diffusion, with papers on financing, firm strategies, the particular case of general purpose technologies and their diffusion, and the role of international trade in diffusing innovation across borders.

The fifth section of the handbook looks at the innovation process and outcomes in agriculture, energy, and environment, as well as the role of innovation in economic development. Then we turn to the problem of measuring innovation input and output, beginning with macroeconomic growth accounting and the microeconomic measurement of the returns to R&D investments. The next two chapters explore two measurement approaches using noneconomic and qualitative data that are specifically tailored to the innovation area: patent data and data from innovation surveys. The final section of the handbook contains three papers on innovation policy, two that look at the system as a whole, and one centering on the considerable impact of defense-related R&D spending on innovation in general.

Acknowledgments

The editors wish to acknowledge the numerous contributions of Rina Gordon Rosenberg in bringing this project to a happy ending. We are also grateful to Jacques Mairesse and Richard Nelson for careful reading of an earlier draft of this introduction. The second editor thanks the staff at the Stanford Institute for Economic Policy Research for valuable assistance.

References

Abramovitz, M. (1956). "Resource and output trends in the United States since 1870". American Economic Review 46 (2), 5–23.

Arrow, K. (1962). "Economic welfare and the allocation of resources for invention". In: Nelson, R.R. (Ed.), The Rate and Direction of Inventive Activity. Princeton University Press, Princeton, NJ.

Balconi, M., Brusoni, S., Orsenigo, L. (2009). "In defence of the linear model: An essay". Research Policy, doi:10.1016/j.respol.2009.09.013.

Griliches, Z. (1990). "Patent statistics as economic indicators: A survey". Journal of Economic Literature 28, 1661–1707.

Hall, B.H., Mairesse, J. (2009). "Measuring corporate R&D returns". In: Presentation to the Knowledge for Growth Expert Group, Directorate General for Research, European Commission, Brussels (January).

Kline, S., Rosenberg, N. (1986). "An overview of innovation". In: Landau, R. (Ed.), The Positive Sum Strategy. Harnessing Technology for Economic Growth. pp. 275–306.

Mansfield, E. (1968). The Economics of Technological Change. W. W. Norton and Co., New York.

Nelson, R.R. (1959). "The economics of invention: A survey of the literature". Journal of Business 32, 101–127.

Pavitt, K. (1984). "Sectoral patterns of technical change: Towards a taxonomy and a theory". Research Policy 13 (6), 343–373.

Röller, L.-H., Mohnen, P. (2005). "Complementarity in innovation policy". European Economic Review 49 (6), 1431–1450.

Romer, P.M. (1990). "Endogenous technological change". Journal of Political Economy 98 (5, pt. 2), S71–S102.

Rosenberg, N. (1982). "How exogenous is science?" In: Rosenberg, N. (Ed.), Inside the Black Box. Cambridge University Press, Cambridge, UK.

Rosenberg, N. (2010). "Schumpeter and history". In: Rosenberg, N. (Ed.), Studies on Science and the Innovation Process. World Scientific Publishing Co., New Jersey.

Schmookler, J. (1966). Invention and Economic Growth. Harvard University Press, Cambridge, MA.

Schumpeter, J.A. (1937). Preface to the Japanese edition of *Theorie der Wirschaftlichen Entwicklung*, translated by I. Nakayama and S. Tobata, Iwanami Shoten, Tokyo. Reprinted in Essays of J. A. Schumpeter, R.V. Clemence (Ed.), Addison-Wesley, Cambridge, MA (1951), p. 158.

Schumpeter, J.A. (1976). Capitalism, Socialism, and Democracy (third ed.; first ed. published 1942). Harper and Row, Publishers, New York.

Schumpeter, J.A. (1951). "Capitalism in the postwar world". In: Clemence, R.V. (Ed.), Reprinted in Essays of J. A. Schumpeter. Addison-Wesley, Cambridge, MA, p. 158.

Solow, R. (1957). "Technical change and the aggregate production function". Review of Economics and Statistics 39, 312–320.

Townsend, J., Henwood, F., Thomas, G., Pavitt, K., Wyatt, S. (1981). Innovations in Britain since 1945. Science Policy Research Unit, University of Sussex, Occasional Paper No. 16.

Verspagen, B., Werker, C. (2003). "The invisible college of the economics of innovation and technological change". Estudios De Economía Aplicada 21 (3), 203–220.

Chapter 2

THE CONTRIBUTION OF ECONOMIC HISTORY TO THE STUDY OF INNOVATION AND TECHNICAL CHANGE: 1750–1914

JOEL MOKYR

Departments of Economics and History
Northwestern University
Evanston, Illinois, USA
Eitan Berglas School of Economics
Tel Aviv University
Tel Aviv, Israel

Contents

Abstract

This chapter surveys the history of modern economic growth and suggests a number of mechanisms that drove the unprecedented technological thrust that account for the discontinuities of economic modernity. The Industrial Revolution and the subsequent developments did not just raise the *level* of technological capabilities; they changed the entire dynamics of how innovation comes about and the speeds of both invention and diffusion. For much of human history, innovation had been primarily a byproduct of normal economic activity, punctuated by periodical flashing insight that produced a macroinvention, such as water mills or the printing press. The mechanisms that account for innovation becoming a routine activity in terms of the production of useful knowledge are reviewed and linked to the "Baconian program" advocated by the eighteenth-century Enlightenment.

Handbooks in Economics, Volume 01

DOI: 10.1016/S0169-7218(10)01002-6

Keywords

economic growth, industrial enlightenment, industrial revolution, innovation

JEL classification: N13, O31

1. Introduction: Technology and economic modernity

There are widely different interpretations on the significance of "economic modernity." Most scholars coming from economics summarize the preindustrial experience by the somewhat casual observation that growth before 1800 was essentially nonexistent, and that modern economic growth in those economies which came to constitute the "convergence club" took off some time after 1830 (Aghion and Durlauf, eds., 2005; Lucas, 2002). Another tradition, older but with equally venerable lineage, views economic modernity as the expansion of goods and factor markets and the rising interdependence of households and firms (Polanyi, 1944; Toynbee, 1884). A third view focuses on industrial organization and places the factory at the center and considers the growing concentration of workers and their subjection to discipline and top-down coordination (Mantoux, 1928; Weber, 1923) as the essence of economic modernity. Yet none of those interpretations would be convincing without the *fundamental* change that underlay all others, namely the changes in technology that characterized the Industrial Revolution and led to modern economic growth. It were these changes that made the factory possible that allowed the creation of transportation networks and communications, the growth in life expectancy and access to information, the urbanization and changes in the quality and variety of goods and services that we associate with modernity.

How does technology advance? Modern endogenous growth theory has postulated that innovation is "produced" within the system, subject to economic incentives, and should be regarded as an output, resulting from inputs, where physical capital, human capital, R&D, and economies of scale all play major roles. The economic agents who brought this about were motivated mostly by selfish considerations of advancement, including the natural human drives of greed and ambition. The greatest technological sea change in history, which is being discussed here, supposedly constitutes a ringing affirmation of this view. Technology does not descend down on us like "manna," or better perhaps, is not given to us like the ten commandments. It was produced within the system by men and (rarely) women whose purpose was normally to achieve some kind of improvement to the process or product they were interested in. Yet the neo-neoclassical view of technological progress needs to cope with the historical parameters of technological progress, which govern a phenomenon unlike anything else in history.

In part this is for reasons quite well understood. Technology, like all forms of knowledge, is nonrivalrous (i.e., by sharing it with another person the original owner does not have less), so that the social marginal cost of sharing it is zero. Since the social marginal product is positive, the optimal static solution is one in which it is made accessible freely to all able and willing to use it. Yet under these conditions no one has much of an incentive to engage in the costly and risky R&D in the first place. The resulting dilemma has led to a debate that is now a quarter of millennium old on how best to establish optimal incentives in innovative activity. Patents and other forms of private property on useful knowledge played a role in the Industrial Revolution, but were not as essential to it as was once supposed. Instead, it has become increasingly clear that useful knowledge is often produced under conditions of "open source," that is, each person who adds to the pool of knowledge does not require or expect to receive some monetary compensation proportional to the social savings of the innovation. He or she insists, however, on receiving credit and recognition for the contribution as part of a signaling game in which the goal is to establish a reputation. Much innovation in the past functioned very similarly. The dichotomy according to which science operated according to open-source systems whereas technology was subject to private property constraints is seriously exaggerated.

Equally important in making innovation a unique topic in economic history is the fact that technology is produced under the kind of uncertainty that can be characterized as a combination of unintended consequences and unknown outcomes (Rosenberg, 1996). In large part this is the case because technology is normally developed when the exact *modus operandi* of the physical, biological, or chemical processes on which it is based are at best understood very partially. Many inventions have unforeseen and unforeseeable spillover effects on the environment, human health, or the social fabric. Moreover, many innovations are often combined with other techniques in ways not originally intended, to produce wholly novel hybrid techniques that do far more than the simple sum of the components. As a consequence, inventors are often surprised by the eventual outcomes of what seems successful innovation. Such surprises can be, of course, positive or negative.

The progress of technology has been explained by both internalist and externalist theories. Internalists see an autonomous logic, an evolutionary process in which one advance leads to another, in which contingency plays a major role, in which the past largely determines the future. Externalists think of technological change as determined by economic needs, by necessity stimulating invention, by induced innovation being guided by factor prices and resource endowments. In the same camp, but with a different emphasis are social constructionists who regard technology as the result of political processes and cultural transformations, in which certain ideas triumph in the marketplace because they serve certain special class or group interests and powerful lobbies. The history of technology since the Industrial Revolution provides support as well as problems for all of those approaches. A more inclusive approach would separate the process into interactive components. For instance, there is no question that economic needs serve as a "focusing device" in Rosenberg's (1976) famous simile, but the popular notion that "necessity is the mother of invention" manages to be simultaneously a platitude and a falsehood. Societies tend to be innovative and creative for reasons that have little to do with pressing economic need; our own society is a case in point. Modern Western society is by and large wealthy enough to not feel any pressing "need," yet it is innovative and creative beyond the wildest dreams of the innovators of the eighteenth century. There was no "necessity" involved in the invention of ipods or botox. The social agenda of technology is often set by market forces or national needs, but there is nothing ever to guarantee that this agenda will be successful and to make sure what it will lead to.

Technology moves at a certain speed and in certain directions, and the study of innovation helps us understand these laws of motion. Moreover, to come to grips with why technology changes the way it does, we need to be clearer about the way in which prescriptive knowledge (technology) and propositional knowledge (science and general knowledge about nature) affect one another. Knowledge about the physical environment creates an epistemic base for techniques in use. Technology, in turn, sets the agenda for scientists, creating a feedback mechanism. Why, for instance, do high-pressure engines work at higher thermal efficiency than low-pressure ones? Why does heating fresh food in tins and then vacuum-closing them prevent putrefaction? Why does injecting people with cowpox pus provide them with protection against the much nastier smallpox? These and similar issues came up during the period under discussion here, and their resolution led to further technological advances.

Technological change, like all evolutionary processes, was often wasteful, inefficient, and frequently wrong-headed. It was inevitably so, because by definition the outcome of the project was unknown, and so mistakes were made, duplicatory efforts took place, blind alleys were entered. Moreover, a great deal of what seems to us successful innovation was not adopted, often for reasons that *ex post* seem hard to fathom and at times frivolous. But the *degree* of inefficiency of the innovative process was not constant over time. As I have argued in Mokyr (2002), the amount of wastefulness in innovation can be substantially reduced if more is known about the underlying process. In that regard, the process has become hugely more efficient in the past quarter millennium. If innovation requires to "try every bottle on the shelf," an improved epistemic base of the technology can at least reduce the number of shelves. It can avoid looking for things known to be blind alleys like perpetual motion machines and processes that convert base metals into gold. It reduces the amount of intellectual energy spent on occult and other activities that the age of Enlightenment increasingly dismissed as "superstition." More and better knowledge of what is used elsewhere can also reduce duplicatory research and avoid reinventing some wheels.

This essay will not be an exercise in technological determinism. Technology does not "drive" History. Improvements in technological capabilities will only improve economic performance if and when they are accompanied by complementary changes in institutions, governance, and ideology. It is never enough to have clever ideas to liberate an economy from an equilibrium of poverty. But it is equally true that unless technology is changing, alternative sources of growth such as capital accumulation or improved allocations of resources (due, for instance, to improved institutions such as law and order and a more commerce-friendly environment) will ineluctably run into diminishing returns. Only a sustained increase in useful knowledge will in the end allow the economy to grow, and to keep growing without limit as far as the eye can see. I have explored the relation between useful knowledge and technology in Mokyr (2002).

The basic proposition of this essay will be that the technological component of economic modernity was created in the century before the Industrial Revolution, not through the growth of foreign trade, the emergence of an urban bourgeoisie, or the growing use of coal (as has often been argued) but by a set of intellectual and ideological changes that profoundly altered the way Europeans interacted with their physical environment. By that I mean both how they related to and studied the physical world in which they lived and the ways they manipulated that knowledge to improve the production of goods and services.

The net result has been that the technological constraints to which premodern societies were subject simply because they did not know enough were slowly lifted. Modern economic growth has been driven by increasing useful knowledge, which is not, as far as is known, subject to decreasing returns. What makes this possible, as was already realized in the eighteenth century, was the growing "division of knowledge" or specialization, in which each person controlled an ever-declining slice of a rapidly increasing total amount of knowledge. Smith (1757, p. 570) argued outright that "speculation in the progress of society. . .like every trade, is subdivided into many different branches. . . and the quantity of science is considerably increased by it." Because total social knowledge equals the *union* of all individual pieces of knowledge, the knowledge available for technological advances was increasing, provided that those who could make best use of it were able to access it. Hence the centrality of what I have called *access costs* (Mokyr, 2005). What has assured the decline in access costs is that the technology of access itself has been improving through such discrete leaps as the invention of the printing press and the internet, as well as through many other advances, both institutional and technological in the creation of open science and the placement of useful knowledge in the public realm and its codification in languages that can be understood or translated easily.

2. Technology in a "Malthusian economy"

Pre-1800 society, both in Europe and in other parts of the world, was able to develop many extremely useful techniques without, usually, understanding why and how they worked. Ignorance did not prevent these societies from making steel without an understanding of metallurgy, brewing beer without understanding the importance of yeasts, to breed animals without genetics, to inoculate against smallpox without immunology, and to practice crop rotations and apply fertilizer without soil chemistry. Technology could change even when the underlying support in propositional knowledge (the epistemic base) was not widening. Traditional societies had developed a "culture of improvement" as Friedel (2007) has recently termed it, and were quite successful in making considerable improvements in communications, transportation, the use of materials and energy, and to enhance their control on the plants and animals that constituted the "organic economy" (Wrigley, 2004). There was more of a "mineral economy" before the Industrial Revolution than is sometimes believed. On the eve of the Industrial Revolution, both home-heating and many industrial processes in Western Europe depended heavily on coal and peat, and iron and other metals found many uses. The transition that took place in the eighteenth century was not primarily one from an organic to a mineral economy, but from a world in which useful knowledge was empirical, unsystematic, more often than not little more than a tacit set of "understandings" of how nature worked and how materials behaved and reacted to heat and motion, to a technological paradigm in which this kind of knowledge was collected and analyzed in a systematic and organized fashion and useful knowledge increasingly became the dynamic agent that changed the economy.

The informal techniques of the preindustrial age were in the end limited in their ability to affect productivity because major new insights from the outside had to be brought to bear on technology. It would be hard to see, for instance, how the bottleneck of bleaching in the late eighteenth-century textile industry would have been overcome, had it not been for the work of Karl Wilhelm Scheele who discovered chlorine in 1756 and that of Claude Berthollet who discovered its bleaching properties. Moreover, the main productivity-enhancing effects of technology take place when it is fitted and stretched to suit local needs and constraints, when it is tweaked to satisfy a somewhat different purpose and when it can be adapted to be hybridized with other techniques to constitute something entirely different. Yet it is exactly such fitting and stretching that is complementary with a more precise knowledge of the nature of the processes, no matter how partial.

The lack of basic understanding of natural processes was not the only reason why premodern Europe grew so slowly. It has been argued repeatedly that these societies were subject to Malthusian regimes, in which even if technological changes took place, they would be undone because mankind in Wells's (1923) words "spent the great gifts of science as rapidly as it got them in a mere insensate multiplication of the common life." Recent writings such as Galor and Weil (2000) and Clark (2007) have re-emphasized this feature of premodern society. Many problems remain with this interpretation (Mokyr and Voth, 2010), and it needs to be complemented by an alternative negative feedback mechanism, namely the fact that economic growth often was undone by the greed of poorer but more violent predatory neighbors and that of tax collectors, guild members, priests, monopolists, and other seekers of exclusionary rents.

The paradox of a Malthusian economy is that, in its most fundamentalist interpretation, any productivity growth fails to lead to long-term improvements in living standards, and that in the long run the "iron law" of wages rules. Yet it is clearly inconsistent with much of evidence on the preindustrial economies. There was growth before the Industrial Revolution, and even if it was relatively slow, over the centuries it was compounded. It simply will not do to argue, as implied by these arguments, that by any set of measurements the standard of living in Western Europe at the time of the Glorious Revolution was comparable to that during the Norman Conquest. Snooks (1994) and Britnell (1996) have both pointed to substantial growth over the long run before 1700. Smith (1776, pp. 365–366) was certain that by the time of *Wealth of Nations* "the annual produce of the land and labour of England" was higher than a century ago, and that it had been growing steadily since the Norman Conquest and even before. The rates of growth, to be sure were low, and progress uneven and at times reversible. Yet in the span of six centuries, even low rates will compound. To what extent was technology responsible for this? The consensus is that before the Industrial Revolution most gains in output and income can be attributed to the growth of commerce and markets. This "Smithian growth" might explain the dynamic characteristics of pre-1800 growth. After all, advances due to commercial expansion were more easily undone and reversed through pointless violence, predatory neighbors, and greedy rent-seekers than technological advances.

Technology, however, was not stagnant. Advances in farming, textiles, shipbuilding, communications, metallurgy, and energy usage were cumulative, and given that British population in 1700 was not more than 50% higher than at its medieval peak, it stands to reason that Smith's view held true. All the same, these advances were limited, as they were based on a combination of serendipity and patient experimentation, and not on anything that a modern economist would ever recognize as research and development.

In the Malthusian economy, most inventions were made by artisans. Artisans were usually organized in craft guilds. Guilds have had a bad reputation in the history of innovation and are often depicted as conservative organizations. In many cases they were, but it has been argued in recent years that guilds were not inevitably conservative but often permitted and even encouraged innovation and were instrumental in its diffusion (Epstein, 1998; Epstein and Prak, 2008). Whatever the role of the guilds in their training and organization, there seems to be little doubt that the presence of a large number of well-trained skilled craftsmen was one of the great advantages that Britain enjoyed in the eighteenth century. Their capabilities made it possible for the most creative minds of the time to actually have their ideas carried out and the devices they designed built according to specifications, not just once but over and over again. They were mechanics, highly skilled clock and instrument makers, metalworkers, woodworkers, toymakers, glasscutters, and similar specialists, who could accurately produce the parts using the correct dimensions and materials, who could read blueprints and compute velocities, understood tolerance, resistance, friction, lubrication, and the interdependence of mechanical parts. These were the applied chemists who could manipulate laboratory equipment and acids, the doctors whose advice sometimes saved lives even if nobody yet quite understood why, the expert farmers who experimented with new breeds of animals, fertilizers, drainage systems, and fodder crops. This level of knowledge is different from the kind of knowledge needed to make scientific discoveries or inventions, and I have used the term *competence* to denote it. Yet the question remains whether skilled artisans *alone* were capable of generating something like the Industrial Revolution. On that matter there should be serious doubt.

Hilaire-Pérez (2007) and Berg (2007) have argued that "an economy of imitation" based on skilled craftsmen led to a self-sustaining process of improvement. This is certainly not a self-evident statement.

Artisans normally reproduced existing technology and in that process incremental microinventive sequences could lead to some improvements, but eventually will fizzle out. Many societies we associate with technological stasis were full of highly skilled artisans, not least of all Southern and Eastern Asia. A purely artisanal-knowledge society will eventually settle down in a technological equilibrium, in contrast to a society where the world of artisans is constantly shocked with infusions of new knowledge from outsiders. To be sure, some of the more famous "great inventors" of the age—starting with Newcomen and his assistant John Calley and the clockmaker John Harrison—were artisans themselves. Yet artisans, unless they were as prodigiously gifted and as well educated as James Watt or the French gunmaker–inventor Edme Régnier, were good at making incremental improvements to existing processes, not in expanding the epistemic base of the techniques they used or applying state of the art knowledge to their craft. Artisans were also normally not well positioned to rely on the two processes of analogy and recombination, in which technology improves by adopting or imitating tricks and gimmicks from other, unrelated, activities. If all that were needed for the Industrial Revolution had been creative artisans, it could have occurred centuries earlier. Artisans, after all, had been around for centuries, and relying on their innovativeness without the infusion of more formalized and systematic useful knowledge for an explanation of the Industrial Revolution would make it difficult to understand why things moved so rapidly after 1750. In textiles, the technical problems were on the whole less complex than in the chemical industry or in power engineering, but even there, as Jacob (2007) shows, mechanical science found its way soon enough to the shopfloor with important consequences for productivity and efficiency. Moreover, France too had skilled artisans, yet for decades it seemed unable to build the steam engines and develop the iron-processing improvements that Britain did on its own. Not all artisans were friendly and conducive to technological progress, as Hilaire-Pérez points out. The armourers' resistance to Honoré Blanc and interchangeable parts in musket making helped derail a potentially promising advance. The Lyon weavers' resistance to the Jacquard loom failed, but only after the innovators were given military protection.

3. The first Industrial Revolution: A new approach

The absence of long-term growth in most societies is thus clearly overdetermined. The real miracle is not that these Malthusian societies grew so slowly, but that they were, in the end, replaced by a society in which rapid growth became the norm. At the core stood something I have called the Industrial Enlightenment (Mokyr, 2002). The Industrial Enlightenment was an attempt to carry out Bacon's dream that useful knowledge would become "a rich storehouse, for the Glory of the Creator and relief of Man's estate" (Bacon, 1996, p. 143). In the *New Organon* Bacon explained what became almost axiomatic to his followers in the eighteenth century: "If Man endeavor to establish and extend the power and dominion of the human race itself over the universe, his ambition…is without doubt a wholesome thing and …noble… Now the empire of man over things depends wholly on the arts and sciences. For we cannot command Nature except by obeying her" (aphorism 129, cited in Bacon, 1999, p. 147).

The influence of Bacon on subsequent generations was enormous. Clearly, he had expressed a sentiment that was already in the air at his time, but by expressing it with precision and impeccable logical reasoning, he became, with Adam Smith, Karl Marx, and John Maynard Keynes, one of those intellectuals whose thinking affected actual economic outcomes. The so-called invisible college that formed in England after his death (and which included such notables as Christopher Wren, Robert Boyle, and Robert Moray) was formalized into the Royal Society, whose declared purpose it was to increase useful knowledge, and to build bridges between formal science and the actual practical applications of the "useful arts." The great experimentalist Robert Boyle expanded the ideas of the Master, pointing out that Lord Verulam (Bacon) had made a distinction between "luciferous" (enlightening) and "fructiferous" (useful) experiments, but that in fact the one led to the other. "There is scarce any physical truth which is not, as it were, teeming with profitable inventions and may not by human skill and industry be made the fruitful mother of diverse things" (Boyle, 1744, vol. 3, p. 155). The Royal Society was explicitly patterned after Bacon's Solomon's House. The Royal Society started off with boundless enthusiasm for practical technical matters. "The business and design of the Royal Society is to improve the knowledge of natural things, and all useful Arts, Manufactures, Mechanic practices, Engines, and Inventions by Experiments" (Lyons, 1944, p. 41). Robert Hooke added in his preface to his *Micrographia* that [the Fellows of the Royal Society] "have one advantage peculiar to themselves, that very many of their number are men of converse and traffick, which is a good omen that their attempts will bring philosophy from words to action, seeing men of business have had so great a share in their first foundation."

The Royal Society eventually lost interest in practical knowledge, but the spirit of Bacon lived on in many other organizations that came to the fore in eighteenth-century Britain. Thus the Society of Arts, founded by William Shipley in 1754, viewed its purpose as follows "Whereas the Riches, Honour, Strength and Prosperity of a Nation depend in a great Measure on Knowledge and Improvement of useful Arts, Manufactures, Etc... several [persons], being fully sensible that due Encouragements and Rewards are greatly conducive to excite a Spirit of Emulation and Industry have resolved to form [the Society of Arts] for such Productions, Inventions or Improvements as shall tend to the employing of the Poor and the Increase of Trade." The second half of the eighteenth century witnessed a veritable explosion of formal societies and academies dedicated to combine natural philosophy with the "useful arts," by bringing together entrepreneurs and industrialists with scientists and philosophers. In 1799, two paradigmatic figures of the Industrial Enlightenment, Sir Joseph Banks and Benjamin Thompson (Count Rumford), founded the Royal Institution, devoted to research and charged with providing public lectures of scientific and technological issues. In the first decade of the nineteenth century, these lectures were dominated by the towering figure of Humphry Davy, in many ways a classic figure of the Industrial Enlightenment.

Did all this lead to the Industrial Revolution? The paradoxical point is that for most of the eighteenth century, the Baconian program had but meager results to report. Many (though not all) of the central inventions of the Industrial Revolution, above all in textiles, had little to do with advances in science or propositional knowledge more widely defined. While the debate between those who feel that modern science played a pivotal role in the Industrial Revolution and those who do not is still ongoing, it is more than the hackneyed discussion whether a glass is half-full or half-empty. The glass started from almost empty and slowly filled in the century and half after 1750. The argument is thus in large part about the *rate* at which this glass filled.

Moreover, the exact delineation of what part of Bacon's luciferous knowledge was supposed to stimulate and enhance the "useful arts" should be defined with some care. Galileo, Newton, Descartes, and Huygens represented a rigorous and analytical science, but in the eighteenth century much of natural philosophy consisted of the three Cs: counting, cataloguing, and classifying. By describing in detail natural phenomena they did not really understand (including technological practices), experimentalists and natural historians provided a huge information base. To be sure, scientists and science (not quite the same thing) had a few spectacular successes in developing new production techniques, above all the chlorine bleaching technique, Leblanc's soda-making process, the lightning rod, and the mining safety lamp. However, the majority of the path-breaking innovations we associate with the Industrial Revolution did not depend much on this knowledge. It did broaden the epistemic base of some techniques that had been in use for centuries, explaining—in part—why the things that were known to work actually did so, paving the road for even more significant advances to come.

The Malthusian and epistemic constraints were broken not only because propositional knowledge got better at informing technology, but also because there was feedback from improved technology into more knowledge that created the virtuous circles that broke the negative feedbacks of preindustrial society. This mechanism, stressed by Rosenberg (1976, 1982) and de Solla Price (1984), has not been fully recognized by economic historians and is worth stressing. Improved technology, broadly defined, made better science possible. While the discovery of the moons of Jupiter thanks to the early telescopes is common knowledge, the phenomenon is wide and broad. The great advances made by Lavoisier and his pupils in debunking phlogiston chemistry were made possible by the equipment manufactured by his colleague Laplace, who was as skilled an instrument maker as he was brilliant a mathematician. The invention of the first battery-like device that produced a steady flow of direct current at a constant voltage, Alessandro Volta's pile of 1800, made it possible to separate elements in the newly proposed chemistry filled in the details of the landscape whose rough contours had been outlined by Lavoisier and his students. Volta's invention made it possible to separate elements in the newly proposed chemistry filled in the details of the landscape whose rough contours had been outlined by Lavoisier and his students. As Humphry Davy, perhaps the most accomplished practitioner of the new electrochemistry put it, Volta's pile acted as an "alarm bell to experimenters in every part of Europe" (cited by Brock, 1992, p. 147).

Improved instruments and research tools thus played important roles in a range of "Enlightenment projects" that might be seen as technological improvements with poetic license. One such improvement was the use of geodesic instruments for surveying. Jesse Ramsden designed a famous theodolite that was employed in the Ordnance Survey of Britain, commenced in 1791. A comparable tool, the repeating circle, was designed by the French instrument-maker Jean-Charles Borda in 1775, and was used in the famed project in which the French tried to establish with precision the length of the meridian. Time, too, was measured with increasing accuracy, which was as necessary for precise laboratory experiments as it was for the solution to the stubborn problem of determining longitude at sea, one of the age of Enlightenment's proudest successes. Experimental engineering also made methodological advances. John Smeaton was one of the first to realize that improvements in technological systems can be tested only by varying components one at a time holding all others constant (Farey, 1827, p. 168). Smeaton's improvements to the water mill and steam engine increased efficiency substantially even if his inventions were not quite as spectacular as those of James Watt. Much experimental work was carried out in the more progressive early factories, often on the shopfloor.

The new technology created factories (Mokyr, 2001). Factories were many things, one of them the repositories of useful knowledge, the sites where techniques were executed through a growing process of specialization. But they were also places in which experimentation, in the best traditions of the Baconian Enlightenment, took place. Of course, only a minority of the great mills carried out such experimentation, but they were the ones that counted. Some of the more famous early mill owners were deeply involved in experimentation. James Watt and Josiah Wedgwood, led the pack, but other such as textile manufacturers Benjamin Gott, John Marshall, and George Lee followed a similar course. They were often in touch with the best scientific minds of their day, but there were limits on what could be learned. The best-practice propositional knowledge of the time was inadequate to guide the industrialists in their technical choices. When the exact natural processes underlying a technique are poorly understood, the best way to advance through systematic trial and error. James Watt wrote in 1794 that even in mechanics theory was inadequate and thus experiment was the only answer. "When one thing does not do, let us try another" (cited by Stewart, 2007, p. 172). Experiments, once the realm of gentlemen scientists, had by the late eighteenth century become a shopfloor activity. In such systems, progress tends to be piecemeal and cumulative rather than revolutionary, yet without such microinventions, the process of innovation would have ground to a halt. Macroinventions and microinventions are inherently complementary, but their capacity to stimulate one another was itself improving in an age that believed deeply in improvement and was learning how to bring it about.

Just as important, technological bottlenecks and issues set the agenda for scientists, just as Bacon and his followers had suggested, and many of them set their minds to solve real-world problems. Among them were the greatest minds of the scientific Enlightenment. Leonhard Euler was concerned with ship design, lenses, the buckling of beams, and (with his less famous son Johann) contributed a great deal to theoretical hydraulics. The great Lavoisier worked on assorted applied problems as a young man, including the chemistry of gypsum and the problems of street lighting. Gottfried Wilhelm Leibniz, William Cullen, Joseph Black, Benjamin Franklin, Gaspar Monge, Joseph Priestley, Humphry Davy, Claude Berthollet, Tobern Bergman, Count Rumford, and Johann Tobias Mayer were among the many first-rate scientific minds who unabashedly devoted some of their efforts to solve mundane problems of technology: how to design calculating machines, how to make better and cheaper steel, increase agricultural productivity and improve livestock, how to build better pumps and mills, how to determine longitude at sea, how to heat and light homes and cities safer and better, how to prevent smallpox, and similar questions.

Of the many examples one could elaborate upon here, the career of René Réaumur (1683–1757) is telling as the epitome of the Enlightenment ideals. Although one of the most recognized scientists of his day (he was a distinguished mathematician and president of the French Académie Royale), his reputation today has been eclipsed by others. Yet in his day he worked on a variety of problems concerning the nature of iron and steel (he was first to suggest the chemical properties of steel), on problems of porcelain and glazing; he showed the feasibility of glass fibers and suggested that paper could be made from wood; carried out a huge research program on entomology and farm pests, egg incubation, and worked on meteorology and temperature measurement (hence the now defunct temperature scale still named after him).

Part of the problem with understanding the Industrial Revolution is the literature's focus on textile industries, which is quite understandable given the central importance of the technological revolution in cotton. Yet as Temin (1997) has argued, the Industrial Revolution was spread over many industries and sectors and technological progress spread to many sectors even if these constituted at first a small part of the British economy. What was unique in the second half of the eighteenth century was not the advances in one industry or another, but the push for progress on a wide front. We tend to be biased in our thinking toward the cases in which success was attainable, such as cotton textiles, steam, iron, and engineering.

Yet there was a similar "push" for improvement in a range of other goods and services, where progress was slower or even barely existent at first simply because nature offered more resistance, that is, the problems were harder.

This resistance is especially notable in agriculture and medicine. The reason economic historians do not speak much about an agricultural revolution anymore is that many of the problems in increasing productivity in farming were beyond the scientific capabilities of the time. But this was not for lack of trying. What is striking about them is the increasingly tight connections agricultural innovators sought with natural philosophers. Arthur Young himself sought the help of the leading British scientist of the 1780s, Joseph Priestley, in preparing his experiments. Many leading scientists were deeply interested in farming. The eminent chemist Humphry Davy was commissioned to give a series of lectures on soil chemistry resulting in his *Elements of Agricultural Chemistry* (published in 1813), which became the standard text until replaced by Von Liebig's work in 1840. The creative Scottish chemist Archibald Cochrane, the ninth earl of Dundonald, published in 1795 a treatise entitled *Shewing the Intimate Connection that Subsists between Agriculture and Chemistry*. Most of these writings were empirical or instructional in nature. Davy had to admit that the field was "still in its infancy" (p. 4), although he realized that it was scarcely possible to do any investigations in agriculture without depending on chemistry. At the time he was writing, the work in organic chemistry carried out in Giessen and which eventually unleashed the agricultural revolution that the Baconian program had promised was another half-century in the future. The same was true mutatis mutandis for medicine. Physicians and public health officials in the eighteenth century launched a massive assault on the main diseases ravaging the population of the day, and while, much like in farming, they scored a few local victories, the main objectives—understanding the nature of infectious diseases—were beyond them.

The central conclusion to take home about the first Industrial Revolution is that its historical importance as the fountainhead of modern economic growth was not so much in the transformations in cotton and steam that occurred between 1760 and 1800, but in the ability of the Western economies to sustain technological progress and somehow managed to avoid the negative feedbacks and hard constraints that had prevented a similar breakthrough after the great macroinventions of the fifteenth century (iron casting, printing, and three-masted shipping, among others). While much of the action in the first 40 years of *Sturm und Drang* of the Industrial Revolution took place in Britain, this was clearly a multinational effort. French, German, North American, and Italian knowledge, as well as that emanating from the Low Countries and Scandinavia were all more or less freely shared in the "Republic of Letters," an international "invisible college" of men and (a few) women of science who shared their knowledge through correspondence and publications and (more rarely) personal contact and travel. This community had already emerged in the sixteenth century (Collins, 1998, pp. 523–569), and by the eighteenth century it had extended to mechanical and technical knowledge (Darnton, 2003; Daston, 1991).

Such a collaboration between scholars and engineers was necessary, because it took the ingenuity and intelligence of people beyond the British boundaries to create the growing knowledge base that was the big difference between the Industrial Revolution and earlier "efflorescences." Contemporaries were aware of that, and while Britain's scientists and mathematicians at times made contributions of substance (one thinks of the work of Priestley, Hale, Cavendish, Black, Faraday, and many others), it is also quite clear that they were quick and ready to adopt and adapt new ideas wherever they came from. John Farey, an eminent engineer, testified in 1829 before a Parliamentary committee that "the prevailing talent of English and Scotch people is to apply new ideas to use, and to bring such applications to perfection, but they do not imagine as much as foreigners" (Great Britain, 1829, p. 153). He provided a long list of such inventions, not entirely accurate, but also omitting some of the most important ones. The Industrial Revolution was a collaborative effort of most of the Western economies, and the British may have had a comparative advantage in competence and microinventions and thus exported skilled craftsmen and mechanics (laws on the books prohibiting such movements notwithstanding), but they imported many of the best ideas.

Precisely because they could draw on a much larger knowledge base than what was produced in Britain alone, British engineers and inventors were able to keep the wheels of innovation turning. This was especially true in chemistry, in which the British, by their own admission, fell behind their European counterparts, such as Lavoisier and his student Claude Berthollet, Berthollet's student J.-L. Gay-Lussac, and many others. What it also meant, however, that Britain's advantage as "the first Industrial Nation" was inherently ephemeral, and the much-discussed British decline in the second half was little more than an equilibrating process, in which the technological capabilities of the other Western nations roughly caught up with Britain, even if differences in national styles and nuances can be readily discerned.

4. The transition to modern growth, 1830–1880

The economies of the West, by 1830, had more or less committed to progress and economic development. The political economy here is fairly complex. On the one hand, the 1815 restoration had reinstated conservative regimes throughout Europe. Yet the impact of the Enlightenment could not be undone. The influence of liberal political economy, the Enlightenment's proudest offspring, soon became too powerful to ignore as the reactionary regimes learned the hard way in 1830 and again in 1848. Moreover, nations were aware of the impact of economic performance on military and political power, and as a consequence increasingly reformed their economy and supported the creation and dissemination of useful knowledge in its various forms.

The most spectacular development of this period in economic history was the growth in transport technology. The railroad was almost exclusively a British invention, and was led by British engineers. Following the development of the high-pressure engine in the first years of the nineteenth century, it at first relied little on formal scientific breakthroughs. The technological history of the railroad is typical of a "hybrid" technology. It combined a number of elements, the most important of which were the flat rail and the high-pressure steam engine. The use of wooden tracks to minimize the friction created by pulling heavy cargoes on wheeled vehicles can be tracked down to the early middle ages, and were quite widely used by British mines in the late eighteenth century. At Coalbrookdale, a cast iron cover was used to reinforce wooden rails in 1767. By the first decade of the nineteenth century, decades before the first successful locomotives, Britain was estimated to have 300 miles of (horse-drawn) railway track (Bagwell, 1974, p. 90). The first "general-purpose" railroad was the Surrey horse-drawn iron railway completed in 1805, built by William Jessop, one of Britain's prime engineers and John Smeaton's star pupil. While no financial success, it indicated what this form of transport could do.

The high-pressure steam engine was a logical extension of the machines of the eighteenth century but its progress was slow, in part because of the deliberate resistance of James Watt and in part because they were difficult to build and unlike low-pressure engines prone to explosions. Its lighter weight made it, however, ideal for transportation purposes. The possibilities of this engine were explored in the first years of the century by Richard Trevithick, and Arthur Woolf, but brought to perfection by a remarkable engineer, George Stephenson, a man of no formal training, barely literate, yet with unfailing technical intuition. The first use of steam power was the Stockton and Darlington railroad (mixed horse and steam power) in 1825. The conventional start of the railway age, however, is taken as the opening of the Liverpool–Manchester route in 1830, and the triumph of Stephenson's famed *Rocket* in the Rainhill competition.

The railroad was the ultimate achievement of British engineering competence. British technicians often laid the foundations of railroad construction and rolling stock design elsewhere. The first locomotives put in service in France were built by Murray and Jackson of Leeds and Bury of Liverpool and even the famed French engineer Marc Séguin purchased engines from Stephenson's shop in Newcastle "to be used as models by French builders" (Daumas and Gille, 1979, pp. 348, 366). It was mostly designed and built by people with little or no formal education, but who had mastered a profound if informal understanding of what did and did not work through a combination of natural talent and access to the right masters.

It is striking how strong the connection was between the railways and the mining sector. Many of the railroad pioneers came from a mining background. The first models were built by Richard Trevithick, whose education was in the Cornish mines, mostly provided by his own father and uncle (Burton, 2000, p. 28). George Stephenson had even less informal education, and both he and William Hedley, the designer of an intermediate proto-model of the locomotive known as "puffing Billy" was trained in the mining sector. Another railroad pioneer, Timothy Hackworth, similarly, was apprenticed to his father (a blacksmith) and he too worked at a colliery. Even when advances were made by non-Englishmen, such as Séguin's fire-tube boiler design, it was made not by a formally trained polytechnician but by a self-made engineer, ignorant of advanced mathematics. The technical problems in the railroad were often hard and perplexing, but they were still mostly of the kind that could be overcome with the traditional empiricist engineering skills that had stood British mining and manufacturing in such good service during the Industrial Revolution. It was, however, not a promising strategy for future technological advances.

The same was true for mechanical engineering. The period after 1815 in Britain was a period of major consolidation, and with it came a huge drive toward the rationalization of manufacturing. As in other industries, Britain was well served by the high skills and broad practical knowledge of its mechanical engineers, in an age in which dexterity and experience could still substitute for a formal training in mathematics and physics. Mechanical engineering, as MacLeod and Nuvolari (2007a) stress, was a core activity of the Industrial Revolution, generating a disproportional share of innovations. The operators of lathes and cutting machines learned to make power-driven machinery that could then be applied in other industries by workers with fewer skills than themselves. Much of this equipment was becoming standardized. The key was special-purpose tools; much like the division of labor, mass production required a specialization in the design of machine tools. Presses, drills, pumps, cranes, and many other forms of mechanical equipment were produced in large series. The idea of a high degree of accuracy, both in measurement and in manufacturing, which had become increasingly prominent in the eighteenth century, was finally becoming operationalized.

Manchester, close to the best customers of many of these machines, became a center of this industry competing with London's. Perhaps the paradigmatic examples of a British engineer in this tradition were Henry Maudslay (in London) and his one-time apprentice Joseph Whitworth (who moved back to Manchester), who helped modernize mechanical production by standardizing screw threads and thus laid a foundation of modern mass production through the modularity of parts. The influence of the machine-tool industry on the advance of manufactures, in the admittedly somewhat biased opinion of one of its leaders, had been comparable to that of the steam engine (Nasmyth, 1841, p. 397). They did so by replacing the human hand in holding the tools of cutting metal by "mechanical contrivances," thereby achieving an accuracy hitherto unimaginable, using far less-skilled labor.

Over the long haul, the emergence of these prosaic devices proved to be one of the most radical innovations of all time. Mass production, based on large batch manufacture of perfectly identical and hence interchangeable parts, has turned out to be one of the unsung heroes of technological history. Less sudden than cotton, less dramatic that steam, less spectacular than gaslighting, mass production was just as much a child of the first Industrial Revolution as cotton and steam and one of the chief causes of how a set of localized technological advances after 1760 turned into a cascade of economic progress. The famous Portsmouth block-making machines, devised by Henry Maudslay and Marc Brunel around 1801 (a project directed by Samuel Bentham, Jeremy's brother) to produce wooden gears and pulleys for the British Navy, were automatic. In their close coordination and fine division of labor they resembled a modern mass-production process, in which a strongly interdependent labor force of 10 workers produced a larger and far more homogeneous output than the traditional technique that had employed more than 10 times as many. As Musson (1975) and others have argued, the widespread belief that Britain fell behind in this area of technology and eventually ceded mass production to the United States, is simply inaccurate. By 1841, a Parliamentary committee could proudly report that the implements after 1820 were "some of the finest inventions of the age" and that by their means the machinery produced by these tools is better as well as cheaper "tools have introduced a revolution in machinery and tool-making" (Great Britain, 1841, p. vii).

The railroad and mechanical engineering notwithstanding, after 1830 the ever-widening epistemic basis of technology was becoming a central factor in technological progress. This process was far from balanced, much less linear and even. But in a number of industries, the importance of scientific understanding became too important to ignore. Inventors did not need to be schooled themselves; it was often sufficient for them to have access to others who were. In iron and steel for instance, the accumulation of useful knowledge played a role in the development of the work of James Neilson. Neilson's "hot blast" was perfected in 1829 and reduced the fuel consumption of blast furnaces by two-thirds. Neilson was not a trained scientist but a practicing and experienced engineer, and his invention was the result of trial and error far more than of logical inference. Yet he was inspired and informed by the courses in chemistry he took in Glasgow, where he learned of the work of the French chemist Gay-Lussac on the expansion of gases he utilized in his invention (Clow and Clow, 1952, p. 354). In steel, a famous paper by Berthollet, Monge, and Vandermonde "Mémoire sur le fer considéré dans ses différents états métalliques" published in France in 1786 explaining the scientific nature of steel may have been above the heads of British steelmakers. The immediate impact of the paper was not large. It was "incomprehensible except to those who already knew how to make steel" (Harris, 1998, p. 220).

But five years later, the British chemist and physician Thomas Beddoes published a paper that relied on it and by 1820 the paper was well known enough to make it into an article in the *Repertory of Arts, Manufactures and Agriculture* (Boussingault, 1821, p. 369), who noted that idea had been adopted by all chemists who have turned their attention to the subject. Further work by scientists, such as Michael Faraday's on the crystalline nature of wootz steel (high-quality steel made directly from ores), increased the understanding of the characteristics of ferrous materials. As Smith (1964, p. 174) noted, "with carbon understood, Bessemer found control of his process easy, though its invention was not a deduction from theory, as the Martins' probably was."

Similar developments can be discerned in some less well-known sectors. In the cement industry, an article in *Rees's Encyclopedia* in 1819 described in detail the chemical processes involved in the hardening of cement, a description deemed "remarkably acute" by a modern expert (Halstead, 1961–1962, p. 43). To be sure, the full explanation of cement's hydraulicity was not put forward until the 1850s, but this was an area on which the new chemistry had a lot to say. The same was true in a different area of chemistry: fatty acids, the raw materials used in candles and soap. Michel Eugène Chevreul, the director of dyeing at the *Manufacture des Gobelins*, who discovered their nature, turned the manufacture of soap and candles from an art into a science. His discovery of stearine served as the basis of improved candles that burned longer and more brightly, with little smoke or smell. The real cost of candle light is estimated to have declined from £15,000 per million lumens-hour in 1760 to below £4000 in constant prices in the 1820s (Fouquet and Pearson, 2006, p. 153). Interestingly enough, Chevreul did not succeed in manufacturing artificial dyes, despite his obvious interest in them.

We should not exaggerate the immediate impact of the Lavoisier revolution in chemistry on industrial practices. Certainly, the full impact of scientific chemistry such as it was on industry did not begin to be felt until 1820 (Daumas, 1979, p. 564), and some modern historians have expressed skepticism whether Lavoisier and his pupils really established a "modern chemistry." Much of the new science remained quite untight—experts still disagreed about basic topics such as the atomic structure of matter and the nature of heat. Some scholars feel that "early nineteenth-century chemists did not regard their practice as having been reformed decisively. They were still in the process of reforming it" (Bowler and Morus, 2005, p. 76). This, of course, is an extreme view; no matter what the underlying philosophy, the establishment of chemical elements and the relations between them, the notation proposed by Berzelius, and the discoveries made by Dalton, Berthollet, Davy, Gay-Lussac, and others did establish new concepts, a new language, and a new set of laboratory tools. The growth of the epistemic base of existing technology made a steady expansion of useful knowledge possible.

In the 1830s, furthermore, the many decades of research in electricity started to see their first payoff: the research of scientists such as Oersted and Joseph Henry led to the development of the electrical telegraph, a breakthrough of truly momentous economic and social consequences later in the nineteenth century. It was a truly international effort. Oersted was Danish and Henry an American, but the research involved Germans and Frenchmen too. All the same, it was two Englishmen, Charles Wheatstone and William Cooke who turned an experiment into an enterprise. It took another decade to convince business interests and bureaucrats that this was indeed a useful technique, but in 1846 Cooke founded the Electric Telegraph Company, and installed 4000 miles of cable in its first 6 years. The first successful submarine cable was laid by Thomas Crampton's Company between Dover and Calais in 1851, and became a technological triumph that lasted 37 years. By 1857, most British cities were linked, and an operating line to the Continent had been established. In telegraph, as elsewhere, the give-and-take between scientists and inventors in the nineteenth century was complex. Before the telegraph could become truly functional, the physics of transmission of electric impulses had to be understood. Physicists, and above all William Thomson (later Lord Kelvin), made fundamental contributions to the technology. Thomson invented a special galvanometer, and a technique of sending short reverse pulses immediately following the main pulse, to sharpen the signal (Headrick, 1989, pp. 215–218).

In steam technology, the books of John Farey and François-Marie Pambour in the 1820s and 1830s summarized the best-practice knowledge of their time, but they were still clearly thinking of steam engines as propelled by the steam rather than heat engines. Oddly enough, it fell to an engineer to suggest for the first time the true nature of the steam engine, namely Sadi Carnot's 1825 *Reflexions*. It took a few decades for the insights to sink in. As one scholar has sighed, "The application of Carnot's explicitly stated results could have been of assistance in some of the problems with which the engineers were wrestling such as the merits of fluids other than water as the working medium or a quantitative estimate of the benefits derived from using high pressure engines. Certainly, the use of Carnot's theory would have, at the very least, prevented many engineers from spending time on hopeless projects" (Kerker, 1960, p. 258).

The transnational and semi-cooperative nature of Western useful knowledge, revived after 1815, was a direct continuation of the Enlightenment *Republica Litteraria*. The development of thermodynamics is another good example of this feature. The 1825 breakthrough paper by Carnot was published by his compatriot Emile Clapeyron in France, but remained unknown in France. It was translated into English in 1837 and into German in 1843, and thus in a position to influence James Joule and William Thomson in Britain and Hermann von Helmholtz and Rudolf Clausius in Germany (Cardwell, 1971, 1972). A young Scottish engineer named William Rankine, more than anyone else, made the new science of thermodynamics part of practical engineering. Rankine insisted that engineering knowledge must have its roots in scientific principles. His style was to state a general problem, solve it if he could, and only then to treat the special cases encountered in practice. Yet his work remained deeply empirical at heart. Much as had been the case for much of the Baconian program, where no general principles were yet accessible, he was much like a "natural historian of an artefactual world," provisionally collecting the empirical regularities and data of engineering with the ultimate intention of subsuming them under scientific law in the best of Baconian traditions (Marsden, 2004).

It is hard to know, exactly, how much the subsequent development of engines owed precisely to Carnot or Rankine specifically, but clearly thermodynamics formed the basis for the continued improvement of engines in the ensuing decades. The Glaswegian engineer John Elder, who receives most of the credit for building the compound marine steam engines that made the final victory of steam over sail possible, worked closely with Rankine and his quadruple compound engine made long sea voyages on steam-driven ships an economic reality (Day and McNeil, 1996, p. 237). As Smith (1990, p. 329) has noted, in Britain the great engineering firms of Manchester and Glasgow required more than just trial-and-error methods to resolve issues of economy and engine efficiency. Rankine's *Manual of the Steam Engine* published in 1859 made thermodynamics accessible to engineers, and the new steam engines made explicit use of the Carnot principle that the efficiency of a steam engine depends on the temperature range over which the engine operates. Rankine has been judged to have developed "a new relationship between science and technology" (Channell, 1982, p. 42).

This is not to say that good, old-fashioned intuition and practical skills were right away relegated to a secondary role. Thus, the famed American Corliss engine was built by a man with little formal education

(1848) and before the revolution of thermodynamics was widely disseminated. It was based on the idea of a shuttle-type valve which gave the engine an automatic variable cutoff capability, which brought a huge improvement in the efficiency with which the engine exploited the expansive power of steam and saved a third of the fuel costs, as well as delivered a much more smooth and responsive delivery of power. It was of central importance in cotton spinning where achievement of higher and constant speeds was central to productivity improvement (Rosenberg and Trajtenberg, 2004, p. 74).

Perhaps, the best way to summarize the kind of useful knowledge that served Britain best in the first half of the nineteenth century was the concept of "mechanical science," almost an oxymoron in our own time (Jacob, 2007; Marsden and Smith, 2005, p. 145). Within the hierarchy of useful knowledge, it was low in the pecking order. The British Association for the Advancement of Science, established in 1831 relegated it to its Section G, founded a few years later, as a bridge between the theoretical sections such as Section A (mathematics and physics) and practical engineers. It was an applied area, and throughout the period attracted some of Britain's most illustrious engineers such as William Fairbairn and the naval architect and engineer John Scott Russell. It provided respectability to the area that, as it seems to us now, Britain was best at, namely to use empirical methods and competence to apply ideas from science to practical engineering issues. Yet the BAAS was not a narrow, national organization, and it made serious efforts to bring foreign scientists to its meetings (Morrell and Thackray, 1981, pp. 372–386). It, too, was a product of the institutions of the eighteenth-century Enlightenment.

5. The second Industrial Revolution

By 1860, the Western world had experienced a revolution in textiles, materials, transportation, and energy. Yet daily life had been affected but little for most of the populations, except that travel had become faster and cheaper, people were wearing cotton clothes, and a number of large industrial towns had sprung up, such as Manchester and Glasgow in Britain, St. Etienne and Mulhouse in France, Ghent and Liège in Belgium, Essen in the Ruhr, and a few budding centers elsewhere. It is quite possible to imagine a counterfactual world in which innovation would have fizzled out at that stage, a world of steam and large cotton mills, of wrought iron and hybrid ships (sailing ships with auxiliary steam engines), of homes illuminated by gas and communications confined to telegraph lines, and in which growth would have slowed down and settled on a set of dominant designs vintage 1860. Such a world would have been different in some visible ways from the world of 1800, but not nearly as spectacularly different as the world of 1860 turned out to differ from that of 1914. The wave of innovations that occurred roughly between those two dates was more radical and spectacular in its technical and conceptual advances than perhaps any era in human history. The period 1859–1873 has been characterized as one of the most fruitful and dense in innovations in history (Mowery and Rosenberg, 1989, pp. 22–23). Vaclav Smil has gone further and characterized this period as a whole as the most revolutionary and innovative in history. It is hard to precisely quantify and test such statements, of course, but almost every new technique developed during the first half of the twentieth century, and many beyond, originated during the period commonly identified as the second Industrial Revolution.

The impact of cheap steel has been hard to overestimate, simply because no other material remotely competitive with it could be made at that time. Steel had been known since the middle ages and before, but its high cost prohibited its use in all but the most demanding uses. Benjamin Huntsman, a Sheffield clockmaker, perfected in 1740 the so-called crucible process, which made it possible to make high-quality steel in reasonable quantities. Huntsman used coke and reverberatory ovens to generate sufficiently high temperatures to enable him to heat blister steel (an uneven material obtained by heating bar iron with layers of charcoal for long periods) to its melting point. In this way, he produced a crucible (or cast) steel that was soon in high demand. Huntsman's process was superior not only in that it produced a more homogeneous product (important in a product such as steel, which consisted of about 2% carbon mixed in with the iron) but also removed impurities better because it created higher temperatures. His product remained too expensive for many industrial uses, however, and attempts to make steel not only good but also cheap, had to wait until the second half of the nineteenth century. Nevertheless, Huntsman's process, one of the early path-breaking inventions of the eighteenth century, is worth mentioning as an important advance. Steel was essential in the production of machine parts, cutting tools, instruments, springs, and anything else that needed a material that was resilient and durable. Crucible steel may have been one critical catalyst to innovation that economic historians have tended to overlook. The quality of crucible steel was such that it was produced in considerable quantities in Sheffield long after the nineteenth-century methods of producing cheap bulk steel had been introduced. Huntsman worked in a world of tacit knowledge, of an instinctive feel for what worked based on experience and intuition, data-driven rather than based on a scientific analysis. As noted, by the 1820s and 1830s, the chemical nature of steel as an alloy of pure iron small quantities of carbon was known, and it is hard to envisage the subsequent advances in steelmaking without it. The two breakthroughs, Bessemer's converter (1856) and the Siemens–Martin process (1865), happened fairly close to one another. Neither of them was built from scratch on purely theoretical reasoning (Bessemer admitted to being surprised by his success), but neither of them would have developed further without the support of an epistemic base that made it possible. At first, Bessemer steel was of very poor quality, but then a trained British metallurgist, Robert Mushet, discovered that the addition of *spiegeleisen*, an alloy of carbon, manganese, and iron, into the molten iron as a recarburizer solved the problem. Scientists such as Henry Clifton Sorby turned new tools (microscopes) on the question of the nature of steel and should be regarded as the founder of what we call today metallography.

Chemistry also helped straighten out problems in both techniques, including the removal of phosphorus from ores, which spoiled the quality of steel. The leader of Britain's Cleveland steel district was Isaac Lowthian Bell, himself a distinguished scientist, who pleaded incessantly for a greater emphasis on science in British steel industry. "The way in which he combined business and science was unusual in Victorian Britain: nevertheless, his abilities as chemist, mineralogist, and metallurgist challenge the view that the economy at that time was run only by empiricists" (Tweedale, 2004). Britain's concern with losing its technological leadership here was to a great deal misplaced. German iron and steel remained dependent on British innovations, and the first Kaiser Wilhelm Institute for Iron and Steel Research was established only in 1917 (Weber, 2003, p. 340). Apart from the Siemens–Martin process, most of the major breakthroughs such as stainless steel came from Britain.

The Bessemer and Siemens–Martin processes produced bulk steel at rapidly falling prices, and were able to use all ores after the Gilchrist-Thomas basic Bessemer process (1878). High-quality steel continued for a long time to be produced in Sheffield using the old crucible technique. However, the steel revolution was brought about by lower prices, not by a novel product. Cheap steel soon found many applications beyond its original spring-and-dagger demand; by 1880 buildings, ships, and railroad tracks were increasingly made out of steel. Steel allowed economies of scale in areas that had until then run into serious constraints: much larger ships and taller buildings. It revolutionized international trade, urban locational patterns, and warfare. It became the fundamental material from which machines, weapons, and implements were made, as well as the tools that made them. The conclusion that cheap steel "created" modern industrial society would be oversimplified and sound like technological determinism. But without it, the morphology of the modern economy would have been dramatically different.

Iron and steel were informed by science, but it remained primarily science of an empirical, descriptive sort. In chemistry, a wider epistemic base turned out to be essential, even if there, too, a full understanding of the principles involved coevolved with the exploitation of new techniques, many of them derived through trial and error. The development of organic chemistry in the late 1820s by two Germans, Friedrich Wöhler and Justus von Liebig, must count as a revolution equal to (and complementing) the insights of Lavoisier and his followers four decades earlier. The novelty was not one compound or another, but the fundamental realization that four elements (oxygen, carbon, nitrogen, and hydrogen) could combine together in almost infinitely many different ways, to produce millions of different compounds (Brock, 1992, p. 201), and that organic compounds could be created through man-made techniques and not just by some mysterious "vital force." Again, this was an international and collaborative effort taking place in the European Republic of Science. Liebig studied in Paris with Gay-Lussac, and Wöhler in Stockholm with Berzelius. The critical insight that soil fertility depended in great measure on nitrogen content was due to a French chemist, Jean-Baptiste Boussingault and two British experimentalists, John Bennet Lawes and Joseph Henry Gilbert. Organic chemistry opened the door for manufacturing in major areas which are often regarded as a core of the second Industrial Revolution: artificial dyes, fertilizers, explosives, and pharmaceutics.

And yet, even here, science went hand in hand with serendipity and patient trial-and-error experimentation. The famous tale of William Perkin, much like the young would-be king Saul, setting out to find one thing and discovering another has often been told. The 18-year-old Perkin searched for a chemical process to produce artificial quinine. While pursuing this work, he accidentally discovered in 1856 aniline purple, or as it became known, mauveine, which replaced the natural dye mauve. The discovery set in motion what was to become the modern chemical industry. Perkin, however, was trained by the German von Hofmann, who was teaching at the Royal College of Chemistry at the time, and his initial work was inspired and instigated by him. Three years later a French chemist, Emanuel Verguin, discovered aniline red, or magenta, as it came to be known. In 1869, after years of hard work, a group of German chemists synthesized alizarin, the red dye previously produced from madder roots, beating Perkin to the patent office by one day. The discovery of alizarin in Britain marked the end of a series of brilliant but unsystematic inventions, whereas in Germany it marked the beginning of a process in which the Germans established their hegemony in chemical discovery (Haber, 1958, p. 83). German chemists succeeded in developing indigotin (synthetic indigo, perfected in 1897) and a series of other dyes. Outside artificial dyes, the most noteworthy discoveries were soda-making, revolutionized by the Belgian Ernest Solvay in the 1860s and explosives, where dynamite, discovered by Alfred Nobel, was used in the construction of tunnels, roads, oil wells, and quarries. If ever there was a labor-saving invention, this was it.

The alleged German advantage in chemicals was based on the scientific lead they had enjoyed since the path-breaking work in Giessen and Göttingen in the 1820s and 1830s. At a range of German universities, chemists slowly unraveled the mysteries of organic compounds. The most famous breakthrough was that of August Kekulé at Bonn, who realized that organic chemistry was the study of carbon compounds and suggested the structure of the benzene compound. But most German chemistry consisted of normal science, cumulative advances by men such as Heinrich Caro (chief researcher at BASF) and Adolf von Baeyer (Professor of Chemistry at Strasbourg and Munich) that added up to a better understanding leading to a flow of innovations that created an industry. British and French contemporaries bewailed the rise of Germany as the chemical giant of the time, but the knowledge on which chemical technology was based was, like all Western science, an open-source endeavor. The techniques themselves, of course, were not, and patent protection was increasingly a factor in this industry, as R&D was costly and often slow. German patent protection was more effective than the British laws, in large part because the 1877 law was shaped by the manufacturers (Murmann and Landau, 1998, pp. 41–42). Germany became the dominant producer of artificial dyes, accounting for as much as 85–90% of the world market.

The German advantage in chemicals in the second Industrial Revolution was, in that respect, comparable to the British advantage in the early cotton industry. Although it made many of the advances itself, its chemists and its chemical knowledge were internationally mobile. If Germany had any advantage that was hard for other countries to replicate it was *competence*. Its polytechnic universities produced a steady stream of well-trained and able midlevel chemists, who were able to implement and execute the new processes, and in the process introduce the stream of microinventions and adaptations that accounted for most gains in productivity and the successful new products. Unlike the early cotton industry, however, the chemical industry required a "scientifically literate workforce" as Murmann (2003, p. 56) has put it, and the German higher education was far better in producing this resource. Such advantages, much like the early British advantage in the eighteenth-century techniques, were inherently ephemeral, and Germany's much-feared industrial superiority in chemicals dissipated after World War I.

Just before the War German chemists produced one of the most spectacular innovations of all times. To be sure, as Vaclav Smil has noted in his brilliant book on the topic (Smil, 2001), major discoveries rarely arise *de novo*, and what seems to us a breakthrough was only the last step in a long intellectual journey. Yet the ability to synthesize ammonia (NH_3) from the atmosphere at reasonable cost, the Haber–Bosch process in 1912, must be counted as one of the most momentous breakthroughs in history. The logic is one of social savings, popularized in the railroad literature of the 1960s. A counterfactual world without nitrates would have been a world in which World War I might have been considerably shorter, but also one in which a human race of 6 billions would have been doomed to a Malthusian disaster: Smil (2001, p. 160) estimates that only half of the current world population could have been fed without nitrogen fertilizers and that with diets that would have been considerably more vegetarian-based.

The story of the invention is clearly another combination of scientific understanding of the process, yet never sufficient to dispense of a large amount of experimental work, trial and error, and innumerable dead alleys and the frantic search in Alvin Mittasch's (Fritz Haber's assistant at BASF) lab for a catalyst that would work well ended up involving 20,000 runs of 4000 different substances, clearly an example of an old-fashioned "try-every-bottle-on-the-shelf" scientific method. Yet in chemistry, much like in other fields, science and formal training prepared the minds that Fortune favored.

This was equally true in electricity, the other spectacular advance of the age of the second Industrial Revolution. Electricity had fascinated many of the best minds of the eighteenth century, and the early nineteenth century, but despite growing understanding of how to generate and control electrical power through the work of, among others, Ampère and Faraday, economically significant applications beyond the telegraph were difficult to bring about for many decades. From the day Faraday and Hippolyte Pixii built the first dynamos (1831), a multiple of scientists and engineers were occupied in a research effort to tame this phenomenon, which promised so much.

Research in electricity shared the three characteristics of nineteenth-century technological change. First, it was multinational, carried out within a community of scholars that had little interest in national identity but only cared about pressing forward. Second, the epistemic base of the techniques that were being developed was emerging more or less hand in hand with the techniques themselves. Formal mathematics was used successfully next to experiments, and the two reinforced one another. It is a field in which multiple discovery was common, simply because access to the best-practice propositional knowledge was available to all participants. International exhibitions and a rapidly growing periodical literature in electrical engineering were central to the easy and cheap access to knowledge. It was also an area in which patenting was common, in part because the costs of experimentation were often high but above all because it was believed that the economic possibilities were indeed promising. The classic multipurpose technology, electricity, could transform production, transportation, and consumption, as many foresaw.

The innovation that made it all possible came in the late 1860s, when the principle of self-excitation could be applied to generate electricity on a large scale. Many could make a claim to being the discoverer of the dynamoelectric generator (as Werner Siemens called it), more than all perhaps the Englishmen Samuel Alfred Varley and Henry Wilde, but Zénobe Théophile Gramme, a Belgian working in Paris, built the first practical generator in 1870. From there, on a cascade of innovations took place, in which more famous names such as Tesla and Edison were able to build devices that could take advantage of the new form of energy. The impact of electricity on both firms and households was profound above all because it allowed energy to be consumed in infinitesimal quantities at constant cost.

Much like the railway and the telegraph before it, electric power involved network externalities, and the possibilities for coordination failures were many—until the present day different currents, frequencies, and even electrical outlets are still not standardized, as the annoyed traveler knows all too well. The mother of all standardization issues was the "battle of the systems" between AC and DC currents, fought in the 1880s, eventually won by Westinghouse and the AC people in 1890. An electric polyphase motor using alternating current was built by the Croatian-born American Nikola Tesla in 1889, and improved subsequently by Westinghouse. Of equal importance was the transformer originally invented by the Frenchman Lucien Gaulard and his British partner John D. Gibbs and later improved by the American William Stanley who worked for Westinghouse (Hughes, 1983, pp. 86–92; Smil, 2005, pp. 68–74). Tesla's polyphase motor and the Gaulard–Gibbs transformer solved the technical problems of alternating current and made it clearly preferable to direct current, which could not overcome the problem of uneconomical transmission. But electricity also required a great deal of systems building, it was "technology not of concentration but of distribution" (Friedel, 2007, p. 458). It required close cooperation between three kinds of experts: pure scientists and mathematicians, practical inventors without necessarily much theoretical knowledge but with a good "feel" for what worked, and entrepreneurs and organizers such as Emil Rathenau and Samuel Insull.

The impact of electricity on industrial productivity was relatively slow in coming, but there can be little doubt that its consumption transformed society. The story of lighting has been told many times (Bowers, 1998), but the impact of streetcars on daily life and the pattern of urbanization was just as dramatic. In the household, within about 15 year, electricity showed how technology could change cooking, heating, entertainment, cleaning, and the cooling of food and the environment in ways that had never been possible. Social savings logic can be applied to any of those advances, but the problem is that they came in a cluster and interacted with one another. Moreover, because it was a democratic form of energy, electricity allowed the survival of small-scale units who could draw the energy they needed from the networks. In an age in which most technological developments were scale-augmenting and pointed to large size, this development pulled in the other direction.

The development of the internal combustion engine shared some characteristics with steel and electricity, but not its social savings aspect. The world would have managed easily without internal combustion for many decades. External combustion, that is to say, steam engines, were being constantly improved, and much of what was done by gasoline-burning engines could have been carried out by ever more efficient and lighter steam cars. There is no reason why steam-driven tractors, while never quite successful, would not have been perfected to the point where they would have been adopted more generally (although one wonders if engines could have ever been built light enough to fly airplanes). But the internal combustion engine outperformed steam for a variety of uses, and in the long-run doomed steam power. It, too, was an international effort, the first internal combustion engine built by a Belgian, Jean-Etienne Lenoir, and the first theoretical paper pointing out the advantages of a four-stroke engine written by a Frenchman, Alphonse Beau de Rochas. Yet most of the critical components of what we would recognize today as "automobile technology" were developed by Germans, above all Nicolaus August Otto who developed the practical four-stroke engine. Otto was anything but a trained scientist. He was an inspired amateur, without formal technical training. Initially, he saw the four-stroke engine as a makeshift solution to the problem of achieving a high enough compression and only later was his four-stroke principle, which is still the heart of most automobile engines, acclaimed as a brilliant break-through (Bryant, 1967, pp. 650–657). Among the other pioneers were Wilhelm Maybach, a Daimler engineer, who invented the modern float-feed carburetor, and finally Gottfried Daimler and Karl Benz who put it all together. Other technical improvements added around 1900 included the radiator, the differential, the crank starter, the steering wheel, pneumatic tires, and pedal-brake control.

Interestingly enough, the French and the Americans adopted these techniques faster than the Germans ever did, and by 1914 they had far more automobiles per person than in Germany. While the four-stroke engine thus had a complex parenthood, its competitor was a one-man production. Rudolf Diesel was a trained engineer, a good specimen of the "new inventor," trained in science, a "rational" engineer, in search of efficiency above all else. Rather than tinkering and tweaking, Diesel started from first thermodynamic principles He began searching for an engine that incorporated the theoretical Carnot cycle, in which maximum efficiency is obtained by isothermal expansion so that no energy is wasted, and a cheap, crude fuel can be used to boot (originally Diesel used coal dust in his engines). Isothermal expansion turned out to be impossible in practice, and the central feature of Diesel engines today has remained compression-induced combustion, which Diesel had at first considered to be incidental (Bryant, 1969), which created a more efficient if more dirty and noisy engines. These Diesel engines powered German submarines during the First World War, and in the following decades gradually replaced steam engines on ships and locomotives and Otto-type engines on trucks, a classic example of the long-run coexistence of two competing techniques.

Changes in ship design were equally dramatic. As happened in power technology, the push to improve and augment efficiency led to a simultaneous advance of both the old and the new techniques. Despite major improvements in sailing ships in the years 1815–60 culminating in the famous clipper ships, wind power as a propulsive force at sea was eventually relegated to niches in sports and leisure boats. First, the materials of which ships were made changed. In the nineteenth-century shipbuilders like Isambard K. Brunel built ships out of iron. The ultimate technological (if not economic) achievement here was the vast *Great Eastern*, completed in 1859 by Brunel. The ship was six times larger than the largest ship built before and the largest ship built in the nineteenth century. Since the maximum speed of a ship increases with its water line, and iron and steel ships could be made much larger than wooden ships, ships grew bigger, more powerful, and faster at unprecedented rates.

To those advances two critical inventions should be added. One was the screw propellor, another example of multiple invention. The propellor's optimal design was a difficult problem, and it took many years until the most efficient propellors emerged after which the clumsy paddle wheels (which still helped move the *Great Eastern*) disappeared. The nature of technological change at this time is illustrated with a little anecdote: in 1837 a British engineer, Francis Pettit Smith launched a steam ship with a screw propellor made out of wood; in one of the trials half of it broke off. It was noted with amazement that this accident actually *increased* the speed of the vessel (Spratt, 1958, p. 147). The propellor had to move at very high speeds, which required complex and heavy gearing. The reinvention of the steam turbine was critical here (it had first been identified by Hero of Alexandria and mentioned repeatedly in earlier technological writings). It was first suggested in its modern form by the Swede Gustav de Laval (who had intended it to be used in butter and cream production) and Charles Parsons in 1884. Its subsequent improvement led to a revolution at sea: the rotary motion of the turbine could develop enormous speed (the prototype that Parsons built in 1884 ran at 18,000 rpm and had to be geared down), was far more efficient, faster, cleaner, and quieter, than the old reciprocating marine steam engines, and their adoption after 1900, when most of the bugs had been removed, was led by naval ships. Parsons, even more than Diesel, personified the second Industrial Revolution. The first turbine was built simply because "thermodynamics told him that it could be done" (Smil, 2005, p. 16). His steam turbine supplied power to both fast moving ships and electrical generators, both of which depended on high speed power. Like Diesel, Parsons was a trained scientist, with 5 years of mathematics under his belt (at Cambridge), then was trained as an apprentice at various engineering firms. His realization that to make a turbine work well the whole expansion of the steam must be subdivided into a number of stages, so that only comparatively moderate velocities have to be dealt with, still forms the basis of all efficient turbine design. At the famous grand Naval review in celebration of Queen Victoria's 60th anniversary in 1897, his ship the *Turbinia* developed speeds never seen before and ran circles around the naval vessels trying to catch up with it, to the delight of the bigwigs present. It was masterful engineering combined with brilliant public relations. Six years later, the new *Dreadnoughts* adopted direct-drive turbines, as did the gigantic passenger ships built before World War I.

While the typical ship of 1815 was not much different from the vessels of 1650, by 1910 both merchant ships and men-of-war (to say nothing of submarines) had little in common with their steam-operated hybrid predecessors half a century earlier. The result was a sharp decline in transportation costs. In the first half of the nineteenth-century freight rates fell by 0.88% a year, which reflected mostly improvements in sailing ships. The decline after 1850 accelerated to 1.5% a year, rates that are all the more impressive in view of persistently rising labor costs. Despite some organizational improvements, there can be little doubt that the decline in transatlantic freight rates was the result of technological improvements (Harley, 1988).

The drive toward improvement in the second Industrial Revolution affected consumers directly in many ways that had never been anticipated let alone experienced. One effect of the new technology was improved diets. Part of the reason was, of course, improved transportation, which allowed far cheaper agricultural imports from nations with a comparative advantage in food production to reach Europe. European farmers responded by specializing in high-end product lines. Dairy products, fresh meat, and fruits and vegetables became increasingly available. These products, too, were exposed to world competition. The efficient method of preserving beef in transit was by deep freezing it at about 14 °F. In 1876, the French engineer Charles Tellier built the first refrigerated ship, the *Frigorifique*, which sailed from Buenos Aires to France with a load of frozen beef. By the 1880s, beef, mutton, and lamb from South America and Australia were supplying European dinner tables.

Of special interest to the historian interested in economic welfare is the development of food preparation and preservation. Much human suffering has been caused over the ages by nutritional deficiencies and by the unwitting consumption of spoiled foods. Food canning had been invented as early as 1795, but because the process was not understood, the food was overprocessed and tasted poorly. Canned food was already consumed at the Battle of Waterloo, played an important role in provisioning the armies in the American Civil War, and led to more consumption of vegetables, fruit, and meat in the rapidly growing cities. Only when Louis Pasteur's path-breaking discoveries showed why canning worked and the epistemic base of food canning widened in the late nineteenth century, did it become clear that the optimal cooking temperature was about 240 °F and quality improved noticeably. Other food preservation techniques were also coming into use. Gail Borden invented condensed milk powder in the 1850s and helped win the Civil War for the Union and made a fortune in the process. By the end of the century his dehydration idea was also successfully applied to eggs and soups.

In terms of economic welfare, it is hard to argue that any technological development can match the impact of the dramatic improvements in health that took place during the second Industrial Revolution. The statistical evidence from demography seems to bear this out without any question. Between 1870 and 1914 infant mortality in the West declined by about 50%: in France, which was fairly typical, the rate fell from 201 per thousand in 1870 to 111 in 1914. In Germany the corresponding numbers were 298 and 164. Life expectancy at birth increased accordingly, in Britain it went from about 40 to 50 years. This decline, it has been argued, was in part simply due to rising incomes: as people enjoyed higher incomes, they could buy more and better food, live in less congested and better heated dwellings, own better clothes, and had access to running water, sewage, and medical care.

But there was more. The eighteenth century had made a huge effort to fight the many scourges that afflicted people, to the point that some scholars have called the age "a medical enlightenment" (Porter, 1982). Before 1850, the results were disappointing but not altogether absent. Over the entire period, medical progress followed a strange and unbalanced path. It was especially significant in preventive measures. The main advances of this period were the discovery that fresh fruits and vegetables could prevent scurvy, the use of cinchona bark (quinine) to fight off the symptoms of malaria, the prescription of foxglove (now known as digitalis) as a treatment for edemas (first recommended by Dr. William Withering, a member of the Lunar Society, in 1785), the consumption of cod liver to prevent rickets, and above all the miraculous vaccination against smallpox discovered by Edward Jenner in 1796. Jenner's discovery, in many ways, epitomizes the huge changes that had occurred in Europe in the preceding century, which made the application of new useful knowledge an effective tool in improving the material conditions of life. By that time, statistics and probability calculations were already becoming part and parcel of scientific discourse, and Jenner's discovery had to be verified by more systematic minds. Yet on the whole, medical progress was constrained by the narrow epistemic base of the medical profession, and especially the failure to understand the nature of infectious diseases, including their etiologies and modes of transmission.

Clinical treatment made but little progress, and most advances were through the abandonment of useless or harmful practices such as bleeding, purging, and obsessive ventilation. Medical practices before 1914 improved only in isolated areas, such as obstetrics, surgery, and better diagnostic tools like stethoscopes, and had only local effect. The main route to progress in this area before 1860 was through the careful collection of data on the occurrence of diseases and the search for empirical regularities, without much understanding of the mechanisms involved. The development of statistical methods to test for the efficacy of curative technology owed most to Pierre C.A. Louis who developed a "numerical method" for evaluating therapy and in about 1840 provided statistical proof that bloodletting was useless, leading to the gradual demise of this technique (Hudson, 1983, p. 206). A few years later, Ignaz Semmelweis observed on the basis of significant difference in the mortality rate that postnatal puerperal fever was caused by contaminated hands and could be reduced by delivery-room doctors and attendants washing their hands in antiseptic solution. In Britain, the use of statistics in the nineteenth century was heavily relied on by William Farr, superintendent of the statistical department of the Registrar General (Eyler, 1979). After 1850, the use of statistics in public health became almost a rage: between 1853 and 1862 a quarter of all papers read at the Statistical Society of London were on public health and vital statistics (Wohl, 1983, p. 145). The most famous triumph of the "empirical" approach to preventive medicine was the discovery of the water-borne sources of cholera in 1854 by John Snow and William Farr through the quantitative analysis of the addresses of the deceased. At about the same time, William Budd demonstrated the contagious nature of typhoid fever and its mode of transmission and successfully stamped out a typhoid epidemic in Bristol.

The epistemic base of medical care was rapidly augmented, thanks to the sudden growth in the understanding in the nature of infectious disease due to the work of Pasteur, Koch, and their associates. Within a few decades, the medical profession managed to work out a more or less complete theory of infectious disease in which many of the causative agents were identified and their modes of transmission established. The main impact that the new bacteriology had was again in preventive medicine, both public and private. The advances in public medicine in separating drinking water from sewage and preventing other epidemics have been well documented (Szreter, 1988).

Households, too, increasingly realized that by following certain simple "recipes" that involved minor redeployments of resources, they could reduce the incidence of infectious disease. Germs could not be seen, but they could be fought by simple household techniques, available at relatively low cost. Once water was established as a carrier of certain diseases, people began to realize the importance of filtering, boiling, and later chlorination. When insects were identified as a carrier of malaria and yellow fever, a war against insects erupted. Food-borne diseases could be reduced by proper cooking, cleaning, and preservation. All this had to be taught and the teaching took time. Many mistakes were made, wrong turns made, causal mechanisms misidentified, and false recommendations made. Yet when all is said and done, the effects of this technological revolution on human welfare are the most unequivocal: the sharp declines in mortality and morbidity rates in this period speak for themselves.

6. A suggested interpretation

The explosion of modern technology and the concomitant economic modernization has been the subject of a huge multidisciplinary literature, and will continue to fascinate scholars as one of the central issues in long-term analysis. Technological creativity seems to be a uniform and ubiquitous feature of the human species, and yet just once in history has it led to a sea change comparable to a phase transition in physics or the rise of *Homo sapiens sapiens* in evolutionary biology. The Industrial Revolution and the subsequent developments did not just raise the *level* of technological capabilities; they changed the entire dynamics of how innovation comes about and the speeds of both invention and diffusion. For much of human history, innovation had been primarily a byproduct of normal economic activity, punctuated by a periodical flashing insight that produced a macroinvention, such as water mills or the printing press. But sustained and continuous innovation resulting from systematic R&D carried out by professional experts was simply unheard of until the Industrial Revolution.

To create the new dynamic, a lot of things had to come together. This is precisely what happened in the West in the eighteenth century. The Baconian program of the eighteenth century that led into the Industrial Revolution came on top of the heritage of the Renaissance and the Scientific Revolutions of the seventeenth century. To thrive beyond occasional flashes, innovation needed a society with an urban sector and a middle class that produced a level of national income that was sufficiently above subsistence to sustain a class of people whom we would call "professionals"—merchants, engineers, scientists, artists, and professors. A nation that consisted primarily of peasants struggling to survive and soldiers and priests keeping them in check was not likely to create a flow of innovations, although even the European early middle ages, mistakenly dubbed a "dark age," was still capable of creating some remarkable innovations.

But generating a few bursts of technological advances was one thing, creating a world in which sustained progress becomes the rule rather than the exception was another. What happened is that in the slightly amorphous region known as the "West" the dynamic of innovation began to change in the eighteenth century. The notion that uncovering nature's secrets and understanding its regularities and laws were the key to economic progress ripened slowly in the minds of a growing number of influential scientists and *philosophes* in the eighteenth century. In the 1760s, without having as yet much of sense of what was in store, Joseph Priestley reflected in purely Baconian terms on the history of knowledge that it is here that "we see the human understanding to its greatest advantage. . .increasing its own powers by acquiring to itself the powers of nature. . .whereby the security and happiness of mankind are daily improved" (Priestley, 1769, p. iv).

Beyond the gradual expansion of useful knowledge surveyed above, what made the success of the West possible was a set of institutional developments. The new institutional economics has concentrated on constraints on the executive, to make sure that government enforced the rules but did not abuse them. Secure property rights, and limits on the predatory behavior of people in power are seen as the taproot of economic growth (North, 1990, 1995). The historical problem is, of course, that such favorable institutions explain first and foremost the kind of Smithian growth in which the expansion of commerce, credit, and more labor mobility were the main propulsive forces. But the exact connection between institutional change and the rate of innovation seems worth exploring, precisely because the Industrial Revolution marked the end of the old regime in which economic expansion was driven by commerce and the beginning of a new Schumpeterian world of innovation.

The growth of useful knowledge occurred in an institutional context that has been called "a market for ideas" (Mokyr, 2007). The market for ideas is not a real market in the literal sense, but it is a useful metaphor. In it, people with ideas and beliefs tried to sell them to others, acquiring influence and through it prestige. Just as commodity markets can be judged by their efficiency if they, for instance, observe the law of one price, we can define yardsticks for the efficiency of a market for ideas. Three criteria should be emphasized here: *consensus*, *contestability*, and *cumulativeness*.

Markets for ideas can be assessed as to whether there is a built-in tendency to converge to a consensus. Knowledge can be characterized as *tight* when it is held by a wide consensus and with high confidence, in which case it is more likely to lead to applications. Much of the knowledge in the areas crucial to modern economic growth in chemistry, biology, medicine, and physics, and which is held with a reasonably high degree of tightness in modern society, was the subject of heated debates in the seventeenth and eighteenth centuries, the resolution of which was sometimes difficult. Consensus was achieved when there was a widely accepted set of criteria by which hypotheses are accepted or rejected and when the selection environment was relatively stringent, so that there was no room for what is known as "occasionalism." In other words, laws of nature were viewed as firm and there was little interest in miracles or magic. In a sense tightness is self-referential, that is, there has to be consensus about how to achieve consensus. An efficient market for ideas has widely-accepted rhetorical tools by which it assesses experiments, observation, and logical analysis. In the eighteenth century, the assessment of experimental data, mathematical logic, and improved observation tools to bear on deciding what was right matured, even if the rhetorical tools were often social as well as epistemological (Shapin, 1994).

Contestability was political in nature, implying limits on intellectual authority, and is the flip side of consensus. It is the equivalent of the notion of "free entry" in markets. Free entry in preindustrial societies was often blocked by reactionary political forces. The notions of "heresy" and "black magic" were still raised against such eminent scientists such as Jan Van Helmond and Giambattista Della Porta as late as the seventeenth century. Physics and metaphysics were still too closely bound up to be free of coercion, by either religious or secular authorities. Between the execution of Jan Hus in 1415 and the expulsion of the Huguenots in 1685, a great deal of senseless violence and suppression thwarted the contestability of new ideas. By the eighteenth century, such coercion was increasingly abandoned, although the political tensions engendered by the French Revolution and the subsequent wars brought it back for a few decades.

Cumulativeness means an effective means of passing knowledge from generation to generation. Knowledge resides in people's minds and is thus subject to depreciation. Without some mechanism that preserved knowledge and made it available in the future, each generation would have to reinvent a few wheels, and worse, some important knowledge might have been lost. Cumulativeness thus depended on the efficacy of the institutions in charge of passing knowledge from generation to generation, and their technological support in knowledge-storage devices such as books and artifacts (Lipsey et al., 2005,

p. 260). Codifiable knowledge was accumulated through the publication of books and periodicals storing useful knowledge. Here, too, the eighteenth century represented a change in degree more than in essence, but degree is everything in these matters. The age of Enlightenment took a special delight in compiling books that summarized existing knowledge, added sophisticated and detailed drawings that elaborated the operation of technical devices, and placed these books in public libraries. The growth of scientific books and periodicals in the eighteenth century was impressive. An analysis of the topics of the books published in the eighteenth century presented shows that the proportion books published on "Science, Technology and Medicine" increased from 5.5% of the total in 1701–1710 to 9% in 1790–1799. As the absolute number of books published in the British Isles tripled over this period, this implies a quintupling of the total number of such books (Mokyr, 2009).

The growth of technical books, dictionaries, compendia, and encyclopedias were a typical eighteenth-century phenomenon. An example is Thomas Croker's three-volume *Complete Dictionary* published in 1764–1766, which explicitly promised to its readers that in it the "the whole circle of human learning is explained and the difficulties in the acquisition of every Art, whether liberal or mechanical, are removed in the most easy and familiar manner." These works were perhaps the prototype of a device meant to organize useful knowledge efficiently: weak on history and biography, strong on brewing, candle-making, and dyeing. They contained hundreds of engravings, cross-references, and an index. These books and journals circulated widely and the growth of libraries made access to them easier and easier.

Yet cumulativeness could become an encumbrance if it degenerated into orthodoxy, and therefore the third component of efficient knowledge markets, contestability, is critical. No social system of knowledge can work without some notion of authority, but in a well-functioning market for ideas there should be no sacred cows and no belief should be beyond challenge. Market theory teaches us that free entry creates on the whole a more salutary outcome than a monopolist. Hence, it was the combination of cumulativeness with contestability that created the unique environment for the rapid growth of useful knowledge.

The cumulativeness of tacit knowledge operated through different channels, and depended on both formal and informal intergenerational transmission mechanisms. Universities had existed in Europe since the middle ages, and not all of them were concerned with useful knowledge. In the eighteenth century, Oxford and Cambridge had little impact in this regard, but the Scottish Universities taught many useful topics and trained many of the key figures in the Industrial Revolution. But artisanal skills and that mixture of intuition and experience that may best be called a technological *savoir faire* were accumulated and passed on from generation to generation by means of personal relationships, usually father–son or master–apprentice.

To achieve consensus, contestability, and cumulativeness, the intellectual and technological communities that produced useful knowledge needed to be integrated and closely knit. Knowledge had to be distributed and shared, so that it could be compared to existing notions, tested, vetted, and accepted, rejected, or left as undecided. Once absorbed and accepted, it could form the basis of new techniques of production. Integration of that nature required, above all, a freedom from coercion and repression by interest groups with a stake in the intellectual status quo. Yet the main historical phenomenon that made the improvement of these features of useful knowledge possible was the sharp decline in *access costs* (Mokyr, 2005). Access costs are the cost incurred by anyone seeking knowledge from another person or storage device. Access costs consisted of physical costs, affected by such technological advances as the printing press, cheaper paper, postal services, cheaper personal transportation, and of institutional changes such as the development of schools and universities, and the establishment of academies and scientific societies. It was also strongly affected by the emergence of open science and the decline of secrecy in the generation of new useful knowledge (Eamon, 1994).

The decline in access costs had momentous consequences for the characteristics of useful knowledge. It increased the tightness of much knowledge because it became understood that any experiment could be replicated and any theorem's proof thoroughly checked simply because others could easily access any result of interest. Potentially productive ideas were first made accessible to other intellectuals, who could peer-review and criticize them. If found to be acceptable by the rhetorical conventions of the time, they could be extended, recombined with other ideas, and applied. For nonexperts, this setup, at least in theory, increased the reliability of useful knowledge. People in the fields and the workshops might be more likely to make use of ideas that could be trusted because experts had presumably vetted them. In reality, for many decades, this idea remained more wishful thinking than reality, but just the shared ideal resulted in more trust and cooperation than there would have been in its absence.

What emerged in Europe in the century before the Industrial Revolution was an open-source system of knowledge creation, based on priority credit, in which the participants essentially placed their knowledge in the public domain, to be accessed by others, those who could verify it and those who could use it. This system rewarded its most successful participants through a signaling or reputation game, in which the most successful participants were rewarded with patronage and cushy jobs, pensions, or tenured positions in universities (David, 2004). No wonder that the late seventeenth century witnessed the beginnings of some of the fiercest priority fights between scientists.

The institutional context of access also requires consideration. Useful knowledge was organized, as it had to be, in formal and informal organizations. Informal organizations such as the invisible college that preceded the Royal Society or the Lunar Society of Birmingham a century later are justly famous. As noted, formal educational institutions played but a modest role in the growth of useful knowledge, and most of the leading engineers and inventors of the Industrial Revolution were self-taught or were educated privately. Many of the leading figures in the eighteenth century were still enjoying patronage employment, often through the national academies: Euler in St. Petersburg, Reaumur in Paris, and Aepinus in Berlin. Local scientific societies, founded in provincial towns, emerged everywhere in Enlightenment Europe. They were small-scale marketplaces for ideas, where knowledge was exchanged, lectures given, libraries utilized, and many conversations took place. They were a prime example of Habermas's "public sphere" but one that had momentous consequences for the economies in the long run. Scotland occupied a disproportionately large place in the evolution of these institutions. The flourishing of Scottish applied science and engineering in the eighteenth century have prompted some scholars to think of the Scottish tail wagging the English dog (Herman, 2001).

Access costs were, in part, supply-determined by the technology of communication and mobility. As Eisenstein (1979) has stressed, the invention of the printing press played a major role in the intellectual development of Europe, as did the improvements in shipping, navigation, and overall increase in the level of commerce and flow of people and objects after 1492. Less well-known but equally important was the improvement in the continent-wide flow of mail, thanks to the organizational abilities of de Tasso family, led by Francisco de Tasso (later known as Franz von Taxis) and his brothers who established regular postal services in Italy, Germany, and the Habsburg lands in the early sixteenth century. Their postal system covered much of the Continent by the middle of the sixteenth century and created one of the most durable business dynasties in history. Postal rates remained quite high, in part because they were a convenient revenue-raising device for the State. The establishment of the famous London penny post in 1683 and its gradual extension in the eighteenth century meant that by 1764 most of England and Wales received mail daily (Headrick, 2000, p. 187). Postal rates depended, in part, on the cost of internal transportation, and as roads were improved, canals dug, and carriages made faster and reliable, the effectiveness of internal communications increased greatly in the age of Enlightenment.

Consensus and contestability demanded new rhetorical tools. The centuries before the Industrial Revolution could be characterized by a growing irreverence toward to classical authorities and great canons that rules intellectual life in the later middle ages. The Age of Reformation brought significant changes. The iconoclastic physician Paracelsus (a contemporary of Martin Luther) delighted in ridiculing the great medical books, and the French philosopher Pierre de la Ramée wrote a dissertation that roughly translates as "everything Aristotle has taught is mistaken." Instead, consensus was to be attained by new criteria: mathematical logic, careful observation, and experimentation. When knowledge is untight, coercion can play an important role in deciding outcomes in the market for ideas. Part of the platform of the Enlightenment of the eighteenth century was to leave no stone unturned in its efforts to make knowledge tighter by confronting hypotheses with evidence and by allowing more and different evidence as admissible. In this effort, it failed more often than it succeeded before 1800, but the effort itself was significant. By the eighteenth century, these rhetorical rules had themselves become more or less incontestable, but everything else was continuously expected to be challenged.

Contestability, of course, depended on more than access costs, and it is here that Europe's unique position as a politically fragmented yet intellectually coherent region paid off handsomely (Mokyr, 2007). New knowledge had powerful enemies, both those with a strong stake in the status quo and those who feared that intellectual changes would upset carefully calibrated coalitions in an age in which physics and metaphysics were still closely related areas and religious affiliation had significant political ramifications. Yet the astonishing historical fact is that the reaction against innovation was on the whole unsuccessful, and that a mind set of toleration slowly pushed out bigotry and repression in the late seventeenth and eighteenth centuries. Of course, intolerance did not disappear completely, and there was always a risk associated with coming up with radical innovations, both in useful knowledge and in its applications. But it seems fair to say that by the time of the Industrial Revolution, such risks were significantly smaller that they had been in 1600.

How did this come about? It could be argued that in a highly fragmented world of independent political units, many of which were regionally decentralized, freedom and progress could take advantage of a massive coordination failure. Whereas Hus could still be burned through a treacherous alliance between the Pope and the Emperor, after Luther such cases were rare. Unless the reactionary forces resisting innovation could coordinate, it would be very difficult to quell innovation everywhere. Rebellious and unconventional thinkers could and did move around the Continent a great deal, and many intellectuals skillfully played one power against another. Among the many peripatetic intellects on the sixteenth and seventeenth centuries, those of Paracelsus, Comenius, and Descartes stand out (Mokyr, 2006). Moreover, the difficulties of coordination that plagued the conservative powers (even between those ostensibly on the same side, like the Catholic Bourbons and Habsburgs) meant that even if intellectual progress and the Enlightenment could successfully be suppressed in some areas (as it was, e.g., in southern Italy), it could always proceed elsewhere, and leave the repressive countries at a disadvantage.

The net result was that in the eighteenth century, coercion and repression were relegated to marginal roles in the market for ideas. Some Enlightenment figures such as Rousseau and Helvétius published controversial books and had to flee to more tolerant environments, but normally such events were fleeting, and they popped up back in their countries after a few years. It almost seems that after 1750 many of the formerly repressive regimes in Europe adopted a "if you can't beat them join them" attitude toward innovation. To be sure, when these so-called enlightened despots tried to reform their institutions in the 1770s and 1780s, they ran into a great deal of resistance (Scott, 1990). The triumph of Enlightenment forces was sealed by Napoleon's artillery, demonstrating that all innovation contains elements of violence and persuasion.

In any event, as long there were some tolerant environments in Europe, intellectual and technological innovation could not be suppressed effectively and nations that tried to crush novelties found themselves at a disadvantage relative to their neighbors. Enlightenment thinkers realized this well. Edward Gibbon wrote that "Europe is now divided into twelve powerful, though unequal, kingdoms, three respectable commonwealths, and a variety of smaller, though independent, states: the chances of royal and ministerial talents are multiplied, at least, with the number of its rulers... In peace, the progress of knowledge and industry is accelerated by the emulation of so many active rivals; in war, the European forces are exercised by temperate and undecisive contests." (Gibbon, 1789, vol. 3, p. 636). The contradiction between the cosmopolitan and pacifist elements in Enlightenment thought and the realization that political fragmentation and interstate rivalry could be beneficial for innovation because it created a more efficient market for ideas remained one of the unresolved issues of the time.

The eighteenth century saw a number of other fundamental changes in the way the market for ideas worked reach their final stage. One of those was the accommodation between religion and the search for useful knowledge. The Enlightenment cannot be characterized as a purely secular, much less an atheistic, movement. Especially in England, many of the leaders of what we think of as the Enlightened community were quite committed to their religious principles and communities. But religion was optimistic, with a faith in progress and a belief in the benefits of useful knowledge. Jacob (2000, p. 277) has argued that it provided the belief of a rational and enlightened God, "not Calvin's inscrutable and judgmental one." The point was not that religion was irrelevant or even secondary, but rather that the investigation of nature for material purposes was to proceed unencumbered by religion, wherever it may lead. Darwin might not have had it so easy had there been no Enlightenment.

Institutions, of course, mattered. Formal institutions, such as the tax and legal systems, have been widely credited for the economic success of modern Europe, but their connection with technological change is actually anything but transparent. In Britain, the connection of the state with the technological changes in the Industrial Revolution was rather tenuous. In a few instances the government initiated innovation. The best-known case was the Board of Longitude, established in 1714 after a naval disaster caused by faulty navigation. It inspired the perfection of Harrison's chronometer, one of the epochal innovations of the eighteenth century. Another was the Portsmouth dockyards, mentioned above. Demand for military supplies, especially the boring of cannon by John Wilkinson, was obviously a factor in the iron industry. Decades later Henry Bessemer, too, was prompted to his foray into steelmaking by his attempt to make ordnance for the military.

Intellectual property rights and their enforcement have been an important theme in the institutional analysis of innovation. Britain had a patent system that in principle was supposed to encourage and incentivize innovation, but its net effects are still much in dispute and probably were of secondary magnitude (MacLeod and Nuvolari, 2007b). Parliament voted pensions and prizes to a few inventors it deemed to have particularly valuable to society, such as smallpox vaccination, the power loom, and the mule. But in most of the areas where we think government may contribute a great deal to innovation, such as education, government-sponsored research, and investment in transport overhead, the British government did little. Continental governments were more interventionist and *dirigiste*, but even there the government was secondary to private enterprise.

Perhaps the most important thing that British institutions, and after 1815 much of the Western World, contributed to the progress of innovation was what they did *not* do: they did not expropriate the profits of innovators and entrepreneurs. While entrepreneurial activities in the Industrial Revolution were exceedingly risky, it was quite obvious to all that they were reasonably safe from predatory rulers who might have taxed away the rents produced by successful innovation. Britain in the eighteenth century was heavily taxed, but most of the revenues came from excise taxes on middle-class goods such as sugar, tobacco, alcoholic beverages, candles, and so on. Successful industrialists used their gains to purchase country estates, and their children moved into the highest circles. Money bought not only comfort and beauty, but also social prestige (Perkin, 1969, p. 85). If money could be made by innovating, as became increasingly the case, the hope for social advancement created the most powerful incentive of all.

It is worth stressing that there was nothing inexorable about this outcome. It is easy to imagine the survival of short-sighted, bellicose, autocratic regimes, guided by militant mercantilism. To win conflicts over the distribution of resources and the gains from trade, they slaughtered the geese that lay the golden eggs by taxing the wealth of innovators and entrepreneurs or by allowing others (such as tax farmers) to redistribute wealth away from them. Arguably, the Napoleonic wars moved Britain in that direction and perhaps briefly threatened the progress of innovation by threatening personal freedoms. Fortunately, the regime remained committed to innovation and never wavered in its support for entrepreneurs and industrialists at the cutting edge of the Industrial Revolution.

Innovation needed venture capital, and eighteenth-century investors, it seems, were highly risk-averse. Formal capital markets typically invested in government securities and a few public projects such as canals and turnpikes (later railways). Yet oddly, there is not that much evidence that fixed capital goods, embodying the new technology, were especially difficult to accumulate and that their high cost constituted a serious bottleneck on the growth of the early industrializing economies. Economists have been prone to look at the roots of accumulation through the prism of perfect capital markets, summarized by "the" rate of interest. Yet nothing of the sort existed in Britain or elsewhere. Governments and certain public overhead projects such as canals and turnpikes, as well as mortgage borrowers could indeed access capital markets, but innovators were usually excluded from them and needed to rely on their own resources except for short-term credit. These resources constituted, first and foremost, ploughed-back profits, as has been pointed out many times (Cottrell, 1980; Crouzet, 1972). But beyond that, entrepreneurs had access to private informal networks that supplied them with credit on the basis of personal relations and trust. These private networks, often established between members of local associations, freemason lodges, or churches, allowed businessmen to diversify their portfolios by investing in the projects of their relatives, coreligionists, or acquaintances (Pearson and Richardson, 2001). The existence of these networks, dubbed the "associational society" by one historian (Clark, 2000), was another facilitating institutional element of the Industrial Revolution. These networks, of course, imposed *informal* institutions, rules by which people behaved although they were not enforced by third-party organizations such as courts. But they allowed the creation of partnership relationships between innovators and businessmen based on trust, both of whom were expected to behave like "gentlemen." This meant above all that they professed not to be so greedy as to engage in opportunistic behavior. Concerns about reputation assured that the bulk of industrialists, merchants, bankers, professionals, skilled workers, and substantial farmers kept their word and paid their debts.

The typical entrepreneur in the Industrial Revolution, then, was hardly the ferocious, unscrupulous, merciless money grabber that some of the more sentimental accounts make him out to be. Far more important were "to be known and trusted in the locality" and to have "standing in the community" in addition to some form of property (Hudson, 1986, p. 262). Trust created successful partnerships between innovators and entrepreneurs. The classic association of Watt the engineer and Boulton the businessman is the standard example here. There were many others: John Marshall, the Leeds flaxspinner who could rely on his technical manager Michael Murray, and the great engineer Richard Roberts, notorious for having poor business management skills, who had able partners such as Thomas Sharpe and Benjamin Fothergill. Other famous teams were that of Cook and Wheatstone, one the businessman, the other the scientist, as were John Kay and Richard Arkwright, the railroad engineer George Stephenson and his partner and promoter Henry Booth, and the rubber pioneers Thomas Hancock and Charles Macintosh. Equally effective were the partnerships of William Woollett and his brother-in-law Jedediah Strutt, the inventors of the improved knitting frame that could produce ribbed stockings (1758), and that of James Hargreaves (inventor of the spinning jenny) and his employer Robert Peel.

The other mechanism through which informal institutions affected innovation was through training. Britain, as is well known, lagged behind other Continental economies in most areas of human capital formation. Apart from a few enclaves in which a useful education was provided, such as the Scottish universities and dissenting academies, Britain had no engineering schools during the Industrial Revolution, the first ones being established in the 1830s. Yet by all indications it had an advantage in the number and quality of its skilled artisans. These craftsmen provided Britain with the "competence" needed for the advances discussed earlier. In part, this advantage can be explained by geography. As a mining country, there clearly was a need to solve certain well-defined problems such as the design of better pumps. As a sea-faring nation, it needed carpenters, sailcloth makers, ropemakers, and the makers of navigational instruments. Britain also was fortunate to have a large clockmaking industry, many of them Huguenot refugees and their descendants. It had a relatively substantial middle class, people with the means to purchase consumer durables that required skills such as high-quality ceramics, musical instruments, watches, and fancy toys. In the century before the Industrial Revolution that middle-class demand expanded considerably (De Vries, 2008).

On the supply side, human capital in the forms of skilled craftsmen was created primarily through apprenticeships, in which adolescents were trained by accomplished craftsmen. The contract between master and pupil was notoriously incomplete, and lent itself to a great deal of opportunistic behavior (Humphries, 2003). On the Continent, these contracts were normally enforced and administered by guilds, which helped reproduce skills, but often crystallized existing technology. In Britain, craft guilds were weak, and by the eighteenth century did little to enforce apprenticeship relationships. Yet the institution survived and indeed lasted deep into the nineteenth century. The main reason was that the relationship was typically not established between strangers, but between neighbors, friends, business relations, and coreligionists. Mistreating one's apprentice or cheating one's master could bring long-term damage to reputations, on which credit and commerce depended. Information about such behavior was readily accessible in a nation of clubs, societies, and other associations, in which people networked and traded information.

Outside Britain, innovation had to rely on more powerful institutional friends. Continental Europe, as Gerschenkron famously pointed out, relied on large investment banks and government guarantees and subsidies to create the capital required by firms on the technological frontier. Different countries followed different recipes on the way to financing innovation, and it is clear that there was more than one way to skin this cat. In subsidizing research, too, the economies of the Continent faced more activist governments, which subsidized what they believed to be important research, from agricultural research stations in Saxony and Polytechnics in Prussia to the Kaiser Wilhelm Society for the promotion of sciences to the *Grandes Écoles* of France and the Pasteur Institute in Paris which, while in part established through private subscriptions, was under close government supervision. A typical Continental figure was Friedrich Althoff, the powerful undersecretary of the Prussian Ministry of Education who firmly established state control over the system of higher education in Germany between 1882 and 1907 (e.g., vom Brocke, 1991).

A venerable theory linking innovation to the economic environment is the one that relates the direction of innovation to pre-existing factor prices. The theory of induced innovation was famously applied to economic history by H.J. Habakkuk and has been the subject of a considerable literature in the 1960s and 1970s about the impact of high wages on the rate of technological change (Habakkuk, 1962; Ruttan, 2001). A recent attempt by Allen (2008) to revive this approach looks at the cost of labor relative to that of fossil energy, and postulates that labor-saving innovation in Britain was the direct result of the high cost of labor and the cheapness of coal. Coal-using machinery that saved labor made sense in such an economy. Innovation induced by factor prices seems a convincing story, as Ruttan and others have pointed out, when it comes to adoption and diffusion, but much less so when it comes to invention itself. What complicates matters, however, is that adoption and diffusion themselves often involve a great deal of local learning, and generate a flow of specific microinventions that may be crucial to ultimate effects on productivity.

There are both empirical and theoretical difficulties in the argument. One empirical difficulty is that coal was cheap near the mines, but much less so than in Cornwall and in London, where it had to be shipped in from a distance, yet Cornwall and London used steam engines and large amounts of coal. While the steam engine may appear the ultimate labor-saving machine, subsequent improvements were primarily aimed at saving fuel (i.e., capital). In its earliest forms, steam power often was intended to save horse- or water-power rather than labor. Coal mining itself, moreover, was highly labor intensive because until the introduction of compressed air in the late nineteenth century, there was no way to actually introduce labor-saving devices down the shaft. The most remarkable invention in deep-mining technology was Davy's mining lamp, a safety device that saved lives and prevented accidents but did not save labor costs. The other empirical difficulty is that most patentees, when asked the purpose of their invention, failed to mention the saving of labor (MacLeod, 1988, pp. 160–171). Interpreting this evidence is difficult not least because in eighteenth-century Britain, "labor saving" was still a fighting word for many militant artisans who feared that their employment would be jeopardized. Yet even after adjusting for this bias, the proportion of patents that can be classified as labor saving was about 21%.

The theoretical problems are not less. A rather obvious one is that high wages by themselves do not mean that labor was expensive and that labor costs were high. If labor productivity in Britain was higher simply because the quality of labor was higher, for instance because it was physically stronger and/or because it was more skilled, better motivated, or even better drilled and disciplined, then Britain's higher wages, stressed by Allen, would at least in part reflect not the high cost of labor but its quality. Some contemporaries had no qualms about this: Arthur Young, writing in the late 1780s, noted that "labour is generally in reality the cheapest where it is nominally the dearest" (Young, 1790, p. 311).

But the main argument against factor prices (or indeed any kind of initial endowment) affecting technological progress is that it often conflates the *rate* and the *direction* of technological change. The rate or intensity of innovation, much like a car's engine, determined the power of society's technological thrust, while endowments may have steered innovation into a particular direction. Resources are, in Rosenberg's (1976) terminology, a classic "focusing device" but they themselves do not determine the rate of innovation. A coal-intensive economy, given that the other factors facilitating technological creativity were present, may well direct its creativity in some way into coal-using techniques, but the large deposits of coal in, say, the Donetsk region of Ukraine or the Kemerovo region in Russia were not sufficient to create a coal-using technology until it had been developed elsewhere.

The fundamental difficulty with biased or induced innovation, then, is that it makes an inference from a static model about a dynamic setting. Economic theory suggests that factor prices determine technological choices from a given menu of techniques. It does not give us much insight on how the menu was written in the first place, and why some items are on it and others are not (Rosenberg, 1976, pp. 108–125). An ingenious attempt to link the two relies on learning-by-doing and localized innovation (David, 1975). The logic is that at first factor, prices dictate the technique chosen, and as the technique is being used, people acquire experience and make further innovations in the area of the technique being used, rather than in those that "exist" somewhere but are not actually implemented anywhere. This contrasts with the model suggested by Hansen and Prescott (2002), for example (and implied by other endogenous growth models), in which techniques that are "known" but not in use can experience an increase in productivity that switches producers to their use. What is not made explicit in these models is the role of the underlying epistemic base of the technological frontier. If the knowledge base permits developing techniques that do not currently exist but are feasible given the knowledge this society controls, such sudden switches triggered by factor prices or other stimuli may well induce technological change in a particular direction. But if a society does not know how to prospect for coal, how to dig the shafts, how to pump out the water, bring the coal to the surface and then transport it to users at a reasonable price, no amount of abundant coal—no matter how dear labor may be.

Were the great technological advances of this age the result of discrete quantum leaps in knowledge or of small incremental and cumulative microinventions? Putting the question in this form is a kin to asking if a bicycle moves thanks to it front or its rear wheel. The two types of processes were highly complementary. Historically, the path of innovation thus contained elements of both. At times, macroinventions occurred which switched some industries (and in the case of General-Purpose Technologies, entire sectors) onto a wholly new path. Such was the case, as we showed earlier, with sectors such as the chemical industry and electricity in the later nineteenth century. In many others, the technique in use was improved gradually through cumulative microinventions, through extensions and combinations with other techniques, as in textiles and iron. Such distinctions are not easy to make empirically because of the continuous spillovers between techniques and the pervasive effect of GPTs, and so it makes little sense to speak of "new-technology" sectors and "old-technology" sectors. In either case, what really counted was the underlying knowledge that created the technological opportunities and allowed the items on the technological menu to be written. To take advantage of them, other factors had to be available. A car with a powerful engine and flat tires or no coolant is of little use. Yet in the end, it is the engine that determines the car's power. The engine of growth in the economic history of the West was the international cooperative agenda for the accumulation of useful knowledge. It is in its dynamic that the key to economic growth is to be found.

References

Aghion, P., Durlauf, S. (Eds.), (2005). Handbook of Economic Growth, 2 vols. Elsevier, Amsterdam.

Allen, R.C. (2008). The British Industrial Revolution in Global Perspective. Unpublished manuscript, Nuffield College, Oxford University.

Bacon, F. (1996). In: Vickers, B. (Ed.), The Major Works. Oxford University Press, Oxford.

Bacon, F. (1999). In: Sargent, R.-M. (Ed.), Selected Philosophical Works. Hackett Publishing, Indianapolis.

Bagwell, P.S. (1974). The Transport Revolution from 1770. Barnes & Noble, New York, NY.

Berg, M. (2007). "The genesis of useful knowledge". History of Science 45 (Pt. 2, 148), 123–134 (June).

Boussingault, J.B. (1821). "On the presence of silicia platina and its presence in steel". Repertory of Arts, Manufactures, and Agriculture 39 (2nd series), 366–373.

Bowers, B. (1998). Lengthening the Day: A History of Lighting Technology. Oxford University Press, Oxford.

Bowler, P.J., Morus, I.R. (2005). Making Modern Science. University of Chicago Press, Chicago, IL.

Boyle, R. (1744). The Works of the Honourable Robert Boyle, (5 vols). A. Millar, London.

Britnell, R.H. (1996). The Commercialization of English Society, 1000–1500. Manchester University Press, Manchester.

Brock, W.H. (1992). The Norton History of Chemistry. W.W. Norton, New York, NY.

Bryant, L. (1967). "The beginnings of the internal combustion engine". In: Kranzberg, M., Pursell Jr., C.W. (Eds.), Technology in Western Civilization, vol. 1. Oxford University Press, New York, NY, pp. 648–663.

Bryant, L. (1969). "Rudolf Diesel and his rational engine". Scientific American 221, 108–117 (August).

Burton, A. (2000). Richard Trevithick: Giant of Steam. Aurum Press, London.

Cardwell, D. S. L. (1971). From Watt to Clausius: The Rise of Thermodynamics in the Early Industrial Age. Cornell University Press, Ithaca, NY.

Cardwell, D. S. L. (1972). Turning Points in Western Technology. Neale Watson/Science History Publications, New York, NY.

Channell, D.F. (1982). "The harmony of theory and practice: The engineering science of W.J.M. Rankine". Technology and Culture 23 (1), 39–52 (January).

Clark, P. (2000). British Clubs and Societies, 1580–1800: The Origins of an Associational World. Clarendon Press, Oxford.

Clark, G. (2007). Farewell to Alms: A Short Economic History of the World. Princeton University Press, Princeton, NJ.

Clow, A., Clow, N.L. (1952). The Chemical Revolution: A Contribution to Social Technology. Batchworth Press, London Reprinted in 1992, Gordon and Breach, New York, NY.

Collins, R. (1998). The Sociology of Philosophies: A Global Theory of Intellectual Change. Harvard University Press, Cambridge, MA.

Cottrell, P.L. (1980). Industrial Finance, 1830–1914. Methuen, London.

Crouzet, F. (1972). "Capital formation in Great Britain during the Industrial Revolution". In: Crouzet, F. (Ed.), Capital Formation in the Industrial Revolution. Methuen, London.

Darnton, R. (2003). "The unity of Europe". George Washington's False Teeth. W.W. Norton, New York, NY.

Daston, L. (1991). "The ideal and reality of the Republic of Letters in the Enlightenment". Science in Context 4 (2), 367–386.

Daumas, M. (1979). "The rise of the heavy chemical industry". In: Daumas, M. (Ed.), A History of Technology & Invention, vol. III. Crown, New York, NY.

Daumas, M., Gille, P. (1979). "Transportation and communications". In: Daumas, M. (Ed.), A History of Technology & Invention, vol. III. Crown, New York, NY.

David, P.A. (1975). Technical Choice, Innovation and Economic Growth. Cambridge University Press, New York, NY.

David, P.A. (2004). Patronage, Reputation, and Common Agency Contracting in the Scientific Revolution. Unpublished manuscript, Stanford University (August).

Day, L., McNeil, I. (Eds.), (1996). Biographical Dictionary of the History of Technology. Routledge, London.

de Solla Price, D.J. (1984). "Notes towards a philosophy of the science/technology interaction". In: Laudan, R. (Ed.), The Nature of Knowledge: Are Models of Scientific Change Relevant? Kluwer, Dordrecht.

De Vries, J. (2008). The Industrious Revolution: Consumer Behavior and the Household Economy, 1650 to the Present. Cambridge University Press, Cambridge (January).

Eamon, W. (1994). Science and the Secrets of Nature. Princeton University Press, Princeton, NJ.

Eisenstein, E. (1979). The Printing Press as an Agent of Change. Cambridge University Press, Cambridge.

Epstein, S.R. (1998). "Craft guilds, apprenticeship and technological change in pre-industrial Europe". Journal of Economic History 58 (3), 684–713 (September).

Epstein, S.R., Prak, M. (Eds.), (2008). Guilds, Innovation, and the European Economy, 1400–1800. Cambridge University Press, Cambridge.

Eyler, J.M. (1979). Victorian Social Medicine: The Ideas and Methods of William Farr. Johns Hopkins University Press, Baltimore, MD.

Farey, J. (1827). A Treatise on the Steam Engine, Historical, Practical and Descriptive. Longman, Rees, Orme, Brown, and Green, London.

Fouquet, R., Pearson, P. J. G. (2006). "Seven Centuries of Energy Services: The Price and Use of Light in the United Kingdom, 1300–2000". The Energy Journal 27 (1), 139–177.

Friedel, R. (2007). A Culture of Improvement: Technology and the Western Millennium. MIT Press, Cambridge, MA.

Galor, O., Weil, D. (2000). "Population, technology, and growth". American Economic Review 90 (4), 806–828 (September).

Gibbon, E. (1789). The History of the Decline and Fall of the Roman Empire (new ed.). A. Strahan and T. Cadell, London.

Great Britain. (1829). British Parliamentary Papers, vol. III (332), ("Select Committee on State of Law and Practice relative to Granting of Patents for Inventions").

Great Britain. (1841). British Parliamentary Papers, vol. VII (201), ("Select Committee on Operation of Laws affecting the Exportation of Machinery, first report").

Habakkuk, H.J. (1962). American and British Technology in the Nineteenth Century. Cambridge University Press, Cambridge.

Haber, L.F. (1958). The Chemical Industry during the Nineteenth Century. Clarendon Press, Oxford.

Halstead, P.E. (1961–1962). "The early history of Portland cement". Transactions of the Newcomen Society 34, 37–54.

Hansen, G.D., Prescott, E.C. (2002). "Malthus to Solow". American Economic Review 92 (4), 1205–1217 (September).

Harley, C.K. (1988). "Ocean freight rates and productivity 1740–1913: The primacy of mechanical invention reaffirmed". Journal of Economic History 48, 851–876 (December).

Harris, J.R. (1998). Industrial Espionage and Technology Transfer: Britain and France in the Eighteenth Century. Ashgate, Aldershot.

Headrick, D.R. (1989). The Invisible Weapon: Telecommunications and International Politics, 1851–1945. Oxford University Press, New York, NY.

Headrick, D.R. (2000). When Information Came of Age: Technologies of Knowledge in the Age of Reason and Revolution, 1700–1850. Oxford University Press, New York, NY.

Herman, A. (2001). How the Scots Invented the Modern World. Crown, New York, NY.

Hilaire-Pérez, L. (2007). "Technology as public culture". History of Science 45 (Pt. 2, 148), 135–153 (June).

Hudson, R.P. (1983). Disease and Its Control: The Shaping of Modern Thought. Greenwood Press, Westport, CT.

Hudson, P. (1986). The Genesis of Industrial Capital: A Study of West Riding Wool Textile Industry, c. 1750–1850. Cambridge University Press, Cambridge.

Hughes, T.P. (1983). Networks of Power: Electrification in Western Society, 1880–1930. Johns Hopkins Press, Baltimore, MD.

Humphries, J. (2003). "English apprenticeship: A neglected factor in the first Industrial Revolution". In: David, P.A., Thomas, M. (Eds.), The Economic Future in Historical Perspective. Oxford University Press, Oxford, pp. 73–102.

Jacob, M.C. (2000). "Commerce, industry, and the laws of Newtonian Science: Weber revisited and revised". Canadian Journal of History 35 (2), 275–292 (August).

Jacob, M.C. (2007). "Mechanical science of the shopfloor". History of Science 45 (2), 197–221 (June).

Kerker, M. (1960). "Sadi Carnot and the steam engine engineers". Isis 51 (3), 257–270 (September).

Lipsey, R.G., Carlaw, K.I., Bekar, C.T. (2005). Economic Transformations: General Purpose Technologies and Long-term Economic Growth. Oxford University Press, Oxford.

Lucas, R.E. (2002). Lectures on Economic Growth. Harvard University Press, Cambridge, MA.

Lyons, H.G. (1944). The Royal Society, 1660–1940, a History of Its Administration under Its Charters. Cambridge University Press, Cambridge.

MacLeod, C. (1988). Inventing the Industrial Revolution: The English Patent System, 1660–1880. Cambridge University Press, Cambridge.

MacLeod, C., Nuvolari, A. (2007a). "Glorious Times: The Emergence of Mechanical Engineering in Early Industrial Britain, c. 1700–1850". Unpublished manuscript.

MacLeod, C., Nuvolari, A. (2007b). "Inventive Activities, Patents, and Early Industrialization: A Synthesis of Research Issues". Unpublished manuscript.

Mantoux, P. (1928). The Industrial Revolution in the Eighteenth Century. Harper Torchbooks, New York, NY. Revised ed., 1961.

Marsden, B. (2004). Rankine, William John Macquorn (1820–1872). Oxford Dictionary of National Biography. Oxford University Press, Oxford.

Marsden, B., Smith, C. (2005). Engineering Empires: A Cultural History of Technology in Nineteenth-Century Britain. Palgrave MacMillan, Houndmills, Basingstoke.

Mokyr, J. (2001). "The rise and fall of the factory system: Technology, firms, and households since the Industrial Revolution". Carnegie–Rochester Conference Series on Public Policy, vol. 55, pp. 1–45 (December).

Mokyr, J. (2002). The Gifts of Athena: Historical Origins of the Knowledge Economy. Princeton University Press, Princeton, NJ.

Mokyr, J. (2005). "The intellectual origins of modern economic growth [Presidential address]". Journal of Economic History 65 (2), 285–351 (June).

Mokyr, J. (2006). "Mobility, creativity, and technological development: David Hume, Immanuel Kant and the economic development of Europe. Prepared for the session on "Creativity and the Economy"" In: Abel, G. (Ed.), Kolloquiumsband of the XX. Deutschen Kongresses für Philosophie, Berlin, German Association of Philosophy, Berlin, pp. 1131–1161.

Mokyr, J. (2007). "The market for ideas and the origins of economic growth in eighteenth century Europe". Tijdschrift voor Sociale en Economische Geschiedenis 4 (1), 3–38.

Mokyr, J. (2009). The Enlightened Economy. Yale University Press, London.

Mokyr, J., Voth, J. (2010). "Understanding growth in Europe, 1700–1870: Theory and evidence". In: Broadberry, S., O'Rourke, K. (Eds.), The New Cambridge Economic History of Europe, Volume 1, (in press).

Morrell, J., Thackray, A. (1981). Gentlemen of Science: Early Years of the British Association for the Advancement of Science. Oxford University Press, Oxford.

Mowery, D., Rosenberg, N. (1989). Technology and the Pursuit of Economic Growth. Cambridge University Press, Cambridge.

Murmann, J.P. (2003). Knowledge and Competitive Advantage. Cambridge University Press, Cambridge.

Murmann, J.P., Landau, R. (1998). "On the making of competitive advantage: The development of the chemical industries of Britain and Germany since 1850". In: Arora, A., Landau, R., Rosenberg, N. (Eds.), Chemicals and Long-term Economic Growth. John Wiley & Sons, New York, NY.

Musson, A.E. (1975). "Joseph Whitworth and the growth of mass-production engineering". Business History 17, 109–149 (July).

Nasmyth, J. (1841). Remarks on the introduction of the slide principle in tools and machines employed in the production of machinery Appendix B in Robertson Buchanan (3rd. ed., with additions by George Rennie). Practical Essays on Mill Work and Other Machinery, vol. 2. J. Weale, London.

North, D.C. (1990). Institutions, Institutional Change, and Economic Performance. Cambridge University Press, Cambridge.

North, D.C. (1995). Understanding the Process of Economic Change. Princeton University Press, Princeton, NJ.

Pearson, R., Richardson, D. (2001). "Business networking in the Industrial Revolution". Economic History Review 54 (4), 657–679.

Perkin, H.J. (1969). The Origins of Modern English Society, 1780–1880. Routledge & Kegan Paul, London.

Polanyi, K. (1944). The Great Transformation. Beacon Press., Boston, MA Reprinted in 1985.

Porter, R. (1982). "Was there a medical Enlightenment?" British Journal for Eighteenth-Century Studies 5, 49–63.

Priestley, J. (1769). The History and Present State of Electricity, with Original Experiments (second ed., corrected and enlarged). J. Dodsley, London.

Rosenberg, N. (1976). Perspectives on Technology. Cambridge University Press, Cambridge.

Rosenberg, N. (1982). "How exogenous is science?" Inside the Black Box: Technology and Economics. Cambridge University Press, Cambridge.

Rosenberg, N. (1996). "Uncertainty and technological change". In: Fuhrer, J.C., Little, J.S. (Eds.), Technology and Growth. Conference Series, no. 40. Federal Reserve Bank of Boston, Boston, MA, pp. 91–110.

Rosenberg, N., Trajtenberg, M. (2004). "A general-purpose technology at work: The Corliss steam engine in the late-nineteenth-century United States". The Journal of Economic History 64 (1), 61–99 (March).

Ruttan, V.W. (2001). Technology, Growth and Development: An Induced Innovation Perspective. Oxford University Press, New York, NY.

Scott, H.M. (Ed.), (1990). Enlightened Absolutism: Reform and Reformers in Later Eighteenth-Century Europe. Palgrave MacMillan, Houndmills, Basingstoke.

Shapin, S. (1994). The Social History of Truth. University of Chicago Press, Chicago, IL.

Smil, V. (2001). Enriching the Earth: Fritz Haber, Carl Bosch and the Transformation of World Food Production. MIT Press, Cambridge, MA.

Smil, V. (2005). Creating the Twentieth Century: Technical Innovations of 1867–1914 and Their Lasting Impact. Oxford University Press, Oxford.

Smith, A. (1757). Lectures on Jurisprudence. Clarendon Press, Oxford Reprinted in 1978.

Smith, A. (1776). Cannan, E. (Ed.), The Wealth of Nations. University of Chicago Press, Chicago, IL Reprinted in 1976.

Smith, C.S. (1964). "The discovery of carbon in steel". Technology and Culture 5 (2), 149–175 (Spring).

Smith, C. (1990). "Energy". In: Olby, R.C., Cantor, G.N., Christie, J. R. R., Hodge, M. J. S. (Eds.), Companion to the History of Science. Routledge, London.

Snooks, G. (1994). "The Industrial Revolution revisited". In: Snooks, G.D. (Ed.), Was the Industrial Revolution Necessary? Routledge, London.

Spratt, H.P. (1958). "The marine steam-engine". In: Singer, C., Holmyard, E.J. (Eds.), A History of Technology, vol. V, pp. 141–156.

Stewart, L. (2007). "Experimental spaces and the knowledge economy". History of Science 45 (148), 155–177 (June).

Szreter, S. (1988). "The importance of social intervention in Britain's mortality decline, c. 1850–1914: A re-interpretation of the role of public health". Social History of Medicine 1, 1–37.

Temin, P. (1997). "Two views of the British Industrial Revolution". Journal of Economic History 57 (1), 63–82 (March).

Toynbee, A. (1884). Toynbee's Industrial Revolution: A Reprint of Lectures on the Industrial Revolution. David & Charles, New York, NY Reprinted in 1969.

Tweedale, G. (2004). "Bell, Sir (Isaac) Lowthian, first baronet (1816–1904)". Oxford Dictionary of National Biography. Oxford University Press, Oxford.

vom Brocke, B. (1991). "Friedrich Althoff: A great figure in higher education policy in Germany". Minerva 29 (3), 269–283 (September).

Weber, M. (1923). General Economic History. Collier Books, New York, NY. Reprinted in 1961.

Weber, W. (2003). "Science, Technology, and Society in Germany from 1800 to the Present". In: Ogilvie, S., Overy, R. (Eds.), Germany: A New Social and Economic History, vol. III. Arnold, London.

Wells, H.G. (1923). Men Like Gods. Cassell, London.

Wohl, A.S. (1983). Endangered Lives: Public Health in Victorian Britain. Harvard University Press, Cambridge.

Wrigley, E.A. (2004). Poverty, Progress, and Population. Cambridge University Press, Cambridge.

Young, A. (1790). In: Maxwell, C. (Ed.), Travels in France during the Years 1787, 1788, and 1789. Cambridge University Press, Cambridge. Reprinted in 1929.

Chapter 3

TECHNICAL CHANGE AND INDUSTRIAL DYNAMICS AS EVOLUTIONARY PROCESSES

GIOVANNI DOSI* *AND* RICHARD R. NELSON†

**Laboratory of Economics and Management*

Sant'Anna School of Advanced Studies

Pisa, Italy

†*Institute for Social and Economic Research and Policy*

Columbia University, New York

New York, USA

Contents

Handbooks in Economics, Volume 01

DOI: 10.1016/S0169-7218(10)01003-8

Abstract

This chapter reviews and integrates much of what has been learned on the processes of technological evolution, their main features, and their effects on the evolution of industries.

First, we map and integrate the various pieces of evidence concerning the nature and structure of technological knowledge, the sources of novel opportunities, the dynamics through which they are tapped, and the revealed outcomes in terms of advances in production techniques and product characteristics. Explicit recognition of the evolutionary manners through which technological change proceeds has also profound implications for the way economists theorize about and analyze a number of topics central to the discipline.

One is the theory of the firm in industries where technological and organizational innovation is important. Indeed a large literature has grown up on this topic, addressing the nature of the technological and organizational capabilities which business firms embody and the ways they evolve over time. Another domain concerns the nature of competition in such industries, wherein innovation and diffusion affect growth and survival probabilities of heterogeneous firms. The processes of knowledge accumulation and diffusion involve winners and losers, changing distributions of competitive abilities across different firms, and, with that, changing industrial structures. Both the sector-specific characteristics of technologies and their degrees of maturity over their life cycles influence the patterns of industrial organization—including size distributions, degrees of concentration, relative importance of incumbents and entrants, etc. This is the second set of topics which we address.

Finally, in the conclusions we briefly flag some fundamental aspects of economic growth and development as an innovation-driven evolutionary process.

Keywords

innovation, technological paradigms, technological regimes and trajectories, evolution, learning, capability-based theories of the firm, selection, industrial dynamics, emergent properties, endogenous growth

JEL classification: O30, O31, O32, O33, L16, L20

1. Introduction

A wide ensemble of scholars, in both economics and several other disciplines, have been studying technological advance, viewed as an evolutionary process. This perspective on technological change is closely linked to recent research on industrial dynamics and on economic growth as processes intertwined with and driven by technological and organizational innovation. In this chapter, we lay out the basic premises of this research and review and integrate much of what has been learned on the processes of technological evolution, their main features, and their effects on the evolution of industries.[1]

The proposition that technology advances through an evolutionary process is not a new idea. Nearly 300 years ago, Bernard de Mandeville, pointing to what he regarded as one of the most complex and sophisticated artifacts of his era, the (then) modern Man of War [the warship], explained how its design came about this way:

> *"What a Noble as well as Beautiful, what a glorious Machine is a First-Rate Man of War. ...We often ascribe to the Excellency of Man's Genius, and the Depth of his Penetration, what is in reality owing to the length of Time, and the Experience of many Generations, all of them very little differing from one another in natural Parts of Sagacity." (Mandeville, 1714, vol. II, pp. 141–142)*[2]

Note also that Adam Smith begins *The Wealth of Nations* by highlighting the importance of technological advance to economic growth, and discusses the processes involved in a way that anticipates modern evolutionary analyses. In his interpretation of the factors behind the enormous improvements in workers' productivity—in general, and in his pin making example, in particular—Smith proposes that a key driving force has been:

> *"...the invention of a great number of machines which facilitate and abridge labour, and enable one man to do the work of many." (Smith, 1776, p. 17)*

In turn,

> *"a great part of the machines 'made use of' in those manufactures in which labour is most subdivided were originally the invention of common workmen, who being each of them employed in some very simple operation, naturally turned their thoughts toward finding easier and readier methods of performing it." (Smith, 1776, p. 20)*

Together,

> *"many improvements have been made by the ingenuity of the makers of the machine...and some by that of those who are called philosophers or men of speculation, whose trade it is, not to do anything but to observe everything; and who, upon that account are often capable of combining together the powers of the most distant and dissimilar objects." (Smith, 1776, p. 21)*

[1] Earlier reviews and discussions in a germane spirit upon which we build are Dosi (1988, 1991, 1997), Cimoli and Dosi (1995), Dosi and Nelson (1994), Dosi et al. (2005b), Nelson (1981, 1996, 1998, 2005), Freeman (1982, 1994), Nelson and Winter (1977, 2002), and Dosi and Winter (2002); more specifically on evolutionary theories of economic growth, see also Silverberg and Verspagen (2005b), and on evolutionary models within an ACE modeling perspective, see the detailed survey in Dawid (2006).

[2] On Mandeville as a precocious evolutionary economist, see Rosenberg (1963).

The processes through which "modern" warship design came to be and productivity was improved, both via "learning by doing"—we would say nowadays—and through the development of new machines, that Mandeville and Smith are suggesting clearly are "evolutionary," in the broad sense of the term that we will develop shortly.

To return to Mandeville's discussion of the evolution of the design of the modern battleship, he does not deny the purpose and competence of those who are designing warships at any time. On the other hand, he clearly is denying that the state of the art in this arena at his time was the result of great sagacity and creativity on the part of a small number of individuals, much less coherent rational planning, and proposing rather that it was the product of many minds and many generations of designers, each working somewhat myopically, with later generations building on the achievements and learning from the mistakes of earlier ones.

That is, Mandeville, as most contemporary scholars analyzing technological advance as an evolutionary process, departs from any assumption of strong "rationality," in the sense either of a fully informed global scan of alternatives made by inventors at any time, or accurate forward-looking technological expectations. The ubiquitous presence of drivers of behavior distinct from strong rationality in the above sense will be indeed a first recurring evolutionary theme in the interpretations of technological and economic change that follow.

A second theme well in tune with evolutionary ideas which will repeatedly appear in our discussion is the emphasis on disequilibrium dynamics as a general feature of "restless capitalism," as Stan Metcalfe put it. As in the case of Smith's "practical men" and "philosophers", the search for new techniques of production and new products (as well as many other economic behaviors—including investment, pricing, production decisions) most often entail trials and errors, gross mistakes, and unexpected successes. This applies also to industrial organization and industrial change: also at this level of analysis, an evolutionary perspective focuses upon the processes by which firms persistently search for and adopt new technologies as well as new organizational forms and new behavioral patterns as means of gaining advantages over their competitors, and upon the features of the competitive process driving the growth, the decline and possibly the disappearance of various firms.

A third theme regards the identification of the regularities in the processes of technological and industrial change, notwithstanding the lack of an *ex ante* commitment to any equilibrium notion. For example, can we identify some relatively invariant patterns in the processes of innovation? How are innovations selected? What are the relationships between technologies and forms of corporate organization? And between technical change and forms of competition? How can one characterize the ways through which relatively orderly processes of industrial change emerge out of underlying "disequilibrium" behaviors? What is the relative role of "chance" and "necessity" in evolutionary processes, and relatedly, to what extent is techno-economic evolution path-dependently shaped by events occurring along its historical unfolding? In which ways do institutions and policies embed the processes of technological and economic change?

Come as it may, as Freeman (1982) already noted, since the classics not much progress had been made for almost two centuries in our understanding of the ways new technical knowledge is generated and its impact works through the economy. Karl Marx and Joseph Schumpeter stand out as major exceptions, but they were rather lonely voices.[3] The importance of technological change reappeared, almost

[3] Alfred Marshall too offered rich insights into the evolution of industries even if the subsequent systematization of his contribution builds on an equilibrium skeleton.

by default, in Robert Solow's growth analysis in the 1950s, but it is only over the last 40 years that one has systematically started looking—using the felicitous expression of Nate Rosenberg—inside the "blackbox of technology," investigating the sources of novel opportunities, the dynamics through which they are tapped and the revealed outcomes in terms of advances in production techniques and product characteristics. The first part of this chapter maps and integrates such pieces of evidence. Explicit recognition of the evolutionary manners through which technological change proceed has also profound implications for the way economists theorize about and analyze a number of topics central to the discipline.

One is the theory of the firm in industries where technological innovation is important. Indeed a large literature has grown up on this topic, addressing the nature of the technological and organizational capabilities which business firms embody and the ways they evolve over time.

Another domain concerns the nature of competition in such industries, wherein innovation and diffusion affect growth and survival probabilities of heterogeneous firms, and, relatedly, the determinants of industrial structure. The processes of knowledge accumulation and diffusion involve winners and losers, changing distributions of competitive abilities across different firms, and, with that, changing industrial structures. Both the sector-specific characteristics of technologies and their degrees of maturity over their life cycles influence the patterns of industrial organization—including size distributions, degrees of concentration, relative importance of incumbents and entrants, etc. This is the second set of topics which we shall address below.

Third, the full acknowledgment of technical change as an evolutionary process bears distinct implications also for the understanding of the processes of economic growth, fuelled as they are by technological and organizational innovation. The "physiology" of modern capitalism rests on the evolution of multiple technologies and industries coupled with each other via input–output and knowledge flows. Some sectors shrink, others expand, yet other new ones appear generally associated with the emergence of radically new technologies. Overall, the patterns of growth of modern economies—with both their secular increase in per-capita productivity and incomes and their fluctuations and discontinuities—are deeply shaped by the underlying patterns of technological and organizational evolution. In Section 5, we shall offer some comments on these points.

The foregoing domains of analysis define also the structure of this work, which will start from some basic notions on the nature of technologies (Section 2) and the analysis of how technologies evolve (Section 3) together with a brief discussion of how technologies are embedded into business organizations and of the implications of all that for the theory of the firm (which is discussed from the angle of strategic management in Chapter 16). Next, we will explore the coupled dynamics of technological change and industrial evolution (Section 4). Finally, in Section 5, we shall briefly flag some fundamental aspects of economic growth and development as an innovation-driven evolutionary process.

First of all, to set the stage we need to briefly discuss what we mean by "technology."

2. On the nature of "technology"

In the most general terms, a technology can be seen as a human designed means for achieving a particular end—being it a way of making steel like the oxygen process, a device to process information such as a computer, or the ensemble of operations involved in heart surgery. These means most often entail particular pieces of *knowledge*, *procedures*, and *artifacts*. These different aspects offer different but complementary ways of describing technologies.

2.1. Technology and information

What are the characteristics of technological knowledge?

It is useful to take as starting points some very basic features shared by technological knowledge and *information*, in general.[4]

First, technological knowledge (even when taken to be equal to information) is nonrivalrous in use. Use by one economic agent in no way by itself reduces the ability of other economic agents to use that same knowledge.

Second, there is an intrinsic indivisibility in the use of information (half of a statement about whatever property of the world or of a technology is not worth half of the full one: most likely it is worth zero).

Third, both technological knowledge and sheer information involve high up-front generation cost as compared with lower cost in their repeated utilization, when the technology is "in place" (with "being in place" roughly meaning "with practitioners and organizations actually mastering and using it"). Moreover, information *stricto sensu* typically displays negligible cost of reproduction, which closely relates (but is not identical) to the proposition that information can be used on any scale (greater or equal than one). In fact, there is something genuinely special of information in general and also of technical knowledge in that they share a sort of notional *scale-free* property. So, in a first approximation (not to be taken too literally: see below), an "idea" when fully developed does not imply any intrinsic restriction on the scale of its implementation. In a language which we do not particularly like, were there a "production function" with information as the only input, it would display an output equal to zero for an information below "one unit" and a vertical line for information equal one.[5]

Fourth, as a consequence, there is a fundamental increasing returns property to the use of information and technological knowledge. The use of standard economic goods, ranging from shoes to machine tools, implies that use wears them out. This does not apply either to information or to technological knowledge. On the contrary, the persistent use of either implies at the very least its nondepreciation, at least in technical terms (their economic value is a different matter).

Indeed, important branches of contemporary economic theory are finally beginning to take on board the implications of having information as a fundamental input in all economic activities: other chapters in this Handbook address the advances in the fields such as "new growth" and "new trade" theories, informational externalities, and standard setting, incorporating increasing returns implications which the economic use of information intrinsically imply.[6] And such exploration is far from over.

[4] For the basics and several ramifications of the economics of information, see Arrow (1962a), Nelson (1959), Simon (1962), Akerlof (1984), Greenwald and Stiglitz (1986), and Radner (1992, 1993) among others.

[5] Compare with Romer (1994) for a discussion of the implications for (new) growth theories.

[6] The properties of information and its distribution—most likely imperfect, incomplete and asymmetric—across a multiplicity of economic agents bears also fundamental macroeconomic consequences which cannot be explored here: however the interested reader may appreciate the intuitive compatibility between analyses such as Greenwald and Stiglitz (1986) and Stiglitz (1994), on the one hand, and the microeconomics of production, competition, and economic change put forward in this chapter, on the other.

Notice also that even neglecting the features of technologies which are different from pure "information" (on which more below), the nonrival use, upfront generation cost, and indivisibility characteristics of the latter bear far-reaching implications for any theory of economic coordination and change. As Arrow (1996) emphasizes:

"[c]ompetitive equilibrium is viable only if production possibilities are convex sets, that is do not display increasing returns," but ... "with information constant returns are impossible" (p. 647). "The same information [can be] used regardless of the scale of production. Hence there is an extreme form of increasing returns." (p. 648)

Needless to say, a fundamental consequence of this statement is the tall demand of providing accounts of economic coordination which do not call upon the properties of competitive equilibria. We shall see later the progress done by evolutionary-inspired theories.

Granted the foregoing properties of technology/information, technological knowledge has important characteristics of its own, highlighted by a body of interpretation pioneered in the 1960s and 1970s by Christopher Freeman in the United Kingdom and a few scholars in the United States, which could be called the "Stanford–Yale–Sussex (SYS) synthesis" (cf. Dosi et al., 2006b) based on the locations where at the time most of the major contributors were based. In brief, such an interpretation takes on board the basic intuitions on the economics of information already present in Arrow (1962a) and Nelson (1959), and further refinements (cf. David, 1993, 2004 among a few others), together with works focusing on the specific features of technological knowledge (including Dosi, 1982, 1988; Freeman, 1982, 1994; Freeman and Soete, 1997; Mowery and Rosenberg, 1989; Nelson, 1962, 1981; Nelson and Winter, 1977, 1982; Pavitt, 1987, 1999, 2005; Rosenberg, 1976, 1982; Winter, 1982, 1987, 2005, 2006a). In such a synthesis, one fully acknowledges some common features of information and knowledge—in general, and with reference to scientific and technological knowledge in particular. Together, however, one also distinguishes the specific characteristics of technological knowledge and of the ways it is generated and exploited in contemporary economies.

In the case of technology, it may well be that even if a body of knowledge might be *notionally* utilizable on any scale (say, a production process which can be applied ten or a million times), this does not imply that replication or imitation is necessarily easy and cheap (see Winter, 2005, 2006a; Winter and Szulanski, 2001, 2002). As we shall see at greater detail below, in the case of technological knowledge the "scale-free reproduction property" is subject to three major qualifications.

Certainly, *first*, the nonrivalry in use implies *nondepletability by reproduction or by transfer* of both scientific and technological knowledge: of course Pythagoras' theorem is depleted neither by repeated use by Pythagoras himself nor by learning on the part of his disciples. This property, however, is quite distinct from the easiness and costs of replication: this applies to the costs of teaching the theorem itself and, more so, to technological knowledge, concerning, say, the fine working of a plant *even within the same firm.*

Second, scientific and, even more so, technological knowledge share, to different extents, some degrees of *tacitness* (more on it below). This applies to the pre-existing knowledge leading to any discovery and also to the knowledge required to interpret and apply even codified information after it is generated. As Pavitt (1987) puts it with regards to technological knowledge:

"most technology is specific, complex...[and] cumulative in its development... It is specific to firms where most technological activity is carried out, and it is specific to products and processes, since most of the expenditures is not on research, but on development and production

engineering, after which knowledge is also accumulated through experience in production and use on what has come to be known as 'learning by doing' and 'learning by using'." (p. 9)

Moreover,

"the combination of activities reflects the essentially pragmatic nature of most technological knowledge. Although a useful input, theory is rarely sufficiently robust to predict the performance of a technological artefact under operating conditions and with a high enough degree of certainty, to eliminate costly and time-consuming construction and testing of prototype and pilot plant." (p. 9)

Notice that given these features of technological knowledge, equating it to a *pure* "public good" might be quite misleading. While the characteristic of being nonrivalrous in use means that there are significant benefits to society as a whole if developed technologies were open for all to try to master and employ, even when there are no explicit barriers to use, there usually are non trivial costs to acquiring the relevant capabilities (see below on technological heterogeneity among firms, bearing far-reaching implications also in terms of growth and development theories).

The easiness and cost of replication across diverse economic actors is generally positive, often quite significant, and varies a lot too. In fact, as we shall see, the conditions and costs for replicability and imitation are important distinguishing marks of different technologies. Hence, in the technological domain the "scale-freeness" should not be taken too literally: "scaling-up" is by itself a challenging learning activity, often associated with the quest for economies of scale (see Section 3 on technological trajectories, and Winter, 2008).

Knowledge differs from sheer information in its *modes and costs of replication* (see Winter and Szulanski, 2001, 2002; for insightful discussions). While the metaphor of "reproduction of ideas" is just pushing a button on the computer with the instruction "copy" and possibly "send," the replication of technological knowledge concerning processes, organizational arrangements, and products is a painstaking and often quite expensive business (see Mansfield et al., 1981 among others). The bottom line is that even when there is an *Arrow core*, as Winter and Szulanski (2002) put it, in the sense of an informationally codifiable template, the actual process of reproduction involves significant efforts, costs, and degrees of uncertainty about the ultimate success—all linked also with the tacit elements involved in technological know-how.

All this bears important consequences also in terms of the theory of production.[7] The divisibility axiom is certainly not on the cards as a plausible assumption, in that even "ideas"—let alone "technologies"—bear the mark of "indivisibility": "half an idea"; to repeat, is certainly not of half the usefulness of a whole idea. And, together, technologies are ridden with indivisibilities of machines, plants, headquarters, etc. Conversely, "additivity"—under some important caveats—may stand (much more in the insightful discussion by Winter, 2008).

As Winter (1987) suggests, taxonomies based on different degrees of tacitness together with other dimensions provide a useful interpretative grid by which to classify different types of knowledge.

Tacitness refers to the inability by the actor(s), or even by sophisticated observers, to explicitly articulate the sequences of procedures by which "things are done," problems are solved, behavioral patterns are formed, etc. (see Dosi et al., 2005a; Nelson and Winter, 1982, especially Chapter 4; Polanyi, 1967, and the

[7] For more details, see Winter (1982, 1987, 2005), Nelson (1981), and Dosi and Grazzi (2006) among others.

references therein). In a nutshell, tacitness is a measure of the degree to which "we know more than we can tell."[8] In turn, the different degrees of tacitness of particular bodies of knowledge and the dynamics of knowledge codification bear ramified implications in terms of patterns of innovation, division of labor and presence/absence of "markets for technology." For example, interorganizational division of labor often requires a good deal of codification of "who does what," and even more codification is needed for the existence of a market for technologies, if by that we mean a market for pieces of knowledge which can be put to use by someone other than the originator of the technology itself, and which can be an object of negotiation and exchange (Arora and Gambardella, 1994; Arora et al., 2002; Granstrand, 1999; Chapter 15 in this handbook).

More generally, technological activities draw upon specific elements of knowledge, partly of the know-how variety and partly of a more theoretical kind. In fact as we shall see below, important advances have been made over the last quarter of a century in the identification across different technologies of (a) the *characteristics* of such knowledge—for example, to what extent is it codified and openly available in the relevant professional communities as distinct from the tacit skills of the actors themselves—and (b) its *sources*—does it come from external institutions such as universities and public laboratories, from other industrial actors such as suppliers and customers, or is it endogenously accumulated by the people and organizations who actually use it.[9]

Regarding the sources of technological knowledge, the reconstruction of the diverse institutional origins of novel learning opportunities helps also in going beyond any first, very rough, representation of "endogenous" versus "exogenous" technical progress. For the time being, let us stick to the basic notion that in no technological activity "knowledge drops for the sky." Even in the most science-based sectors, a good deal of technological advances are endogenously generated by more "applied," task-focused organizations. At the same time most if not all of the activities which have experienced the highest rates of technological progress, at least over the last half-century, are also those which have been also fuelled by "exogenous" scientific advances.

To understand both the nature and the dynamics of technological knowledge, a crucial step regards the understanding of *where technological knowledge resides* and how it is expressed, stored, improved upon (see Section 3). In that, the account of technology in terms of pieces of knowledge, their combinations and their changes has to be complemented by a more operational representation of *technology in action*.

2.2. *Technologies as recipes*

The conception, design, and production of whatever artifact or the completion of whatever service generally involves (often very long) sequences of cognitive and physical acts. Hence, it is useful to think of a technology also like a "recipe" entailing a design for a final product, whenever there is a final

[8] On the possibilities, obstacles and determinants of "tacitness reduction" via knowledge codification, in general and with reference to contemporary technologies, see Cowan (2001), Cowan et al. (2000), Nelson (2003), Nightingale (2003), and Pavitt (1987, 1999). More specifically on the contemporary patterns of codification of manufacturing technologies based on ICT instrumentation and computing, see Becker et al. (2005), Balconi (2002), and Lazaric and Lorenz (2003) among others. A more specific illustration in the case of the software industry is in Grimaldi and Torrisi (2001).

[9] A further distinction still largely unexplored, regards the *codification of learning processes* as distinct from the codification of search *outcomes*: see the insightful discussion in Prencipe and Tell (2001).

physical artifact—such as in the cookbook case—together with a *set of procedures* for achieving it. The recipe specifies a set of actions that need to be taken to achieve the desired outcome, and identifies the inputs that are to be acted on, and any required equipment (if sometimes implicitly). Where a complex physical product or artifact is the end of the procedure or a basic element of it, that artifact itself may be considered a technology, a view we will consider later in this section. Thus Mandeville's Man of War can be considered as a piece of technology. By the recipe view, so would be the way of building that ship. And quite sophisticated technologies, in the sense of the required procedures, might be involved also in sailing and using it effectively as a "Man of War."

The recipe specifies the sequence of procedures that are "legal," at the very least in the sense that they are technically feasible and apt to allow the desired outcome. In that respects, acts like "break the eggs smashing them with the pan over the sink" are not "legal" in the cake-making procedures in that they will never yield eventually a cake. As such, (well-constructed) recipes obey to sorts of *grammars* which prescribe what can or cannot be done on the ground of particular knowledge bases. Recipes are *coded programs* instructing on the sequential combinations of physical and cognitive acts, along the sequence involving various material inputs and machine services.[10]

The *technologies as recipes* view offers an enormous progress in the understanding of what technological knowledge is all about as compared to the blackboxing entailed by any representation of the kind cake $= f$(list of ingredients). Moreover, as we shall see below, the recipe view offers promising angles also to the formal representation of the dynamics of problem-solving procedures involved in any technological activity, However, recalling our earlier discussion of technologies as knowledge it is important to recognize that recipes have tacit aspects as well as articulated ones, and that the written-down recipe, what we call the codified recipe, is far from the whole story. Tacit knowledge is precisely what is not (or, sometimes cannot even in principle) be conveyed in the codified recipe itself, but—in the example of the cake recipe—remains in the head (or better in the practice) of grandmothers and French cooks, and is transmitted more by example than by instruction. There is a general principle here: *no good artifact or service comes out of codified recipes alone* (for a detailed discussion, see Winter, 2006a). Or, putting the other way round, there is much more knowledge in technological procedures than any codified recipe can reveal.

In some cases, like the literal example of cooking recipes, one single person embodies the whole set of skills necessary to lead from the raw inputs to the final output, involving, say, how to break the eggs, mix them with flour, put the butter in the pan, etc., all the way to the final production of a cake. However, in the domain of industrial technologies this is not generally the case: the various pieces of knowledge and skills are distributed across many individuals and a crucial issue concerns when and how they are called for. Such a procedural, know-how centered, interpretation of technologies brings into sharp view the blurry lines between, or, better, the intertwining of technology, division of labor, organization, and management: more below. Thus if one considers the "recipe" for building a Man of War, or for sailing it, or for designing it, generally more than one person is involved, and this is so

[10] On the representation of "technologies as codes," see Baldwin and Clark (2000). It also worth mentioning the *funds-flow* theory of production which, while falling short of an explicit procedural representation of production activities, attempts to nest the use of inputs into an explicit temporal sequence flagging when the inputs themselves are used (i.e., when the flows of their services are called upon): cf. Georgescu-Roegen (1970) and the reappraisal, refinements, and applications in Morroni (1992).

regardless of whether complex artifacts are employed as production inputs: no matter how mechanized (as it is in contemporary times), the building of a ship is a team operation. Different people, and groups, are assigned different parts of the process. In fact technologies very rarely are just individual activities of sheer manipulation of physical objects. Rather, they involve intrinsic social elements, nested in particular organizations, and ensembles of them, which have led one of us to suggest the notion of *social technologies* (Nelson and Sampat, 2001), meant to capture the system of norms, beliefs, and social practices shaping the "ways of doing things." In turn, how Mandeville's ship turns out will depend not only on the overall ship design and recipe that nominally is being followed, but also on "social technologies" governing how the work is divided, the match up of the skills and understandings of what is to be done under that division of labor with what actually needs to be done, and how effectively the work is coordinated and managed.

2.3. Technologies as routines

The term "routines" has been proposed to recognize and denote the multiperson nature of the way organizations "make or do things": see Nelson and Winter (1982), Cohen et al. (1996), Teece et al. (1997), Dosi et al. (2000), the special issues of *Industrial and Corporate Change* edited by Augier and March (2000) and by Becker et al. (2005), Montgomery (1995), Becker and Lazaric (2009), and Foss and Mahnke (2000). A routine that is commanded by an organization is "an executable *capability* for repeated performance in some *context* that has been *learned* by an organization" (Cohen et al., 1996, p. 683). Routines, as thoroughly argued in Nelson and Winter (1982), (i) embody a good part of the memory of the problem-solving repertoires of any one organization; (ii) entail complementary mechanisms of governance for potentially conflicting interests (for more detailed discussions, see Cohen et al., 1996; Coriat and Dosi, 1998); and (iii) might well involve also some "meta-routines," apt to painstakingly assess and possibly modify "lower-level" organizational practices (the more incremental part of R&D activities, and recurrent exercises of "strategic adjustment," are good cases to the point).

Routines involve multiple organizational members who "know" how to appropriately elicit an action pattern or a signal in response to the specific environmental circumstances:

> *"Each individual is constantly engaged in receiving signals from other members of the organization or from the environment, responding to the signal with some operation from his repertoire, and thereby creating a signal for other members of the organization, or an effect in the environment. Here, the incoming signal might be the appearance of a partially finished automobile on a production line, the operation may be tightening particular screws and the outgoing 'signal' is the slightly-more-finished automobile going down the line. Or, the incoming signal may be a report summarizing last month's expense account submissions from the sales force, the operation may be a comparison with standards and past experience, and the outgoing signal a letter of protest." (Winter, 2006a, p. 134)*
>
> *"'Knowing your job' in [the] organization is partly a matter of having the necessary repertoire of actions, and partly knowing which actions go with which incoming signals. Each individual has some ability to perform a considerably larger set of actions than are called for in his job, but to the extent that 'practice makes perfect' he will acquire superior skill in the ones actually called for." (Winter, 2006a, p. 134)*

Note that the "program" built into routines generally involves, at the same time, recipes which tend to be silent regarding the division of labor, together with particular divisions of labor, plus specific modes of coordination: in the language introduced earlier, the former aspect primarily captures the "physical" technology involved, while the latter entails specific "social technologies" (Nelson and Sampat, 2001).

In turn, ensembles of organizational routines are the building blocks of distinct organizational competences and capabilities. In the literature, the two terms have often been used quite liberally and interchangeably. In the introduction to Dosi et al. (2000, 2008a), it is proposed that the notion of capability ought to be confined to relatively purposeful "high-level" tasks such as, for example, "building an automobile" with certain characteristics, while "competences," for sake of clarity, might be confined to the ability to master specific knowledge bases (e.g., "mechanical" or "organic chemistry" competences). Clearly, such notion of competences/capabilities largely overlaps with what has come to be known as the "competence view of the firm" (cf. Helfat et al., 2007; see also below and Chapter 19).

2.4. Technologies as artifacts

The *procedure-centered* representation of technology is highly complementary to what we could call an *artifact-centered* account of what technologies are and their dynamic over time (see Arthur, 2007; Baldwin and Clark, 2000; Basalla, 1988; Frenken and Nuvolari, 2004 among others). Indeed, recipes often involve *designs* of what it is there to be achieved as a final output. (Although not always: think of services such as airline booking system or a surgical operation.) Even when the procedure involves a notion of design, the latter is in general only one of the many possible configurations which can be achieved on the grounds of any one knowledge base. In fact, when outputs are physical artifacts, it is useful to study their dynamics in the *design space* (Bradshaw, 1992; Frenken and Nuvolari, 2004), defined by the properties of the components which make up the final output and their combinations. So, in the case of the warship, the technology—seen as a complex product system (Helfat, 2003; Prencipe et al., 2003)—is made in turn of components—the hull, the sailing apparatus, the guns, etc., held together by binding technical consistency conditions.[11] Further, dynamically, innovation can be fruitfully studied in terms of modifications and improvements of the performance characteristics of each components and the system as a whole. After all, the numerous discontinuities in naval history from the "Man-of-War" of Mandeville's times to the contemporary USS air carrier *Ronald Reagan* map into the dynamics of both "incremental" change and more radical ruptures in the structure and functionalities of the artifacts: these are precisely two central concerns of evolutionary theories of innovation.

The artifact angle on technologies is in fact useful for a rather general purpose, namely the identification of the techno-economic characteristics of specific final products on the one hand, and of machines, components, intermediate inputs, on the other. Hence, as we shall see, the history of technologies can be usefully tracked, from one angle, through the dynamics of outputs in their appropriate characteristics space. This is also the "hedonic" dimension of product innovations. Symmetrically technological advances are reflected by the specific performances of particular pieces

[11] Visitors of Stockholm can still admire a beautiful seventeenth-century warship, the *Vasa*, immaculately conserved because it almost immediately sunk, due to the King's interventions on the design which made it violate precisely those conditions.

of equipment (e.g., how fast can this cutting machine cut? What is the tolerance of that boring machine? How many bits of information can this computer process per second? etc.).

2.5. Knowledge, procedures, and input/output relations

Note that in a *procedural view* of technology, the orienting focus is not immediately the list of inputs and equipment used to produce, say, a semiconductor with certain properties, but rather it rests in the design of the devices, and the procedures used in the transformation of the raw silicon into a microprocessor; not on the quantities of iron, plastic, and copper that go into an automobile of specified characteristics, but rather on the design of the automobile and the procedures used to produce it. Concerning technological advance, modifications and refinements of procedures and designs are "where the action is," while changes in input/output relations are in a way the byproduct of successful attempts to achieve effective procedures and designs with certain performances and to change them both in desired directions. Thus, students of the theory of production should notice that what comes under the heading of "production functions" of whatever kind, is basically just the *ex post* descriptions of what appears in the "quantity part" of the recipe—in the foregoing cooking example, the amount of eggs, butter, flour, pans, electricity, human labor, that goes into the production of a cake—but such quantities themselves derive quite strictly from the nature of the recipe and the characteristics of the final product one is meant to obtain. So, for example, procedures involving 90% eggs and 10% flour are not "legal" (they are not part of an admissible procedure), because they will yield at most an omelette, and not a cake, *irrespectively of relative prices.*

Note also that, dynamically, in most cases efforts to change recipes directly entail changes in input characteristics and "intensities" and, conversely, attempts to substitute one input for another involve changes in production procedures. Good examples of the former are, in economic history, the changes in "capital intensity" associated with the "taylorist" and "fordist" transformation of business firms—roughly a century ago—as such an attempt of major proportions to change the "ways of doing things" within organizations. Symmetrically, attempts to "substitute more expensive inputs"—so easy when seen from the angle of some "production function"—often require the painstaking search of new recipes and effective procedures.

A question with crucial ramification for any theory of production regards precisely the mappings between *procedure-centered* and *input/output-centered* representations of technologies. Suppose one has some metrics in the input/output space, and one is also able to develop, some (albeit inevitable fuzzy) metrics in the high-dimensional "problem-solving space."[12] Granted that, how do the latter map into the former? In particular, were one able to put together all the notional recipes known at a certain time apt to yield a cake (or for that matter a microprocessor or a car) what would the distribution look like in terms of input/output coefficients? In particular, would one find very many recipes which could be ordered in such a way as to be approximately described by a homogeneous function (possibly of degree one)? Indeed, there is nothing a priori in the nature of technological knowledge and in the nature

[12] As we shell briefly survey below, in the literature formal representations of technologies as recipes are quite rare. One of such exception is Auerswald et al. (2000). There the "distance" between any two recipes is the minimum number of operation that must be changed in order to convert one into the other (p. 397). Such a definition is well in tune also with the formalization in Marengo et al. (2000) and Marengo and Dosi (2006).

of recipes and routines which suggests this to be the case (the evidence below will just reinforce the point). In fact, nothing excludes the possibility of recipes that are quite "near" in terms of sequences of procedures which they entail, but quite far in the input/output space. *Vice versa*, it is equally possible to have recipes regarding, say, the production of steel, chemicals, or semiconductors, which might appear at a first look "near" in terms of input intensities but are in fact quite far away from the point of view of underlying knowledge and procedures.

Issues of the same kind regard the relationship between changes in the recipes and routines, on the one hand, and changes in the nature and relative intensities in the use of the various inputs, on the other. Do "small" changes in procedures correspond to "small" changes in input/output relations? And, *vice versa*, do major technological revolutions affecting "the way of doing things" imply also major changes in the proportions in which different artifacts and types of labor enter into the recipes for whatever output? In fact, the existence of possible regularities in the dynamic of procedures, artifact characteristics, and input intensities will be one of the central topics of the next section.[13]

Another implication is that the foregoing view of technologies focused on the procedures involved in, say, designing and manufacturing cars, software, chemical compounds, etc., rather than on the (derived) input/output relations allows a straightforward account for the ample variance in revealed performances across firms which one observes within each industrial sector. Especially if procedures are long, complex and possibly only partly understood by the organizations implementing them, one is likely to expect that (a) each organization knows only one or very few of them, (b) even for apparently similar recipes, any two organizations might master them with very different degrees of effectiveness. Heterogeneity across firms is, thus, the rule, even in presence of identical relative prices: more on all this below.

3. How technologies evolve

As we suggested above, scholars from a wide variety of disciplines who have studied technological advance in some detail have converged on the proposition that technological advance needs to be understood as proceeding through an evolutionary process. (Among economists and economic historians, the list includes many contributors to the SYS synthesis, cited earlier and also Chandler, 1992; Chandler and Galambos, 1970; Metcalfe, 1994, 1998, 2005b; Mokyr, 1990, 2002; Ziman, 2000.)[14] In a broad sense, the process is evolutionary meaning at least that at any time there generally are a wide variety of efforts going on to advance the technology, which to some extent are in competition with each other, as well as with the prevailing practices. The winners and losers in this competition are determined to a good extent through some *ex post* selection mechanisms. At no instance the interpretation of the process gains much by trying to rationalize it either in terms of consistent "gambles" by forward-looking players or by efficient "market processing" over *ex ante* blind ones. As such, the processes through which technologies evolve are also different in important respects from evolutionary processes in

[13] Germane discussions are in Nelson and Winter (1982), Nelson (1981), Auerswald et al. (2000), Winter (2006a), and Dosi and Grazzi (2006). A somewhat similar problem in biology is the mapping between genotypic and phenotypic structures: see Stadler et al. (2001).

[14] Quite a few others, without explicitly calling themselves "evolutionary" have expressed largely overlapping views, *in primis* Landes (1969, 1998) and David (1985, 1989, 2005) among others.

biology. In particular, the proposition that technology evolves in the above sense in no way denies, or plays down, the role of human purpose in the process, or the sometimes extremely powerful body of understanding and technique used to guide the efforts of those who seek to advance technology. Thus efforts at invention and innovation are by no means totally blind, or strictly random, as often is assumed to be the case regarding biological "mutation." At the same time, as we shall discuss below, purposefulness of search does not mean at all any accurate matching between forecasts and realized outcomes. Hence also the fundamental role of trials, errors, and *ex post* selection among competing variants of artifacts and processes of production.

Vincenti (1990) has described the kinds of complex knowledge and technique that modern aeronautical engineers possess, and discusses in detail how these focus and give power to their efforts at design. This body of knowledge and technique enables engineers to roughly analyze the likely plusses and minuses of various design alternatives through analytic methods or simulations, and thus focus their efforts on particular designs and variants. A portion of the body of understanding that guides problem solving and designing by professionals in a technological field comes often from operating experience. At the same time, in the contemporary world, many technologies are associated with specific fields of applied science or engineering. A good deal of the relevant body of understanding is codified in these fields, and serves as the basis for the training of new technologists and applied scientists. And these fields also are fields of research. In modern "high-tech" industries, research in the underlying scientific disciplines is an important source of new understandings and techniques that become part of the kit used by designers (Cohen et al., 2002; Rosenberg and Nelson, 1994; see also below).

Whenever efforts at inventing and designing are oriented by relatively strong professional understanding, *part* of the relevant variation and of the selection which is involved in the evolution of technologies occurs in the human mind, in thinking and analysis, in discussion and argument, in exploration and testing of models, as contrasted with being tried out there in practice. A good deal of the effort to advance technology proceeds "off-line," as it were. Research and development (R&D) is the term customarily given to such off-line efforts, particularly when they involve groups of scientists and engineers working within a formal organization who have such work as their principal activity. Technologies and industries vary in regards to the amount of funds invested in R&D, and the extent to which R&D is the principal source of technological advance, as contrasted with learning by doing and by using (the intersectoral evidence discussed in Dosi, 1988 and Pavitt, 1984 broadly applies also nowadays; see also below). However, even in fields where the science base is strong and the lion's share of efforts to advance a technology proceeds off-line, learning by doing and by using still plays an important role (cf. Freeman, 1994; Rosenberg, 1982, Chapter 6). Pavitt's foregoing point holds throughout past and contemporary technologies: *ex ante* well-codified knowledge, no matter how important, does not suffice to establish the detailed properties of any production process or artifact. There are three reasons.

First, even where the underlying sciences are strong, a good part of the know-how that professionals bring to bear in their efforts to advance a technology is acquired through operating experience, rather than through formal training in the sciences.

Second, in any case, as Vincenti argues, efforts at inventing and solving technological problems inevitably reach beyond the range of options that are perfectly understood. Ultimately what works and what does not, and what works better than what, must be learned through actual experience.

Third, as we will highlight later, firms in an industry tend to differ from one another in the details of the products and processes they produce and employ, in the set of customers and suppliers they know well, and in their past history of successes and failures, all of which influences how they focus and undertake search activities. Such differences in knowledge and practice hardly come from either science or engineering principles, but rather form idiosyncratic experience.

We have been sketching so far some quite general characteristics of technological advance that hold across fields and across countries, often driven by diverse behaviors of multiple agents searching and competing with each other. Pushing further, let us ask whether there are some *invariances in the knowledge structure and in the ways technological knowledge accumulates* and, together, *what distinguishes different fields and different periods of technological advance, if any.*

3.1. Technological paradigms and technological trajectories

From the earlier discussion it should be clear that each technology needs to be understood as comprising (a) a specific body of practice—in the form of processes for achieving particular ends—together of course with an ensemble of required artifacts on the "input side"; (b) quite often some distinct notion of a design of a desired "output" artifacts; and (c) a specific body of understanding, some relatively private, but much of it shared among professionals in a field. These elements, together, can be usefully considered as constituent parts of a *technological paradigm* (Dosi, 1982, 1988), somewhat in analogy with Kuhn's (1962) scientific paradigm.[15]

A paradigm embodies an *outlook*, a definition of the relevant problems to be addressed and the patterns of enquiry in order to address them. It entails a view of the purported needs of the users and the attributes of the products or services they value. It encompasses the scientific and technical principles relevant to meeting those tasks, and the specific technologies employed. A paradigm entails *specific patterns of solution to selected techno-economic problems*—that is, specific families of recipes and routines—based on highly selected principles derived from natural sciences, jointly with specific rules aimed at acquiring related new knowledge. Together, the paradigm includes a (generally imperfect) understanding about just how and (to some extent) why prevailing practice works.

An important part of paradigmatic knowledge takes the form of *design concepts* which characterize in general the configuration of the particular artifacts or processes that are operative at any time. Shared general design concepts are an important reason why there often is strong similarity among the range of particular products manufactured at any time—as the large passenger aircrafts produced by different aircraft companies, the different television sets available at the electronics stores, etc. Indeed, the establishment of a given technological paradigm is quite often linked with the emergence of some *dominant design* (see Abernathy and Utterback, 1978; Henderson and Clark, 1990; Rosenbloom and Cusumano, 1987; Suarez and Utterback, 1995; Utterback and Suarez, 1993; and the critical review of the whole literature in Murmann and Frenken, 2006). A dominant design is defined in the space of artifacts and is characterized both by a set of core design concepts embodied in components that

[15] Here as well as in Dosi (1982), we use the notion of paradigm in a *microtechnological* sense: for example, the semiconductor paradigm, the internal combustion engine paradigm, etc. This is distinct from the more "macro" notion of "techno-economic paradigm" used by Perez (1985, 2010) and Freeman and Perez (1988) which is a constellation of paradigms in our *narrow* sense: for example, the electricity techno-economic paradigm, ICTs, etc. The latter broader notion overlaps with the idea of "general purpose technologies" from Bresnahan and Trajtenberg (1995) (see also the remarks below, Section 5). Moreover, the notion of paradigm used here bears a good deal of overlapping with that of "regimes" put forward in Nelson and Winter (1977).

correspond to the major functions performed by the product and by a product architecture that defines the ways in which these components are integrated (Murmann and Frenken, 2006; drawing upon Henderson and Clark, 1990). However, sometimes the establishment of a dominant paradigm is *not* associated with a dominant design. A revealing case to the point are pharmaceutical technologies which *do* involve specific knowledge basis, specific search heuristics, etc.—that is, the strong mark of paradigms—without however any hint at any dominant design. Molecules, even when aimed at the same pathology, might have quite different structures: in that space, one is unlikely to find similarities akin those linking even a Volkswagen Beetle 1937 and a Ferrari 2000. Still, the notion of "paradigm" holds in terms of underlying features of knowledge bases and search processes.[16] Whether the establishment of a dominant paradigm entails also the established of a dominant design or not bears a lot of importance also in terms of dynamics of industry structure along the life cycle of the industries to which a particular paradigm is associated. We shall come back to that in Section 4.

Technological paradigms identify the operative constraints on prevailing best practices and the *problem-solving heuristics* deemed promising for pushing back those constraints. More generally, they are the cognitive frames shared by technological professionals in a field that orient what they think they can do to advance a technology (Constant, 1980). Technological paradigms also encompass normative aspects, like criteria for assessing performance, and thus provide ways of judging what is better than what, and goals for the improvement of practice. Each paradigm involves a specific "technology of technical change," that is specific *heuristics of search*. So, for example in some sectors, such as organic chemicals these heuristics relate to the ability of coupling basic scientific knowledge with the development of molecules that present the required characteristics, while in pharmaceutical field the additional requirement is the ability to match the molecular knowledge with receptors and pathologies. In microelectronics search concerns methods for further miniaturization of electrical circuits, the development of the appropriate hardware capable of "writing" semiconductor chips at such a required level of miniaturization and advances in the programming logic to be built into the chip. The examples are very many: a few are discussed in Dosi (1988). Here notice in particular that distinct (paradigm-specific) search and learning procedures, first, imply as such diverse modes of creating and accessing novel technological opportunities, and, second, entail also different organizational forms suited to such research procedures.[17] Both properties will turn out to be central when trying to characterize distinct "regimes" of technological and industrial evolution (see below).

Together, the foregoing features of technological paradigms both provide a focus for efforts to advance a technology and channel them along distinct *technological trajectories*, with advances (made by many different agents) proceeding over significant periods of time in certain relatively invariant directions, in the space of techno-economic characteristics of artifacts and production processes. As paradigms embody the identification of the needs and technical requirements of the users, trajectories may be understood in terms of the progressive refinement and improvement in the supply responses to such potential demand requirements. A growing number of examples of technological trajectories include

[16] A notion quite akin to "dominant design" is that of "technological guideposts" (Sahal, 1981, 1985), a guidepost being the basic artifact whose techno-economic characteristics are progressively improved over time.

[17] Note also that there seems to be major differences between science-driven and technology-driven search (cf. Nightingale, 1998), with heuristics that in one case focus on "puzzles further ahead"—given what one knows—while in the technological domain, heuristics typically address "how can one solve this problem," irrespectively of the underlying theoretical knowledge.

aircrafts, helicopters, various kinds of agricultural equipment, automobiles, semiconductors, and a few other technologies (Dosi, 1984; Gordon and Munson, 1981; Grupp, 1992; Sahal, 1981, 1985; Saviotti, 1996; Saviotti and Trickett, 1992). So, for example, technological advances in aircraft technologies have followed two quite distinct trajectories (one civilian and one military) characterized by log-linear improvements in the tradeoffs between horsepower, gross takeoff weight, cruise speed, wing load, and cruise range (Frenken and Leydesdorff, 2000; Frenken et al., 1999; Giuri et al., 2007; Sahal, 1985; and more specifically on aircraft engines Bonaccorsi et al., 2005). Analogously, in microelectronics, technical advances are accurately represented by an exponential trajectory of improvement in the relationship between density of electronic chips, speed of computation, and cost per bit of information (see Dosi, 1984, but the trajectory has persisted since then). As an illustration consider Figure 1 on computers, from Nordhaus (2007) highlighting also the changing trajectories associated with paradigm changes, and Figure 2 pointing out the long-term trajectory-like patterns in semiconductors and the ways they have

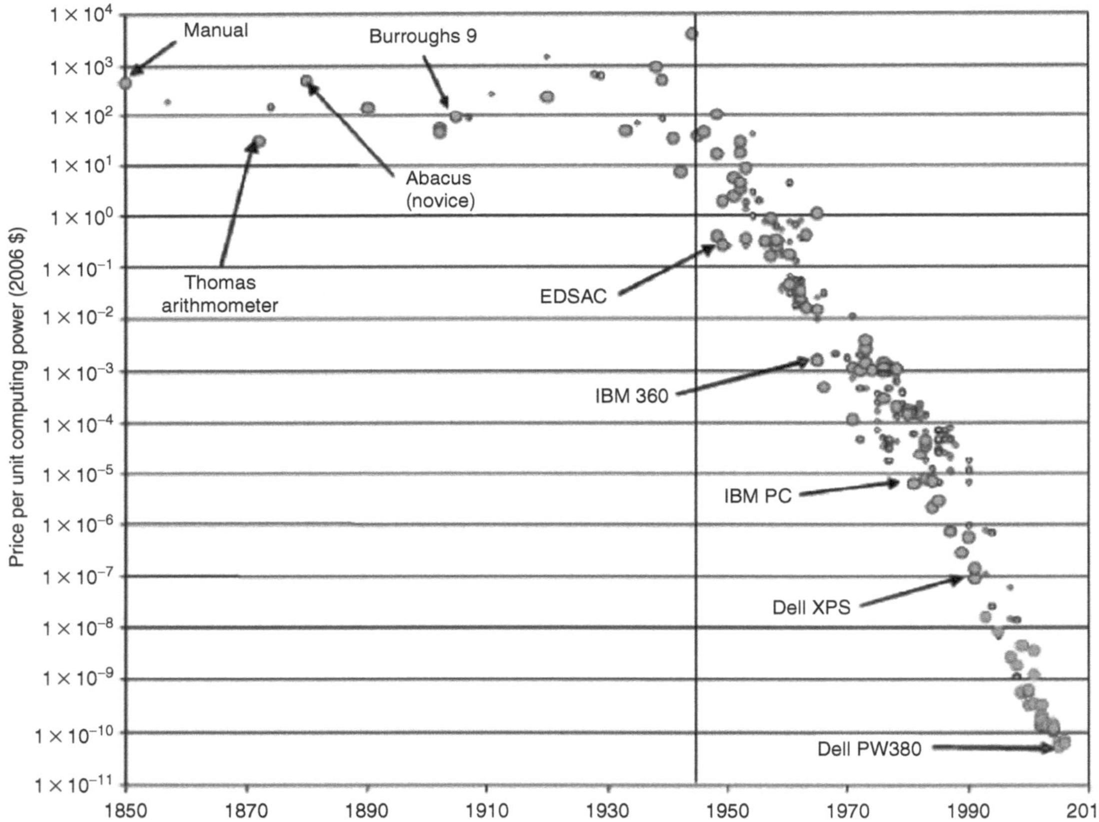

Figure 1. The progress of computing measured in cost per computation per second deflated by the price index for GDP in 2006 prices. Source: Nordhaus (2007).

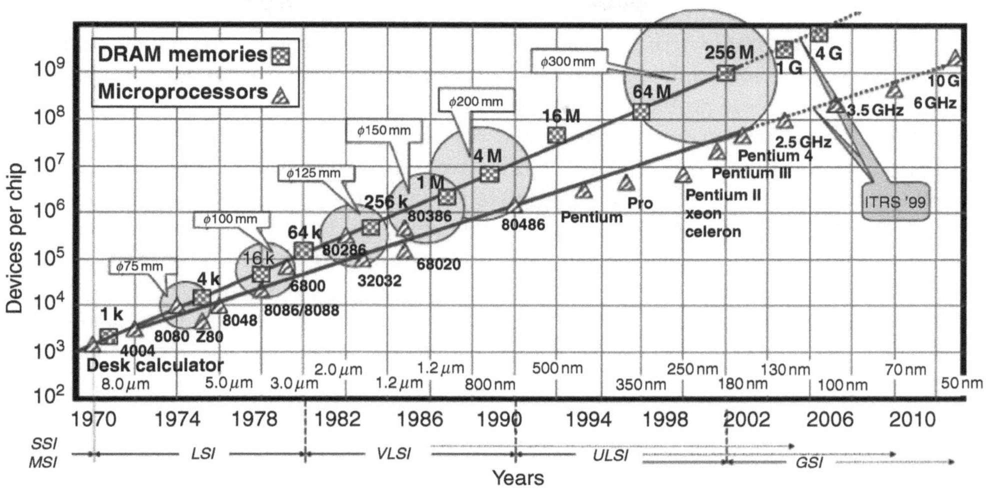

Figure 2. Moore's law and technology scaling. Source: Zheng (2008).

been punctuated by different families of devices. In fact, it is fair to say that trajectory-like patterns of technological advance have been generally found so far whenever the analyst bothered to plot over time the fundamental techno-economic features of discrete artifacts or processes, say from the DC3 to the Airbus 380, among aircrafts, or from crucible to Bessemer to basic oxygen reduction among steel-making processes. (Admittedly, trajectories in the space of processes and related input intensities have been studied much less than trajectories in the output characteristic space, and this is indeed a challenging research area ahead.)

The emergence of relatively ordered trajectories, as already hinted, sometimes is and sometimes is not associated with the emergence of dominant designs. When it does, the trajectories appear to be driven by "hierarchically nested technological cycles" entailing both relatively invariant core components improving over time and a series of bottlenecks and "technological imbalances" (Rosenberg, 1976) regarding the consistency among all the components of the systems (cf. Murmann and Frenken, 2006). Come as it may, some properties of trajectories are important to notice here.

First, trajectories *order* and *confine* but do not at all eliminate the persistent *generation* of *variety*, in the product and process spaces, which innovative search always produces. The paradigm defines proximate boundaries of feasibility and together shapes the heuristics of search. However, there continues to be plenty of possible tradeoffs between output characteristics which different producers explore (Saviotti, 1996) and which will be eventually the object of (imperfect and time-consuming) market selection.

Second, by the same token, trajectories so to speak "extrapolated forward"—in so far as their knowledge is shared by the community of firms, practitioners, engineers—are a powerful *uncertainty reducing representations* of what the future is likely to yield in technological terms. However, this remains a far cry from any unbiased expectation on the time and costs involved in "getting there"—whatever "there" means—and, even more so, of the probability distributions of individual actors over both technological and economic success. That is trajectories are not means to reduce Knightian uncertainty into probabilizable risk.[18] Indeed, notwithstanding roughly predictable trajectories of advance, both *substantive uncertainty*—concerning future states of the world—and *procedural uncertainty*—regarding yet to come problem-solving procedures—continue to be ubiquitous.[19]

Note that there is no a priori economic reason why one should observe limited clusters of technological characteristics at any one time and ordered trajectories over time. On the contrary, as we already argued in Dosi (1988)—given consumers with different preferences and equipment users with different technical requirements, and different relative prices in different countries, if technologies were perfectly "plastic" and malleable—as standard economic representations are implicitly suggesting—one would tend to observe sorts of "isoquants" with the familiar shape in the space of techniques and of techno-economic characteristics of products. And, over time, if technological recipes—in both the procedural aspects and their input contents—could be freely added, divided, recombined, substituted, etc., one would also tend to observe an increasingly disperse variety of technical and performance combinations in products, production inputs, and available techniques (even if not necessarily in their use, given relative prices). The ubiquitous evidence on trajectories, on the contrary, suggests that technological advances are circumscribed within a quite limited subset of the techno-economic characteristics space. We could say that the paradigmatic, cumulative, nature of technological knowledge provides *innovation avenues* (Sahal, 1985) which channel technological evolution, while major discontinuities tend to be associated with changes in paradigms. Indeed, here and throughout we shall call "normal" technical progress those advances occurring along a given trajectory—irrespectively of how "big" they are and how fast they occur—while we reserve the name of "radical innovations" to those innovations linked with paradigm changes.

A change in the paradigm generally implies a change in the trajectories. Together with different knowledge bases and different prototypes of artifacts, the techno-economic dimensions of innovation also vary. Some characteristics may become easier to achieve, new desirable characteristics may emerge, some others may lose importance. Relatedly, the engineers' vision of future technological advances changes, together with a changing emphasis on the various tradeoffs that characterize the new artifacts. So, for example, the technological trajectory in active electrical components based on thermionic valves had as fundamental dimensions heat-loss vacuum parameters, miniaturization and reliability over time. With the appearance of solid-state components (the fundamental building block of the microelectronic revolution) heat loss became relatively less relevant, while miniaturization increased enormously in importance. Similar examples of change in the dimensions of the design space can be found in most transitions from one paradigm to another. Of course, one does not always observes clear-cut paradigmatic "revolutions". It is sometimes the case that "normal" advances on established knowledge bases is intertwined with new sources of knowledge. This appears to be the case in electronics-based industrial automation and might apply also to drugs and biotech: cf. Hopkins et al. (2007).

Are there some features which most technological trajectories share?

[18] Such persistent uncertainty is also reflected by systematic forecasting errors concerning costs of innovative search, future demand and future profitabilities of new products and processes: see Starbuck and Mezias (1996), Beardsley and Mansfield (1978), Freeman and Soete (1997), Dawid (2006), and Gary et al. (2008) among others. Indeed, all evidence points in a direction opposite to any assumption of "rational technological expectations"!

[19] More on the notions of substantive and procedural uncertainty in Dosi and Egidi (1991). For a discussion of the related modeling efforts, Dawid (2006).

A common feature which characterizes trajectories in process technologies and in the related equipment-embodied advances is a powerful trend toward mechanization and/or automation of production activities. Recent pieces of evidence are in Klevorick et al. (1995), but the phenomenon has been noticed since the classics and plays an important role in the analyses of the dynamics of capitalist economies by Adam Smith and Karl Marx. Note that such a tendency holds across sectors and across countries characterized by different capital intensities and broadly occurs irrespectively of variations in relative prices.[20] Due to its generality, in another work (Nelson and Winter, 1977) we called it a "natural trajectory": of course there is nothing "natural," strictly speaking, but it is indeed a general reflection of a long-term trend toward the substitution of inanimate energy to human and animal efforts, and more recently also of inanimate information processing to human cognition and control.

There is another relatively common feature of trajectories of innovation (even if we still do not know how common—a task indeed for empirical research ahead), namely *learning curves*. Chapter 10 is devoted to learning by doing and its different formalizations. Here, let us just mention some basic regularities and their bearing on the properties of technological trajectories. It has been found that costs fall according to a power law of the kind:

$$p = \alpha X^{\beta}, \tag{1}$$

where X is the cumulated production, α and β are two (technology-specific) constants, and p generally stands for unit costs but sometimes represents unit labor inputs or also some indicator of product performance. This original statement of the "law" comes from Wright (1936)[21] based on aircraft manufacturing (see also Alchian, 1963). Similar regularities appear in various energy producing technologies, in computers, light bulbs, and many other artifacts and processes: for technology-specific evidence, surveys, and discussions see Conoway and Schultz (1959), Conley (1970), Baloff (1971), Dutton and Thomas (1984), Gritsevskyi and Nakicenovic (2000), MacDonald and Schrattenholzer (2001), Neij (1997), Yelle (1979), Argote and Epple (1990), and Chapter 10 in this Handbook. Semiconductors offer an archetypical example of a trajectory driven by miniaturization efforts yielding the so-called Moore's law involving the doubling of the density of elementary transistor per chip and later microprocessors every 2–3 years (cf. Figure 2; more details in Dosi, 1984; Gordon and Munson, 1981; Jovanovic and Rousseau, 2002; Nordhaus, 2007).[22]

Interestingly, a steady fall in unit labor inputs seems—at least in some circumstances—seems to appear even when holding the equipment constant. It is the so-called "Horndahl effect," named after a Swedish steel mill (Lundberg, 1961), which contributed to inspire Arrow (1962b) on learning by doing.[23] Notice that learning effects are present at the levels of industry, firms and plants, even if rates and intertemporal variabilities are different, with micro-learning displaying higher irregularities over time than industry-level rates of progress (for some discussion of the evidence, see Auerswald et al., 2000). The interpretation

[20] For some more detailed discussion, see also Dosi et al. (1990).

[21] Who, somewhat confusingly, calls "performance" the left-hand variable and "prevalence" the right-hand one.

[22] Moore's law, technically, is formulated in terms of time rather than cumulated output such as in Equation (1). However, it can be easily reformulated accordingly, noticing that output flows exhibit an exponential growth profile over time.

[23] Strictly speaking, the Horndahl effect showed around 2% per year growth in productivity, and thus, again, linked performance with time and experience rather than accumulated output, but see footnote 22.

of the learning mechanisms underlying the observed performance trajectories and of their variations across different paradigms are indeed important tasks ahead for evolutionary analyses of innovation.[24]

Together with differences across paradigms in the rates of technological advance, one observes major differences in the processes through which such advances occur. In fact, significant progress has been made in the conceptualization of what different technological paradigms have in common and how they differ in terms of the sources of knowledge upon which they draw—that is, the *technological opportunities* which they tap, the mechanisms through which such opportunities are seized, and the possibilities they entail for innovators to extract economic benefit from their technological advances—that is, the *appropriability conditions*.

Let us consider these properties.

3.2. Technological opportunities, the processes of knowledge accumulation, and their cumulativeness

Prevailing technological paradigms differ over time and across fields regarding the nature of the knowledge underlying the opportunities for technical advances. Relatedly, they differ in the extent to which such knowledge has been gained largely through operating experience, as contrasted to scientific research.

While in most fields there is a mix, in the fields generally thought of as "high tech" a more significant contribution is nowadays grounded in specialized fields of science or engineering.

Where operating experience and learning by doing and using are the primary basis for professional understanding, as was the case with Mandeville's example of eighteenth-century ship design, the learning trajectory is going to advance paced by experience with actual new designs (and nowadays with the advances incorporated into new vintages of capital equipment and ability of using it). In the other hand, understanding can advance rapidly when there are fields of science dedicated to that objective. Several recent studies (see, e.g., Klevorick et al., 1995; Nelson and Wolff, 1997) have shown that the fields of technology that, by a variety of measures, have advanced most rapidly are associated with strong fields of applied science or engineering. Moreover, firms operating in these fields also tend to have higher than average levels of R&D intensity. In fact, in a secular perspective, the evidence is in tune with Mokyr's general conjecture that the "epistemic" elements of technological knowledge—that is, those elements associated with an explicitly casual knowledge of natural phenomena—have had a crucial (and increasing) importance in modern technological advances (Mokyr, 2002, 2010; Nelson, 2003; Nelson and Nelson, 2002; Nelson and Wolff, 1997).

Since the Industrial Revolution, the relative contribution of sciences to technology has been increasing, and in turn such a science base has been largely the product of publicly funded research, while the knowledge produced by that research has been largely open and available for potential innovation to use (more in David, 2001a,b, 2004; Nelson, 2004; Pavitt, 2001).

This, however, is not sufficient to corroborate any simple "linear model" from pure to applied science, to technological applications.

First, the point made elsewhere by Rosenberg (1982), Kline and Rosenberg (1986), Pavitt (1999), and Nelson (1981) continues to apply: scientific principles help a lot but are rarely enough. An enlightening case

[24] For more evidence on the characteristics of specific paradigms and trajectories, see also Consoli (2005), Chataway et al. (2004), Mina et al. (2007), Possas et al. (1996), Dew (2006), and Castaldi et al. (2009) among others.

to the point, indeed in a "science-based" area—medical innovation—is discussed in Rosenberg (2009). Semiconductors technology is another good example. For many decades, efforts to advance products and process technology—crucially involving the ability to progressively make circuits smaller and smaller—have taken advantage of the understandings in material science and the underlying solid-state physics. However, much more pragmatic and tacit elements of technological knowhow have persistently been crucial.

Second, it is quite common that scientific advances have been made possible by technological ones, especially in the fields of instruments: think of the example of the electronic microscope with respect to the scientific advances in life sciences (more in Rosenberg, 1982, 1994).

Third, it is not unusual that technologies are made to work before one understands why they do: the practical (steam) engine was developed some years before science modeled the theoretical Carnot engine; even more strikingly, the airplane was empirically proved to work few decades before applied sciences "proved" that it was theoretically possible![25] In fact, the specificities of the links between technological advances and advances in applied sciences are a major discriminating factor among different technological paradigms and different sectors (see below on sectoral taxonomies).

Generally speaking, while it usually holds that technological advance tends to proceed rapidly where scientific understanding is strong and slowly where it is weak, the key has often been the ability to design controllable and replicable practices that are broadly effective around what is understood scientifically[26] (for a more detailed discussion, see Nelson, 2008a).

Given whatever potential opportunities for innovation, what are the properties of the processes through which they are tapped? An important feature distinguishing different paradigms has to do with the *cumulativeness* of innovative successes. Intuitively, the property captures the degrees to which "success breeds success," or, in another fashionable expression, the measure to which innovative advances are made by dwarfs standing on the shoulders of past giants (as such, possibly, the integral of many dwarfs). Cumulativeness captures the incremental nature of technological search, and, crucially, varies a lot across different innovative activities (Breschi et al., 2000; Malerba and Orsenigo, 1996b; see also below). More formally, a way to capture cumulativeness is in terms of *future probabilities* of success conditional on *past realizations* of the stochastic process. In that respect, it is a widespread instance of *knowledge-based dynamic increasing returns.*

Quite a few technological paradigms embodying knowledge generated to a large extent *endogenously* tend to display dynamics of knowledge accumulation which are more cumulative than trajectories of advance which are, so to speak, fuelled "from outside" (e.g., via the acquisition of new pieces of equipment generated in other industrial sectors). A further distinction concerns the *domain* at which cumulative learning tends to occur. It is at the level of individual firms or is it at the level of the overall

[25] In fact, history quite often offers examples of a coevolutionary kind with the main arrow of causation running in one direction or the other depending also on the period and stage of development of knowledge. Take the case of the steam engine. While it is true that practical advances in the first half of the eighteenth century preceded subsequent advancement in classical thermodynamics and the theory of heat engines, it also holds that earlier attempts to exploit the power of steam were palpably influenced by the scientific investigations of Torricelli, Pascal, Boyle, and Hooke on the existence and properties of atmospheric pressure (Kerker, 1961). This may also explain why the steam engine was not invented in China, even if all constituent parts (piston, cylinder, etc.) were available also there (Needham, 1962–1963) (we thank A. Nuvolari for pointing it out to us).

[26] Note that this property does not bear any direct implication in terms newness of the scientific understanding itself. Moreover, high rates of advance often occur when new pieces of knowledge (new paradigms) are applied to older, much less science-based technologies. ICT applications to industrial machinery used in "traditional" industries are a good case to the point.

community of firms, would-be entrepreneurs, technical communities associated with each paradigms, etc.? In Teece et al. (1994), one points at examples such as Intel where cumulativeness applies at both paradigm and firm level (see also Breschi et al., 2003). At the opposite extreme, several instances point at patterns of technological change which are anticumulative in that they imply *competence destruction* at the level of individual incumbents (cf. Tushman and Anderson, 1986). Yet other historical examples highlight discontinuities engendered by firm-specific *organizational diseconomies of scope* even under largely cumulative industry-level patterns of accumulation of technological knowledge: Bresnahan et al. (2008) offer a vivid illustration concerning the introduction of the PC and the browser in the case of IBM and Microsoft, respectively.

3.3. Demand and other socioeconomic factors shaping the direction of technological advance

The tendency of the advance of a technology to follow a particular trajectory is not an indication that user needs and preferences and economic conditions such as relative prices do not affect the path of technological development. While the nature of technological opportunities does limit the range of directions along which a technology can advance, there generally is still significant scope for variation, and, as mentioned above, built into the paradigms that guide technological development are also a set of understandings about users' would-be requirements.

Let us consider in more detail the interplay between knowledge-driven venues of search and mechanisms of economic inducement.

A widespread view is that, in fields where the underlying science is strong, efforts to advance the technology generally are triggered by new scientific knowledge, and are directed to taking advantage of that new knowledge. While there certainly are quite a few circumstances where new science has directly stimulated new inventive efforts, several studies suggest that usually this is not the case, with the science being applied in industrial R&D usually not being particularly new. Conversely, these same studies show that firm level efforts to advance practice are very strongly influenced by perceptions of what users' value or at least by the perception of a problem with clear practical applications (cf. the evidence collected in the still classic Sappho project, comparing innovative successes and failures across otherwise similar firms: cf. Freeman, 1982; similar findings of the importance of perceived user needs are reported in Cohen et al., 2002). At the same time, considerations of technological feasibility tend to influence how these perceived demands are addressed.

An important aspect of the technological regime that shapes progress in a field is the character of the user community, their wants and constraints, more generally the (perceived) market for the new products and services that efforts to advance the technology might engender. User markets differ greatly both in the nature of the needs and preferences they reflect, and in the sophistication of the purchasers. Thus to sell their wares to the airlines, the producers of large passenger aircraft know that their designs have to meet a long list of quite precise requirements which the airlines have the technical sophistication to assess quite accurately. There also are regulatory safety standards that a new aircraft must pass before airlines can purchase and use it. Hence, the market for large commercial aircraft is far more tuned to technical characteristics of the product, far less moveable by advertising aimed to influence tastes, than say the market for automobiles. The market for operating system software mostly consists of the designers and producers of computers for whom various technical qualities are important, while the market for software games is mainly individuals who are attracted by different sorts of product quality. Indeed, there have been several studies that have explored the reasons why certain technological innovations were successful commercially while other ones, similar in many technical respects, were not. The principal factor often turned out to be understanding of the needs and desires of users by the successful innovator (see again Freeman, 1982).

Granted such broad and widespread interactions between users' demands and technological advances, it holds also that each body of knowledge specific to particular technologies, that is, each paradigm shapes and constrains the notional opportunities of future technical advance and also the boundaries of the set of input coefficients which are feasible on the grounds of that knowledge base (so that, e.g., irrespectively of the relative price of energy, it is difficult to imagine, given our current knowledge base, a technology for the production of hyperpure silicon which would not be very energy-intensive...).

Within such boundaries, change in the orientation of the new technologies created and developed can be induced by changes in demand-side factors in three analytically different ways.[27]

First, within a particular paradigm changes in relative prices and demand or supply conditions may well affect the orientation of search heuristics. This is what Rosenberg (1976) has called *focusing devices*, and historically documented in a few cases of supply shocks and technological bottlenecks (recall also the similar notion of "reverse salients" by Hughes, 1983), from the continental blockade during Napoleonic wars to various instances of technical bottlenecks in mechanical technologies. The mid-nineteenth-century history of machine tools provides indeed a fascinating example. Users always wanted tools that would cut faster, and inventors and designers responded. As higher cutting speeds were achieved, this put stress on the metals used in the machine blades. New blade materials were invented. And higher speeds also increased the temperatures at which blades had to operate; better cooling methods were invented and developed. (Bounded rationality and lack of "rational" technological expectations stand behind the relevance of these behaviorally mediated inducements effects. But, as already mentioned, evolutionary theories—quite in tune with empirical evidence—are at ease with these assumptions.)

Other powerful and quite general inducement factors have to do with industrial relations and industrial conflict. As analyzed by Rosenberg (1976), the resistance of nineteenth-century English labor, especially skilled labor, to factory discipline and terms of employment, has acted as a powerful stimulus to technical change. Karl Marx vividly put it:

> *"In England, strikes have regularly given rise to the invention and application of new machines. Machines were, it may be said, the weapon employed by the capitalists to equal the result of specialized labour. The self-acting mule, the greatest invention of modern industry put out of action the spinners who were in revolt. If combinations and strikes had no other effect than of making the efforts of mechanical genius react against them, they would still exercise an immense influence on the development of the industry." (Marx, 1847, p. 161; also cited in Rosenberg, 1976)*

Similarly, industrial conflict has been a powerful driver of the trajectories of mechanization of production based on taylorist principles (Coriat and Dosi, 1998).

[27] For important discussions of "inducement effects," see Binswanger and Ruttan (1978) and Ruttan (1997).

Symmetrically, on the demand side, along with obvious feasibility conditions, users' requirements have a major influence on the ensuing trajectories in the products characteristic space. As illustrations, think of the role of the requirements of the space and military industry on the early (United States and world) trajectories in semiconductor devices, or the influence of the characteristics of the US market on the trajectories of product innovation in automobile (in this case, largely specific to North America). And of course the extreme case of users' requirements influencing the patterns of innovation is when users themselves are innovators (von Hippel, 2005).

In all these instances, "inducement" stands for the influences that the actual or perceived environmental conditions exert upon the problem-solving activities which agents decide to undertake.

The earlier caveat that knowledge bases constrain the directions of search is crucial as well, and this applies to both single technologies and broad technological systems (or "techno-economic paradigms" in the sense of Perez, 1985; Freeman and Perez, 1988) which dominate in the economy over particular phases of development (e.g., steam power, electricity and electromechanical technologies, microelectronics and information technologies, etc.). Consider for example, Moses Abramovitz's proposition that:

> *"In the nineteenth century, technological progress was heavily biased in a physical capital-using direction [and] it could be incorporated into production only by agency of a large expansion in physical capital per worker...[while]... in the twentieth century... the bias weakened [and] may have disappeared altogether." (Abramovitz, 1993, p. 224)*

As we read it, it is a proposition on the nature of the knowledge available at a certain time in the society and the ways it constrains its economic exploitation, irrespectively of relative prices. That is, the proposition concerns the boundaries of the opportunity set attainable on the grounds of the available paradigms[28] and the limits to possible "inducement effects."

Second, inducement may also take the form of an influence of market conditions upon the *relative allocation of search efforts* to different technologies or products, that is in the allocation of inventive efforts across different paradigms. Note that while the former inducement process concerned the *directions* of search within a paradigm (e.g., in the inputs space or in terms of product characteristics), this second form regards the *intensity of search* and, other things being equal, the rates of advance, between paradigms. In the literature, it has come to be known as "Schmookler's hypothesis" (Schmookler, 1966), suggesting that cross-product differences in the rates of innovation (as measured by patenting) could be explained by differences in the relative rates of growth of demand. While it is no *a priori* reason why the perception of demand opportunities should not influence the allocation of technological efforts, the general idea of "demand-led" innovation has been criticized at its foundation for its theoretical ambiguities. (Does one talk about observed demand? Expected demand? And how are these expectations formed? More in Dosi, 1982; Freeman, 1982; Mowery and Rosenberg, 1979). The empirical evidence is mixed. Schmookler's empirical research has shown how changes over time in the sales of different kinds of products tend to be followed, with a short lag, by changes in patenting in the same direction. Thus the rise in the sales of automobiles and motorized tractors in the first half of the twentieth century, and the fall off in the use of horses for transportation and farm work,

[28] A pale image of all that appear even after blackboxing the whole process into aggregate production functions, via different elasticities of substitution and factor saving biases. A pertinent discussion is the cited work by Abramovitz (1993). Relatedly, see also Nelson (1981).

was accompanied by a large increase in patenting relating to the first two products, and a fall of patenting relating to horse shoes. However, the review in Freeman (1994) concludes that "the majority of innovation characterized as 'demand led'. . .were actually relatively minor innovations along established trajectories," while as shown by Walsh (1984) and Fleck (1988), "counter-Schmookler-type patterns was [the] characteristic of the early stage of innovation in synthetic material, drugs, dyestuff,. . ." and robotics (Freeman, 1994, p. 480). As emphasized by Freeman himself and by Kline and Rosenberg (1986), the major analytical step forward here (mentioned already) is the abandonment of any "linear" model of innovation (no matter whether driven by demand or technological shocks) and the acknowledgment of a coevolutionary view embodying persistent feedback loops between innovation, diffusion, and endogenous generation of further opportunities of advancement.

Both mechanisms of "inducement" discussed so far ultimately rest on the ways production and market conditions and their change influence "cognitive foci" and incentives, and in turn, the way the latter affect behavioral patterns—in terms of both search heuristics and allocation rules of those working to created new technology. However, changing relative prices can easily "induce" changes in the directions of the technical changes brought to practice by users/adopters of new technologies, even holding search behavior constant, via the selection of the (stochastic) outcomes of search itself. This is the *third* inducement process. Suppose the allocation of resources dedicated to search were invariant to changing relative prices. Even in this case, however, would-be innovations—being they new production techniques or new machines to be sold to a user firm—will be implemented/selected only if they will yield total costs lower than those associated with the incumbent techniques/machines. But the outcome of the comparison obviously depends on relative prices (a formalization of the process is sketched out below, Section 3.8).

To summarize, one ought to disentangle three sources of "inducement" related to (a) changes in microeconomic rules of search, affecting the direction of exploration in the notional opportunity space and the pattern of adoption of machine-embodied technical change within paradigms; (b) changes in the allocation of resources to search efforts (irrespectively of their "directions") *across paradigms and lines of business*; and (c) market-induced changes in the selection criteria by which some techniques or products are compared with alternative varieties. An evolutionary interpretation of such processes easily allow for endogenous interactions (i.e., "coevolution") between the incentive structure (stemming from relative prices and demand patterns), on the one hand, and learning capabilities, on the other. In this respect, Wright (1997) is an excellent illustration of the point. Even in the case of mineral resources—that is, the nearest one can get to a "naturally" determined opportunity set—Wright shows that opportunities themselves have been the outcome of both public and private search efforts (see also David and Wright, 1997 and more generally, Mowery and Nelson, 1999, Mowery and Rosenberg, 1982, and Nelson, 1999). Conversely, more conventional views of inducement, by making stronger commitments to both optimizing rationality and equilibrium, obscure—in our opinion—the distinctions between behavioral effects and system-level ("selection") effects, and, together, render very difficult any account of the sector-specific and period-specific patterns of knowledge accumulation. The blackboxing under unobservable constructs like "elasticities of substitution" in aggregate or sectoral production functions just helps to rationalize *ex post* the dynamic outcome while obscuring the process driving it.

Of course, in the longer term major changes in the patterns of innovation are associated with the emergence of new technological paradigms. Thus the shift in inventive efforts from horse-driven carriages to automobiles and motor tractors can be regarded as the result of successful efforts to advance an ensemble of new technological paradigms associated with the successful development of, for example, gasoline engines, cheaper steel, electromechanical machine tools, etc. From this point of view, over such longer time-scale it is the emergence and development of new technological paradigms that molds the direction as well as the rate of technological advance, rather than "inducement" in any strict sense of such a notion.

3.4. Means of appropriation

Most researchers at universities and public laboratories do their work, which on occasion may result in a significant technological advance, without expectation of benefiting directly from it financially. Some inventors invent because of the challenge of it, and the sense of fulfillment that comes with solving a difficult problem. And, more important, as already mentioned, in contemporary societies most scientific knowledge—of both the "pure" and "applied" nature—has been generated within a regime of *open science*. The fundamental vision underlying and supporting such a view of publicly supported open science throughout a good part of the twentieth century entailed (i) a sociology of the scientists community largely relying on self-governance and peer evaluation, (ii) a shared culture of scientists emphasizing the importance of motivational factors other than economic ones, and (iii) an ethos of disclosure of search results driven by "winner takes all" precedence rules.[29] In Nelson (2006), David and Hall (2006), and Dosi et al. (2006b), one discusses the dangers coming from the erosion of Open Science institutions. We cannot get into details here. We have already mentioned above the importance of (free flowing) advances in pure and applied sciences as a fundamental fuel for technological advances—albeit with significant variation across technologies, sectors, and stages of development of each technological paradigm. However, the major share of inventive activities finalized to economically exploitable technologies that go on in contemporary capitalist societies is done in profit-seeking organizations with the hope and expectation of being economically rewarded, if that work is successful. In turn, the very existence of a relation between economically expensive search efforts by private agents, and (uncertain) economic rewards from successful innovations, entails the fundamental incompatibility—originally pointed out by Marx and Schumpeter—between any sort of zero-profit general equilibrium and any incentive to *endogenous* innovation (i.e., endogenous to the private, "capitalist," sector of the economy).

Granted that, however, two major sets of questions arise.

First, how profound is such a tradeoff between monopolistic departures from competitive (zero-profit) conditions and incentives to innovate?[30] More precisely, what is the evidence, if any, on some monotonic relation between (actual and expected) returns from innovation, on the one hand, and innovative efforts, on the other?

Such a monotonic relation is in fact built-in as one of the core assumptions within most "neo-Schumpeterian" models of growth, while the limited ability to appropriate returns to invention and innovation often is offered as the reason why the rate of technological progress is very slow in some

[29] On these points, following the classic statements in Bush (1945), Polanyi (1962), and Merton (1973), see the more recent appraisals in Dasgupta and David (1994), David (2004), Nelson (2004), and the conflicting views presented in Geuna et al. (2003).

[30] Note that the possible "tradeoff" discussed here is distinct from the purported, and somewhat elusive ("Schumpeterian"), tradeoff referred to in the literature between propensity to innovate and market structure: more on the theoretical side in Nelson and Winter (1982) and, on the empirical evidence, in Soete (1979) and Cohen and Levin (1989).

industries. The aforementioned studies on the nature and sources of technological opportunities suggest that this is unlikely to be the primary reason. Rather, it is far more likely that the reason for the highly uneven rates of progress among industries lies in differences in the strength and richness of technological opportunities. More generally, let us suggest that the widespread view that the key to increasing technological progress is in strengthening appropriability conditions, mainly through making patents stronger and wider, is deeply misconceived. Obviously, inventors and innovators must have a reasonable expectation of being able to profit from their work, where it is technologically successful and happens to meet market demands. However, in most industries this already is the case. And there is no evidence that stronger patents will significantly increase the rate of technological progress. (more in Granstrand, 1999, 2005; Jaffe, 2000; Mazzoleni and Nelson, 1998; Dosi et al., 2006c; and the growing literature cited therein). In fact, in many instances the opposite might may well be the case. We have noted that in most fields of technology, progress is cumulative, with yesterday's efforts- both the failures and the successes- setting the stage for today's efforts and achievements. If those who do R&D today are cut off from being able to draw from and build on what was achieved yesterday, progress may be hindered significantly. Historical examples, such as those presented in Merges and Nelson (1994) on the Selden patent around the use of a light gasoline in an internal combustion engine to power an automobile or the Wright brothers patent on an efficient stabilizing and steering system for flying machines, are good cases to the point. They show how the IPR regime probably slowed down considerably the subsequent development of automobiles and aircrafts, due to the time and resources consumed by lawsuits against the patents themselves. The current debate on property rights in biotechnology suggests similar problems, whereby granting very broad claims on patents might have a detrimental effect on the rate of technical change, insofar as they preclude the exploration of alternative applications of the patented inventions.

This is particularly the case when inventions concerning fundamental techniques or knowledge are concerned, for example, genes or the Leder and Stewart patent on a genetically engineered mouse that develops cancer. This is clearly a fundamental research tool. To the extent that such techniques and knowledge are critical for further research that proceeds cumulatively on the basis of the original invention, the attribution of broad property rights might severely hamper further developments. Even more so, if the patent protects not only the product the inventors have achieved (the "oncomouse") but also all the class of products that could be produced through that principle, that is, "all transgenic nonhuman mammals," or all the possible uses of a patented invention (say, a gene sequence), even though they are not named in the application. In this respect, Murray et al. (2009) offer a striking illustration of how "opening up upstream" (again, in the case of the mouse)—in such an instance, a discrete change in the IPR regime in the United States—yielded more search/more diverse rates of exploration of "downstream" research paths.[31]

In general, today's efforts to advance a technology often need to draw from a number of earlier discoveries and advances which painstakingly build upon each other. Under these circumstances, IPRs are more likely to be a hindrance than an incentive to innovate (more in Heller and Eisenberg, 1998 and Merges and Nelson, 1994). If past and present components of technological systems are patented by different parties, there can be an *anticommons* problem (the term was coined by Heller and Eisenberg).

[31] It is not possible to discuss here the underlying theoretical debates: let us just mention that models range from "patent races" equilibrium models (cf. the discussion in Stoneman, 1995) to much more empirically insightful "markets for technologies" analyses (Arora et al., 2002), all the way to evolutionary models of appropriability (Winter, 1993).

While in the standard commons problem (such as an open pasture) the lack of proprietary rights is argued to lead to overutilization and depletion of common goods, in instances like biotechnology the risk may be that excessive fragmentation of IPRs among too many owners may well slow down research activities because each owner can block each other. Further empirical evidence on the negative effects of strong patent protection on technological progress is in Mazzoleni and Nelson (1998); and at a more theoretical level, see the insightful discussion in Winter (1993) showing how tight appropriability regimes in evolutionary environments might deter technical progress (cf. also the formal explorations in Marengo et al., 2009). Conversely, well before the contemporary movement of "open-source" software, one is able to document cases in which groups of competing firms or private investors, possibly because of some awareness of the anticommons problem, have preferred to avoid claiming patents and, on purpose, to operate in a *weak* IPR regime somewhat similar to that of open science, involving the free disclosure of inventions to one another: see Allen (1983) and Nuvolari (2004) on blast furnaces and the Cornish pumping engine, respectively. Interestingly these cases of "collective invention" have been able to yield rapid rates of technical change. Similar phenomena of free revelation of innovation appear also in the communities of users innovators (see von Hippel, 2005).

The *second* set of questions regards the characteristics of the regimes with respect to *how* inventors appropriate returns. The conventional wisdom long has been that patent protection is the key to being able to appropriate them. But this is the case only in few fields of technology. Pharmaceuticals is an important example. However, a series of studies (Cohen et al., 2002; Levin et al., 1985; Mansfield et al., 1981 among others) has shown that in many industries patents are not the most important mechanism enabling inventors to appropriate returns. Thus Levin et al. (1985) find that for most industries:

> *"lead time and learning curve advantages, combined with complementary marketing efforts, appear to be the principal mechanisms of appropriating returns to product innovations." (p. 33)*

Patenting often appears to be a complementary mechanism for appropriating returns to product innovation, but not the principal one in most industries. For process innovations (used by the innovator itself) secrecy often is important, patents seldom so. These findings were largely confirmed by a follow-on study done a decade later by Cohen et al. (2002). Teece (1986) and a rich subsequent literature (cf. the Special Issue of *Research Policy*, 2006; taking stock on the advancements since his original insights) have analyzed in some detail the differences between inventions for which strong patents can be obtained and enforced, and inventions where patents cannot be obtained or are weak, and in the firm strategies needed for reaping returns to innovation. A basic and rather general finding is that in many cases building the organizational capabilities to implement the new technology, also by means of complementary assets such as manufacturing capabilities, enables returns to R&D to be high, even when patents are weak. Thus, despite the fact that patents were effective in only a small share of the industries considered in the study by Levin et al. (1985), some three quarters of the industries surveyed reported the existence of at least one effective means of protecting process innovation, and more than 90% of the industries reported the same regarding product innovations (Levin et al., 1985). These results have been confirmed by a series of other subsequent studies conducted for other countries (see, e.g., the PACE study for the European Union; cf. Arundel et al., 1995).

If there are some bottom lines so far to this broad area of investigation, they are that, *first*, there is no evidence on any monotonic relation between degrees of appropriability and propensity to undertake innovative search, above some (minimal) appropriability threshold; *second*, appropriability mechanisms currently in place are well sufficient (in fact, probably overabundant); *third* the different rates of innovation across sectors and technological paradigms can be hardly explained by variations in the effectiveness of appropriability mechanisms, and, *fourth*, even less so by differences in the effectiveness if IPR protection.

3.5. Technological advance and the theory of the firm

As mentioned, another chapter of this Handbook is devoted to the management of innovating firms. Here let us just sketch telegraphically some links between the theory of corporate organization and the evolution of technological knowledge and artifacts: related discussions are in Nelson and Winter (1982), Winter (1987, 2006b), Dosi et al. (2000, 2008a), Marengo and Dosi (2006), and Helfat et al. (2007), a few chapters of Fagerberg et al. (2005) and Granstrand (1998).

While in earlier eras much of inventing was done by self-employed individuals, under modern capitalism business firms have become a central locus of efforts to advance technologies. And firms long have been the economic entities that employ most new technologies, produce and market the new products, operate the new production processes. As already mentioned, most modern firm operates in environments that are changing over time in ways that cannot be predicted in any detail. Technological advances are one of the primary forces causing continuing uncertainty, but other causes concern the nature of markets and of competition regardless of whether these are associated with technological advance. That is, to recall, Knightian uncertainty obtains, both of the "substantive" and the "procedural" kinds. In these circumstances there is no way that a truly optimal policy can be even defined (among other things the choice set is not well specified), much less achieved. Rather, firms ought to be seen as "behavioral entities," largely characterized by routinized patterns of action, modified in the longer term by more explicit "strategic" orientations. In turn, as already sketched above, organizational routines and capabilities stemming from ensembles of them represent to a large extent the procedural counterpart of what we have discussed so far largely in terms of knowledge and its dynamics over time. In this respect, possibly one of the most exciting, far from over, intellectual enterprises over the last two decades has involved the interbreeding between the evolutionary research program, largely evolutionary-inspired technological innovation studies, and an emerging competence/capability-based theory of the firm, with complementary roots drawing back to the pioneering organization studies by March, Simon, and colleagues (Augier and March, 2000, 2002; Cyert and March, 1992; March, 1988; March and Simon, 1958; Simon, 1957). Deeply complementary to the analyses of innovative activities focused on dynamics of knowledge, artifact characteristics and input coefficients, organizational analyses have began addressing the behavioral meaning of statements such as "firm X is good at doing Y and Z. . . ." Relatedly, what are the mechanisms that govern how organizational knowledge is acquired, maintained, and sometimes lost?

Organizational knowledge is in fact a fundamental link between the social pool of knowledge, skills, opportunities for discoveries, on the one hand, and the micro efforts aimed at of their *actual* exploration, on the other.

Distinctive organizational capabilities bear their importance also in that they persistently shape the destiny of individual firms—in terms of, for example, profitability, growth, probability of survival. Equally important, their distributions across firms shape the patterns of change of broader aggregates such as particular sectors (see Section 4) and whole countries.

Over time, organizational capabilities change, partly as a result of deliberate search: the ongoing stream of research on *dynamic capabilities* (Helfat et al., 2007; Teece et al., 1997; Winter, 2003) addresses precisely the criteria and processes by which capabilities evolve at least partly steered by the effort of strategic management. But this fact in no way diminishes the significance of the limits on what particular firms are capable of doing at any time, and the constraints on the range of new things that they can learn to do in a reasonable period of time. In fact one often notices the apparent inability of established firms to cope with changes in paradigms associated with the development of alternative technologies based on different design principles and requiring different skills for their mastery and advancement, and the tendency for periods where regimes are changing to be marked by the entry of new firms which may come to dominate the industry in coming years. These limits and constraints on existing firms, and the consequent openness of an industry to entry under conditions when technologies are changing radically are a central aspect of a capability-based theory and also straightforwardly link with the analysis of the drivers of industrial evolution (more in Section 4).

3.6. The dynamic of productive knowledge, and the dynamics of production coefficients

It is a fundamental consequence of the foregoing view of technology and innovation and of the related knowledge-based theory of the firm that firms themselves ought to be expected to generally differ in the techniques they master. They are likely to differ in both the broad "recipes" they use, and even when they use the same nominal recipe (i.e. with the same codified elements) they almost certainly will differ in the tacit aspects of those recipes. The ways work on a particular technique is organized and managed almost never is the same across firms in the same nominal industry. Firms command and use different routines. Some important implications which are indeed quite at odds with traditional thinking in economics are the following:

(a) In general, there is at any point in time one or very few best-practice techniques which dominate the others irrespectively of relative prices.
(b) Different firms are likely to be characterized by persistently diverse (better and worse) techniques.
(c) Over time the observed aggregate dynamics of technical coefficients in each particular activity is the joint outcome of the process of imitation/diffusion of existing best-practice techniques, of the search for new ones, of the death of some others and of the changing shares of the incumbent ones over the total (these processes of course might or might not correspond to a similar dynamics in terms of *firms* which are so to speak the carriers of these techniques: see below).
(d) Changes over time of the best-practice techniques themselves are likely to display rather regular paths (i.e., trajectories) in the space of input coefficients.

Let us further illustrate the previous points with a graphical example.

Suppose that, for the sake of simplicity, we are considering here the production of a homogeneous good under constant returns to scale with two variable inputs only, x_1 and x_2.[32]

[32] Note that fixed inputs, vintage effects, and economies of scale would just strengthen the argument.

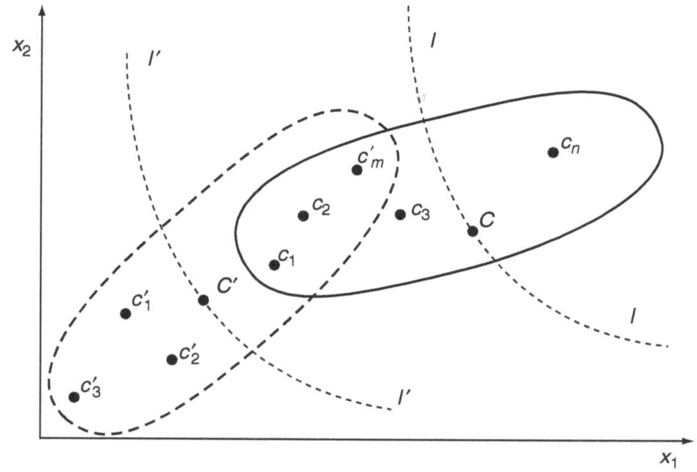

Figure 3. Microheterogeneity and technological trajectories.

A paradigm-based theory of production predicts that, in general, in the space of unit inputs, micro-coefficients are distributed somewhat as depicted in Figure 3. Suppose that at time t the coefficients are $c_1, \ldots, c_n$, where $1, \ldots, n$ are the various techniques labeled in order of decreasing efficiency at time t. It is straightforward, for example, that technique c_1 is unequivocally superior to the other ones no matter what relative prices are: it can produce the same unit output with less inputs of both x_1 and x_2. The same applies to the comparison between c_3 and c_n, etc.

A rapidly expanding evidence robustly supports the existence of *wide* and *persistent* inter-firm and inter-plant asymmetries in production coefficients at all levels of disaggregation (cf. Baily et al., 1992; Baldwin, 1995; Bartelsman and Doms, 2000; Bottazzi et al., 2007; Dosi, 2007; Jensen and McGuckin, 1997; Nelson, 1981; Power, 1998; Rumelt, 1991; Syverson, 2004).

Typically the support of inter-firm/inter-plant distributions of both labor productivities and "total factor productivity"[33] are strikingly wide even at relatively high levels of sectoral disaggregation. So, for example, Syverson (2004) finds that at a four-digit disaggregation, "the average 90-10 and 35-5 percentile [labour] productivity ratios within industries are over 4 to 1 and 7 to 1 respectively" (p. 535). Similar interfirm dispersion at three-digit disaggregation are found in the Italian industry by Bottazzi et al. (2007) and Dosi (2007). Moreover, such productivity differentials are quite stable over time with just some mild regression-to-the-mean tendency (cf. Dosi, 2007). A similar picture emerges from all micro longitudinal data banks we are aware of. It is also important to notice that inter-firm/inter-plant differences in labor productivities are not accounted for by differences in relative factor intensities (cf. Syverson, 2004; preliminary elaborations by one of us on the Italian industry show that the within-industry/cross-firm correlations between labor productivities and output/capital ratios are basically nil). Interestingly, such widespread differences in production efficiency *across firms and across plants*

[33] Notwithstanding the ambiguities of such latter measure, discussed in Dosi and Grazzi (2006).

continue to apply irrespectively of the degrees of sectoral disaggregation of the data. As Griliches and Mairesse (1997) put it,

> *"we. . .thought that one could reduce heterogeneity by going down from general mixtures as 'total manufacturing' to something more coherent, such as 'petroleum refining' or 'the manufacture of cement'. But something like Mandelbrot's fractal phenomenon seem to be at work here also: the observed variability-heterogeneity does not really decline as we cut our data finer and finer. There is a sense in which different bakeries are just as much different form each others as the steel industry is from the machinery industry."*

For evolutionary perspective, heterogeneity in the degrees of innovativeness and production efficiencies should not come as a surprise. A non-negligible part of the differences in production efficiencies must be due to different distributions of capital equipment of different vintages (the early intuition about the phenomenon is from Salter, 1962). However, broader differences are what one ought to expect to be the outcome of idiosyncratic capabilities (or lack of them), mistake-ridden learning and path-dependent adaptation.

Let us call this property *technological dominance*, and call some measure of the distribution of the coefficients across heterogeneous firms as the *degree of asymmetry* of that industry (e.g., in Fig. 3 the standard deviation around the mean value C).

The first question is why doesn't the firm using the nth technique adopt instead technique c_1? The simplest answer based on the foregoing argument is "because it does not know how to do it." That is, even if it is informed about the existence of c_1, it might not have the capabilities of developing or using it. Remarkably, this might have little to do with the possibility for c_1 to be legally covered by a patent. The argument is much more general: precisely because technological knowledge is partly tacit, also embodied in complex organizational practices, etc., technological lags and lead may well be persistent even without legal appropriation. The opposite also holds: if the two firms have similar technological capabilities, imitation might occur relatively quickly, patent protection notwithstanding, by means of "inventing around" a patent, reverse engineering, etc.

We are prepared to push the argument further and suggest that even if all firms were given the codified part of the recipe *for* technique c_1 (or, in a more general case, also all the pieces of capital equipment associated with it), performances and thus revealed input coefficients might still widely differ. It is easy to illustrate this by means of the foregoing cooking example: despite readily available cooking recipes, one obtains systematically asymmetric outcomes in terms of widely shared standards of food quality. Note that this has little to do even in the domain of cooking with "variety of preferences": indeed, we are ready to bet that most eaters randomly extracted from the world population would systematically rank samples of English cooks to be "worse" than French, Chinese, Italian, Indian,. . . ones, even when performing on identical recipes!!. If one accepts the metaphor, this should apply, much more so, to circumstances whereby performances result from highly complex and opaque organizational routines. (Incidentally, Leibenstein's X-efficiency rest also upon this widespread phenomenon).

Suppose now that at some subsequent time t' we observe the changed distribution of microcoefficients $c'_3, \ldots, c'_m$. How do we interpret such a change?

The paradigm-based story would roughly be the following. At time t, all below-best-practice firms try with varying success to imitate technological leader(s). Moreover, firms change their market shares, some may die and other may enter: all this obviously changes the weights (i.e., the relative frequencies) by which techniques appear. Finally, at least some of the firms try to discover new techniques, prompted by the perception of innovative opportunities, irrespectively of whether relative prices change or not (for the sake of illustration, in Figure 3, the firm which mastered the technique labeled three succeeds in leapfrogging and becomes the technological leader while m is now the marginal technique). Conversely, does one gain much by adding on the two "isoquants" I and I' passing through the respective means and by calling their shift "technical progress"? In our view, not much: rather it is going to blur the true underlying dynamics just described.

As discussed at greater length in Cimoli and Dosi (1995), and in several contributions to Cimoli et al. (2009), this interpretation of the distributions of techniques of production bears fundamental implications also in terms of international growth patterns. Consider again the illustration of Figure 3 and suppose that the evidence does not refer to two distributions of technical microcoefficients over time within the same country, but instead to two countries at the same time: after all, paraphrasing Robert Lucas, we only need informed tourists to recognize that most countries can be ranked in terms of unequivocal average technological gaps. The explanation of such international differences fundamentally rest upon the processes of accumulation of technological capabilities. Indeed, the economic discipline has undertaken far too few exercises at the highest available disaggregation on international comparisons among micro technical coefficients. Our conjecture is that less developed countries may well show higher utilization of all or most inputs per unit of output and perhaps even higher relative intensity of those inputs that conventionally would be consider more scarce (i.e., some loose equivalent of what euphemistically the economic profession calls in international trade the Leontief "paradox"). An evolutionary interpretation is straightforward: unequivocal technological gaps account for generalized differences in input efficiencies. Moreover, if technical progress happens to involve also high rates of saving in physical capital and skilled-labor inputs, one may observe less developed countries which do not only use more labor per unit of output but more capital input as compared to technological leaders (Figure 3 illustrates a similar case: compare, e. g., techniques c_3' and c_1).[34]

3.7. *Technological regimes: Sectoral specificities in patterns of technological advance, and the characteristics of innovative actors*

An important area of investigation has concerned over the last couple of decades the identification of different patterns of industrial evolution conditional on specific regimes of technological learning. By "regimes" here we mean distinct ensembles of technological paradigms with their specific learning modes and equally specific sources of technological knowledge. One of the aims of the well-known taxonomy by Pavitt (1984) is precisely to capture such relations mapping "industry types" and industry dynamics (see also Marsili, 2001 for important refinements). To recall, Pavitt taxonomy comprises four groups of sectors, namely:

(i) "Supplier-dominated" sectors, whose innovative opportunities mostly come through the acquisition of new pieces of machinery and new intermediate inputs (textile, clothing, metal products belong to this category)
(ii) "Specialized suppliers," including producers of industrial machinery and equipment

[34] The models in Nelson (1968) and Nelson and Pack (1999) are congenial formalizations of productivity differences across nations that have these features. Dosi et al. (1990) and Cimoli and Soete (1992) present also formalizations of international trade flows driven by technology gaps across countries.

(iii) "Scale-intensive" sectors, wherein the sheer scale of production influence the ability to exploit innovative opportunities partly endogenously generated and partly stemming from science-based inputs.[35]

(iv) "Science-based" industries, whose innovative opportunities coevolve, especially in the early stage of their life with advances in pure and applied sciences (microelectronics, informatics, drugs, and bioengineering are good examples).

Other, rather complementary, taxonomic exercises have focused primarily on some characteristics of the innovation process, distinguishing between a "Schumpeter Mark I" and a "Schumpeter Mark II" regime, dramatizing the difference between the views of innovative activities from Schumpeter (1911) and Schumpeter (1942); see Dosi et al. (1995), Breschi et al. (2000), Malerba and Orsenigo (1995, 1997), and Marsili (2001). The Mark I regime is characterized by innovations carried to a good extent by innovative entrants and by relatively low degrees of cumulativeness of knowledge accumulation, at least at the level of individual firms. Conversely under the Mark II regime innovative activities are much more cumulative and undertaken to a greater extent by a few incumbents which turn out to be "serial innovators."

In our view, such taxonomic exercises are important in their own right in that they identify discretely different modes through which innovation occurs in contemporary economies. And they are also important because they allow a link between such modes of innovative learning, the underlying sources of knowledge, the major actors responsible for the innovative efforts, and the ensuing forms of industrial organization. See Table 1 from Pavitt (1984) for one of such empirical attempts.

Note also that different technological regimes are supported by distinct institutions governing public research and training and, at the market end, by different forms of organization of the interactions among producers. Such institutions, together with the corporate actors involved contribute to define distinct *sectoral systems of innovation and production*: see Malerba (2002, 2004).

3.8. Formal models of search and technological evolution

The dichotomy between knowledge-ridden recipes and routines, on the one hand, and more "black-boxed" input/output representations is also reflected by two quite different styles of modeling, still in search for systematic links with each other.

The newer, and less developed, procedure-centered modeling *genre* builds on the notion that a technology is made of a discrete set of operations or components (Auerswald et al., 2000; Dosi et al., 2003; Levinthal, 1997; Levinthal and Warglien, 1999; Marengo and Dosi, 2006). Whatever name is chosen they stand for physical or cognitive acts eventually leading to the solution of whatever "problem," being it, for example, the construction of an automobile or the design of a piece of software. Different notional sequences of operations on components are associated with different degrees of efficiency in the solution of such problems (or no solution at all). One way of synthetically capturing these formalizations, represented over a relatively simple topology, is by nesting them over a *fitness landscape*. The notion was originally developed in biology as a way of mapping configurations of possibly interrelated traits into their fitness values (see Kauffman, 1993; Kauffman and Levin, 1987). Within this modeling style central

[35] Here one should in fact distinguish between "discontinuous" complex-product industries such as automobiles, white goods and other consumer durables versus "continuous" flow industries such as oil refining or steel making.

Table 1

Sectoral technological trajectories: Determinants, directions, and measured characteristics

Category of firm	Typical core sectors	Determinants of technological trajectories			Technological trajectories	Measured characteristics			
		Sources of technology	Type of user	Means of appropriation		Source of process technology	Relative balance between product and process innovation	Relative size of innovating firms	Intensity and direction of technological diversification
(1)	(2)	(3)	(4)	(5)	(6)	(7)	(8)	(9)	(10)
Supplier dominated	Agriculture; housing; private services traditional manufacture	Suppliers research extension services; big users	Price sensitive	Nontechnical (e.g., trademarks, marketing, advertising, aesthetic design)	Cost-cutting	Suppliers	Process	Small	Low vertical
Scale intensive	Bulk materials (steel, glass); assembly (consumer durables and autos)	PE suppliers; R&D	Price sensitive	Process secrecy and know-how, technical lags, patents, dynamic learning economies, design know-how, knowledge of users,	Cost-cutting (product design)	In-house; suppliers	Process	Large	High vertical
Specialized suppliers	Machinery; instruments	Design and development users	Performance sensitive	patents	Product design	In-house; customers	Product	Small	Low concentric
Science based	Electronics/ electrical; chemicals	R&D public science; PE	Mixed	R&D know-how, patents, process secrecy and know-how, dynamic learning economies	Mixed	In-house; suppliers	Mixed	Large	Low vertical High concentric

Source: Pavitt (1984, p. 12).
PE, Production Engineering Department.

questions regard the characteristics and efficacy of different ways of "decomposing" the overall problem, the implications of different search/adaptation strategies (e.g., whether involving "local" vs. "global" exploration), and the conditions under which "lock-in" into suboptimal outcomes occurs.

A domain of analysis to which such a modeling enterprise seem to straightforwardly apply is the theory of organization and its boundaries, and this is in fact where most of the attention has gone so far (more in Marengo and Dosi, 2006; see the discussion in Dawid, 2006, with reference to a large ensemble of agent-based—ACE—models: more on the latter in Tesfatsion and Judd, 2006). However, to repeat, not much effort has gone so far into the mapping between the recipe dynamics and the input/output dynamics.[36] In a rare exception, Auerswald et al. (2000) assume that the labor requirement associated with each "operation" is a random variable (so that the labor requirement of each recipe is a random field). Indeed, a quite challenging modeling frontier regards the explicit representation of evolving problem-solving procedures, constrained by paradigm-shaped "grammars" and their ensuing dynamics in the more familiar space of input/output coefficients.

As things stand now, even in the evolutionary camp, formal representations of technologies tend to "blackbox" the procedural part. As a result, most of the representations of techniques are in terms of quantities of inputs per units of output, with the output itself being often assumed homogeneous or sometimes defined by specific performance characteristics. Hence, the innovative dynamics is characterized by the evolution of the input vector (and, possibly, the output characteristics vector) over time. At this level of analysis, important modeling questions regard the form, and the support of the probability distribution of "innovative draws" agents may access, whether access is conditioned upon expensive investment ("R&D") and whether innovations are embodied or not in particular pieces of equipment. One feature, however, is common to most evolutionary representations of techniques in that they assume at any given time that firms are characterized by *fixed coefficients* of production (in the jargon they are endowed with Leontief techniques). In our view, this is a quite natural representation of the (degenerate) "production possibility set" firms are able to access in the short term: in fact, agents essentially know how to master the recipe actually in use while it is quite far-fetched to postulate that they have, so to speak, cupboards full of notional recipes which they *could instantaneously adopt* were relative prices different. Rather, any attempt to change technique has to be considered as a time-consuming, innovative effort, most often subject to uncertain outcomes.

Well supported by the microeconomic evidence discussed above, the basic unit of analysis of many evolutionary models are *heterogeneous techniques which at any point in time coexist and compete with each other, and evolve over time according to some search/learning process*. Straightforwardly, each technique can be pictured as a vector $\boldsymbol{x}_{(\cdot,\cdot)}(t)$ specifying, in the simplest case, the quantities of inputs per unit of homogeneous output. Each technique may or may not be labeled also in terms of agents which embody and hence master them. As reviewed in Silverberg and Verspagen (2005a), a family of models sticks to the "technique-as-the-primitive" representation (cf. Conlisk, 1989; Silverberg and Lehnert, 1993, 1994). The postulated "search" under this assumption is blackboxed within some random arrival process, drawing from a time-drifting normal distribution (Conlisk, 1989) or either time invariant or drifting Poisson distributions (Silverberg and Lehnert, 1993, 1994). Think for simplicity of a one-

[36] To our knowledge, the only attempt to link also at a formal level a dynamic in the space of recipes yielding learning-curve-type trajectories in the space input efficiencies is Auerswald et al. (2000) (see also Muth, 1986, albeit for a much more "blackboxed" perspective).

dimensional process, whereby one draws, say, in the space of labor productivities. The process for sound empirical reasons is assumed to be *multiplicative* on the techniques already in use (as witnessed, e.g., by the observed dynamics in labor productivities: cf. Dosi, 2007).

In another style of modeling, the technique is also tagged to specific firms, trying to capture the idiosyncratic features of innovative (and imitative) search. A model to that effect is presented in Iwai (1984a,b), where the distribution of techniques is taken to correspond to a distribution of firms which both innovate and imitate each other (with probabilities that are a function of the frequencies of the particular firms/techniques in the industry).

In quite a few modeling exercises, in tune with Nelson and Winter (1982), firm-level search is represented as a two stages stochastic process. In the first stage firms draw from a Bernoulli process the event "access to innovation" (or to imitation), with a probability dependent on the amount of resources invested in search. A successful draw yields access to a second stochastic process determining the actual "innovation" (or imitation) defined by the input coefficients of a new technique (which in fact might turn out to be inferior to the incumbent one, and in that case the firm sticks to the latter).

The whole family of models typically assumes a process whereby advances are likely to occur in the neighborhood of the techniques already in use within any one firm: this is also a straightforward representation of the *cumulativeness and locality* of technological advances.[37]

It follows also from the foregoing discussion that the ways opportunities are tapped and degrees of success in doing so depend to a good extent upon the capabilities and past achievements of economic agents. So, more technically, think of "opportunities" as some measure on the set of input coefficients which are reachable at time t, with positive probability, conditional on the vector $\boldsymbol{x}_j(t)$ of coefficients that agent j ($j = 1,\ldots, n$) masters at that time. And, straightforwardly, the transition probabilities can be seen as capturing both paradigm-specific opportunities and capabilities, specific to each j for any given search effort.[38] Differing opportunities can be straightforwardly captured by different width of the support of the probability distribution of possible draws, as well as by the shape of the distribution itself.[39]

It is also relatively easy to formalize the "inducement mechanisms" discussed in Section 3.3. Effects on the *direction of search* formally imply that market shocks induce different partitions of the notional search space attainable at t, and focus search in those regions where one is more likely to find, say, savings on the inputs which are perceived as scarce and more expensive. Note that, for example, part of the (highly convincing) interpretation of inducements to mechanization in the American nineteenth-century economy suggested by David (1975) can be rephrased in this way.[40]

[37] Related formalizations of "local" technical learning are in Atkinson and Stiglitz (1969) and Antonelli (1995).

[38] This is to make things simple: in more complicated but more realistic accounts, allowing for imitation, transition probabilities of each j should depend also on the states achieved by all other agents and some metrics on their distances: see, for example, Chiaromonte and Dosi (1993), Dosi et al. (1994a), and Fagiolo and Dosi (2003).

[39] For example, in Dosi et al. (2006a), one assumes a β-distribution which, depending on the parameterization, may attribute the major mass to "bad draws" (in the case of scarce opportunities) and *vice versa*. The opportunities actually tapped depend crucially also on the agents' ability to explore and exploit them. In Nelson (1982), we sketch a model with a two-stage stochastic process ("study and test" and next "design/blueprint drawing") wherein agents' knowledge influences the "quality" of the choice set of new techniques—in terms of expected cost for achieving an advance of a given magnitude or expected magnitude of advances for a given R&D investment.

[40] Without any analytical loss, except the dubious commitments to rational choice with reference to a mysterious "innovation possibility frontier."

As already mentioned, relative prices may induce changes in the *revealed* directions of technological change even when the micro directions of search remain invariant.

Let us illustrate it by recalling the very basics of the Markov model of factor substitution from Nelson and Winter (1982, pp. 175–192).

It has been mentioned earlier that "innovative opportunities," when talking about process innovations, can be represented as the (bounded) set of states in the space of inputs (per unit of output) attainable starting from an arbitrary technique in use at time t. Suppose that search is a random process invariant in t (this implies that one excludes both decreasing returns to innovative efforts and those inducement effects upon search rules, discussed earlier). As already sketched in Section 3.3 when a new technique is drawn, it is compared with the one currently in use, given the prevailing input prices, and the minimum cost one is obviously chosen. The sequence of factor ratios displayed by a firm can be described by a Markov process characterized by the transition probability matrix $\boldsymbol{F} = [f_{ik}]$, where f_{ik} is the probability that state i follows state k.[41] Note that the transition matrix is time invariant but actual transition probabilities depend on relative input prices. This is because of the "comparison check": holding constant the initial technique and the one drawn, whether the latter will be adopted or not might depend on relative prices,[42] and such a choice will set different initial conditions for the next draw, etc. The intuition on dynamic-choice-of-technique inducement suggests that if the relative price of some input increases, the transition probabilities, loosely speaking of "getting away" from the techniques which intensively use that input will also increase. And in fact, Nelson and Winter (1982, pp. 180–192) establish the result, in a two-input case, that, with the appropriate ordering in terms of relative input intensities, the transition matrix $\hat{\boldsymbol{F}}$ (based on the new relative prices) stochastically dominates the "old" one, $\boldsymbol{F}$. It is an appealing result, resting so far on many formal qualifications, but certainly worth further exploration.[43] The bottom line is the following. Even if opportunities do not change and agents do not change their search rules, it is enough that relative prices enter into the criteria of choice between what has been found by search and what is already in use, to determine—in probability—"induced" changes in the patterns of factor use, at the level of individual firms and whole industries.[44]

[41] Nelson and Winter (1982), quite in tune with the general idea that there are "paradigm-based" constraints to the scope of factor substitution, assume that factor ratios can take only N possible values; thus, $i, k = 1, \ldots, N$.

[42] It obviously does not whenever the newly discovered technique is more efficient in terms of every input—a case which evolutionary interpretations easily allow.

[43] Among other points, the clarity of representation in terms of a time-invariant finite-state Markov process has its inevitable downside in that—taking seriously the question of "what happens as time goes to infinity?"—all persistent states return infinitely often in the limit (see also below on path dependency). However, it should not be formally impossible to make transition probabilities phase-space dependent, thus giving also more persistence to the weight of past "inducements." However, more down to the earth, does the fact that in the mathematical limit, say, Honduras will interchange with Sweden an infinite number of times weakens the (indeed, formally, transient-bound) proposition that both Sweden and Honduras are likely to display path-dependent technical coefficients over any reasonable, finite, window of observation?

[44] We do not dare extend this conjecture to whole economies, since not much has been done toward the exploration of multi-sectoral systems, linked by input–output relations, checking also the empirical plausibility of phenomena like reswitching of techniques, etc.—which appeared prominently in the theoretical debates in the 1970s and disappeared by magic later on. A few evolutionary formalizations are multisectoral, including Verspagen (1993) and some include also an admittedly rudimentary input/output structure such as Chiaromonte and Dosi (1993), Fagiolo and Dosi (2003), and Dosi et al. (2008b), but, to our knowledge none has addressed the dynamics of technique in a multisector "general disequilibrium" framework.

Evolutionary formalizations of search, innovation and imitation abhor any assumption of "rational technological expectations," and thus deny the possibility, in the actual world and in theory, of deriving the amount of resources devoted to search from unbiased expectations about probabilities of innovation/imitation the future returns from them. Rather, the somewhat extreme opposite assumption is generally made: propensity to invest in R&D are time-invariant behavioral routines possibly changed only if performances fell below a certain "satisfying" threshold (with few exceptions: see Silverberg and Verspagen, 1996 for a model with adaptive variations of such propensities to invest in search; Kwasnicki and Kwasnicka, 1992; Yildizoglu, 2002 for a model wherein R&D rules evolve stochastically by means of a genetic algorithm-based search).

Clearly firm-specific dynamics of innovation nurture a persistent heterogeneity across firms in terms of production efficiencies (and, too rarely in the models but most often in reality, product characteristics) curbed only to a partial extent by the processes of imitation. In turn, as we shall discuss in Section 4, such interfirm differences underlie different competitive abilities and contribution to shape the evolution of industrial structures.

3.9. Invention, innovation, and diffusion

Innovation diffusion is the subject of Chapter 17, and we refer to it for a more detailed survey of the evidence.[45] However, as that chapter is explicitly confined to equilibrium analyses of such an evidence, let us offer some basic elements of distinct interpretations more in tune with the evolutionary view outlined so far.

One of the contributions of J. Schumpeter's work that is often cited with reference to technological change concerns his distinction between invention, innovation, and diffusion. According to his definition, invention concerns the original development of some novel would-be process of production or product while innovation entails its actual introduction and tentative economic exploitation. Diffusion describes its introduction by buyers or competitors. It is a rough and "heroic" conceptual distinction, which can hardly be found in practice, since the empirical processes are usually never precisely like this. The invention is often introduced from the start as an innovation by economically minded research establishments. Diffusion entails further innovation on the part of both developers and users. All three activities are often associated with changes in the characteristics of, and incentives for, potential innovators/adopters. However, Schumpeter's distinction between invention, innovation, and diffusion is still a useful theoretical point of departure. For example, invention is suggestive of the sort of unexploited potential for technological progress whose sources we discussed above, while innovation and diffusion hint at the economically motivated efforts aimed at the incorporation of technological advances into economically exploitable products and processes.

The three major stylized facts already highlighted by early classic analyses including Mansfield (1961), Griliches (1957), Nabseth and Ray (1974), and Rosenberg (1972, 1976) are, *first* that diffusion is a time-consuming process, *second* that the speed varies widely across technologies and across countries, *third*, that diffusion of successful innovations most often follows S-shaped, but asymmetric, profiles (Figure 4 illustrates all three points). However, *fourth*, a good percentage of innovations, even when

[45] See also Hall (2005), Nakicenovic and Gruebler (1991), Geroski (2000) and Stoneman (2007), and the discussions from an evolutionary angle in Metcalfe (1988, 2005a).

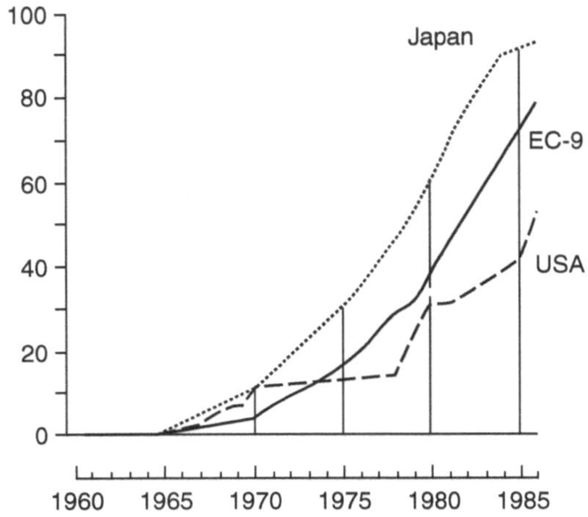

Figure 4. The diffusion of continuous casting of steel, as a percentage of total crude steel production. Source: Ray (1989, p. 4).

introduced by a small number of initial adopters, never diffuses and thus ultimately fail (hence also the sample selection bias stemming from considering only successful ones).

There are few basic ingredients which evolutionary analyses (of both the empirical and the theoretical kinds) share in the interpretation of diffusion dynamics. An obvious building block is the acknowledgment of the ubiquitous heterogeneity across would-be adopters on nearly every dimension which might think of as influencing adoption—ranging from sheer size all the way to different "absorptive capacities" (Cohen and Levinthal, 1990) and abilities to use the new techniques, pieces of equipment, and even consumption goods. Indeed, if one adds to adopters' heterogeneity also some dynamics in the characteristics of the good to be diffused, one goes a long way in accounting for the observed *retardation factors* in innovation diffusion (cf. David, 1990). The copious empirical literature estimating probit models of diffusion is well in tune.

On the supply side, heterogeneity is amply endogenous to the dynamics of learning, innovation, imitation, and selection among producers (see the next section): product characteristics and their prices change and with that also the market shares and the very identity of producers themselves.[46]

On the demand side, especially when the artifact to be diffused is a production good, learning by using is a powerful driver of diffusion. And, indeed, in evolutionary worlds, the ability to learn how to use and exploit new technologies is likely to be subject to unexpected bonanzas as well as dire delusions (the model in Silverberg et al., 1988 highlights the point; discussions of the related "cognitive biases" are in Dosi and Lovallo, 1997; Gary et al., 2008). Conversely, the frequent requirements of organizational changes associated with the adoption of innovations, especially when the latter are producer

[46] In fact, diffusion in production is intimately intertwined with the process of imitation, generally ridden with improvements in the initial artifact and in the techniques to produce it: an illustration of the point in the case of the steam engines in Rosenberg (1996).

goods, represent a powerful retardation factor, both with respect to adoption as such and to the reaping of its economic benefits (Brynjolfsson and Hitt, 2000 convincingly illustrate the point).

The process involves important collective dimensions as well, including knowledge spillovers, network externalities, endogenous evolution of preferences, as well as sheer herd behaviors.

How does one formally represent such dynamics? In a nutshell, full-fledged evolutionary models of innovation, imitation, and selection basically entail diffusion dynamics as a corollary of the whole process (Silverberg et al., 1988 is an early example). Interestingly, evolutionary models are capable of generating the major "stylized facts" of diffusion dynamics recalled earlier as *emergent properties* of the evolutionary process whereby the system collectively "self-organize" around the use of a new technology. However, an interesting family of "reduced-form" models compresses the interfirm competition dynamics while offering a succinct account of diffusion nested into heterogeneous populations, and driven by dynamic increasing returns, network effects, and endogenous preferences. A powerful and versatile formal instrument are *generalized Pólya urns* (cf. Arthur et al., 1987; Dosi and Kaniovski, 1994; Bassanini and Dosi, 2001, 2006). Let us just recall here that such formal machinery is well apt to account for (a) the influences of stochastic events along the evolutionary dynamics upon the long-term outcomes (and thus the related path dependency of technology selection); (b) the widespread importance of dynamic increasing returns (possibly intertwined with forms of decreasing returns within "badly behaved" dynamics: cf. Dosi and Kaniovski, 1994); and (c) the possibility that technological evolution "gets it wrong" (in the sense of convergence to the dominance of a technology which is "inferior" to other ones available in some form from the start, which however the collective dynamics of adoption did not reinforce: see below).

Can one identify different families of evolutionary processes of diffusion? An attempt to do so is in Nelson et al. (2004) where one distinguishes four "archetypes" of diffusion patterns conditional on the presence/absence of dynamics increasing returns and of sharp persuasive feedbacks on the returns to adoption itself. Phenomena like fads belong to one extreme (absence on both dimensions), while QWERTY-type diffusion (David, 1985) belongs to the opposite one. To recall, the QWERTY keyboard became dominant, as David argued, through path dependent externalities in production and use, notwithstanding its intrinsic inferiority to other configurations.

3.10. The path dependence of the processes of technological evolution

Two quite general features of the processes of technological innovation discussed so far are dynamic increasing return, path dependency and their interaction. Since other chapters of this Handbook are devoted to these two topics, we need not address the details of such phenomena. However, again, let us flag their role in technological evolution. (We shall come back to some of the issues below when addressing industrial evolution.)

Let us consider the relationship between evolutionary success, intrinsic "fitness," and chance (i.e., unpredictable historical events) in the development and diffusion of innovations.

Students of technical advance long have noted that, in the early stages of a technology history, there usually are a number of competing variants or even competing paradigms. This was the case of vehicles, some driven by the combustion engines, some by steam engines, and some by batteries. As we know, gasoline-fuelled engines came to dominate and the other two possibilities were mostly abandoned. The standard interpretation for this is that gasoline engines were potentially superior and with time, trial and error and learning such superiority became manifest. There is, however, an alternative explanation grounded in the interaction between dynamic increasing returns of some kind, network externalities, and path dependency (cf. Arthur, 1988, 1989; David, 1985, 1988, 2001b; Dosi and Kaniowski, 1994; and a few contributions to Antonelli et al., 2006). In this second interpretation, the internal combustion engine need not have been innately superior. All that would have been required was that, because of a run of luck, it became heavily used or bought, and this started a rolling snowball mechanism fuelled by some sort of collective positive feedback.

What might be behind an increasing returns rolling snowball? Arthur, David, and other authors suggest several different possibilities. One of them is that the competing technologies involved are strongly cumulative technologies. In a cumulative technology, today's technical advances draw from and improve upon the technology that was available at the start of the period, and tomorrow's in turn build on today's. So, in the case of the history of automobile engine technology—according to the cumulative technology interpretation—gasoline engines, steam engines, and electrical engines, all were plausible alternative technologies for powering cars, and it was not clear which of these means would turn out to be superior. Reflecting this uncertainty, different inventors tended to make different technological bets. Assume, however, that simply as a matter of chance (or marginal choice or political decision), a large share of these efforts just happened to focus on one of the variants—for example, the internal combustion engine—and as a result, over this period there was much more overall improvement in the design of internal combustion engines than in the design of the two alternative power sources. Or, alternatively, assume that while the distribution of inventive efforts were relatively even across the three potential paradigms simply as a matter of chance significantly greater advances were made on internal combustion engines that on the other ones. But then, at the end of the first period, if there were a rough tie before, gasoline-powered engines now are better that steam or electric engines. Cars embodying internal combustion engines will sell better. More inventors thinking about where to allocate their efforts now will be deterred from allocating their attention to steam or electric engines because large advances in these need to be achieved before they would become competitive even with existing internal combustion engines. Thus, there are many strong incentives for the allocation of inventive efforts to be shifted toward the variant of the technology that has been advancing most rapidly. The process is cumulative. The consequences of increased investment in advancing internal combustion engines, and diminished investment in advancing the other two power forms, are likely to be that the former pulls even further ahead. Relatively shortly, a clear dominant paradigm has emerged. And all the efforts to advance technology further in this broad area come to be concentrated on improving that particular paradigm.

There are two other largely complementary dynamic increasing returns stories. One stresses network externalities or other advantages to consumers or users if what different individuals buy are similar, or compatible, which lends advantage to a variant that just happened to attract a number of customers already. The other stresses systems aspects where a particular product has a specialized complementary product or service, whose development lends that variant special advantage. Telephone and computer networks, in which each user is strongly interested in having other users have compatible products, are commonly employed examples of the first case. Video cassette recorders which run cassettes that need to be specially tailored to their particular design, or computers that require compatible programs, are often used examples of the second. David's (1985) story of the reasons why the seemingly inefficient "QWERTY" typewriter keyboard arrangement has persisted so long as a standard involves both its familiarity to experienced typists and the existence of typewriter training programs that teach

QWERTY. As in the QWERTY story, the factors leading to increasing returns often are intertwined, and also linked with the processes involved in the development of cumulative technologies. Thus, to return to our automobile example, people who learned to drive in their parents' or friends' car powered by an internal combustion engine naturally were attracted to gas-powered cars when they themselves came to purchase one, since they knew how they worked. At the same time the ascendancy of automobiles powered by gasoline-burning engines made it profitable for petroleum companies to locate gasoline stations at convenient places along highways. It also made it profitable for them to search for more sources of petroleum, and to develop technologies that reduced gasoline production costs. In turn, this increased the attractiveness of gasoline-powered cars to car drivers and buyers.

Note that, for those who consider gas engine automobiles, large petroleum companies, and the dependence of a large share of the nation's transportation on petroleum, a complex that spells trouble, the story spun out above indicated that "it did not have to be this way." If the toss of the die early in the history of automobiles had come out another way, we might today have had steam or electric cars. A similar argument recently has been made about the victory of AC over DC as the "system" for carrying electricity (David, 1992). The story also invites consideration of possibly biased professional judgments and social or political factors as major elements in the shaping of long-run economic trends. After all, in these stories all it takes may be just a little push.

It is difficult to precisely assess the importance and frequency of such path-dependent processes, since of course counterfactuals involving "running the tape of history another time" are impossible (in social sciences but also in biology). Come as it may, evolutionary interpretations of technological change—and as we shall see of industrial dynamics and development—are deeply skeptical of any view of evolution as the inevitable unfolding of a process leading from the good to the better. Such a view tries to justify and explain any end state of the system as being the best possible outcome given the (perceived) constraints by imperfectly informed but fully "rational" agents along the whole path. The view emphatically illustrated in Liebowitz and Margolis (1995) basically aims at rationalizing whatever one observes as an equilibrium and, at the same time, at attributing rational purposefulness to all actions which led to any present state.

On all that, David (2001b) and Dosi (1997) coincide in the rejection of any Panglossian interpretations of history as "the best which could have happened," mainly "proved" by the argument that "rational agents" would not have allowed anything short of the optima to happen (compare the amazing similarities with Dr. Pangloss' remarks in Voltaire's *Candide* on the optimizing virtues of Divine Providence).

4. Schumpeterian competition and industrial dynamics

The evidence discussed in the previous section highlights both the general characteristics that technological knowledge displays and at the same time the widespread diversity in the mode and efficacy by which individual firms access and exploit such knowledge even when undertaking very similar activities and operating in the same lines of business.

Idiosyncratic capabilities and, dynamically, idiosyncratic patterns of learning by individual firms are the general rule. In turn, such persistently heterogeneous firms are nested in competitive environments which shape their individual economic fate and collectively the evolution of the forms of industrial organization. In the following, we shall first offer an overview of some broad features of such competitive environments. Next, we shall consider at greater detail a few properties of the processes of industrial evolution, trying to distinguish those elements which are common to all industries and others which are regime-specific. Finally, we shall discuss the modeling efforts which try to interpret the patterns of industrial evolution.

Differences in products, and in processes of production—and as a consequence in costs and prices—are central features of the competitive process in which firms are involved at multiple levels. Let us call *Schumpeterian competition* the process through which heterogeneous firms compete on the basis of the products and services they offer and get selected, with some firms growing, some declining, some going out of business, and some new ones always entering on the belief that they can be successful in this competition. Such processes of competition and selection are continuously fuelled by the activities of innovation, adaptation, and imitation by incumbent firms and by entrants. Such processes involve both selection *across firms*, and learning and selection among techniques, organizational practices, and product attributes *within* the firms themselves.

In all that user selection of particular technological variants over others, together with firms' selection in financial markets, are central drivers of competition, industrial demographics, and changing industry structures. It is important to consider both users and suppliers. It is reasonable to start from the observation that the production and adoption of "superior" consumption goods, capital goods, and intermediate inputs often underlies the competitive advantage of particular firms. And, indeed, a major analytical question bears on the precise drivers and mechanisms of the competition process. Another one regards how long "competitive advantages," of whatever kind, last. In industries where a company which introduces a very attractive innovation is able to prevent rapid imitation by competitors, and also is able to expand its own market share rapidly, the result may be a highly concentrated industry. This certainly has been the result in some well-known cases, for example, IBM's long domination of the mainframe computer industry, and Intel's continuing domination of the market for microprocessors. However, in many other instances successful innovators have not been able to develop and hold on to a dominant market position, in the face of continuing efforts at innovation by their competitors. Joseph Schumpeter employed the term "creative destruction" to refer both to the nature of technological advance, and to what often happens to leading firms in industries where technological advance is rapid and incumbents are unable to seize novel opportunities. In fact, significant changes in industrial structure as a result of innovation are more likely when the success of a particular new product or process is associated with the ascendancy of new technological paradigms. Successful innovations in these cases are associated with different design concepts, or different ways of doing things, than what it replaced. Continued viability of firms in this area of activity then may require learning to work effectively with the (partly) new knowledge bases and new organizational routines. In such a context, an industry structure that had been stable for a considerable period of time may be ripe for the success of new entrants.

If we step back from the details of particular industry patterns, there are a few general properties that stand out from industry studies. *First*, as Schumpeter, and Marx before him, argued long ago, competition in industries where innovation is central has little to do with the idea that such process generates results that are economically "efficient" in the standard static sense of that concept in economics. What is driving the process is the striving by some firms to get an economic advantage over their competitors. As discussed in Section 3, both the cross section and the time profiles of modern industrial sectors inevitably show considerable variation across firms in measures of economic efficiency and in profitability: in short, industries are characterized by considerable and persistent "inefficiency" in the standard allocative sense of that term. *Second*, in industries marked by continuing innovation, competitive conditions may be fragile. This applies particularly to the cases whereby firms who have been successful innovators are able to hold off imitation or other effective competitive responses, and their profitability enables them to stretch their advantage further. *Third*, this notwithstanding, while the evolutionary notion of "competition" differs from competition of the economic textbooks in fundamental respects, it does serve a related function. To the extent competition is preserved, a significant share of the benefits of technological progress go to the customers/users of the technology. And on the supply side, over industrial evolution, competition tends to roughly keep prices moving in line with costs (including R&D costs).

This is the bird eye interpretation of innovation-driven competition and the ensuing industrial evolution. How well does it hold against the evidence? Are there some finer regularities in such processes? What are the distinct characteristics of firms and their distribution which systematically persist over time, if any? How do such characteristics within the population of competing firms affect their relative evolutionary success? And, moreover, among the foregoing properties and relation between them which ones are invariant across industries, and, conversely, which ones depend on the technological and market characteristics of particular sectors?

Let us begin with the evidence concerning some features of the *dynamics* in (i) *industrial structures* and *firms characteristics*, broadly understood to cover variables such as size, productivity, innovativeness, and their intraindustry distributions; (ii) *performances*—including individual profitabilities, growth profiles, and survival probabilities, together, again, with their aggregate distributions; and (iii) their mapping into *regimes of learning*—for example, modes of innovative search, etc. (cf. Section 3.7).[47]

4.1. *Microeconomic heterogeneity: Size, to begin with*

We have repeatedly emphasized it already: firms persistently differ over all dimensions one is able to detect.

A first, extremely robust, "stylized fact" regards the quite wide variability in firm sizes. More precisely, one observes—throughout industrial history and across all countries—right-skewed distributions of firm sizes: within a large literature see Steindl (1965), Hart and Prais (1956), Ijiri and Simon (1977), Hall (1987), Bottazzi et al. (2007), Lotti et al. (2003), Bottazzi and Secchi (2005), and Dosi (2007).

Irrespectively of the precise form of the density function, the intuitive message is the coexistence of many relatively small firms with quite a few large and very large ones—indeed in a number much higher than the one would predict on the ground of any Gaussian shape. In turn, all this militates against any naive notion of some "optimal size" around which empirical distributions should be expected to fluctuate. Notice that, as a consequence, also any theory of production centered around invariant U-shaped cost curves, familiar in microeconomic theory, looses a lot of plausibility: were they the

[47] See more on all this in Dosi et al. (1995, 1997) and Dosi (2007), where one can find also a more detailed discussion of the literature.

rule, one ought to reasonably expect also a tendency to converge to the corresponding technologically optimal equilibrium size. On the contrary, plausible candidates to the representation of the empirical size distributions are the log-normal, Pareto, and Yule ones. Certainly, the full account of the distributions suffers from serious problems in offering also an exhaustive coverage for the smallest firms. Recent attempts to do that, such as Axtell (2001) on the population of US firms, lend support to a *power law* distribution linking firm size probability densities with the size ranking of firms themselves.

All this primarily concerns *aggregate manufacturing* firm size distributions. Are these properties robust to disaggregation? Size differences are. However an increasing body of finer sectoral data suggest that in fact invariances in the distributions are not. Corroborating a conjecture put forward in Dosi et al. (1995) and further explored in Marsili (2001), aggregate "well-behaved" Pareto-type distributions may well be a puzzling outcome of sheer aggregation among diverse manufacturing sectors, characterized by diverse regimes of technological learning and market interactions, which do not display Paretian distributions. While some sectors present distributions rather similar to the aggregate ones, others are almost log-normal and yet others are bimodal or even multimodal. (More evidence is summarized in Dosi, 2007). Together, admittedly circumstantial evidence hints at a plausible oligopolistic core versus fringe firms separation in several sectors—indirectly supported by the mentioned bimodality of size distributions.[48]

Finally, note that even relatively stable industrial structures—as measured in terms of stability of size distributions—hide a much more turbulent microeconomics. Incumbents change their relative share and ranking[49] with a lot of "churning" of new firms: roughly half disappear before they get to the age of 5,[50] but a subset of the survivors grows to significant share of most industries, and is also an important carrier of innovation and productivity growth.[51]

Come as it may, industrial structures—in this case proxied by size distributions—are the outcomes of the growth dynamics undergone by every entity in the industrial population (jointly, of course, with birth and death processes). What about such growth processes?

4.2. Corporate growth rates and corporate profitabilities

There are many studies that have explored empirically the extent to which Gibrat's law, which proposes that firm growth rates are multiplicative and statistically independent of size, is a good first approximation of actual industrial dynamics. Lotti et al. (2003) provides a rich review. The evidence suggests that:

(i) Most often, smaller firms that survive over the period under analysis on average grow faster than larger firms. However, most studies do not count firms in existence at the start of the period that disappear somewhere over the period, and many small firms are young firms that generally have high mortality rates.

[48] Indeed, an important research task ahead concerns the transition probabilities between "core" and "fringe."

[49] Cf. with Louça and Mendonça (2002) on long-term patterns in the upper tail of the size distributions over the whole industrial sector. However, within-industry rankings seem to be rather inertial: on German evidence Cantner and Krüger (2004). See also the comments in Dosi et al. (2008c).

[50] For comparative evidence of the OECD countries, compare with Bartelsman et al. (2005).

[51] Converging pieces of evidence are in Audretsch (1997), Baldwin and Gu (2006), and Foster et al. (2008).

(ii) No strikingly robust relationship appears between size and average rates of growth (cf. Bottazzi and Secchi, 2006; Bottazzi et al., 2003; Coad, 2008; Hall, 1987; Kumar, 1985; Mansfield, 1962; Sutton, 1997 among others). The relationship between size and growth is modulated by the age of firms themselves—with age, broadly speaking, exerting *negative* effects of growth rates, but *positive* effects on survival probabilities, at least after some post-infancy threshold (cf. Evans, 1987).[52]

Such pieces of evidence are easily consistent with evolutionary theories of industrial change. Indeed an evolutionary interpretation would be rather at odds with a notion of convergence to some invariant "optimal" size, with decreasing returns above it. Conversely, it is rather agnostic on the precise specification of *non-decreasing* returns. In particular, it does not have any difficulty in accepting a world characterized by *roughly constant returns to scale*, jointly with drivers of firm growth uncorrelated on average with size itself. Conversely, precious clues on the basic characteristics of the processes of market competition and corporate growth are offered by the statistical properties of the "error term." Note in this respect that the absence of any structure in the growth processes would be very damaging indeed to evolutionary theories of industrial change. In fact, if one were to find corroboration to any "strong Gibrat" hypothesis according to which growth would be driven by a multiple, small "atomless" uncorrelated shocks, this would come as bad news to evolutionary interpretations whose basic building blocks—to recall—comprise the twin notions of (i) persistent heterogeneity among agents and (ii) systematic processes of competitive selection among them. What properties in fact do the statistics on firm growth display?

One of the most important pieces of evidence able to throw some light on the underlying drivers of corporate growth regards the distribution of growth rates themselves. The evidence suggests an extremely robust stylized fact: growth rates display distributions which are *at least exponential (Laplace) or even fatter in their tails*.[53] This property holds across (i) levels of aggregation, (ii) countries, (iii) different measures of size (e.g., sales, employees, value added, assets), even if (iv) one observes some (moderate) variations across sectors with respect to the distribution parameters. Such statistical properties are indeed good news for evolutionary interpretations. The generalized presence of fat tails in the distribution implies much more structure in the growth dynamics than generally assumed. More specifically, ubiquitous fat tails are a sign of some underlying correlating mechanism which one would rule out if growth events were normally distributed, small, and independent. In Bottazzi et al. (2003) and Dosi (2007), one conjectures that such mechanisms are likely to be of two types. First, the very process of competition induces correlation. Market shares must obviously add up to one: someone's gain is someone else's loss. Second, in an evolutionary world one should indeed expect "lumpy" growth events (of both positive and negative sign) such as the introduction of new products, the construction/closure of plants, entry to and exit from particular markets.[54]

Together with corporate growth, profitability is another crucial measure of revealed corporate performances. Concerning the variable, there is indeed a robust literature on the *persistent profitability*

[52] Moreover, the statistical relationships between size and growth rates appear to be influenced by the stage of development of particular industries along their life cycles: cf. Geroski and Mazzucato (2002).

[53] See Stanley et al. (1996) and Bottazzi and Secchi (2003) on US data, Bottazzi et al. (2001a) on the international pharmaceutical industry, Bottazzi et al. (2002, 2003) on the Italian industry, and the discussion in Dosi (2007).

[54] A suggestive attempt to model increasing-return dynamics yielding the observed fat-tailed distribution is in Bottazzi and Secchi (2005).

differences across firms: see, among others, Mueller (1986, 1990), Cubbin and Geroski (1987), Geroski and Jacquemin (1988), Geroski (1998), Goddard and Wilson (1999), Cefis (2003a), Gschwandtner (2004), and Dosi (2007). Moreover, the autocorrelation over time in profit margins is extremely high in all manufacturing sectors, with just a relatively mild tendency to mean reversion, while, interestingly, the rates of change in profit margins display distributions which are again fat-tailed (at least exponential, or even fatter-tailed). That is, we find again here the mark of powerful underlying correlation mechanisms which tend to induce "coarse-grained" shocks upon profitabilities.

Indeed, the bottom line is that core indicators of corporate performances such as growth and profitability confirm the already familiar widespread multifaceted *heterogeneity* across firms notwithstanding the competition process. Given all that, a natural question concerns the roots of such heterogeneity itself.

4.3. Behind heterogeneous performances: Innovation and production efficiency

Straightforward candidates for the explanation of the differences in corporate performances are in fact (i) differences in the ability to innovate and/or adopt innovation developed elsewhere regarding product characteristics and production processes, (ii) different production efficiencies, (iii) different organizational arrangements, and (iv) different propensities to invest and grow conditional on the foregoing set of variables. Plausibly the former three ensembles of variables may be expected to be related with each other (the behavioral aspects are a distinct matter). For example, technological innovations typically involve also changes in the organization of production; different ways of searching for innovations imply distinct organizational arrangements regarding the relationships among different corporate tasks (e.g., R&D, production, sales, etc.). And, intuitively, technological and organizational innovations ultimately shape the degrees of efficiency in which inputs happens to generate outputs.

What is the evidence concerning the patterns of technological innovation, on the one hand, and production efficiencies on the other? (We are forced to neglect here the role of organizational variables. In fact, *organizational capabilities* are intimately linked with the very process of technological innovation and with production efficiencies: cf. the insightful evidence in Brynjolfsson and Hitt, 2000.)

We have discussed at length in Section 3.7 the evidence on asymmetries in production efficiencies—no matter how measured, for example, in terms of labor productivities or TFPs: widespread and persistent asymmetries are the general rule.

Together, the literature on the economics of innovation surveyed in Section 3 primarily from the angle of knowledge dynamics, indeed suggests widespread differences across firms in their ability to innovate:

(i) Innovative capabilities appear to be highly asymmetric, with a rather small number of firms in each sector responsible for a good deal of innovatios even among highly developed countries.
(ii) Somewhat similar considerations apply to the *adoption* of innovations, in the form of new production inputs, machinery, etc. (see Section 3.9 on "diffusion") revealing asymmetric capabilities of learning and "creative adaptation."
(iii) Differential degrees of innovativeness are generally persistent over time and often reveal a small "core" of systematic innovators (cf. Bottazzi et al., 2001a; Cefis, 2003b; Cefis and Orsenigo, 2001; Malerba and Orsenigo, 1996a among others).

(iv) Relatedly, while the arrivals of major innovations are rare events, they are not independently distributed across firms. Rather, recent evidence suggests that they tend to arrive in firm-specific "packets" of different sizes.[55]

In fact, all the evidence on wide asymmetries in the abilities to innovate and imitate is consistent with the interpretation of the patterns of knowledge accumulation put forward in Section 3. And so is the evidence on micro correlations of innovative events, well in tune with an evolutionary notion of few, high-capability, persistent innovators.

On a much larger scale, the persistent asymmetries across countries, even within the same lines of business, cry out in favor of profound heterogeneities in learning and searching capabilities.[56]

4.4. Corporate capabilities, competition, and industrial change

Differences in innovative abilities and efficiencies (together with differences in organizational setups and behaviors) ought to make up the distinct corporate "identities" which in turn should somehow influence those corporate performances discussed above.

But do they? How? And how are these relations influenced by behavioral (partly "strategic") considerations on the side of individual firms?

Let us consider first the impact of different degrees of innovativeness and different efficiencies upon profitability, growth, and survival probabilities.

In several studies, firms that are identified as innovators tend to be more profitable than other firms: see Geroski et al. (1993), Cefis (2003a), Cefis and Ciccarelli (2005), Roberts (1999), and Dosi (2007) among others. Production efficiency also shows a systematic positive influence upon profitability (cf. Bottazzi et al., 2009; Dosi, 2007).

The impact upon growth is much less clear cut. Certainly, there are some serious questions about how both superior innovative performance and superior production efficiency are identified and measured.[57] Even if the measurements are taken at face value, the impact of both measured innovativeness and production efficiency upon growth performances appear to be quite uncertain. Mainly North American evidence, mostly at *plant* level, does suggest that increasing output shares in high-productivity plants and decreasing shares of output in low-productivity ones are important drivers in the growth of sectoral productivities, even if the process of displacement of lower efficiency plants is rather slow (cf. the evidence discussed in Ahn, 2001; Baily et al., 1992; Baldwin, 1995; Baldwin and Gu, 2006). *Firm-level* data are less straightforward. For example, Italian and French data (cf. Bottazzi et al., 2009; Dosi, 2007)

[55] On the statistical properties of the discrete innovations, in general, cf. Silverberg (2003) showing a secular drifting Poisson-type process. However, at a much finer level of observation the firm-specific patterns of innovation do not happen to be Poisson-distributed. Rather, as one shows in Bottazzi et al. (2001a) in the case of the pharmaceutical industry, few firms "draw" relatively large "packets" of innovations well described by Bose–Einstein (rather than Poisson) statistics.

[56] Much more on that in Dosi et al. (1990), Verspagen (1993), Fagerberg (1994), Nelson (1996), and Cimoli et al. (2009).

[57] An important caveat here is that there might be an intrinsic sample selection bias in the data in favor of *successful* innovations: firms that try to innovate and do badly are not adequately counted as innovative firms. Another caveat, is that generally "efficiency" is measured, due to data availability, in terms of deflated value added or deflated sales, folding together price and volume levels, and dynamics. A rare exception is Foster et al. (2008) who are able to draw upon microdata separating the two at microlevel.

show a weak or nonexistent relationship between relative (labor) productivities and growth: more efficient firms do *not* grow more. Moreover even when some positive relation between efficiency and growth appears, this is almost exclusively due to the impact of few *outliers* (the very best and the very worst).

Concerning the impact of innovation the evidence from some industry-specific data sets such as the international pharmaceutical industry shows that more innovative firms do *not* grow more (Bottazzi et al., 2001a; for some qualifications of the statement still on the drugs industry cf. Demirel and Mazzucato, 2008; and concerning a few high-tech sectors cf. Coad and Rao, 2008). Rather the industry constantly displays the coexistence of heterogeneous types of firms (e.g., innovators vs. imitators). There is a sort of a puzzle here awaiting further research in that such statistical evidence appears to be somewhat at odds with more qualitative reconstructions of industrial evolution whereby technological advances appear to be at the centre of competitive advantages and ultimately the drive toward corporate leadership: cf. among others Dosi (1984) on semiconductors and Murmann (2003) on chemicals.

In complementary efforts, a growing number of scholars has indeed began doing precisely what we could call *evolutionary accounting* (even if most do not call it that way; however for an early example of the *genre*, cf. Nelson and Winter, 1982). The fundamental evolutionary idea is that distributions (including, of course, their means, which end up in sectoral and macro statistics!) change as a result of (i) learning by incumbent entities, (ii) differential growth (i.e., a form of selection) of incumbent entities themselves, (iii) death (indeed, a different and more radical form of selection), and (iv) entry of new entities. Favored by the growing availability of micro longitudinal panel data, an emerging line of research (see Baily et al., 1996; Baldwin and Gu, 2006; Bottazzi et al., 2009; Brown et al., 2006; Foster et al., 2001 among others, and the discussion in Bartelsman and Doms, 2000) investigates the properties of decompositions of whatever mean sectoral performance variable, typically productivity of some kind, of the following form, or variations thereof:

$$\begin{aligned}\Delta\Pi_t = &\textstyle\sum_i s_i(t-1)\Delta\Pi_i(t) + \sum_i \Pi_i(t-1)\Delta s_i(t) \\ &+ \textstyle\sum_e s_e(t)\Pi_e(t) + \sum_f s_f(t-1)\Pi_f(t-1) \\ &+ \text{some interaction terms},\end{aligned} \tag{2}$$

where Π are the productivities (or, for that matter, some other performance variables), s are the shares[58] of each firm in the industry total, while i is an index over incumbents, e over entrants, and f over exiting entities.

The first term stands for the contribution of firm-specific changes holding shares constant (sometimes called the *within* component), the second one captures the effects of the changes in the shares themselves, holding initial firm productivity levels constant (also known as the *between* component) and the last two take up the effect of entry and exit, respectively.

Of course, there is a considerable variation in the evidence depending on countries, industries and methods of analysis. However, some patterns emerge. *First*, the *within* component generally is significantly larger than the *between* one: putting it another way improvement of productivity by existing firms *dominates selection* across firms as a mode of industry advancement—at least concerning productivity (both labor and TPF). This emerges both from the foregoing "evolutionary accounting" exercises and from estimates of the relationship between efficiency and subsequent growth, allowing for firm fixed effects. And, it holds in both the short and the medium term. So, for example, in the analyses of Bottazzi et al. (2009) on Italy and France, firm-specific factors generally account for almost an order of

58 Shares in terms of what is a delicate issue: in terms of output? Value added? Or, conversely, employment? Relocation of resources and output across firms involves both changes in inputs and market shares.

magnitude more than "selection" of the variance in firm growth rates. *Second*, relative efficiencies do influence survival probabilities, and it may well turn out that selective mechanisms across the population of firms operate much more effectively in the medium–long term at this level rather than in terms of varying shares over the total industry output.

We have focused so far upon the linkages between admittedly rough proxies for innovativeness and productivity, on the one hand, and growth and survival, on the other. What about the relationships between profitability and the latter two variables? The evidence we are familiar with strikingly shows little or no link between profitability and firm growth of incumbents (cf. again Bottazzi et al., 2009 on Italian and French longitudinal data). However, other pieces of evidence suggest also systematic effects of profitability upon survival probabilities (cf. the discussion in Bartelsman and Doms, 2000; Foster et al., 2008).

The implications of all these empirical regularities are far-reaching.

Certainly, the recurrent evidence at all levels of observation of *interfirm heterogeneity* and its persistence over time is well in tune with an evolutionary notion of idiosyncratic learning, innovation (or lack of it) and adaptation. Heterogeneous firms compete with each other and, given (possibly firm-specific or location-specific) input and output prices, obtain different returns. Putting it in a different language, they obtain different "quasi rents" or, conversely, losses above/below the notional "pure competition" profit rates. Many firms enter, a roughly equivalent number of firms exits. In all that, the evidence increasingly reveals a rich structure in the processes of learning, competition and growth. As mentioned, various mechanisms of correlation—together with the "sunkness" and indivisibilities of many technological events and investment decisions—yield a rather structured process of change in most variable of interest—for example, size, productivity, profitability—also revealed by the "fat-tailedness" of the respective growth rates. At the same time, market selection among firms—the other central mechanism at work together with firm-specific learning in evolutionary interpretations of economic change- does not seem to be particularly powerful, at least on the yearly or multiyearly timescale at which statistics are reported (while the available time series are not generally long enough to precisely assess what happens in the long run, say, decades). Conversely, diverse degrees of efficiencies and innovativeness seem to yield primarily relatively persistent profitability differentials. That is, contemporary markets do not appear to be too effective selectors delivering rewards and punishments in terms of relative sizes or shares—no matter how measured—according to differential efficiencies. Moreover, the absence of any strong relationship between profitability and growth militates against the "naively Schumpeterian" (or for that matter "classic") notion that profits feed growth (by plausibly feeding investments). Selection among different variants of a technology, different vintages of equipment, different lines of production does occur and is a major driver of industrial dynamics. However, it seems to occur to a good extent *within* firms, driven by the implementation of "better" processes of production and the abandonment of older less productive ones.

Finally, the same evidence appears to run against the conjecture, put forward in the 1960s and 1970s by the "managerial" theories of the firm on a tradeoff between profitability and growth with "managerialized" firms trying to maximize growth subject to a minimum profit constraint.[59]

[59] In fact, the absence of such a tradeoff had been already noted by Barna (1962). Note also that this proposition is orthogonal to the finding that current growth appears to be correlated with *future* long-term profitability (cf. Geroski et al., 1997).

In turn, the (still tentative) observation that market selection that winnows directly on firms may play less of a role than that assumed in many models of evolutionary inspiration (see below) demands further advances in the understanding of how markets work (or do not), and of the structure of demand (broadly in the perspective of this work, cf. Nelson, 2008b, and Aversi et al., 1999). Here note the following. *First*, one measures "efficiency"—supposedly a driver of differential selection—very imperfectly: we have already mentioned, as emphasized by Foster et al. (2008), that one ought to disentangle the price component of "value added" (and thus the "price effect" upon competitiveness) from "physical efficiency" to which productivity strictly speaking refers. This applies to homogeneous products and even more so when products differ in their characteristics and performances: as this is often the case in modern industries, one ought to explicitly account for the impact of the latter upon competitiveness and revealed selection processes. *Second*, but relatedly, the notion of sharp boundaries between industries and generalized competition within them is too heroic to hold. It is more fruitful in many industries to think of different submarket of different sizes as the locus of competition (cf. Sutton, 1998). The characteristics and size of such submarkets offer also different constraints and opportunities for corporate growth. Ferrari and Fiat operate in different submarkets, face different growth opportunities and do not compete with each other. However, the example is interesting also in another respect: Fiat can "grow," as it actually happened, by acquiring Ferrari. *Third*, a growing microevidence highlights the intertwining between technological and organizational factors as determinants of Schumpeterian competition: Bresnahan et al. (2008) illustrate the point in the case of IBM and Microsoft facing the introduction of the PC and the browser, respectively. Both firms, the work shows, faced organizational diseconomies of scope precisely in the corporate activities where they were stronger. *Fourth*, in any case, the links between efficiency and innovation, on the one hand, and corporate growth, on the other, are mediated by large degrees of *behavioral freedom*, in terms, for example, of propensities to invest, export, expand abroad; pricing strategies; patterns of diversification; etc.

4.5. Industry-specific dynamics and industry life cycles

So far, we have discussed some properties of industrial evolution which appear to hold broadly across all industrial sectors. Conversely, are there sectoral specificities in the patterns of industrial evolution? And do they map into those different technological and production regimes discussed above? Moreover, different sectors happen to be at different stages of their life cycles. How does that influence the characteristics of the processes of industrial evolution?

In fact, significant industry-specific differences emerge from the data. The finding that variables like capital intensity, advertising intensity, R&D intensity—along with structural measures like concentration and performance measures like profitability—differ widely across sectors is at the very origin of the birth of industrial economics as a discipline. Longitudinal microdata add further evidence. So for example, Jensen and McGuckin (1997) observe that industry-specific effects also significantly influence firms' heterogeneity, even if most of the observed variance in plants and firms characteristics is *within* industries.[60] Thus, it

[60] Other studies (e.g., Geroski and Jacquemin, 1988; Mueller, 1990) showed that the persistence of profit also appears to depend on industry-specific characteristics as well as on firm-specific ones. In particular, industry-specific features such as the intensity of advertising and of R&D appear to be highly correlated with the persistence of higher than average profits.

should not come as too big a surprise that phenomena like entry, exit, and survival, persistence in firms attributes and performances, innovative activities and firms' growth also exhibit significant interindustry variability. Audretsch (1997) reports on the relationships between entry, exit, and survival entrants on the one hand, and industry characteristics like the rate of innovation and capital intensity on the other. This evidence suggests, in particular, that survival is easier in those industries in which small firms are important sources of innovation, and that new surviving firms tend to grow faster in innovative industries and as a function of the gap between minimum efficient scale of output and actual firm size. At the same time, however, the likelihood of survival decreases as a function of that gap. The same happens in terms of innovation rates.

Can one move a step further and link at least some characteristics of evolutionary patterns with the underlying technological regimes? It is a conjecture put forward in Winter (1984) and Dosi et al. (1995), explored in both circumstances via simulation models, which the empirical evidence begins to corroborate (Marsili, 2001; Marsili and Verspagen, 2002), even if probably more disaggregate classification of the regimes themselves are needed beyond the "Schumpeter Mark I" versus "Schumpeter Mark II" distinction. Together *market regimes* variables have to be introduced (Marsili and Verspagen, 2002).

Do different industrial regimes correspond also to different innovation strategies of business firms? The issue is still largely underexplored; however, Srholec and Verspagen (2008) suggest that within a sector, strategic heterogeneity dominates upon sectoral effects: indeed, a challenging puzzle crossing over economics and strategic management.

Thus far our discussion has been concerned with differences that exist across industries at any time. Now we shift our attention to changes that occur over time within an industry.

No matter the technological regime in which they are embedded, individual industries evolve since their emergence all the way to their maturity, and frequently decline.

Klepper (1997) offers a broad fresco of many *industry life cycle dynamics*:

> *"Three stages of evolution are distinguished. In the initial exploratory or embryonic stage, market volume is low, uncertainty is high, the products design is primitive, and unspecialized machinery is used to manufacture the product. Many firms enter and competition based on product innovation is intense. In the second, intermediate or growth stage, output growth is high, the design of the product begins to stabilize, product innovation declines, and the production process becomes more refined as specialized machinery is substituted for labour. Entry slows and a shakeout of producers occurs. Stage three, the mature stage, corresponds to a mature market. Output growth slows, entry declines further, market shares stabilize, innovation are less significant, and management, marketing and manufacturing techniques become more refined. Evidence on first mover advantages [...] and the link between market shares and profitability [...] suggests that the firms that ultimately capture the greater share of the market and earn the greatest returns on investment tend to be those that enter earliest." (Klepper, 1997, p. 148)*

Moreover, the surviving and often dominant firms tend to be those characterized by distinct innovative capabilities (Klepper and Simons, 2005; Bergek et al., 2008; Cantner et al., 2009) which often were there at the start of the firms themselves.

There are now a large number of studies exploring the explanatory power of technology/product cycle theory in a wide range of industries. For many industries major parts of the story hold up pretty well.[61] Figure 5A–C regarding cars, tires, and TVs is a good illustration. However, there is a range of industries where economies of scale in production never become so great, or the advantages of learning by doing so significant, that only large firms can survive, and entry is blocked. Many "supplier-dominated" sectors (cf. Pavitt's taxonomy above) such as textiles and clothing are good examples. In other cases, while the large economies of scale predicted by product life cycle (PLC) theory in fact have emerged, the nature of the demand for a product class is sufficiently varied so that a single dominant design cannot emerge and take a large share of the overall market. As surveyed in detail by Klepper (1997), alternatives to the canonic PLC template include *first*, industries wherein the dominant trend is toward "Smithian" specialization across components along the overall production chain. *Second*, and relatedly, the requirements by end users may well be sufficiently diverse to define technologically diverse market niches. When, together, knowledge maintains a significant tacit cumulative and niche-specific component, such submarkets are likely to be supplied by different firms throughout the history of the industry. As we discuss in Dosi et al. (2008c) this is the case of most producer good industries including machine tools and instruments and several "complex product systems" (cf. Figure 5D for an illustration concerning lasers; the case of jet engines is discussed in Bonaccorsi and Giuri, 2000).

Equally interesting deviations from (or complications of) the technology cycle theory are industries where, while something like a product cycle dynamic seems to hold in particular eras, from time to time significantly new technologies arise, which upset the old order, and start off a new product cycle. Striking cases include the dramatic changes in aircraft systems technology, and together the identity of the dominant firms, set in train when the turbojet engine became preferred to the older gasoline reciprocating engines; the change in the dominant players in electronic circuitry when transistors and later integrated circuits replaced vacuum tubes; the rise of biotechnology as a vehicle for drug discovery and design. Note that these are essentially cases associated, at least partly, to *paradigm discontinuities*. In these and other cases when a radically new technology has replaced an older mature one, as we have noted, old dominant firms often have difficulty in making the adjustments. In such circumstances, technological change has been what Tushman and Anderson (1986) have called "competence destroying." The industry may experience a renewal of energy and progress, but often under the drive of a new set of firms.

4.6. Models of industrial dynamics

How does one formally represent the processes of industrial evolution? Evolutionary models of industrial dynamics—and economic change more generally—rest on the representation of multiple "boundedly rational" heterogeneous agents interacting with each other (Nelson and Winter, 1982; Bottazzi et al., 2001b; Dosi et al., 1994a, 1995, 2006a; Iwai, 1984a,b; Malerba et al., 1999, 2007, 2008; Silverberg and Lehnert, 1993; Silverberg and Verspagen, 1996; Winter, 1984; Winter et al., 2003; see also the early insights in Winter, 1971).

[61] In such industries, the transition between the initial to the "mature" phase appears to be associated also with different degrees of instability of market shares (cf. Mazzucato, 2002 on the PC industry) and departures from Gibrat-type properties of growth (which seems to be higher in the post-shakeout phase: cf. Geroski and Mazzucato, 2002).

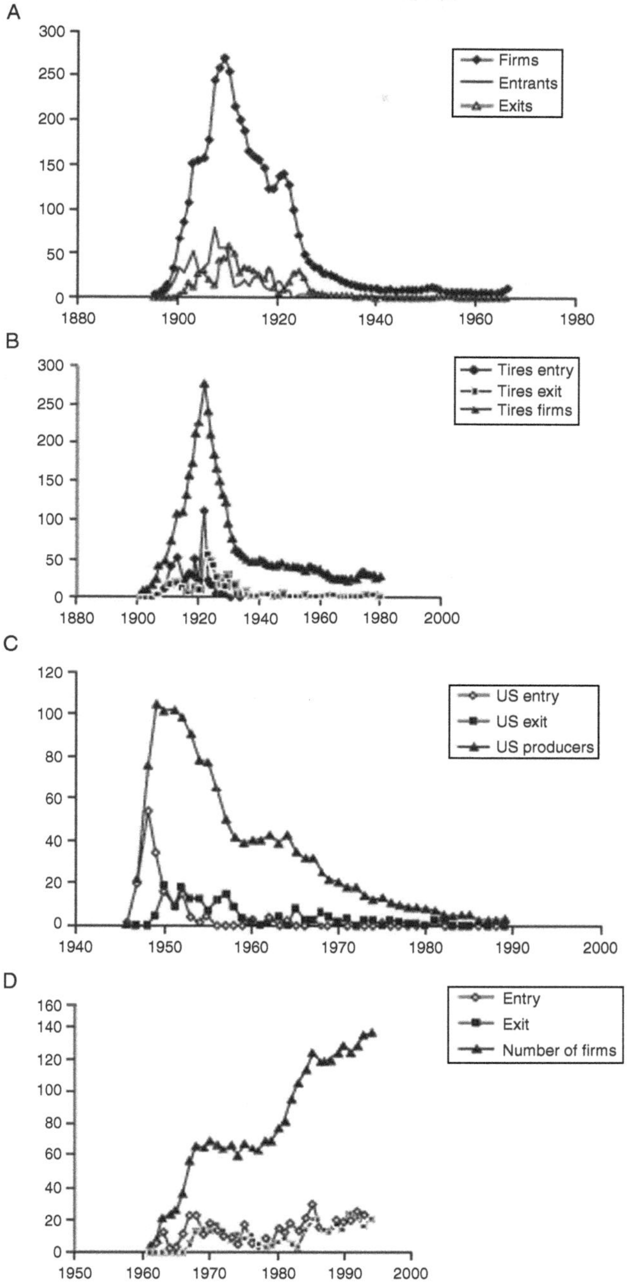

Figure 5. Entry, exit, and number of (A) automobile producers, 1985–1996; (B) tire producers, 1901–1980; (C) television producers, 1946–1989; and (D) laser producers, 1961–1994. Source: Arora et al. (2006).

"Bounded rationality" also takes the form of limited understanding by the agents of the causal structure of the environment in which they are embedded and a limited ability to think through future contingencies, while behavioral patterns are often described in terms of relatively invariant routines. On the other hand, in this approach agents are capable of learning and thus improve their performance over time by changing their technologies and organizational practices.[62]

The symmetric complement of the assumptions on what agents know, learn, and do concerns how markets (and other interaction environments) operate. Observed industrial dynamics are obviously the joint outcome of both. But it makes a lot of difference (except for some rather peculiar circumstances), in terms of the properties of the dynamics themselves, whether and to what extent individual entities can figure out, so to speak "in their heads," *ex ante*, what is going to happen to them, at least in probability, because they also know (and possibly collectively share) a common "model" of their environment and shape their decision accordingly. In that respect, evolutionary models are far from that extreme view whereby everyone knows *ex ante* everything that is relevant to know—about, for example, technologies, distribution of "talents" or other causes of heterogeneity across the population of agents, strategies, etc.—and thus markets operate essentially as collective arrangements setting incentive-compatible schemes. In that, since agents "work it out" beforehand, not much happens through the markets themselves—the consistency of individual plans being guaranteed by the (certainly "hyper-rational") assumptions on micro knowledge.[63] Evolutionary interpretations are nearer the opposite interpretation whereby agents hold quite different views on what is going to happen to them (or to the same effect that they hold a rather wild distribution of beliefs largely uncorrelated with what economists call the "fundamentals") and, together, operate a diverse array of both physical and "social" technologies. This applies notwithstanding the fact that firms in any one industry share a similar body of technological knowledge, that is the same paradigm.[64] Under these circumstances, markets operate first of all as selection devices, determining, *ex post*, profitabilities, survival probabilities, and rates of growth.[65] Short of any belief in full micro rationality and collective equilibrium, the challenge for evolutionary models is to understand how joint processes of micro learning and collective selection yield the observed dynamic patterns. And, indeed, this is a central task for evolutionary interpretations.[66] There, as already mentioned, the commitment to individual rationality is much lower and, symmetrically, the explanatory burden placed upon some combination of idiosyncratic innovative learning and

[62] Broadly defined "bounded rationality" applies—even more so—in models of organizational ecologies (for surveys and discussions, see Carroll, 1997; Carroll and Hannan, 2000) whereby firms carry with them their idiosyncratic features at birth.

[63] Of course, this view implies also that empirical observations—such as those presented above—should in principle be interpreted as sequences of equilibrium outcomes, nested into collectively consistent, highly sophisticated, plans of intertemporally maximizing agents (and this is indeed the spirit by which Hopenhein, 1992; Lucas, 1978, e.g., try to account for the evidence on skewed distributions of firms' sizes, positive rates of entry and exit, etc.).

[64] And in fact it happens that the effective entry of technologies based on a new paradigm often requires also the entry of new firms (a formalization of this idea is in Malerba et al., 2007).

[65] Interpretations based on "pure selection" and "pure *ex ante* rationality" happen to be equivalent whenever the underlying equilibria coincide, and, together, each empirical observation might be understood to be a rather close approximation to the "limit" (in a mathematical sense) of some adjustment process operating at a timescale of order of magnitude faster than that at which empirical observations themselves are collected. Frankly, we find this possibility rather awkward, at best, as a general interpretative framework.

[66] Including Nelson and Winter (1982), Winter (1984), Silverberg et al. (1988), Dosi et al. (1995), Bottazzi et al. (2001b, 2007), Winter et al. (2000, 2003), and Silverberg and Verspagen (1996).

market selection is correspondingly higher. An explicit market dynamics is assumed. Innovation is the main engine of dynamics and evolution. As biologists would say, the "evolutionary landscape" upon which evolution occurs is not fixed, but is continuously deformed by the endogenous learning activities of agents. Relatedly, one ought to interpret the aggregate regularities that are observed in the data as emerging from disequilibrium interactions among heterogeneous agents on the basis of some well-specified dynamic process.

We have reviewed above (Section 3.8) a few evolutionary approaches to modeling the *learning* part of the dynamics, that is the formal representation of stochastic innovation and imitation by individual firms. Conversely, the *selection* part of the process is basically captured by different instantiations of some *replication dynamics*—in a closer or looser analogy with the biological counterpart.[67] The bottom line is a relation between some corporate characters—that is, technological, organizational, or behavioral traits—which the particular interactive environment "favors," on the one hand, and the rate of variation of the frequencies in the *carriers* of such characters in the relevant populations on the other (more in Andersen, 2004; Metcalfe, 1998, 2005b; Silverberg, 1988; Silverberg and Verspagen, 2005b). A basic formulation in discrete time is

$$\Delta s_i = f(E_i(t) - \bar{E}(t))s_i(t), \tag{3}$$

where $s_i(t)$ is the market share of firm i at t, $E_i(t)$ is a sort of (blackboxed) measure of its "competitiveness" in turn determining the relative "fitness" (with $\bar{E}(t) = \sum_i E_i(t)s_i(t)$). Of course, *first*, the $E_i(\cdot)$ may well change over time, and indeed the learning dynamics is precisely about such changes. Moreover, *second*, E_i is most likely a vector capturing multiple corporate features influencing the revealed "competitiveness" of each firm. *Third*, the $f(\cdot,\cdot,\cdot)$ function is most likely nonlinear (hence a further reason for a "rugged selection landscape"). *Fourth*, needless to say, one may add varying degrees of stochastic noise to the selection process, apart from the inherent stochasticity of firm-specific processes of change. In the basic linear case with fixed micro characteristics it is possible to derive analytically also some important properties of the dynamics of *industrial means* as a function of the *variances* across the micro $E_i(\cdot)$ variables.[68]

Many evolutionary models explicitly represent the selection process entailed by market interactions via variants of a replicator equation: see, for example, Silverberg et al. (1988), Verspagen (1993), and Dosi et al. (1995, 2006). In other models the "replication process" is implicit into the rates of expansion/contraction of heterogeneous firms as a result of their differential efficiencies. Nelson and Winter (1982) is an exemplar of this modeling approach. Different production efficiencies imply different firm-specific unit costs. The latter (possibly modulated by some behavioral rules governing output) determine different unit profit margins for each firm. If there is some monotonic relation between profit margins and investments in future production capacity, higher efficiency yields higher investment which entails higher relative shares into the $(t + 1)$ overall output.

[67] The original biological formulation comes from Fisher (1930).

[68] More in Metcalfe (2005b). Incidentally note that the whole field of *evolutionary games*, which we cannot discuss at any detail here, fundamentally studies the process of (deterministic or, more often, stochastic) adaptation/selection across a population of *given* traits/trait-carrying agents by analyzing its asymptotic properties (a little more discussion congenial to our argument here is in Dosi and Winter, 2002).

A replication process similar in spirit involves equipment-embodied technological advances and rates of adoption of particular vintages proportional to their profitabilities: see for example, Soete and Turner (1984) analyzing technological diffusion and Silverberg and Lehnert (1993) for a model addressing the microeconomics of long-term growth.

For the most part the models considered above are highly abstract and general. The recent modeling of Malerba et al. (1999, 2007, 2008) is guided by another theoretical strategy: that of trying to explain particular patterns of evolution observed in certain industries.[69]

One has only begun to systematically link evolutionary models with the "stylized facts" of industrial dynamics discussed earlier, and, together of macrodynamics and growth. Here the big challenge regards the ability of the models of generating—and in that sense "explaining"—rich ensembles of observed empirical regularities, both those that are generic, holding across sectors, countries and phases of the industry life cycles, and those that are regime-specific. Indeed, what the analytical perspective has achieved so far is highly encouraging: it has contributed, in our view, important insights on the nature and drivers of industrial dynamics, highlighting also the ways different patterns of learning and market selection influence variables such as the degrees of industrial concentration, turbulence in market shares, the dynamics of asymmetries across firm in production efficiency, and firm mortality.[70]

One major field of exploration has been indeed the mapping between regimes of learning and the ensuing industrial dynamics—from Nelson and Winter (1982) on the "Schumpeterian tradeoffs"; to Winter (1984) on the properties of different innovative regimes; to Dosi et al. (1995), Marsili (2001), Winter et al. (2000, 2003), and Bottazzi et al. (2001b). More precisely, Dosi et al. (1995) and Marsili (2001) study the ways differences in the processes by which innovative opportunities are tapped (e.g., by entrants vs. incumbents, with or without cumulative learning) affect the evolution of industry structures, the degrees of turbulence of the latter, and the statistical properties of corporate growth. Conversely, Bottazzi et al. (2001b) and Winter et al. (2000, 2003) focus on the properties of the "churning" process characterizing industrial evolution, and on the ensuing dynamics in costs and prices.

Another major area of analysis has focused upon more aggregate statistical phenomena. After all, one of the major questions addressed in Nelson and Winter (1982) and earlier Nelson (1968) was indeed whether the model was able to *generate* as an *emergent property* (at the time this was not the language but in fact the meaning) macro-time series analogous to those analyzed by Robert Solow in his pioneering growth accounting and modeling efforts. And the answer was gloriously positive. A good deal of work has gone on in the area. In fact all evolutionary models naturally generate innovation-driven endogenous growth resting on underlying industrial dynamics of the type discussed above. Some models of evolutionary growth have studied the features of the micro dynamics and the interaction

[69] The authors call their style of modeling "history friendly." As the name suggests, it is meant to be much nearer to the phenomenology of particular industry dynamics, their technological and market characteristics, and the actual chronology of events (e.g., the introduction of the PC in the history of computers or that of integrated circuits in the history of semiconductors) and symmetrically try to account for relatively detailed features of the actual evolution of particular industries.

[70] Incidentally note in this respect that evolutionary modes have abundantly vindicated the proposition that market structures, rather than being a determinant of innovative patterns, are—at least in a first instance—the outcome of innovation-driven industrial evolution.

patterns underlying the long-term properties of growth (cf. Chiaromonte and Dosi, 1993; Silverberg and Lehnert, 1994; Silverberg and Verspagen, 1994). Other has focused upon the convergence/divergence dynamics among trading economies (cf. Dosi et al., 1994b; Verspagen, 1993 among others). More recently, one has begun to explore the properties of growth dynamics jointly with an ensemble of "cyclical" macro properties (e.g., fluctuations in macro demand, employment rates, investment, etc.) grounded upon the same evolutionary industrial foundations (cf. Dosi et al., 2006a, 2008b and Dawid, 2006 which offers a broad survey of the general family of agent-based models).

More generally, the reader is invited to refer to Dosi et al. (1988), Dopfer (2005), Malerba and Brusoni (2007), and Hanusch and Pyka (2007) to grasp the progress that has been made since Nelson and Winter (1982), both empirically and theoretically, toward a full fledged evolutionary theory of economic change, and also the gaps that are still there. As we see it, there is a very promising and very challenging future ahead for evolutionary/agent-based formalizations. The ambition, not out of reach, is to offer a relatively unified interpretation of a large ensemble of phenomena at different level of aggregation—ranging from the "industrial stylized facts" discussed above to phenomena concerning the properties of growth and fluctuations (and crises). Concerning the theoretical tools, if we were to pick just one major challenge to *formal* evolutionary modeling, we would name the following.

More work certainly is needed on selection processes and dynamics. A major step forward in this respect would involve a detailed analysis of *how markets work*. Surprisingly enough, we still have very few empirical works of the kind pioneered by Kirman and colleagues (Delli Gatti et al., 2001; Kirman, 2001; Weisbuch et al., 2000), studying the institutional architectures, the actual mechanisms of exchange, and the ensuing dynamics of prices and quantities. And, symmetrically, we have still very few models—most likely of the "agent-based" kind (cf. again the critical review in Dawid, 2006)—exploring the same phenomena from the side of the theory. Needless to say, the analysis of how markets work is crucial to understand what are the main dimensions of the "selection" landscape and how market selection operates.

As we have noted, to date most formal evolutionary modeling has presumed that a large share of "selection" occurs through the selection on firms—and through that on the technologies and practices of which firms are carriers (i.e., the equivalent of their "genotypes"), while the empirical evidence suggests that this is not the main part of the selection story. At least over the short and medium run a good deal of selection of techniques and practices goes on *within* firms. Moreover, the generality of evolutionary models so far has assumed some monotonicity in the relations between "fundamental" determinants of competitiveness/revealed "fitness," and subsequent relative growth.[71] However, as we have seen above, the evidence on these selective processes suggests that selection forces, on practices as well as on firms, are weaker than those theorized. In turn, these persistent asymmetries may well be the consequence of various forms of market "imperfections"—including informational ones—which, together with endemic "satisficing" behaviors, allow firms characterized by diverse degrees of efficiency and product qualities to coexist without too much selective pressure. On the modeling side such evidence entails two complementary challenges. *First*, one ought to pay more attention to the workings of diffusion processes into the evolutionary dynamics (one of the few incumbent examples is Silverberg et al., 1988). *Second*, the models ought to be able to account for evolution occurring over "fitness landscapes" which for a good portion are roughly *flat*.

[71] Note that the same considerations apply, *just much more so*, to "equilibrium evolutionary dynamics" such as those in Jovanovic (1982) and Ericson and Pakes (1995).

5. Innovation, industrial evolution, and economic growth: Some conclusions

In this chapter, we have led the reader from the investigation of the nature and dynamics of *technological knowledge* all the way through the analysis of how technological (and organizational) innovation and imitation drive the evolution of industries. The understanding of the *structure* of technological knowledge and its diversity across different technological paradigms, together with the understanding of the ways such knowledge is generated, augmented, and diffused—we have argued—are fundamental also for the understanding of the rates and directions of innovative activities, well beyond the incentive economic agents face.

Different abilities to innovate and imitate are central aspects and drivers of industrial evolution, shaping the patterns of growth, decline and exit over populations of competing firms, as well as the opportunities of entry of new firms. In this chapter, we have discussed such dynamics as evolutionary processes driven by the twin forces of (often mistake-ridden) idiosyncratic learning by persistently heterogeneous firms, on the one hand, and (imperfect) market selection delivering prizes and penalties—in terms of profits, possibilities of growth, and survival probabilities—across such heterogeneous corporate populations, on the other. In that, we argued, firm-specific learning processes appear to be relatively more powerful than between-firms selection dynamics.

The learning going on in an economy has a collective as well as an individual element. While their capabilities and actions remain far from identical, firms in the same industry often learn similar things about how to operate the technological developments that are emerging. And firms learn from each other, sometimes as a result of deliberate communication, sometimes because at least a portion of what is going on in individual firms becomes public knowledge. As we stressed, the broad elements of technological paradigms are common property for technical people in a field. As a consequence, even while selection on firms often is relatively weak, there generally is significant selection on new technological variants that are being introduced to the field, with advances that tend to get into the general practice, although, as the diffusion studies we described earlier attest, the process may take considerable time.

We have also here the basic ingredients of an evolutionary interpretation of economic growth and development. Such an evolutionary account, which we cannot discuss in detail here, would highlight the significant differences in the rates of progress at any time across different technologies and industries, which we alluded to in our earlier discussion. There is a developing body of research and writing that aims to explain such differences (see, e.g., Nelson, 2003, 2008a). As mentioned earlier, an important underlying variable seems to be the strength of the scientific fields that illuminate the technologies used in an area of practice. However, there clearly are a number of factors at work. And as we have noted, while there are exceptions, progress within a field of technology tends to become more narrowly focused and to slow down as the technology matures. While repressed in neoclassical growth theory, the process of economic growth as we have historically experienced it has been driven by the continuing introduction of new products and new technologies, and the continuing shifting of resources from older industries where the rate of advance had slowed down to emerging new industries. The continuing growth of output per worker and per-capita incomes that industrialized economies have experienced would not have been possible without this kind of an evolutionary process.[72]

A full evolutionary account of economic growth would also take into account that the historical time path of growth tends to be punctuated by "eras" characterized by the development and diffusion of specific constellations of "general-purpose" technologies (Bresnahan and Trajtenberg, 1995; Rosenberg and Trajtenberg, 2004), that is broad techno-economic paradigms in the sense of Perez (1985), Freeman and Perez (1988), and Freeman and Louça (2001). During a particular economic era, much of the economic growth is accounted for by innovation and productivity growth in the industries that produce the goods that directly incorporate the driving technological paradigms and also in the downstream industries that are able to use these goods as inputs (historically, this was the case of steam power, later electricity and the internal combustion engine, and today it is the case of ICT technologies).[73]

Evolutionary processes of economic growth are embedded in a rich structure of institutions. There is now an extensive empirical literature concerned with the institutions of what have been called innovation systems (see Freeman, 1993; Freeman and Louça, 2001; Lundvall, 1992; Nelson, 1993). That literature has been concerned with matters like cooperative arrangements among firms, the role of universities in technological progress and modes of university–industry interaction in different industries, the variety of government programs supporting technological advance, and other supporting institutions. Others relevant institutions pertain to the "political economy" of socio-economic arrangements governing how firms are organized and managed, labor markets, finance/industry relations, corporate laws, etc. In fact, a general conjecture here is that economic growth is driven by the coevolution of technologies and institutions (Freeman, 2008; Nelson, 2008c; Boyer and Saillard, 2002; Hodgson, 1999).

Detailed analysis of macroeconomic growth as an innovation-driven evolutionary process, however, is beyond the scope of this chapter. Consider the foregoing discussion as a sketch of its underlying building blocks.

[72] In fact, an important link between the evolution of individual sectors and aggregate dynamics rests upon their changing shares of output and employment—intertwined as they are by evolving input/output profiles and final demand patterns. The analysis of the dynamics of sectoral structures has been pioneered long ago by Kuznets (1972), Burns (1934), Mitchell (1925), and Svennilson (1954) among others, but unfortunately largely neglected in more recent times. However, those structural changes—which have been formally discussed by Pasinetti (1981) and more recently Saviotti and Pyka (2008a,b)—are a crucial link between changes in individual industries, the primary locus of innovation, diffusion and competition, and broader aggregates. (See also Metcalfe et al., 2005). In this respect, incidentally note how the bad habit common to a good deal of the contemporary economic discipline to compress interagent intrasectoral relations as well as intersectoral ones into some dynamics driven by a purported "representative agent," has obfuscated both the characteristics of industrial dynamics, and also the drivers and properties of macro growth and fluctuations.

[73] Granted that, the relationship between techno-economic paradigms (and even more so individual general-purpose technologies thereof), on the one hand, and growth patterns, on the other, continues to be a challenging area of investigation. In this respect note that chronology of diffusion of general-purpose innovations is far from smooth (a good illustration in the case of the steam engine is in Nuvolari and Verspagen, 2009). Moreover, the application of the same technology in different sectors is characterized by quite uneven rates of technical change (a point already noted by Pavitt, 1986, concerning the impact of microelectronic technologies). Broad discussions on such a relationship are in von Tunzelmann (1995), Freeman and Louça (2001), and Perez (2002). A critical discussion of the very notion of "General Purpose Technologies" is in Field (2008).

Acknowledgments

Comments by Marianna Epicoco, Bronwyn Hall, Stan Metcalfe, Alessandro Nuvolari, Carlota Perez, Nate Rosenberg, and Sid Winter on previous drafts or direct antecedents have helped a lot to shape the work. Mario Cimoli, Marco Grazzi, Luigi Orsenigo and Mauro Sylos Labini coauthored earlier papers upon which we draw. Giorgia Barboni and Matteo Barigozzi have been competent research assistants. Last but not least, we thank Laura Ferrari who helped the work along with utmost competence and endless patience. Support to one of us (G. Dosi) has been granted by the European Community's 6th FP (Contract CIT3-CT-2005-513396), Project: DIME—Dynamics of Institutions and Markets in Europe 7th FP (FP7/2007-2013) under Socioeconomic Sciences and Humanities, grant agreement no. 217466, and MIUR, PROT 2007HA3S72_003, PRIN 2007.

References

Abernathy, W.J., Utterback, J. (1978). "Patterns of innovation in industry". Technology Review 80 (7), 40–47.

Abramovitz, M. (1993). "The search for the sources of growth: Areas of ignorance, old and new". Journal of Economic History 53 (2), 217–243.

Ahn, S. (2001). Firm dynamics and productivity growth: A review of micro evidence from OECD countries. Working Paper No. 297, OECD, Paris.

Akerlof, G.A. (1984). An Economic Theorist's Book of Tales. Cambridge University Press, Cambridge.

Alchian, A. (1963). "Reliability of progress curves in airframe production". Econometrica 31 (4), 679–693.

Allen, R.C. (1983). "Collective invention". Journal of Economic Behavior & Organization 4, 1–24.

Andersen, E.S. (2004). "Population thinking, Price's equation and the analysis of economic evolution". Evolutionary and Institutional Economics Review 1, 127–148.

Antonelli, C. (1995). The Economics of Localized Technological Change and Industrial Dynamics. Springer, New York.

Antonelli, C., Foray, D., Hall, B., Steinmüller, W.E. (Eds.), (2006). New Frontiers in the Economics of Innovation and New Technology, Essays in Honor of Paul A. David. Edward Elgar Publishing, Cheltenham, UK.

Argote, L., Epple, D. (1990). "Learning curves in manufacturing". Science 247 (4945), 920–924.

Arora, A., Gambardella, A. (1994). "The changing technology of technological change: General and abstract knowledge and the division of innovative labour". Research Policy 23, 523–532.

Arora, A., Fosfuri, A., Gambardella, A. (2002). Markets for Technology. MIT Press, Cambridge, MA.

Arora, A., Gambardella, A., Klepper, S. (2006). "Organizational capabilities and the rise of the software industry in the emerging economies: Lessons from the history of some US industries". In: Arora, A., Gambardella, G. (Eds.), From Underdogs to Tigers. Oxford University Press, Oxford, pp. 171–206.

Arrow, K.J. (1962a). "Economics of welfare and the allocation of resources for invention". In: Nelson, R.R. (Ed.), The Rate and Direction of Inventive Activity. Princeton University Press, Princeton, NJ, pp. 609–625.

Arrow, K. (1962b). "The economic implications of learning by doing". Review of Economic Studies 29, 155–173.

Arrow, K. (1996). "Technical information and industrial structure". Industrial and Corporate Change 5 (2), 645–652.

Arthur, W.B. (1988). "Competing technologies: An overview". In: Dosi, G., Freeman, C., Nelson, R., Silverberg, G., Soete, L. (Eds.), Technical Change and Economic Theory. Pinter Publishing, London, pp. 590–607.

Arthur, W.B. (1989). "Competing technologies, increasing returns, and lock-in by historical events". Economic Journal 99 (394), 116–131.

Arthur, W.B. (2007). "The structure of invention". Research Policy 36 (2), 274–287.

Arthur, W.B., Ermoliev, Y., Kaniovski, Y. (1987). "Path-dependent processes and the emergence of macro-structure". European Journal of Operational Research 30 (3), 294–303.

Arundel, A., van de Paal, G., Soete, L. (1995). Innovation strategies of Europe's largest firms. Results of the PACE survey. European Innovation Monitoring System, Report No. 23, European Commission, Brussels.

Atkinson, A.B., Stiglitz, J.E. (1969). "A new view of technological change". Economic Journal 79 (315), 573–578.

Audretsch, D.B. (1997). "Technological regimes, industrial demography and the evolution of industrial structures". Industrial and Corporate Change 6 (1), 49–82.

Auerswald, P., Kaufmann, S., Lobo, J., Shell, K. (2000). "The production recipe approach to modelling technological innovation: An application to learning by doing". Journal of Economic Dynamics and Control 24, 389–450.

Augier, M., March, J. (2000). "Roots and branches of organizational economics". Industrial and Corporate Change 9 (Special Issue), 555–788.

Augier, M., March, J. (2002). The Economics of Choice, Change and Organizations: Essays in Memory of Richard M. Cyert. Edward Elgar Publishing, Cheltenham, UK.

Aversi, R., Dosi, G., Fagiolo, G., Meacci, M., Olivetti, C. (1999). "Demand Dynamics with Socially Evolving Preferences". Industrial and Corporate Change 8 (2), 353–408.

Axtell, R.L. (2001). "Zipf distribution of U.S. firm sizes". Science 293, 1818–1820.

Baily, M., Hulten, C., Campbell, D. (1992). The distribution of productivity in manufacturing plants. Brookings Papers on Economic Activities: MicroeconomicsBrookings Institution, Washington, DC pp. 187–249.

Baily, M., Bartelsman, E., Haltiwanger, J. (1996). "Downsizing and productivity growth: Myth or reality?" Small Business Economics 8 (4), 259–278.

Balconi, M. (2002). "Tacitness, codification of technological knowledge and the organisation of industry". Research Policy 31 (3), 357–379.

Baldwin, C.Y., Clark, K.B. (2000). Design Rules. MIT Press, Cambridge, MA.

Baldwin, J.R. (1995). The Dynamics of Industrial Competition: A North American Perspective. Cambridge University Press, New York, NY.

Baldwin, J.R., Gu, W. (2006). "Plant turnover and productivity growth in Canadian manufacturing". Industrial and Corporate Change 15 (3), 417–465.

Baloff, N. (1971). "Extension of the learning curve. Some empirical results". Operation Research Quarterly 22, 329–340.

Barna, T. (1962). Investment and Growth Policies in British Industrial Firms. Cambridge University Press, Cambridge.

Bartelsman, E., Doms, M. (2000). "Understanding productivity: Lessons from longitudinal microdata". Journal of Economic Literature 38, 569–594.

Bartelsman, E., Scarpetta, S., Schivardi, F. (2005). "Comparative analysis of firm demographics and survival: Evidence from micro-level sources in OECD countries". Industrial and Corporate Change 14, 365–391.

Basalla, G. (1988). The Evolution of Technology. Cambridge University Press, Cambridge.

Bassanini, A., Dosi, G. (2001). "Heterogenous agents, complementarities and diffusion: Do increasing returns imply convergence to international technological monopolies?" In: Delli Gatti, D., Gallegati, M., Kirman, A. (Eds.), Market Structure, Aggregation, and Heterogeneity. Cambridge University Press, Cambridge.

Bassanini, A., Dosi, G. (2006). "Competing technologies, technological monopolies and the rate of convergence to a stable market structure". In: Antonelli, C., Foray, D., Hall, B.H., Steinmueller, E.W. (Eds.), New Frontiers in the Economics of Innovation and New Technology, Essays in Honor of Paul A. David. Edward Elgar Publishing, Cheltenham, UK, pp. 23–59.

Beardsley, G., Mansfield, E. (1978). "A note on the accuracy of industrial forecasts of the profitability of new products and processes". Journal of Business 51, 127–135.

Becker, M.C., Lazaric, N. (Eds.), (2009). Recent Empirical Advances on Organizational Routines. Edward Elgar Publishing, Cheltenham, UK.

Becker, M.C., Lazaric, N., Nelson, R.R., Winter, S.G. (2005). "Applying organizational routines in analyzing organizations". Industrial and Corporate Change 14 (5), 775–791.

Bergek, A., Tell, F., Berggren, C., Watson, J. (2008). "Technological capabilities and late shakeouts: Industrial dynamics in the advanced gas turbine industry, 1987–2002". Industrial and Corporate Change 17, 335–392.

Binswanger, H., Ruttan, V. (1978). Induced Innovation. Johns Hopkins University Press, Baltimore, MD.

Bonaccorsi, A., Giuri, P. (2000). "When shakeout doesn't occur. The evolution of the turboprop engine industry". Research Policy 29, 847–870.

Bonaccorsi, A., Giuri, P., Pierotti, F. (2005). "Technological frontiers and competition in multi-technology sectors. Micro evidence from the aero-engine industries". Economics of Innovation and New Technology 14, 23–42.

Bottazzi, G., Secchi, A. (2003). "Why are distributions of firm growth rates tent-shaped?" Economics Letters 80, 415–420.

Bottazzi, G., Secchi, A. (2005). "Growth and diversification patterns of the worldwide pharmaceutical industry". Review of Industrial Organization 26, 195–216.

Bottazzi, G., Secchi, A. (2006). "Explaining the distribution of firms growth rates". RAND Journal of Economics 37, 234–263.

Bottazzi, G., Dosi, G., Lippi, M., Pammolli, F., Riccaboni, M. (2001a). "Innovation and corporate growth in the evolution of the drug industry". International Journal of Industrial Organization 19, 1161–1187.

Bottazzi, G., Dosi, G., Rocchetti, G. (2001b). "Modes of knowledge accumulation, entry regimes and patterns of industrial evolution". Industrial and Corporate Change 10, 609–638.

Bottazzi, G., Cefis, E., Dosi, G. (2002). "Corporate growth and industrial structure. Some evidence from the Italian manufacturing industries". Industrial and Corporate Change 11, 705–723.

Bottazzi, G., Dosi, G., Cefis, E., Secchi, A. (2007). "Invariances and diversities in the evolution of manufacturing industries". Small Business Economics 29 (1), 137–159.

Bottazzi, G., Dosi, G., Jacoby, N., Secchi, A., Tamagni, F. (2009). Corporate performance and market selection. LEM Working Paper Series 2009/13, Sant'Anna School of Advanced Studies, Pisa, Italy, forthcoming on Industrial and Corporate Change (2010).

Boyer, R., Saillard, Y. (Eds.) (2002). Regulation Theory. The State of the Art. Routledge, London and New York.

Bradshaw, G.F. (1992). "The airplane and the logic of invention". In: Giere, R.N. (Ed.), Cognitive Models of Science. The University of Minnesota Press, Minneapolis, MN, pp. 239–250.

Breschi, S., Malerba, F., Orsenigo, L. (2000). "Technological regimes and Schumpeterian patterns of innovation". Economic Journal 110, 388–410.

Breschi, S., Lissoni, F., Malerba, F. (2003). "Knowledge-relatedness in firm technological diversification". Research Policy 32 (1), 69–87.

Bresnahan, T.F., Trajtenberg, M. (1995). "General purpose technologies: 'Engines of growth'?" Journal of Econometrics 65 (1), 83–108.

Bresnahan, T., Greenstein, S., Henderson, R. (2008). Schumpeterian competition and diseconomies of scope; illustrations from leading historical firms in computing. Kellogg Institute WP.

Brown, J.D., Earle, J.S., Telegdy, A. (2006). "The Productivity Effects of Privatization: Longitudinal Estimates from Hungary, Romania, Russia, and Ukraine". Journal of Political Economy 114 (1), 61–99.

Brynjolfsson, E., Hitt, L.M. (2000). "Beyond Computation: Information technology, organizational transformation and business performance". Journal of Economic Perspectives 14, 3–48.

Burns, A.F. (1934). Production Trends in the United States Since 1870. NBER, New York, NY.

Bush, V. (1945). Science: The Endless Frontier. Government Printing Office, Washington, DC.

Cantner, U., Krüger, J. (2004). "Geroski's stylized facts and mobility in large German manufacturing firms". International Journal of Industrial Organization 24, 267–283.

Cantner, U., Krüger, J., Rhein, K.V. (2009). "Knowledge and creative destruction over the industry life cycle. The case of the German automobile industry". Economica 76, 132–148.

Carroll, G.R. (1997). "Long-term evolutionary changes in organizational populations: Theory, models and empirical findings". Industrial and Corporate Change 6 (1), 119–143.

Carroll, G.R., Hannan, M.T. (2000). The Demography of Corporations and Industries. Princeton University Press, Princeton, NJ.

Castaldi, C., Fontana, R., Nuvolari, A. (2009). "'Chariots of fire': The evolution of tank technology, 1915–1945". Journal of Evolutionary Economics 19 (4), 545–566.

Cefis, E. (2003a). "Persistence in innovation and profitability". Rivista Internazionale di Scienze Sociali 110, 19–37.

Cefis, E. (2003b). "Is there persistence in innovative activities?" International Journal of Industrial Organization 21, 482–515.

Cefis, E., Ciccarelli, M. (2005). "Profit differentials and innovation". Economics of Innovation and New Technologies 14 (1–2), 43–61.

Cefis, E., Orsenigo, L. (2001). "The persistence of innovative activities: A cross-country and cross-sector comparative analysis". Research Policy 30, 1139–1158.

Chandler, A. (1992). "Organizational capabilities and the economic history of the industrial enterprise". Journal of Economic Perspectives 6 (3), 79–100.

Chandler, A., Galambos, L. (1970). "The development of large-scale economic organizations in modern America". Journal of Economic History 30 (1), 201–210.

Chataway, J., Tait, J., Wield, D. (2004). "Understanding company R&D strategies in agro-biotechnology: Trajectories and blind spots". Research Policy 33 (6–7), 1041–1057.

Chiaromonte, F., Dosi, G. (1993). "Heterogeneity, competition, and macroeconomic dynamics". Structural Change and Economic Dynamics 4 (1), 39–63.

Cimoli, M., Soete, L. (1992). "A generalized technology gap trade model". Economie Appliquée, vol XLV, no. 3, pp. 33–54.
Cimoli, M., Dosi, G. (1995). "Technological paradigms, patterns of learning and development: An introductory roadmap". Journal of Evolutionary Economics 5 (3), 243–268.
Cimoli, M., Dosi, G., Stiglitz, J.E. (Eds.), (2009). Industrial Policy and Development: The Political Economy of Capabilities Accumulation. Oxford University Press, Oxford.
Coad, A. (2008). The Growth of Firms: A Survey of Theories and Empirical Evidence. Edward Elgar Publishing, Cheltenham, UK.
Coad, A., Rao, R. (2008). "Innovation and firm growth in high-tech sectors: A quantile regression approach". Research Policy 37 (4), 633–648.
Cohen, W., Levin, R. (1989). "Empirical studies of innovation and market structure". In: Schmalensee, R., Willig, R. (Eds.), Handbook of Industrial Organization, vol. 2. Elsevier, Amsterdam.
Cohen, W.M., Levinthal, D.A. (1990). "Absorptive capacity: A new perspective on learning and innovation". Administrative Science Quarterly 35, 128–152.
Cohen, M.D., Burkhart, R., Dosi, G., Egidi, M., Marengo, L., Warglien, M., Winter, S. (1996). "Routines and other recurring action patterns of organizations: Contemporary research issues". Industrial and Corporate Change 5 (3), 653–698.
Cohen, W., Nelson, R.R., Walsh, J.P. (2002). "Links and impacts: The influence of public research on industrial R&D". Management Science 48 (1), 1–23.
Conley, P. (1970). "Experience curves as a planning tool". IEEE Spectrum 7 (6), 63–68.
Conlisk, J. (1989). "An aggregate model of technical change". Quarterly Journal of Economics 104, 787–821.
Conoway, R., Schultz, A. (1959). "The manufacturing progress function". Journal of Industrial Engineering 10, 39–53.
Consoli, D. (2005). "The dynamics of technological change in UK retail banking services: An evolutionary perspective". Research Policy 34 (4), 461–480.
Constant, E. (1980). The Origins of the Turbojet Revolution. Johns Hopkins University Press, Baltimore, MD.
Coriat, B., Dosi, G. (1998). "Learning how to govern and learning how to solve problems: On the co-evolution of competences, conflicts, and organizational routines". In: Chandler, A., Hagström, P., Sölvell, O. (Eds.), The Dynamic Firm. Oxford University Press, Oxford, pp. 103–134.
Cowan, R. (2001). "Expert systems: Aspects of and limitations to the codifiability of knowledge". Research Policy 30, 1355–1372.
Cowan, R., David, P.A., Foray, D. (2000). "The explicit economics of knowledge codification and tacitness". Industrial and Corporate Change 9 (2), 211–253.
Cubbin, J., Geroski, P. (1987). "The convergence in profits in the long-run: Inter-firm and inter-industry comparisons". Journal of Industrial Economics 35, 427–442.
Cyert, R.M., March, J.G. (1992). A Behavioral Theory of the Firm (second ed.). Basil Blackwell Publishing, Cambridge.
Dasgupta, P., David, P.A. (1994). "Towards a new economics of science". Research Policy 23, 487–521.
David, P.A. (1975). Technical Choice, Innovation and Economic Growth. Cambridge University Press, Cambridge.
David, P.A. (1985). "Clio and the economics of QWERTY". American Economic Review 75, 332–337.
David, P.A. (1988). Path dependence: Putting the past into the future of economics. Technical Report No. 533, Institute for Mathematical Studies in the Social Sciences, Stanford, CA.
David, P.A. (1989). A paradigm for historical economics: Path dependence and predictability in dynamic systems with local network externalities. Working PaperStanford University Press, Stanford, CA.
David, P.A. (1990). "The dynamo and the computer: An historical perspective on the modern productivity paradox". American Economic Review 80 (2), 355–361.
David, P.A. (1992). "Heroes, herds and hysteresis in technological history". Industrial and Corporate Change 1 (1), 129–179.
David, P.A. (1993). "Knowledge property and the system dynamics of technological change". In: Summers, L., Shah, S. (Eds.), Proceedings of the World Bank Conference on Development Economics. The World Bank, Washington, DC, pp. 215–248.
David, P.A. (2001a). From keeping nature's secrets to the institutionalization of open science. Discussion Papers in Economic and Social History, University of Oxford, Oxford.
David, P.A. (2001b). "Path dependence, its critics and the quest for 'historical economics'". In: Garrouste, P., Ioannides, S. (Eds.), Evolution and Path Dependence in Economic Ideas: Past and Present. Edward Elgar Publishing, Cheltenham, UK.
David, P.A. (2004). "Understanding the emergence of 'open science' institutions: Functionalist economics in historical context". Industrial and Corporate Change 13 (3), 571–589.
David, P.A. (2005). "Path dependence in economic processes: Implications for policy analysis in dynamical system contexts". In: Dopfer, K. (Ed.), The Evolutionary Foundations of Economics. Cambridge University Press, Cambridge.

David, P.A., Hall, B. (2006). "Property and the pursuit of knowledge: IPR issues affecting scientific research". Research Policy 35 (6), 767–771.

David, P.A., Wright, G. (1997). "Increasing returns and the genesis of American resource abundance". Industrial and Corporate Change 6 (2), 203–245.

Dawid, H. (2006). "Agent-based models of innovation and technological change". In: Tesfatsion, L., Judd, K. (Eds.), Handbook of Computational Economics, vol. 2. Elsevier, Amsterdam, pp. 1235–1272.

Delli Gatti, D., Gallegati, M., Kirman, A. (Eds.), (2001). Market Structure, Aggregation and Heterogeneity. Cambridge University Press, Cambridge.

Demirel, P., Mazzucato, M. (2008). Does market selection reward innovators? R&D, patents and growth in the U.S. pharmaceutical industry. IKD Working Paper No. 33, Open University, London.

Dew, N. (2006). "Incommensurate technological paradigms? Quarreling in the RFID industry". Industrial and Corporate Change 15 (5), 785–810.

Dopfer, K. (Ed.), (2005). Principles of Evolutionary Economics. Cambridge University Press, Cambridge.

Dosi, G. (1982). "Technological paradigms and technological trajectories. A suggested interpretation of the determinants and directions of technical change". Research Policy 11, 147–162.

Dosi, G. (1984). Technical Change and Industrial Transformation. Macmillan, London.

Dosi, G. (1988). "Sources, procedures and microeconomic effects of innovation". Journal of Economic Literature 26 (3), 1120–1171.

Dosi, G. (1991). "The research on innovation diffusion: An assessment". In: Nakicenovic, N., Grübler, A. (Eds.), Diffusion of Technologies and Social Behavior. Springer, Laxenburg.

Dosi, G. (1997). "Opportunities, incentives and the collective patterns of technical change". Economic Journal 107 (444), 1530–1547.

Dosi, G. (2007). "Statistical regularities in the evolution of industries. A guide through some evidence and challenges for the theory". In: Malerba, F., Brusoni, S. (Eds.), Perspectives on Innovation. Cambridge University Press, Cambridge.

Dosi, G., Egidi, M. (1991). "Substantive and procedural uncertainty. An exploration of economic behaviours in changing environments". Journal of Evolutionary Economics 1 (2), 145–168.

Dosi, G., Grazzi, M. (2006). "Technologies as problem-solving procedures and technologies as input–output relations: Some perspectives on the theory of production". Industrial and Corporate Change 15 (1), 173–202.

Dosi, G., Kaniovski, Y. (1994). "On "Badly Behaved" Dynamics. Some applications of generalized urn schemes to technological and economic change". Journal of Evolutionary Economics 4 (2), 93–123.

Dosi, G., Lovallo, D. (1997). "Rational entrepreneurs or optimistic martyrs? Some considerations on technological regimes, corporate entries and the evolutionary role of decision biases". In: Garud, R., Nayyar, P.R., Shapira, Z.B. (Eds.), Technological Innovation Oversights and Foresights. Cambridge University Press, Cambridge.

Dosi, G., Nelson, R.R. (1994). "An introduction to evolutionary theories in economics". Journal of Evolutionary Economics 4 (3), 153–172.

Dosi, G., Winter, S.G. (2002). "Interpreting economic change: Evolution, structures and games". In: Augier, M., March, J. (Eds.), The Economics of Choice, Change, and Organizations. Edward Elgar Publishing, Cheltenham, UK.

Dosi, G., Freeman, C., Nelson, R.R., Silverberg, G., Soete, L. (Eds.), (1988). Technical Change and Economic Theory. Pinter Publishing, London.

Dosi, G., Pavitt, K., Soete, L. (1990). The Economics of Technical Change and International Trade. Wheatsheaf/New York University Press, Brighton, UK/New York, NY.

Dosi, G., Fabiani, S., Aversi, R., Meacci, M. (1994a). "The dynamics of international differentiation: A multi-country evolutionary model". Industrial and Corporate Change 2 (3), 225–241.

Dosi, G., Freeman, C., Fabiani, S. (1994b). "The process of economic development. Introducing some stylized facts and theories on technologies, firms and institutions". Industrial and Corporate Change 3 (1), 1–45.

Dosi, G., Marsili, O., Orsenigo, L., Salvatore, R. (1995). "Learning, market selection and the evolution of industrial structures". Small Business Economics 7, 411–436.

Dosi, G., Malerba, F., Marsili, O., Orsenigo, L. (1997). "Industrial structures and dynamics: Evidence, interpretations and puzzles". Industrial and Corporate Change 6, 3–24.

Dosi, G., Nelson, R.R., Winter, S. (Eds.), (2000). The Nature and Dynamics of Organizational Capabilities. Oxford University Press, Oxford.

Dosi, G., Levinthal, D., Marengo, L. (2003). "Bridging contested terrain: Linking incentive-based and learning perspectives on organizational evolution". Industrial and Corporate Change 12 (2), 413–436.

Dosi, G., Marengo, L., Fagiolo, G. (2005a). "Learning in evolutionary environments". In: Dopfer, K. (Ed.), The Evolutionary Foundations of Economics. Cambridge University Press, Cambridge, pp. 255–328.

Dosi, G., Orsenigo, L., Sylos Labini, M. (2005b). "Technology and the economy". In: Smelser, N.J., Swedberg, R. (Eds.), The Handbook of Economic Sociology. (second ed.). Princeton University Press/Russell Sage Foundation, Princeton, NJ.

Dosi, G., Fagiolo, G., Roventini, A. (2006a). "An evolutionary model of endogenous business cycles". Computational Economics 27 (1), 3–34.

Dosi, G., Llerena, P., Sylos Labini, M. (2006b). "Science–technology–industry links and the 'European paradox': Some notes on the dynamics of scientific and technological research in Europe". Research Policy 35 (10), 1450–1464.

Dosi, G., Marengo, L., Pasquali, C. (2006c). "How much should society fuel the greed of innovators? On the relations between appropriability, opportunities and rates of innovation". Research Policy 35 (8), 1110–1121.

Dosi, G., Faillo, M., Marengo, L. (2008a). "Organizational capabilities, patterns of knowledge accumulation and governance structures in business firms: An introduction". Organization Studies 29 (8), 1165–1185.

Dosi, G., Fagiolo, G., Roventini, A. (2008b). Schumpeter meeting Keynes: A policy-friendly model of endogenous growth and business cycles. LEM Working Paper Series, 2008/21, Sann't Anna School of Advanced Studies, Pisa, Italy, forthcoming in Journal of Economic Dynamics and Control, 2010.

Dosi, G., Gambardella, A., Grazzi, M., Orsenigo, L. (2008c). "Technological revolutions and the evolution of industrial structures: Assessing the impact of new technologies upon the size and boundaries of firms". Capitalism and Society 3 (1).

Dutton, J.M., Thomas, A. (1984). "Treating progress functions as a managerial opportunity". Academy of Management Review 9 (2), 235–247.

Ericson, R., Pakes, A. (1995). "Markov perfect industry dynamics: A framework for empirical work". Review of Economic Studies 62 (1), 53–82.

Evans, D.S. (1987). "The relationship between firm growth, size and age: Estimates for 100 manufacturing industries". Journal of Industrial Economics 35 (4), 567–581.

Fagerberg, J. (1994). "Technology and international differences in growth rates". Journal of Economic Literature 32 (3), 1147–1175.

Fagerberg, I., Mowery, D.C., Nelson, R.R. (Eds.), (2005). The Oxford Handbook of Innovation. Oxford University Press, Oxford.

Fagiolo, G., Dosi, G. (2003). "Exploitation, exploration and innovation in a model of endogenous growth with locally interacting agents". Structural Change and Economic Dynamics 14, 237–273.

Field, A.J. (2008). "Does Economic History Need GPTs?" Available at http://ssrn.com/abstract=1275023.

Fisher, R.A. (1930). The General Theory of Natural Selection. Clarendon Press, Oxford.

Fleck, J. (1988). "Innofusion or diffusation? The nature of technological developments in robotics". Programme on Information and Communications Technologies (PICT) Working Paper, Edinburgh.

Foss, N.J., Mahnke, V. (Eds.), (2000). Competence, Governance, and Entrepreneurship: Advances in Economic Strategy Research. Oxford University Press, Oxford.

Foster, L., Haltiwanger, J., Krizan, C.J. (2001). "Aggregate productivity growth: Lessons from microeconomic evidence". In: Dean, E., Harper, M., Hulten, C. (Eds.), New Developments in Productivity Analysis. University of Chicago Press, Chicago, IL.

Foster, L., Haltiwanger, J., Syverson, J. (2008). "Reallocation, firm turnover, and efficiency: Selection on productivity or profitability?" American Economic Review 98 (1), 394–425.

Freeman, C. (1982). The Economics of Industrial Innovation (second ed.). Frances Pinter Publishing, London.

Freeman, C. (1993). "The 'National System of Innovation' in historical perspective". Cambridge Journal of Economics 19 (1), 5–24.

Freeman, C. (1994). "The economics of technical change". Cambridge Journal of Economics 18, 463–514.

Freeman, C. (2008). Systems of Innovation: Selected Essays in Evolutionary Economics. Edward Elgar Publishing, Cheltenham, UK.

Freeman, C., Louça, F. (2001). As Time Goes By: The Information Revolution and the Industrial Revolutions in Historical Perspective. Oxford University Press, Oxford.

Freeman, C., Perez, C. (1988). "Structural crises of adjustment: Business cycles and investment behavior". In: Dosi, G., Freeman, R., Nelson, R., Silverberg, G., Soete, L. (Eds.), Technical Change and Economic Theory. Pinter Publishing, London.

Freeman, C., Soete, L. (1997). The Economics of Industrial Innovation (third ed.). Pinter Publishing, London.

Frenken, K., Leydesdorff, L. (2000). “Scaling trajectories in civil aircraft (1913–1997)”. Research Policy 28, 469–488.

Frenken, K., Nuvolari, A. (2004). “The early development of the steam engine: An evolutionary interpretation using complexity theory”. Industrial and Corporate Change 13, 419–450.

Frenken, K., Saviotti, P.P., Trommetter, M. (1999). “Variety and niche creation in aircraft, helicopters, motorcycles and microcomputers”. Research Policy 28 (5), 469–488.

Gary, M.S., Dosi, G., Lovallo, D. (2008). “Boom and bust behavior: On the persistence of strategic decision biases”. In: Hodgkinson, G.P., Starbuck, W.H. (Eds.), The Oxford Handbook of Organizational Decision Making. Oxford University Press, Oxford.

Georgescu-Roegen, N. (1970). “The economics of production”. American Economic Review 60 (2), 1–9.

Geroski, P.A. (1998). “An applied econometrician’s view of large company performance”. Review of Industrial Organization 13, 271–293.

Geroski, P.A. (2000). “Models of technology diffusion”. Research Policy 29, 603–625.

Geroski, P.A., Jacquemin, A. (1988). “The persistence of profits: A European comparison”. Economic Journal 98, 357–389.

Geroski, P.A., Mazzucato, M. (2002). “Learning and the sources of corporate growth”. Industrial and Corporate Change 11, 623–644.

Geroski, P.A., Machin, S., Van Reenen, J. (1993). “The profitability of innovating firms”. RAND Journal of Economics 24, 198–211.

Geroski, P.A., Machin, S., Walters, C. (1997). “Corporate growth and profitability”. Journal of Industrial Economics 45 (2), 171–189.

Geuna, A., Salter, A., Steinmuller, W.E. (Eds.), (2003). Science and Innovation: Rethinking the Rationale for Funding and Governance. Edward Elgar Publishing, Cheltenham, UK.

Giuri, P., Tomasi, C., Dosi, G. (2007). L’industria aerospaziale. Innovazione, tecnologia e strategia economica. Il Sole 24 Ore e Fondazione Cotec, Milan.

Goddard, J.A., Wilson, J. O. S. (1999). “The persistence of profit: As new empirical interpretation”. International Journal of Industrial Organization 17, 663–687.

Gordon, T.J., Munson, T.R. (1981). Research into Technology Output Measures. The Futures Group, Glastonbury, CT.

Granstrand, O. (1998). “Toward a theory of the technology-based firm”. Research Policy 27, 465–489.

Granstrand, O. (1999). The Economics and Management of Intellectual Property. Edward Elgar Publishing, Cheltenham, UK.

Granstrand, O. (2005). “Innovation and intellectual property rights”. In: Fagerberg, J., Mowery, D.C., Nelson, R.R. (Eds.), The Oxford Handbook of Innovation. Oxford University Press, Oxford, pp. 266–290.

Greenwald, B., Stiglitz, J. (1986). “Externalities in economies with imperfect information and incomplete markets”. Quarterly Journal of Economics 101 (2), 229–264.

Griliches, Z. (1957). "Hybrid Corn: An Exploration in the Economics of Technological Change". Econometrica, 25, 501–522.

Griliches, Z., Mairesse, J. (1997). “Production function: The search for identification”. In: Strøm, S. (Ed.), Econometrics and Economic Theory in the Twentieth Century: The Ragnar Frisch Centennial Symposium. Cambridge University Press, Cambridge.

Grimaldi, R., Torrisi, S. (2001). “Codified-tacit and general-specific knowledge in the division of labour among firms. A study of the software industry”. Research Policy 30 (9), 1425–1442.

Gritsevskyi, A., Nakicenovic, N. (2000). “Modeling uncertainty of induced technological change”. Energy Policy 28, 907–921.

Grupp, H. (1992). Dynamics of Science-Based Innovation. Springer, Berlin.

Gschwandtner, A. (2004). “Profit persistence in the “very” long run: Evidence from survivors and exiters”. Vienna Economics Papers 0401, University of Vienna, Department of Economics.

Hall, B.H. (1987). “The relationship between firm size and firm growth in the U.S. manufacturing sector”. Journal of Industrial Economics 3 (4), 583–606.

Hall, B.H. (2005). “Innovation and diffusion”. In: Fagerberg, J., Mowery, D.C., Nelson, R.R. (Eds.), The Oxford Handbook of Innovation. Oxford University Press, Oxford.

Hanusch, H., Pyka, A. (2007). Elgar Companion to Neo-Schumpeterian Economics. Edward Elgar Publishing, Cheltenham, UK.

Hart, P.E., Prais, J.S. (1956). “The Analysis of Business Concentration”. Journal of the Royal Statistical Society 119, 150–191.

Helfat, C. (Ed.), (2003). The SMS Blackwell Handbook of Organizational Capabilities Emergence, Development, and Change. Blackwell Publishing, London.

Helfat, C., Finkelstein, S., Mitchell, W., Peteraf, M.A., Singh, H., Teece, D.J., Winter, S.G. (2007). Dynamic Capabilities: Understanding Strategic Change in Organizations. Blackwell Publishing, Oxford.

Heller, M., Eisenberg, R. (1998). "Can patents deter innovation? The anti-commons in biomedical research". Science 280, 698–701.

Henderson, R.M., Clark, K.B. (1990). "Architectural innovation: The reconfiguration of existing product technologies and the failure of established firms". Administrative Science Quarterly 35, 9–30.

Hodgson, G.M. (1999). Evolution and Institutions: On Evolutionary Economics and the Evolution of Economics. Edward Elgar Publishing, Cheltenham, UK.

Hopenhein, H.A. (1992). "Entry, exit and firm dynamics in long run equilibrium". Econometrica 60 (5), 1127–1150.

Hopkins, M.M., Martin, P.A., Nightingale, P., Kraft, A., Mahdi, S. (2007). "The myth of the biotech revolution: An assessment of technological, clinical and organizational change". Research Policy 36 (4), 566–589.

Hughes, T.P. (1983). Networks of Power: Electrification in Western Society. Johns Hopkins University Press, Baltimore, MD.

Ijiri, Y., Simon, H.A. (1977). Skew Distributions and the Sizes of Business Firms. North-Holland, Amsterdam.

Iwai, K. (1984a). "Schumpeterian dynamics, Part I". Journal of Economic Behavior & Organization 5, 159–190.

Iwai, K. (1984b). "Schumpeterian dynamics, Part II". Journal of Economic Behavior & Organization 5, 321–351.

Jaffe, A.B. (2000). "The U.S. patent system in transition: Policy innovation and the innovation process". Research Policy 29 (4–5), 531–577.

Jensen, B., McGuckin, H. (1997). "Firm performance and evolution: Empirical regularities in the US micro data". Industrial and Corporate Change 6 (1), 25–47.

Jovanovic, B. (1982). "Selection and the evolution of industry". Econometrica 50 (3), 649–670.

Jovanovic, B., Rousseau, P.L. (2002). "Moore's law and learning by doing". Review of Economic Dynamics 5 (2), 346–375.

Kauffman, S.A. (1993). The Origins of Order: Self-Organization and Selection in Evolution. Oxford University Press, Oxford.

Kauffman, S.A., Levin, S. (1987). "Toward a general theory of adaptive walks on rugged landscapes". Journal of Theoretical Biology 128, 11–45.

Kerker, M. (1961). "Science and the steam engine". Technology and Culture 2, 381–390.

Kirman, A.P. (2001). "Market organization and individual behavior: Evidence from fish markets". In: Rauch, J.E., Casella, A. (Eds.), Networks and Markets. Russell Sage Foundation, New York, NY, pp. 155–195.

Klepper, S. (1997). "Industry life cycles". Industrial and Corporate Change 6 (1), 145–181.

Klepper, S., Simons, K.L. (2005). "Industry shakeouts and technological change". International Journal of Industrial Organization 23, 23–43.

Klevorick, A.K., Levin, R.C., Nelson, R.R., Winter, S.G. (1995). "On the sources and significance of interindustry differences in technological opportunities". Research Policy 24, 185–205.

Kline, S.J., Rosenberg, N. (1986). "An overview of innovation". In: Landau, R., Rosenberg, N. (Eds.), The Positive Sum Strategies: Harnessing Technology for Economic Growth. National Academic Press, Washington, DC, pp. 275–305.

Kuhn, T. (1962). The Structure of Scientific Revolutions. University of Chicago Press, Chicago, IL.

Kumar, M.S. (1985). "Growth, acquisition activity and firm size: Evidence from the United Kingdom". Journal of Industrial Economics 33, 171–196.

Kuznets, S. (1972). Modern Economic Growth. Yale University Press, New Haven, CT.

Kwasnicki, W., Kwasnicka, H. (1992), "Market, innovation, competition: An evolutionary model of industrial dynamics", Journal of Economic Behavior & Organization, 19 (3), 343–368.

Landes, D.S. (1969). The Unbound Prometheus. Cambridge University Press, Cambridge.

Landes, D.S. (1998). The Wealth and Poverty of Nations: Why Are Some So Rich and Others So Poor? W.W. Norton, New York, NY.

Lazaric, N., Lorenz, E. (Eds.), (2003). Knowledge, Learning and Routines. Edward Elgar Publishing, Cheltenham, UK.

Levin, R.C., Cohen, W.M., Mowery, D.C. (1985). "R&D appropriability, opportunity and market structure: New evidence on some Schumpeterian hypotheses". American Economic Review Proceedings 75, 20–24.

Levinthal, D.A. (1997). "Adaptation on rugged landscapes". Management Science 43, 934–950.

Levinthal, D.A., Warglien, M. (1999). "Landscape design: Designing for local action in complex worlds". Organization Science 10 (3), 342–357.

Liebowitz, S.J., Margolis, S.E. (1995). "Path dependence, lock-in and history". Journal of Law, Economics, and Organization 11, 205–226.

Lotti, F., Santarelli, E., Vivarelli, M. (2003). "Does Gibrat's law hold in the case of young, small firms?" Journal of Evolutionary Economics 13 (2), 213–235.
Louça, F., Mendonça, S. (2002). "Steady change: The 200 largest U.S. manufacturing firms throughout the 20th century". Industrial and Corporate Change 11 (4), 817–845.
Lucas, R.E. (1978). "On the size distribution of business firms". Bell Journal of Economics 9, 508–523.
Lundberg, E. (1961). "Produktivitet och Räntabilitet". P.A. Norstedt & Söner, Stockholm. Quoted in Arrow (1962b).
Lundvall, B.A. (1992). National Systems of Innovation: Towards a Theory of Innovation and Interactive Learning. Pinter Publishing, London.
MacDonald, A., Schrattenholzer, L. (2001). "Learning rates for energy technologies". Energy Policy 29, 255–261.
Malerba, F. (2002). "Sectoral systems of innovation and production". Research Policy 31 (2), 247–264.
Malerba, F. (Ed.), (2004). Sectoral System of Innovation: Concepts, Issues, and Analyses of Six Major Sectors in Europe. Cambridge University Press, Cambridge.
Malerba, F., Brusoni, S. (Eds.), (2007). Perspectives on Innovation. Cambridge University Press, Cambridge.
Malerba, F., Orsenigo, L. (1995). "Schumpeterian patterns of innovation". Cambridge Journal of Economics 19 (1), 47–65.
Malerba, F., Orsenigo, L. (1996a). "Schumpeterian patterns of innovation are technology-specific". Research Policy 25 (3), 451–478.
Malerba, F., Orsenigo, L. (1996b). "The dynamics and evolution of industries". Industrial and Corporate Change 5, 51–87.
Malerba, F., Orsenigo, L. (1997). "Technological regimes and sectoral patterns of innovative activities". Industrial and Corporate Change 6, 83–117.
Malerba, F., Nelson, R., Orsenigo, L., Winter, S. (1999). "History friendly models of industry evolution: The case of the computer industry". Industrial and Corporate Change 8 (1), 3–40.
Malerba, F., Nelson, R., Orsenigo, L., Winter, S. (2007). "Demand, innovation, and the dynamics of market structure: The role of experimental users and diverse preferences". Journal of Evolutionary Economics 17 (4), 371–399.
Malerba, F., Nelson, R., Orsenigo, L., Winter, S. (2008). "Vertical integration and dis-integration of computer firms: A history friendly model of the co-evolution of the computer and semiconductor industries". Industrial and Corporate Change 17 (2), 197–231.
Mandeville, B. (1714). The Fable of the Bees, or, Private Vices, Public Benefits. Clarendon Press, Oxford Reprinted in Kaye, F.B. (Ed.) (1924). Mandeville's Fable of the Bees: Or, Private Vices, Public Benefits. Clarendon Press, Oxford.
Mansfield, E. (1961). "Technical change and the rate of imitation". Econometrica 29 (4), 741–766.
Mansfield, E. (1962). "Entry, Gibrat's law, innovation and the growth of firms". American Economic Review 52, 1023–1051.
Mansfield, E., Schwartz, M., Wagner, S. (1981). "Imitation costs and patents: An empirical study". Economic Journal 91 (364), 907–918.
March, J. (1988). Decision and Organizations. Basil Blackwell Publishing, Oxford.
March, J., Simon, H. (1958). Organizations. Basil Blackwell Publishing, New York, NY (second ed., 1993).
Marengo, L., Dosi, G. (2006). "Division of labor, organizational coordination and market mechanisms in collective problem-solving". Journal of Economic Behavior & Organization 58, 303–326.
Marengo, L., Dosi, G., Legrenzi, P., Pasquali, C. (2000). "The structure of problem-solving knowledge and the structure of organizations". Industrial and Corporate Change 9, 757–788.
Marengo, L., Pasquali, C., Valente, M., Dosi, G. (2009). Appropriability, patents, and rates of innovation in complex products industries. LEM Working Paper Series, 2009/05, Sant' Anna School of Advanced Studies, Pisa, Italy.
Marsili, O. (2001). The Anatomy and Evolution of Industries: Technological Change and Industrial Dynamics. Edward Elgar Publishing, Cheltenham, UK.
Marsili, O., Verspagen, B. (2002). "Technology and the dynamics of industrial structures: An empirical mapping of Dutch manufacturing". Industrial and Corporate Change 11 (4), 791–815.
Marx, K. (1847). The Poverty of Philosophy (English version). Marx-Engels Institute, Moscow.
Mazzoleni, R., Nelson, R.R. (1998). "The benefits and costs of strong patent protection: A contribution to the current debate". Research Policy 27 (3), 273–284.
Mazzucato, M. (2002). "The PC industry: New economy of early life cycle". Review of Economic Dynamics 5, 318–345.
Merges, R.P., Nelson, R.R. (1994). "On limiting or encouraging rivalry in technical progress: The effect of patent scope decisions". Journal of Economic Behavior & Organization 25 (1), 1–24.

Merton, R.K. (1973). The Sociology of Science: Theoretical and Empirical Investigations. University of Chicago Press, Chicago, IL.
Metcalfe, J.S. (1988). "The diffusion of innovations: An interpretative survey". In: Dosi, G., Freeman, C., Nelson, R., Silverberg, L., Stole, L. (Eds.), Technical Change and Economic Theory. Pinter Publishing, London, pp. 560–589.
Metcalfe, J.S. (1994). "Evolutionary economics and technology policy". Economic Journal 104 (425), 931–944.
Metcalfe, J.S. (1998). Evolutionary Economics and Creative Destruction. Routledge, London.
Metcalfe, J.S. (2005a). "Ed Mansfield and the diffusion of innovation: An evolutionary connection". Journal of Technology Transfer 30, 171–181.
Metcalfe, J.S. (2005b). "Evolutionary concepts in relation to evolutionary economics". In: Dopfer, K. (Ed.), The Evolutionary Foundations of Economics. Cambridge University Press, Cambridge.
Metcalfe, J.S., Foster, J., Ramlogan, R. (2005). "Adaptive economic growth". Cambridge Journal of Economics 30 (1), 7–32.
Mina, A., Ramlogan, R., Tampubolon, G., Metcalfe, J.S. (2007). "Mapping evolutionary trajectories: Applications to the growth and transformation of medical knowledge". Research Policy 36 (5), 789–806.
Mitchell, W.C. (1925). "Quantitative analysis in economic theory". American Economic Review 15 (1), 1–12.
Mokyr, J. (1990). The Lever of Riches: Technological Creativity and Economic Progress. Oxford University Press, Oxford.
Mokyr, J. (2002). The Gifts of Athena: Historical Origins of the Knowledge Economy. Princeton University Press, Princeton, NJ.
Mokyr, J. (2010). The contribution of economic history to the study of innovation and technical change: 1750–1914, this volume.
Montgomery, C. (1995). Resource-Based and Evolutionary Theories of the Firm. Kluwer, Boston, MA.
Morroni, M. (1992). Production Process and Technical Change. Cambridge University Press, Cambridge.
Mowery, D.C., Nelson, R.R. (Eds.), (1999). Sources of Industrial Leadership. Cambridge University Press, Cambridge.
Mowery, D.C., Rosenberg, N. (1982). Paths of Innovation, Technological Change in 20th Century America. Cambridge University Press, Cambridge.
Mowery, D.C., Rosenberg, N. (1989). Technology and the Pursuit of Economic Growth. Cambridge University Press, New York, NY.
Mowery, D.C., Rosenberg, N. (1979). "The influence of market demand upon innovation: A critical review of some recent empirical studies". Research Policy 8 (2), 102–153.
Mueller, D. (1986). Profits in the Long-Run. Cambridge University Press, Cambridge.
Mueller, D. (1990). The Dynamic of Company Profits. An International Comparison. Cambridge University Press, Cambridge.
Murmann, J.P. (2003). Knowledge and competitive advantage. Cambridge University Press, Cambridge.
Murmann, J.P., Frenken, K. (2006). "Toward a systematic framework for research on dominant designs, technological innovations, and industrial change". Research Policy 35 (7), 925–952.
Murray, F.E., Aghion, P., Dewatripont, M., Kolev, J., Stern, S. (2009). "Of mice and academics: Examining the effect of openness on innovation". NBER Working Paper Series, No. 14819.
Muth, J.F. (1986). "Search theory and the manufacturing progress function". Management Science 32, 948–962.
Nakicenovic, N., Grübler, A. (Eds.), (1991). Diffusion of Technologies and Social Behaviour. Springer, Heidelberg.
Needham, J. (1962–1963). "The pre-natal history of the steam engine". Transactions of the Newcomen Society 35, 3–58.
Nasbeth, L., Ray, G.F. (Eds.), (1974). The diffusion of new industrial processes. Cambridge University Press, Cambridge.
Neij, L. (1997). "Use of experience curves to analyse the prospects for diffusion and adoption of renewable energy technology". Energy Policy 25 (13), 1099–1107.
Nelson, R.R. (1959). "The simple economics of basic scientific research". Journal of Political Economy 67 (3), 297–306.
Nelson, R.R. (Ed.), (1962). The Rate and Direction of Inventive Activity. Princeton University Press, Princeton, NJ.
Nelson, R.R. (1968). "A diffusion model of international productivity differences in manufacturing industry". American Economic Review 58 (5), 1219–1248.
Nelson, R.R. (1981). "Research on productivity growth and productivity differences: Dead ends and new departures". Journal of Economic Literature 19 (3), 1029–1064.
Nelson, R.R. (1982). "The role of knowledge in R&D efficiency". Quarterly Journal of Economics 97 (3), 453–470.
Nelson, R.R. (1993). National Systems of Innovation: A Comparative Analysis. Oxford University Press, New York, NY.
Nelson, R.R. (1996). The Sources of Economic Growth. Harvard University Press, Cambridge, MA.
Nelson, R.R. (1998). "The agenda for growth theory: A different point of view". Cambridge Journal of Economics 22, 497–520.
Nelson, R.R. (1999). "The sources of industrial leadership: A perspective on industrial policy". De Economist 147, 1–18.
Nelson, R.R. (2003). "On the uneven evolution of human know-how". Research Policy 32 (6), 909–922.

Nelson, R.R. (2004). "The market economy, and the scientific commons". Research Policy 33 (3), 455–471.
Nelson, R.R. (2005). Technology, Institutions, and Economic Growth. Harvard University Press, Cambridge, MA.
Nelson, R.R. (2006). "Reflections on "The Simple Economics of Basic Scientific Research": Looking back and looking forward". Industrial and Corporate Change 15, 145–149.
Nelson, R.R. (2008a). "Factors affecting the powers of technological paradigms". Industrial and Corporate Change 17, 485–497.
Nelson, R.R. (2008b). "Why do firms differ and how does it matter? A revisitation". Seoul Journal of Economics 28, 607–619.
Nelson, R.R. (2008c). "What enables rapid economic progress? What are the needed institutions?" Research Policy 37 (1), 1–11.
Nelson, R.R., Nelson, K. (2002). "On the nature and evolution of human know-how". Research Policy 31, 719–733.
Nelson, R.R., Pack, H. (1999). "The Asian miracle and modern growth theory". Economic Journal 109 (457), 416–436.
Nelson, R., Sampat, B. (2001). "Making sense of institutions as a factor shaping economic performance". Journal of Economic Behavior & Organization 44, 31–54.
Nelson, R.R., Winter, S.G. (1977). "In search of a useful theory of innovation". Research Policy 6, 36–76.
Nelson, R.R., Winter, S.G. (1982). An Evolutionary Theory of Economic Change. Harvard University Press, Cambridge, MA.
Nelson, R.R., Winter, S.G. (2002). "Evolutionary theorizing in economics". Journal of Economic Perspectives 16 (2), 23–46.
Nelson, R.R., Wolff, E.N. (1997). "Factors behind cross-industry differences in technical progress". Structural Change and Economic Dynamics 8 (2), 205–220.
Nelson, R.R., Peterhausl, A., Sampat, B. (2004). "Why and how innovations get adopted: A tale of four models". Industrial and Corporate Change 13, 679–699.
Nightingale, P. (1998). "A cognitive model of innovation". Research Policy 27, 689–709.
Nightingale, P. (2003). "If Nelson and Winter are only half right about tacit knowledge, which half? A Searlian critique of "codification"" Industrial and Corporate Change 12, 149–183.
Nordhaus, W.D. (2007). "Two centuries of productivity growth in computing". Journal of Economic History 67 (1), 128–159.
Nuvolari, A. (2004). "Collective invention during the British industrial revolution: The case of the Cornish pumping engine". Cambridge Journal Economics 28, 347–363.
Nuvolari, A., Verspagen, B. (2009). "Technical choice, innovation and British steam engineering, 1800–1850". Economic History Review 63, 685–710.
Pasinetti, L.L. (1981). Structural Change and Economic Growth. A Theoretical Essay on the Dynamics of the Wealth of Nations. Cambridge University Press, Cambridge.
Pavitt, K. (1984). "Sectoral patterns of technical change: Towards a taxonomy and a theory". Research Policy 13, 343–373.
Pavitt, K. (1986). "'Chips' and 'Trajectories': How does the semiconductor influence the sources and directions of technical change?" In: MacLeod, R.M. (Ed.), Technology and the Human Prospect. Pinter Publishing, London.
Pavitt, K. (1987). "The objectives of technology policy". Science and Public Policy 14, 182–188.
Pavitt, K. (1999). Technology, Management and Systems of Innovation. Edward Elgar Publishing, Cheltenham, UK.
Pavitt, K. (2001). "Public policies to support basic research: What can the rest of the world learn from US theory and practice? (and what they should not learn)". Industrial and Corporate Change 10 (3), 761–779.
Pavitt, K. (2005). "Innovation processes". In: Fagerberg, J., Mowery, D.C., Nelson, R.R. (Eds.), The Oxford Handbook of Innovation. Oxford University Press, Oxford, pp. 86–114.
Perez, C. (1985). "Microelectronics, long waves and world structural change: New perspectives for developing countries". World Development 13, 441–463.
Perez, C. (2002). Technological revolutions and financial capital: The dynamics of bubbles and golden ages. Edward Elgar Publishing, Cheltenham, UK.
Perez, C. (2010). "Technological revolutions and techno-economic paradigms". Cambridge Journal of Economics 34, 185–202.
Polanyi, M. (1962). Personal Knowledge: Towards a Post-Critical Philosophy. University of Chicago Press, Chicago, IL.
Polanyi, M. (1967). The Tacit Dimension. Anchor Books, New York, NY.
Possas, M.L., Salles-Filho, S., Silveira, J.M. (1996). "An evolutionary approach to technological innovation in agriculture: Some preliminary remarks". Research Policy 25 (6), 933–945.
Power, L. (1998). "The missing link: Technology, investment and productivity". Review of Economics and Statistics 80, 300–313.
Prencipe, A., Tell, F. (2001). "Inter-project learning: Processes and outcomes of knowledge codification in project-based firms". Research Policy 30, 1373–1394.
Prencipe, A., Davies, A., Hobday, M. (Eds.), (2003). The Business of Systems Integration. Oxford University Press, Oxford.
Radner, R. (1992). "Hierarchy: The economics of managing". Journal of Economic Literature 30 (3), 1382–1415.

Radner, R. (1993). "The organization of decentralized information processing". Econometrica 61 (5), 1109–1146.
Ray, G. F. (1989). "Full circle: The diffusion of technology", Research Policy, Elsevier, 18 (1), 1–18.
Roberts, P.W. (1999). "Product innovation, product-market competition and persistent profitability in the U.S Pharmaceutical industry". Strategic Management Journal 20, 655–670.
Romer, P.M. (1994). "The origins of endogenous growth". Journal of Economic Perspectives 8 (1), 3–22.
Rosenberg, N. (1963). "Mandeville and Laissez-Faire". Journal of the History of Ideas 24 (2), 183–196.
Rosenberg, N. (1972). "Factors affecting the diffusion of technology". Explorations in Economic History 10, 3–33.
Rosenberg, N. (1976). Perspectives on Technology. Cambridge University Press, Cambridge.
Rosenberg, N. (1982). Inside the Black Box: Technology and Economics. Cambridge University Press, Cambridge.
Rosenberg, N. (1994). Exploring the Black Box: Technology, Economics, and History. Cambridge University Press, Cambridge.
Rosenberg, N. (1996). "Uncertainty and technological change". In: Landau, R., Taylor, T., Wright, G. (Eds.), The Mosaic of Economic Growth. Stanford University Press, Stanford, CA.
Rosenberg, N. (2009). "Some critical episodes in the progress of medical innovation: An Anglo-American perspective". Research Policy 3 (2), 234–242.
Rosenberg, N., Nelson, R.R. (1994). "American universities and technical advance in industry". Research Policy 23 (3), 323–348.
Rosenberg, N., Trajtenberg, M. (2004). "A general-purpose technology at work: The Corliss steam engine in the late-nineteenth-century United States". Journal of Economic History 64, 61–99.
Rosenbloom, R.S., Cusumano, M.A. (1987). "Technological pioneering and competitive advantage: The birth of the VCR industry". California Management Review 29 (4), 51–76.
Rumelt, R. (1991). "How much does industry matter?" Strategic Management Journal 12, 167–185.
Ruttan, V.W. (1997). "Induced innovation, evolutionary theory and path dependence: Sources of technical change". Economic Journal 107 (444), 1520–1529.
Sahal, D. (1981). Patterns of Technological Innovation. Addison-Wesley, New York, NY.
Sahal, D. (1985). "Technological guideposts and innovation avenues". Research Policy 14 (2), 61–82.
Salter, W. (1962). Productivity and Technical Change. Cambridge University Press, Cambridge.
Saviotti, P.P. (1996). Technological Evolution, Variety and the Economy. Edward Elgar Publishing, Cheltenham, UK.
Saviotti, P.P., Trickett, A. (1992). "The evolution of helicopter technology, 1940–1986". Economics of Innovation and New Technology 2, 111–130.
Saviotti, P.P., Pyka A. (2008a). "Micro and macro dynamics: Industry life cycles, inter-sector coordination and aggregate growth". Journal of Evolutionary Economics, 18 (2), 167–182.
Saviotti P.P., Pyka A. (2008b). "Product variety, competition and economic growth". Journal of Evolutionary Economics, 18 (3–4), 323–347.
Schmookler, J. (1966). Invention and Economic Growth. Harvard University Press, Cambridge, MA.
Schumpeter, J.A. (1911). The Theory of Economic Development. Harvard University Press, Cambridge, MA, English Edition, 1934.
Schumpeter, J.A. (1942). Capitalism, Socialism and Democracy. Harper, New York, NY.
Silverberg, G. (1988). "Modelling economic dynamics and technical change: Mathematical approaches to self-organisation and evolution". In: Dosi, G., Freeman, C., Nelson, R., Silverberg, G., Soete, L. (Eds.), Technical Change and Economic Theory. Pinter Publishing, London, pp. 531–559.
Silverberg, G. (2003). "Breaking the waves: A Poisson regression approach to Schumpeterian clustering of basic innovations". Cambridge Journal of Economics 27 (5), 671.
Silverberg, G., Lehnert, D. (1993). "Long waves and 'evolutionary chaos' in a simple Schumpeterian model of embodied technical change". Structural Change and Economic Dynamics 4 (1), 9–37.
Silverberg, G., Lehnert, D. (1994). "Growth fluctuation in an evolutionary model of creative destruction". In: Silverberg, G., Soete, L. (Eds.), The Economics of Growth and Technical Change. Edward Elgar Publishing, Aldershot.
Silverberg, G., Verspagen, B. (1994). "Learning, innovation and economic growth: A long-run model of industrial dynamics". Industrial and Corporate Change 3, 199–223.
Silverberg, G., Verspagen, B. (1996). "From the artificial to the endogenous: Modelling evolutionary adaptation and economic growth". In: Helmstadter, E., Perlman, M. (Eds.), Behavioral Norms, Technological Progress and Economic Dynamics: Studies in Schumpeterian Economics. University of Michigan Press, Ann Arbor, MI, pp. 331–371.

Silverberg, G., Verspagen, B. (2005a). "A percolation model of innovation in complex technology spaces". Journal of Economic Dynamics and Control 29, 225–244.
Silverberg, G., Verspagen, B. (2005b). "Evolutionary theorizing on economic growth". In: Dopfer, K. (Ed.), The Evolutionary Principles of Economics. Cambridge University Press, Cambridge.
Silverberg, G., Dosi, G., Orsenigo, L. (1988). "Innovation, diversity and diffusion: A self-organising model". Economic Journal 98 (393), 1032–1054.
Simon, H. (1957). Administrative Behavior (second ed.). Macmillan, New York, NY.
Simon, H. (1962). "The architecture of complexity". Proceedings of the American Philosophical Society 106 (6), 467–482.
Smith, A. (1776). An Inquiry into the Nature and Causes of the Wealth of Nations. Liberty Fund, Indianapolis, IN Reprinted in 1981.
Soete, L. (1979). "Firm size and inventive activity: The evidence reconsidered". European Economic Review 12 (4), 319–340.
Soete, L., Turner, R. (1984). "Technology diffusion and the rate of technical change". Economic Journal 94 (75), 612–623.
Srholec, M., Verspagen, B. (2008). The Voyage of the Beagle in innovation systems land. Explorations on sectors, innovation, heterogeneity and selection. Working Paper No. 2008-008, UNU-Merit, Maastricht.
Stadler, B.M., Stadler, P.F., Wagner, G.P., Fontana, W. (2001). "The topology of the possible: Formal spaces underlying patterns of evolutionary change". Journal of Theoretical Biology 213, 241–274.
Stanley, M. H. R., Buldyrev, S.V., Havlin, S., Mantegna, R., Salinger, M.A., Stanley, H.E. (1996). "Zipf plots and the size distribution of firms". Economics Letters 49, 453–457.
Starbuck, W., Mezias, J.M. (1996). "Opening Pandora's Box: Studying the accuracy of managers' perceptions". Journal of Organizational Behavior 17 (2), 99–117.
Steindl, J. (1965). Random processes and the growth of firms. Griffin, London.
Stiglitz, J. (1994). Whither Socialism? MIT Press, Cambridge, MA.
Stoneman, P. (Ed.) (1995). Handbook on the Economics of Innovation and Technical Change. Blackwell Publishing, Oxford.
Stoneman, P. (2007). "Technological diffusion: aspects of self-propagation as a neo-Schumpeterian characteristic". In: Hanush, A., Pyka, A. (Eds.), Elgar Companion to Neo-Schumpeterian Economics. Edward Elgar Publishing, Cheltenham, UK.
Suarez, F.F., Utterback, J.M. (1995). "Dominant designs and the survival of firms". Strategic Management Journal 16, 415–430.
Sutton, J. (1997). "Gibrat's legacy". Journal of Economic Literature 35 (1), 40–59.
Sutton, J. (1998). Technology and Market Structure: Theory and Evidence. MIT Press, Cambridge, MA.
Svennilson, I. (1954). Growth and Stagnation in the European Economy. UN Economic Commission for Europe, Geneva.
Syverson, C. (2004). "Product substitutability and productivity dispersion". Review of Economics and Statistics 86 (2), 534–550.
Teece, D. (1986). "Profiting from technological innovation: Implications for integration, collaboration, licensing and public policy". Research Policy 15 (6), 285–305.
Teece, D., Rumelt, R., Dosi, G., Winter, S.G. (1994). "Understanding corporate coherence: Theory and evidence". Journal of Economic Behavior & Organization 23, 1–30.
Teece, D., Pisano, G., Shuen, A. (1997). "Dynamic capabilities and strategic management". Strategic Management Journal 18 (7), 509–533.
Tesfatsion, L., Judd, K. (Eds.), (2006). Handbook of Computational Economics, vol. 2: Agent-Based Computational Economics. North-Holland, Amsterdam.
Tushman, M.L., Anderson, P. (1986). "Technological discontinuities and organizational environments". Administrative Science Quarterly 31, 439–465.
Utterback, J.M., Suarez, F.F. (1993). "Innovation, competition, and industry structure". Research Policy 22 (1), 1–21.
Verspagen, B. (1993). Uneven Growth Between Interdependent Economies: A Evolutionary View on Technology Gaps, Trade and Growth. Ashgate, Aldershot.
Vincenti, W. (1990). What Engineers Know and How They Know It. Johns Hopkins University Press, Baltimore, MD.
Von Hippel, E. (2005). Democratizing innovation. MIT Press, Cambridge, MA.
Von Tunzelmann, G.N. (1995). Technology and Industrial Progress: The Foundations of Economic Growth. Edward Elgar Publishing, Cheltenham, UK.
Walsh, V. (1984). "Invention and innovation in the chemical industry: Demand pull or discovery push". Research Policy 13, 211–234.
Weisbuch, G., Kirman, A., Herreiner, D. (2000). "Market organization and trading relationships". Economic Journal 110, 411–436.

Winter, S.G. (1971). "Satisficing, selection and the innovative remnant". Quarterly Journal of Economics 85 (2), 237–261.
Winter, S.G. (1982). "An essay on the theory of production". In: Hymans, S.H. (Ed.), Economics and the World Around It. University of Michigan Press, Ann Arbor, MI.
Winter, S.G. (1984). "Schumpeterian competition in alternative technological regimes". Journal of Economic Behavior & Organization 5, 287–320.
Winter, S.G. (1987). "Knowledge and competence as strategic assets". In: Teece, D.J. (Ed.), The Competitive Challenge: Strategies for Industrial Innovation and Renewal. Ballinger Publishing, Cambridge, MA, pp. 159–184.
Winter, S.G. (1993). "Patents and welfare in an evolutionary model". Industrial and Corporate Change 2, 211–231.
Winter, S.G. (2003). "Understanding dynamic capabilities". Strategic Management Journal 24 (10), 991–995.
Winter, S.G. (2005). "Towards an evolutionary theory of production". In: Dopfer, K. (Ed.), The Evolutionary Foundations of Economics. Cambridge University Press, Cambridge, pp. 223–224.
Winter, S.G. (2006a). "Toward a neo-Schumpeterian theory of the firm". Industrial and Corporate Change 15 (1), 125–141 (Original Working Paper, 1968).
Winter, S.G. (2006b). "The logic of appropriability: From Schumpeter to Arrow to Teece". Research Policy 35 (8), 1100–1106.
Winter, S.G. (2008). "Scaling heuristics shape technology! Should economic theory take notice?" Industrial and Corporate Change 17 (3), 513–531.
Winter, S.G., Szulanski, G. (2001). "Replication as strategy". Organization Science 12, 730–743.
Winter, S.G., Szulanski, G. (2002). "Replication of organizational routines: Conceptualizing the exploitation of knowledge assets". In: Choo, C.W., Bontis, N. (Eds.), The Strategic Management of Intellectual Capital and Organizational Knowledge. Oxford University Press, Oxford.
Winter, S., Kaniovski, Y., Dosi, G. (2000). "Modeling industrial dynamics with innovative entrants". Structural Change and Economic Dynamics 11 (3), 255–293.
Winter, S., Kaniovski, Y., Dosi, G. (2003). "A baseline model of industry evolution". Journal of Evolutionary Economics 13 (4), 355–383.
Wright, T.P. (1936). "Factors affecting the costs of airplanes". Journal of Aeronautical Sciences 10, 302–328.
Wright, G. (1997). "Towards a more historical approach to technological change". Economic Journal 107 (444), 1560–1566.
Yelle, L.E. (1979). "The learning curve: Historical review and comprehensive survey". Decision Sciences 10, 302–308.
Yildizoglu, M. (2002). "Competing R&D strategies in an evolutionary industry model". Computational Economics 19 (1), 51–65.
Zheng, L. (2008). System-on-Chip Applications, Lecture Notes—Electronics, Computer and Software Systems. Royal Institute of Technology (KTH), Stockholm.
Ziman, J.M. (Ed.), (2000). Technological Innovation as an Evolutionary Process. Cambridge University Press, Cambridge.

Chapter 4

FIFTY YEARS OF EMPIRICAL STUDIES OF INNOVATIVE ACTIVITY AND PERFORMANCE

WESLEY M. COHEN
Fuqua School of Business
Duke University, Durham
North Carolina, USA

Contents

Abstract

This chapter reviews the empirical literature on the determination of firms' and industries' innovative activity and performance, highlighting the questions addressed, the approaches adopted, impediments to progress in the field, and research opportunities. We review the "neo-Schumpeterian" empirical literature that examines the effects of firm size and market concentration upon innovation, focusing

Handbooks in Economics, Volume 01

DOI: 10.1016/S0169-7218(10)01004-X

on robust findings, questions of interpretation, and the identification of major gaps. We also consider the more modest literature that considers the effect on innovation of firm characteristics other than size. Finally, we review the literature that considers three classes of factors that affect interindustry variation in innovative activity and performance: demand, appropriability, and technological opportunity conditions.

Keywords

innovation, market structure, R&D, technological change

JEL classification: O30, O31, O32, O34, O38, L1, L2

1. Introduction

For much of the twentieth century, industrial organization economists examined the determinants of market structure and its effect on price competition and allocative efficiency, largely disregarding technological change. The writings of Joseph Schumpeter in the first half of the century pushed economists to appreciate the fundamental role of technological progress in affecting economic growth and social welfare. Since that time, economists have increasingly appreciated the economic significance of technological progress, and it is now common to hear that a firm's, an industry's, or even a nation's capacity to progress technologically underpins its long-run economic performance. Stimulated by Schumpeter's writings and Solow's (1957) subsequent "discovery" of the contribution of technological change to economic growth, industrial organization economists have conducted numerous empirical studies on the determination of innovative activity and performance. In this chapter we review the empirical literature and highlight the questions addressed, the approaches adopted, and impediments to progress in the field.

Some of these impediments are ironically due to Schumpeter himself. In making his case for the importance of innovation broadly construed, Schumpeter rejected the antitrust orthodoxy of his day. He argued that the large firm operating in a concentrated market had become the locus of technological progress, and, therefore, an industrial organization of large monopolistic firms offered decisive welfare advantages.

Provoked by these claims, industrial organization economists (e.g., Mason, 1951) became preoccupied with the effects of firm size and market concentration on innovation and neglected other, perhaps more fundamental determinants of technological progress. This review will briefly examine this literature on the relationship between innovation and market structure and firm size. We will then, however, review more recent research that has both recast the "neo-Schumpeterian" preoccupations and moved beyond them to study the determinants of technical advance more broadly.

This review updates and draws heavily from the survey written by the author (Cohen, 1995), which in turn drew extensively from a prior survey written by Richard Levin and the author (Cohen and Levin, 1989). As in the prior surveys, we review the empirical literature on the characteristics of industries and firms that influence industrial innovation. In addition to the empirical literature, we also selectively review the case study and institutional literature that often provides richer, more subtle interpretations of the relationships among innovation, market structure, and industry and firm characteristics. Although the literature considered here is extensive, this survey examines only studies of innovation that fall under the rubric of industrial organization economics. Moreover, given a rapid growth in this literature since the 1995 review, the review of the more recent empirical literature will be selective. Section 2 reviews the "neo-Schumpeterian" literature that examines the effects of firm size and market concentration upon innovation. After a brief synopsis of the empirical findings of the Schumpeterian literature, Section 2.3 focuses on questions of interpretation and the identification of major gaps in the empirical literature. Section 3 discusses the more modest literature that considers the effect on innovation of firm characteristics other than size. Section 4 covers recent literature that considers three classes of factors that affect interindustry variation in innovative activity and performance: demand, appropriability, and technological opportunity conditions. We conclude in Section 5.

2. Empirical studies in the Schumpeterian tradition

This section considers the scores of studies on the relationship between market concentration or firm size and innovative effort elicited by Schumpeter's controversial claims about the key role of large monopolistic firms in advancing technology. As a foundation for reviewing this literature, it is useful to understand the questions with which Schumpeter himself was centrally concerned—and not. He wanted to understand the impact of capitalist competition on economic growth and how competition might critically affect growth through its impact on innovation, which he viewed as central to the growth process, and, in turn, long-run improvements in social welfare. Though obviously, therefore, concerned with the different ways that competition might drive innovation, and, more specifically, how different forms of competition might affect the incentives and capacities of firms (and individuals) to innovate, Schumpeter was not centrally concerned with the determinants of innovation more generally—which is the subject of this review.

Prior to Cohen and Levin (1989) and Cohen (1995), three other literature surveys (Baldwin and Scott, 1987; Kamien and Schwartz, 1982; Scherer, 1980) ably summarized findings concerning the two "Schumpeterian" hypotheses and related propositions. Most recently, Gilbert (2006) has reviewed both the theoretical and empirical literatures treating the "Schumpeterian" hypotheses linking, respectively, market structure and firm size to innovation. For this reason, the summary of specific results will be brief. Rather, we focus instead on identifying robust empirical patterns and on questions of interpretation and methodology raised by this substantial body of work.[1]

2.1. *Firm size and innovation*

A literal reading of Schumpeter's (1942) classic discussion suggests that he was primarily impressed by the qualitative differences between the innovative activities of small, entrepreneurial enterprises and those of large, modern corporations with formal R&D laboratories. Nonetheless, the empirical literature has interpreted Schumpeter's claim for a large firm advantage in innovation as a proposition that innovative activity increases more than proportionately than firm size.[2] With some exceptions (e.g., Gellman Research Associates, 1976; Nelson et al., 1967; Pavitt et al., 1987; Scherer, 1965a) the Schumpeterian hypothesis about firm size has been tested by regressing some measure of innovative activity (input or output) on a measure of size.

While Schumpeter confounded his discussions of the impacts on innovation of firm size and market concentration, Galbraith (1952) explicitly argued that large firm size confers an advantage in innovation. Over the years, several justifications (only some of which were suggested by Schumpeter) for a positive effect of firm size on innovative activity have been offered. One claim is that capital market

[1] Also see Scherer (1992) for a discussion of a number of the issues raised in this section.

[2] Markham (1965) and Nelson et al. (1967) have argued, however, that Schumpeter never claimed a continuous relationship between R&D and firm size, but only that innovation no longer depended upon the initiative and genius of independent entrepreneurs, as he had previously suggested (Schumpeter, 1934). Rather, R&D had come to be conducted largely by the professional laboratories of large, bureaucratic corporations that had become the principal source of innovation in modern capitalist societies.

imperfections confer an advantage on large firms in securing finance for risky R&D projects because size is correlated with the availability and stability of internally generated funds. A second claim is that there are scale economies in the R&D function itself. Another is that the returns from R&D are higher where the innovator has a larger volume of sales over which to spread the fixed costs of innovation, particularly process innovation. R&D is also alleged to be more productive in large firms as a result of complementarities between R&D and other nonmanufacturing activities (e.g., marketing and financial planning) that may be better developed within large firms. Finally, it is sometimes suggested that large, diversified firms provide economies of scope or reduce the risk associated with the prospective returns to innovation.

Counterarguments to the proposition have also been suggested (cf. Scherer and Ross, 1990, pp. 652–653). Perhaps the most prominent are that, as firms grow large, efficiency in R&D is undermined either through the loss of managerial control or, alternatively, through excessive bureaucratic control which diverts the attention of the firm's bench scientists and technologists. Also, as firms grow large, the incentives of individual scientists and entrepreneurs may be blunted as either their ability to capture the benefits from their individual efforts diminishes or their creative impulses are frustrated by the conservatism characteristic of the hierarchies of large corporations.[3] Indeed, Schumpeter (1942) himself speculated that the bureaucratization of inventive activity inhering in the large capitalist enterprise would ultimately contribute to capitalism's decline.[4]

Over five decades of empirical research on the relationship between firm size and innovation have spawned a number of robust empirical patterns. Although the establishment of most of these patterns has been subject to some degree of controversy, the profession has tended to arrive at consensus views. Of rather more dispute was the interpretation of these patterns, although even here a consensus emerged, at least for a period of time. Our characterization of these empirical patterns and their interpretation closely follows Cohen and Klepper (1996b).

Based on National Science Foundation (NSF) data from the 1950s and early 1960s, early research established that the likelihood of a firm conducting R&D increases with firm size and approaches unity among the largest firms (Hamberg, 1964; Nelson et al., 1967; Villard, 1958; Worley, 1961). Bound et al. (1984) and Cohen et al. (1987) noted a similar pattern while controlling for industry effects. Notwithstanding early skepticism expressed by Schmookler (1959), the positive relationship between the likelihood of performing R&D and firm size was interpreted as revealing an advantage to large size in the conduct of R&D.

In their search for a "Schumpeterian" advantage to firm size, economists focused most of their attention on the continuous relationship between R&D and firm size. This relationship was typically estimated on cross-sectional samples restricted to R&D performers and was specified in log–log form, linearly or with R&D intensity (i.e., R&D effort divided by a measure of firm size, usually sales) as the dependent variable and a measure of firm size as a regressor. As noted in Cohen and Klepper (1996b), the overwhelming evidence accumulated across different samples, specifications, and estimation

[3] See Sah and Stiglitz (1986, 1988) for a theoretical argument for why hierarchy may dampen innovation.

[4] Rosenberg (1994) argues that Schumpeter both overestimates the degree to which capitalism "automatizes" innovation, and underestimates the degree to which commercial success depends upon the more mundane and rationalized processes associated with downstream R&D and related activities.

methods was that R&D rises monotonically with firm size, and proportionately beyond some modest firm size threshold. Moreover, R&D was found to vary closely with firm size within industries, with size typically explaining over half of its variation.

The earliest studies examining the R&D–size relationship that used samples spanning multiple industries (e.g., Hamberg, 1964; Horowitz, 1962) concluded that R&D rose somewhat more than proportionately with firm size. These early studies, however, omitted controls for industry effects. Since differences in the size distribution of firms across industries may well reflect industry-level differences in technological opportunities, and economies in production and/or distribution and other factors, one would expect industry effects to be correlated with firm size. As a consequence, the omission of such industry effects will likely bias estimates of the effects of size on innovation (cf. Baldwin and Scott, 1987; Nelson et al., 1967).

Most of the subsequent studies conducted using samples spanning multiple industries have controlled for industry effects by including either relatively crude measures of industry-level variables such as technological opportunity, or industry fixed effects. Although some of these studies (e.g., Comanor, 1967; Meisel and Lin, 1983) found R&D to rise more than proportionately than firm size, Scherer (1965a,b) observed a more subtle relationship—that inventive activity, whether measured by input (personnel) or output (patents), increased more than proportionally with size up to a threshold, whereupon the relationship became basically proportional. Confirmed to varying degrees by other investigators (e.g., Link, 1981; Malecki, 1980; Phlips, 1971), Scherer's findings became the profession's tentative consensus by the early 1980s (cf. Kamien and Schwartz, 1982; Scherer, 1980).

Using a larger and more comprehensive sample of American firms than any previously employed to study the size–innovation relationship at the firm level, Bound et al. (1984) found that R&D intensity falls slightly with size among the very smallest firms and then rises somewhat with firm size among the very largest firms. Using data from the Federal Trade Commission's (FTC) Line of Business Program combined with data from the Levin et al. (1987) survey of appropriability and technological opportunity conditions in industry, Cohen et al. (1987) showed that once care was taken to control for industry effects and distinguish between the size of the firm and that of the business unit, neither business unit nor overall firm size significantly affected business unit R&D intensity in the (selected) sample of R&D performers.

Although offering the advantage of exploiting many more observations than studies featuring industry-level samples, the aggregate studies that pool observations across industries restrict the elasticity of R&D with respect to size to be the same across industries. Not subject to this limitation, the industry-level analyses (Link, 1981; Link et al., 1988; Mansfield, 1964; Scherer, 1965b, 1984b; Soete, 1979) suggested that among R&D performers, in most industries, the null hypothesis of proportionality between R&D and firm size could not be rejected, and Scherer (1965b, 1984b) found this to be true of patent counts as well as R&D.[5] In the few industries where it was rejected, there was no single pattern. R&D was found, for example, to rise more than proportionately than size in chemicals (Mansfield, 1964; Scherer, 1965b), less than proportionately in drugs (Grabowski, 1968; Mansfield, 1964), and both less than and greater than proportionately in a number of other industries (Soete, 1979). One weakness, however, of the industry-level studies is that due to the small

[5] As noted by Griliches (1990) and Cohen and Levin (1989), raw patent counts are likely a better measure of innovative input than output, and are closely correlated with R&D.

sample size for most individual industries, there is a statistical presumption in favor of the null hypothesis of proportionality (Cohen and Klepper, 1996b).

Both the studies based on individual industries and the studies pooling observations across industries are subject to several limitations apart from those already noted. First, most of the samples used in the regression studies are nonrandom, and, with a few exceptions (Bound et al., 1984; Cohen et al., 1987; Crepon et al., 1998), no attempt has been made to study the presence or the effects of sample selection bias. Many of the earlier firm-level studies confined attention to the 500 or 1000 largest firms in the manufacturing sector, and firms that reported no R&D were typically excluded from the sample.

Second, the studies vary in the degree to which they control for characteristics of firms (other than size), despite the demonstrated importance of firm effects in explaining R&D intensity (Scott, 1984), and their likely colinearity with firm size. The absence of controls for firm characteristics highlights the related point that many of the studies that hypothesize a relationship between firm size and R&D do so by appealing to the influence of what are claimed to be firm characteristics correlated with size, such as cash flow, degree of diversification, complementary capabilities, economies of scale and scope in the R&D function, and the ability to spread R&D costs over output. Yet, most of these studies have not directly examined whether the observed relationship between size and R&D is indeed due to the influence of any of these hypothesized factors.

Third, although most of the studies of the R&D–firm size relationship attempted to control for industry effects, it is not a simple matter to control properly for industry effects in a sample of firm-level data because most larger firms are aggregations of business units engaged in a variety of industries. Most attempts to control for industry effects have assigned each sample firm to a primary industry and then used either a fixed effects model or specific industry characteristics as covariates. Such assignments are typically made at the two-digit SIC level, a procedure that introduces measurement error to the extent that relevant industry characteristics vary substantially across the constituent four-digit industries. On the other hand, when industry assignments are made at the three- or four-digit level, there is also systematic mismeasurement because many firms (and most large ones) conduct the bulk of their business outside their designated primary industry.

Speaking to one of the rationales offered in the literature for expecting a relationship between firm size and R&D, Henderson and Cockburn (1996) conduct a careful analysis of scale and scope effects of R&D in the pharmaceutical industry. Unique among economic studies of R&D, Henderson and Cockburn (1996) employ program-level data, where programs reflect pharmaceutical firms' research activities within a therapeutic class. They employ approximately 20 years of program-level data on 10 firms, providing 4930 observations. Featuring a count of "important" patents (defined as patents granted in two of the three major jurisdictions: the United States, Europe, and Japan) as their dependent variable, and focusing exclusively on research (not development) expenditures, they find strong scale and scope effects; those research programs that are pursued by firms with larger research budgets and a larger number of distinct research programs are significantly more productive. They did not obtain such clear results when analyzing their data at the firm level; the results were only obtained when they disaggregated to the level of the research program as the unit of analysis. Although unsurprising given the sharp reduction in the number of observations from almost 5000 to about 200 when they move to the firm level, the lack of results at the firm level does suggest the general point that empirical analysis is often much more revealing when conducted at an appropriate, and often finer, level of aggregation. Cockburn and Henderson (2001) conduct a follow-through study of the relationship between their measures of scale and scope with the productivity of the firms' drug *development* efforts, using a logit analysis and defining development success as whether the drug obtained regulatory approval. They again find a strong effect of scope, but little impact of scale. All the effects of their scale and scope variables, defined at the therapeutic class level, become small and insignificant when firm dummies are included, undoubtedly reflecting colinearity as well as other possible effects (e.g., management, firm structure, etc.) operating at the firm level. Although constrained by a limited number of firm-level observations, it is worth highlighting that these studies do not analyze the degree to which scope or scale economies may account for any relationship that might be observed between firm size and R&D.

It is useful to reconsider briefly the Schumpeterian hypothesis in light of the fact that most large firms operate business units in numerous industries. Although some arguments advanced to rationalize Schumpeter's hypothesis refer to the overall size of the firm (e.g., the ability to overcome capital market imperfections), others are more plausible at the level of the business unit (e.g., R&D cost spreading). Although the great majority of the studies we have discussed examine the effect of firm size on firm-level R&D, the FTC's Line of Business data made it possible to separate the effects of business unit and firm size. Scherer (1984b) and Scott (1984) studied the effects of business unit size on business unit R&D, while Cohen et al. (1987) and Cohen and Klepper (1996b) examined the effects of both business unit and overall firm size on business unit R&D intensity. Cohen et al. (1987) found that the size of the business unit rather than the size of the firm as a whole affects the likelihood of performing R&D. Cohen and Klepper (1996b) found in two simple regressions of business unit R&D against, respectively, business unit and overall firm size for each of 75 industries, that business unit size alone explained an average of 65% of the variance in business unit R&D, and the relationship was typically proportionate. In contrast, overall firm size explained an average of only 15%. Moreover, the coefficient of business unit size was positive and significant for almost 90% of the 75 industries, whereas the coefficient of overall firm size was rarely significant, and, indeed, was actually negative for 26 of the 75 industries. Both Cohen et al. (1987) and Cohen and Klepper (1996b) also found that, controlling for business unit size, overall firm size exercised no independent influence on business unit R&D. These results together suggest that it is the size of the business unit (or its correlates) rather than the size of the firm as a whole (or its correlates) that accounts for the close relationship between firm size and R&D.

Notwithstanding the various challenges in evaluating the R&D–firm size relationship, the consensus is that either in the majority of industries, or when controlling for industry effects in more aggregate samples, R&D rises proportionately with firm size among R&D performers (e.g., Baldwin and Scott, 1987; Scherer and Ross, 1990). Although the source of this relationship had not been determined, the finding was widely interpreted through the mid-1990s as indicating that, contrary to Schumpeter, large size offered no advantage in the conduct of R&D. The intuition behind this interpretation is that, if the relationship is proportional, then, holding industry sales constant, the same amount of R&D will be conducted whether an industry is comprised of large firms or a greater number of smaller firms. Fisher and Temin (1973) argued, however, that to the extent that Schumpeter's hypothesis can be given a clear formulation, it must refer to a relationship between innovative output and firm size, not to a relationship between R&D (an innovative input) and firm size, which is the one most commonly tested in the literature. They demonstrated, among other things, that an elasticity of R&D with respect to size in excess of one does not necessarily imply an elasticity of innovative output with respect to size greater than one.[6]

[6] Kohn and Scott (1982) established the conditions under which the existence of the former relationship does imply the latter.

Both before and subsequent to Fisher and Temin's critique, however, several studies exploiting measures of innovative output reinforced the earlier consensus of no advantage to size. Scherer (1965a), Gellman Research Associates (1976, 1982), The Futures Group (1984), Pavitt et al. (1987) and Acs and Audretsch (1988, 1990, 1991b) have shown that, in either panel or cross-sectional data spanning a broad range of firm sizes, smaller firms tend to account for a disproportionately large share of innovations relative to their size, and that R&D productivity (e.g., innovations per unit of R&D) tends to decline with firm size. Bound et al.'s (1984) analysis of patenting activity similarly found that patents produced per R&D dollar for smaller firms (i.e., less than 1 million dollars in sales) is considerably higher than that for larger firms. Acs and Audretsch (1990, 1991b) provide evidence that this pattern varies, however, across industries. Also, Pavitt et al.'s (1987) findings, based on the SPRU data set that counts the successful introduction of "significant" new products or processes, suggest that the relationship may be somewhat U-shaped, with the very largest firms displaying relatively high R&D productivity, defined as simply the number of innovations per R&D dollar. Also drawing upon the SPRU data set, Geroski (1994, Chapter 2) highlights the clear negative correlation between firm size and R&D productivity. Using information on financial service innovations drawn from the *Wall Street Journal* over the period, 1990–2002, Lerner (2006) also observes that smaller firms account for a disproportionate share. Thus, the predominant pattern is that R&D productivity appears to decline with size.

Though less explored than the relationship between firm size and innovations per R&D dollar due to limited availability of data (see below), scholars have also examined the relationship between firm size and the types of innovation pursued, focusing on the degree to which firm size is related to process versus product R&D, or to the generation of incremental versus more significant or "radical" (variously defined; see below) innovations. The key findings are that larger, incumbent firms tend to pursue relatively more incremental (Henderson, 1993; Mansfield, 1981; Wilson et al., 1980) and relatively more process innovation (Cohen and Klepper, 1996a; Link, 1982a; Pavitt et al., 1987; Scherer, 1991) than smaller firms. Whether new ventures and entrants (as opposed to small firms more generally) are chiefly responsible for "radical" innovation—though often talked about—suffers from a dearth of rigorous empirical study. One exception is Prusa and Schmitz (1991), who provide evidence from the personal computer software industry that new firms tend to create new software categories, while established firms tend to develop improvements in existing categories.

Thus, the robust empirical patterns relating to R&D and innovation to firm size are that R&D increases monotonically—and typically proportionately—with firm size among R&D performers within industries, the number of innovations tends to increase less than proportionately than firm size, and the share of R&D effort dedicated to more incremental and process innovation tends to increase with firm size. These patterns, however, raise a number of questions. Why should there be such a close positive, monotonic—no less proportional—relationship between R&D and firm size to begin with? Also, how can this relationship be reconciled with the apparent decline in R&D productivity with firm size, no less with the apparent association between incremental and process innovation with firm size?

The apparent decline in R&D productivity with firm size has been explained in a number of ways. For example, some have argued that smaller firms, especially new ventures, are more capable of innovating than larger firms (e.g., Acs and Audretsch, 1990, 1991b; Cooper, 1964), or, similarly, are more capable of spawning more significant or distinctive innovations than larger incumbents (e.g., Baumol, 2002; Henderson, 1993). Bound et al. (1984) and Griliches (1990) suggest two other explanations. One is selection bias; only the most successful small firm innovators tend to be included in the samples that have been examined, perhaps because greater firm size increases the likelihood of survival, and thus surviving smaller firms likely manifest some compensating advantage such as greater innovative capability. Griliches (1990) also suggests the possibility that measurement error may account for the seemingly greater R&D productivity of small firms due to the systematic underestimation of formal R&D for small firms (cf. Kleinknecht, 1987; Schmookler, 1959; Sirilli, 1987). These explanations for a decline in R&D productivity with size leave open, however, the question of why we should also observe a strong positive relationship between firm size and R&D.

Cohen and Klepper (1996a,b) proposed that the idea that the returns to R&D increase with the level of output over which the fixed costs of innovation may be spread can reconcile the close positive monotonic and typically proportional relationship between R&D and firm size with both declining R&D productivity in firm size, and larger firms' greater propensity to pursue incremental and process innovation.[7] They argue that cost spreading allows the return to R&D to increase with contemporaneous firm size as a consequence of two common features of innovation within industries. First, to profit from their innovations, firms typically rely on appropriability mechanisms such as secrecy or first-mover advantages that entail embodiment of their innovations in their own output. Second, firms expect their growth due to innovation to be limited by their existing size. Together these two conditions imply that the larger is a business unit's output level at the time it conducts its R&D, the greater the expected future output over which the fixed costs of innovation (i.e., its R&D expenditures) can be spread. Consequently, larger business unit size yields a higher expected return per R&D dollar which, in turn, induces more R&D effort. Moreover, assuming diminishing productivity of R&D, if firms of different sizes within industries face similar diminishing R&D productivity schedules (i.e., their R&D efforts are comparably efficient), the larger firms will realize lower average innovative output per unit of R&D because they can profitably undertake R&D projects further down the diminishing marginal productivity of R&D schedule than smaller firms.[8] In this conception, lower R&D productivity does not reflect any relative inefficiency in large firms' ability to generate innovations, but, rather, larger firms' superior ability to profit from their R&D due to their cost-spreading advantage. Moreover, larger firm size should induce relatively greater investment in process and incremental R&D because, relative to product R&D or more breakthrough innovations, the former derive a proportionately greater cost-spreading advantage from the firm's current sales. In a theoretical paper, Rosen (1991) similarly rationalizes a link between incremental innovation and firm size with an appeal to incentives rather than capabilities by arguing that larger firms can gain relatively more from safer, more incremental R&D projects that build on existing technologies because, when successful, such projects magnify their existing competitive advantage as

[7] Such R&D cost spreading had been previously offered as a possible explanation for why R&D might rise more than proportionately than R&D (e.g., Scherer, 1980). R&D cost spreading has also often been used in other models of R&D (e.g., Lunn, 1982; Pakes and Schankerman, 1984; Rosen, 1991).

[8] Tether (1998) shows that the sales due to significant innovations tend to be higher for larger firms. He interprets this finding as suggesting that it is larger firms that tend to generate the most valuable innovations, arguing that larger firms are more capable at generating important innovations. One might entertain an alternative interpretation. Perhaps what it shows is that large firms tend to have more output in which they can embody their innovations, and thus, consistent with Cohen and Klepper (1996b), the same innovation is likely to be tied to more sales for larger firms. Thus, the finding may say little about the abilities of firms of different sizes to generate more or less technologically significant innovations.

well as the advantage that arises from spreading a fixed per unit cost savings over a larger level of output. In contrast, the returns to more revolutionary (i.e., substitute) innovations are less tied to a firm's prior market position.

Cohen and Klepper (1996a,b) empirically test and confirm the R&D cost-spreading model's key predictions. First, as implied by their R&D cost-spreading model, and as noted above, R&D is much more closely tied to business unit rather than overall firm size. More revealing, they confirm the prediction that R&D cost spreading, and, hence the link between R&D and firm size, will be weakest for just those industries, such as pharmaceuticals, where innovations are most saleable in disembodied form, or where the prospects for rapid growth due to innovation are greatest. Of comparable note, the cost-spreading model also predicts that the relationship between firm size and types of R&D will be stronger (weaker) for those *types* of R&D and innovation that lend themselves less (more) readily to rapid, discontinuous growth, such as either process or incremental innovation, and, thus, the share of R&D dedicated to either incremental or process innovation should rise with firm size. Consistent with the literature cited above (Link, 1982a; Pavitt et al., 1987; Scherer, 1991), Cohen and Klepper (1996a) confirm that the share of R&D dedicated to process innovation indeed rises with firm size. And the implication that larger firms pursue relatively more incremental innovation is consistent with previously cited findings (Henderson, 1993; Mansfield, 1981; Wilson et al., 1980).

An important implication of the argument that R&D cost spreading conditions both the rate and composition of R&D is, first, that little can be inferred about the relative R&D capabilities of large versus small firms from simple comparisons of R&D productivity across firms of different sizes. Similarly, large firms' apparent disproportionate pursuit of more incremental innovation also may not be linked to some relative disadvantage in larger firms' capabilities to pursue more significant or breakthrough innovation, contrary to Henderson (1993) or Baumol (2002). Even if disproportionate pursuit of incremental innovation by large firms does reflect investment incentives rather than capabilities, it is also plausible to argue, however, that such an investment strategy may lead over time to larger firms' becoming less capable at the conduct of R&D leading to more significant innovation.

Although Cohen and Klepper suggest that large firms have a cost-spreading advantage in realizing returns from their R&D, this advantage is not innate to size, but arises from two common conditions: (1) appropriability conditions that typically confine firms to exploiting their innovations chiefly through their own output and (2) limited firm growth due to innovation.[9] Both imperfections in the market for information highlighted by theorists such as Arrow (1962a) and Nelson (1959), as well as our empirical understanding of appropriability conditions (Cohen et al., 2000; Levin et al., 1987) rationalize the first claim, at least in the majority of industries. The second claim, that expected growth due to innovation tends to be limited by existing firm size, is empirically plausible in light of widely observed (though not understood) limits on firm growth (e.g., Hart and Prais, 1956; Mansfield, 1962), and the fact that most innovation tends to be incremental (Nelson et al., 1967; Rosenberg and Steinmueller, 1988).[10]

[9] Acs and Audretsch (1987) and Dorfman (1987) found that the relative contributions of small and large firms to innovation may depend on other industry conditions. Acs and Audretsch observed that large firms are more innovative in concentrated industries with high barriers to entry, while smaller firms are more innovative in less concentrated industries that are less mature. In a comparative study of five electronics industries, Dorfman (1987) reached a similar conclusion.

[10] Some theoretical models of R&D (e.g., Dasgupta and Stiglitz, 1980a,b) adopt the opposite assumption, that growth due to innovation is limited only by the market as a whole.

The observation that the tight, monotonic relationship between R&D and firm size and related patterns depend on other industry-level conditions poses, however, a question. If these relationships depend on industry-level conditions, why are these relationships—particularly the monotonic relationship between R&D and size—robust across industries and to the inclusion of industry fixed effects? A simple candidate explanation is that the conditions in question are so pervasive that they cut across the vast majority of industries.

Thus, there are clear, robust results indicating that R&D expenditures are closely linked to firm—or, more accurately—business unit size. Moreover, the source of the relationship appears to be R&D cost spreading, which seems to be attenuated to the extent that technology licensing or rapid growth due to innovation can be expected. An implication of this argument—and one to be qualified in a more detailed discussion below—is that, contrary to a consensus view of decades in duration, is that greater output confers an advantage in realizing returns to R&D. Thus, larger firms appear to be better positioned to profit from the innovations they have in hand, which, in turn implies that the rate of technical advance may depend not only upon the total amount of R&D conducted within an industry, but also its distribution across firms of different sizes. The literature, however, provides less guidance about whether large or small firms are more capable at generating innovations, despite studies suggesting that small firms or entrants may be.

Finally, an area where this literature on the tie between R&D and firm size is relatively mute is the endogeneity of firm size with respect to R&D and innovation. This is surprising given that the simultaneity between market structure and R&D is now widely recognized (as discussed below), and also in light of Hall's (1987) finding of a strong relationship between R&D and firm growth, with growth rising more than proportionately than R&D, especially for smaller firms. Mowery (1983b) also found that R&D contributed to firm survival over the period 1921–1946. Sutton (1998), discussed below, is an exception when he features the importance of R&D for firm growth.

2.2. Monopoly and innovation

In Schumpeter's discussion of the effects of market power on innovation, there are two distinct themes. First, Schumpeter recognized that firms require the expectation of some form of transient market power to have the incentive to invest in R&D. This is, of course, the principle underlying patent law; it associates the incentive to invent with the expectation of *ex post* (i.e., post-innovation) market power tied to the innovations originating from R&D. Second, Schumpeter argued that the possession of *ex ante* market power, linked to an *ex ante* oligopolistic (or monopolistic) market structure, also favored innovation. An oligopolistic market structure, for example, made rival behavior more stable and predictable, he claimed, and thereby reduced the uncertainty associated with excessive rivalry that tended to undermine the incentive to invent. Implicitly assuming that capital markets are imperfect, he also suggested that the profits derived from the possession of *ex ante* market power provided firms with the internal financial resources necessary to invest in innovative activity. Finally, he also appeared to argue that *ex ante* market power would tend to confer *ex post* market power.

The empirical literature has focused principally on the effects of market concentration on innovative behavior. The literature has thus directly tested Schumpeter's conjectures about the effects of *ex ante* market structure. Over the past two decades scholars have also begun to test Schumpeter's claims about *ex ante* market power. In the empirical work exploring the effects of expected *ex post* market power on innovation, traditional measures of market structure have not been employed. Rather, the potential for achieving *ex post* market power through innovation has been characterized under the general heading of appropriability conditions and measured by survey-based indicators of appropriability, as discussed below.

Economists have offered an array of theoretical arguments yielding different and conflicting predictions about the effects of market structure on innovation. Some have supported Schumpeter's position that firms in concentrated markets have a stronger incentive to invest in innovation. Others have demonstrated that a firm's gains from innovation at the margin are larger in an industry that is more competitive *ex ante*. The latter argument has been made, for example, by Arrow (1962a) who highlighted what has come to be known as the "replacement effect"—that, assuming perfect *ex post* appropriability of profits due to innovation, an incumbent monopolist's introduction of, say, a new manufacturing process would partially displace the monopoly rents it was earning beforehand. Therefore, its returns to the innovation are only the increment beyond the monopoly profits that it had previously earned. In contrast, a firm operating in a competitive market would not be displacing any monopoly profit, and could therefore realize the full return to its process innovation. Thus, in Arrow's model, the firm in the competitive industry has the greater incentive to invest in R&D.[11] Also assuming perfect appropriability, but focusing on an incumbent monopolist facing a prospective entrant, Gilbert and Newbery (1982) argue, in contrast, that the monopolist should have the greater incentive to invest in innovation. Assuming the incumbent monopolist considers the avoided cost of losing its monopoly to an innovating entrant, it has an incentive to pre-empt entry by investing more aggressively in innovation than the entrant. On the other hand, Reinganum (1983) shows this result can flip once the uncertainty of the R&D process is considered, assuming discovery conforms to an exponential process. Assuming symmetric Cournot competitors, and dropping the unrealistic assumption of perfect appropriability, Dasgupta and Stiglitz's (1980b) analysis of cost-reducing R&D also suggests that, as an industry becomes more competitive, firms' R&D intensities will decline.

More institutionally grounded arguments have also been offered. Contrary to Schumpeter's original argument, Scherer (1980), for example, argued that insulation from competitive pressures may breed bureaucratic inertia and discourage innovation. Also arguing from a behavioral perspective, in his study of national competitive advantage, Porter (1990, p. 118), states that "active pressure from rivals stimulates innovation as much from fear of falling behind as the inducement of getting ahead," and, in his view, more intense rivalry in national markets contributes to the emergence of more capable, innovative firms.

In all this to-and-fro, theorists have highlighted a number of important points neglected by applied economists. As articulated, for example, in Aghion and Griffith (2005), the way to think about firms' decisions to invest in R&D is to consider their expectations regarding the *difference* between profits earned prior to introducing an innovation and the future profits earned once the innovation is commercialized. Though, as suggested above, the *ex post* market power tied to the innovation is one determinant of this difference, the difference is also partly a function of the position that the firm is starting from, reflecting the firm's *ex ante* market power, which was Arrow's point. Thus, in this logic, the firm's

[11] Gilbert (2006) points out that Arrow's argument does not, however, generalize to product innovation, assuming some degree of horizontal differentiation in the product market. On the other hand, Arrow's replacement effect could dominate even in this instance if the new product renders the old one obsolete.

ex ante market power should matter for R&D investment—though not in the fashion proposed by Schumpeter—and to the degree that market structure is tied to market power, *ex ante* market structure should matter as well. Gilbert and Newbery (1982) also provide an important perspective when they extend Arrow's logic by suggesting that firms consider the difference in *ex ante* and *ex post* profits in anticipation of the possibility that a rival may introduce the innovation instead, suggesting that, under some circumstances, current monopoly profit may represent the opportunity cost of not innovating.

Aside from these general points, the particular theoretical models—those briefly reviewed here and a large number of others—offer conflicting conclusions and depend heavily on a range of assumptions regarding appropriability conditions, the type of innovation (e.g., product vs. process), the importance of the innovations in question ("drastic" vs. not), and the change in the intensity of rivalry associated with innovation. A question for empirical scholars is whether and which of these theories offer testable insights. Another question—one of relevance to both the modern theoretical treatments and the more casually motivated empirical inquiries in this area—is just how important is the nature and intensity of competition relative to other industry- and firm-level determinants of R&D and innovation. We will return to both questions below, but will first review the key empirical findings of the Schumpeterian literature on market structure.

The majority of studies that examine the relationship between market concentration and R&D have found a positive relationship. First among many were Horowitz (1962), Hamberg (1964), Scherer (1967a), and Mansfield (1968). A few have found evidence that concentration has a negative effect on R&D (e.g., Bozeman and Link, 1983; Mukhopadhyay, 1985; Williamson, 1965). Rather than examine the relationship between market structure and R&D—an input, Geroski and Pomroy (1990) and Geroski (1990) consider the relationship between market structure and innovation, the output of innovative activity, which they measure with counts of commercially significant innovations drawn from the SPRU database (cf. Geroski, 1994, Chapter 2; Robson and Townsend, 1984). Geroski (1990) also departs from the prior literature by employing a number of measures of market structure, including market concentration, but also measures of entry, exit, import penetration, and the number of small firms. Geroski (1990) finds a positive relationship between competition and innovation, a qualitative reversal of the majority of prior findings that he attributes to his inclusion of a control for technological opportunity.

In contrast to the studies cited above that focus on *ex ante* market structure, Blundell et al. (1999) probe the relationship between a firm's *ex ante* market power and innovation. Measuring the former with market share and the latter with the number of commercially significant innovations (again drawn from the SPRU data set) in a panel of 340 firms spanning 1972–1982, they find market share to have a positive effect on innovation, while that of overall market concentration is negative, suggesting that, although market share—and perhaps market power—stimulates innovation, concentrated industries may innovate less. They also exploit their time series on the innovation count variable by including the prior stock of innovations on the RHS to control for unobserved, permanent firm effects on the propensity to innovate.

A finding that long ago captured the imagination of numerous theorists was that of Scherer (1967a), who found evidence of a nonlinear, "inverted-U" relationship between R&D intensity and concentration. Using data from the Census of Population, Scherer found that R&D employment as a share of total employment increased with industry concentration up to a four-firm concentration ratio between 50% and 55%, and declined with concentration thereafter. This inverted-U result, in the context of a simple regression of R&D intensity against market concentration and a quadratic term, has been replicated by Scott (1984) and Levin et al. (1985) using the FTC Line of Business data. Using a 21-year panel, spanning 1973–1994, for 17 two-digit industries, Aghion et al. (2005) observe a similar "inverted-U" between industry-level market power, measured with an averaged Lerner index—an arguably better measure of the intensity of competition than a concentration ratio—and industry innovation, measured with the average number of citation-weighted patents.[12]

Few of the studies examining the Schumpeterian hypothesis of a tie between innovation and either *ex ante* market structure or market power have been designed to test any of the several game-theoretic models that might account for such a link. Exceptions include Blundell et al. (1999) and Aghion et al. (2005). Blundell et al. (1999) claim to test the Gilbert and Newbery (1982) argument that firms with more market power have a greater incentive to innovate in order to pre-empt entry that would otherwise dilute their above-normal profits. Their test is, however, indirect, relying on the elimination of alternative explanations for the positive relationship they observe between market share and innovation. One is that cash flow might account for their finding, and another is that more permanent firm or industry effects on innovative performance might account for their finding. They successfully eliminate the considered alternatives. There remain, however, unexamined competing explanations. For example, their analysis does not control for the cost-spreading incentive effects on R&D of business unit (as opposed to firm) size—a potentially important omission in light of a likely correlation between market share and business unit size, and the very tight relationship noted above between business unit size and R&D. Gilbert (2006) also suggests that the authors' use of lagged variables to control for endogeneity is only valid to the extent that omitted firm- and industry-level variables that may drive market share, innovation, and financial returns, such as technological opportunity, appropriability, or even exogenous firm capabilities, are stable over time, which he questions.

To probe Aghion et al.'s (1997) theoretical model, Aghion et al. (2005) offer an explanation for the inverted-U relationship between the intensity of competition and innovation. Assuming innovation occurs in a step-by-step fashion (precluding large advances leading, e.g., to leapfrogging), Aghion et al. (2005) formulate a model in which the qualitative character of rivalry can take one of two forms—either rivals are "neck-and-neck" or there is a leader and a laggard. In the former case, Aghion et al. (2005) show that increasing the intensity of competition (represented as the extent of collusion between the firms) strengthens the incentive to innovate because it leads to a greater difference between the pre- and post-innovation profits that comes from an innovating firm's "escape from competition." In contrast, greater intensity of competition dampens a laggard's incentive to innovate because there is little to gain since, with only a one-step gain from innovation, post-innovation rents remain low due to competition with the leader, and the laggard therefore has little prospect of recovering—no less profiting from—its investment in innovation.[13] Thus, at the level of the economy, competition may either increase or diminish innovation, depending on the mix in the economy of these two forms of rivalry. The authors argue that due to the effect of competition on the steady-state distribution of technology gaps across rivals, the composition of industries is such that both types of rivalry exist in the economy and one

[12] In a study examining the relationship between market concentration and innovation counts, Acs and Audretsch (1990) found that the average number of innovations per unit of sales declines with concentration. This finding is consistent with the finding, discussed in the prior section, that the number of innovations tends to increase less than proportionately than firm size.

[13] This argument seems to assume, however, that the leader in this setting has little or no incentive to innovate.

should consequently observe the inverted-U. Other theoretical models and behavioral explanations have also been offered for the inverted-U.[14] The theoretical model of Scott (2009), for example, suggests that the pattern can be explained if we assume that more concentrated industries display the kind of aggressive, noncooperative interaction where firms pay attention to one another's behavior, while, in more competitive industries, there is no interdependent decision making; firms only see the negative appropriability incentives that might accompany a more perfectly competitive market structure.

Although the Aghion et al. (2005) model successfully predicts a number of other observed empirical patterns, the model is highly stylized, assuming, for example, that rivalry is of one of the two types and that innovation is step by step. As Gilbert (2006) notes, given the disincentive for a laggard to advance just one step, why not allow it to leapfrog the leader? Moreover, the empirical predictions also assume that, where a laggard is facing a leading firm, the leading firm does not invest in R&D. Aghion et al. (2005), however, does usefully suggest that, in thinking about the relationship between market structure and innovation, one might usefully distinguish different forms of rivalry as well as how rivalry may differentially impact firms depending on their competitive position and their capabilities. Consistent with Aghion et al.'s (2005) argument, Lee (2009), using World Bank survey data for nine industries across seven countries, observes that intensity of competition may stimulate more capable firms to invest more heavily in R&D, while less capable firms may invest less.[15] Aghion et al.'s analysis of laggards' incentives is also consistent with Koeller's (1995) finding of a negative effect of concentration on small firms' innovative output, which also raises the question of differential effects of market structure on small versus large firms' innovative activity (cf. Van Dijk et al., 1997).

To understand the link between market competition and innovation, empirical analysis must go beyond a consideration of the behavior of incumbents alone to consider entry. Motivated partly by Blair's (1948, 1974) claim that innovation in the twentieth century has been principally a deconcentrating force, Geroski (1989, 1990, 1991b,c, 1994), Acs and Audretsch (1991a), and others have considered the relationship among innovation, entry, and market structure. Featuring only observable endogenous variables reflecting entry and innovation rates, Geroski (1991b) finds that innovation and entry are positively related, and, in the short run, entry appears to Granger-cause innovation rather than *vice versa*. He interprets the overriding role of industry fixed effects—separately estimated for innovation and entry—as suggesting that more permanent features of markets drive both innovation and entry and thus induce the positive relationship between these two factors in any cross section. Interpreting these fixed effects as reflecting the influence of technological opportunity and entry barriers, he attributes the relationship between entry and innovation to a close positive relationship between technological opportunity and low entry barriers.[16] Geroski (1994) speculates (per a suggestion attributed to Klepper) that high technological opportunity presents opportunities for innovation to both incumbents

[14] Scherer (1967b) developed one of the first detailed theoretical models of R&D rivalry; its implications, like those deduced by Kamien and Schwartz (1976), were consistent with the empirical finding that an "inverted-U" characterized the relationship between R&D investment and market concentration.

[15] However intriguing, Lee's results should be subjected to more scrutiny given the self-reported character of much of the data, and the sharp differences in the institutions and market environments across the sample countries that included, among others, China, India, Canada, and Japan.

[16] Breschi et al.'s (2000) finding of a strong link between entry and technological opportunity partially confirms Geroski's interpretation of his findings.

and potential entrants alike, but given the difficulty of selling innovations in disembodied form (e.g., licensing) in most industries, innovating firms must enter most markets themselves to capitalize on their inventions. Gans et al. (2002) confirm a portion of this intuition when they find that in industries where technology markets function better due to effective patents, startups are more likely to monetize their discoveries through licensing and alliances rather than through entry. Thus, technological opportunity may have little effect on—or possibly dampen—competition where patents are sufficiently strong that outsiders can monetize their innovations by licensing them to industry incumbents. Alternatively, where patents are weaker, technological opportunity will intensify competition because outsiders need to enter a market themselves to profit from their innovations. More generally, Geroski's studies of entry do not suggest that market concentration spawns innovation, but, rather, that entry, innovation, and the intensity of competition are codetermined by technological opportunity and appropriability, where the former creates the potential for innovation and the latter conditions the effect of that innovation on the intensity of competition by affecting whether a potential entrant enters the industry or licenses to an incumbent.

Although not focused on the relationship between competition and R&D or innovation, Nickell (1996) falls within the purview of the Schumpeterian empirical tradition by examining the relationship between competition and total factor productivity growth. Employing a panel of more than 600 publicly listed firms spanning 17 two-digit SIC industries, Nickell finds a significant, positive and economically important relationship between competition and productivity growth, which departs from the spirit of the Schumpeterian hypothesis, but is consistent with Porter's (1990) qualitative analysis and Geroski's (1994) observation of a positive link between industry entry and productivity growth. Nickell employs two measures of the intensity of competition—one is a survey-based measure of the number of competitors faced by each responding firm (available for 147 firms), and the second is a ratio, resembling the Lerner index, of earnings to value added for each firm (available for over 600 firms). He also includes on the RHS measures of concentration and market share, and finds negative, significant effects of each, reinforcing the basic finding. His analysis controls for the endogeneity of both capital and labor, and includes fixed effects for SIC two-digit industries. Nickell acknowledges that he does not know why greater competition is associated with total factor productivity growth. Also, although we know that R&D contributes importantly to productivity growth, it is not clear that this particular result reflects a mediating effect of competition on R&D or innovation, especially in light of the inconsistency between his finding and Blundell et al.'s (1999) finding of a positive relationship between market share and innovation. For example, competition may mainly push firms to operate closer to their production frontiers, consistent with Leibenstein's (1966) notion of "X-efficiency," or may similarly stimulate the adoption of innovations. In any event, Nickell's intriguing result leaves us with the tasks of probing its robustness, and identifying the mechanism(s) behind it—and hopefully reconciling it with what we know—however mixed—about the relationship between competition, R&D, and innovation.

There have been two central challenges facing empirical studies on the relationship between competition and innovation, as suggested by our discussion of entry and innovation. First, it is likely that competition and innovation are simultaneously determined, either with causality running in both directions, or with both innovation and competition codetermined by other exogenous factors. Second, there is a question of the sensitivity of the relationship to industry-level factors, and what that sensitivity might imply about the nature and importance of the influence of competition on innovation.

Phillips (1966) was among the first to propose that causality might run from innovation to market structure, rather than the reverse. Although Schumpeter envisioned that the market power accruing from successful innovation would be transitory, eroding as competitors entered the field, Phillips argued that, to the extent that "success breeds success," concentrated industrial structure would tend to emerge as a consequence of past innovation. Phillips' (1971) monograph on the manufacture of civilian aircraft illustrates how market structure can evolve as a consequence of innovation, as well as how it can affect the conditions for subsequent innovation.

Theoretical support for the proposition that a rapid rate of innovation leads to concentration can be found in the literature on stochastic models of firm growth, notably in the simulation models of Nelson and Winter (1978, 1982b). Klepper's (1996) analytic model also highlights the contribution of innovation to market concentration over time for industries and types of innovation where R&D fixed cost spreading applies, as does Sutton's (1998) model that highlights the importance of endogenous sunk costs. Most analytic results concerning this and related propositions, however, are asymptotic (see Rothblum and Winter, 1985). By contrast, in the short run, the presence of long-lived capital and costly adjustment by firms and consumers implies that innovation, even dramatic innovation, can make a market more or less concentrated, a proposition for which Mansfield (1983) finds empirical support. The short-run effect of innovation on market structure depends, in part, on whether established leaders or new entrants commercialize the innovation.[17]

Recognizing the potential simultaneity between innovation and concentration, some investigators (Howe and McFetridge, 1976; Levin et al., 1985) have instrumented for concentration in regression studies of the effects of market structure on innovative activity. Similarly, Blundell et al. (1999) also instrument for market share, and Aghion et al. (2005) instrument for the Lerner index that is intended to reflect *ex ante* market power. Others (Connolly and Hirschey, 1984; Farber, 1981; Levin, 1981; Levin and Reiss, 1984, 1988; Wahlroos and Backstrom, 1982) have used industry-level data to estimate multiequation models in which concentration and R&D are both treated as endogenous.[18] There is a suggestion that such techniques are appropriate. Levin (1981), Connolly and Hirschey (1984), Levin and Reiss (1984), and Levin et al. (1985) all find that Wu-Hausman tests reject the hypothesis (maintained in the O–L–S specification) that the concentration variables are orthogonal to the error term. This result, however, may well arise from misspecification or omitted variables. In any event, Howe and McFetridge (1976) found that, relative to ordinary least squares, two-stage least squares produced little change in the coefficient on the concentration term in the R&D equation.

Perhaps the most persistent finding concerning the effect of concentration on R&D intensity is that it depends upon other industry-level variables. Scherer (1967a) found that the statistical significance of

[17] Innovation can also affect market structure by increasing or decreasing the efficient scale of production. If technological change causes the efficient scale of a firm to grow more rapidly than demand, concentration tends to increase over time. For a theoretical treatment in which such changes in scale and concentration are both endogenous, see Levin (1978). For evidence that technical change has increased efficient scale in various industries, see Hughes (1971) on electric power generation, Levin (1977) on several chemical industries, and Scherer et al. (1975) on steel, cement, brewing, refrigerators, paints, and batteries.

[18] Data limitations have made it convenient to treat concentration and R&D intensity as simultaneously determined variables, but this is inconsistent with the underlying Schumpeterian theory, as interpreted by Phillips (1966, 1971). Contemporaneous concentration, in this view, should influence R&D spending, but current concentration is the consequence of past innovative activity. Levin (1981) estimates a model in this form, where a distributed lag of past R&D investment, not the current R&D intensity, appears on the right-hand side of the concentration equation.

concentration was attenuated with the addition of dummy variables classifying the industry's technology (chemical, electrical, mechanical, and traditional) and its products (durable/nondurable, consumer/producer goods). The dummy variables, especially those representing technology classes, were highly significant, and explained considerably more variance in the dependent variables than did concentration. Wilson (1977) attained similar results, and Lunn and Martin (1986), splitting their sample into two technology classes, found that concentration had a significant effect on R&D intensity only in "low opportunity" industries. Geroski (1990) observed that dropping the industry fixed effects—interpreted as reflecting technological opportunity—largely reversed his basic finding of a positive relationship between competition and innovation.

Scott (1984) and Levin et al. (1985) provide strong evidence that results concerning the effect of concentration on R&D intensity are sensitive to industry conditions. Using the FTC data on R&D intensity at the business unit level, Scott found that the addition of fixed company and two-digit industry effects rendered statistically insignificant the coefficients on concentration and its square. Using the FTC data at the line of business level (a level of aggregation between the three- and four-digit SIC level), Levin et al. found that the addition of a set of measures representing technological opportunity and appropriability conditions replicated Scott's result in equations for both R&D intensity and innovative performance. With the new variables added, the coefficient and the t-statistic on concentration dropped by an order of magnitude in the R&D intensity equation.

Among others who have found the validity of the Schumpeterian hypothesis to depend on industry characteristics, Comanor (1967) found that the degree of product differentiation conditioned the relationship between concentration and R&D intensity, but he used advertising intensity, presumably a codetermined decision variable, to represent what should more properly have been represented by a set of predetermined product characteristics. Somewhat more defensibly, Shrieves (1978) obtained a similar result by classifying industries according to the nature of the final product market.[19] Angelmar (1985) suggested that the effect of concentration on innovation might depend on the degree of technological uncertainty, but the appropriateness of his measure of uncertainty—the average lag between initiating the development of a new product and its market introduction—is subject to serious doubt. Wedig (1990) probed this hypothesis more directly by examining the financial uncertainty associated with R&D. Employing adjusted β values for a sample of 214 manufacturing firms as the dependent variable, Wedig first confirms Schumpeter's (and others') assumption that R&D is especially risky by finding that the systematic risk associated with R&D is substantially higher than that associated with non-R&D assets. He then provides only modest evidence, however, that this risk is partially offset by market concentration as well as firm size, as Schumpeter had claimed.

Thus, numerous, typically cross-sectional, studies underscore the dependence of the relationship between market structure and R&D or innovation on other industry-level factors. Although several theoretical explanations have been offered for why appropriability or technological opportunity may either condition or account for the relationship (e.g., Nelson and Winter, 1982a; Scherer and Ross, 1990), cross-sectional analyses, even those employing long panels, have offered little insight into the actual role of these industry-level factors. Nonetheless, the key role of these other industry-level factors suggests that market structure does not play—at least in any straightforward way—an important,

[19] Shrieves classified industries on the basis of a factor analysis that took account of the composition of industry demand and the durability of the product.

independent role in affecting innovation. Skepticism about the importance of market structure for innovation is further invited by assessments of its empirical power. Simple tests of the explanatory power of market concentration, for example, find that it contributes little to an explanation of the variance in R&D intensity. Scott found that line of business concentration and its square explained only 1.5% of the variance in R&D intensity across 3388 business units, whereas fixed two-digit industry effects explained 32% of this variance. Similarly, a re-examination of the data used in Levin et al. (1985) revealed that concentration and its square explained only 4% of the variance in R&D intensity across 127 lines of business (Cohen and Levin, 1989). In contrast, Cohen et al. (1987) reported that demand, opportunity, and appropriability measures explained roughly half of the between-industry variance, where industries were defined at roughly the three- to four-digit SIC level.

In his landmark work, Sutton (1998) proposes a different approach to thinking about the mixed theoretical and empirical results on the link between market structure and innovation.[20] Making only two core assumptions about firm behavior, namely "that firms avoid loss-making strategies, and that if a gap appears that can be profitably filled by an entrant then it will be filled," (1998, p. 9) Sutton proposes a class of "reasonably" specified game-theoretic models that can be validated with the use of observables both across and—in a more detailed fashion—within industries. He allows the assumptions of his models to vary, recognizing that there may be little a priori basis to discriminate across them, to bound a set of outcomes describable by the observables of R&D intensity, market concentration, and submarket homogeneity. This "bounds approach" implies that the standard regression analyses commonly employed to estimate the relationship between market structure and R&D intensity are inappropriate if the best you can do is to broadly characterize an admissible space.

In his theory, two key market-level factors affect R&D investment: (1) the degree to which R&D investment can increase a firm's unit price–cost margin by increasing consumers' willingness to pay for a product (or by reducing its unit variable cost of production) and (2) the market reach of a new or improved product, where market reach is not just a function of the overall demand schedule for a market, but also the degree to which that market may be disaggregated into submarkets exogenously distinguished by the degree to which products are substitutes. The first of these two factors conforms to what the literature now refers to as technological opportunity, characterized by Spence (1984), for example, as the elasticity of quality unit cost with respect to R&D, or by Nelson and Winter (1982a) as the magnitude of the enhancement to productivity growth that may be expected from an R&D success (p. 311). Sutton is the first, however, to apply the second factor—the degree to which an industry can be divided into submarkets characterized by imperfect substitutability of products between them—to the analysis of R&D and innovation. Thus, both of these factors—what we will term technological opportunity and submarket homogeneity—increase the returns to R&D spending. Technological opportunity allows the firm to increase its margin per unit of output, and greater submarket homogeneity further increases the returns to R&D by allowing the firm to realize greater R&D cost spreading. And, in his model, with endogenously determined R&D increasing, a firm increases its share of the overall market by increasing the desirability and reach of its offerings. But, this still leaves open the question of how R&D intensity and market concentration might be linked.

Sutton's characterization of R&D as an endogenous sunk cost suggests that, as R&D rises in response to both technological opportunity and market reach, the number of high R&D spending firms that a

[20] See Sutton (2007) for a synthesis and review of his own work and related contributions.

given market can support becomes more limited, *ceteris paribus*, given some overall demand, assuming that firms are to remain profitable (and, accordingly, realize a return on their R&D). Thus, where technological opportunity is high and submarkets are more homogeneous, markets will be more concentrated and R&D intensity will be high. But what if technological opportunity is high, but submarket homogeneity is low, constraining the firm's ability to benefit from R&D cost spreading? In this case, R&D spending may rise relative to sales, but the sales that can be captured through an innovation will be limited due to submarket heterogeneity. R&D intensity (i.e., R&D expenditures divided by the firm's sales) may be high, but concentration, defined in terms of the overall market, will be low. Thus, the overall market will be able to support a larger number of R&D-intensive firms, and market concentration can be quite low to the extent that there are distinct, separate submarkets that are relatively small. In sum, high R&D intensity can occur in industries which with low or high concentration, depending on submarket homogeneity. For industries with high technological opportunity and high R&D intensity, this implies a bounded region of outcomes relating submarket homogeneity and market concentration such that market concentration will exceed some threshold that increases with the homogeneity of the submarkets. In contrast, in industries where technological opportunity is low, there is little incentive to invest in R&D, and thus little sunk cost that the market needs to support, implying a diffuse relationship between market concentration and submarket heterogeneity for more R&D-intensive industries.

Sutton tests—and finds support—for his theory employing cross-sectional data on R&D intensity, market concentration, and constructed measures on submarket homogeneity drawn from the US FTC's Line of Business Program and the US Census of Manufactures for the mid-1970s. He also finds confirmatory patterns by considering what he characterizes as the natural experiments offered by the experience of selected industries (e.g., color film, digital switches, flowmeters). Matraves (1999) tested and largely confirmed Sutton's model in the global pharmaceutical industry, and Marin and Siotis (2007) did the same with plant-level data for the US and European chemical industries.

Sutton's analysis deepens our understanding of the fundamental role of technological opportunity, and introduces us to the importance of considering submarket homogeneity and, in turn, horizontal product differentiation and associated demand conditions, as fundamental drivers of R&D investment. But what are the implications of Sutton's analysis for the analysis of the effect of *ex ante* market structure or market power, on R&D and innovation? His theory and findings again offer little support for the view that market concentration is an independent, significant, and important determinant of innovative behavior and performance. In essence, it suggests that overall market demand, the homogeneity of submarkets, and technological opportunity drive R&D spending and, in R&D-intensive industries, market structure as well, but that neither *ex ante* market structure nor power exercise any influence—even indirect—on R&D.

2.2.1. *Market structure, innovation, and industry dynamics*

Analyses of the long-run evolution of firms, competition, and technology within industries put into play the question of the degree to which technological opportunity or even submarket homogeneity should be assumed exogenous with respect to R&D—in Sutton's analysis, or in the cross-sectional empirical analyses reviewed above. More generally, applied economists, who have tended to focus on cross-sectional empirical patterns, can learn a great deal from analyses of the dynamics of industries and of the patterns linking the evolution of innovative activity and technology to features of the evolution of markets, and to regularities characterizing entry, exit, prices and market structure observed by Abernathy (1978), Abernathy and Utterback (1978), Utterback (1979), Gort and Klepper (1982), Klepper and Graddy (1990), and Klepper and Simons (2005). To the extent that empirical scholars acknowledge underlying dynamics, it has typically been in the form of collapsing a simultaneous relationship between innovation and market structure into their static models, or simply employing instruments to control for possible endogeneity. Such simultaneous structures obviously do not do justice to the richer dynamic relationships that they are trying to capture. In an important exception to the approach, Sutton (2007) employs cross-sectional analysis in tandem with case histories of selected industries to test his models. An understanding of the ways in which industry-level factors, such as appropriability, technological opportunity and demand, as well as firm-specific capabilities might affect both R&D and market structure will, however, be best illuminated by the study of the interaction of firm growth, competition, and technological change that unfolds over time.[21]

Mueller and Tilton (1969) long ago offered evidence that the role of market structure is related to an industry's stage in the product life cycle, which reflects the idea that product markets experience a life cycle over which the nature of innovation changes in a predictable manner (Abernathy and Utterback, 1978; Utterback, 1979). In the early years of an industry's evolution, the emphasis is on product innovation, as numerous small firms compete to establish a market position. New product ideas are tested, and eventually a "dominant design" emerges. With the dominant design comes product standardization and a new emphasis on process innovation. In this phase, process innovation is pursued; effort is concentrated on realizing the benefits of large-scale production, mechanization, improving production yields, etc. The industry becomes more concentrated, the potential for further process innovation is eventually exhausted, and the industry becomes subject to external threats from competing products that eschew the dominant design. Although this life-cycle model may provide a coherent interpretation of the history of the US automobile industry (Abernathy and Utterback, 1978), its generality is limited. For example, the model fits the experience of some segments of the worldwide semiconductor industry (memory, devices) but not the experience of others (logic devices and microprocessors).[22] Moreover, in a detailed study of the history of the automobile, tire, penicillin, and television industries, Klepper and Simons (2005) suggest that it is difficult to identify the emergence of a dominant design in these industries as a watershed event precipitating an industry shakeout of the quality suggested by Abernathy and Utterback.

Geroski (1991b) suggests that the product life cycle is related to patterns in entry, exit, market structure, and innovation, arguing that early in the product life cycle, high technological opportunity stimulates entry, but as the product matures, entry barriers rise, entry falls off, concentration increases, and innovation shifts to more incremental and process innovation. Klepper's (1996) model suggests that the shift in the nature of innovation over the life cycle to more process and incremental innovation is endogenously determined by the growth of dominant firms over time, which then not only invest more in R&D due to their larger size, but invest more in those types of R&D that disproportionately capitalize

[21] See Chapter 3 for a review of the literature on technological change and industrial dynamics from an evolutionary perspective.

[22] While particular features of this life-cycle model do not fit many chemical industries, the evolutionary patterns in firms' innovative activity from 1930 to 1982 in the chemical process industry identified by Achilladelis et al. (1990) indeed suggest a movement away from radical innovation to more incremental innovation over time.

on existing output, namely process and incremental R&D. Thus, the features of the evolution of technological change that are implicitly characterized as exogenous either in the product life-cycle model (e.g., Abernathy and Utterback, 1978) or in Nelson and Winter's (1977) or Dosi's (1982) characterizations of "natural" or technological trajectories may have an important endogenous component. This raises the broader, fundamental question of the sources of technological opportunity. Specifically, does technological opportunity originate largely from public research and other extra-industry sources that may be reasonably characterized as exogenous? Or from incumbents, in which case the assumption of exogeneity becomes suspect? In most instances, the sources will be mixed, and one would want to consider how that mix varies both across industries and over time.

In their seminal treatment of industry evolution, Nelson and Winter's (1978, 1982a) simulation model of the evolution of technology and market structure provides a clear, yet nuanced sense of how technological opportunity and appropriability conditions can account for a cross-sectional link between a concentrated market structure and R&D intensity. In their stochastic model, starting with a set of similarly sized firms, the greater is technological opportunity (reflected in the set of latent opportunities for realizing productivity growth), the greater the technical advance and associated sales growth that some firm will achieve. With that growth, the firm will conduct more R&D (assuming, in their model, that R&D is some fixed percentage of sales or is funded out of cash flow), which advantages that firm relative to the competition for the next round, and so on. In their model, weaker (stronger) appropriability, reflected in rivals' ability to imitate one another's advances, can attenuate (increase) market concentration. Also, to the degree that the technical advances themselves tend to be both indivisible and larger, concentration will tend to be greater.

Those analyzing cross-sectional or panel data would be well advised to pay careful attention to the models and empirical analyses of industry dynamics. First, these analyses should inform the development of any theories speaking to cross-sectional patterns, as well as the empirical methods employed. Second, the outcome of dynamics should be considered when trying to interpret cross-sectional empirical relationships. For example, while cross-sectional studies might tell us that industry-level conditions, such as technological opportunity, matter for the link between market structure and innovation, the dynamic studies are much better at providing insight into how and why they matter.

A moment when the evolution of technology can be dramatically tied to that of market structure is when incumbent firms exit *en masse* in the face of a radical change in the technology underlying an industry's products or processes. As highlighted long ago by Schumpeter (1942), and reflected succinctly in his notion of "creative destruction," such moments occur with some regularity and they can have important consequences. In the transitions from steam to diesel locomotives, from propeller to jet aircraft engines, and from vacuum tubes to transistors, leading firms changed technological regimes too late or with too little commitment. Also, established market structures were entirely overturned in each of these cases. Scherer and Ross (1990) suggest that the number of instances in which industry entrants introduce revolutionary product or process innovations, often with grave consequences for industry incumbents, are legion.

Although reviewing the full breadth of this literature is beyond the scope of the current chapter, this literature highlights several important issues bearing on the relationship between innovation and market structure. There is first, however, the question of what exactly constitutes "radical" or "drastic" innovation, and few studies are clear about that, with the notable exceptions of Arrow (1962a), Henderson and Clark (1990), Henderson (1993), Ehrnberg and Sjöberg (1995), Christensen (1997),

Rosenbloom and Christensen (1994), and Tripsas (1997b). In studies of what may drive revisions in market structure or leadership, scholars also need to be attentive to not defining radical innovation in such a way to presuppose the dependent variable (e.g., an innovation associated with the failure of leading industry incumbents). In turn, this raises the issue of just how often "radical" innovation—however defined—does lead to such a change in the market. For example, one might argue that, despite a dramatic change in the science underlying drug discovery over the past three decades, most of the major incumbents in pharmaceuticals retain their dominance (albeit perhaps now as merged entities). This observation calls for consideration of the circumstances under which dominant firms fail and why in the face of significant innovation. The literature has offered a number of suggestions.

In a series of case studies on the shifts from manual to numerically controlled metal cutting machine tools, from stand-alone to flexible manufacturing systems, and from noncellular to cellular mobile telephony, Ehrnberg and Sjöberg (1995) suggest that more radical innovations may overturn existing market structures when there is a technological transition involving a new generic technology which both substitutes for current technology and diffuses rapidly. Tushman and Anderson (1986) focus on the changes in the technical expertise required. In her analysis of the photolithographic equipment industry, Henderson (1993) also considers the sort of organizational capabilities that a new technology may warrant, highlighting their role in explaining why industry incumbents had considerable difficulty even in exploiting significant but not revolutionary technological changes (see above). Examining the disk drive industry, Christensen (1997) argues that such transitions have little to do with a firm's capabilities—that firms, if motivated, would acquire them or invest in their development in a timely fashion. Rather, he focuses attention on incumbents' investment incentives, suggesting that incumbents will invest in new technologies—even when "radically" new—as long as they appeal to current customers, but will not invest in new technologies if they appeal only to new or marginal buyers. Moreover, it is these technologies, and the initially smaller firms that pursue them, that end up displacing the dominant players. In her study of the typesetting industry, Tripsas (1997a) notes, however, that many incumbents did invest in the new technology in a timely fashion, but, due to a failure in execution, were not able to offer competitive products. Building on Mitchell's (1989) early analysis of the medical imaging industry, Tripsas (1997a) focuses on whether the innovation in question undermines the value or utility of the extant specialized complementary capabilities of the incumbent.[23] Finally, Tripsas and Gavetti (2000), on the basis of a detailed examination of Polaroid's experience in digital imaging, argue that the culprit is managerial cognition; that even where top management may be able to acquire or develop a new technology, they may not perceive the need to adopt a different commercialization strategy to support it. They are cognitively stuck in their existing "schema" linked to their existing business model.

For our purpose, a key insight from this literature on industry dynamics is that the links between innovation and market dominance are much more complex and multifaceted than what cross-sectional studies typically convey. Moreover, there are likely a range of reasons for significant turnover among dominant incumbents in the face of the introduction of significantly new technologies, and that some of these reasons are best illuminated by disciplinary perspectives outside of economics, including

[23] Also taking exception to the generality of Christensen's argument, Chesbrough (1999) shows that leading incumbents in the disk drive industry in Japan retained their market dominance, largely due to their ability, through their extended keiretsu structure, to entertain and develop new technologies, and due to less threat from entrants. The latter reflects less developed venture finance in Japan as well as norms and employment practices that discourage the labor mobility that often lies behind new venture creation.

organizational science and social psychology. Consider, for example, Henderson's (1993) provocative finding that Gilbert and Newbery's (1982) attention to incumbents' desire to avoid losses due to entry was only confirmed once controls were included for the new technologies' organizational demands. Finally, one important question that this varied literature does not address in any systematic fashion is whether the number and size distribution of firms within the industry (i.e., market structure)—as opposed to the identity of the leading firms—changes with radical innovation.[24]

2.3. Evaluation of empirical research in the Schumpeterian tradition

In this section, after briefly synopsizing the key findings reviewed in Sections 2.1 and 2.2, we focus on questions of interpretation, as well as identify gaps in the empirical literature to highlight opportunities for future research.

2.3.1. Firm size and innovation

The most robust finding from the empirical research conducted in the Schumpeterian tradition is that, in cross-sectional data, there is a close, positive monotonic relationship between size and typically contemporaneous R&D which appears to be roughly proportional among R&D performers in the majority of industries or when controlling for industry effects in more aggregate samples. In addition, innovative output, variously measured, appears to increase less than proportionately than firm size. A consensus of decades' duration was that these relationships implied that size has little effect on innovation, and that large size confers no advantage for innovation, and perhaps a disadvantage.[25] More recent research, however, suggests that the close, positive monotonic relationship between firm (or, more precisely, business unit) size and R&D likely signals a fixed cost-spreading advantage to larger firm size in R&D. Klepper's (1996) and Klepper and Simons' (2005) analyses also suggest that this advantage to firm size can be self-reinforcing over time. As a firm grows large, its returns to R&D, and, in turn, its R&D investment, increase, and, assuming some amount of success in its R&D, the firm is likely to grow larger still and so on.

Moreover, the widely observed decline in R&D output with size does not necessarily imply that larger firms are less efficient in the generation of innovation. Employing the Schumpeterian distinction between invention and innovation, the empirical patterns signal a large firm size advantage in innovation, while implying little about the effect of firm size on the efficiency of invention. Thus, the question of the relationship between firm size and R&D efficiency remains open, with the measurement and selection challenges associated with the assessment of R&D productivity of small versus large firms only strengthening the point.

An important feature of the cost-spreading advantage of business unit size for R&D, according to Cohen and Klepper (1996b), is its dependence on two conditions, namely appropriability conditions that commonly limit firms to profiting from their innovations by embodying them in their own output rather

[24] One partial exception is King and Tucci (2002) who show that the technological changes in the US disk drive industry examined by Christensen (1997) indeed led to entry, but did not typically precipitate the exit of the previously dominant firms.

[25] Indeed, this was the author's own view expressed in Cohen and Levin (1989), though revised in Cohen (1995).

than by, for example, licensing them, and that growth due to innovation is typically limited by the size of the firm. Sutton's (1998) analysis of the link between market structure and R&D suggests a third factor conditioning the firm size relationship with R&D—namely submarket homogeneity, and, in turn, the nature of buyer preferences.

If there is a private advantage to larger size due to R&D cost spreading in the majority of industries, is there also a social welfare advantage? This is not apparent for at least two reasons. First, the dynamic effects of the R&D cost-spreading advantage of larger firm size raise a concern for social welfare. If R&D cost spreading implies that as firms grow larger, their incentives to pursue more incremental innovation intensify—independent of the technological opportunities that are available—then, over time, depending on entry conditions, there may well be technological opportunities that are foregone, to society's detriment. Second, as firms grow large within a market, the number of firms supported by that market declines, *ceteris paribus*. If one assumes—and this is a strong assumption—that technological diversity (e.g., the variety of approaches adopted to address a technological challenge) both promotes technical advance and is associated with a larger number of firms within an industry, then, as suggested by Cohen and Klepper (1992b), larger firm size may come at the cost of the benefits of technological diversity.[26] Although numerous scholars have suggested that the greater is technological diversity within an industry, the greater the rate of an industry's technical advance (e.g., Jewkes et al., 1958; Metcalfe, 1988; Nelson, 1981; Porter, 1990; Scott, 1991), neither the relationship between the number of firms within an industry and technological diversity, nor that between technological diversity and innovation have, however, been empirically examined in any depth. Whatever we know about these relationships is suggestive, at best. For example, using a coarse measure of technological diversity within industries drawn from the Levin et al. (1987) survey, Cohen and Malerba (2001) observe a significant, positive relationship between a survey-based measure of technological diversity (cf. Klevorick et al., 1995) and a subjective measure of industries' rates of technical advance.[27] More generally, whether due to the role of diversity, or due to a systematic, self-reinforcing link between larger firm size and more incremental innovation, it does not follow that an industry composed of larger firms is more innovative over the long run, notwithstanding a cost-spreading advantage to size. Moreover, these offsetting effects imply a social welfare tradeoff associated with firm size and challenge us to understand the factors that might condition it. In any event, the link between market structure and technological diversity, as well as the latter's impact on technical advance warrants further study.

2.3.2. Market structure and innovation

Moving on to our consideration of the relationship between market structure and R&D, the empirical patterns are mixed, and not terribly informative. Even before one controls for industry effects, the variance in R&D intensity explained by market concentration is small. Moreover, whatever relationship that exists in cross sections becomes imperceptible with the inclusion of controls for industry characteristics, whether expressed as industry fixed effects or in the form of survey-based and other measures

[26] One could also argue that, beyond some point, a plethora of technologies and approaches to technical challenges could become excessive and inefficient from a social welfare perspective (Gilbert, 2006).

[27] Moreover, the link observed between competition and productivity growth by Nickell (1996) could well reflect a mediating effect of technological diversity due to larger numbers of competitors.

of industry characteristics such as technological opportunity, appropriability conditions, and demand. In parallel to a decades-long accumulation of mixed results, theorists have also spawned an almost equally voluminous and equivocal literature on the link between market structure and innovation.

How should we think about the sensitivity of the link between market structure and innovation to other industry-level variables, no less the modest explanatory power of market structure observed in cross-sectional data? And how should we consider this relationship in light of the lack of theoretical consensus on the question? As noted above, in response to the indeterminacy of the theory and the inconclusiveness of the empirical results, Sutton (1998) develops a class of game-theoretic models yielding a range of possible Nash equilibria that bound a set of possible measurable outcomes. The substantive conclusion of his analysis is consistent with the earlier conclusion of Cohen and Levin (1989), namely that, though market concentration and R&D intensity may be correlated, market concentration is not an independent, important driver of innovation. Although a question of time horizon, the argument that market structure is not a fundamental determinant of innovation turns on the empirical case that market structure may be at least partly a function of other, more plausibly exogenous variables, and possibly innovation itself. Thus, the correlation observed between market structure and R&D intensity reflects either their codetermination or the impact of innovation on market structure.[28] Such a conclusion calls into question the "Schumpeterian tradeoff"—at least that between the allocative efficiency tied to *ex ante* market structure and the dynamic efficiency associated with the pace of technical advance.[29]

In his review of the "Schumpeterian" (neoclassical) theoretical and empirical literatures, Gilbert (2006) suggests that we should not conclude from the mixed empirical record that market structure does not matter for innovation. Rather, he argues that one should consider that there is no one theory of the relationship, but many. And his response to the sensitivity of the empirical relationship between market structure and R&D to industry effects and the inconclusiveness of the findings is (1) that industry effects "mask" the relationship; (2) that empirical scholars have not controlled for the contingencies highlighted by theorists; and (3) that empirical scholarship suffers from limited data, measures, and methods. Regarding the need to consider the contingencies, empiricists should work harder at understanding the key features of industries and technologies that may condition the relationship, and at least a couple of these contingencies are highlighted by theorists (as well as empiricists), including firms' abilities to protect their innovations, and the type of innovation in question—particularly process versus product innovation. The challenge for testing the game-theoretic models of R&D rivalry, however, is that only in their most stripped down and simplified version do they provide clear, testable empirical implications (see below), partly because they analyze behavior in highly stylized and counterfactual settings. For example, many models focus on the interaction of a single incumbent and a single prospective entrant, or, alternatively, symmetric competitors. Moreover, many of the results obtained in this literature depend upon typically unverifiable assumptions concerning the distribution of information, the identity of the decision variables, and the sequence of moves.

[28] Indeed, in light of the mixed empirical and theoretical results, one might entertain Griliches' (1962, p. 353) remark of almost 50 years ago: "I do not deny that the relation between the form of industrial organization and inventiveness may be of interest to the industrial organization man, I only doubt its importance to the invention and economic growth oriented researcher.... Even if there were some relationship between the ... degree of market control and the rate of inventive activity ... it could at best have only a second order effect."

[29] Recall, however, that Schumpeter also argued that the expectation of *ex post* market power acquired by successful innovation provides an important incentive to undertake inventive activity.

Regarding measures, there can be little disagreement with Gilbert's contention that the commonly employed measure of market structure, market concentration, does not accurately reflect the nature or intensity of competition. Progress has been made here, however, particularly with the use of measures of market power, such as the modified Lerner index employed by Nickell (1996) and Aghion et al. (2005), although even those measures are limited. Also promising is the strategy of using multiple measures of competition, as employed by Geroski (1990), and subsequently Artes (2009), with the idea that one wants to look for robust results across them.

For the study of the relationship between competition and innovation, just as fundamental as the absence of appropriate measures of the intensity of competition, is our limited understanding of exactly how firms compete with respect to innovation. Grabowski and Baxter (1973) conducted the first empirical study of strategic interaction, offering weak evidence that firms in the chemical industry engage in competitive matching of R&D investment. Little empirical attention was devoted to the subject until the 1990s (e.g., Cockburn and Henderson, 1994; Khanna, 1995; Lerner, 1997; Meron and Caves, 1991), when a number of scholars looked for evidence of strategic interaction in R&D spending patterns. Due, however, to the contingent quality of the theoretical literature's results, these studies focus on the simplest form thereof, namely competitive matching in either R&D investment or new product introductions. Henderson and Cockburn (1994) look for evidence of positively correlated project-level research investments within therapeutic drug classes in the ethical pharmaceutical industry. Meron and Caves (1991) try to discern matching in firms' overall R&D expenditures in 28 US manufacturing industries. Khanna's (1995) study of market segments within the high-end computer industry and Lerner's (1997) study of Winchester disk drives search for evidence of matching behavior with respect to product introductions. The difficulty facing all these analyses is that there are numerous explanations for positively correlated R&D or product introduction behaviors other than strategic behavior. These include common changes in industry-level technological opportunity and demand conditions, spillovers from leading firms which increase the marginal productivities of rival R&D, or simply a catchup phenomenon where equally capable firms involved in similar activities all move in the same direction at roughly similar rates with it being a matter of chance that any one firm moves before the others.

Considering the difficulties in controlling for these alternative factors, it is not surprising that the studies, considered together, are inconclusive. Cockburn and Henderson (1994) find that purposive matching does not appear to characterize "the bulk of research investment" in ethical pharmaceuticals. Despite qualitative evidence suggesting some purposive matching, Lerner (1997) cannot reject a simpler probabilistic "catchup" explanation. In contrast, on the basis of quantitative and as well as qualitative evidence, Khanna (1995) concludes that there is competitive matching within segments of the computer industry. For most of the industries in their sample, Meron and Caves (1991) found some evidence of strategic matching within groups of what they identified as core firms within each industry.[30]

[30] The Carnegie Mellon Survey administered by Cohen et al. (2000) also speaks to an informational premise of some models of strategic behavior in R&D, namely that firms know what R&D projects their rivals are pursuing at a relatively early stage. Having asked their respondents (R&D lab managers and directors) at what stage in the innovation process did they first learn of a major R&D project of a rival, only 15% of over 1000 respondents indicated that they were aware of the project at project inception or during the research stage, and 85% reported that they did not learn of the project until either the development stage, or subsequent to product introduction, with the 85% evenly divided between the two. The implication is that firms tend to learn what their rivals are doing rather late in the game, calling into question assumptions about the timeliness of firms' awareness of rivals' R&D activities.

Thus, some of the studies suggest that strategic interaction affects innovative activity and some suggest that it does not. Also, it is hard to know from these results whether the absence of a common result across the different industries examined signifies that strategic interaction matters more or differently in some industries than in others, but, if that is true, it would be interesting to know what conditions its character and importance. Even in the studies that suggest that competitive interaction matters, however, we have little sense of how important it is relative to other factors. Indeed, Geroski (1991c) has speculated that strategic rivalry may be of second-order importance when compared to the influence of factors such as technological opportunity, and Cockburn and Henderson (1994) suggest that, in addition to opportunity, heterogeneous firm capabilities also appear to be much more important. These comments do not, however, suggest that we should dismiss the impact of strategic interaction on innovation, but, rather, suggest that we need to devote more study to the issue.

2.3.3. Crosscutting considerations

The research on firm size, market concentration and R&D suggests that either interpreting observed relationships or searching for their sources is a hazardous venture always, but even more so without the discipline imposed by at least a simple theoretical model. More generally, evaluation of the Schumpeterian hypotheses should take place within the context of more complete models of the determination of technological progress. Only with such models will we be able to understand the basis for the robust relationships that do exist, as between firm size and R&D, or the source of the fragility of other relationships, as between market concentration and R&D intensity. One important qualification to the admonition to ground empirical analysis in theory is that a good deal of what we know empirically about the links among firm size, market structure, and innovation originates from exercises that are either descriptive or only casually motivated. Moreover, these empirical patterns, and judgments about their robustness, in turn, have informed and guided subsequent theorizing, such as that by Sutton (1998), Nelson and Winter (1982a), Klepper (1996), and Klette and Kortum (2004), and have, in this way, deepened our understanding.

Although the Schumpeterian empirical tradition is divided between studies of firm size and market structure, Sutton's (1998) and others' work show that it can be productive to bridge across these two levels of analysis. The simple reason is that whatever forces associated with innovation may contribute to a single firm's dominance within an industry obviously bear on the determination of market structure. Moreover, as illustrated by Blundell et al.'s (1999) analysis, to the degree that market power impacts innovation, its effect may be clearest at the level of the firm and business unit.

As suggested above, the analyses of cross-sectional and panel data that dominate empirical work in this area can benefit from the lessons of industry studies, as they did in the case of Phillips' (1971) influential study of the aircraft manufacturing industry. They should also attend to the lessons of the growing body of literature dedicated to the dynamics of innovation, entry, exit, and market structure (e.g., Nelson and Winter, Jovanovic, Klepper, Utterback, etc.). Indeed, the work of Geroski, Phillips, and others highlights the difficulty of inferring possibly important dynamic relationships from cross-sectional or short panel data. It is, however, just these dynamic relationships that account for many of the relationships that this literature is trying to explain.

An important gap where empirical scholars working in the Schumpeterian tradition have paid little attention to analyses of industry dynamics is the now large body of work dedicated to the study of network externalities, pioneered by Katz and Shapiro (1985), David (1985), and Arthur (1989), and reviewed in Farrell and Klemperer (2007) in this handbook series, where the development of selected technologies have obviously had profound implications for the emergence of concentrated markets and dominant firms. This neglect reflects a broader point. However important, network externalities are only one illustration of the phenomenon of dynamic increasing returns where market dynamics lend themselves to self-reinforcing feedback. In addition to network externalities, other sources of dynamic increasing returns with implications for technical advance and its links to market structure include learning by doing (cf. Chapter 10), and learning by using (cf. Rosenberg, 1982). R&D cost spreading constitutes another source of dynamic increasing returns, but one whose implications for industry evolution and technical advance has been examined in some detail, as discussed above. As noted below in our consideration of technological opportunity, to link these different sources of dynamic increasing returns to innovation and market structure, one might usefully distinguish among the sources on the basis of the degree to which they are tied to specific firms (e.g., learning by doing, or R&D fixed cost spreading), versus those which are tied to technologies that can potentially stand apart from the firms that may have first introduced them (e.g., network externalities or learning by using). In this latter case, the nature of the innovation, and possibly its complementarity with other technologies, will tend to drive market structure rather than the reverse.

A limitation of the majority of empirical studies of firm size, market structure, and R&D is an implicit assumption about the innovation process: that firms are autarkic with respect to innovation; the entire process of innovation, from invention through commercialization is housed within a given firm. The discussion above on firm size and R&D indicates that the ability to out-license inventions to others can have important consequences for the advantages of firm size for innovation. Thus, simply recognizing the possibility of out-licensing informs our understanding of the role of firm size. We have not, however, addressed the implications for the study of effects of firm size or market structure of in-licensing, which has become more pervasive over the past two decades (Arora et al., 2001). In-licensing and the related proliferation of R&D alliances (cf. Ahuja et al., 2009) raise challenges for measurement and empirical analysis. For example, what does R&D intensity signify if inventions are in-licensed and the licensing fees are not accounted for in a firm's R&D budget? Illustrating the importance of considering technology markets for this literature, Gilbert (2006) describes Czarnitzki and Kraft's (2004, 2005) tests of Gilbert and Newbery's (1982) hypothesis that dominant incumbents should invest more in R&D than entrants in order to pre-empt entry that would eliminate their monopoly rents. Focusing on only R&D expenditures, and employing data on German firms, Czarnitzki and Kraft (2004) found that entrants invested proportionately more than incumbents. In contrast, Czarnitzki and Kraft (2005) found that the incumbents spent more on in-licensing. More generally, although it has long been true that firms regularly draw on extramural sources of knowledge (cf. Jewkes et al., 1958, second ed., 1969; Mueller, 1962), it is now becoming more common for these relationships to be market-mediated, at least in some industries, and that in-licensing, out-licensing, and associated relationships should be considered as we think about the relationship between firm size and innovation (cf. Chapter 15).

In conclusion, obtaining a better understanding and evaluation of the Schumpeterian hypotheses is a good reason to move toward better data as well as more complete models of technological change. There are, however, reasons to move the profession's agenda beyond the Schumpeterian hypotheses and to focus attention on more fundamental determinants of technological progress. First, the welfare gains associated with technological progress are large. Second, we have, at present, only a limited understanding of the primary economic forces driving innovation and how they differ across industries, and particularly across firms within industries.

3. Firm characteristics

This section considers firm-level determinants of innovative effort and performance other than firm size.[31] As noted above, applied economists often control for firm size in their studies of innovative effort by using R&D intensity as the dependent variable measuring innovative effort. Employing the FTC Line of Business Program data that distinguishes between the firm and its constituent business units, Scott (1984) found that once one controls for size by expressing R&D intensity as R&D normalized by business unit sales, fixed firm effects explain about 50% of the variance in R&D intensity. Economists have made, however, only modest headway in explaining interfirm differences in R&D intensity and performance since Cohen and Levin (1989) observed that the most widely considered measures of firm characteristics up to that time, namely cash flow and the degree of diversification, jointly explained less than 10% of the variance in business unit R&D intensity explained by fixed firm effects.

Variables reflecting firm characteristics may be interpreted as affecting the cost of innovation per quality unit of output (cf. Spence, 1984). Some firm characteristics, such as cash flow, may be captured by single variables. Others, such as firms' R&D capabilities, are better conceptualized as multidimensional.

As a possible determinant of R&D, cash flow may be the most thoroughly examined firm characteristic in this literature (e.g., Antonelli, 1989; Armour and Teece, 1981; Branch, 1974; Caves et al., 1980; Elliot, 1971; Grabowski, 1968; Hall, 2002; Hamberg, 1966; Hao and Jaffe, 1993; Himmelberg and Peterson, 1994; Johannisson and Lindstrom, 1971; Kamien and Schwartz, 1978; Kraft, 1989; Link, 1981; Mueller, 1967; Smyth et al., 1972; Switzer, 1984; Teece and Armour, 1977). Indeed, the claim that cash flow affects R&D constitutes one of Schumpeter's (1942) arguments for an advantage to large firm size.

The rationale for considering cash flow as a determinant of R&D effort typically assumes (1) that capital markets are imperfect; (2) that those imperfections especially constrain investments whose outcomes are more uncertain; and (3) that the returns to R&D, as a class of investment expenditure, are more uncertain than the returns to, say, investment in PP&E.[32] The possibility that R&D expenditure is liquidity constrained due to capital market imperfections is important for policy makers because it represents another rationale for public support for R&D beyond the widely recognized market failure affecting R&D investment associated with R&D spillovers.

[31] See Ahuja et al. (2009) for a review of the management literature on this subject.

[32] Using a sample spanning large and small Italian firms, Antonelli (1989) entertains a behavioral rationale for the effect of cash flow on R&D. Consistent with Cyert and March's (1962) notion of "problemistic" search (that hypothesizes that firms engage in search for better ways of doing things—which is one way to characterize R&D activity in general—when their performance falls below some minimally acceptable performance threshold), he finds that firms invest more in innovation when performance falls below a minimum threshold, but that the relationship between cash flow and innovative effort appears to be positive otherwise.

Many, but not all, of the studies, have found that a firm's cash flow is associated with higher levels of R&D intensity or unnormalized R&D effort. Scholars have disagreed over the interpretation of this finding. Some have argued that it is difficult to distinguish cash flow as a measure of liquidity from its possible function as a signal of the future profitability of R&D investment (Elliot, 1971). Indeed, the likelihood that R&D and cash flow both respond to the same demand or other shocks is likely the most challenging of the empirical issues facing researchers on this topic. Others question whether cash flow encourages R&D or whether it simply reflects the profitability of past R&D expenditures which tend to be stable over time (Branch, 1974).[33] Moreover, a positive relationship between cash flow and R&D expenditures may simply reflect their codetermination by a firm's underlying capabilities. The intertemporal smoothing characteristic of R&D spending also makes a relationship more difficult to discern.

Exploiting a simple lag structure, Hao and Jaffe (1993) provide evidence that causality runs from liquidity to R&D, at least among smaller firms. Consistent with this finding, Himmelberg and Peterson (1994) find that the R&D spending of a 5-year (1983–1987) panel of 179 small firms in four high-tech industries rises with cash flow, where both R&D and cash flow are scaled by beginning-of-period asset values. Himmelberg and Peterson control for firm effects, and find within-firm and between-firm effects of cash flow to be significant in specifications that include sales and Tobin's q as controls for profitability. They also find evidence of downward bias due to the dampening of transitory effects in the between-firm estimates. Using panel data from the United States, Israel, and Japan, Hall et al. (1999) observe that R&D (as well as investment in PP&E) appears to be more sensitive to cash flow in the United States than in either Japan or Israel.

One line of research that has examined the effect of cash flow on R&D has considered R&D expenditures as simultaneously determined with other forms of investment (e.g., Mueller, 1967), or as one type of investment which is simultaneously determined with other investment and financial decisions (e.g., Guerard et al., 1987; Switzer, 1984). Switzer (1984) provides mild support for the notion that cash flow positively conditions R&D, and Guerard et al. (1987) provide some evidence that dividends may constitute an alternative use of funds, and both studies find some evidence of a positive relationship between R&D and capital expenditures.[34] Embedding her consideration of cash flow in the context of the relationship between corporate finance and R&D investment, Hall (1990) finds that firms that increase their leverage reduce their R&D intensities, sometimes considerably. While Acs et al. (1991) obtain a similar result for larger firms, they find that the smaller firms with more debt actually increase their R&D effort and suggest that larger firms appear to finance their R&D effort principally through equity while smaller firms do so through debt. Economists have also considered the determinants of the cost of financing R&D, reflecting more broadly on the mix of means that a firm may use to finance innovation, including debt, equity, venture funding, as well as internal funds (e.g., Acs et al., 1991; Hall, 1990, 2002).

After reviewing a number of other studies (e.g., Bhagat and Welch, 1995; Bond et al., 1999; Bougheas et al., 2003; Brown, 1997; Hall et al., 1999; Harhoff, 1998), Hall (2002) summarizes the

[33] There is evidence, from case studies (Mansfield et al., 1971) and from econometric work (Ravenscraft and Scherer, 1987), that the mean lag in returns from R&D expenditure is on the order of 4–6 years.

[34] In a carefully structured analysis of the relationship between R&D and capital investment for a sample of 191 firms in science-based industries, Lach and Schankerman (1989) also find a close relationship between capital investment and R&D, but suggest that while R&D appears to Granger-cause investment, this relationship arises principally because the same persistent factors determine both R&D and investment.

more recent literature as suggesting that, "debt is disfavored as a source of finance for R&D investment," and that "'the Anglo-Saxon' economies, with their thick and highly developed stock markets and relatively transparent ownership structures, typically exhibit more sensitivity and responsiveness of R&D to cash flow than Continental economies." Thus, while the methodological challenges that plague this literature are not entirely addressed, the weight of findings suggests that, at least in the United States and Great Britain, cash flow likely contributes to R&D spending. A more comprehensive assessment of the role of cash flow and the other means of financing R&D is provided in Chapter 14.

The other widely studied corporate attribute is diversification. The influence of product diversification upon basic research spending was first suggested by Nelson (1959), who argues that, because the results of basic research tend to be unpredictable, the diversified firm possesses more opportunities for exploiting the new knowledge. A link between diversification and innovative effort is also suggested by the argument that large, diversified firms are better positioned to exploit complementarities among the diverse activities and the economics of scope that may be associated with R&D. This second argument, however, does not necessarily imply a positive relationship between diversification and R&D intensity for any given line of business. Nonetheless, both arguments implicitly assume what Arrow (1962a) expressed clearly: the market for information is imperfect and appropriability may be better achieved by the internal application of knowledge than by its sale.

In the first study to examine the relationship between degree of diversification and R&D expenditures, Scherer (1965a) found that an index of diversification was highly significant and explained considerable variance when introduced into simple cross-section regressions of patents and R&D intensity on firm size. The effect of diversification, however, was barely discernible in separate regressions at the two-digit industry level, which suggests that Scherer's diversification measure may have reflected the influence of omitted two-digit industry effects in the full cross section.

Subsequent results have been mixed. For example, Grabowski (1968) found that diversification encourages R&D spending in chemicals and drugs, but not in petroleum; McEachern and Romeo (1978) obtained precisely the opposite results. Link and Long (1981) explicitly tested Nelson's hypothesis suggesting a relationship between diversification and basic research in particular. Using tobit estimation for a cross-sectional sample of 250 Fortune 1000 firms responding to a survey for the one year, 1977, they found a significant, positive relationship between the firms' degree of diversification and basic research intensity. They acknowledge, however, that, given the cross-sectional character of the exercise, causality could run from basic research to diversification. Scott and Pascoe (1987) examined the hypothesis that R&D expenditures depend on the particular pattern of a firm's diversification. They found that when a firm diversifies into technologically related industries, its pattern of R&D expenditures differs from the case where diversification is not so "purposive." In particular, such a firm tends to allocate a large share of R&D to industries in which appropriability is high.

MacDonald (1985) looked at the reverse direction of causation, attempting to explain a firm's direction of diversification as the effect of accumulated intangible R&D capital in its primary industry, and found the R&D intensity of the home and target industries of diversifying firms to be similar. Montgomery and Harihan (1991) found R&D-intensive firms more likely to diversify, and Ravenscraft and Scherer (1987) observed that more R&D-intensive industries tended to spawn more diversification activity. In her review of the literature on corporate diversification, Montgomery (1994, p. 174) interprets these findings as suggesting, "that existing organizational capabilities, particularly in R&D and marketing, often guide diversified expansion."

The absence of robust findings concerning an effect of diversification on R&D is unsurprising given the problems that beset this research. First, it is just as likely, if not more, that the direction of the relationship runs opposite to that posited in the early work. Moreover, with the exception of Link and Long (1981), Doi (1985), and Scott and Pascoe (1987), little attention has been paid to controlling for the influence of industry-level variables.

Perhaps the most revealing work on the relationship between R&D and diversification has focused on the relationship between innovative performance and firm diversification across submarkets within one industry, namely drugs. As described above, Henderson and Cockburn (1996) observed a significant, robust positive relationship between the number of research programs pursued by drug firms (where a program is distinguished by therapeutic area), and the productivity of each program's R&D, measured by the number of "important" patents produced (i.e., patents granted in two of the three major jurisdictions). This result suggests the presence of scope economies where diverse R&D activities are related across submarkets.[35]

Differences in R&D activities and innovative performance across firms have focused economists' attention on the influence on R&D of differences in firms' R&D-related capabilities. Before discussing this literature, it is important, however, to take heed of Scott's (1991) point that one should not presume that heterogeneity in firms' R&D activities or performance necessarily signal differences in capabilities rather than differences in the incentives faced by comparably capable firms. Indeed, in Cohen and Klepper's (1996b) R&D cost-spreading model, firms of different sizes may pursue different types of R&D not because of differences in capabilities, but due to the different R&D incentives conditioned by size. Obviously, however, differences in capabilities may contribute to differences in the level and direction of R&D effort.

Anticipating more contemporary work, Rothwell et al. (1974), in their "Project SAPPHO," try to identify the features of the management and organization of firms that might account for innovative success. In a detailed study of 43 matched pairs of successful and unsuccessful innovations, Rothwell et al. (1974) found that the most important determinants of success were (1) close attention to user needs, (2) effective marketing, (3) efficient management of the development process, (4) ability to utilize outside technology and communicate with the external scientific community in areas specifically relevant to the innovation, and (5) project management in the hands of a relatively senior individual who could serve effectively as a "product champion" within the organization.[36]

The firm characteristic that economists have been long believed to affect R&D performance is the degree of integration between R&D and other functions within the firm. Starting with the work of Mansfield (1968), economists such as Teece (1986, 1987) and Mowery and Rosenberg (1989) have highlighted the importance of the links across marketing, manufacturing, and R&D in conditioning innovative success. Indeed, as noted below in our discussion of appropriability conditions, Teece considers the possession of hard-to-develop or acquire complementary capabilities as critical to the successful commercialization of innovation across a wide range of industries. Teece (1986) builds on the line of inquiry developed by Williamson (1975) that suggests that in the presence of asset specificity, uncertainty and opportunism, differences in internal

[35] Some case studies suggest the presence of economies of scope to R&D in vertically related industries. For example, Malerba's (1985) work on the semiconductor industry suggests that the advantages of vertical integration for innovative activity vary over the life cycle of the technology.

[36] Additional discussion of the SAPPHO project and related research on innovation is found in Freeman (1982).

organization and interfirm contractual relationships may have important implications for innovative behavior and performance. Teece applies this transaction cost framework in arguing that, to be a successful innovator, a firm typically requires a set of in-house, complementary capabilities, including manufacturing and marketing capabilities, which he terms "cospecialized assets," to commercialize new products and processes.[37] Scholars such as Teece have provided numerous qualitative accounts affirming the importance of these links. Few direct tests of Teece's arguments regarding the importance of complementary capabilities have been offered.[38] An important but unexamined implication of Teece's and others' argument is, to the extent that the possession of complementary capabilities is a determinant of R&D performance, investment in such capabilities—including, for example, some portion of marketing expenditures—should be treated as codetermined with R&D.[39] Indeed, an important development in this area has been a more sophisticated consideration of complementarities broadly, and their implications for innovative performance. We will return to this point below.

Economists have been slow to take up the call to consider the role of capabilities within the R&D function itself. Although one could treat such capabilities as observable, the construction of measures of R&D-related capabilities suitable for analyzing innovation in large samples spanning multiple industries is a formidable challenge. Construction of such measures demands detailed knowledge of the nature of the technologies and other activities associated with innovation in each sample industry. Moreover, it is not even clear how such capabilities would be defined or distinguished in a broad cross section. Partly reflecting these difficulties and partly reflecting economists' inattention to organizational processes and the nature of organizations more generally, only modest progress has been made in conceptualizing differences in firm R&D-related capabilities in ways that lend themselves to hypothesis formulation and testing.

A small number of empirical studies on the role of capabilities in determining innovative performance within industries date from the late 1980s and early 1990s.[40] Clark et al. (1987) and Clark and Fujimoto (1991) identify wide disparities in firms' R&D productivities in the international automobile manufacturing industry, and Iansiti (1995a,b) identifies similar disparities in the mainframe computer industry. These studies all offer evidence suggesting that these disparities result from differences in the organization of product development in these firms, including the degree to which tasks are subdivided, how the different phases of the development process are coordinated, how technical problems get

[37] See Teece's discussion of technological innovation and firm capabilities in Chapter 16. Also see more discussion of Teece's framework in our consideration of appropriability conditions in Section 4.3 below.

[38] In one of the few attempts to test directly Teece's prediction that the possession of complementary capabilities should favor R&D spending, Helfat (1997) "...found that firms with more complementary assets in the form of coal reserves undertook more R&D in synthetic fuels derived from coal" (Teece, 2006, pp. 1134–1135). In a counterexample to Teece's claim, Khazam and Mowery (1991) describe, however, the experience of the Sun company in commercializing its RISC chip without possessing in-house the complementary production capabilities. The example of Sun's early history is instructive in suggesting the possibility that where standard components are available and markets for tangible and intangible inputs abound, as was the case in Silicon Valley at the time, complementary capabilities may be less critical to the commercialization of R&D.

[39] Even Comanor (1964, 1965, 1967)—who studied both R&D and advertising extensively, and argued that the two activities were indeed linked in their contributions to product differentiation (Comanor, 1964)—never considered the two activities to be codetermined in his empirical analyses.

[40] Although going beyond the study of innovation *per se*, Chandler's (1990) historical analysis of the growth of large manufacturing firms claims that the development of organizational capabilities in the domains of production, marketing, and R&D underpinned the development and success of the large manufacturing firm.

solved, the relative autonomy of the project leader, and the nature of the links between product development and upstream applied research. In a study of the effects of R&D-related capabilities on innovative performance in the photolithographic alignment equipment industry, Henderson (1988, 1993) and Henderson and Clark (1990) compare entrants' and incumbents' abilities to exploit significant changes in the underlying technology. Henderson (1993) finds that entrants are typically at an advantage, despite the absence of evidence in the small sample employed that entrants invested more than incumbents in radical or major innovation or that their R&D productivity was greater. From interview-based studies of the firms in the industry, Henderson (1988) speculates that the result may reflect differences in information processing capabilities.[41]

Reflecting the greater feasibility of measuring firms' R&D capabilities within selected industries, Henderson and Cockburn (1994) and Cockburn and Henderson (1998) examine the impact of selected firm characteristics on the generation of important patents (see above) both at the therapeutic program level (Cockburn and Henderson, 1994) and at the firm level (Cockburn and Henderson, 1998) in the pharmaceutical industry. The authors suggest several classes of variables may affect a firm's R&D productivity, including (1) the firm's domain expertise, including its disciplinary-based knowledge as well as the knowledge bearing on specific areas of technology, which, for drug firms, bears on specific disease areas and (2) the firm's ability to acquire and use knowledge, which is a function of its ability to integrate knowledge across disciplines and across units within the firm, as well as its ability to integrate extramural knowledge. Even though benefiting from rich program-level data, data constraints still limit the authors' ability to construct measures that map to these two classes of variables; the measures employed only loosely correspond to the variables that they think should matter.

In the Cockburn and Henderson studies, the measure that robustly influences R&D productivity both at the program level and at the firm level is the degree to which the firm rewards its scientists for publishing. The authors claim that this measure reflects the firm's integration with the broader scientific community. The measure may also, however, reflect a more direct impact of the incentive on their dependent variable, namely patents, since patents in biomedicine are often linked to publications. A related variable that is positive and significant in their firm-level analysis is the fraction of papers that are coauthored with academics, which more defensibly reflects the role of the strength of ties to public research. In the two papers, the authors also find that greater authority of the project leader tends to be associated with lower R&D productivity, perhaps reflecting the effects of a more autocratic decision-making process. The authors are careful to characterize their results as descriptive, acknowledging that their featured variables, "may be measures of symptoms as much as they are measures of causes..." (Cockburn and Henderson, 1994, p. 79) As discussed below, the possible endogeneity of variables characterizing firm R&D-related capabilities is a pervasive concern for this literature.

Another capability that may affect firms' innovative activity and its success is the way in which a firm's management structures the incentives provided its R&D personnel. Empirical work on this feature of firms has only just begun. In a sample of over 300 publicly traded firms observed from 1987 to 1998, Lerner and Wulf (2007) find that firms with centralized labs whose top R&D managers are offered greater long-term incentives as a fraction of total compensation produce more patents, more

[41] On the basis of the experience of three of the larger firms in the industry, Henderson (1988) suggests that the result may partly reflect differences between incumbents and entrants with regard to communication channels, information filters and problem-solving strategies developed in the course of pursuing incremental innovation on the prior generation of the technology.

heavily cited patents and more original patents. In this analysis, they consider the possibility of spurious correlation by controlling for the long-term compensation offered to other senior executives as well. Their analysis is consistent with the notion that R&D management makes better decisions regarding project selection and management when it is offered long-term financial incentives. As the authors acknowledge, however, these results may also be driven by a selection effect; better quality managers are attracted by firms that offer such compensation packages. Also consistent with the importance of individual-level incentives is Cockburn and Henderson (1998) and Henderson and Cockburn's (1994) result that pharmaceutical firms that promote their scientists at least partly on the basis of their publications generate more important patents.[42]

This modest economics literature on the role of firm-specific capabilities in affecting innovation raises the question of exactly what kinds of capabilities matter for innovative activity and performance. The research broadly distinguishes two sorts. Clark, Henderson, Mansfield, Mowery, Nelson (1991), and others emphasize the role of organizational or procedural capabilities in conditioning the R&D productivity of firms. In this view, firms may pursue similar innovative activities, but some firms are more successful than others in either generating or profiting from innovation due to superior organization of internal processes. In contrast, Jewkes et al. (1958), Cohen and Klepper (1992a,b), Lerner (1997), and others focus on differences in firms' areas of substantive technological or domain expertise which leads them to pursue different innovative activities with different degrees of success.[43]

Notwithstanding the type of capability that is thought to matter, many studies of firm capabilities assume, often implicitly, that key R&D-related capabilities are exogenously given. This literature, however, does not make a strong theoretical case for this assumption, nor does it, with the exception of the limited evidence offered by Henderson (1993), indicate whether or under what circumstances it may be true. Although partly a question of time horizon, case study and anecdotal evidence support to some degree the assumption that selected R&D-related capabilities are exogenous by highlighting the fact that a firms' "core" capabilities are hard to change. The question of whether R&D capabilities are endogenous or not suggests the utility, however, of trying to address the fundamental questions raised by Clark et al. (1987, p. 781) about why such differences exist, how they evolve, and why they persist. One would suspect that the answers to these questions will vary across types of R&D activities as well as across industries depending upon the nature of the knowledge that underpins innovation and how that knowledge accumulates over time. In this regard, Winter's (1987) discussion of the degree to which an underlying knowledge or competence is tacit or articulable, and Merges and Nelson's (1990) consideration of the cumulative quality of some types of expertise may provide some guidance as to how to think about how distinct capabilities may persist and their roles in conditioning innovative activity and performance. The answers to Clark's questions, however, would also benefit from more

[42] Although examining the innovative performance of academic rather than industrial scientists, Azoulay et al. (2009), building on the work of Manso (2006) and Amabile (1996), consider whether the way in which financial incentives are structured can affect innovative performance. The particular proposition tested is that an incentive system that provides more time and opportunity for search and feedback stimulates greater creativity because it encourages researchers to become less instrumental and more exploratory in their thinking, and thus more willing to adopt higher risk, higher payoff approaches to technological challenges. Exploiting what they characterize as a "natural experiment," Azoulay et al. (2009) find what may be a treatment effect associated with the "admission" to a regime characterized by very patient capital and encouraging of bold initiatives, namely being named and supported as a Howard Hughes Medical Institute investigator.

[43] Henderson and Cockburn (1994), discussed above, consider both sorts of capabilities to be relevant.

eclectic studies of the sort conducted by Holbrook et al. (2000) on the early history of the semiconductor industry that combine economics, organizational science and technology, and business history.[44]

Holbrook et al. (2000) highlight the enduring impact of founding conditions, among other factors, in affecting both the domain expertise and organizational attributes of firms. Aside from founding conditions, however, it would be useful to consider other factors that may affect the incentives to develop or acquire important capabilities, including, for example, the availability of specialized talent, locational attributes that may affect the character and density of knowledge flows, and the degree of uncertainty that might affect firms' estimation of the returns to the acquisition of such capabilities.[45]

As suggested above, the idea that there may exist complementarities between R&D and other functions within the firm complementarities or across different activities within the R&D function itself poses a challenge for both methods and measurement. The issue is how can one estimate such complementarities given the possibility that an apparent complementarity across activities or capabilities may rather reflect the effect of unobserved firm heterogeneity; that, for example, there is some unobserved firm characteristic that is driving what is believed to be complementary activities as well as R&D performance itself. For example, in an exploratory analysis using European Community Innovation Survey (CIS)[46] data, Kremp and Mairesse (2004) observe a suggestive relationship between innovative performance and four different knowledge management practices within firms, including the promotion of a "knowledge sharing culture," the adoption of incentive policies to retain R&D professionals, the employment of alliances for knowledge acquisition and the adoption of a written knowledge management policy by the firm. They observe positive effects of virtually all of these on innovative performance, but the question is whether these factors as well as the higher performance with which they are associated are themselves reflect the influence of some other firm characteristic. Work that tries to control for the possibility of unobserved heterogeneity includes, for example, Cassiman and Veugelers' (2006) examination of the relationship between firms' acquisition of extramural knowledge and internal R&D. The authors underscore the challenge of estimating the impact of a complementary capability—in their case, one associated with acquiring external knowledge—on internal R&D, and the difficulty in finding appropriate instruments critical to such an exercise. Also building on the methodological insights of Arora (1996) and Athey and Stern (1998) that consider the challenge of estimating complementarities using cross-sectional data, Leiponen (2005) tries to demonstrate the

[44] Holbrook et al. (2000) showed important differences across four key early competitors: Schockley, Motorola, Sprague, and Fairchild in their technical goals and their approaches to solving similar problems. Another key difference across the firms was the degree to which top management was able to, "integrate the activities and information flows across the different functional areas of R&D, manufacturing, and sales." The authors conclude that there were exogenous sources of these differences stemming from "the pre-entry and early experiences of the firms. . .or their principals, and these experiences had lasting effects on management's beliefs about what was worth doing, as well as their ability to do it." The authors also note that the effects of these experiences endured, "not because the firms did not see the need for change, but because, when a need for change was recognized. . .they found change difficult. Constraints on change stemmed in part from an inability to secure the necessary kind of expertise and to integrate it across functions sufficiently rapidly to remain at the technological frontier" (Holbrook et al., 2000, pp. 1037–1038).

[45] The impact of such uncertainty on firm's investment in the development of its capabilities is subject, however, to a bit of a "chicken-and-egg" problem, in that the expectations of the return to a capability may well be conditioned, in turn, by the firm's already existing expertise in an area (Cohen and Levinthal, 1994).

[46] See a brief discussion of the Community Innovation Survey data below in Section 5.

complementarities that exist between firms' technical skills and firm innovation, where the latter is measured by whether firms innovate or not, based on the Finnish CIS data from 1992.[47]

With the exception of this recent work entertaining the role of unobserved firm heterogeneity, economists have tended to focus on the role of more readily measured firm characteristics in their attempts to explain interfirm differences in innovative activity and performance. Cohen and Klepper (1992a) adopt a different approach that does not assume that the key sources of heterogeneity are observable. They start their analysis by examining the within-industry distributions of firm R&D intensities for a sample of 99 manufacturing industries. They found that the industry R&D intensity distributions reveal a regular pattern—unimodal with either an internal or external mode, positively skewed with a long tail to the right, and include a large number of R&D nonperformers. In their view, the existence of such distributional regularities suggests the possibility of a common, underlying probabilistic process, which operates by conditioning an unobservable determinant of R&D intensity.

To probe this conjecture, Cohen and Klepper (1992a) developed a probabilistic model where firms' R&D activities and, in turn, R&D intensities within an industry are determined by unobserved, firm-specific, R&D-related capabilities which are allocated across firms via a simple Bernoulli process.[48] Coupled with business unit size conditioning the returns to any given R&D effort (cf. Cohen and Klepper, 1996b), this process yields a binomial distribution of business unit R&D intensities. Cohen and Klepper (1992a) showed that this distribution could account for the initially observed patterns in the industry R&D intensity distributions. More importantly, it also successfully predicted the cross-industry correlations of the first three moments and the coefficient of variation of the industry R&D intensity distributions. In the course of an additional distribution fitting exercise, however, Cohen and Klepper found that the actual distributions departed somewhat from a binomial distribution. Nonetheless, the evidence suggesting that industry R&D intensity distributions can be characterized as the outcome of a common probabilistic process is strong.[49] One useful next step would be to develop measures of the parameters governing the probabilistic process and evaluate their explanatory power.

Related to the observation that firms differ in their innovative activities and capabilities is the notion that no one firm may pursue all the activities necessary to generate and commercialize a given

[47] Also employing CIS data, Mohnen and Roller (2005) similarly probe the effects of complementarities between firm R&D and different dimensions of the policy environment that may affect innovation.

[48] Cohen and Klepper's (1992a) suggestion that probabilistically allocated, distinct firm capabilities affect innovative activity and performance is consistent with Nelson and Winter's (1982a) earlier suggestion that, under conditions of bounded rationality, differences among firms in technological capabilities, accumulated in part by experience and in part by good draws from a stochastic environment, may be sources of interfirm differences in behavior and performance. (The corporate strategy literature (e.g., Porter, 1980) also implicitly assumes that such capabilities are exogenously given when it prescribes that firms should exploit their core competences.) Like Nelson and Winter, Cohen and Klepper break with convention by suggesting that the allocation of capabilities across firms may be characterized at any point in time as the outcome of a probabilistic rather than optimizing process. Cohen and Klepper's model departs, however, from Nelson and Winter's perspective by assuming that capabilities, once possessed, are optimally exploited. Expressing a different view of the origins of cross-firm differences in innovative activities, Scott (1991) develops a model in which firm capabilities are identical within industries, but firms choose to pursue different R&D activities in order to gain some degree of a monopolistic advantage.

[49] Using World Bank data on seven industries for six countries, Lee (2002) confirms the existence of a common distributional pattern across industries and nations, but finds that a lognormal rather than a binomial distribution fits the distributions better. He also finds these patterns to conform to a distribution of firms' self-reported "technological competence" from World Bank survey data.

innovation. This observation should push economists to embrace the argument that Jewkes et al. (1958; second ed., 1969) made in the context of their 61 innovation case histories: that technical advance within an industry is commonly realized through the interactions of firms that are distinguished by size, expertise, capabilities, and other attributes. For example, large firms often buy out small ones to bring an innovation to market, or enter into contracts with small firms or independent inventors to acquire critical skills or knowledge. In a view subsequently echoed by Nelson et al. (1967), Scherer (1980), and Dorfman (1987), Jewkes et al. (1958, second edn. 1969) suggested that, "It may well be that there is no optimum size of firm but merely an optimal pattern for any industry, such a distribution of firms by size, character and outlook as to guarantee the most effective gathering together and commercially perfecting of the flow of new ideas" (p. 168). One may build upon this insight to argue, first, that, for the achievement of technical advance, there is no optimal set of characteristics for any one firm, and, moreover, that there is no optimal set of distributed firm characteristics that stands apart from key features of the technology and the industry.

The conjecture of Jewkes et al. should be interpreted as an invitation to pursue the inquiry begun by Nelson (1986, 1989, 1993) and Freeman (1987) to explore the complementarities and relationships across firms and between firms and other institutions (e.g., universities, technical societies, government) that facilitate innovation. We need to consider the circumstances under which a division of labor between the institutions generating new knowledge and the firms engaged in its generation and commercialization occurs and is efficient, and how such a "division of innovative labor" may vary with industry conditions, such as appropriability and technological opportunity, and with the features of knowledge that may affect how readily it may be transmitted across organizations (cf., Malerba and Torrisi, 1992; Pavitt, 1987; von Hippel, 1994). In Chapter 15, Arora and Gambardella review the literature that considers the nature and effects of cross-firm relationships—especially market transactions—that may support innovation within an industry.

4. Industry characteristics

In seeking to understand why industries differ in the degree to which they engage in innovative activity, empirical researchers such as Pakes and Schankerman (1984) have come to classify explanatory variables under three headings: product market demand, technological opportunity, and appropriability conditions. Although the importance of each of these classes of variables has been illustrated in the historical literature and in case studies as well as in some pioneering empirical studies, their study and measurement had been relatively neglected until the mid-1980s. One reason for this relative neglect has been the profession's preoccupation with the effects of firm size and market structure. Another reason is the absence of a clear and precise understanding of how the forces classified under the headings of technological opportunity and appropriability should be conceptualized and given operational definitions. Finally, even where a particular variable is well defined and a clear hypothesis is formulated regarding its influence, the data necessary for empirical work are often unavailable or unreliable.

The sections that follow summarize and interpret what is known about how demand, technological opportunity, and appropriability conditions vary across industries and how they contribute to an explanation of interindustry differences in innovative activity. We also offer suggestions to guide further exploration of the roles of these variables.

4.1. Demand

In his seminal work on technological change in various capital goods industries, Schmookler (1962, 1966) made two arguments. The first and more general claim was that the direction and rate of technological change could be explained as the outcome of the behavior of purposive, profit-seeking firms, and was thus amenable to economic analysis. To demonstrate the importance of economic incentives in general, he highlighted the role of demand in particular. To substantiate the claim that demand determined the rate and direction of inventive activity, he marshaled evidence indicating that cycles in the output of capital goods and in capital expenditures by downstream industries "led" cycles in the time series on relevant capital goods patents.

Schmookler's focus on the role of demand sparked a lively debate among economic historians and other economists concerning whether "demand-pull" or "technology-push" was the primary force behind technological change. In the terminology that has since come into use in industrial organization, the debate was about the relative importance of demand versus that of technological opportunity.[50] In arguing for the primacy of demand, Schmookler claimed that scientific knowledge and technological capability were applicable to a wide range of industrial purposes and were readily available. Although he recognized that generic knowledge and capability tend to grow, he argued that at any point in time a common pool was uniformly available for industrial application. The industries that made use of this common resource by making their own complementary investments in applied research and development were those induced to do so by large and growing markets. Though he presented an impressive array of data to support the view that demand matters, Schmookler never attempted to test the maintained hypothesis that the supply conditions for innovation (technological opportunities) were uniform across industries.

Schmookler's proposition that demand almost alone determines the rate and direction of technical change has not survived empirical scrutiny. The consensus, after dozens of case studies, is that the Marshallian scissors cuts with two blades. Perhaps the most persuasive refutation of Schmookler's proposition is offered by Parker (1972) and Rosenberg (1974), who document several important historical examples (e.g., the mechanization of hand operations in agriculture, the use of coal as an industrial fuel) in which the sequence of particular applications of a "generic" technological idea was determined not by demand, but by the state of knowledge and inherent technological complexity of particular industrial applications. Offering statistical evidence that both blades matter, Stoneman (1979) showed that the cost of, as well as the demand for inventions conditioned the level of innovative effort. Scherer (1982c) offered additional evidence, finding that dummy variables classifying industries by technology (chemical, electrical, mechanical, etc.) and variables representing demand conditions were statistically significant in a regression analysis of line of business patenting activity, but the technology variables explained considerably more variance. Employing aggregate time series data on an index of manufacturing output, and innovation and patent counts over the period 1948–1983 for the United Kingdom, Geroski and Walters (1995) provide evidence that demand matters when they find that output Granger causes both the innovation count and patents. They also conclude, however, that the effect of

[50] The early debate has been thoroughly reviewed by Mowery and Rosenberg (1979) and does not require detailed attention here.

aggregate demand on innovation is small relative to what they estimate to be stochastic determinants that they interpret as supply side shocks, pointing to a role for technological opportunity.

A particularly interesting perspective on the demand-pull/technology-push debate is offered by Walsh (1984), who combined the case-study approach with the time series methods of Schmookler. Walsh found that, in several chemical industries, the production series indeed leads the patent series, but growth in production also tends to follow one or several major innovations. An interpretation of this pattern is that relatively exogenous major innovation induces growth in demand, which in turn creates the incentive for subsequent incremental innovation. The suggestion that major technological innovations may induce changes in demand, obvious as it may seem to the historian, gives pause to the economist, who typically models tastes as given and immutable. Consistent with Walsh's argument, Kleinknecht and Verspagen's (1990) reanalysis of Schmookler's own data and analysis of more disaggregated industry-level Dutch data suggest mutual causation between demand conditions and innovative activity.

There are two principal respects in which interindustry differences in demand conditions might be expected to affect the incentives to engage in innovative activity. First, as Schmookler himself emphasized, there is the size of the market, which might be represented in static terms by a scale parameter and in dynamic terms by a rate of growth. The argument is straightforward. The (expected) investment required to produce a given reduction in unit cost or a given improvement in product quality is independent of the level of output that will be produced once the innovation is made. The benefits realized by such investment, however, are proportional to the size of the market in which the innovation is used. More inventive activity would therefore be expected in the larger of two markets, holding constant the cost of innovation; in two markets of equal size, more inventive activity would be expected in the market that is expected to grow more rapidly.

Second, Kamien and Schwartz (1970) suggested that the price elasticity of demand will also affect the marginal returns to investment in R&D. They demonstrated that the gains from reducing the cost of production (process innovation) are larger the more elastic is demand. On the other hand, Spence (1975) demonstrates that the gains from improvement in product quality (product innovation) will, under many circumstances, be larger the more inelastic is demand, since inelastic demand tends to magnify the gains from a rightward shift in the demand curve. Thus, the effect of price elasticity will be ambiguous in empirical studies that do not distinguish between process and product innovation.

The distinction between process and product innovation raises the subtle conceptual and operational question of how to characterize the demand conditions relevant to product innovation. In the case of intermediate products, such as those studied by Schmookler, the demand function for inputs of higher quality can in principle be derived from estimates of final product demand and the downstream production technology. It is more difficult to characterize and estimate the demand for consumer product innovation. A variety of econometric techniques can be used to estimate the demand for routine improvements in some measurable dimension of product quality (e.g., hedonic price models, Lancasterian demand models, and discrete choice models). Such techniques, though useful in particular applications, are unlikely to be fruitful in cross-industry analysis, since they require very detailed data and special modeling efforts for each specific product. A vastly more difficult problem is posed by major innovations that introduce an entirely new product (e.g., the television, the automobile). In such cases, there is no straightforward way to characterize latent demand from data on existing products, particularly if one acknowledges that tastes themselves may change as a consequence of a major innovation. A similar though less daunting challenge is also faced in the more common situation where firms add significantly new performance attributes to an existing product. Indeed, it is an interesting empirical question how firms actually assess consumers' willingness to pay or market size when the product or product attribute in question is entirely new to the market.[51]

In regression studies of R&D investment, demand conditions, although rarely featured, have often been considered. A variety of categorical variables have been used, presumably as proxies for interindustry differences in price elasticity. Most common are those distinguishing durables from nondurables, as well as those distinguishing consumer goods from material inputs or investment goods. Some researchers have used input–output data on the disposition of industry output (i.e., the shares of output destined for personal consumption, intermediate use, exports, the government sector, etc.). Although these categorical and input–output variables are sometimes statistically significant in regressions explaining a measure of innovative activity, there are no notably robust findings.

In an attempt to develop industry-level demand measures, Levin (1981) calculated, from a set of estimated constant elasticity demand functions for consumer goods and the input–output table, three demand parameters for each four-digit industry: a price elasticity, an income elasticity, and an exponential shift parameter. Although these parameters were significant as explanatory variables in simple regressions of R&D intensity on size and other industry characteristics (Cohen et al., 1987), their contamination with measurement error may have hampered their usefulness in the estimation of more complex specifications (Levin and Reiss, 1984, 1988).[52]

To capture market size and growth effects, sales and the rate of growth of sales are typically used, despite the obvious problem that these variables do not measure demand conditions, but the interaction of demand and supply. Moreover, since improved products will have larger markets, sales revenue and growth are endogenous with respect to innovation. Measures of market size that are defensibly exogenous and stand apart from supply conditions have just recently been incorporated into studies of innovation and R&D, though only for the pharmaceutical industry, by Acemoglu and Linn (2004) and Cerda (2007). The dependent variable of the former is the number of new drug approvals by the US FDA in broad categories distinguished by therapeutic effects, and the authors distinguish between nongeneric and generic drugs. The latter paper, in contrast, focuses exclusively on what are called "new molecular entities," which exclude generics. In both papers, the exogenous component of potential market size is derived from Current Population Survey data on demographic trends combined with differences in the population age profiles of expenditures and use for different types of drugs. Both studies find the effect of market size on product introductions to be economically important. Acemoglu and Linn (2004), for example, estimate that a 1% increase in market size is associated with a 4% increase in the introduction of new nongeneric drugs. They also find that current market size has the strongest effect generally, though 5-year leads of market size also strongly affect new molecular entities and generics. An important issue considered by Acemoglu and Linn is the need to control for possible time varying changes in technological opportunity specific to therapeutic classes. Acemoglu and Linn's introduction

[51] As pointed out in Urban et al. (1996), the question of how firms should assess market demand under such conditions also challenges the marketing literature.

[52] The quality of industry-level price elasticities did not inspire confidence. Mueller (1986) could not reject the hypothesis that Levin's price elasticities were uncorrelated with another set of estimates provided by Ornstein and Intriligator.

of a range of controls for technological opportunity has little effect on the magnitude of the effects of market size.[53]

As discussed above, Sutton (1998) highlights submarket homogeneity as another dimension of demand that conditions R&D spending. In his model, submarket heterogeneity, and the interaction between the nature of products and the buyer preferences that give rise to it, conditions the degree to which firm size (or, more properly, business unit size) offers an R&D cost-spreading advantage, and, thus, the degree to which firm size increases the incentive to conduct R&D. One might argue that submarket heterogeneity simply begs the question of how one defines a market, though, as discussed at length by Sutton, market definitions need to be empirically implementable. Moreover, product differentiation can be such that submarket heterogeneity is truly a matter of degree.

A limitation of economists' consideration of demand is the assumption, highlighted by Malerba (2007), that buyers are passive participants in the innovation process. This is far from true. Cohen et al.'s (2002b) survey research shows, for example, that, of all the sources of ideas for new R&D projects outside the R&D lab itself, including suppliers, rivals, university and government labs, or even a firm's own manufacturing operations, customers are far and away the most important. Von Hippel (1976, 1977, 1988) shows that in semiconductors and scientific instruments, some customers—what he calls "lead users"—may not only suggest ideas, but may themselves act on them, modifying firms' products, and, in von Hippel's view, anticipate or even shape the future direction of product innovation.[54] Economists would surely benefit from a more detailed understanding of the role of buyers in affecting the rate and direction of technical advance. Indeed, the notion that innovation may commonly be the outcome of an interaction between firms and their customers raises a number of questions. What is the impact of these interactions on expected demand? To what extent are firms' marketing and sales functions involved in these interactions? Bearing on this last point, and as noted above, it may be fruitful to try to begin to understand the relationship between firms' R&D investment and their investments in marketing and sales.

Finally, economists have not empirically analyzed the way in which demand dynamics might affect R&D incentives, particularly where demand is itself a function of the nature and diffusion of innovation. Obvious examples of such include products that lend themselves to network externalities of a direct sort as well as network externalities of a more indirect sort for which the development of complementary technologies are essential.

4.2. Technological opportunity

Much of the empirical literature takes for granted that technical advance, at prevailing input prices, is "easier" (less costly) in some industries than in others. Although it is widely accepted that industries differ in the opportunities they face for technical advance, there is no consensus on how to make the concept of technological opportunity precise and empirically operational. In the framework of the standard neoclassical theory of production, technological opportunity can be regarded as the set of

[53] Although Acemoglu and Linn (2004) find market size to significantly affect new product introductions, they find no effect of market size on the number of patents filed by their sample firms.

[54] Riggs and von Hippel (1994) report that 44% of scientific instruments used to study the chemistry of solid surfaces were developed by users. See von Hippel's discussion of user innovation in Chapter 9.

production possibilities for translating research resources into new techniques of production that employ conventional inputs. Some theoretical treatments have thus represented technological opportunity as one or more parameters in a production function relating research resources to increments in the stock of knowledge, with the stock of knowledge entering in turn as an argument, along with conventional inputs, in the production function for output (Griliches, 1979; Pakes and Schankerman, 1984). Related approaches treat technological opportunity as the elasticity of unit cost or unit price–cost margin with respect to R&D spending (Dasgupta and Stiglitz, 1980a; Spence, 1984; Sutton, 1998), as a shift parameter determining the location of an innovation possibility frontier representing the tradeoffs in the direction of technical change (Levin, 1978), and as a shift parameter determining the location of a frontier describing the tradeoff between the time and cost of an R&D project (Scherer, 1984b).

These formulations lend themselves in principle to direct econometric estimation, if only adequate data were available to identify the technological opportunity parameter(s) and other relevant parameters for each industry. Only Pakes and Schankerman (1984) have attempted this type of structural estimation. The panel data they used did not permit identification of the parameter representing technological opportunity or its contribution to the explanation of variance in R&D intensity. They were, however, able to identify the fraction of variance explained jointly by opportunity and appropriability, which they found to be substantial.

Perhaps the greatest empirical challenge associated with operationalizing technological opportunity as technical advance per unit of R&D effort is the measurement of the technical advance associated with changes in product quality, variety, and the introduction of altogether new products (cf. Griliches, 1979). Trajtenberg's (1989, 1990a,b) study of the introduction of computerized tomography scanners confronts similar challenges when it assesses the social welfare gains associated with a major product innovation. It may be fruitful to build on Trajtenberg's (1990b) method of using patents weighted by their citations in other patent applications to develop retrospective measures of the technological opportunity associated with significant product innovations (see Carpenter et al., 1981; Narin, 1983; Narin and Wolf, 1983).

Most attempts to represent technological opportunity as a determinant of innovative activity in regression studies have followed the practice introduced by Scherer (1965a), who classified industries on the basis of the scientific or technological field with which each was most closely associated. Scherer's initial classificatory scheme (chemical, electrical, mechanical) was refined in his subsequent work (Scherer, 1967a, 1982c), and variants have been used by numerous investigators. Although Scherer intended to capture interindustry differences in the vigor of advance of underlying scientific and technological knowledge, he recognized that statistical results obtained with the use of such crudely defined categorical variables would also likely reflect the influence of unspecified industry practices or demand effects not captured by other regressors. Nonetheless, the simple classification of industries into a small number of technology groups has powerful statistical consequences; it has explained a substantial fraction of variance in patenting activity (Scherer, 1965a, 1982c) and R&D intensity (Scott, 1984).

Several investigators have used proxy variables thought to be associated with technological opportunity to explain innovative activity. Shrieves (1978) performed factor analysis on the distribution of scientific and technological employees by field across 411 firms to develop several technology factors; these constructed variables fared poorly in a regression analysis of R&D expenditures. Jaffe (1986, 1988, 1989b) used data on the distribution of patents across patent classes to assign firms to 21 "technological opportunity clusters." The vector of cluster dummies was statistically significant in regressions explaining interfirm differences in R&D, patents, total factor productivity, profits, and Tobin's q. Jaffe typically found, however, that conventional industry dummy variables performed equally well. Moreover, except in the regression explaining R&D, it was difficult to distinguish between the effects of the clusters and industry dummies. Geroski (1990) observed a highly significant effect of technological opportunity using innovation counts in an earlier period as a proxy variable representing industry-level technological opportunity in a subsequent period.

In the optimization model of Levin and Reiss (1984), specific parameters of the cost function were interpreted as unobservable measures of technological opportunity and appropriability conditions. Each parameter was then formally treated as a function of observable variables. To represent technological opportunity, Levin and Reiss augmented a set of technology class dummy variables with measures of industry age (intended to capture the effects of technological life cycles), the fraction of R&D devoted to basic research (intended to capture an industry's "closeness" to science), and government R&D (intended to capture externally generated opportunities for privately funded R&D). Each of these variables was statistically significant in an equation for R&D intensity.

The survey of R&D executives in 130 manufacturing industries described by Levin et al. (1987) and Klevorick et al. (1995) measured several variables thought to represent an industry's technological opportunity. Among these are the contribution of various basic and applied sciences to each industry's technological advance and the contribution of external sources of technical knowledge, such as upstream suppliers of the industry's materials, production, and research equipment, downstream users of the industry's product, universities, government agencies and labs, professional and technical societies, and independent inventors. Although these survey variables, constructed from responses along a semantic scale, are contaminated with considerable measurement error, a number of them have performed well in regression studies of innovative activity. Levin et al. (1985), Cohen et al. (1987), and Cohen and Levinthal (1989) all found opportunity variables representing closeness to science and the sources of extraindustry knowledge to be jointly significant and to explain a substantial fraction of interindustry variance in R&D intensity.[55] The survey variables performed less well in estimates of the more structured optimization model of Levin and Reiss (1988), no doubt reflecting the shortcomings of the highly stylized model as well as the imprecision of the data. Although the Levin et al. survey-based measures of technological opportunity perform reasonably well in less structured empirical models, Geroski (1990) suggests that technological opportunity is perhaps best treated as an unobservable variable given the difficulty of constructing technological opportunity measures for samples spanning numerous industries.

For a fuller account of the role of technological opportunity, it is useful to consider the rich institutional and historical literature, as well as a few interesting theoretical conjectures. Consider first the role of science. Among economists, Rosenberg (1974) has argued most strongly for a close link between scientific and technological advance, offering a convincing account of why certain

[55] To the extent that the relevance of science and the contribution of extraindustry knowledge sources reflect an industry's technological opportunity, one would expect a positive relationship between these variables and innovative output. Greater opportunity, however, need not imply greater expenditure on R&D. Thus, although Levin et al. (1985) found that each measure of opportunity derived from the R&D survey had a positive coefficient in an equation to explain each industry's self-reported rate of innovation, Levin et al. (1985) and Cohen et al. (1987) found that an increase in the contribution to R&D of equipment suppliers reduced an industry's own R&D intensity.

technological innovations could not have occurred without certain foundational scientific advances. In a case study of the invention of the transistor, Nelson (1962) demonstrated that the contribution of science to invention is by no means simple. He explained, first, that the essential scientific knowledge required and utilized by the inventors of the transistor was in place more than 15 years before the invention. He also illustrated how scientific knowledge directed and structured the thinking of the Bell Lab's research team at various steps along the way to the ultimate discovery. Most remarkably, however, the invention of the device itself preceded and actually triggered the inquiry leading to a full scientific understanding of how it worked. Rosenberg (1982) and Kline and Rosenberg (1986) elaborate the point that technological developments and challenges may stimulate and focus basic scientific research, and highlight more generally the importance for upstream research of feedback from downstream research, development, and commercialization activity.

Rosenberg (1974) also suggested a simple mechanism by which the growth in scientific knowledge encourages innovation; he claimed that "as scientific knowledge grows, the cost of successfully undertaking any given, science-based invention declines..." (p. 107). Conceptualizing R&D as a stochastic search process, Evenson and Kislev (1976) and Nelson (1982a) similarly suggested that "strong" science affects the cost of innovation by increasing the productivity of applied research. Nelson in particular argued that a strong science base narrows the set of research options and focuses attention on the most productive approaches. The consequence is that the research process is more efficient. There is less trial and error; fewer approaches need to be evaluated and pursued to achieve a given technological end. From this perspective, the contribution of science is that it provides a powerful heuristic guiding the search process associated with technological change. Evenson and Kislev (1976) and Nelson (1982a) enrich this characterization of the role of science by suggesting that stronger science may also provide a larger pool of candidate approaches to achieving some technological objective, thereby increasing the expected payoff, although at the cost of broadening the search. Cohen and Klepper (1992b) offer a complementary view in which a more vital underlying science and technology may also increase the number of technological objectives to be pursued rather than simply increase the number of approaches to pursuing any given objective.

Although the effect of the advance of science on firms' and industries' incentives to invest in R&D surely constitutes an important dimension of an industry's technological opportunity, cross-industry empirical research on the subject is still undeveloped. Using the survey data described in Levin et al. (1987), Klevorick et al. (1995) examine across a broad range of manufacturing industries the general relevance of 11 fields of basic and applied science to each industry's technological progress, as well as the relevance of university-based research for those same fields. Klevorick et al. find that the reported relevance of science to technical advance varies considerably across industries and identify the industries to which science appears to be most relevant. For science in general, they found drugs and semiconductors closest to a field of science among the industries for which they received 10 or more responses. They found that university-based research to be much less relevant than science more generally except in the case of the agricultural and medical applications of the biological sciences.[56] While these survey results represent an important first step in examining an important dimension of

[56] Klevorick et al. (1995) accounted for this difference in the relevance of science in general versus that of each scientific field's university-based research by arguing that the former may also reflect the relevance of training in these fields and that recent scientific discoveries, most of which originate in universities, are less relevant in any immediate way.

technological opportunity, empirical economists need to build on these initial efforts and develop data—survey and other data—that are more readily interpretable and speak more directly to some of our theoretical notions of the role of science in promoting technical advance.

Highlighting another aspect of technological opportunity, the historical and case-study literatures also suggest that the development of technology may follow a course that is relatively independent of market influences. At any given time, innovative efforts within an industry or a complex of related industries tend to be concentrated on a limited number of distinct, identifiable problems. A breakthrough in one area typically generates new technical problems, creating imbalances that require further innovative effort to realize fully the benefits of the initial breakthrough. Rosenberg (1969) identifies this phenomenon as a "compulsive sequence," citing examples from the history of technology in the machine tool industry. The development of high speed steel, for instance, improved cutting tools and thus stimulated the development of sturdier, more adaptable machines to drive them. Similar "bottleneck-breakthrough" sequences have been described in nineteenth-century textile manufacture, iron and steel, and coal and steam technology (Landes, 1969), in twentieth-century petroleum refining (Enos, 1962), and in other technologies (Ayres, 1987).

A related phenomenon is the tendency for technologies to develop along what Nelson and Winter (1977) termed "natural trajectories."[57] The notion is that in certain instances technological development proceeds along a relatively clear path, as if moving toward some physical limit. Engineers do not move myopically from one bottleneck to the next; they repeatedly focus on a particular class of engineering problems, drawing upon and strengthening a familiar method of solution. A good example of a natural trajectory is the progressive extension of the range of output over which scale economies are attainable, which has been documented for electric power by Hughes (1971) and for several chemical industries by Levin (1977). For a period that lasted approximately 25 years in both cases, engineers understood that lower production costs were possible if they could solve the design problems associated with building bigger plants. Another example is the progressive miniaturization of semiconductor devices (e.g., Braun and Macdonald, 1982; Levin, 1982). In this instance, engineers have understood for almost five decades that a tighter packing of circuit elements would lead to higher speeds for performing logical or data storage operations, but a host of related technological problems—such as obtaining sufficiently pure materials and etching ever-finer lines in silicon—have required solution with each successive generation of devices.

Just as a strong science base narrows the set of approaches that a researcher must pursue to achieve a given technological objective, it might be argued that working within a particular technological regime narrows the set of objectives to be pursued, and hence the range of specific technological problems to be investigated. Linkages to science and natural trajectories can thus both be understood as ways of coping with, and reducing, the enormous uncertainty inherent in the complex decision problem of formulating an optimal R&D strategy.

The presence of identifiable "technological regimes" or trajectories in at least some industries suggests two potentially fruitful and complementary directions for empirical research. First, in such industries, the participants in the R&D process probably have a relatively clear idea about how to characterize technological opportunities and the constraints on technical advance. Thus, interview and questionnaire methods may be a particularly appropriate way to gather useful data. Second, where a

[57] The idea has been further developed by Sahal (1981) and Dosi (1982).

particular trajectory or other technological regime is present, careful modeling on an industry-specific basis may permit identification and estimation of the technological opportunity parameters that have proven elusive in cross-industry econometric work.

Although the case-study literature provides many examples of the evolution of the technologies, we know little about the degree to which phenomena such as natural trajectories, compulsive sequences, and other patterns are representative of the manufacturing sector as a whole. In one analysis of the subject spanning a broad number of industries, Klevorick et al. (1995) examine the degree to which firms' R&D units "consistently and repeatedly" pursue each of 11 different broadly defined classes of technological activity at a point in time—Nelson and Winter's (1977) "natural trajectories." Examples of these activities include changes in the scale of production, improvement of process yields, improvement in inputs materials, improvement in product characteristics, and design for market segments. With only a couple of exceptions, each of these 11 broad classes of technological activity was viewed as important by at least 30% of respondents to their survey. Leiponen and Drejer (2007) build on Pavitt's (1984) argument that industries' innovative activities can be characterized as supplier-dominated, scale-intensive, science-based, or specialized-supplier. Using CIS data for Denmark and Finland for the period, 1994–1996, Leiponen and Drejer (2007) find for the majority of industries that, across firms within industries, multiple technological activities tend to be pursued, with little evidence that any one type of activity tends to dominate innovation.

To the degree that "natural trajectories" and other patterns of technical advance are typical, we have limited understanding of the forces that spawn them. Theoretical work in the area of dynamic increasing returns (e.g., Arthur, 1989; Katz and Shapiro, 1985) offers some promise of illuminating the sources of some of these patterns, and, more generally, has deepened economists' appreciation for the role of history in affecting technical advance. As applied to technological change, a central idea of this work is that the development or use of some technologies may be subject to self-reinforcing, positive feedback cycles that, once set in motion by what may be considered small, random events, may become "locked-in" to a particular time path of development. In this framework, since there are multiple dynamic paths that may be followed, the particular path that emerges need not be socially optimal *ex post*, suggesting that Adam Smith's hidden hand does not necessarily work its magic in such settings. Economists' early research on this theme considered the emergence and impact of technical standards that permit the realization of external economies among users (e.g., a railroad gauge, a color television standard, a programming language). David (1985), for example, has provided an account of how the QWERTY typewriter keyboard became "locked-in" despite the presence, *ex post*, of the demonstrably superior alternative Dvorak keyboard, and Arthur (1990) and David (1988), among others, have provided additional examples.[58]

[58] Among others, Farrell and Saloner (1985, 1986) and Katz and Shapiro (1985, 1986) have developed theoretical models of dynamic increasing returns in which the choice of a Pareto-inferior standard is possible. Also, Cowan (1990) has illustrated the emergence of "lock-in" in the nuclear power reactor industry. Liebowitz and Margolis (1990), however, adopt a skeptical posture regarding "lock-in" due to network externalities when they employ the historical record to argue that the Dvorak keyboard was not a demonstrably superior standard to QWERTY. Spulber (2002) endorses this skepticism, suggesting that other examples of lock-in do not hold up to scrutiny, and that, "The challenge for historians of technology is to fully explore the complex effects of consumer choice and producer competition on technology standards" (p. 14).

In addition to external economies among users, other sources of increasing returns with important dynamic implications have been identified in the theoretical and empirical literatures. While some of these are associated exclusively with innovation, others involve innovation in an important way. Such sources include increasing returns due to learning and the development of expertise (cf. Arrow, 1962b; Cohen and Levinthal, 1994), the self-reinforcing positive externalities that accrue to geographic proximity of innovating firms, buyers, suppliers, and universities and other institutions (sometimes called "agglomeration economies"), and the increasing returns to innovation due to the nonrivalrous character of the application of innovations to output (cf. Romer, 1990). Although these sources of increasing returns have been recognized for some time, their implications for the patterns in the advance of technology over time within industries have not been considered until recently.

One intriguing suggestion regarding the origins of patterns of technical advance is that these patterns may have little to do with the character of a given technology or exogenous drivers of its advance, but may be driven by evolving firm economic incentives reflecting the evolution of the number and size distribution of firms within an industry. As discussed above, for example, Klepper (1996) suggests that in those industries where firms can enjoy R&D cost-spreading advantages to size, R&D incentives shift endogenously over time as firms grow large, away from product and more significant innovation to more incremental and process innovation. Thus, the evolutionary pattern in some industries of innovation becoming more incremental and process oriented may not reflect a depletion of opportunities, but the shifting of economic incentives as firms grow large.

Linked to the progress of underlying science and technology that may affect an industry's technological opportunity is the contribution of technical knowledge from sources external to the industry: suppliers, customers, universities, technical societies, government, and independent inventors. A voluminous institutional literature documents the contribution of such extraindustry sources to technological progress.[59] The case studies of Jewkes et al. (1958) contain instances of virtually every type of external influence. A notable example of institutional–empirical work on this subject is von Hippel's (1976, 1977, 1988) treatment of the contributions of users to technological development in a variety of industries, including scientific instruments and semiconductor process equipment. Klevorick et al. (1995) offer the first broad, cross-industry empirical examination of the contributions to technical advance made by extraindustry sources of embodied and disembodied knowledge. Not surprisingly, they found that what they call sources "within the industrial chain," such as buyers and materials and equipment suppliers, contribute more to most industries' technical advance than nonindustrial sources such as universities and government labs.

Bresnahan and Trajtenberg (1995) propose that a key source of technological opportunities that reflects the influence that innovation in one industry may have on downstream and complementary technologies are "general purpose technologies"—technologies possessed of "generality of purpose" that may enable and stimulate applications across a range of markets, and whose prospective

[59] To cite a few examples, Brock (1975) indicates that most of the computer industry's innovations could be traced to technological developments outside the industry. Peck (1962) makes the same point in his study of innovation in the aluminum industry. Mueller (1962) traces the origin of the majority of DuPont's product and process innovations over the period, 1920–1950, to extramural sources, including independent inventors.

applications typically elicit follow-on R&D, thus exhibiting "innovational complementarities."[60] In other words, there are technologies that, once developed, provide significant, and often pervasive technological opportunities for innovation in other industries. As elaborated by Lipsey et al. (1998), such GPT technologies are also dynamic in the sense that the technologies as well as their various applications lend themselves to improvement and revision over time, opening up yet further opportunities. Examples include the dynamo, the transistor, the internal combustion engine, the internet, and so on. Lipsey et al. (1998) provide a nuanced characterization of the origins of GPTs when they claim that, depending upon the particular GPT, it may be endogenous to the economic system (e.g., Newcomen's atmospheric engine), or exogenous to the economic system but endogenous either to science and/or to the political–military system (pp. 35–36). Given a GPT, one may consider systematically the nature and magnitude of its application innovations in downstream and related industries, and, in this sense, evaluate its contributions to technological opportunity.

An extensively studied extraindustry influence on technological opportunity is that of government. In numerous sectors, notably agriculture, aircraft, electronics, and medicine, government has contributed to reducing the private cost of innovation and has influenced the direction of industrial research by its own research, by its support of academic research, by subsidizing and sponsoring private sector research and by disseminating technological knowledge developed in its own labs and elsewhere.[61] The distribution of government expenditures on R&D across industries is highly skewed, especially in the United States, where industries supplying the military, and, more recently, universities conducting research in the life sciences, are the principal recipients of R&D support.[62] Although its direct role in creating and disseminating knowledge is substantial in some sectors, its indirect influence is also felt through a variety of other channels that have different impacts across industries. Most important is the impact of government demand on the rate and direction of innovation.[63]

Over the past 20 years, scholars have also studied intensively the effect of university research on industrial innovation. Since this area of study is reviewed in the chapter written by Foray and Lissoni in Chapter 6, we only selectively review the contributions. On the basis of the Levin et al. (1987) survey of appropriability and technological opportunity conditions in the US manufacturing sector, Nelson (1986) and Klevorick et al. (1995), for example, find university research to be an important source of

[60] Klevorick et al. (1995) highlight the same phenomenon when, in their discussion of the sources of technological opportunity represented by technological advances outside an industry, they state: "The creation of new general purpose components, for example, power sources or electronic components, quite often opens new technological opportunities in a variety of industries that use that kind of component. Thus, the light internal combustion engine made possible both a viable automobile design and machine powered flight..." (Klevorick et al., 1995, pp. 190–191).

[61] A good introduction to the role of government in the United States is the collection of case studies edited and summarized by Nelson (1982b). A particularly nice case study of the impact of government policy on the development of the computer industry is provided by Flamm (1988). For a survey of international differences in the contribution of government to technological development in the major OECD countries, see Nelson (1984). Also, Nelson (1993) offers an impressive collection of studies on the interactions between government, firms, and the range of other institutions affecting technical advance for 17 nations. A number of papers also find that government R&D and, particularly, government procurement expenditures, have had a significant impact on private R&D spending (e.g., Levin and Reiss, 1984; Levy and Terleckyj, 1983; Lichtenberg, 1987, 1988). See Steinmueller's review of the literature on technology policy in Chapter 28.

[62] See Chapter 29 for Mowery's review of the influence of defense spending on innovation.

[63] See the case studies of semiconductors, computers, and aircraft in the Nelson (1982b) volume. Also see Temin (1979), Grabowski and Vernon (1982), and Baily (1972) on pharmaceuticals, and Caves (1962) on civilian aircraft.

innovation in only selected industries, especially those associated with biology. In contrast, also relying upon survey data, but collected over a decade later, Cohen et al. (2002b) found the impact of public research (including university and government sources) research to be more pervasive.[64] Blumenthal et al. (1986), Jaffe (1989b), Adams (1990, 1993), Mansfield (1991), Acs et al. (1992), and Jaffe et al. (1993) have all found that university research has substantial effects on innovative activity and performance. For a selected number of largely R&D-intensive industries, Mansfield (1991) suggests that about one-tenth of the new products commercialized during period 1975–1985 could not have been developed without substantial delay in the absence of recent academic research. Using state-level patent, innovation, and patent citation data, respectively, Jaffe (1989b), Acs et al. (1992), and Jaffe et al. (1993) add an important dimension to the discussion by showing that the effects of university research on innovation increase with geographic proximity.[65] Jaffe (1989b) and Acs et al. (1992) also observe that the effect of university spillovers differ across industry groups. Contrary to Link and Rees' (1990) finding that small firms tend to exploit university research more than large firms, Cohen et al. (2002b) found that, with the notable exception of startups, larger firms benefit disproportionately from university research. Another finding of relevance to policy is that the main channels through which university research reaches industrial R&D labs are the traditional channels of open science, including publications and public meetings and conferences, as well as informal interaction and consulting (Agrawal and Henderson, 2002; Cohen et al., 2002b). Moreover, as reported in Cohen et al. (2002b), these channels are considerably more important than the licensing and formal cooperative efforts that have garnered the attention of policymakers in recent years. Adams et al. (2003) find a different result in their study of US federal R&D laboratory interactions with firms. Using survey and other data on 220 industrial R&D labs collected during the period, 1996–1997, Adams et al. (2003) find that cooperative research and development agreements (CRADAs) in particular dominate other channels—including personnel exchange, licensing, or citation-based measures of knowledge flows—in their influence on industrial R&D labs, as reflected in labs' patenting and R&D expenditures.[66]

Just as spillovers from extraindustry sources may augment a recipient firm's technological opportunity, so may spillovers within an industry reduce the own R&D required to achieve a given level of technical performance. As discussed below, spillovers may also yield complementarity effects by increasing recipient firms' R&D productivity, as well as by increasing firms' incentive to invest in

[64] It is unclear whether this difference in findings reflected an increase in the relevance of public research in the intervening 11 years between the two surveys, or, more likely, a difference in the way in which the questions and response scales were framed in the two surveys.

[65] Thompson and Kean (2005a), however, challenge the Jaffe et al. (1993) finding, suggesting that it was an artifact of employing an insufficiently fine level of technology classification, and not selecting control patents with sufficient technological similarity. For further discussion, see Henderson, Jaffe and Trajtenberg (2005)and Thompson and Kean (2005b).

[66] While much of the work considering university–industry relationships has focused on the effect of university research and training on industrial innovation, Dasgupta and David (1987, 1994) have suggested that we consider the effect of industrial innovation on university-based scientific research. They argue that as university-based science becomes more tied to industry, the profit incentive that drives industrial innovation may displace the incentives of priority and "communalism" (cf. Merton, 1962) that tend to motivate university-based researchers. In the process, the norms of public disclosure that govern basic scientific research and even the pursuit of basic research itself may be undermined, possibly to the long-run detriment of technological progress. Chapter 5 also reviews this argument and the growing literature that considers the impact of links with industry on academic research.

own R&D to develop what Cohen and Levinthal (1989, 1990) call the "absorptive capacity" to exploit them.[67] Within-industry spillovers, however, also reduce the incentive to engage in R&D, because a firm must share with its competitors the benefits of its investment. We defer further discussion of this incentive effect to the next section.

There have been several econometric attempts to measure the effects of both extraindustry and intraindustry spillovers on firms' productivity. Pursuing a method suggested by Griliches (1979), Jaffe (1986, 1988, 1989a) used data on the distribution of patents by patent class to measure the technological relatedness of every pair of firms in a sample of over 500 firms. For each firm, he constructed a "spillover pool," defined as the sum of all other firms' R&D weighted by the measure of relatedness. He found that the size of the spillover pool had a powerful positive effect on a firm's patents, R&D and total factor productivity. Examining the effects on the total factor productivity of firm in the chemical industry, Adams and Jaffe (1993) also found that R&D labs that are either geographically or technologically more distant have less of a spillover effect on the total factor productivity of plants in the chemical industry. Qualifying Jaffe's earlier results, Geroski (1991a) found that the productivity impact on technologically "neighboring" industries of knowledge spillovers not embodied in innovations is modest.

Bernstein (1988, 1989), Bernstein and Nadiri (1988, 1989), and Nadiri (1993) took a more direct approach to estimating the magnitude of spillover effects by including the R&D capital of other firms or industries in the cost function of the receiving firm or industry. They found evidence of large efficiency gains from both intraindustry and extraindustry spillovers.[68] On the basis of these and other studies, Griliches (1992), in a review of the econometric literature on R&D spillovers, concluded that, "R&D spillovers are present, their magnitude may be quite large, and social rates of return remain significantly above private rates" (p. 24). The more recent literature on R&D spillovers is reviewed in Chapter 24.

Thus, most of the work on technological opportunity over the past decade has focused on particular sources—notably that emerging from public research and other firms. With the exception of Klevorick et al. (1995), there is a paucity of work that probes the nature and sources of technological opportunity, or that operationalizes the concept and examines its role across industries, despite the fact that a number of prominent scholars consider technological opportunity to be not only a key determinant of R&D and technical advance, but foundational to the evolution of market structure and entry, as well as key to the links across market structure, entry, and innovation (e.g., Geroski, 1994; Sutton, 1998). Moreover, despite Griliches et al.'s (1991) creative—though ultimately unsuccessful—attempt to measure technological opportunity by trying to discern an effect of firm patent counts on market value (controlling for demand and R&D), little progress has been made on developing measures of technological opportunity

[67] Cohen and Levinthal (1989, 1990) and Rosenberg (1990) argue that the incentive to exploit extramural knowledge may explain why some firms invest in basic research even though it tends not to yield directly appropriable results. This argument implies that in industries where extramural knowledge is particularly important to innovation, the productivity of applied research and development may be an increasing function of a firm's basic research. Employing annual patent application counts as a measure of innovation for 14 pharmaceutical firms over the period 1973–1986, and the number of published scientific papers as a measure for each firm's basic scientific capability, Gambardella (1992) indeed finds that firms with greater basic scientific capabilities generate more innovations for a given level of R&D spending.

[68] Bernstein and Nadiri (1989) found elasticities of average cost with respect to intraindustry spillovers to be approximately -0.1 in machinery and instruments and approximately -0.2 in chemicals and petroleum. Most of their interindustry elasticities (Bernstein and Nadiri, 1988) fell in the range of -0.05 to -0.1.

beyond the technology and industry dummies first employed decades ago and the survey-based measures developed in the 1980's (Klevorick et al., 1995).

4.3. Appropriability

To the extent that new knowledge is transmitted at relatively low cost from its creator to prospective competitors, and particularly to the extent that such knowledge is embodied in new processes and products that may be copied or imitated at relatively low cost, appropriable rewards may be insufficient to justify innovative effort. Recognition of this problem of appropriability predates classical, let alone neoclassical, economics. Indeed, the notion that monopoly privileges were required to provide economic incentives for inventive activity were reflected in the public policy of fifteenth-century Venice (Kaufer, 1989) and also motivated the Statute of Monopolies, passed by the English Parliament in 1623 (Penrose, 1951). Later, the problem was explicitly recognized by the framers of the Constitution of the United States.[69]

In theory, patents solve the problem of imperfect appropriability; the exclusive right granted by society enhances the incentive to invent by sanctioning restriction of an invention's use. To the would-be inventor, the prospect of a patent represents the expectation of *ex post* market power that Schumpeter claimed was an essential spur to innovation. In fact, however, industries differ widely in the extent to which patents are effective. The evidence suggests that patents are featured as a means of protecting innovation in only a few industries, while, in most industries, firms tend to rely more heavily upon other means of appropriation. In some instances, imitation is costly despite the absence of strong patent protection. In others, investment in complementary capabilities such as marketing and manufacturing can facilitate appropriation when neither strong patents nor technical barriers to imitation are present. In this section, we first review the growing body of evidence on interindustry differences in appropriability conditions, with some discussion of interfirm differences as well. We then discuss the more limited evidence on how appropriability conditions affect innovative activity and performance.

In an early investigation that revealed substantial interfirm differences in patenting behavior, Scherer et al. (1959) suggested that the value of patent protection might differ across industries. The suggestion was pursued by Taylor and Silberston (1973), who examined the use and effectiveness of patents with a small sample of 27 firms in four British industries. They found that 60% of pharmaceutical R&D, 15% of chemical R&D, 5% of mechanical engineering R&D, and a negligible amount of electronics R&D, was dependent upon patent protection. Mansfield et al. (1981), using data on 48 product innovations, said that, across his respondents, 90% of pharmaceutical innovations and about 20% of chemical, electronics, and machinery innovations would not have been introduced without patents.

Mansfield (1986) provided more comprehensive evidence on the extent to which the value and effectiveness of patents differs across industries. Mansfield asked executives from a random sample of 100 firms from 12 (mostly two-digit) industries to estimate the proportion of inventions developed in 1981–1983 that would not have been developed in the absence of patent protection. Only

[69] In empowering Congress to grant "for limited times to authors and inventors the exclusive rights to their respective writings and discoveries," the express purpose of the framers was "to promote the progress of science and useful arts" (Article 1, Section 8).

pharmaceutical and chemical inventions emerged as substantially dependent on patents; respondents reported that 65% of pharmaceutical inventions and 30% of chemical inventions would not have been introduced without such protection. Patents were judged to be essential for 10–20% of commercially introduced inventions in three industries (petroleum, machinery, and metal products) and for less than 10% in the remaining seven industries (electrical equipment, instruments, primary metals, office equipment, motor vehicles, rubber, and textiles). The last four of these industries reported that patent protection was not essential for the introduction of any of their inventions during the period studied.[70]

Mansfield's findings were reinforced by the results of the Levin et al. (1987) survey, administered in 1983 to a sample of Fortune 1000 firms in 130 more narrowly defined lines of business. As a means of appropriating returns, product patents were regarded as highly effective (scoring six or more on a seven-point semantic scale) in only five industries—including drugs, organic chemicals, and pesticides—and as moderately effective (five to six on the scale) in about 20 other industries, primarily those producing chemical products or relatively uncomplicated mechanical equipment.[71] Only three industries, however, regarded process patents as even moderately effective.

Administered in 1994 to a more representative sample of firms that spanned the firm size distribution, the Cohen et al. (2000) survey confirms the basic finding of Levin et al. (1987) that patents are rarely featured in the manufacturing sector as a means of protection. Cohen et al. (2000) found that only in two industries, drugs and medical equipment, are patents, on average, reported to be effective for more than 50% of product innovations. And in only three industries, are patents reported to be effective for between 40% and 50% of the firms' product innovations.

Quantitative evidence on the costs and lags associated with imitation (cf. Mansfield, 1985) added some understanding about how patents help protect innovations—or not. Both Mansfield et al. (1981) and Levin et al. (1987) found that patents raise imitation cost substantially in the chemical and petroleum industries but only slightly in electronics. Moreover, Levin et al. identified several industries, concentrated in the aerospace and industrial machinery sectors, which reported very high imitation costs and imitation time lags despite very weak patent protection. In these instances, the relative complexity of the products presumably makes reverse engineering difficult even in the absence of patent protection.[72]

[70] Despite the relative inefficacy of patents outside the pharmaceutical and chemical industries, Mansfield (1986) found that all 12 of his sample industries patented at least half of their patentable inventions during the 1981–1983 period. This implies that the benefits of patenting exceed the cost in most cases, suggesting, consistent with the discussion below, that there may have been payoffs to patenting that Mansfield had not fully considered.

[71] Levin et al. (1987) suggested that the most probable explanation for the robust finding that patents are particularly effective in chemical industries is that comparatively clear standards can be applied to assess a chemical patent's validity and to defend against infringement. The uniqueness of a specific molecule is more easily demonstrated than the novelty of, for example, a new component of a complex electrical or mechanical system. Similarly, it is easy to determine whether an allegedly infringing molecule is physically identical to a patented molecule; it is more difficult to determine whether comparable components of two complex systems, in the language of the patent case law, "do the same work in substantially the same way."

[72] More than 85% of the industries covered by the Levin et al. (1987) survey reported that the cost of imitating an unpatented major innovation was at least 50% of the innovator's R&D cost. More than 40% of the responding industries indicated that imitation costs were in excess of 75% of innovation costs. Evidence that imitation (a noncooperative endeavor) is quite costly even in the absence of patent protection is reinforced by findings in the literature on technology transfer (a cooperative endeavor), where it has been found that firms must make substantial investments to utilize technology licensed from other firms, or even technology transferred from another plant operated by the same firm (see, e.g., the studies contained in Mansfield et al., 1982).

Though probed in somewhat different ways, both the Levin et al. (1987) and Cohen et al. (2000) surveys also considered why firms believe patents to be limited in their effectiveness. In the Levin et al. (1987) survey, the principal reason cited for the limited effectiveness of patents was that competitors can legally "invent around" patents, with concern over the disclosure that comes with patenting playing a small role. Cohen et al. (2000) probed the question in a slightly different way. Asking respondents to report the reasons why they did not patent the most recent invention that they decided not to patent, Cohen et al. (2000) found—like Levin et al. (1987)—"inventing around" to be very important. However, unlike in Levin et al., respondents were just as concerned over the information disclosed in the patent. It makes sense that these two reasons would be comparably cited given that concern over disclosure would be especially merited where inventing around was feasible.

The Levin et al. survey revealed that firms in many industries tend to regard mechanisms other than patents as quite effective in appropriating the returns from innovation. In contrast to the 4% of industries that regarded product patents as highly effective, 80% regarded investments in complementary sales and service efforts as highly effective in capturing competitive advantage from their R&D activities. In numerous lines of business outside the chemical and pharmaceutical industries, firms reported that the advantages of a head start and the ability to move quickly down the learning curve were more effective means of appropriation than patents. In their survey administered 11 years later in 1994, Cohen et al. (2000) also found that firms typically believed means other than patents to be most effective in protecting innovation. In contrast to the Levin et al. results, however, secrecy was ranked as the most effective mechanism in 17 of 34 three-digit SIC code-level manufacturing industries, and thus at least comparable in importance to lead time, which is ranked first in 13 industries. Interpreting the reports of the effectiveness of patents and the other appropriability mechanisms require, however, some care. For example, Cohen et al. (2000) show that these different mechanisms are not mutually exclusive in their use.[73] Thus, Cohen et al. (2000) interpret the different evaluations of effectiveness or strength as reflecting the centrality of the different means of protection to firms' appropriability strategies, and argue that such data do not transparently reflect the economic return to their use, as discussed below.

Controlling for differences in the firm size distributions between the Levin et al. (1987) and Cohen et al. (2000) samples, Cohen et al. conducted an intertemporal comparison of the responses for the subset of the manufacturing industries surveyed by both studies. A comparison of the rank order of the different mechanisms in the 33 "comparison industries" that have at least four observations in each survey showed secrecy ranked as the first or second most effective mechanism in 24 of the 33 industries in the Cohen et al. study versus zero in the Levin et al. study. Similarly, even though secrecy figured much more prominently in the protection of process than product innovation in the Levin et al. survey—ranking first or second in 12 of the 33 comparison industries—it is ranked first or second in 31 of the 33 comparison industries in the Cohen et al. study.

Perhaps reflecting the pro-patent change in the legal and policy environment between 1983 when the Levin et al. (1987) survey was administered, and 1994, when the Cohen et al. (2000) survey was

[73] For example, secrecy is invariably employed prior to the filing for a patent, and, often, prior to the disclosure of the patent application. Moreover, patents can often contribute to a lead time advantage. Finally, complementary capabilities may also enable firms to achieve a lead time advantage.

administered, the relative ranking of patents among the appropriability mechanisms increased modestly, from being ranked first or second for seven of the 33 comparison industries in the Levin et al. survey, to 12 times in the Cohen et al. survey.

Although representing the first comprehensive survey of appropriability conditions in the US manufacturing sector, the Levin et al. survey data are restricted to the largest firms in the manufacturing sector, and one might expect the evaluation of the effectiveness of different appropriability mechanisms to differ between large and small firms. With its more representative sample, Cohen et al. explored the partial correlations, controlling for industry, between the reported effectiveness of the different appropriability mechanisms and firm size. Surprisingly, the reported effectiveness of all the appropriability mechanisms except patents (i.e., complementary capabilities, lead time, and secrecy) was not significantly correlated with firm size. Patent effectiveness, in contrast, was positively correlated with business unit size ($r = 0.23$) and with overall firm size ($r = 0.18$), suggesting that larger firms believe patents to be more effective. A significant negative correlation of -0.23 between firm size and whether the cost of defending a patent motivated firms' decisions not to patent in turn suggests that larger firms may consider patents to be more effective due to superior access to legal resources.

In addition to differences in appropriability conditions over time and across firms of different sizes, appropriability conditions may also differ across regions. Though not comparable in their detail, Arundel's (2001) analysis of data from Europe's Community Innovation Surveys suggests similarities to the United States. Lead time advantage is the dominant mechanism by a considerable margin, and patents are clearly subordinate. A survey, essentially identical to that administered in the United States by Cohen et al. (2000), was administered in parallel in Japan by NISTEP under the direction of Akira Goto. As discussed in Cohen et al. (2002a), differences in responses were striking. First, the effectiveness of patents in protecting product innovations in Japan were at least comparable to that of the other major mechanisms, and equivalent to the most effective mechanism, lead time. Even for process innovation, patents were the most effective mechanism, behind the use of complementary manufacturing capabilities. What is also notable is that the absolute score on patents was close to that observed for the United States, perhaps suggesting that patents are not more effective in Japan, but that other mechanisms are less. If true, this in turn implies that appropriability in Japan may be less overall. Cohen et al.'s (2002a) comparison of imitation lags suggests this to be the case, with the imitation lags for new products and processes more compressed in Japan by about a third. This cross-national comparison underscores the point that appropriability conditions can differ across regions for the same technologies and markets. For our purpose, an important question is the impact of these differences in appropriability conditions on innovation and technical advance. Indeed, despite firms' apparent lower ability to appropriate their profits due to innovation in Japan, during this same period, the R&D intensity of the Japanese manufacturing sector was higher. How can we reconcile this observation with the notion that appropriability plays an important role in stimulating R&D? We return to this question below.

Despite the evidence on interindustry as well as cross-national differences in appropriability conditions, there is no clear empirical consensus about whether greater appropriability encourages innovative activity. The simplest hypothesis, derived from the standard argument supporting the patent system, is that innovative activity will increase monotonically with appropriability because spillovers create a disincentive to innovative effort. By this argument, the more effective are the means of appropriation, or the less extensive are intraindustry spillovers, the greater will be industry R&D investment. However, when the "efficiency effect" of spillovers is considered (reflected by less duplicative R&D), some simple models (e.g., Spence, 1984) predict that although industry R&D intensity will rise with appropriability (fall with spillovers), industries' innovative output may decline (increase with spillovers).

In the model of Cohen and Levinthal (1989) that builds on Spence's earlier formulation, the simple "disincentive effect" of spillovers remains, but there is an offsetting incentive to invest in "absorptive capacity"-building R&D to make use of them, suggesting that R&D spillovers are not as much of a public good as previously supposed. Levin and Reiss (1988) have suggested yet another countervailing incentive effect. To the extent that own and rival R&D are different but complementary, the knowledge produced by a firm's competitor may raise the marginal product of own R&D. Thus, the same knowledge flows that can diminish appropriability can increase the productivity of other firms' R&D and offer a positive incentive to invest in R&D (cf. Cassiman and Veugelers, 2002). In either instance, an increase in spillovers (decrease in appropriability) has an ambiguous effect on industry R&D. If we broadly consider the Levin and Reiss and Cohen and Levinthal models as together positing a complementarity effect of R&D spillovers, then a key question for understanding R&D incentives and innovation at the industry level is, in addition to considering the efficiency effect of spillovers identified by Spence, what factors condition the tradeoff between spillovers' negative appropriability incentive effect and their positive complementarity effects.

The empirical findings to date do not establish whether the net effect of appropriability on R&D incentives is positive or negative, nor do we yet know the extent to which the net effect varies across industries. Although Bernstein and Nadiri (1989) found that intraindustry spillovers have a negative effect on R&D in each of four US industries, Bernstein (1988) found a positive effect in three R&D-intensive industries in Canada. Levin et al. (1985) and Levin (1988) found that various survey-based measures of appropriability were individually and jointly insignificant in regressions that explain R&D intensity at the industry level. Using business unit data, however, Cohen et al. (1987) found some of these measures to have positive and significant effects on R&D intensity in pooled regressions, although the results were not robust across separate two-digit industry regressions. For example, they found a negative effect of appropriability within the electrical equipment sector, a result that Cohen and Levinthal (1989) replicated and interpreted as possibly reflecting a high payoff to investment in absorptive capacity. A fuller understanding of the empirical consequences of imperfect appropriability will require tests that distinguish more sharply among the various mechanisms by which spillovers affect the incentives for R&D directed toward innovation, investment in underlying technological capabilities as well as imitation.

Studies on the overall effects of appropriability have, however, stalled over the past two decades as economists became more concerned with the narrower (though important) question of the effects on R&D and innovation of patents in particular.[74] One reason for this interest was the pro-patent shift in policy and the courts witnessed in the United States since the early 1980s (cf. Jaffe, 2000; National Research Council (NRC) (2004)). Economists have argued that the impact of patents—and stronger patents—on innovation is not as straightforward as popularly believed. Also, as noted in Arora et al. (2008), there are both theoretical and empirical reasons to question whether patents stimulate R&D. For example, although the expectation of monopoly rents should induce inventive effort, the prospect of patent disclosures can offset the private gains from patenting (cf. Horstmann et al., 1985). Also, "stronger" patents mean that, not only any given firm's patents, but also those of its rivals are stronger (cf. Gallini, 2002; Jaffe, 2000), possibly dissuading innovation. Merges and Nelson (1990), Scotchmer (1991), and subsequently Bessen and Maskin (2009) argue that, where technologies progress cumulatively and patents are broad, the licensing

[74] See Rockett's review of the theoretical literature on intellectual property rights in Chapter 7.

or enforcement decisions of upstream inventors may retard downstream innovation. Heller and Eisenberg (1998), Shapiro (2000), and Hunt (2006) also argue that, in some instances, the fragmentation or proliferation across different stakeholders of potentially overlapping patent rights for a given technology can raise transactions costs sufficiently to retard or even block otherwise worthwhile innovation.[75]

In light of both the Levin et al. (1987) and Cohen et al. (2000) survey findings that patents are among the least "effective" of the means firms use to protect their innovations, one might question why firms patent as much as they do, especially in those industries where patents were not reported to be especially effective such as semiconductors or communications equipment, where the R&D-weighted average patent propensities for product innovations—defined as the percentage of product innovations patented[76]—were 49% and 59%, respectively.[77] More generally, Cohen et al. found that, for their sample of manufacturing firms, the R&D-weighted product and process patent propensities overall were 49% and 31%, respectively, with 30% of their respondents reporting never having patented in the prior 3 years.[78] The question of why firms patent is reinforced by the threefold growth in patents issued, and a growth of about 50% in patents per R&D dollar between the early 1980s and late 1990s considered by Kortum and Lerner (1998) and Hall (2004). Moreover, the preponderance of this growth of patenting occurred in industries such as semiconductors and telecommunications equipment, where patents are reported to be less effective than in, say, drugs or medical devices (Hicks et al., 2001).

Addressing the question of why firms patent to begin with, the Cohen et al. (2000) report that, for product innovations, 96% of their survey respondents said they patented to prevent copying. Arguing that this particular response may reflect a social desirability response bias (i.e., firms offer this response because they think this is what patents are supposed to do), the authors expressed greater interest in the next most important reason—blocking other firms from patenting or commercializing their innovations. Probing this reason, and one of the other more common responses—to use patents for negotiations—the authors uncovered systematic differences in the reasons for patenting across industries in the US manufacturing sector. To understand the source of these differences, the authors distinguished what they called "complex product industries," where commercializable products are comprised of large numbers (e.g., hundreds) of patentable elements, from "discrete product industries," where products are comprised of relatively few patentable elements—perhaps only a handful or even one. In the former, they argue that any one firm would be unlikely to amass all the patents they needed to commercialize a product, and thus—consistent with Grindley and Teece (1997) and Levin (1982)—is more likely to patent strategically to gain bargaining leverage in cross-licensing negotiations to access rivals' patents,

[75] Encaoua et al.'s (2006) review of the theoretical literature concludes that patents "often contribute to enhancing incentives to invent, to disclosing and trading technology, but they also generate costs to society in terms of monopoly rents and barriers to access and use of knowledge."

[76] This definition of patent propensity differs from much of the literature that defines patents propensity as the number of patents per R&D dollar.

[77] Arundel and Kabla (1998) report the similarly defined but sales-weighted average patent propensities for 604 large industrial firms in Europe were 36% and 25% for product and process innovations, respectively, ranging from 8% in textiles to 79% in pharmaceuticals.

[78] The drug and medical device industries patented their product innovations intensively, with patent propensities of 96% and 67%, respectively. The industries patenting new processes most intensively included petroleum products and chemicals, both of which patented just over 60% of their process innovations. In the Cohen et al. (2000) sample, product patent propensities ranged from an average of 3% in metals to 96% in pharmaceuticals.

or to achieve freedom of operation and design. Hall and Ziedonis (2001) also conclude from both field study and empirical analysis that in the semiconductor industry—a complex product industry—such strategic interaction likely spawned patent portfolio races and the consequent rapid rise in industry patenting. After coarsely distinguishing between "complex" and "discrete" product industries, Cohen et al. (2000) found that the preponderance of firms in the former indeed patented to use those patents in negotiations such as cross-licensing, while in "discrete product" industries, where firms typically have less need to access rivals' technologies to develop a product, they tend to patent to protect against rivals' development of copies or substitute products.[79]

A clear conclusion that emerges from these studies is that firms commonly patent for reasons other than simply to protect the commercialization of a product or process or enable a licensing transaction. For our purpose, the key question is the effect of patents on R&D—even where patents are used for purposes of cross-licensing, freedom of operation and design, etc. The antecedent question, however, is whether the patenting of inventions indeed yields a return, presumably by increasing the value of whatever inventions that are patented. The work of Pakes, Schankerman, Lanjouw, and colleagues in their examinations of European firms' patent renewal decisions (see, e.g., Deng, 2007; Lanjouw, 1998; Pakes, 1986; Schankerman, 1998; Schankerman and Pakes, 1986), suggest, for Europe, that patent protection does yield a return, sometimes substantial, with estimates of the implied subsidy to R&D varying between 10% (Lanjouw, 1998) and 25% (Schankerman, 1998).

Using the Carnegie Mellon Survey data for the US manufacturing sector, Arora et al. (2008) estimate the "patent premium"—the proportional increment to the value of an invention conferred by patenting it—by estimating a structural model linking a firm's R&D effort with its decision to patent, recognizing that R&D and patenting affect one another and are both driven by many of the same factors. Their results indicate that, even though most innovations are not worth patenting, patents are valuable for a subset of innovations and thus do provide incentives for R&D. Moreover, conditional on patenting, the expected premium is substantial, with firms earning an average 50% premium, with this conditional premium ranging from an average of 60% in the health related industries to about 40% in electronics. Employing the estimated coefficients, the authors also simulate the effect of changing the premium on R&D. They find that an increase in the mean of the patent premium distribution for a typical manufacturing firm in their sample would significantly stimulate R&D, especially in industries where the patent premium is high, such as drugs, biotech, and medical instruments. But, even in industries where the patent premium is lower and firms rely more heavily upon means other than patents to protect their inventions, such as electronics and semiconductors, the estimates imply that patents stimulate R&D, though less so.

Thus, consistent with the descriptive findings of Levin et al. (1987) and Cohen et al. (2000), Arora et al. (2008) find that patents are not as central to the protection of inventions as other mechanisms except in a few, selected industries; their estimates confirm that in most industries, patenting the *typical* innovation is indeed not profitable. However, even in these industries, some innovations are profitable

[79] Consistent with the intuition that this difference in the way patents are used is driven by the number of patents per product, Cohen et al. (2002a) found that in Japan—where it takes many more patents to protect a product given fewer claims per patents and a narrower judicial interpretation of those claims—all firms used patents similarly to the way firms in US complex industries used them. Ordover (1991) provides insight into the different roles that patents play in the United States versus Japan.

to patent, thus explaining why firms may patent even though they report patents to be less effective than other appropriability mechanisms.[80]

The Arora et al. (2008) study offers several methodological implications. Although the survey research studies discussed above indicate that firms in most industries do not feature patents among their various means of protection, those studies should not be interpreted as showing that patent protection either does not add to the value of the underlying inventions or provides no incentive for R&D. The Arora et al. (2008) analysis also shows how self-reported and often subjective measures from a survey may be used in structural modeling and estimation to generate quantitative estimates of returns and impacts on behavior. Indeed, the estimated patent premia, as well as the elasticity of output with respect to R&D were well within the estimates of the prior literature (e.g., Pakes and Griliches, 1984; Schankerman, 1998). The embedding of such survey responses in a rigorous, structural model also imposes a discipline that offers a more grounded, coherent interpretation of the underlying survey data than that sometimes drawn directly from descriptive analyses of the same.

An important feature and limitation of both the patent renewal and Arora et al. (2008) studies is their focus on the direct, private returns to patenting. By restricting R&D spillovers or potentially creating complex thickets, patents may, however, produce aggregate, more system-wide effects not easily discerned by simulating the responses of individual firms to changes in the strength of their own patents. Indeed, empirical studies of the relationship between patenting and innovation at the aggregate levels of nations or industries have provided rather more ambiguous results about the impact of patents on innovation, although this ambiguity may be partly due to difficulties in controlling for the endogeneity of patent policy or the joint determination of R&D and patenting.[81] Thus, neither the results of the patent renewal studies nor those of Arora et al. (2008) imply that patents necessarily yield a net social welfare benefit overall.

Another question not addressed by any of the studies of the impact of patenting is, given the availability of other means of protection, what would happen to R&D spending if the option of patenting were removed. Providing historical insight, Moser's (2005) analysis of the invention records associated with two World's Fairs in the second half of the nineteenth century shows that, in countries without patent laws,

[80] Building on the model developed in Schankerman and Pakes (1986), Schankerman (1998) also provides econometric evidence that patents are typically less effective than other mechanisms in appropriating the rents due to innovation. In an analysis of French patent renewal data for the period 1969–1972, Schankerman concludes that while patents may be a significant source of returns to innovative effort, they are not the major one.

[81] As summarized in Arora et al. (2008), the more aggregate studies analyzing the impact of patents on innovation and growth have yielded mixed and, at times, difficult-to-interpret results. Most studies using aggregate cross-national data find a positive and significant effect (e.g., Kanwar and Evenson, 2003; Lederman and Maloney, 2003; Park and Ginarte, 1997). A limitation of most of these studies, however, is that policy may be endogenous with respect to innovation. Lerner (2002) employs an instrumental variables approach to address this endogeneity in his examination of the impact of 177 policy changes on innovation over a 150-year period and across 60 countries. He finds, however, that strengthening patent protection appears to have few positive effects on the patent applications by domestic entities in the country undertaking the policy change.

In their general equilibrium model of the impact of R&D, innovation, and diffusion, Eaton and Kortum (1999) consider, among other questions, the impact of patents on R&D and growth. Estimating key parameters, and relying upon the literature to specify others (notably the difference in imitation rates for patented vs. unpatented innovations), they conclude that eliminating patent protection would reduce R&D and economic growth. Like Arora et al. (2008), and in contrast to other empirical studies of patent protection and R&D, Eaton and Kortum model the patenting and R&D decisions as simultaneously determined, with the value of the invention and the strength of patent protection conditioning both.

inventors tended to focus their effort on technologies where other means of protection were available. This finding suggests that, if patents were eliminated, firms would not necessarily continue inventing in the same markets using alternative means of protection, but may redirect their innovative efforts.

In addition to studies that consider the impact of patenting on the returns to R&D, a small number of studies attempt to examine the effect of patenting on R&D directly. A few empirical studies have considered the effect of patent strength or policies on R&D at the firm level. In one, Sakakibara and Branstetter (2001) exploit the 1988 change in Japanese patent policy from a policy of one claim per patent to one allowing multiple claims per patent. Interpreting this as an increase in patent strength, Sakakibara and Branstetter find only a small positive effect using a reduced-form model estimated with a panel data set of Japanese firms. Contrary to Bessen and Maskin's (2009) conjecture that patent protection offered no inducement for R&D or innovation in software in the 1980s and 1990s, Lerner and Zhu (2007) find that software firms' increased reliance on patents due to a reduction of software copyright protection in the early 1990s was associated with higher R&D investments.

One key lesson of the Levin et al. (1987) and Cohen et al. (2000) surveys is that firms use a range of appropriability mechanisms—and patents are not the central means of protection in most US manufacturing industries. Economists have not, however, devoted the energy to the empirical study of the use of the other mechanisms as they have to patents, nor, in their analyses of the effects of patents, have they explicitly considered patents as just one of several possible mechanisms in any given industry. A likely reason for this inattention is the difficulty of finding suitable data and formulating precise tests to distinguish among competing hypotheses concerning the deployment of the different means to achieve appropriability. An important exception is provided by Teece's (1986) qualitative framework that characterizes conditions under which firms might employ different means to achieve appropriability, highlighting that such conditions and the consequent strategies can differ importantly, not only across industries, but also across firms within industries. Teece argues that two conditions will affect firms' abilities to profit from their innovations: (1) whether they possess or can readily acquire or develop the complementary manufacturing, marketing, sales, and other capabilities that commercialization requires and (2) the "appropriability regime," reflecting whether a new product or process can be readily imitated, which, in turn, is a function of whether patents are effective in protecting a particular technology, or whether a new product or process is simply difficult to copy due to its complexity, the tacit quality of the underlying knowledge, etc.

Teece's framework has been embraced by scholars of innovation, particularly those concerned with management. For economists, the framework offers several implications. First, it suggests that appropriability conditions have a firm-level component since appropriability depends upon the distribution of hard-to-acquire complementary capabilities. Second, this framework provides another rationale for a relationship between firm size and R&D; firms should invest more in R&D to the extent they possess complementary capabilities that are essential for commercialization. Teece's framework also provides insight into why innovating firms may not end up successfully commercializing their own innovations. If property rights are weak and an innovating firm does not possess or cannot acquire the requisite complementary capabilities to commercialize an innovation, other more capable firms may be better positioned to do so.

Although Teece and others provide numerous qualitative accounts that are consistent with this framework, little broad-based econometric work has directly tested the implications of Teece's framework for R&D spending, as noted above in Section 3. Gans et al. (2002), however, test an extension of Teece's framework. They argue that when technology markets work due to strong patents and the possession of complementary capabilities are consequently less essential to appropriation, incumbents are at lower risk of being unseated by successful startups because the latter can monetize their innovations through licensing or being acquired rather than through entry. Employing a sample of 118 startup projects, they find that those startups possessing patents are 23% more likely to either license or become acquired than to enter a market. Arora and Ceccagnoli (2006) test yet another extension of Teece's framework. Using the Carnegie Mellon Survey data, they find that, as patent effectiveness increases, firms lacking complementary capabilities are more likely to license, while firms possessing those capabilities become less likely to license. The latter case shows that, even as stronger patents may make licensing more viable, they also increase a capable firm's return to commercializing an innovation itself all the more.

Although Cohen et al. (2000) and Arundel (2001) highlight the importance of secrecy to the appropriability strategies of firms in, respectively, the United States and Europe, there has been little study of the implications of firms' use of secrecy for R&D incentives. Of all the appropriability mechanisms, secrecy entails the clearest suppression of knowledge flows and thus its use may entail the sharpest tradeoff between the appropriability incentive effect on R&D versus the complementarity and efficiency benefits of spillovers, pitting the private incentives of firms most clearly against the innovative performance of an industry as a whole. Indeed, in an exploratory study at the industry level, Cohen and Walsh (2000) find that the only appropriability mechanism associated with both a reduction of industry-level knowledge flows and an increase in its associated appropriability incentive effect is secrecy.

Gilson's (1999) discussion of the unenforceability of noncompete agreements in California is consistent with the idea that secrecy may not only suppress knowledge flows—those associated specifically with labor mobility—but that suppression may undercut innovation, at least at the regional level. He conjectures that the vitality of Silicon Valley reflected in venture creation and innovation is partly a function of the longstanding unenforceability of noncompete agreements in California, which increases interfirm mobility of personnel and, in turn, the kind of knowledge sharing that may stimulate innovative performance. Testing the premise of Gilson's conjecture, Fallick et al. (2006) use data from the Current Population Survey to measure the rate of labor mobility across firms in Silicon Valley and elsewhere. Largely consistent with Gilson's conjecture, they discern a positive "California effect" on mobility relative to other states, though they cannot rule out sources other than employee noncompete statutes.[82] Marx et al. (2009) are better able to discern the effect of such noncompete agreements on the mobility of inventors listed on patents by taking advantage of a "natural experiment" in Michigan where the state legislature inadvertently reversed the state's noncompete enforcement policy in 1985. They indeed found noncompetes attenuate mobility, especially in very specialized technical areas.

Although Fallick et al.'s and Marx et al.'s results largely substantiate the first element of Gilson's argument, there is limited data showing a tie between such mobility and venture creation and innovation within an industry. Stuart and Sorensen (2003) do observe more startups, however, in states that do not enforce noncompete agreements. That does not, however, establish a clear tie between innovation and noncompetes. Indeed, there is the possibility that in areas with enforceable employee noncompete agreements such as Massachusetts, firms may invest more in innovation due to greater appropriability offsetting the otherwise dampened complementarity effects of the spillovers associated with labor mobility.

[82] Moreover, this effect seems to be specific to the computer industry, suggesting that the effect is not one strictly of location, but an interaction between industry and location.

Thus, the study of the use of secrecy has only begun, but is quite important, not only to help us understand the determinants of innovative activity and performance, but also for policy. Policy discussions on the strength of patents, for example, should proceed in light of firms' other options for protecting their innovations. To illustrate, a weakening of patents may induce firms to rely more heavily on secrecy, yielding fewer R&D spillovers, not more.

Moreover, the examination of secrecy, and particularly its expression in the form of noncompetes and their impact on personnel flows, highlight a broader point regarding the study of appropriability and R&D spillovers. While useful to have measures of the effectiveness and use of different appropriability mechanisms, the study of appropriability conditions and their impact on innovation also require that we develop better measures of spillovers that reflect flows of knowledge that are sometimes embodied in technology, sometimes disembodied, and sometimes contained in the heads of mobile personnel.[83] Moreover, when considering the impacts of such knowledge flows, one needs to be attentive to the associated tradeoffs for R&D and innovation between the appropriability incentive effects of such flows, on the one hand, and their complementarity and efficiency effects, on the other.

One of the most important reasons to know which appropriability mechanisms tend to be used within industries is their indirect effects on R&D incentives. As suggested in this review, different mechanisms can differentially affect the nature and extent of knowledge flows within an industry, the way in which inventions are monetized, the viability of innovation-based entry, the relationship between market structure and innovation, and, in turn, the rate and direction of inventive activity within an industry (cf. Arora and Ceccagnoli, 2006; Cohen and Klepper, 1996b; Gans and Stern, 2003; Gans et al., 2002, 2008; Teece 1987 among others). As just discussed, consider that more pervasive use of secrecy, rather than, say, patents tends to dampen knowledge flows, possibly diminishing the contribution of R&D spillovers to technical advance. Or consider that in industries where appropriability is best achieved through the use of complementary capabilities rather than patents, larger firms will tend to have an advantage in profiting from innovation, conferring a self-reinforcing advantage on size. Possible consequences include the squeezing out of less capable rivals as well as the discouragement of innovation-based entry, with dampening effects on the technological diversity and technical advance that may accompany entry (cf. Teece, 1986). We also know from Gans et al. (2008) and Arora and Ceccagnoli (2006) that the effectiveness of patents can importantly affect whether smaller firms tend to sell their innovations in disembodied form (as opposed to exploiting them through their own output). This, in turn, will affect the advantages to firm size in R&D (Cohen and Klepper, 1996b)[84] and, in turn, the nature of technical advance (Cohen and Klepper, 1996a). Attention to the effects of the dominance of different means of protection on innovation begs the question, however, of what factors determine

[83] Supporting the point that important knowledge that spills out is not necessarily detailed knowledge of a tangible product or process, Levin et al. (1987) report that conversations with R&D managers suggest that they find it very valuable to know what technical problem a competitor is trying to solve, what technical approach has been adopted or what approach has succeeded. This suggests that the problem of appropriability is not limited to protecting successful innovations. Knowledge that a project has failed may save a competitor money or help a competitor succeed.

[84] Per the discussion above, where patents are not effective, firms may rely upon secrecy or first-mover advantages, which require the embodiment of their innovations in their own output.

which of these mechanisms tend to be favored for a given technology or industry.[85] Also, a challenge in assessing the determinants and effects of different appropriability mechanisms is the absence of data on their use, with the exception of patents.

Although most of the literature has focused on how appropriability conditions within a single industry affect innovative activity within that industry, von Hippel (1982, 1988) suggests that the appropriability of profits due to innovation across vertically related industries affects the locus of innovative effort. In an attempt to specify the conditions under which, for example, process machinery is developed by machinery manufacturers rather than users of the machinery, von Hippel emphasizes considerations such as the extent to which new knowledge is embodied in the machinery, the relative efficacy of patents or secrecy, whether the machinery is used in one industry or many, the market structures of the manufacturing and using industries, and, more recently, the degree to which information required by problem solvers is "sticky"—that is, specific to the organizational setting in which it was developed (von Hippel, 1994). These factors, hypothesized to determine the locus of innovation between vertically related industries, may also affect the amount of innovation.[86]

5. Conclusion

Since the first version of this survey, coauthored with Richard Levin, appeared over 20 years ago (Cohen and Levin, 1989), there has been enormous growth in economists' writing on the economics of technological change and innovation. Indeed, the fact that the subject has warranted three entire handbook volumes including the present one (Fagerberg et al., 2005; Stoneman, 1995), and not simply the two chapters[87] that appeared in the 1989 Handbook of Industrial Organization (Schmalensee and Willig, 1989), makes the point. Yet, the progress in advancing our empirical understanding of the subject of this review—the determination of firms' and industries' innovative activity and performance—has been uneven. This review attempts to highlight important and robust empirical findings on the determination of R&D, and, to a lesser extent, R&D performance. We have divided our consideration across three broad areas of inquiry: (1) the Schumpeterian hypotheses relating innovation to market structure and firm size; (2) the role of firm characteristics; and (3) the role of industry-level variables broadly characterized as reflecting demand, technological opportunity, and demand conditions.

Our review of the two Schumpeterian hypotheses highlights one longstanding, robust finding: a monotonic relationship between firm size and R&D. There is also reason to believe that the source of

[85] Levin et al. (1987) offers a cogent argument why patents were particularly effective in pharmaceuticals—namely that there is typically a one-to-one mapping between chemical structure and the action of a given drug that makes inventing around difficult.

[86] Although these issues have not yet been thoroughly explored in the econometric literature, Farber (1981) introduced and found some support for the hypothesis that concentration on the buyer's side of the market influences R&D spending on the seller's side. Suggesting additional factors affecting the locus of R&D activity across vertically related industries, Harhoff (1996) argued that suppliers with considerable monopoly power, such as aluminum manufacturers in the past, may have an incentive to do R&D on downstream applications for their products in order to increase the demand for their output by both increasing the demand for the downstream products and making the downstream industry more competitive.

[87] In addition to the Cohen and Levin (1989) chapter on the empirical literature on R&D, Reinganum (1989) reviewed the theoretical literature.

this relationship is the R&D cost-spreading incentive effects of firm size. This cost-spreading effect of size itself, however, reflects underlying appropriability conditions that confine firms to exploiting their innovations in their own output, typically limited growth due to innovation, and market segmentation. The relationship between market structure and R&D remains, however, problematic. No clear theoretical rationale for the relationship has emerged. And empirical results remain weak, with measures—albeit poor ones—explaining little of the variance in R&D spending. More importantly, the work of Scherer (1967a), Scott (1984), Levin et al. (1985), Sutton (1998), and others suggests it likely reflects the influence of other more fundamental determinants of technical advance, specifically technological opportunity and appropriability conditions, as well as the degree of market segmentation. Thus, while the Schumpeterian tradeoff may apply to firm size (to the degree that size is tied to market power), whether it in fact applies to market structure (controlling for firm size) is less apparent. We need to develop a deeper understanding, however, of the ways in which industry-level factors might condition the relationship between market structure and innovation.

We have documented a movement, beginning about 50 years ago, away from a preoccupation with the roles of firm size and market concentration to embrace a broader research agenda laid out by Schmookler, Arrow, Nelson, Griliches, Rosenberg, Mansfield, Scherer, and other pioneers to consider what Cohen and Levin (1989) characterized as more "fundamental" determinants of industrial R&D, including demand, appropriability, technological opportunity, and key firm characteristics. It is surely a question of judgment whether a determinant of innovative activity is fundamental. Although, for example, tastes, technological opportunity, and appropriability conditions themselves are subject to change over time in response to some radical innovations that alter the technological regime, these conditions may be reasonably assumed to determine interindustry differences in innovative activity over relatively long periods. It is in this sense that such industry-level conditions may be considered fundamental. There is much less consensus about whether variables that distinguish firms within industries are comparably fundamental. For example, are there types of firm-specific R&D-related expertise that are so difficult or costly to acquire at any point in time that they should be legitimately considered exogenous?

Despite the consensus regarding the importance of demand, appropriability, and technological opportunity conditions in affecting innovative activity and performance, consideration of the role of these variables still has a considerable distance to go. These three key classes of industry-level variables appear to explain a good deal of the variance in firms' innovative activities. Yet, our understanding of just how this influence is exercised is limited. We have ideas, some clearer than others, of how different dimensions of demand, technological opportunity, and appropriability may influence innovative activity and performance. Our tests for the presence and importance of these particular effects are, however, often indirect. Although some descriptive evidence has begun to accumulate on how the nature and effects of demand, opportunity, and appropriability differ across industries, there have been few efforts to collect original data on these variables—Levin et al. (1987), Cohen et al. (2000), and the Community Innovation and related survey efforts being the exceptions. Moreover, the absence of suitable data still constrains more detailed examinations of the roles of these industry-level factors. For example, we still have little empirical understanding of the tradeoff for industries' R&D incentives of the tradeoff between the negative appropriability incentive effect and the positive complementarity effects of R&D spillovers. In addition, empirical studies of such effects would benefit from more rigorous models which lend themselves to empirical testing.

In comparison to our understanding of the influence of industry-level variables, our understanding of the role of firm-level variables is more primitive still. Economists reemphasized the study of the influence of firm characteristics from the mid-1980s to the mid-1990s, but there has been little work on the topic since, perhaps again reflecting the challenge of collecting suitable data. Moreover, while it is difficult to collect data on many firm characteristics, an analysis of their role in affecting innovation is especially challenging due to the possible endogeneity of many of them. The consequence is the additional requirement of identifying and measuring appropriate instruments. Particularly problematic in this regard is the study of R&D-related capabilities. Efforts to probe the nature and role of R&D-related capabilities will probably require more detailed industry studies, like those of Henderson and Cockburn (1996), which build on close knowledge of the firms and technologies involved.

The role of R&D-related firm capabilities raises an issue with regard to the analysis of industry-level variables. The proposition that firms differ in the sorts of expertise they possess, the approaches to innovation they pursue, and how well they pursue them suggests that the effect of industry-level factors, whether technological opportunity, demand, or R&D spillovers and appropriability, may vary across firms within industries. Closely linked to the importance for innovation of firm capabilities is the increasingly appreciated fact that technological information is costly to process, and how costly at a point in time depends upon the nature of the information and the ability of the firm to evaluate and use it. This perspective suggests that the influence on innovative activity and performance of all three classes of industry-level variables depends upon firms' assimilation and use of knowledge, whether it be knowledge of demand conditions, knowledge flows from competitors, or information about new scientific or technical advances; yet how capable firms are at learning will depend upon prior investments in that capacity. Indeed, considering the implications of the observation that firms that invest in their own R&D are more capable of assimilating and exploiting externally generated knowledge (Baldwin, 1962; Evenson and Kislev, 1973; Mowery, 1983a), Cohen and Levinthal (1989) formulated and successfully tested a model in which firms deliberately invest in R&D with two purposes: to generate new knowledge and, as discussed above, to develop "absorptive capacity"—the ability to evaluate, assimilate, and exploit outside knowledge.[88] Thus, while one may assume that there is some latent technological opportunity that is uniform across firms within an industry, its effect for any given firm may be conditioned by whether the firm possesses the ability to evaluate, assimilate, and exploit the relevant knowledge. More generally, to understand how and to what degree industry-level factors exercise their effects, we need to know more about what conditions firm learning and information processing within and across industries. Malerba (1992) has provided a taxonomy of firm learning, distinguishing among learning that emerges from experience in production or use, the learning that allows firms to exploit extramural knowledge, and the learning that is focused on internal problem solving, and shows that these different sorts of learning actually affect the kinds of innovative activities pursued by firms.

Another issue to which the empirical literature has devoted more attention since the mid-1970s is the dynamics of innovation, firm growth, and market structure, where the simulation models of Nelson and

[88] Using the FTC's Line of Business data and the Levin et al. (1987) survey data, Cohen and Levinthal (1989) found strong support for the model—and thus, the existence of an endogenous absorptive capacity that itself depended upon, among other factors, the extent of technological opportunity measured using a survey-based measure (Klevorick et al., 1995). Employing survey data on over 200 corporate R&D laboratories belonging to 115 firms, Adams (2006) estimates the shares of R&D actually devoted to learning about the knowledge flows originating both from other firms and from universities.

Winter (1982a) have led the way. Since then, the analytic model and empirical analysis of Klepper (1996) and colleagues have provided illuminating theoretical and empirical treatments of the issues. Once again, however, the availability of data—in this case, historical data—represents an important brake on the advance of empirical testing.

Long ago, in making the case for the primacy of profit as a driving force behind technical change, economists such as Schmookler (1962), Griliches (1957), Nelson (1959), and Arrow (1962a) sensibly argued that the rate and direction of technological change could be understood as the outcome of *firms'* rational, profit-driven investment in innovation. In doing so, they subordinated consideration of the impact of individuals and their motives on technical advance. Recent work has, however, suggested that the motives and incentives of individual R&D employees within firms—including their nonpecuniary motives—may be usefully considered to advance our understanding of innovative performance.[89] Stern's (2004) empirical analysis of new biomedical Ph.D.'s consideration of job offers from drug firms shows, for example, that nonpecuniary motives can importantly affect the cost of industrial research. Specifically, Stern estimates that job candidates accepted a 25% salary cut for jobs that allowed them to do more academic-like science; they had a "taste for science." Employing a sample of almost 2000 Ph.D. respondents to NSF's SESTAT survey who work in industrial R&D, Sauermann and Cohen (2008) find that, controlling for firm effects and for individuals' abilities, training and effort levels, the intensity of individuals' desire for intellectual challenge is strongly associated with greater innovative output, measured by the number of their patent applications or commercialized patents.[90] The suggestion that individuals' motives may matter for industrial innovation, even after controlling for firm effects, suggests that economists could fruitfully expand their consideration of the determinants of industrial innovation beyond the features of firms and industries to consider individual-level motives and incentives, as well as other characteristics of the individuals who work in industrial R&D.

There are several overarching lessons that may be drawn from the empirical literature on innovative activity and performance. First, as repeatedly suggested, this is a field in need of more and better data on the range of independent variables considered—industry-level variables, firm attributes, and, as suggested above, even individual-level variables. Better measurement of innovative activity itself is also essential. Consider, for example, the lack of current R&D data collected in the United States at a

[89] Indeed, Schumpeter (1934, 1942) himself suggested a critical role of a range of motives for entrepreneurship and innovative activity. However plausible the claim, why should the motives and incentives of individual scientists and engineers—as opposed to the profit incentive of the firm as a whole—matter for innovation? As pointed out by Sauermann and Cohen (2008), first, a firm's R&D employees are able to exercise more autonomy than most employees since there is typically uncertainty about how to tackle technical challenges and the technologists themselves are often more expert about the technologies in question. As a consequence, it is in management's interest that the technologists retain some significant degree of autonomy. Moreover, inventive or "creative" effort is hard to observe by outsiders, and given the uncertainty endemic to the outcomes of R&D projects, observable outcomes are not very informative of effort expended by employees. Thus, R&D labs are settings where there is often significant delegation of authority to the individual employee, where the opportunity for bureaucratic control is limited (Prendergast, 1999), and the innovative performance of firms thus depend importantly on the motives of its scientists and engineers.

[90] The question for this study is why, controlling for the level of individual effort, should love of challenge be associated with greater R&D productivity? Drawing from the social and cognitive psychology literatures, Sauermann and Cohen (2008) conjecture that such intrinsic motivation may, for example, be associated with respondents' cognition and, in turn, their ability to solve problems. The authors acknowledge, however, that they cannot altogether rule out the possibility of reverse causality; that technologists' success in their innovative efforts may elicit greater appreciation of intellectual challenge.

sufficiently disaggregated (e.g., four-digit) line of business level, no less industry-level data on the composition of R&D, distinguishing between process and product R&D, R&D dedicated to incremental improvement versus the development of altogether new products and processes, as well as the division of R&D expenditures between basic and applied research versus development (cf. NRC, 2005).[91] As emphasized by Lunn (1986), the common assumption that R&D activities are qualitatively homogeneous makes it difficult to specify an empirical model; some variables expected to influence process innovation, for example, may be thought to have no influence on product innovation. The importance of particular firm- or industry-level explanatory variables may also differ across types of activities. The availability of patent protection would be expected to have a stronger effect on product R&D than on process R&D (Levin et al., 1987). Or, a firm's degree of diversification would be expected to have a stronger effect on basic research than on applied research and development (Nelson, 1959), and, as noted above, firm size is more closely tied to process and incremental R&D than it is to more significant product innovation. Moreover, these relationships may well vary across industries. Another data limitation in the United States is the absence of measures of innovative activity that occurs outside of the formal labs and accounting categories encompassed by R&D. For a small, informal sample of firms, Mansfield (1968) observed that R&D expenditures reflected only about 50% of manufacturing firms' investments in product and process innovation. Almost all the studies to date proceed on the implicit and untested assumption that R&D expenditures are a reliable index of this broader investment activity. Moving to remedy this situation over the past fifteen years or so, data collection efforts in Canada, Europe and elsewhere, under the heading in Europe of the "Community Innovation Surveys," have identified innovating, not just R&D performing firms, and collected data on the extent, features and correlates of innovative activity broadly considered (see Arundel, 2007; Arundel et al., 2006; Gault, 2003; Smith, 2005; as well as the review of innovation surveys by Mairesse and Mohnen in Chapter 26).

Yet another long-recognized data limitation, and perhaps the most serious, is the absence of accurate measures of innovation itself—the output of innovative activity (cf. Cohen and Levin, 1989). Although progress has been made on measuring innovation with citation-weighted patent counts, it is understood that such a measure is still flawed given differences in the propensity to patent across industries and firms, and given our still limited understanding of citing behavior (cf. Alcacer et al., 2009).[92]

[91] The US National Science Foundation is in the process of significantly expanding and improving its collection of R&D and related data, and is moving to collect detailed R&D data at the line of business level. In the past, individual scholars have, on occasion, collected data on the composition of R&D. For example, Mansfield (1981) and Link (1982b) collected and analyzed data distinguishing basic research from applied research and development at the firm level. Link (1982a) collected firm-level data on process and product R&D, and Scherer's (1982a, 1984a) classification of all US patents granted within a 10-month period in the mid-1970s by industry of origin and industry of use permits this distinction to be made for a much broader sample. In Scherer's framework, process innovations are those represented by patents used in their industry of origin and product innovations represent the balance. In addition to providing an interesting picture of interindustry flows of technology (Scherer, 1982b, 1984a,b), Scherer's data have been used to divide an industry's or business unit's R&D expenditures between process and product R&D (Cohen and Klepper, 1996a; Levin and Reiss, 1988; Lunn, 1986) by assuming that each industry or business unit devotes to processes a percentage of R&D equal to the percentage of patents assigned to processes.

[92] The Community Innovation and related surveys have also usefully collected data on the percentage of firm sales accounted for by products, processes, and services that are new to the market, as well as only new to the enterprise. Although not a direct measure of innovation, such a measure reflects at least one dimension of the commercial importance of innovation for any given firm.

A major lacuna in our understanding of the determinants of innovative activity and performance is our virtual ignorance of innovation in the service sector, with the exception of a modest literature on financial services.[93] This gap is not simply a matter of a lack of data. It speaks to our very notion of innovation, and how to capture innovation when a good deal of it occurs outside of R&D labs, and especially in the context of client-specific relationships. This is, however, also associated with the way R&D is measured—and not. For example, if a firm innovates in the course of providing service to a client, accounting conventions require that that be counted as cost of goods sold; it cannot be considered R&D for reporting purposes. And a good deal of innovation in the service sector occurs in just this fashion.

Perhaps one of the more basic lessons to emerge from the empirical literature is that, although testing loosely motivated hypotheses may yield empirical results, even robust ones, their interpretation can be challenging, and the insight that can be gleaned from such findings is often limited in the absence of underlying theory. However banal, the prescription that modeling can serve as a useful handmaiden to empirical analysis has often been overlooked in studies of innovation. Indeed, even very simple theory can radically revise the interpretation of what may appear to be the most straightforward of empirical relationships, such as that between R&D productivity and firm size.

There is also, however, an important role to be played by more inductive efforts. Consider, for example, that much of our empirical understanding of innovation derives not from the estimation of econometric models, but from the use of other empirical methods. As we have illustrated, the historical and case-study literatures provide a rich array of insights and factual information, and often constitute a source of hypotheses and inspiration for more rigorous approaches. More strikingly, many of the most credible empirical regularities have been established not by estimating and testing elaborate models with published data but by the painstaking collection of original data, often in the form of responses to simple questions. Even as econometric methods advance and the quality of published data improves, it will be important to remain catholic in the application of empirical techniques.

Acknowledgments

The author wishes to acknowledge the research assistance of James Melnick, Marcelo Ochoa, Fabrice Smieliauskas and Ulya Tsolmon, as well as the helpful comments of the editors of this volume.

References

Abernathy, W.J. (1978). "The Productivity Dilemma". Johns Hopkins University Press, Baltimore, MD.

Abernathy, W.J., Utterback, J.M. (1978). "Patterns of industrial innovation". Technology Review 41–47.

Acemoglu, D., Linn, J. (2004). "Market size in innovation: Theory and evidence from the pharmaceutical industry". Quarterly Journal of Economics 119, 1049–1090.

Achilladelis, B., Schwarzkopf, A., Cities, M. (1990). "The dynamics of technological innovation: The case of the chemical industry". Research Policy 19, 1–34.

[93] See Lerner (2006) for an empirical study of the determinants of financial services innovation, and Frame and White (2004) for a review of the modest literature on innovation in the financial services sector.

Acs, Z.J., Audretsch, D.B. (1987). "Innovation, market structure, and firm size". Review of Economics and Statistics 71, 567–574.
Acs, Z.J., Audretsch, D.B. (1988). "Innovation in large and small firms". American Economic Review 78, 678–690.
Acs, Z.J., Audretsch, D.B. (1990). Innovation and Small Firms. MIT Press, Cambridge, MA.
Acs, Z.J., Audretsch, D.B. (1991a). "Innovation as a means of entry: An overview". In: Geroski, P., Schwalbach, J. (Eds.), Entry and Market Contestability: An International Comparison. Basil Blackwell, Oxford.
Acs, Z.J., Audretsch, D.B. (1991b). "R&D, firm size, and innovative activity". In: Acs, Z.J., Audretsch, D.B. (Eds.), Innovation and Technological Change: An International Comparison. Harvester Wheatsheaf, New York, NY.
Acs, Z.J., Audretsch, D.B., Isberg, S.C. (1991). "Innovation, firm size and corporate finance". Economics Letters 35, 323–326.
Acs, Z.J., Audretsch, D.B., Feldman, M. (1992). "Real effects of academic research: Comment". American Economic Review 82 (1), 363–367.
Adams, J.D. (1990). "Fundamental stocks of knowledge and productivity growth". Journal of Political Economy 98 (4), 673–702.
Adams, J.D. (1993). "Science, R&D, and invention potential recharge: U.S. evidence". American Economic Review: Proceedings 83, 458–462.
Adams, J.D. (2006). "Learning, internal research, and spillovers". Economics of Innovation and New Technology 15, 5–36.
Adams, J.D., Jaffe, A.B. (1993). "On the microeconomic structure of R&D spillovers". Mimeo, University of Florida.
Adams, J.D., Chiang, E.P., Jensen, J.L. (2003). "The influence of federal laboratory R&D on industrial research". Review of Economics and Statistics 85, 1003–1020.
Aghion, P., Griffith, R. (2005). Competition and Growth: Reconciling Theory and Evidence. MIT Press, Cambridge, MA.
Aghion, P., Harris, C., Vickers, J. (1997). "Competition and growth with step-by-step innovation: An example". European Economic Review 41, 771–782.
Aghion, P., Bloom, N., Blundell, R., Griffith, R., Howitt, P. (2005). "Competition and Innovation: An inverted U-relationship". Quarterly Journal of Economics 120, 701–728.
Agrawal, A., Henderson, R. (2002). "Putting patents in context: Exploring knowledge transfer from MIT". Management Science 48, 44–60.
Ahuja, G., Lampert, C.M., Tandon, V. (2009). "Moving beyond Schumpeter: Management research on the determinants of technological innovation". The Academy of Management Annals 2, 1–98.
Alcacer, J., Gittelman, M., Sampat, B. (2009). "Applicant and examiner citations in U.S. patents: An overview and analysis". Research Policy 38, 415–427.
Amabile, T. (1996). Creativity in Context. Westview Press, Boulder, CO.
Angelmar, R. (1985). "Market structure and research intensity in high technological-opportunity industries". Journal of Industrial Economics 34, 69–79.
Antonelli, C. (1989). "A failure-inducement model of research and development expenditure". Journal of Economic Behavior & Organization 12, 159–180.
Armour, H.O., Teece, D.J. (1981). "Vertical integration and technological innovation". Review of Economics and Statistics 62, 470–474.
Arora, A. (1996). "Testing for complementarities in reduced-form regressions: A note". Economics Letters 50, 51–55.
Arora, A., Ceccagnoli, M. (2006). "Profiting from licensing: The role of patent protection and commercialization capabilities". Management Science 52, 293–308.
Arora, A., Fosfuri, A., Gambardella, A. (2001). Markets for Technology: Economics of Innovation and Corporate Strategy. MIT Press, Cambridge, MA.
Arora, A., Ceccagnoli, M., Cohen, W.M. (2008). "R&D and the patent premium". International Journal of Industrial Organization 26, 1153–1179.
Arrow, K.J. (1962a). "Economic welfare and the allocation of resources for invention". In: Nelson, R.R. (Ed.), The Rate and Direction of Inventive Activity. Princeton University Press, Princeton, NJ.
Arrow, K.J. (1962b). "The economic implications of learning by doing". Review of Economic Studies 29, 155–173.
Artes, J. (2009). "Long-run versus short-run decisions: R&D and market structure in Spanish firms". Research Policy 38, 120–132.
Arthur, W.B. (1989). "Competing technologies, increasing return, and lock-in by historical events". Economic Journal 99, 116–131.
Arthur, W.B. (1990). "Positive feedbacks in the economy". Scientific American 262, 92–99.
Arundel, A. (2001). "The relative effectiveness of patents and secrecy for appropriation". Research Policy 30, 611–624.

Arundel, A. (2007). Innovation survey indicators: What impact on innovation policy? Science, Technology and Innovation Indicators in a Changing World: Responding to Policy Needs. OECD Publishing, Paris, pp. 49–64.

Arundel, A., Kabla, I. (1998). "What percentage of innovations are patented? Empirical estimates for European firms". Economic Journal 27, 127–141.

Arundel, A., Bordoy, C., Mohnen, P., Smith, K. (2006). "Innovation surveys and policy: Lessons from the CIS". Mimeo, MERIT, University of Maastricht .

Athey, S., Stern, S. (1998). An empirical framework for testing theories about complementarity in organizational design. Working paper No. 6600. National Bureau of Economic Research, Cambridge, MA.

Ayres, R.U. (1987). "Barriers and breakthroughs: An "expanding frontiers" model of the technology–industry life cycle". Technovation 7, 87–115.

Azoulay, P., Manso, G., Zivin, J. (2009). "Incentives and creativity: Evidence from the academic life sciences". Mimeo, MIT Sloan School of Management.

Baily, M.N. (1972). "Research and development costs and returns: The U.S. pharmaceutical industry". Journal of Political Economy 80, 70–85.

Baldwin, W.L. (1962). "Contracted research and the case for big business". Journal of Political Economy 70, 294–298.

Baldwin, W.L., Scott, J.T. (1987). Market Structure and Technological Change. Harwood, Chichester.

Baumol, W.J. (2002). The Free-Market Innovation Machine. Princeton University Press, Princeton, NJ.

Bernstein, J.I. (1988). "Costs of production, intra- and inter-industry R&D spillovers: Canadian evidence". Canadian Journal of Economics 21, 324–347.

Bernstein, J.I. (1989). "The structure of Canadian inter-industry R&D spillovers, and the rates of return to R&D". Journal of Industrial Economics 37 (3), 315–328.

Bernstein, J.I., Nadiri, M.I. (1988). "Interindustry R&D spillovers, rates of return, and production in high-tech industries". American Economic Review: Proceedings 78, 429–439.

Bernstein, J.I., Nadiri, M.I. (1989). "Research and development and intraindustry spillovers: An empirical application of dynamic duality". Review of Economic Studies 56, 249–267.

Bessen, J., Maskin, E. (2009). "Sequential innovation, patents, and imitation". RAND Journal of Economics 40, 611–635.

Bhagat, S., Welch, I. (1995). "Corporate research and development investments: International comparisons". Journal of Accounting and Economics 19, 443–470.

Blair, J.M. (1948). "Technology and size". American Economic Review 38, 121–152.

Blair, J.M. (1974). Economic Concentration. Harcourt Brace Jovanovich, New York, NY.

Blumenthal, D., Gluck, M., Louis, K., Wise, D. (1986). "Industrial support of university research in biotechnology". Science 231, 242–246.

Blundell, R., Griffith, R., Van Reenen, J. (1999). "Market share, market value and innovation in a panel of British manufacturing firms". Review of Economic Studies 66, 529–554.

Bond, S., Harhoff, D., Van Reenen, J. (1999). Investment, R&D, and financial constraints in Britain and Germany. Working paper No. 99/4. Institute of Fiscal Studies, London.

Bougheas, S., Goerg, H., Strobl, E. (2003). "Is R&D financially constrained? Theory and evidence from Irish manufacturing". Review of Industrial Organization 22, 159–174.

Bound, J., Cummins, C., Griliches, Z., Hall, B.H., Jaffe, A. (1984). "Who does R&D and who patents?" In: Griliches, Z. (Ed.), R&D Patents and Productivity. University of Chicago Press for the NBER, Chicago, IL.

Bozeman, B., Link, A.N. (1983). Investments in Technology: Corporate Strategies and Public Policy Alternatives. Praeger, New York, NY.

Branch, B. (1974). "Research and development activity and profitability: A distributed lag analysis". Journal of Political Economy 82, 999–1011.

Braun, E., MacDonald, S. (1982). Revolution in Miniature. Cambridge University Press, Cambridge, MA.

Breschi, S., Malerba, F., Orsenigo, L. (2000). "Technological regimes and Schumpeterian patterns of innovation". Economic Journal 110, 388–410.

Bresnahan, T., Trajtenberg, M. (1995). "General purpose technologies—Engines of growth?" Journal of Econometrics 65, 83–108.

Brock, G.W. (1975). The U.S. Computer Industry. Ballinger, Cambridge, MA.

Brown, W. (1997). "R&D intensity and finance: Are innovative firms financially constrained?" Mimeo, London School of Economics Financial Market Group .

Carpenter, M.P., Narin, F., Wolf, P. (1981). "Citation rates to technologically important patents". World Patent Information 3, 160–163.
Cassiman, B., Veugelers, R. (2002). "R&D cooperation and spillovers: Some empirical evidence from Belgium". American Economic Review 92, 1169–1184.
Cassiman, B., Veugelers, R. (2006). "In search of complementarity in innovation strategy: Internal R&D and external knowledge acquisition". Management Science 52, 68–82.
Caves, R. (1962). Air Transport and Its Regulators. Harvard University Press, Cambridge, MA.
Caves, R., Porter, M., Spence, M. (1980). Competition in the Open Economy: A Model Applied to Canada. Harvard University Press, Cambridge, MA.
Cerda, R. (2007). "Endogenous innovations in the pharmaceutical industry". Journal of Evolutionary Economics 17, 473–515.
Chandler, A. (1990). Scale and Scope. Harvard University Press, Cambridge, MA.
Chesbrough, H. (1999). "Arrested development: The experience of European hard disk drive firms in comparison to U.S. and Japanese firms". Journal of Evolutional Economics 9, 287–330.
Christensen, C. (1997). The Innovator's Dilemma: When New Technologies Cause Great Firms to Fail. Harvard Business School Press, Boston, MA.
Clark, K.B., Fujimoto, T. (1991). Product Development Performance. Harvard Business School Press, Boston, MA.
Clark, K.B., Chew, W.B., Fujimoto, T. (1987). "Product development in the world auto industry". Brookings Papers on Economic Activity, pp. 729–771.
Cockburn, I., Henderson, R. (1994). "Racing to invest? The dynamics of competition in ethical drug discovery". Journal of Economics & Management Strategy 3, 481–519.
Cockburn, I., Henderson, R. (1998). "Absorptive capacity, coauthoring behavior, and the organization of research in drug discovery". Journal of Industrial Economics 46, 157–182.
Cockburn, I., Henderson, R. (2001). "Scale and scope in drug development: Unpacking the advantages of size in pharmaceutical research". Journal of Health Economics 20, 1033–1057.
Cohen, W.M. (1995). "Empirical studies of innovative activity". In: Stoneman, P. (Ed.), Handbook of the Economics of Innovation and Technical Change. Basil Blackwell, Oxford.
Cohen, W.M., Klepper, S. (1992a). "The anatomy of industry R&D intensity distributions". American Economic Review 82, 773–788.
Cohen, W.M., Klepper, S. (1992b). "The tradeoff between firm size and diversity in the pursuit of technological progress". Small Business Economics 4, 1–14.
Cohen, W.M., Klepper, S. (1996a). "Firm size and the nature of innovation within industries: The case of process and product R&D". Review of Economics and Statistics 78, 232–243.
Cohen, W.M., Klepper, S. (1996b). "A reprise of size and R&D". Economic Journal 106, 925–951.
Cohen, W.M., Levin, R.C. (1989). "Empirical studies of innovation and market structure". In: Schmalensee, R., Willig, R. (Eds.), Handbook of Industrial Organization. North-Holland, Amsterdam.
Cohen, W.M., Levinthal, D.A. (1989). "Innovation and learning: The two faces of R&D—Implications for the analysis of R&D investment". Economic Journal 99, 569–596.
Cohen, W.M., Levinthal, D.A. (1990). "Absorptive capacity: A new perspective on learning and innovation". Administrative Science Quarterly 35, 128–152.
Cohen, W.M., Levinthal, D.A. (1994). "Fortune favors the prepared firm". Management Science 40, 227–251.
Cohen, W.M., Malerba, F. (2001). "Is the tendency to variation a chief cause of progress?" Journal of Industrial and Corporate Change 10, 587–608.
Cohen, W.M., Levin, R.C., Mowery, D.C. (1987). "Firm size and R&D intensity: A re-examination". Journal of Industrial Economics 35, 543–563.
Cohen, W.M., Nelson, R.R., Walsh, J.P. (2000). Protecting Their Intellectual Assets: Appropriability Conditions and Why U.S. Manufacturing Firms Patent (or Not). Working paper No. 7552. National Bureau of Economic Research, Cambridge, MA Revised 2004 (February).
Cohen, W.M., Goto, A., Nagata, A., Nelson, R.R., Walsh, J.P. (2002a). "R&D spillovers, patents and the incentive to innovate in Japan and the United States". Research Policy 31, 1349–1367.
Cohen, W.M., Nelson, R.R., Walsh, J.P. (2002b). "Links and impacts: The influence of public research on industrial R&D". Management Science 48, 1–23.

Cohen, W.M., Walsh, J.P. (2000). "R&D Information Flows, Appropriability and R&D Intensity". In: Spivack, R.N. (Ed.), Papers and proceedings of the Advanced Technology Program's International Conferences on the Economic Evaluation of Technological Change, Washington, DC. NIST Special Publication 952. USGPO, Washington, pp. 22–29.

Comanor, W.S. (1964). "Research and competitive product differentiation in the pharmaceutical industry in the United States". Economica 31, 372–384.

Comanor, W.S. (1965). "Research and technical change in the pharmaceutical industry". Review of Economics and Statistics 47, 182–190.

Comanor, W.S. (1967). "Market structure, product differentiation, and industrial research". Quarterly Journal of Economics 81, 639–657.

Connolly, R.A., Hirschey, M. (1984). "R&D, market structure; and profits: A value-based approach". Review of Economics and Statistics 66, 682–686.

Cooper, A.C. (1964). "R&D is more efficient in small companies". Harvard Business Review 42 (3), 75–83.

Cowan, R. (1990). "Nuclear power reactors: A study in technological lock in". Journal of Economic History 50, 541–567.

Crepon, B., Duguet, E., Mairesse, J. (1998). "Research, innovation and productivity: An econometric analysis at the firm level". Economics of Innovation and New Technology 7, 115–158.

Cyert, R., March, J. (1962). A Behavioral Theory of the Firm. Wiley-Blackwell, New York, NY.

Czarnitzki, D., Kraft, K. (2004). "An empirical test of the asymmetric models on innovative behavior: Who invests more into R&D, the incumbent or the challenger?" Journal of Economic Behavior & Organization 54, 153–173.

Czarnitzki, D., Kraft, K. (2005). "License expenditures of incumbents and potential entrants: An empirical analysis on German firm behavior". ZEW—Centre for European Economic Research Discussion Paper No. 05-035. Available at SSRN: http://ssrn.com/abstract=732303, http://papers.ssrn.com/sol3/papers.cfm?abstract_id=732303#.

Dasgupta, P.S., David, P.A. (1987). "Information disclosure and the economics of science & technology". In: Feiwel, G.R. (Ed.), Arrow and the Ascent of Modern Economic Theory. , pp. 519–542

Dasgupta, P.S., David, P.A. (1994). "Toward a new economics of science". Policy Research 23, 487–521.

Dasgupta, P.S., Stiglitz, J.E. (1980a). "Industrial structure and the nature of innovative activity". Economic Journal 90, 266–293.

Dasgupta, P.S., Stiglitz, J.E. (1980b). "Uncertainty, industrial structure and the speed of R&D". Bell Journal of Economics 11, 1–28.

David, P.A. (1985). "Clio and the economics of QWERTY". American Economic Review: Proceedings 75, 332–337.

David, P.A. (1988). Path-dependence: Putting the past into the future of economics. IMSSS Technical Report No. 533. Stanford University, Stanford, CA.

Deng, Y. (2007). "The effects of patent regime changes: A case study of the European patent office". International Journal of Industrial Organization 25, 121–138.

Doi, N. (1985). "Diversification and R&D activity in Japanese manufacturing firms". Managerial and Decision Economics 6, 47–52.

Dorfman, N.S. (1987). Innovation and Market Structure: Lessons from the Computer and Semiconductor Industries. Ballinger, Cambridge, MA.

Dosi, G. (1982). "Technological paradigms and technological trajectories: A suggested interpretation of the determinants and directions of technical change". Research Policy 11, 147–162.

Eaton, J., Kortum, S. (1999). "International technology diffusion, theory and measurement". International Economic Review 40, 537–570.

Ehrnberg, E., Sjöberg, N. (1995). "Technological discontinuities, competition and firm performance". Technology Analysis & Strategic Management 7, 93–107.

Elliot, J.W. (1971). "Funds flow versus expectational theories of research and development expenditures in the firm". Southern Economic Journal 37, 409–422.

Encaoua, D., Guellec, D., Catalina, M. (2006). "Patent systems for encouraging innovation: Lessons from economic analysis". Research Policy 35, 1423–1440.

Enos, J.L. (1962). Petroleum Progress and Profit. MIT Press, Cambridge, MA.

Evenson, R.E., Kislev, Y. (1973). "Research and productivity in wheat and maize". Journal of Political Economy 81, 1309–1329.

Evenson, R.E., Kislev, Y. (1976). "A stochastic model of applied research". Journal of Political Economy 84, 265–281.

Fagerberg, J., Mowery, D.C., Nelson, R.R. (2005). The Oxford Handbook of Innovation. Oxford University Press, Oxford.

Fallick, B., Fleischman, C., Rebitzer, J. (2006). "Job-hopping in Silicon Valley: Some evidence concerning the micro-foundations of a high technology cluster". Review of Economic Statistics 88, 472–481.
Farber, S. (1981). "Buyer market structure and R&D effort: A simultaneous equations model". Review of Economics and Statistics 62, 336–345.
Farrell, J., Klemperer, P. (2007). "Coordination and lock-in: Competition with switching cots and network effects". In: Armstrong, R.H., Porter, R.H. (Eds.), Handbook of Industrial Organization, vol. 3. Elsevier, Amsterdam.
Farrell, J., Saloner, G. (1985). "Standardization, compatibility, and innovation". RAND Journal of Economics 16, 70–83.
Farrell, J., Saloner, G. (1986). "Installed base and compatibility: Innovation, product preannouncements, and predation". American Economic Review 76, 940–955.
Fisher, F.M., Temin, P. (1973). "Returns to scale in research and development: What does the Schumpeterian hypothesis imply?" Journal of Political Economy 81, 56–70.
Flamm, K. (1988). Creating the Computer. The Brookings Institution, Washington, DC.
Frame, W.S., White, L.J. (2004). "Empirical studies of financial innovation: Lots of talk, little action?" Journal of Economic Literature 42, 116–144.
Freeman, C. (1982). The Economics of Industrial Innovation (second ed.). MIT Press, Cambridge, MA.
Freeman, C. (1987). Technology Policy and Economic Performance: Lessons from Japan. Pinter, London.
Galbraith, J.K. (1952). American Capitalism: The Concept of Countervailing Power. Houghton Mifflin, Boston, MA.
Gallini, N. (2002). "The economics of patents: Lessons from recent U.S. patent reform". Journal of Economic Perspectives 16, 131–154.
Gambardella, A. (1992). "Competitive advantages from in-house scientific research: The US pharmaceutical industry in the 1980s". Research Policy 21, 391–407.
Gans, J., Stern, S. (2003). "The product market and the market for 'ideas': Commercialization strategies for technology entrepreneurs". Research Policy 32, 333–350.
Gans, J., Hsu, D., Stern, S. (2002). "When does start-up innovation spur the gale of creative destruction". RAND Journal of Economics 33, 571–586.
Gans, J., Hsu, D., Stern, S. (2008). "The impact of uncertain intellectual property rights on the market for ideas: Evidence for patent grant delays". Management Science 54, 982–997.
Gault, F. (2003). Understanding Innovation in Canadian Industry. McGill-Queen's University Press, Montreal, QC.
Gellman Research Associates. (1976). Indicators of international trends in technological innovation. Final Report to the National Science Foundation, NTIS Document PB-263-738. Gellman Research Associates, Jenkintown, PA.
Gellman Research Associates. (1982). The relationship between industrial concentration, firm size and technological innovation. Final Report, U.S. Small Business Administration. The Futures Group, Inc., Glastonbury, CT
Geroski, P.A. (1989). "Entry, innovation and productivity growth". Review of Economics and Statistics 71 (4), 572–578.
Geroski, P.A. (1990). "Innovation, technological opportunity, and market structure". Oxford Economic Papers 42, 586–602.
Geroski, P.A. (1991a). "Innovation and the sectoral sources of UK productivity growth". Economic Journal 101, 1438–1451.
Geroski, P.A. (1991b). "Entry and the rate of innovation". Economic Innovation and New Technology 1, 203–214.
Geroski, P.A. (1991c). Market Dynamics and Entry. Basil Blackwell, Cambridge, MA.
Geroski, P.A. (1994). Market Structure, Corporate Performance and Innovative Activity. Clarendon Press, Oxford.
Geroski, P.A., Pomroy, R. (1990). "Innovation and the evolution of market structure". Journal of Industrial Economics 38, 299–314.
Geroski, P.A., Walters, C.F. (1995). "Innovative activity over the business cycle". Economic Journal 105, 916–928.
Gilbert, R. (2006). "Looking for Mr. Schumpeter: Where are we in the competition–innovation debate?" In: Jaffe, A., Lerner, S., Stern, S. (Eds.), Innovation Policy and the Economy, vol. 6. MIT Press/National Bureau of Economic Research, Cambridge, MA, pp. 159–215.
Gilbert, R., Newbery, D. (1982). "Preemptive patenting and the persistence of monopoly". American Economic Review 72, 514–526.
Gilson, R.J. (1999). "The legal infrastructure of high technology industrial districts: Silicon Valley, Route 128, and covenants not to compete". New York University Law Review 74, 575–626.
Gort, M., Klepper, S. (1982). "Time paths in the diffusion of product innovations". Economic Journal 92, 630–653.
Grabowski, H.G. (1968). "The determinants of industrial research and development: A study of the chemical, drug, and petroleum industries". Journal of Political Economy 76, 292–306.

Grabowski, H.G., Baxter, N.D. (1973). "Rivalry in industrial research and development: An empirical study". Journal of Industrial Economics 21, 209–235.
Grabowski, H.G., Vernon, J.M. (1982). "The pharmaceutical industry". In: Nelson, R.R. (Ed.), Government and Technical Progress: A Cross-Industry Analysis. Pergamon Press, New York, NY.
Griliches, Z. (1957). "Hybrid corn: An exploration in the economics of technological change". Econometrica 25, 501–522.
Griliches, Z. (1962). "Comment". In: Nelson, R.R. (Ed.), The Rate and Direction of Inventive Activity. Princeton University Press, Princeton, NJ.
Griliches, Z. (1979). "Issues in assessing the contribution of research and development to productivity growth". Bell Journal of Economics 10, 92–116.
Griliches, Z. (1990). "Patent statistics as economic indicators: A survey". Journal of Economic Literature 28, 1661–1707.
Griliches, Z. (1992). "The search for R&D spillovers". Scandinavian Journal of Economics 94, 29–47.
Griliches, Z., Hall, B.H., Pakes, A. (1991). "R&D, patents, and market value revisited: Is there a second (technological opportunity) factor?" Economics of Innovation and New Technology 1, 183–201.
Grindley, P., Teece, D. (1997). "Managing intellectual capital: Licensing and cross licensing in semiconductors and electronics". California Management Review 39, 8–41.
Guerard, J.B., Jr., Bean, A.S., Andrews, S. (1987). "R&D management and corporate financial policy". Management Science 33 (11), 1419–1427.
Hall, B.H. (1987). "The relationship between firm size and firm growth in the U.S. manufacturing sector". Journal of Industrial Economics 35, 583–606.
Hall, B.H. (1990). The impact of corporate restructuring on industrial research and development. Brookings Papers on Economic Activity. The Brookings Institution, Washington, DC.
Hall, B.H. (2002). "The financing of research and development". Oxford Review of Economic Policy 18, 35–51.
Hall, B.H. (2004). "Exploring the patent explosion". Journal of Technology Transfer 30, 35–48.
Hall, B.H., Ziedonis, R.H. (2001). "The patent paradox revisited: An empirical study of patenting in the U.S. semiconductor industry, 1979–1995". RAND Journal of Economics 32, 101–128.
Hall, B.H., Mairesse, J., Branstetter, L., Crepon, B. (1999). "Does cash flow cause investment and R&D: An exploration using panel data for French, Japanese, and United States firms in the scientific sector". In: Audretsch, D., Thurik, A.R. (Eds.), Innovation, Industry Evolution and Employment. Cambridge University Press, Cambridge, MA.
Hamberg, D. (1964). "Size of firm, oligopoly, and research: The evidence". Canadian Journal of Economics and Political Science 30, 62–75.
Hamberg, D. (1966). R&D: Essays on the Economics of Research and Development. Random House, New York, NY.
Hao, K.Y., Jaffe, A. (1993). "Effect of liquidity on firms' R&D spending". Economics of Innovation and New Technology 2, 275–282.
Harhoff, D. (1996). "Strategic spillovers and incentives for research and development". Management Science 42, 907–925.
Harhoff, D. (1998). "Are there financing constraints for R&D and investment in German manufacturing firms?" Annales d'Economie et de Statistique 49 (50), 421–456.
Hart, P.E., Prais, S.J. (1956). "The analysis of business concentration". Journal of the Royal Statistical Society 119 (2), 150–181 (May–August).
Helfat, C.E. (1997). "Know-how and asset complementarity and dynamic capability accumulation: The case of R&D". Strategic Management Journal 18, 339–360.
Heller, M., Eisenberg, R. (1998). "Can patents deter innovation? The anticommons in biomedical research". Science 28, 698–701.
Henderson, R.M. (1988). "The failure of established firms in the face of technical change: A study of photolithographic alignment equipment". Ph.D. Thesis, Harvard University.
Henderson, R.M. (1993). "Underinvestment and incompetence as responses to radical innovation: Evidence from the photolithographic alignment equipment industry". RAND Journal of Economics 24 (2), 248–270.
Henderson, R.M., Clark, K.B. (1990). "Architectural innovation: The reconfiguration of existing product technologies and the failure of established firms". Administrative Science Quarterly 35, 9–30.
Henderson, R.M., Cockburn, I. (1994). "Measuring competence: Exploring firm effects in pharmaceutical research". Strategic Management Journal 15, 63–84.
Henderson, R.M., Cockburn, I. (1996). "Scale, scope, and spillovers: Determinants of research productivity in the pharmaceutical industry". RAND Journal of Economics 27, 32–59.

Henderson, R.M., Jaffe, A., Trajtenberg, M. (2005). "Patent Citations and the Geography of Knowledge Spillovers: A Reassessment: Comment". American Economic Review 95, 461–464.

Hicks, D., Breitzman, T., Olivastro, D., Hamilton, K. (2001). "The changing composition of innovative activity in the U.S.—A portrait based on patent analysis". Research Policy 30, 681–704.

Himmelberg, C., Peterson, B. (1994). "R&D and internal finance: A panel study of small firms in high tech industries". Review of Economics and Statistics 76, 38–51.

Holbrook, D., Cohen, W.M., Hounshell, D., Klepper, S. (2000). "The nature, sources and consequences of firm differences in the early semiconductor industry". Strategic Management Journal 21, 1017–1041.

Horowitz, I. (1962). "Firm size and research activity". Southern Economic Journal 28, 298–301.

Horstmann, L., MacDonald, J.M., Slivinski, A. (1985). "Patents as information transfer mechanisms: To patent or (maybe) not to patent". Journal of Political Economy 93, 837–858.

Howe, J.D., McFetridge, D.G. (1976). "The determinants of R&D expenditures". Canadian Journal of Economics 9, 57–61.

Hughes, W.R. (1971). "Scale frontiers in electric power". In: Capron, W.M. (Ed.), Technological Change in Regulated Industries. The Brookings Institution, Washington, DC.

Hunt, R. (2006). "When do more patents reduce R&D?" American Economic Review: Papers and Proceedings 96, 87–91.

Iansiti, M. (1995a). "Technology integration: Managing technological evolution in a complex environment". Research Policy 24, 521–542.

Iansiti, M. (1995b). "Science-based product development: An empirical study of the mainframe computer industry". Production and Operations Management Journal 4, 335–359.

Jaffe, A.B. (1986). "Technological opportunity and spillovers of R&D". American Economic Review 76, 984–1001.

Jaffe, A.B. (1988). "Demand and supply influences on R&D intensity and productivity growth". Review of Economics and Statistics 72, 431–437.

Jaffe, A.B. (1989a). "Characterizing the "technological position" of firms, with application to quantifying technological opportunity and research spillovers". Research Policy 18, 87–97.

Jaffe, A.B. (1989b). "Real effects of academic research". American Economic Review 79, 957–970.

Jaffe, A.B. (2000). "The U.S. patent system in transition: Policy innovation and the innovation process". Research Policy 29, 531–558.

Jaffe, A.B., Trajtenberg, M., Henderson, R. (1993). "Geographic localization of knowledge spillovers as evidenced by patent citations". Quarterly Journal of Economics 108, 576–598.

Jewkes, J., Sawers, D., Stillerman, R. (1958). The Sources of Invention. Macmillan, London.

Johannisson, B., Lindstrom, C. (1971). "Firm size and inventive activity". Swedish Journal of Economics 73, 427–442.

Kamien, M.I., Schwartz, N.L. (1970). "Market structure, elasticity of demand, and incentive to invent". Journal of Law and Economics 13, 241–252.

Kamien, M.I., Schwartz, N.L. (1976). "On the degree of rivalry for maximum innovative activity". Quarterly Journal of Economics 90, 245–260.

Kamien, M.I., Schwartz, N.L. (1978). "Self-financing of an R&D project". American Economic Review 68, 252–261.

Kamien, M.I., Schwartz, N.L. (1982). Market Structure and Innovation. Cambridge University Press, Cambridge, MA.

Kanwar, S., Evenson, R. (2003). "Does intellectual property protection spur technological change?" Oxford Economic Papers 55, 235–264.

Katz, M., Shapiro, C. (1985). "Network externalities, competition, and compatibility". American Economic Review 75, 424–440.

Katz, M., Shapiro, C. (1986). "Technology adoption in the presence of network externalities". Journal of Political Economy 94, 822–841.

Kaufer, E. (1989). The Economics of the Patent System. Harwood, Chichester.

Khanna, T. (1995). "Racing behavior technological evolution in the high-end computer industry". Research Policy 24, 933–958.

Khazam, J., Mowery, D. (1991). "RISC: Rewriting the dominant order in the microprocessor industry". Mimeo, University of California, Berkeley .

King, A.A., Tucci, C.L. (2002). "Incumbent entry into new market niches: The role of experience and managerial choice in the creation of dynamic capabilities". Management Science 48, 171–186.

Kleinknecht, A. (1987). "Measuring R&D in small firms: How much are we missing?" Journal of Industrial Economics 36, 253–256.

Kleinknecht, A., Verspagen, B. (1990). "Demand and innovation: Schmookler re-examined". Research Policy 19, 387–394.

Klepper, S. (1996). "Entry, exit, growth, and innovation over the product life cycle". American Economic Review 86, 562–583.
Klepper, S., Graddy, E. (1990). "The evolution of new industries and the determinants of market structure". RAND Journal of Economics 21, 27–44.
Klepper, S., Simons, K.L. (2005). "Industry shakeouts and technological change". International Journal of Industrial Organization 23, 23–43.
Klette, T.J., Kortum, S. (2004). "Innovating firms and aggregate innovation". Journal of Political Economy 112, 986–1018.
Klevorick, A.K., Levin, R., Nelson, R.R., Winter, S. (1995). "On the sources and significance of interindustry differences in technological opportunities". Research Policy 24, 195–205.
Kline, S.J., Rosenberg, N. (1986). "An overview of innovation". In: Landau, R., Rosenberg, N. (Eds.), The Positive Sum Strategy. National Academy Press, Washington, DC.
Koeller, T. (1995). "Innovation, market structure and firm size: A simultaneous equation model". Managerial and Decision Economics 16, 259–269.
Kohn, M.G., Scott, J.T. (1982). "Scale economies in research and development: The Schumpeterian hypothesis". Journal of Industrial Economics 30, 239–249.
Kortum, S., Lerner, J. (1998). "Stronger protection or technological revolution: What is behind the recent surge in patenting?" Carnegie–Rochester Conference Series on Public Policy 48, 247–304.
Kraft, K. (1989). "Market structure, firm characteristics and innovative activity". Journal of Industrial Economics 37, 329–336.
Kremp, E., Mairesse, J. (2004). Knowledge management, innovation and productivity: A firm-level exploration based on French manufacturing CIS3 data. Working paper No. 10237. National Bureau of Economic Research, Cambridge, MA.
Lach, S., Schankerman, M. (1989). "Dynamics of R&D and investment in the scientific sector". Journal of Political Economy 97 (41), 880–994.
Landes, D.S. (1969). The Unbound Prometheus: Technological Change and Industrial Development in Western Europe from 1750 to the Present. Cambridge University Press, Cambridge, MA.
Lanjouw, J.O. (1998). "Patent protection in the shadow of infringement: Simulation estimations of patent value". Review of Economic Studies 65, 671–712.
Lederman, D., Maloney, W.F. (2003). R&D and development. Working paper No. 3024. World Bank, Washington, DC.
Lee, C.Y. (2002). "Industry R&D intensity distributions: Regularities and underlying determinants". Journal of Evolutionary Economics 12, 307–341.
Lee, C.Y. (2009). "Competition favors the prepared firm: Firms' R&D responses to competitive market pressure". Research Policy 38, 861–870.
Leibenstein, H. (1966). "Allocative efficiency vs. "X-efficiency"" American Economic Review 56, 392–415.
Leiponen, A. (2005). "Skills and innovation". International Journal of Industrial Organization 23, 303–323.
Leiponen, A., Drejer, I. (2007). "What exactly are technological regimes? Intra-industry heterogeneity in the organization of innovation activities". Research Policy 36, 1221–1238.
Lerner, J. (1997). "An empirical exploration of a technology race". RAND Journal of Economics 28, 228–247.
Lerner, J. (2002). Patent protection and innovation over 150 years. Working paper No. 8977. National Bureau of Economic Research, Cambridge, MA.
Lerner, J. (2006). "The new financial thing: The origins of financial innovations". Journal of Financial Economics 79, 223–255.
Lerner, J., Wulf, J. (2007). "Innovation and incentives: Evidence from corporate R&D". Review of Economics and Statistics 89, 634–644.
Lerner, J., Zhu, F. (2007). "What is the impact of software patent shifts? Evidence from Lotus v. Borland". International Journal of Industrial Organization 25, 511–529.
Levin, R.C. (1977). "Technical change and optimal scale: Some evidence and implications". Southern Economic Journal 44, 208–221.
Levin, R.C. (1978). "Technical change, barriers to entry, and market structure". Economica 45, 347–361.
Levin, R.C. (1981). Toward an empirical model of Schumpeterian competition. Working paper series A, No. 43. Yale School of Organization and Management, New Haven, CT.
Levin, R.C. (1982). "The semiconductor industry". In: Nelson, R.R. (Ed.), Government and Technical Progress: A Cross-Industry Analysis. Pergamon Press, New York, NY.
Levin, R.C. (1988). "Appropriability, R&D spending and technological performance". American Economic Review: Proceedings 78, 424–428.

Levin, R.C., Reiss, P.C. (1984). "Tests of a Schumpeterian model of R&D and market structure". In: Griliches, Z. (Ed.), R&D Patents and Productivity. University of Chicago Press for the NBER, Chicago, IL.
Levin, R.C., Reiss, P.C. (1988). "Cost-reducing and demand-creating R&D with spillovers". RAND Journal of Economics 19, 538–556.
Levin, R.C., Cohen, W.M., Mowery, D.C. (1985). "R&D appropriability, opportunity, and market structure: New evidence on some Schumpeterian hypotheses". American Economic Review 75, 20–25.
Levin, R.C., Klevorick, A.K., Nelson, R.R., Winter, S.G. (1987). "Appropriating the return from industrial R&D". Brookings Papers on Economic Activity, pp. 783–820.
Levy, D., Terleckyj, N. (1983). "The effects of government R&D on private R&D and productivity: A macroeconomic analysis". Bell Journal of Economics 14, 551–561.
Lichtenberg, F. (1987). "The effect of government funding on private industrial research and development: A reassessment". Journal of Industrial Economics 36, 97–104.
Lichtenberg, F. (1988). "The private R&D investment response to federal design and technical competitions". American Economic Review 78, 550–559.
Liebowitz, S.J., Margolis, S.E. (1990). "The fable of the keys". Journal of Law and Economics 33, 1–25.
Link, A.N. (1981). Research and Development in U.S. Manufacturing. Praeger, New York, NY.
Link, A.N. (1982a). "A disaggregated analysis of industrial R&D: Product versus process innovation". In: Sahal, D. (Ed.), The Transfer and Utilization of Technical Knowledge. Lexington Books, Lexington, MA.
Link, A.N. (1982b). "An analysis of the composition of R&D spending". Southern Economic Journal 49, 342–349.
Link, A.N., Long, J. (1981). "The simple economics of basic scientific research: A test of Nelson's diversification hypothesis". Journal of Industrial Economics 30, 105–109.
Link, A.N., Rees, J. (1990). "Firm size, university based research, and the returns to R&D". Small Business Economics 2, 25–31.
Link, A.N., Seaks, T.G., Woodbery, S.R. (1988). "Firm size and R&D spending: Testing for functional form". Southern Economic Journal 54 (4), 1027–1032.
Lipsey, R.G., Bekar, C., Carlaw, K. (1998). "What requires explanation?" In: Helpman, E. (Ed.), General Purpose Technologies and Economic Growth. MIT Press, Cambridge, MA.
Lunn, J. (1982). "Research and development and the Schumpeterian hypothesis: Alternate approach". Southern Economic Journal 49, 209–217.
Lunn, J. (1986). "An empirical analysis of process and product patenting: A simultaneous equation framework". Journal of Industrial Economics 34, 319–330.
Lunn, J., Martin, S. (1986). "Market structure, firm structure, and research and development". Quarterly Review of Economics and Business 26, 31–44.
MacDonald, J.M. (1985). "R&D and the direction of diversification". Review of Economics and Statistics 47, 583–590.
Malecki, E.J. (1980). "Firm size, location and Industrial R&D: A disaggregated analysis". Review of Business and Economic Research 16, 29–42.
Malerba, F. (1985). The Semiconductor Business: The Economics of Rapid Growth and Decline. University of Wisconsin Press, Madison, WI.
Malerba, F. (1992). "Learning by firms and incremental technical change". Economic Journal 102, 845–859.
Malerba, F. (2007). "Innovation and the dynamics and evolution of industries: Progress and challenges". International Journal of Industrial Organization 25, 675–699.
Malerba, F., Torrisi, S. (1992). "Internal capabilities and external networks in innovative activities: Evidence from the software industry". Economics of Innovation and New Technology 2, 49–71.
Mansfield, E. (1962). "Entry, Gibrat's law, innovation, and the growth of firms". American Economic Review 52, 1023–1051.
Mansfield, E. (1964). "Industrial research and development expenditures: Determinant prospects, and relation of size of firm and inventive output". Journal of Political Economy 72, 319–340.
Mansfield, E. (1968). Industrial Research and Technological Innovation: An Econometric Analysis. Norton, New York, NY.
Mansfield, E. (1981). "Composition of R and D expenditures: Relationship to size, concentration, and innovation output". Review of Economics and Statistics 62, 610–614.
Mansfield, E. (1983). "Technological change and market structure: An empirical study". American Economic Review: Proceedings 73, 205–209.
Mansfield, E. (1985). "How rapidly does new industrial technology leak out?" Journal of Industrial Economics 34, 217–223.

Mansfield, E. (1986). "Patents and innovation: An empirical study". Management Science 32, 173–181.
Mansfield, E. (1991). "Academic research and industrial innovation". Research Policy 20, 1–12.
Mansfield, E., Rapoport, J., Schnee, J., Wagner, S., Hamburger, M. (1971). Research and Innovation in the Modern Corporation. Norton, New York, NY.
Mansfield, E., Schwartz, M., Wagner, S. (1981). "Imitation costs and patents: An empirical study". Economic Journal 91, 907–918.
Mansfield, E., Romeo, A., Schwartz, M., Teece, D., Wagner, S., Brach, P. (1982). Technology Transfer, Productivity, and Economic Policy. Norton, New York, NY.
Manso, G. (2006). "Motivating innovation". Mimeo, MIT Sloan School.
Marin, P., Siotis, G. (2007). "Innovation and market structure: An empirical evaluation of the 'bounds approach' in the chemical industry". The Journal of Industrial Organization 55, 93–111.
Markham, J.W. (1965). "Market structure, business conduct, and innovation". American Economic Review 55, 323–332.
Marx, M., Strumsky, D., Fleming, L. (2009). "Mobility, skills, and the Michigan non-compete experiment". Management Science 55, 875–889.
Mason, E.S. (1951). "Schumpeter on monopoly and the large firm". Review of Economics and Statistics 33, 139–144.
Matraves, C. (1999). "Market structure, R&D and advertising in the pharmaceutical industry". Journal of Industrial Economics 47, 169–194.
McEachern, W.A., Romeo, A. (1978). "Stockholder control, uncertainty, and the allocation of resources to research and development". Journal of Industrial Economics 26, 349–361.
Meisel, J.B., Lin, S. A. Y. (1983). "The impact of market structure on the firm's allocation of resources to research and development". Quarterly Review of Economics and Business 23, 28–43.
Merges, R.P., Nelson, R.R. (1990). "On the complex economics of patent scope". Columbia Law Review 90, 839–916 (May).
Meron, A., Caves, R.E. (1991). "Rivalry among firms in research and development outlays". Mimeo, Harvard University .
Merton, R.K. (1962). "Priorities in scientific discovery: A chapter in the sociology of science". In: Barber, B., Hirsch, W. (Eds.), The Sociology of Science. Free Press, New York, NY.
Metcalfe, J.S. (1988). "Evolution and Economic Change". In: Silberston, A. (Ed.), Technology and Economic Progress. Macmillan, London.
Mitchell, W. (1989). "Whether and when? Probability and timing of incumbent's entry into emerging industrial subfields". Administrative Science Quarterly 34, 208–234.
Mohnen, P., Roller, L.H. (2005). "Complementarities in innovation policy". European Economic Review 49, 1431–1450.
Montgomery, C. (1994). "Corporate diversification". Journal of Economic Perspectives 8, 163–178.
Montgomery, C., Hariharan, S. (1991). "Diversified expansion in large established firms". Journal of Economic Behavior & Organization 15, 71–89.
Moser, P. (2005). "How do patent laws influence innovation? Evidence from 19th-century world's fairs". American Economic Review 95, 1215–1236.
Mowery, D.C. (1983a). "The relationship between intrafirm and contractual forms of industrial research in American manufacturing, 1900–1940". Explorations in Economic History 20, 351–374.
Mowery, D.C. (1983b). "Industrial research and firm size, survival, and growth in American manufacturing, 1921–46: An assessment". Journal of Economic History 43, 953–980.
Mowery, D.C., Rosenberg, N. (1989). Technology and the Pursuit of Economic Growth. Cambridge University Press, Cambridge, MA.
Mowery, D.C., Rosenberg, N. (1979). "The influence of market demand upon innovation: A critical review of some recent empirical studies". Research Policy 8, 102–153.
Mueller, W.F. (1962). "The origins of the basic inventions underlying DuPont's major product and process innovations, 1920 to 1950". In: Nelson, R.R. (Ed.), The Rate and Direction of Inventive Activity. Princeton University Press, Princeton, NJ.
Mueller, D.C. (1967). "The firm's decision process: An econometric investigation". Quarterly Journal of Economics 81, 58–87.
Mueller, D.C. (1986). Profits in the Long Run. Cambridge University Press, New York, NY.
Mueller, D.C., Tilton, J.E. (1969). "Research and development costs as a barrier to entry". Canadian Journal of Economics 2, 570–579.
Mukhopadhyay, A.K. (1985). "Technological progress and change in market concentration in the U.S.: 1963–77". Southern Economic Journal 52, 141–149.

Nadiri, W.I. (1993). Innovations and technological spillovers. Working paper No. 4423. National Bureau of Economic Research, Cambridge, MA (August).
Narin, F. (1983). "Patent citation analysis as an analytical tool in marketing research". Mimeo, CHI Research/Computer Horizons, Inc.
Narin, F., Wolf, P. (1983). "Technological performance assessments based on patents and patent citations". Mimeo, CHI Research/Computer Horizons, Inc.
National Research Council (NRC). (2004). In: Merrill, S.A., Levin, R.C., Myers, M.B. (Eds.), A Patent System for the 21st Century. National Academies Press, Washington, DC.
National Research Council (NRC). (2005). Measuring Research and Development Expenditures in the U.S. Economy. National Academies Press, Washington, DC.
Nelson, R.R. (1959). "The simple economics of basic scientific research". Journal of Political Economy 67, 297–306.
Nelson, R.R. (1962). "The link between science and invention: The case of the transistor". In: Nelson, R.R. (Ed.), The Rate and Direction of Inventive Activity. Princeton University Press, Princeton, NJ.
Nelson, R.R. (1981). "Assessing private enterprise: An exegesis of tangled doctrine". Bell Journal of Economics 12, 93–111.
Nelson, R.R. (1982a). "The role of knowledge in R&D efficiency". Quarterly Journal of Economics 97, 453–470.
Nelson, R.R. (Ed.), (1982). Government and Technical Progress: A Cross-Industry Analysis. Pergamon Press, New York, NY.
Nelson, R.R. (1984). High-Technology Policies: A Five-Nation Comparison. American Enterprise Institute, Washington, DC.
Nelson, R.R. (1986). "Institutions supporting technical advance in industry". American Economic Review: Proceedings 76, 186–189.
Nelson, R.R. (1989). "Capitalism as an engine of progress". Research Policy 19, 193–214.
Nelson, R.R. (1991). "Why do firms differ, and how does it matter?" Strategic Management Journal 12, 61–74.
Nelson, R.R. (Ed.), (1993). Technical Innovation and National Systems. Oxford University Press, New York, NY.
Nelson, R.R., Winter, S. (1977). "In search of useful theory of innovation". Research Policy 6, 36–76.
Nelson, R.R., Winter, S. (1978). "Forces generating and limiting concentration under Schumpeterian competition". Bell Journal of Economics 9, 524–548.
Nelson, R.R., Winter, S. (1982a). An Evolutionary Theory of Economic Change. Harvard University Press, Cambridge, MA.
Nelson, R.R., Winter, S. (1982b). "The Schumpeterian tradeoff revisited". American Economic Review 72, 114–132.
Nelson, R.R., Peck, M.J., Kalachek, E.D. (1967). Technology, Economic Growth, and Public Policy. The Brookings Institution, Washington, DC.
Nickell, S.J. (1996). "Competition and Corporate Performance". Journal of Political Economy 104, 724–746.
Ordover, J. (1991). "A patent system for both diffusion and exclusion". Journal of Economic Perspectives 5, 43–60.
Pakes, A. (1986). "Patents as options: Some estimates of the value of holding European patent stocks". Econometrica 54, 755–784.
Pakes, A., Griliches (1984). "Patents and R&D at the firm level: A first look". In: Griliches, Z. (Ed.), R&D, Patents and Productivity. University of Chicago Press, Chicago, Illinois.
Pakes, A., Schankerman, M. (1984). "An exploration into the determinants of research intensity". In: Griliches, Z. (Ed.), R&D Patents and Productivity. University of Chicago Press for the NBER, Chicago, IL.
Park, W.G., Ginarte, J.C. (1997). "Intellectual Property Rights and Economic Growth". Contemporary Economic Policy 15, 51–61.
Parker, W.N. (1972). "Agriculture". In: Davis, L.E., Easterlin, R.A., Parker, W.N. (Eds.), American Economic Growth: An Economist's History of the United States. Harper & Row, New York, NY.
Pavitt, K. (1984). "Sectoral patterns of technical change: Towards a taxonomy and a theory". Research Policy 13, 343–373.
Pavitt, K. (1987). "The objectives of technology policy". Science and Public Policy 14, 182–188.
Pavitt, K., Robson, M., Townsend, J. (1987). "The size distribution of innovating firms in the UK: 1945–1983". Journal of Industrial Economics 35, 297–316.
Peck, M.J. (1962). "Inventions in the postwar American aluminum industry". In: Nelson, R.R. (Ed.), The Rate and Direction of Inventive Activity. Princeton University Press, Princeton, NJ.
Penrose, E. (1951). The Economics of the International Patent System. Johns Hopkins University Press, Baltimore, MD.
Phillips, A. (1966). "Patents, potential competition, and technical progress". American Economic Review 56, 301–310.
Phillips, A. (1971). Technology and Market Structure. D.C. Heath, Lexington, MA.
Phlips, L. (1971). Effects of Industrial Concentration: A Cross-Section Analysis for the Common Market. North-Holland, Amsterdam.

Porter, M.E. (1980). Competitive strategy: Techniques for analyzing a business, industry and competitors. Free Press, New York, NY.
Porter, M.E. (1990). The Competitive Advantage of Nations. Free Press, New York, NY.
Prendergast, C. (1999). "The provision of incentives in firms". Journal of Economic Literature 37, 7–63.
Prusa, T.J., Schmitz, J.A.Jr. (1991). "Are new firms an important source of innovation?" Economics Letters 35? 339–342.
Ravenscraft, D., Scherer, F.M. (1987). "The lag structure of returns to research and development". Applied Economics 14, 603–620.
Reinganum, J.F. (1983). "A dynamic game of R and D: Patent protection and competitive behavior". Econometrica 50, 671–688.
Reinganum, J.F. (1989). "The theory of innovation: Research, development and diffusion". In: Schmalensee, R., Willig, R.D. (Eds.), Handbook of Industrial Organization. North-Holland, Amsterdam.
Riggs, W., von Hippel, E. (1994). "Incentives to invent and the sources of innovation: The case of scientific instruments". Research Policy 23, 459–469.
Robson, M., Townsend, J. (1984). "Users manual for ESRC archive file on innovations in Britain since 1945: 1984 update". Mimeo, Science Policy Research Unit, University of Sussex.
Romer, P.M. (1990). "Endogenous technological change". Journal of Political Economy 98 (5), 71–102.
Rosen, P.J. (1991). "Research and development with asymmetric firm sizes". RAND Journal of Economics 22, 411–429.
Rosenberg, N. (1969). "The direction of technological change: Inducement mechanisms and focusing devices". Economic Development and Cultural Change 18, 1–24.
Rosenberg, N. (1974). "Science, invention, and economic growth". Economic Journal 84, 90–108.
Rosenberg, N. (1982). Inside the Black Box: Technology and Economics. Cambridge University Press, New York, NY.
Rosenberg, N. (1990). "Why do firms do basic research (with their own money)?" Research Policy 19, 165–174.
Rosenberg, N. (1994). Exploring the Black Box: Technology, Economics and History. Cambridge University Press, Cambridge, MA.
Rosenberg, N., Steinmueller, W.E. (1988). "Why are Americans such poor imitators?" American Economic Review: Proceedings 78, 229–234.
Rosenbloom, R., Christensen, C. (1994). "Technological discontinuities, organizational capabilities, and strategic commitments". Industrial and Corporate Change 3, 655–685.
Rothblum, U.G., Winter, S.G. (1985). "Asymptotic behaviour of market shares for a stochastic growth model". Journal of Economic Theory 36, 352–366.
Rothwell, R., Freeman, C., Horsley, A., Jervis, V. T. P., Robertson, A.B., Townsend, J. (1974). "SAPPHO updated—Project SAPPHO phase II". Research Policy 3, 258–291.
Sah, R.J., Stiglitz, J.E. (1986). "The architecture of economic systems: Hierarchies and polyarchies". American Economic Review 76, 716–727.
Sah, R.J., Stiglitz, J.E. (1988). "Committees, hierarchies and polyarchies". Economic Journal 98, 451–470.
Sahal, D. (1981). Patterns of Technological Innovation. Addison-Wesley, New York, NY.
Sakakibara, M., Branstetter, L. (2001). "Do stronger patents induce more innovation? Evidence from the 1988 Japanese patent law reforms". RAND Journal of Economics 33, 77–100.
Sauermann, H., Cohen, W.M. (2008). What makes them tick? Employee motives and industrial innovation. Working paper No. 14443. National Bureau of Economic Research, Cambridge, MA.
Schankerman, M. (1998). "How valuable is patent protection? Estimates by technology field". RAND Journal of Economics 29, 77–107.
Schankerman, M., Pakes, A. (1986). "Estimates of the value of patent rights in European countries during the post-1950 period". Economic Journal 96, 1052–1076.
Scherer, F.M. (1965a). "Firm size, market structure, opportunity, and the output of patented inventions". American Economic Review 55, 1097–1125.
Scherer, F.M. (1965b). "Size of firm, oligopoly, and research: A comment". Canadian Journal of Economics and Political Science 31, 256–266.
Scherer, F.M. (1967a). "Market structure and the employment of scientists and engineers". American Economic Review 57, 524–531.
Scherer, F.M. (1967b). "Research and development resource allocation under rivalry". Quarterly Journal of Economics 81, 359–394.

Scherer, F.M. (1980). Industrial Market Structure and Economic Performance (second ed.). Rand McNally, Chicago, IL.
Scherer, F.M. (1982a). "Inter-industry technology flows in the United States". Research Policy 11, 227–245.
Scherer, F.M. (1982b). "Inter-industry technology flows and productivity growth". Review of Economics and Statistics 44, 627–634.
Scherer, F.M. (1982c). "Demand-pull and technological innovation: Schmookler revisited". Journal of Industrial Economics 30, 225–237.
Scherer, F.M. (1984a). "Using linked patent and R&D data to measure interindustry technology flows". In: Griliches, Z. (Ed.), R&D Patents and Productivity. University of Chicago Press for the NBER, Chicago, IL.
Scherer, F.M. (1984b). Innovation and Growth: Schumpeterian Perspectives. MIT Press, Cambridge, MA.
Scherer, F.M. (1991). "Changing perspectives on the firm size problem". In: Acs, Z.J., Audretsch, D.B. (Eds.), Innovation and Technological Change: An International Comparison. Harvester Wheatsheaf, New York, NY.
Scherer, F.M. (1992). "Schumpeter and plausible capitalism". Journal of Economic Literature 30, 1419–1436.
Scherer, F.M., Ross, D. (1990). Industrial Market Structure and Economic Performance. Houghton Mifflin, Boston, MA.
Scherer, F.M., Herzstein, S., Jr., Dreyfoos, A., Whitney, W., Bachmann, O., Pesek, C., Scott, C., Kelly, T., Galvin, J. (1959). Patents and the Corporation (second ed.). Privately published, Boston, MA.
Scherer, F.M., Beckenstein, A., Kaufer, E., Murphy, R.D. (1975). The Economics of Multi-plant Operation. Harvard University Press, Cambridge, MA.
Schmalensee, R., Willig, R. (Eds.), (1989). Handbook of Industrial Organization. North-Holland, Amsterdam.
Schmookler, J. (1959). "Bigness, fewness, and research". Journal of Political Economy 67, 628–632.
Schmookler, J. (1962). "Economic sources of inventive activity". Journal of Economic History 22, 1–10.
Schmookler, J. (1966). Invention and Economic Growth. Harvard University Press, Cambridge, MA.
Schumpeter, J.A. (1934). The Theory of Economic Development. Harvard University Press, Cambridge, MA.
Schumpeter, J.A. (1942). Capitalism, Socialism, and Democracy. Harper, New York, NY.
Scotchmer, S. (1991). "Standing on the shoulders of giants: Cumulative research and the patent law". Journal of Economic Perspectives 5, 29–41.
Scott, J.T. (1984). "Firm versus industry variability in R&D intensity". In: Griliches, Z. (Ed.), R&D Patents and Productivity. University of Chicago Press for the NBER, Chicago, IL.
Scott, J.T. (1991). "Research diversity and technical change". In: Acs, Z., Audretsch, D. (Eds.), Innovation and Technical Change. University of Michigan Press, Ann Arbor, MI.
Scott, J.T. (2009). "Competition in research and development: A theory for contradictory predictions". Review of Industrial Organization 34, 153–171.
Scott, J.T., Pascoe, G. (1987). "Purposive diversification of R&D in manufacturing". Journal of Industrial Economics 36, 193–206.
Shapiro, C. (2000). "Navigating the patent thicket: Cross licenses, patent pools, and standard setting". In: Jaffe, A., Lerner, J., Stern, S. (Eds.), Innovation Policy and the Economy, vol. 1. National Bureau of Economic Research/MIT Press, Cambridge, MA.
Shrieves, R.E. (1978). "Market structure and innovation: A new perspective". Journal of Industrial Economics 26, 329–347.
Sirilli, G. (1987). "Patents and inventors: An empirical study". Research Policy 16, 157–174.
Smith, K. (2005). "Measuring innovation". In: Fagerberg, J., Mowery, D.C., Nelson, R.R. (Eds.), The Oxford Handbook of Innovation. Oxford University Press, Oxford.
Smyth, D.J., Samuels, J.M., Tzoannos, J. (1972). "Patents profitability liquidity and firm size". Applied Economics 4, 77–86.
Soete, L. L. G. (1979). "Firm size and innovative activity: The evidence reconsidered". European Economic Review 12, 319–340.
Solow, R. (1957). "Technical change and the aggregate production function". Review of Economics and Statistics 34, 312–320.
Spence, A.M. (1975). "Monopoly, quality, and regulation". Bell Journal of Economics 6, 417–429.
Spence, A.M. (1984). "Cost reduction, competition, and industry performance". Econometrica 52, 101–121.
Spulber, D.F. (2002). Famous Fables of Economics—Myths of Market Failure. Blackwell Publishers, Oxford.
Stern, S. (2004). "Do scientists pay to be scientists?". Management Science 50, 835–853.
Stoneman, P. (1979). "Patenting activity: A re-evaluation of the influence of demand pressures". Journal of Industrial Economics 27 (4), 385–401.
Stoneman, P. (Ed.), (1995). Handbook of the Economics of Innovation and Technical Change. Basil Blackwell, Oxford.
Stuart, T., Sorensen, O. (2003). "Liquidity events and the geographic distribution of entrepreneurial activity". Administrative Science Quarterly 48, 175–201.
Sutton, J. (1998). Technology and market structure. MIT Press, Cambridge, MA.

Sutton, J. (2007). "Market structure: Theory and evidence". In: Armstrong, M., Porter, R.H. (Eds.), Handbook of Industrial Organization, vol. 3. Elsevier, Amsterdam.
Switzer, L. (1984). "The determinants of industrial R&D: A funds flow simultaneous equation approach". Review of Economics and Statistics 67 (1), 163–168.
Taylor, C.T., Silberston, Z.A. (1973). The Economic Impact of the Patent System: A Study of the British Experience. Cambridge University Press, Cambridge, MA.
Teece, D.J. (1986). "Profiting from technological innovation: Implications for integration, collaboration, licensing and public policy". Research Policy 15, 286–305.
Teece, D.J. (1987). "Profiting from technological innovation: Implications for integration, collaboration, licensing, and public policy". In: Teece, D.J. (Ed.), The Competitive Challenge: Strategies for Industrial Innovation and Renewal. Ballinger, Cambridge, MA.
Teece, D.J. (2006). "Reflections on "profiting from innovation"" Research Policy 35, 1131–1147.
Teece, D.J., Armour, H.O. (1977). "Innovation and divestiture in the U.S. oil industry". In: Teece, D.J. (Ed.), R&D in Energy—Implications of Petroleum Industry Reorganization. Institute for Energy Studies, Stanford, CA.
Temin, P. (1979). "Technology, regulation, and market structure in the modern pharmaceutical industry". Bell Journal of Economics 10, 429–446.
Tether, B.S. (1998). "Small and large firms: Sources of unequal innovations?" Research Policy 27, 725–745.
The Futures Group. (1984). Characterization of Innovations Introduced on the U.S. Market in 1982. U.S. Small Business Administration, Office of Advocacy., Washington, DC NTIS No. PB84-212067 (March).
Thompson, P., Kean, M.F. (2005a). "Patent citations and the geography of knowledge spillovers: A reassessment". American Economic Review 95, 450–460.
Thompson, P., Kean, M.F. (2005b). "Patent citations and the geography of knowledge spillovers: A reassessment: Reply". American Economic Review 95, 465–466.
Trajtenberg, M. (1989). "The welfare analysis of product innovations, with an application to computed tomography scanners". Journal of Political Economy 97 (2), 444–479.
Trajtenberg, M. (1990a). Economic Analysis of Product Innovation—The Case of CT Scanners. Harvard University Press, Cambridge, MA.
Trajtenberg, M. (1990b). "A penny for your quotes: Patent citations and the value of innovations". RAND Journal of Economics 21, 172–187.
Tripsas, M. (1997a). "Unraveling the process of creative destruction: Complementary assets and incumbent survival in the typesetter industry". Strategic Management Journal 18, 119–142.
Tripsas, M. (1997b). "Surviving radical technological change through dynamic capability: Evidence from the typesetter industry". Industrial and Corporate Change 6, 341–377.
Tripsas, M., Gavetti, G. (2000). "Capabilities, cognition and inertia: Evidence from digital imaging". Strategic Management Journal 21, 1147–1161.
Tushman, L.M., Anderson, P. (1986). "Technological discontinuities and organizational environments". Administrative Science Quarterly 31, 439–465.
Urban, G.L., Weinberg, B.D., Hauser, J.R. (1996). "Premarket forecasting of really-new products". Journal of Marketing 60, 47–60.
Utterback, J.M. (1979). "The dynamics of product and process innovation in industry". In: Hill, C.T., Utterback, J.M. (Eds.), Technological Innovation for a Dynamic Economy. Pergamon Press, New York, NY.
Van Dijk, B., Den Hertog, R., Menkveld, B., Thurik, R. (1997). "Some new evidence on the determinants of large- and small-firm innovation". Small Business Economics 9, 335–343.
Villard, H. (1958). "Competition, oligopoly and research". Journal of Political Economy 66, 483–497.
von Hippel, E. (1976). "The dominant role of the user in the scientific instrument innovation process". Research Policy 5, 212–239.
von Hippel, E. (1977). "The dominant role of the user in semiconductor and electronic subassembly process innovation". IEEE Transactions on Engineering Management 24, 60–71.
von Hippel, E. (1982). "Appropriability of innovation benefit as a predictor of the source of innovation". Research Policy 11, 95–115.
von Hippel, E. (1988). The Sources of Innovation. Oxford University Press, New York, NY.

von Hippel, E. (1994). ""Sticky information" and the locus of problem solving: Implications for innovation". Management Science 40, 429–439.

Wahlroos, B., Backstrom, M. (1982). "R&D intensity with endogenous concentration: Evidence for Finland". Empirical Economics 7, 13–22.

Walsh, V. (1984). "Invention and innovation in the chemical industry: Demand pull or discovery-push?" Research Policy 13, 211–234.

Wedig, G.J. (1990). "How risky is R and D? A financial approach". Review of Economics and Statistics 74, 296–302.

Williamson, O.E. (1965). "Innovation and market structure". Journal of Political Economy 73, 67–73.

Williamson, O.E. (1975). Markets and Hierarchies. Free Press, New York, NY.

Wilson, R.W. (1977). "The effect of technological environment and product rivalry on R&D effort and licensing of innovations". Review of Economics and Statistics 59, 171–178.

Wilson, R.W., Ashton, P.K., Egan, T.P. (1980). Innovation, Competition, and Government Policy in the Semiconductor Industry. Lexington Books, Lexington, MA.

Winter, S.G. (1987). "Knowledge and competence as strategic assets". In: Teece, D.J. (Ed.), The Competitive Challenge: Strategies for Industrial Innovation and Renewal. Ballinger, Cambridge, MA.

Worley, J.S. (1961). "Industrial research and the new competition". Journal of Political Economy 69, 183–186.

PART II

INVENTION AND INNOVATION

Chapter 5

THE ECONOMICS OF SCIENCE

PAULA E. STEPHAN
Andrew Young School of Policy Studies
Georgia State University
Atlanta, Georgia, USA
NBER, Cambridge
Massachusetts, USA

Contents

Handbooks in Economics, Volume 01

DOI: 10.1016/S0169-7218(10)01005-1

Abstract

This chapter examines the contributions that economists have made to the study of science and the types of contributions the profession is positioned to make in the future. Special emphasis is placed on the public nature of knowledge and characteristics of the reward structure that encourage the production and sharing of knowledge. The role that cognitive and noncognitive resources play in discovery is discussed as well as the costs of resources used in research. Different models for the funding of research are presented. The chapter also discusses scientific labor markets and the extreme difficulty encountered in forecasting the demand for and supply of scientists. The chapter closes with a discussion of the relationship of scientific research to economic growth and suggestions for future research.

Keywords

economics of science, knowledge production, patenting, priority, publishing, research

JEL classification: O31, O34, O43, Z13

1. Introduction

Science commands the attention of economists for at least three reasons. First, science is a source of growth. The lags between research and growth may be long, but the economic impact of science is indisputable. The evidence is quite tangible. Advances in information technology, for example, have contributed significantly to growth in the service sector in recent years. Medical research has done a considerable amount to extend work and life expectancy, first with the introduction of antibiotics and more recently with the introduction of new classes of drugs and medical devices.

Second, scientific research has properties of a public good. It is not depleted when shared and once it is made public others cannot easily be excluded from it. As economists, we have special concerns regarding the failure of economies to produce public goods efficiently. A major reason for studying science is that a reward system has evolved in science that goes a long way toward solving the appropriability problem associated with the production of a public good.

Third, the public nature of research and the spillovers inherent in such a system are fundamental to the concept of endogenous growth theory developed by Paul Romer and others that is now a cornerstone of growth theory in economics.

This chapter attempts to bring together lines of inquiry concerning science and to incorporate into the discussion salient facts about science and scientists that have been observed by colleagues working in other disciplines. We begin by discussing the public nature of knowledge and characteristics of the reward structure. Special attention is given to the recognition that priority of discovery is a form of property rights. We then explore how science is produced, emphasizing not only labor inputs but also the important role that materials and equipment play in scientific discovery and the ways in which discovery is affected by advances in technology. This is followed by a discussion of scientific contests and the character of research. We next discuss outcomes. Included is a discussion of the relationship of gender to productivity and the inequality observed among both publishing and patenting outcomes.

The second half of this essay begins with a discussion of efficiency considerations and funding regimes. Included is a discussion of efficiency considerations related to the reward system in science and whether there are too many contestants in certain scientific contests. This leads to a discussion of how the incentives to disclose information in a timely fashion relate to the type of property right sought. We see that it is not uncommon for scientists in industry to publish, nor for scientists working in the nonprofit sector to "privatize" information. We continue by discussing scientists working in industry, and more generally discuss the market for scientists and engineers. We close with a discussion of empirical studies relating scientific research to economic growth and endogenous growth theory.

2. The public nature of knowledge and the reward structure of science

In his 1962 article concerning the economics of information, Kenneth Arrow discussed properties of knowledge that make it a public good. Others (e.g., Dasgupta and David, 1987, 1994; Johnson, 1972; Nelson, 1959) have also commented on the public nature of knowledge: it is not depleted when shared, and once it is made public others cannot easily be excluded from its use.[1] Moreover, the incremental cost of an additional user is virtually zero[2] and, unlike the case with other public goods, not only is the stock of knowledge not diminished by extensive use, it is often enlarged. This means that the transmission of knowledge is a positive sum game (Foray, 2004, p. 93).[3]

Economists were not the first to note the public nature of knowledge. More than 190 years ago, Jefferson (1967 edition, p. 433, section 4045) wrote:

> *"If nature has made any one thing less susceptible than all others of exclusive property, it is the action of the thinking power called an idea, which an individual may exclusively possess as long as he keeps it to himself; but the moment it is divulged, it forces itself into the possession of every one, and the receiver cannot dispossess himself of it. Its peculiar character, too, is that no one possesses the less, because every other possesses the whole of it. He who receives an idea from me, receives instruction himself without lessening mine; as he who lights his taper at mine, receives light without darkening mine."*[4]

A cornerstone of economic theory is that competitive markets provide poor incentives for the production of a public good. The nonexcludable nature of public goods invites free-riders and consequently makes it difficult for providers to capture the economic returns. Thus, incentives for provision are not present. Moreover, the nonrivalrous nature of public goods means that if and when public goods are produced, the market will fail to provide them efficiently where marginal cost equals marginal revenue since the marginal cost of an additional user is zero. Such observations regarding the provision of public goods, however, relate to incentives that are market based. An important contribution of the sociologists of science and the economists who have extended their work is the demonstration that a nonmarket reward system has evolved in science that provides incentives for scientists to produce and share their knowledge, thus behaving in socially desirable ways. In the sections that follow, we analyze the components of that reward system as well as the behavior it encourages.

2.1. The importance of priority

As economists, we owe a substantial debt to Robert Merton for establishing the importance of priority in scientific discovery. In a series of articles and essays begun in the late 1950s, Merton (1957, 1961, 1968, 1969) argues convincingly that the goal of scientists is to establish *priority of discovery* by being first to

[1] Research findings only become a public good when they are codified in a manner that others can understand. The distinction, therefore, is often drawn between knowledge, which is the product of research, and information, which is the codification of knowledge (Dasgupta and David, 1994, p. 493).

[2] In reality, the marginal cost of use is greater than zero because users must incur the opportunity cost of time as well as the direct cost of access to journals or attendance at meetings. Information, of course, is only of use to those who possess the requisite intellectual framework and know the "code." Michel Callon (1994) argues that the public nature of science is greatly overstated. Tacit knowledge (discussion to follow), which by definition cannot be codified, is more costly to learn than knowledge that is codified.

[3] It is the user value of knowledge that does not diminish with use. The market value of knowledge can fall with dissemination.

[4] Jefferson also noted that ideas are "like fire expansible over all space, without lessening their density in any point." (quoted in David, 1993, p. 226). David stresses the infinite expansibility of knowledge rather than the nonrival characteristics of knowledge.

communicate an advance in knowledge and that the rewards to priority are *the recognition awarded by the scientific community for being first.* Merton further argues that the interest in priority and the intellectual property rights awarded to the scientist who is first are not a new phenomenon but have been an overriding characteristic of science for at least three centuries.

The recognition awarded priority has varied forms, depending upon the importance the scientific community attaches to the discovery. Heading the list is eponymy, the practice of attaching the name of the scientist to the discovery. Haley's comet, Planck's constant, Hodgkin's disease, the Copernican system are all examples.[5] Recognition also comes in the form of prizes. Of these, the Nobel is the best known, carrying the most prestige and the largest purse (approximately $1.4 million in 2009), but hundreds of others exist, a handful of which have purses in excess of $500,000, such as the Lemelson-MIT Prize with a $500,000 (US) purse, the Shaw Prize ($1 million US) and the Spinoza Prize (1.5 million euros).[6] The number of prizes awarded has grown in recent years. Zuckerman (1992) estimates that approximately 3000 prizes in the sciences were available in North America alone in the early 1990s, five times the number awarded 20 years earlier. Although no systematic study of prizes has been done since, anecdotal evidence suggests that the number continues to grow. *Science*, the highly cited journal of the American Association for the Advancement of Science, regularly features recent recipients of prizes, many of which are awarded by companies and recently established foundations and often have purses in excess of $250,000.[7]

Publication is a lesser form of recognition, but a necessary step in establishing priority. While eponymy or the receipt of a prestigious prize is perceived by most to be beyond their reach, the reward of publication is within the reach of most scientists. A common way to measure the importance of a scientist's contribution is to count the number of citations to an article or the number of citations to the entire body of work of an investigator. While this used to be a laborious process, changes in technology, as well as the incentives to create new products, such as Google Scholar, have meant that researchers, and those who evaluate them, can quickly (and sometimes incorrectly) count citations to their work as well as where they stand relative to their peers. Thompson Scientific, for example, markets a product that ranks scientists, within a field, in terms of citations.[8]

It is important to stress that priority is established by being first. The behavior such an incentive structure elicits is one of the themes of this chapter. One consequence is the perceived need to publish quickly. It is not unknown for scientists to write and submit an article in the same day. Neither is it

[5] The Higgs particle is much in the news these days with the construction of the new accelerator at CERN (the LHC) and its associated four colliders. Named for the Scottish physicist Peter Higgs, who first postulated its existence, its existence has been sought at every collider since then.

[6] The Fields Medal is the closest equivalent to the Nobel Prize in math. Awarded every 4 years, to up to four mathematicians under the age of 40, it carries a nominal purse of around $13,000. It garnered considerable attention in 2007 when one of the four recipients of the Medal, Grigory Perelman, honored for his proof of the Poincaré conjecture, refused the prize. In 2002, the Norwegian government established the Abel Prize in mathematics; the 2006 award carried a purse of $920,000, making it the largest prize in mathematics.

[7] By way of example, Johnson&Johnson established the Dr Paul Janssen Award for Biomedical Research in 2005 with a purse of $100,000; the Heinz Foundation awards Heinz Prizes ($250,000); the Peter Gruber Foundation began to award several prizes beginning in 2000, including one in genetics for $250,000; GE partnered with *Science* to create the Prize for Young Life Scientists in 1995 ($25,000); General Motors awards the General Motors Cancer Research Prize ($250,000).

[8] Such lists are not without errors. The presence of common names, especially among the Asian community, means that attribution can be incorrect and thus such rankings must be cautiously used and carefully monitored.

unknown to negotiate with the editor of a prestigious journal the timing of a publication or the addition of a "note added" so that work completed between the time of submission and publication can be reported, thus making the claim to priority all the more convincing (Stephan and Levin, 1992). The time between receipt of a manuscript and publication is considerably shorter in science than in the social sciences. At the extreme is the practice of the journal *Science* to ask that referee reports be returned within 7 days of receipt and to then publish quickly following the editorial decision to accept. Ellison (2002) documents discipline differences and how these have changed over time. The move to electronic publication is quickening the process and may narrow the difference between science and the social sciences.

Another consequence of a priority-based reward system is the energy that scientists devote to establishing priority over rival claims. Moreover, such practices are not new. Merton (1969, p. 8) describes the extreme measures Newton took to establish that he, not Leibniz, was the inventor of the calculus.[9]

Science is sometimes described as a "winner-take-all" contest," meaning that there are no rewards for being second or third. One characteristic of science that contributes to such a reward structure is the difficulty that occurs in monitoring scientific effort (Dasgupta, 1989; Dasgupta and David, 1987). This class of problem is not unique to science. Lazear and Rosen (1981) have investigated incentive-compatible compensation schemes where monitoring is costly. Another factor that contributes to such a reward structure is the low social value of the contributions made by the runner-up. "There is no value added when the same discovery is made a second, third, or fourth time." (Dasgupta and Maskin, 1987, p. 583).

But it is somewhat extreme to view science as a winner-take-all contest. Even those who describe scientific contests in such ways note that it is a somewhat inaccurate description, given that replication and verification have social value and are common in science. It is also inaccurate to the extent that it suggests that only a handful of contests exist. True, some contests are world class, such as identification of the Higgs particle or the development of high-temperature superconductors. But there are many other contests that have multiple components, and the number of such contests appears to be on the increase. By way of example, while for many years it was thought that there would be "one" cure for cancer, it is now realized that cancer takes multiple forms and that multiple approaches are needed to find a cure. There will not be but one winner; there will be multiple winners.

A more realistic metaphor is to see science as following a tournament arrangement, much like tournaments in golf or tennis, where the losers, too, get some rewards. This keeps individuals in the game, raises their skills, and enhances their chances of winning a future tournament. A similar type of competition exists in science. Dr X is passed over for the Lasker Prize, but her work is sufficiently distinguished that she is invited to give an important lecture, consistently receives support for her research and is awarded an honorary degree from her undergraduate institution.

2.2. Financial remuneration and the satisfaction derived from solving the puzzle

Financial remuneration is another component of the reward structure of science. While scientists place great importance on priority and are highly motivated by an interest in puzzle-solving, money clearly plays a role in the reward structure. Rosovsky (1990) recounts how, upon becoming dean of the Faculty

[9] A tension that exists between experimentalists and theorists in physics is the "awkward matter of credit." "Who should get the glory when a discovery is made: the theorist who proposed the idea, or the experimentalist who found the evidence for it?" (Kolbert, 2007, p. 75).

of Arts and Sciences at Harvard, he asked one of Harvard's most eminent scientists the source of his scientific inspiration. The reply (which "came without the slightest hesitation") was "money and flattery." (p. 242).

The tournament nature of the race places much of the risk on the shoulders of the scientists.[10] It is, therefore, not surprising that compensation in science is generally composed of two parts: one portion is paid regardless of the individual's success in races; the other is priority-based and reflects the value of the winner's contribution to science. While this clearly oversimplifies the compensation structure, counts of publications and citations play a significant role in academic promotions and raises, at least in the United States, although empirical work regarding the relationship is considerably dated (Diamond, 1986b; Tuckman and Leahey, 1975).[11] Salaries and resources are based on productivity in other countries, as well. Chinese researchers who place in the top half of their colleagues in terms of bibliometric measures can earn three to four times the salaries of coworkers (Hicks, 2007). The funding for academic departments in the United Kingdom is based in part on published output, as is that in Australia (Hicks, 2007). Unfortunately, we know little about the reward structure for scientists in industry or in government labs, particularly as the reward structure relates to priority.[12]

The flat profile of earnings in science (at least for those employed in academe) is frequently noted. Ehrenberg (1992), for example, calculates that the average full professor in the physical and life sciences earns only about 70% more than the average new assistant professor. In countries where faculties are civil servants, the profiles are also rather flat. The shape of the profile arguably relates to monitoring problems and the need to compensate scientists for the risky nature of their work. On the other hand, if earnings are expanded to include other forms of compensation, the profiles are not as flat as is assumed. The additional monetary awards that await the successful scientist take the form of prize money, speaking and consulting fees, and royalties. A fruitful area for further research would be to investigate the shape of the earnings profile when the definition of income is broadened to include other forms of compensation briefly elaborated below.

Royalties from patents are one form of additional compensation available to certain university faculty. Thursby and Thursby (2007) find that 10.3% of faculty at the highly selective US universities they study disclosed an invention to the technology transfer office in 1999. While many disclosures are not patented and most patents produce a small royalty stream at best, some produce substantial sums and in rare cases extraordinary sums. For example, Emory University in July 2005 sold its royalty interests in emtricitabine, also known as Emtriva®, and used in the treatment of HIV, to Giliad Sciences, Inc. and Royalty Pharma. The university received \$525 million (US). The three Emory University scientists involved received approximately 40% of the sale price, reflecting the university policy that was in place at the time (http://sec.edgar-online.com/2005/08/04/0001193125-05-157811/Section7.asp).

[10] Arrow (1962) noted that it is fortuitous that teaching and research activities are two sides of the same profession since the arrangement provides for researchers to be remunerated not on the basis of research (which would lead to a highly irregular pattern) but on that of teaching.

[11] The relationship between productivity and salary can be enhanced by the awarding of an endowed chair which pays a supplement over and above the scientist's salary. In some US universities the relationship between compensation and productivity is further enhanced through the university's practice of sharing indirect costs with faculty as a way to increase incentives for faculty to submit grant proposals.

[12] There is some evidence that increasing amounts of risk are being shifted to the scientist. For example, in the US university scientists, even those who are tenured, increasingly are expected to raise a portion of their salary from grants and contracts.

Royalty payments received by universities have dramatically increased in recent years, suggesting that faculty royalty payments have increased as well. Within the United States, for example, the amount of annual net royalty payments received by the university went from $195.0 to $866.8 million (US) during the period 1993–2003 (National Science Board, 2006, Table 5-28, vol. 2). University policies vary in terms of how royalties are shared with faculty inventors, but in all cases the inventor receives a portion of the stream of revenues. Lach and Schankerman (2008) have investigated how the structure of the sharing formula relates to invention disclosure and provide empirical support for the view that invention activity, as measured by invention disclosures, is positively related to the share of license income accruing to faculty.[13]

Faculty may also earn income and wealth through their role in start-up companies. In the most extreme case, the faculty member reaps rewards when the company goes public. Sometimes these are of staggering proportions, at least on paper. A case in point is Eric Brewer, a computer scientist at UC Berkeley, who was listed on *Fortune* magazine's list of the 40 richest Americans under 40 in October 1999 with a net worth of $800 million (US), a result of the role he played in founding a company that went public in 1998 (Wilson, 2000). Edwards et al. (2006) document that, in the event a biotechnology firm makes an initial public offering, the median value of equities held by an academic with formal ties to the company, based on the IPO's closing price, ranged from $3.4 million to $8.7 billion, depending upon the period analyzed. The incidence of being on a scientific advisory board (SAB) is nontrivial. Ding et al. (2006b) identify 785 academic scientists who are members of one or more SABs of companies that made an initial public offering in biotechnology in the United States.

The other reward often attributed to science is the satisfaction derived from solving the puzzle. Hagstrom (1965, p. 16), an early sociologist of science, noted this when he said "Research is in many ways a kind of game, a puzzle-solving operation in which the solution of the puzzle is its own reward." The philosopher of science Hull (1988, p. 305) describes scientists as being innately curious and suggests that science is "play behavior carried to adulthood." Feynman (1999), explaining why he did not have anything to do with the Nobel Prize (which he won in 1965), said: "I don't see that it makes any point that someone in the Swedish Academy decides that this work is noble enough to receive a prize—I've already got the prize. The prize is the pleasure of finding the thing out, the kick in the discovery . . ." This suggests that time spent in discovery is an argument in the utility function of scientists. Pollak and Wachter (1975) demonstrate that maximization problems of this type are generally intractable, because implicit prices depend upon the preferences of the producer. While this provides a rationale for excluding the process of discovery from models of scientific behavior, the failure of economists to acknowledge the puzzle as a motivating force makes economic models of scientific behavior lack credibility. Recent work by Sauermann and Cohen (2007) seeks to address this in part for scientists and engineers working in industry.

[13] Not all inventions made by faculty are patented by the university. Thursby et al. (2009) find that 29% of patents by US faculty are assigned to firms. Likewise, the practice of "professor privilege" that exists in several European countries means that inventions made by professors need not be assigned to the university. Crespi et al. (2009) find that the large majority of university-invented patents in their sample are not owned by universities. Instead, most are assigned to firms. We know virtually nothing about the royalties from patents assigned outside the university.

3. How knowledge is produced

> *"Any new idea—a new conceptualization of an existing problem, a new methodology, or the investigation of a new area—cannot be fully mastered, developed into the stage of a tentatively acceptable hypothesis, and possibly exposed to some empirical tests without a large expenditure of time, intelligence, and research resources."*

So Stigler (1983, p. 536) described the "production function" for knowledge in his 1982 Nobel lecture. Here we explore these components in more detail.

3.1. Time and cognitive inputs

Although it is popular to characterize scientists as having instant insight, studies suggest that science takes time. Investigators often portray productive scientists—and eminent scientists especially—as strongly motivated, with the "'stamina' or the capacity to work hard and persist in the pursuit of long-range goals." (Fox, 1983).[14]

Several dimensions of cognitive resources are associated with discovery. One aspect is ability. It is generally believed that a high level of intelligence is required to do science, and several studies have documented that, as a group, scientists have above average IQs.[15] There is also a general consensus that certain people are particularly good at doing science and that a handful are superb.[16] Another dimension of cognitive inputs is the knowledge base the scientist(s) working on a project possesses. This knowledge is used not only to solve a problem but to choose the problem and the sequence in which the problem is addressed.

The importance knowledge plays in discovery leads to several observations. First, it intensifies the race, because the public nature of knowledge means that multiple investigators have access to the knowledge needed to solve a problem. Second, knowledge can either be embodied in the scientist(s) working on the research or disembodied, but available in the literature (or from others). Different types of research rely more heavily on one than the other. The nuclear physicist Leo Szilard, who left physics to work in biology, once told the biologist Sydney Brenner that he could never have a comfortable bath after he left physics. "When he was a physicist he could lie in the bath and think for hours, but in biology he was always having to get up to look up another fact" (Wolpert and Richards, 1988, p. 107).

Third, the knowledge base of a scientist can become obsolete if the scientist fails to keep up with changes occurring in the discipline. On the other hand, the presence of fads in science (such as in particle physics) means that the latest educated are not always the best educated (Stephan and Levin, 1992). Vintage may matter in science but not always in the way that Mincer's "secular progression of knowledge" would lead us to believe (Mincer, 1974, p. 21).

[14] Hermanowicz (2006) reports that slightly over one-half of the physicists in his sample chose persistence from the list of 25 adjectives in response to the question "What do you think are the most important qualities needed to be successful at the type of work you do?" No other quality came close to persistence. Smartness was second, mentioned by 25%.

[15] Harmon (1961, p. 169) reports that PhD physicists have an average IQ in the neighborhood of 140. Catherine Cox, using biographical techniques to estimate the intelligence of eminent scientists, reports IQ guestimates of 205 for Leibnitz, 185 for Galieleo, and 175 for Kepler. Roe (1953, p. 155) summarizes Cox's findings.

[16] Feist (2006) examines the psychological forces at play in the development of an individual's interest, talent and creativity in science.

Fourth, there is anecdotal evidence that "too much" knowledge can be a bad thing in discovery in the sense that it "encumbers" the researcher. There is the suggestion, for example, that exceptional research may at times be done by the young because the young "know" less than their elders and hence are less encumbered in their choice of problems and the way they approach a question.[17]

Finally, the cognitive resources brought to bear on a problem can be enhanced by assembling a research team or, at a minimum, engaging in a collaborative arrangement with investigators in other labs and countries. Research is rarely done in isolation, especially research of an experimental rather than theoretical bent (Fox, 1991). Scientists work in labs. How these labs are staffed varies across countries. For example, in Europe research labs are often staffed by permanent staff scientists, although increasingly these positions are held by temporary employees (Stephan, 2008). In the United States, while positions such as staff scientists and research associates exist, the majority of scientists working in the lab are doctoral students and postdocs. Stephan et al.'s study (2007b) of 415 labs affiliated with a nanotechnology center finds that the average lab has 12 technical staff, excluding the principal investigator (PI). Fifty percent of these are graduate students; 16% are postdocs, and 10% are undergrads.[18] Such patterns mean that labs in the United States are disproportionately staffed by young, temporary workers. The reliance on such a system, with its underlying pyramid scheme, at a time when there has been minimal expansion in faculty positions, has resulted in an increasing supply of scientists trained in the United States (as well as those trained abroad, who come to the United States to take a postdoctoral position) who are less and less likely to find permanent PI positions in the university.[19]

One way of seeing how team size and collaboration have changed is to examine trends in coauthorship patterns. Adams et al. (2005), for example, find that the mean number of authors per paper increased from 2.8 to 4.2 for an 18-year interval, ending in 1999.[20] The rate of growth was greatest during the period 1991–1996 when use of email and the Internet was rapidly accelerating. The growth has been due both to a rise in lab size and to an increase in the number of institutions—especially foreign institution—collaborating on a research project. During the period 1988–2003, for example, the number of addresses on an article with at least one US address grew by 37% while the number of foreign addresses more than tripled (National Science Board, 2006, Table 5-18).[21]

[17] There is a literature suggesting that individuals coming from the margin—"outsiders" if you will—make greater contributions to science than those firmly entrenched in the system (Gieryn and Hirsch, 1983). Stephan and Levin (1992) argue that this is one reason why exceptional contributions are more likely to be made by younger persons. In studying Nobel laureates they conclude that although it does not take extraordinary youth to do prize-winning work, the odds decrease markedly by midcareer.

[18] Approximately a third of the PIs were affiliated with departments of engineering, a third with departments of chemistry and the remainder with departments of physics.

[19] Hollingsworth (2006) argues that the organizational structure of the institution in which the research is being performed also contributes to productivity. He sees extreme decentralization, permitting exceptionally productive scientists a high degree of autonomy and flexibility, to be a key characteristic of organizations where major discoveries occur.

[20] The study is restricted to articles in science and engineering having one or more authors from a top-110 US university.

[21] During the same period, the number of names increased by approximately 50%, suggesting that lab size was growing slightly faster than institutional collaboration growth (National Science Board, 2006, Table 5-18).

Several factors contribute to the increased role that collaboration plays in research.[22] First, the importance of interdisciplinary research and the fact that major breakthroughs often occur in emerging disciplines, encourages collaboration. Systems biology, which involves the intersection of biology, engineering, and physical sciences, is a case in point.[23] By definition, no one has all the requisite skills required to work in the area; researchers must rely on working with others. Second, and related, researchers arguably are acquiring narrower expertise over time in order to compensate for the educational demands associated with the increase in knowledge (Jones, 2005b). Narrower expertise, in turn, leads to an increased reliance on teamwork for discovery. Third, the rapid spread of connectivity, which began in the early 1980s with the adoption of bitnet by a number of universities and accelerated in the early 1990s with the diffusion of the Internet, has decreased the costs of collaboration across institutions (Agrawal and Goldfarb, 2008; Levin et al., 2006). Another factor that fosters collaboration is the vast amount of data that is becoming available, such as that from the Human Genome Project (and the associated GenBank database). Although that is probably the best known, many other large databases have recently come online, such as PubChem, which as of this writing contained 17,655,303 recorded substances, and the Worldwide Protein Data Bank (wwPDB), a worldwide depository of information regarding protein structures.[24] The practice of sharing research materials also leads to increases in the number of authors appearing on an article.

Increased complexity of equipment also fosters collaboration. At the very extreme are the teams assembled to work at colliders. CERN's four colliders have combined team size of just under 6000: 2520 for the Compact Muon Detector (CMS), 1800 for the Atlas, 1000 for ALICE, and 663 for LHCb (Overbye, 2007). Barnett et al. (1988) suggest two other factors that lead persons to seek coauthors. One is the desire to minimize risk by diversifying one's research portfolio through collaboration; the other is the increased opportunity cost of time. An additional factor is quality. The literature on scientific productivity suggests that scientists who collaborate produce "better" science than do individual investigators (Andrews, 1979; Lawani, 1986; Wuchty et al., 2007). Some of the factors encouraging collaboration are new (such as connectivity) but growth in the number of authors on a paper is not. Wuchty et al. (2007) find that team size has grown in all but one of the 171 S&E fields studied during the past 45 years.

Other chapters in this volume will address the role of networks in research. Here we note that governments on both sides of the Atlantic have bought heavily into the importance of funding collaborative research across institutions. The National Institutes of Health (NIH), in an effort to encourage collaborative research, funds P01 grants which support broadly based multidisciplinary research with multiple investigators. On the other side of the Atlantic, the European Union (EU) is committed to funding networks of excellence. While such grants clearly create incentives for individuals to work together, research has yet to show their effectiveness relative to other forms of funding. One possible reason for creating these networks is to improve incentives for labs to share data and material across labs.

[22] Changing patterns in collaboration present certain challenges for organizations. For example, as the number of coauthors grows, it becomes increasingly difficult to evaluate curriculum vitas at tenure and promotion time. Historically, for example, individuals were penalized if they only published with their mentor after completing a postdoctoral appointment. In recent years, however, programs such as the Medical College at the University of Pennsylvania have relaxed this rule and now consider such individuals for promotion.

[23] Systems biology studies the relationship between the design of biological systems and the tasks they perform.

[24] The Large Hadron Collider (LHC) at CERN will create vast amounts of data. According to Kolbert (2007, p. 74), "If all the LHC data were burned onto disks, the stack would rise at the rate of a mile a month."

3.2. Research resources

The production of knowledge also requires resources. In the social sciences this generally translates into a personal computer, access to a database and one or two graduate research assistants. For the experimental physical sciences the resource requirements are considerably more extensive, involving access to substantial equipment. Sometimes this equipment is in the lab, but particle physics experiments require time on an accelerator; astronomers require time on a telescope. Research in nanotechnology requires "clean" labs and specialized equipment such as a scanning tunneling microscope. Super computers increasingly play a role in research, both at the theoretical and at the experimental level. Moreover, as large databases become increasingly available, the use of super computers is accelerating.

Research in the biomedical sciences also increasingly requires access to sophisticated equipment. The DNA gene sequencer and synthesizer and the protein synthesizer and sequencer comprise the technological foundation for contemporary molecular biology. The revolution in proteomics and systems biology relies on analytical tools such as mass spectrometry (Chait, 2006). Robotics technology is becoming increasingly important in sequencing proteins. Research in the biomedical sciences is not only *in vitro*. *In vivo* studies have become progressively more important, especially those involving mice, which are estimated to account for more than 90% of all mammals used in research (Malakoff, 2000).

The increasing sophistication of research tools in the biomedical sciences has dramatically changed the output of a lab. While in 1990 the best equipped lab could sequence 1000 base pairs a day, by January 2000 the 20 labs involved in mapping the human genome were collectively sequencing 1000 base pairs a second, 24/7. The cost per finished base pair fell from $10.00 in 1990 to under $0.05 in 2003 (Collins et al., 2003) and was roughly $0.01 in 2007 (www.biodesign.asu./edu/news/232/) (see Figure 1). Measured in terms of base pairs sequenced per person per day, for a researcher operating multiple machines, productivity increased more than 20,000-fold from the early 1990s to 2007, doubling approximately every 12 months (http://www.bio-era.net/news/add_news_18.html).[25] More recently, next-generation sequencing machines, which first came on the market in 2007 and read millions of sequences at once, have made the earlier technology for sequencing obsolete.

The increasing sophistication of research tools means that the "capital–labor ratio" for research, at least in the biomedical sciences, is changing. In 2008, for example, the Venter Institute eliminated 29 sequencing-center jobs, announcing that the staff reduction "is a direct result of a technology shift and is not a reflection of the tough economic times that we are all facing in the United States today." (http://www.jcvi.org/cms/press/press-releases/full-text/article/j-crai...quencing-staff-positions/?tx_ttnews%5baclkPid%5D=67&Hash=db443577b0).

The substitution of capital for labor in research is an underresearched area which has clear implications for the demand for scientists. The dramatic changes in technology that have occurred have also substantially changed the nature of dissertation research. For example in chemistry, nuclear magnetic

[25] The decline in the cost of sequencing has led to the hope that personal genomes can be sequenced for $1000 or less (www.biodesign.asu.edu/news/232/). In March 2007, the Archon X Prize for Genomics was established with the goal of awarding $10 million to the first group that can "build a device and use it to sequence 100 human genomes within 10 days or less, with an accuracy of no more than one error in every 100,000 bases sequenced, with sequences accurately covering at least 98% of the genome, and at a recurring cost of no more than $10,000 per genome." (http://thepersonalgenome.com/category/sequency-cost/).

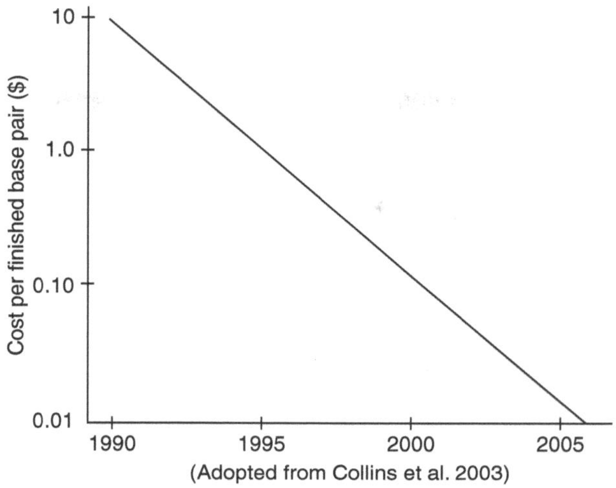

Figure 1. Cost per finished base pair ($).

resonance combined with X-ray crystallography and advanced computing power allows protein structures to be elucidated much more rapidly. As a result, while a PhD thesis used to be focused on defining the structure of a single protein domain, now a thesis in a similar field might examine and compare dozens of structures.

The importance of equipment is one reason to stress the nonlinearity of scientific discovery. Scientific research can lead to technological advance, but technology very much affects advances in science. The history of science is the history of how important resources and equipment are to discovery—a theme in the research of Rosenberg and Mokyr, among others. In some instances, and perhaps what is most efficient, the scientist is both the researcher and the inventor of new technology (Franzoni, 2009). The biologist Leroy Hood, author of more than 500 papers and winner of the 1987 Lasker Award for Basic Medical Research, exemplifies the researcher–inventor. In recognition of his inventions, which include the automated DNA sequencer and an automated tool for synthesizing DNA, he received the 2002 Kyoto Prize for Advanced Technology. In 2003 he was the recepient of the Lemelson-MIT Prize for inventing "four instruments that have unlocked much of the mystery of human biology, including the automated DNA sequencer."[26] (http://web.mit.edu/invent/n-pressreleases/n-press-03LMP.html).

Equipment for research is costly.[27] At the extreme are costs associated with building and running an accelerator. The 27-km-long LHC which is scheduled to come online early in 2008 at CERN will cost $8 billion; the Spallation Neutron Source (SNS) at Oak Ridge National Laboratory in the United States cost

[26] Hood's interest in tools and cutting edge research was instilled in him by his mentor William Dreyer, who reportedly told the then Cal Tech doctoral student "If you want to practice biology, do it on the leading edge and if you want to be on the leading edge, invent new tools for deciphering biological information." (http://web.mit.edu/invent/a-winners/a-hood.html).

[27] US academic institutions spent $1.8 billion (US) in 2003 for research equipment, approximately 2.5 times the amount spent 20 years before in real dollars (National Science Board, 2006, Appendix Table 5-13).

$1.41 billion (*Science*, vol. 312, 5 May 2006; p. 675). A microscope used for nanotechnology research can cost $750,000 (http://www.unm.edu/~market/cgi-bin/archives/000132.html). A sequencer, such as Applied Biosystems' 3730 model costs approximately $300,000. Next-generation sequencers cost between $400,000 and $500,000.

Mice are not free. An inbred off-the-shelf mouse costs between $17 and $60; mutant strains begin around $40 and can go to $500 plus. Prices are for mice supplied from live breeding colonies. Many strains, however, are only available from cyropreserved material. Such mice cost considerably more: in 2009 the cost to recover any strain from cryopreservation (either from cryopreserved sperm or embryos) was $1900. For this, investigators receive at least two breeding pairs of animals in order to establish their own breeding colony.[28] Custom made mice can cost much more. Johns Hopkins University, for example, estimates that it costs $3500 to engineer a mouse to order.

With the large number of mice in use (over 13,000 are already published), the cost of mouse upkeep becomes a significant factor in doing research. US universities, for example, charged from $0.05 to $0.10 per day per mouse (mouse per diem) in 2000 (Malakoff, 2000). This can rapidly add up. Irving Weissman of Stanford University reports that before Stanford changed its cage rates he was paying between $800,000 and $1 million a year to keep the 10,000 to 15,000 mice in his lab.[29] Costs for keeping immune deficient mice are far greater (on the order of $0.65 per day), given their susceptibility to disease.[30]

The importance of equipment and research materials in scientific research means that exchange, which has a long tradition in science (Hagstrom, 1965), plays a considerable role in fostering research and in creating incentives for scientists to behave in certain ways. For example, scientists routinely share information and access to research materials and expertise in exchange for citations and coauthorship.[31] But, as research materials have become increasingly important, exchange has arguably taken on more importance. Walsh et al. (2005, 2007) examine the practice of sharing materials (such as cell lines, reagents, and antigens) among academic biomedical researchers and find that 75% of the academic respondents in their sample made at least one request for material in a 2-year period, with an average of 7 requests for materials to other academics and two requests for materials from an industrial lab (Walsh et al., 2005).[32]

Murray examines how the advent of patenting life-forms has influenced patterns of exchange among mouse researchers during the past 100 years. She argues that although mouse geneticists resisted the imposition of patents, in recent years they accommodated them, incorporating them into their exchange relationship: "Having patents became a signal to other scientists that you were a valuable exchange

[28] The NIH and a number of other foundations have provided long-term support for the Jackson Laboratories which serves as a critical institution for the preservation and upkeep of thousands of mice, making them available to researchers and providing important economies of scale. More than 67% of the strains from the Jackson lab are only available from cryopreserved material.

[29] Given such costs, it is no surprise that "mouse packages" play a role in recruitment. The McLaughlin Research Institute in Great Falls, Montana, for example, successfully recruited a researcher when they offered him a mouse package with a per diem of $0.036 (Vogel, 2000). (Cage costs converted to mice costs at the rate of 5 mice per cage.)

[30] Researchers not only buy mice and equipment to take care of the mice; they also buy equipment to observe and record mouse activity. For example, the titanium dorsal skinfold chamber (which is designed to fit under the back of a mouse) allows the researcher to "nondestructively record and visualize microvascular functions" according to an ad placed in *Science* (June 9, 2006, p. 1439).

[31] LaTour (1987) provides a detailed account of how academics use exchange to nurture their expertise.

[32] This is not to say that scientists always "share" or exchange data and resources. See discussion of "having one's cake and eating it too" in section 6.4.

partner and therefore worthy of coauthorship. Scientific collaboration was never entered into indiscriminately but under the commercial regime, a patent became a way of signaling your value to other scientists and co-opting them in your bid for prestige and reputation." (Murray, 2006, p. 34).

The overwhelming importance of equipment to the research process and the associated costs of equipment mean that in most fields access to resources is a necessary condition for doing research. It is not enough to decide to do research, as a standard human capital model might assume. One must also have access to research inputs. At US universities, equipment, and funding for graduate and postdoc stipends, are generally provided by the dean at the time of hire in the form of start-up packages.[33] Thereafter, equipment, some buyoff for faculty time,[34] and the stipends that graduate students and postdocs receive, become the responsibility of the scientist. Scientists whose work requires access to "big" machines off campus must also submit grants to procure time (e.g., beam time) at the research facility. This means that for a variety of fields funding becomes a necessary condition for doing "independent" research that is initiated and conceived by the scientist. Scientists working in these fields in the United States take on many of the characteristics of entrepreneurs. As graduate students and postdocs they must work hard to establish their "credit-worthiness" through the research they do in other people's labs. If successful in the endeavor, and if a position exists (see discussion of cohort effects), they will subsequently be provided with a lab at a research university. They then have several years to leverage this capital into funding. If they succeed, they face the onerous job of continually seeking support for their lab; if they fail, the probability is low that they will be offered a start-up package by another university. The emphasis on the individual scientist to generate resources is not as strong in many other countries, where researchers are hired into government-funded and government-run laboratories such as CNRS in France. Nevertheless, fits and starts in funding for such programs translate into the possibility that certain cohorts of scientists enter the labor market when conditions are favorable for research while other cohorts do not.

3.3. Serendipity

Serendipity also plays a role in scientific discovery; it is not that uncommon for researchers to find different, sometimes greater, riches than the ones they are seeking. Although serendipity is sometimes referred to as the "happy accident," this is a bit of a misnomer. True, Pasteur "discovered" bacteria while trying to solve problems that were confronting the French wine industry. But his discovery, although unexpected, was hardly "an accident." Distinguishing between the unexpected and the "accidental" is especially difficult when research involves exploration of the unknown. The analogy to discovery makes the point: Columbus did not find what he was looking for—but the discovery of the new world was hardly an accident.[35]

[33] Ehrenberg et al. (2003) survey US universities regarding start-up packages. They find that the average package for an assistant professor in chemistry is $489,000; in biology it is $403,071. At the high-end it is $580,000 in chemistry; $437,000 in biology. For senior faculty they report start-up packages of $983,929 in chemistry (high-end is $1,172,222); and of $957,143 in biology (high-end is $1,575,000).

[34] Universities increasingly expect faculty to write off part of their academic-year salary on grants. This is an absolute necessity for faculty on soft money positions, but also is becoming increasingly common for tenure and tenure-track faculty.

[35] I thank Nathan Rosenberg for this analogy.

Thus, it is perhaps more appropriate to think of serendipity as the act of finding answers to questions not yet posed. Important medical advances, for example, have come from fundamental, nonmission directed, research. A scientist studying marine snails found a powerful new drug for chronic pain. A widely used cancer medication came out of studies of how electricity affects microbes.[36] The discovery of AGM-1470—a drug being tested for an entirely different approach to the treatment of cancer, is described as having started with a "laboratory accident." The narrative: the dish in which Don Ingber was culturing capillary endothelial cells became contaminated with fungus. Ingber noticed that the fungus induced cell rounding, which his previous work had shown to be associated with inhibition of capillary growth.[37] The hope: that the drug will block the growth of blood vessels, which tumors need in order to survive. It may have been an accident, but, to quote Pasteur, "Where observation is involved, chance favors only the prepared mind."

Scientists not only benefit from serendipitous occurrences; they also note them at times, as does Robert Richardson, Nobel laureate in Physics in 1996, in his short bio (National Academies, 2005, p. 148):

> *"He (Richardson) obtained his PhD degree from Duke in 1966. His thesis advisor was Professor Horst Meyer. In the Fall of 1966 he began work at Cornell University in the laboratory of David Lee. Their Research goal was to observe the nuclear magnetic phase transition in solid ^{3}He that could be predicted from Richardson's thesis work with Horst Meyer at Duke. In collaboration with Douglas Osheroff, a student who joined the group in 1967, they worked on cooling techniques and NMR instrumentation for studying low temperature helium liquids and solids.* In the fall of 1971, they made the accidental discovery that liquid ^{3}He undergoes a pairing transition similar to that of superconductors. The three were awarded the Nobel Prize for that work in 1996."

4. Choice of scientific contests and character of research

4.1. Choice of contests

The importance attached to priority of discovery dictates that scientists choose the contests they enter with care. The probability of being scooped is a constant threat. This is particularly true in the case of "normal" science, where the accumulated knowledge and focus necessary for the next scientific breakthrough is "in the air."[38] Young scientists, in particular, must choose their contests with care if they are to successfully signal their ability and "resource worthiness" to receive funding. Young biomedical researchers in the United States must choose a research trajectory that is sufficiently independent from that of their mentors to appeal to funders, yet sufficiently close to signal the effectiveness of their training.

[36] National Institutes of General Medical Sciences: 2008–2012 Strategic Plan (http://publications.nigms.nih.gov/strategicplan/chapter2.htm).

[37] www.aids.org/atn/a-135-04.html

[38] Note the distinction between social and individual risk. Because accumulated knowledge is an important input in the process of discovery, normal science is not especially risky from the social point of view (Arrow, 1962; Dasgupta and David, 1987, p. 526). From the individual investigator's point of view, however, risks can be substantial: being in the air is entirely different from being in scientist X's air.

Scientists can minimize the threat of being scooped by seeking ways to monopolize a line of research. During the seventeenth and eighteenth centuries discoveries in process were sometimes reported in the form of anagrams for the "double purpose of establishing priority of conception and yet not putting rivals on to one's original ideas, until they had been further worked out" (Merton, 1957, p. 654). It was also not uncommon to deposit a sealed and dated manuscript with a learned society to protect both priority and idea. More recently, the ownership of apparatus or strains has proved to be a convenient way to monopolize a line of research. Another strategy for minimizing the threat of being scooped is to develop a particularly novel technique for research and to then collaborate with others in applying the approach or technique to a range of questions. Scientists can also minimize the threat of being scooped by choosing to work on problems that fall outside the mainstream of "normal science" or by working "in the backwaters" of research (Stephan and Levin, 1992). The downside of such a strategy is that, while the low number of competitors increases the probability of being first, the contest that is won may be of little interest to the larger scientific community and hence receive minimal recognition.

Researchers must choose not only a line of research. They must also choose a research strategy, because more than one method can be used to address the same question (Dasgupta and David, 1994). In the life sciences, this involves not only choosing one's research topic but also the approach for one's research. Here, too, uncertainty enters the equation.[39] The use of novel methods, for example, can prove rewarding, but the risk of coming up empty-handed can be quite large when an unorthodox approach is employed or when a difficult problem is approached in a way that is not divisible into intermediate outputs.[40] The uncertainty associated with the *process* of discovery can be substantial. The outcome may not have been envisioned, neither may the outcome relate to the original objective of the researcher. As noted above, in the process of trying to solve some very practical problems concerning fermentation and putrefaction in the French wine industry, Pasteur established the modern science of bacteriology (Rosenberg, 1990).

Research often provides answers to unposed questions.[41] Consequently, the risk associated with such research can be lessened by shifting goals during the course of research. Nelson (1959) argues that this strategy is more appropriate for scientists working in a nonprofit-based environment than for scientists working in the profit sector because the former can more easily capture the rewards regardless of where the research leads. On the other hand, companies having a broad technological base can benefit from research that is not directed to a specific goal. At the time General Electric developed synthetic diamonds, for example, it was the most diversified company in the United States.

A number of institutional arrangements have evolved in science to help minimize risk or provide some insurance against risk. Some of these, such as the ability to monopolize a line of research, have already been noted.[42] Others include the adoption of a research portfolio that contains projects with

[39] Susan Linquist, an HHMI Investigator at MIT who studies protein function, reports the risky choice she made early in her career to change her research focus from fruit flies to yeast (Dreifus, 2007).

[40] A consequence is that rival teams often select highly correlated research strategies. From a social point of view, highly correlated research strategies produce inefficiencies by failing to provide the kind of portfolio diversification that society would choose if it were allocating resources in a way to maximize the probability of success (Dasgupta and David, 1994). The gains to society from sponsoring multiple lines of independent research are examined by Scherer (1966).

[41] The unpredictable nature of scientific discovery is explored by Polanyi (1962).

[42] The ability to monopolize a line of research is being weakened by the increasingly rapid disclosure requirements being placed on researchers by databases as well as rules placed on researchers, such as the Bermuda Rule for gene sequence disclosure.

varying degrees of uncertainty, the formation of research teams and networks and the practice of "gift giving" whereby scientists, by acknowledging intellectual debts to their colleagues (via citations), pay "protection money" to insure that those colleagues "won't deny their grants, spread slander, or—worst of all—ignore their work altogether." (Fuller, 1994, p. 13).

4.2. *The character of research*

It was common practice for many years to classify research as either basic or applied and many government statistical agencies continue to classify research accordingly. Such a classification, while useful for governmental statistical agencies, oversimplifies the research process and reasons for doing research. Stokes (1997) notes that much of today's research is both "use inspired" and inspired by a quest for fundamental understanding. In honor of Louis Pasteur, Stokes classifies such research as falling into "Pasteur's Quadrant." Stokes argues that increasingly scientists work in Pasteur's Quadrant, in part because of the scientific opportunities that have become available in recent years in such areas as molecular biology and, to extend his argument, nanotechnology. Stokes contrasts this to research that falls in "Bohr's Quadrant"—research that is motivated exclusively for fundamental understanding—and research in "Edison's Quadrant"—research inspired exclusively by use.

It is also an oversimplification to assume that research occurring in the public sector is distinct from that occurring elsewhere. The research boundaries between public sector and other sectors are porous, and are becoming increasingly so. Gibbons et al. (1994) see this as one characteristic of what they call Mode 2, a new mode of knowledge production, which they argue is distinct from what they call Mode 1, where research is done within the university, within disciplinary boundaries, and is homogeneous and hierarchical. By contrast, "The new mode operates within a context of application in that problems are not set within a disciplinary framework. It is transdisciplinary rather than mono- or multidisciplinary. It is carried out in nonhierarchical, heterogeneously organized forms which are essentially transient. It is not being institutionalized primarily within university structures." (p. vii).

While there is considerable debate over some of these claims (e.g., the "newness" of Mode 2; Pavitt, 2000), it is clear that university researchers work with researchers outside their own disciplines. It is also clear that university researchers are heavily influenced by the research and technological opportunities that occur outside the academy and that they frequently work with scientists and engineers located outside the university. Moreover, this cross-sectoral work often enhances the research activity of academic scientists and engineers. Zucker et al. (1998a,b), for example, find that the productivity of academic scientists is enhanced when they work with scientists in biotechnology companies; Mansfield (1995) found that academic researchers with ties to firms report that their academic research problems frequently or predominately are developed out of their industrial consulting and that the consulting also influences the nature of work they propose for government-funded research.

4.3. *Production of dual knowledge*

One choice that scientists working in the public sector increasingly must make is whether to disclose their findings exclusively through publication, to seek intellectual property protection, or to both patent and publish. While the presence of time in the production function for knowledge suggests that patenting and publishing may be substitute activities, there are good reasons to argue that complementarity is more likely and that patents can be a logical outcome of research activity that is designed first and foremost with an eye to publication. The reasons for complementarity are threefold. First, the results of research, especially research in Pasteur's Quadrant, can often be both patented and published, having a dual nature. Second, the increased opportunities that academic researchers have to work with industry may enhance productivity and encourage patenting. Third, the reward structure in academe encourages patenting as one outcome of research.

A handful of studies in recent years have examined the relationship of publishing to patenting (Agrawal and Henderson, 2002; Calderini et al., 2007; Carayol, 2007; Wuchty et al., 2007). While various methodological issues arise, such as endogeneity, most find evidence that publishing and patenting are complementary rather than substitute activities. Researchers have also examined the relationship between patenting and publishing. Azoulay et al. (2009), for example, examine the impact of patenting on the publication activity of university researchers working in areas related to biotechnology and find that patenting has a positive effect on publication. Markiewicz nd Di Minin (2004) also find patents to have a positive and significant effect on publication production of university researchers in their sample of US scientists, as do Breschi et al. (2009) in a study of Italian scientists.[43]

5. Outcomes

Articles are a major output of scientific research. Over time, the number of articles written has increased substantially, as well as the distribution of those writing the articles. This is best seen in Figure 2, which shows worldwide article production in science and engineering (measured by fractional counts) for the 16-year period 1988–2003. The numbers are impressive—the most recent data enumerate more than 650,000 articles (fractional counts). The figure also shows how the dominance of the United States has waned in recent years, as counts from the EU-15 and the East-Asia 4 have dramatically increased.

Academics contribute disproportionately to research that is codified through journal publication. This is seen in Table 1, which gives article output (fractional counts) by sector for selected years for the United States. During the period, the academic share rose from about 72% to 74%. That of industry and the Federal government declined, while that of private nonprofits and Federally Funded Research and Development Centers (FFRDCs) remained approximately the same. We will return to this table later in the chapter when we discuss scientists working in industry.

Patents provide another indicator of research output. As in the case of articles, over time the number of patents has increased substantially, as has the number granted to an academic institution. For example, the number of patents granted in the United States almost doubled between 1990 and 2003, going from 90,000 to 169,000 (National Science Board, 2006, Table 6-12).[44] Some of the dramatic increase undoubtedly relates to problems with the patent system (Jaffe and Lerner, 2004). During the same period of time, the number of patents granted to US universities increased by more than 2.5 times, going from about 1300 in 1993 to 3450 in 2003 (National Science Board, 2006, Table 5-28). Similar

[43] Their research suggests that the positive effect is not due to patenting *per se* but to advantages derived by having strong links with industry.

[44] The number of US patents granted to a foreign inventor more than doubled during this same period.

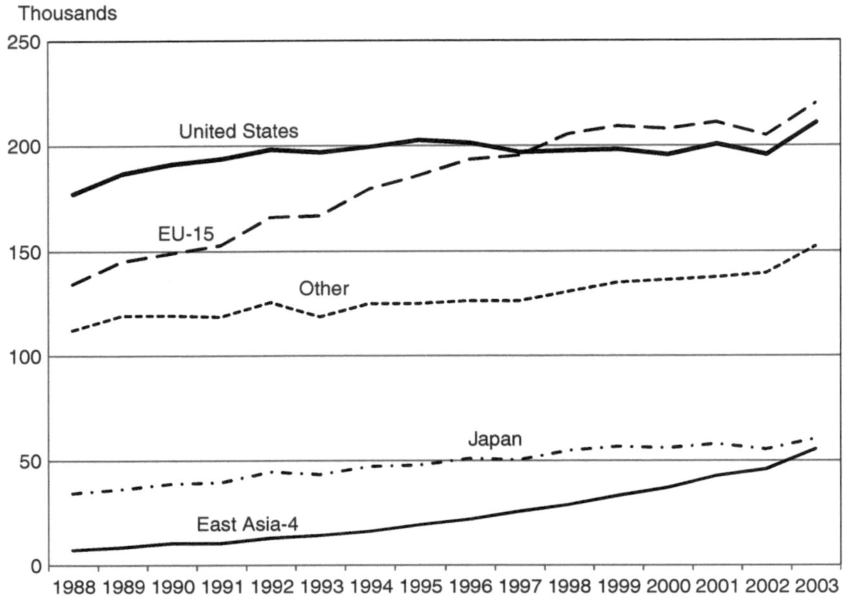

Figure 2. S&E article output (fractional counts) of major S&E publishing centers: 1988–2003. *Notes*: Article counts are on a fractional basis, that is, for articles with collaborating institutions from multiple publishing centers, each publishing center receives fractional credit on the basis of the proportion of its participating institutions. East Asia-4 includes China, Singapore, South Korea, and Taiwan. China includes Hong Kong. Source: National Science Foundation (2007, Figure 6).

Table 1
S&E article output (fractional counts) of US institutions: 1988–2003 (1000s)

Year	FFRDC	Federal government	Other	Private-for-profit	Private nonprofit	Academe	Total
1988	4.9	14.4	3.5	15.1	12.4	127.3	177.6
1991	5.1	15.2	3.8	16.9	13.5	139.3	193.8
1993	4.7	15.3	4.1	16.4	14.6	142.3	197.4
1995	5.4	15.5	3.7	16.4	15.4	146.5	202.9
1997	5.2	14.3	3.9	14.6	15.0	144.6	197.6
1999	5.2	13.9	4.1	14.5	15.4	145.5	198.6
2001	5.2	14.0	3.7	14.2	16.0	147.8	200.9
2003	5.7	14.1	4.0	14.5	16.3	156.6	211.2

Source: National Science Board (2006, Table 5-19 and table underlying Figure 5.51).

trends exist in Europe although academic patents are more difficult to trace because of the practice in many countries of "professor privilege."

The productivity of scientists and engineers, especially those working in academe, has been studied by a number of researchers. While most of this work focuses on the publication of articles, in recent years researchers have also examined the patenting output of faculty. Most of the early work was conducted by sociologists; in more recent years sociologists have been joined by economists and researchers in public policy in their efforts to understand factors related to productivity, especially at the individual level. Issues of interest include: (1) whether science is a young person's game; (2) the extent to which cohort effects are present in science; and (3) the degree to which output is related to gender and underlying reasons for such a differential, if it is found to exist. In addition, there has been considerable interest in the distribution of output across scientists and factors leading to the extreme inequality that is observed.

Data for productivity studies is drawn from a variety of sources. Early studies, for example, generally used survey data collected specifically for the study. Some researchers have matched public data with outcome data (Levin and Stephan, 1991; other researchers have collected data from national funding organizations or institutes (Gonzalez-Brambilia and Veloso, 2007; Turner and Mairesse, 2003),while others rely on data that is available from CVs (Cañibano and Bozeman, 2009). Here we examine several of these studies, organizing the discussion by the type of research question addressed. Two types of outcomes are examined, where relevant: (1) publication measures and (2) patent measures.

5.1. Is science a young person's game?

Einstein once said that "a person who has not made his great contribution to science before the age of thirty will never do so." (Brodetsky, 1942, p. 299). There is a great deal of anecdotal evidence (Stephan and Levin, 1992) that he was right, that science is the domain of the young. However, investigating the veracity of the statement statistically is fraught with problems: measurement issues abound, as do the confounding of aging effects with cohort effects, as well as the availability of appropriate databases. We examine these issues, prefacing them with a discussion of theoretical reasons that one might expect age to be related to productivity.

For economists, the theoretical reason to expect a relationship between age and productivity rests on human capital theory.[45] General models of human capital predict that, due to the finiteness of life, investment behavior declines (eventually) over time. Several authors have adapted the human capital framework to develop life-cycle models of scientists or academics. Like their first cousins, these models are driven by the finiteness of life and investigate the implications this has for the allocation of time to research over the life cycle. The models differ in the assumptions they make concerning the objective function of the scientists but reach somewhat similar conclusions. In its simplest form the objective is the maximization of income, itself a function of prestige capital (Diamond, 1984). In a more complex form, the objective is the maximization of a utility function that includes income as well as research output (Levin and Stephan, 1991).[46] The latter is included given the strong anecdotal evidence that puzzle solving is part of the reward to doing science.[47] The implications of these models are that the stock of prestige capital peaks during the career and then declines and that the publishing profile

[45] Sociologists, psychologists, and neurologists have other reasons regarding why there may be a relationship between age and productivity. See Stephan and Levin (1993) for a summary.

[46] The objective function can also include fame as an end in itself, not only as a means for generating income.

[47] This way of dealing with the puzzle issue is not completely satisfactory because it assumes that it is the *product* of discovery that enters the utility function, not the input of time in discovery. Yet, it is the *process* of discovery that is often reported as giving enjoyment to scientists.

declines over the life cycle. The addition of puzzle solving to the objective function produces the result that research activity is greater at any time, the greater is the satisfaction derived from puzzle solving; it also produces the strong suggestion that the research profile is flatter, the larger is the satisfaction derived from puzzle solving.[48]

Several classes of problems present themselves in studying research productivity in a life-cycle context. These include measurement, the confounding of aging effects with cohort effects, and the availability of an appropriate database.

Publication counts are generally used as a proxy for research activity. This is justified on the grounds of the high acceptance rates—often in excess of 70% (Hargens, 1988) that exist among scientific journals and that publication is a necessary condition for communicating research findings and establishing priority. The question of attribution regarding coauthored articles is sometimes addressed by prorating article counts among coauthors. Article quality is often proxied by weighting article counts by some type of citation measure.

Because scientists of different ages come from different cohorts, aging effects are confounded with cohort effects in cross-sectoral studies. One type of cohort effect is associated with change in the knowledge base of the scientific field. If, for example, there is a secular progression of knowledge (to paraphrase Mincer, 1974, p. 21), the latest educated should be the best educated and hence the most productive, other things being equal. Another factor that affects research productivity and varies by cohort is access to resources that affect research. Variation occurs primarily through fluctuations in the job market that lead certain cohorts to have relatively easy access to jobs rich in resources while others, who graduate during periods when job openings in the research sector are scarce, have considerably less access to the resources that contribute to productivity.

The presence of cohort effects dictates a research design that uses a pooled-cross-section time series database.[49] Such databases are not only costly to create: issues of confidentiality can limit access to the ones that do exist. Diamond (1986a) uses a database he assembled for mathematicians at Berkeley; Levin and Stephan develop a database by matching records from the National Science Foundation's (NSF) biennial 1973–1979 Survey of Doctorate Recipients (SDR) with publishing information from the Science Citation Index. Weiss and Lillard (1982) use a sample of Israeli scientists; Turner and Mairesse examine the productivity of solid-state physicists working at CNRS in France; Bombaradaro and Veloso examine the productivity of scientists supported by the National System of Researchers (SNI) of Mexico.

[48] Thursby et al. (2005) expand the life cycle model to examine the effects of licensing on academic research. Their model builds on that of Levin and Stephan (1991), but divides research into a basic and applied component. The latter component has the potential of producing income through licenses. Work in progress by Doh-Shin Jeon (correspondence) suggests that life-cycle effects can be mitigated by the presence of teams in science, especially for stars. The idea is that the inclusion of a star scientist as a member of the team provides certification value. As the number of individuals working in an area increases, more scientists propose ideas and the star can select among the best. This not only increases the productivity of the star but also increases the probability that the star will remain active over a longer period of the life cycle.

[49] See Hall et al. (2007) for a discussion of the problems arising in identifying age, cohort, and period effects in studying scientific research productivity.

5.1.1. Age and publishing

Levin and Stephan analyze six areas of science. They find that, with the exception of particle physicists employed in PhD-granting departments, life-cycle effects are present in a fully specified model that controls for fixed effects such as motivation and ability.[50] For the fields of solid-state physics, atomic and molecular physics, and geophysics, the evidence suggests that publishing activity initially increases but declines somewhere in mid-career. For particle physicists at FFRDCs, as well as for geologists, the profile decreases throughout the career. The absence of life-cycle effects for particle physicists at PhD-granting institutions is not totally unexpected. Theorists working on unification are often depicted as involved in a "religious quest," handed them by Albert Einstein, or, as is commonly stated in the literature, the "search for the Holy Grail."

Diamond finds that the publishing activity of Berkeley mathematicians declines slightly with age. Weiss and Lillard use a pooled model to estimate the growth rate of publications for 1000 Israeli scientists. They find that the average annual number of publications tends to increase in the early phase of the academic career and then decline. They also find that, along with the mean, the variance of publications increases markedly over the first 10–12 years of the academic career.

Turner and Mairesse find virtually no "aging" effects for their sample of condensed matter physicists working at CNRS during the period 1986–1997. Several reasons may explain the difference between their findings and those of Levin and Stephan. These include their controlling for career stage, which is highly correlated with age, their use of a highly selective sample (by definition all of the sample are "research" scientists) while Levin and Stephan focus their research on university faculty, many of whom may not be doing research. Because their research spans a later period, they may also be picking up the fact that as publications have become more important to careers and, as more coauthors are involved, the incentives and ability to stay productive over longer periods of time have changed. They are also able to control for variables related to the lab in which the researcher works, something that Levin and Stephan could not do. Gonzalez-Brambilia and Veloso examine publication activity of Mexican researchers supported by the Mexican SNI. Their sample is restricted to individuals who have at least one publication during the period of observation. They find that conditional upon being supported by SNI and having published at least one article there is a fairly consistent level of publishing output over time within broad disciplines.

5.1.2. Age and exceptional contributions

Research on Nobel laureates suggests that the relationship between age and exceptional contribution is more pronounced than the relationship between age and productivity for what could be thought of as "journeymen" scientists. Stephan and Levin (1993), for example, find for Nobel laureates during the period 1901–1992 that although it does not require extraordinary youth to do prize-winning work, the odds decrease markedly in mid-life. The relationship is field dependent as well as dependent upon the definition used to measure the age at which the award-winning work was done. But regardless of field, the odds of commencing research for which a Nobel prize is awarded decline dramatically after

[50] Vintage variables (discussion to follow) cannot be included in a fixed-effects model because the vintage variable is invariant over time for an individual. Equations were also estimated that included vintage variables but excluded the fixed-effects.

age 40 and very, very few laureates undertake prize-winning work after the age of 55. To wit, during the period studied less than 2% of the laureates commenced their work after age 55. Jones (2005a) finds that the age at which peak output occurs for Nobel laureates has increased over the past century. When Stephan and Levin extend their analysis to include laureates from 1993 to 2006, they find that the median age at the time laureates began their research increased by 1 year, going from 31 to 32.

5.1.3. Age and patenting

The increase in patenting among faculty has been accompanied by a spate of studies that examine determinants of faculty patenting behavior. The focus of some of these studies (noted in Section 4.3 of this chapter) is the relationship between patenting and publishing, with a special focus on whether the two are substitutes or complements. The focus of others is broader, examining specific determinants of patenting activity. The work of Azoulay et al. (2006) is a case in point and examines the patenting behavior of a panel of 3884 academic life scientists. Each scientist is observed from the year that he or she earned a PhD until 1999, beginning in 1967. Those who exit academia are dropped from the sample at the time they leave. The authors find "pronounced life-cycle effects on the propensity to patent, with mid-career academics being much more likely to patent than younger and older faculty members" (Azoulay et al., 2005, p. 1).[51] Thursby and Thursby (2007) examine the disclosure activity of a panel of scientists and engineers working at six universities over a 17-year period. They find that patenting declines with age; they also find that other things being equal, newer cohorts are less likely to patent than are earlier cohorts.

5.2. The presence of cohort effects

There are various reasons to expect scientific output to be related not only to age but also to the cohort to which the scientist belongs. Levin and Stephan, for example, focus on the relationship between productivity and the "vintage" of the scientist, arguing that certain vintages may be more productive than others. They investigate the hypothesis by identifying changes in each of the six subfields that they study that had the potential of making scientists in the subfield obsolete. Changes were identified through the use of case studies conducted through personal interviews, a small mail survey, and various publications. They then estimate a model that controls for age, time period effects,[52] and vintage. The most striking finding of this aspect of their research is that at conventional levels of significance, in no field are the latest vintages more productive than the earliest, benchmark, vintage. Stated differently, there is no evidence that the latest vintage, with supposedly the most up-to-date knowledge, engages in more research than does the earliest vintage. Furthermore, in several subfields, depending on the output measure used, there is some indication that the latest vintages are less productive than are the earliest vintages.

There is, of course, reason to believe that other types of cohort effects may exist, relating to such factors as variability in job market conditions over time or changes in the "culture" of research. For

[51] The authors also find that patenting is often accompanied by a "flurry" of publication activity in the preceding year. They also create a variable which measures the latent patentability of the scientist's research through the use of keywords and find a positive relationship between this measure and the propensity to patent.

[52] The inclusion of time period effects is desirable given that such things as resources for research vary over time.

example, in recent years it has been particularly difficult for young life scientists in the United States to obtain tenure-track appointments, just as it has been exceedingly difficult for young researchers in Italy, France and Germany to find permanent research positions. This is in contrast to earlier times, when research budgets were growing and universities (in the case of the United States) had healthy budgets to create new positions (Stephan, 2008). Oyer (2006) has examined how variability in initial labor market outcomes affects research over the long term for a sample of economists. Oyer's research shows that "initial career placement matters a great deal in determining the careers of economists." Consistent with the cohort hypothesis, the effect persists holding innate ability constant; that is, initial placement matters independent of ability.

5.3. Gender

The presence of a gender differential in publishing outcomes is well established. Fox (2005), for example, finds that women published or had accepted for publication 8.9 papers in the 3-year period beginning in the early 1990s, compared to 11.4 for men. The difference owes to disparities at both extremes of the productivity distribution. Women are almost twice as likely as men to publish zero or one paper during the period (18.8% compared to 10.5%); men are almost twice as likely as women to publish 20 or more papers during the period (15.8% for men compared to 8.4% for women).[53] Gender differentials have also declined over time. Xie and Shauman (1998) find the female-to-male ratio to have been about 0.60 in the late 1960s, and to have increased to 0.82 by 1993.

The question as to why research output is related to gender has long interested those studying scientific productivity. In economic terms, the question is often examined in terms of supply versus demand characteristics. Stated in these terms, the question is whether women publish less than men because of specific attributes, such as family characteristics, amount of time spent doing research, etc., or whether women publish less than men because they have fewer opportunities to be productive, due to hiring and funding decisions as well as possible network outcomes. This dichotomy is misleading, of course, to the degree that interactions exist between the two. Differential placement opportunities, for example, may lead women to allocate their time to activities that are rewarded (such as teaching) but diminish publishing activity. One of the most in-depth studies to be done on the subject in recent years is that by the sociologists (Xie and Shauman, 1998, 2003, p. 23). After carefully analyzing four datasets that span a 24-year period, they conclude that "women scientists publish fewer papers than men because women are less likely than men to have personal characteristics, structural positions, and facilitating resources that are conducive to publication." In other words, both demand and supply play a role.

The increase in patenting among academic scientists raises the question of whether differential patenting patterns exist by gender. The question is of interest not only because patenting is another indicator of output but also because of the role that royalty payments from patents can play in remuneration as well as the role that patents arguably play in exchange and hence, indirectly, in fostering productivity (see above).

[53] Kelchtermans and Veugelers (2007) find that, relative to men, women faculty at Katholieke Universeteit Leuven are more likely to consistently not publish and slightly less likely to be in the top performance category.

The most thorough study of patenting to date has been Ding et al.'s study (2006a) of life scientists who received their PhDs between 1967 and 1995. Among the 4227 in the sample who had at least a 5-year history of post-PhD publishing at an academic institution, women were found to be far less likely to have at least one patent then men: 5.65% of the women in the sample; 13.0% of the men. The hazard models that they estimate indicate that gender differences cannot be entirely explained away by such things as contact with industry, number of coauthors, past publications, institutional support for patents (as measured by the number of patents the institution has received), or subfield. Although controlling for these measured characteristics reduces gender disparity, the coefficient on women in their proportional hazard model remains positive and significant, indicating that other things being equal women patent at 0.40 times the rate of men. In light of the earlier discussion, it is interesting to note that they also find indication of strong cohort effects. The cumulative hazard for patenting for those who received their PhDs in the earliest period studied (1967–1975) was 4.4 times higher for men than that for women; the differential declined to 2.1 for those who received their PhDs in the middle period (1976–1985) and further declined to 1.8 for those who received their PhDs in the latest period studied (1986–1985). These findings are consistent with the views older women expressed in interviews conducted by the authors that they felt excluded from industry relationships early in their careers and were never able to develop an understanding of how commercial science works.[54]

5.4. Inequality

A defining characteristic of contests that have winner-take-all characteristics such as those that exist in science is extreme inequality in the allocation of rewards. Science, too, has extreme inequality with regard to scientific productivity and the awarding of priority. One measure of this is the highly skewed nature of *publications*, first observed by Lotka (1926) in a study of nineteenth century physics journals. The distribution that Lotka found showed that approximately 6% of publishing scientists produce half of all papers. Lotka's "law" has since been found to fit data from several different disciplines and varying periods of time (Price and Solla, 1986).[55]

Patents are even more highly skewed than are publications. Stephan et al. (2007b), for example, find for a sample of 10,962 US academics studied over the 5-year period 1990–1995, that 90.1% reported zero patents; 8.7% reported 1–5 patents; 0.4% reported 6–10, and 0.1% reported greater than 10. By comparison, only 14.4% reported zero publications, 40.8% reported 1–5 publications, 20.9% reported 6–10, and 23.9% reported over 10 articles.[56]

[54] There is also considerable evidence that a gender gap exists in entrepreneurial activity among university scientists (Stephan and El-Ganair, 2007).

[55] Lotka's law states that if k is the number of scientists who publish one paper, then the number publishing n papers is k/n^2. In many disciplines this works out to some 5% or 6% of the scientists who *publish at all* producing about half of all papers in their discipline. Although Lotka's Law has held up well over time and across disciplines, David (1994) shows that other statistical distributions also provide good fits to observed publication counts.

[56] Inequality appears in other dimensions of science, as well. Terviö (2006) measures departmental influence using a method similar to that used by Google to rank web pages and finds for the fields studied that the distribution of influence is significantly more skewed than the distribution of academic placements.

Inequality in scientific productivity could be explained by differences among scientists in their ability and motivation to do creative research (to have the "right stuff"). But scientific productivity is not only characterized by extreme inequality at a point in time; it is also characterized by increasing inequality over the careers of a cohort of scientists, suggesting that at least some of the processes at work are state dependent. Weiss and Lillard (1982), for example, find that not only the mean but also the variance of publication counts increased during the first 10–12 years of the career of a group of Israeli scientists.

Merton christened his explanation for inequality in science the Matthew Effect, defining it to be the accruing of greater increments of recognition for particular scientific contributions to scientists of considerable repute and the withholding of such recognition from scientists who have not yet made their mark (Merton, 1968, p. 58).

He argues that the effect results from the vast volume of scientific material published each year, which encourages scientists to screen their reading material on the basis of the author's reputation. Other sociologists (e.g., Allison and Stewart, 1974; Cole and Cole, 1973) have argued that additional processes of "cumulative advantage" are at work in science, such as the ability to leverage past success into research funding as well as the "taste" for recognition that success engenders. A funding system such as NIH's that awards grants, at least in part, on past success clearly contributes to cumulative advantage. While the interaction of the "right stuff" and the processes of cumulative advantage are not fully understood, a strong case can be made that a variety of factors are at work in helping able and motivated scientists leverage their early successes and that some form of feedback mechanism is at work (David, 1994). This observation is consistent with other work in winner-take-all contests. Frank and Cook (1992, p. 31) observe that "in all their manifestations, winner-take-all effects translate small differences in the underlying distribution of human capital into much larger differences in the distribution of economic reward."

6. Efficiency considerations and funding regimes

6.1. *Efficient nature of the reward system*

The socially desirable properties attached to a reward system that is priority-based are substantial. Priority solves the monitoring problem. "Since effort cannot in general be monitored, reward cannot be based upon it. So a scientist is rewarded not for effort, but for achievement." (Dasgupta and David, 1987, p. 530). Priority also means that shirking is rarely an issue in science. The knowledge that multiple discoveries are commonplace makes scientists exert considerable effort.[57] A reward structure based on priority requires that scientists share information in a timely fashion if they are to establish priority. Such a process in turn permits peer evaluation, which discourages plagiarism and fraud and builds consensus in science (Dasgupta and David, 1987; Ziman, 1968). The process also provides scientists the reassurance that they have the capacity for original thought (Merton, 1957) and encourages scientists to acknowledge the roots of their own ideas, thereby reinforcing the social process. Reputation also serves as a signal of "trustworthiness" to scientists wishing to use the results of another in their own research without incurring

[57] The prevalence of multiples in science is discussed below. Fox (1983) and Hull (1988) discuss the effort and work patterns of successful scientists.

the cost of reproducing and checking the results. It also serves as a signal of trustworthiness to foundations. As such, reputation provides an answer to the agency problem (Turner, 1994) posed by Coase (1937).[58]

From an economist's point of view, an exceedingly appealing attribute of a reward system that is rooted in priority is that it offers nonmarket-based incentives for the production of the public good "knowledge." (Stephan, 2004). Merton noted the functionality of the reward system in the inaugural lecture of the George Sarton Leerstoel that he delivered October 28, 1986 at the University of Ghent. In the lecture, published 2 years later in *Isis*, Merton spoke of the public nature of science, writing that "...a fund of knowledge is not diminished through exceedingly intensive use by members of the scientific collectivity—indeed, it is presumably augmented..." (Merton, 1988, p. 620). Merton not only recognized this but stood the public–private distinction on its head, proposing that the reward structure in science of priority functioned to make a public good private. "I propose the seeming paradox that in science, private property is established by having its substance freely given to others who might want to make use of it." He continues (1988, p. 620) by saying that "only when scientists have published their work and made it generally accessible, preferably in the public print of articles, monographs, and books that enter the archives, does it become legitimately established as more or less securely theirs" or, as he says elsewhere, "one's private property is established by giving its substance away" (1988, p. 620).

Dasgupta and David (1987, p. 531) express the private–public paradox exceedingly well: "Priority creates a privately owned asset—a form of intellectual property—from the very act of relinquishing exclusive possession of the new knowledge." Arrow (1987, p. 687), commenting on their work, articulates the cleverness of such a system:

> *"The incentive compatibility literature needs to learn the lesson of the priority system; rewards to overcome shirking and free-rider problems need not be monetary in nature; society is more ingenious than the market."*

6.2. Funding regimes

The conventional wisdom holds that because of problems related to appropriability, public goods are underproduced if left to the private sector. Although priority goes a long way toward solving the appropriability problem in science, this ingenious form of compensation does not insure that efficient outcomes will be forthcoming. In addition to problems caused by uncertainty and indivisibilities

[58] This is not to say that the reward structure is without problems. Fraud and misconduct occur with some frequency in science (Kohn, 1986). In recent years there have been several high-profile cases involving misconduct and fraud, including the fabrication of data by Woo Suk Hwang regarding the creation of embryonic stem cells (various online sources) and the University of Wisconsin researcher, Elizabeth Goodwin, who, according to a University of Wisconsin investigation, falsified data in grant applications (Couzin, 2006). In China an "unprecedented number of researchers stand accused of cheating—from fudging resumes to fabricating data—to gain fame or plumb positions" (Xin, 2006, p. 1464). According to Lu Youngxiang, president of the Chinese Academy of Science (CAS), "Too many incentives have blurred the reasons for doing science in some people's minds." (p. 1464). Feigenbaum and Levy (1993) discuss the market for (ir)reproducible results; Fox and Braxton (1994) discuss other issues related to fraud. There is also the considerable issue that the reward structure in science appears to have favored white and Asian men over women and members of underrepresented groups.

(Arrow, 1962), there is the problem that scientific research requires access to substantial resources. Unless priority can be translated into resources, it cannot come close to generating a socially optimal amount of research. Research must be subsidized, by either the government or philanthropic institutions.[59] The government's rationale for supporting scientific research also rests on the importance of research and development to defense, the desire to win what Johnson (1972) calls the "Scientific Olympics"; and the importance of science to economic growth.

Many countries fund scientists indirectly by supporting the research institutes where they work, such as the CNRS in France, the CNR in Italy, and the Institute of Molecular and Cell Biology in Singapore. In some instances this means that scientists are directly employed by the government; in other instances the funds pass from the government to the institute where the hiring arrangements are made. The practice of funding the institute rather than the scientist is less common in the United States, especially in academe, but the practice exists outside academe. The National Institute of Standards and Technology (NIST), for example, which is federally funded, operates on such a model and has been the research home of several Nobel laureates. FFRDC's, of which SLAC (formerly called the Stanford Linear Accelerator) is an example, work on such a model as well and NIH has several large intramural research programs. Notwithstanding the above, competitive processes also exist outside the United States, especially in Europe, for funding researchers. In the United Kingdom, for example, researchers are supported by a grants program administered by various councils such as EPSRC—Engineering and Physical Research Council; in Belgium, the Flemish Science Foundation (FWO) provides a peer-review system for supporting research. The European Union has long supported research through the "Framework Program," which is now in its seventh form (Seventh Framework Program (EP7)). A particular focus in recent years has been the fostering of networks across countries and universities.[60] In an effort to encourage peer-reviewed-investigator-initiated research, the European Research Council (ERC) was established in 2006 with a focus on "cutting-edge" basic research (Vogel, 2006, p. 1371).

As noted earlier, in the United States, scientists working in academe and at certain research institutes are responsible for raising their own funds through the submission of proposals to funding agencies. The largest agencies funding academic researchers are NIH, NSF, Department of Defense (DOD), and Department of Energy (DOE), in that order. While each agency uses a somewhat different approach in evaluating projects, NIH and NSF rely on peer review.[61] The NIH review process puts considerable weight on the presence of preliminary data as well as past accomplishments, in terms of publications as well as of "lineage" as measured by where the scientist trained and did postdoc work. NSF puts less emphasis on reputation, but does require a two-page bio.

Reputation also plays a role in the funding available to academic departments and research institutes. In the United Kingdom, for example, departments receive funding based in part on the quality of their research and the number of students, through the Research Assessment Exercise (RAE) which occurs

[59] Callon (1994) proposes that public support of science is needed to ensure that multiple lines of inquiry remain open.

[60] The FP7, as proposed, has four parts: (1) funds to foster cooperation for applied research projects that require participation from many labs or companies across the continent; (2) funds to support portable Marie Curie grants for young researchers; (3) funds to support new research infrastructure; and (4) funds for the newly created European Research Council (Enserink, 2006).

[61] Not all funds dispersed by government agencies are awarded by a process of evaluation. A common practice in the United States (De Figueiredo and Silverman, 2007) is for universities to receive funds "earmarked" by Congress at the time of the appropriation.

every 5 years. A related system exists in the Netherlands.[62] In Germany, the Wissenschaftsrat evaluates the institutes that are to be placed on the blue list (Blaue Liste). One consequence of such a system is the recruitment of stars to bolster rankings ("just-in-time hiring") and thus funding.

Certain private foundations also support research at universities. The largest in the United States is the Howard Hughes Medical Institute, which funds research in the life sciences. A number of smaller foundations as well as disease-specific foundations (such as the American Cancer Society) also support research. Funding of research by philanthropic organizations also occurs outside the United States. In France, for example, the Association Against Myopathies (AFM) funds a considerable amount of research in the biomedical sciences; the Wellcome Trust in the United Kingdom is the world's largest medical research charity, funding research in human and animal health.

Industry also provides support for academic research; the importance of this source grew during the 1980s and 1990s in the United States as well as in other OECD countries (National Science Board, 2006, Figure 4-44). Moreover, in certain countries the amount of support is substantial. In Germany, for example, industry currently supplies 12% of funding for academic research and in Canada industry supplies close to 8%. By these standards, the percent of academic research supported by industry in the United States (which peaked at around 6% in 1999) is modest.

Governments (and to a much lesser extent nonprofit foundations) also support research by encouraging the study of science and engineering through the provision of fellowships. Such funds can be targeted directly to students (as in the case of NIH training grants, NSF dissertation awards, and Marie Curie awards) or indirectly, through the support of faculty research which includes funds for graduate students and postdoctoral students. The amount of funds provided can vary considerably over time and in response to perceived needs, as occurred when the United States responded to the launching of Sputnik by creating the National Defense Education Act to encourage the study of science in the late 1950s.[63]

Differences in funding regimes raise the question of whether knowledge advances more rapidly under the peer-review grants system or under the "institute" approach. The issue, to the best of our knowledge, has been ignored by the economics profession. It is therefore hoped that the *ad hoc* discussion that follows will stimulate research on this important topic.

The institute approach has its benefits: it insures that scientists can follow a research agenda (with an uncertain outcome) over a substantial period of time, and it exempts scientists from devoting long hours to seeking resources. These benefits are not trivial.

The costs of the institute approach are also substantial. Foremost is the question of the research agenda. In many institutes the agenda is set by the director, and younger scientists are constrained from following leads they consider promising. The guarantee of resources also encourages shirking; consequently, alternative methods of monitoring must be found. The institute approach also enhances stratification in science and hence the possible waste of human resources. Most appointments are made early in the career. If the scientist does not succeed in getting a tenured appointment, the scientist

[62] In late 2006, the Reading University became the 21st university in the United Kingdom since 1997 to announce the closure of its physics department. The reason: not enough new students—or enough research income (Another Physics Department Down, 2006).

[63] The focus of government funding, as well as the amount, is also quite variable. For example, in recent years in the United States, while funds for the life sciences have grown significantly, funds for the physical sciences, earth sciences and mathematics have languished, as have funds more recently for engineering research.

will have minimal access to resources in that country for the rest of his or her career. One effect of such a system is to encourage migration of those who do not obtain such an appointment.

The grants system also has its benefits. It encourages scientists to remain productive throughout the life cycle, because scientists who wish to have a lab must remain active. To the extent that success in the grants system is not completely determined by past success, the system provides some opportunity for last year's losers to become this year's winners. Peer review arguably promotes quality and the sharing of information. The system also encourages entrepreneurship among scientists. Getting money from a venture capitalist is not that much different from getting money from a funding agency. Both require making a "pitch."

Just as some of the benefits of the grants system are costs of the institute system, so, too, some of the benefits of the institute approach are costs of the grants system. Grant applications and administration divert scientists from spending time doing research.[64] A 2006 survey of US scientists found that scientists spend 42% of their research time filling out forms and in meetings; tasks split almost evenly between pregrant (22%) and postgrant work (20%). The tasks cited as the most burdensome were filling out grant progress reports, hiring personnel and managing laboratory finances (Kean, 2006).[65] The grants system also encourages scientists to choose sure(r) bet short-term projects that in the longer run may have lower social value. The system also implicitly encourages scientists to misrepresent their work or the effort required to generate certain outcomes. It is typical, for example, for scientists to apply for work that is nearing completion (yet not acknowledge the degree to which it has been performed) and to use some of the proceeds of funding to support research that may lead to future funding or research of a riskier nature that may not be fundable.

The process used to evaluate proposals is not without its problems. For example, considerable concern exists regarding the peer-review system used to evaluate proposals. At NIH, for example, the increased number of proposals that accompanied the doubling of the budget led to an increasing percentage of proposals being triaged and thus not reviewed. Agencies report problems getting individuals, especially experienced individuals, to be reviewers, and the charge has been made that the quality of reviews is declining. A related issue is the extent to which scientists engage in "gift-giving" by awarding favorable reviews to acquaintances and coauthors.

While the grants system, in theory, should be more open than the institute system, there is evidence that early career scientists are having difficulty at NIH. The average age at which one receives first independent funding increased from 37.3 to 42.0 between 1985 and 2005; the percent of R01 grants awarded to individuals 40 or younger fell from 25.2% in 1995 to 15.0 in 2005, while the percent awarded to individuals 51 or older increased from 29.1 to 45.8.[66]

[64] The Framework programs in the European Union award contracts, rather than grants. The ERC will award research grants instead, which are viewed by ERC leadership as being potentially less burdensome (Vogel, 2006).

[65] The survey was completed by 6083 university scientists. The study was sponsored by the Federal Demonstration Partnership, a coalition of university and federal officials interested in streamlining government research regulations.

[66] The R01 is the basic independent research grant awarded by NIH; more than 50% of the Institute's resources are used to support R01 research. The data come from the Office of Extramural Research (OER), NIH.

6.3. Are there too many contestants in certain contests?

Governments also encourage the production of knowledge by granting property rights to the discoverer. With rare exception, patents have been the primary form of intellectual property rights that economists have examined, arguing that patents provide for appropriability while placing knowledge in the public domain.[67] Moreover, it has been shown (Dasgupta and Stiglitz, 1980) that under a wide array of circumstances social inefficiency results from patent races among rival groups. This inefficiency manifests itself in "excessive duplication of research effort (or) . . . too fast a pace of advance of the frontiers of knowledge" (Dasgupta and David, 1987, p. 532).

The recognition that priority is a form of property rights leads to the question of whether there are "too many" contestants in certain scientific contests. Would the social good be served by having fewer? In a speech delivered at the conference commemorating the 400th anniversary of the birth of Francis Bacon, Merton detailed the prevalence of what he called "multiples" in scientific discovery. And Merton was not the first to note their presence. In what Merton calls a "play within a play," he gives 20 "lists" of multiples that were compiled between 1828 and 1922. Moreover, Merton is quick to point out that the absence of a multiple does not mean that a multiple was not in the making at the time the discovery was made public. This is a classic case of censored data where scooped scientists abandon their research after a winner is recognized. Indeed, Merton argues that "far from being odd or curious or remarkable, the pattern of independent multiple discoveries in science is in principle the dominant pattern rather than a subsidiary one." (Merton, 1961, p. 356).[68]

The presence of multiple discoveries is due in part to the free access scientists have to knowledge and in part to the fact that uncertainty associated with who will make a discovery leads scientists to choose research portfolios that are correlated (Dasgupta and Maskin, 1987).[69] The knowledge that multiples exist keeps scientists from shirking and moves the enterprise of science at a rapid pace. Such observations invite the question of whether science moves at too rapid a pace and whether certain contests attract too many entrants. Dasgupta and David (1987, p. 540) argue that the priority system can create excesses, just as the patent system does, provided the "reward to the discoverer . . . is tempting enough." They make no effort to define the boundary of temptation, but one wonders if the general knowledge that certain contests deserve the Nobel Prize does not attract an excessive number of scientists.[70]

[67] While neither goal is perfectly achieved by the patent process, the goal of disclosure arguably suffers the most. "The imperfections we have examined in the patent as a device for rewarding disclosures of knowledge are not at all surprising; a stone flung at two birds really ought not be expected to make a clean strike on either" (Dasgupta and David, 1987, p. 534).

[68] Stigler (1980) argues that multiples are less common than Merton assumes and that incomplete knowledge of who "is working on a problem and what his achievement will be" is the only reason why full multiples should occur.

[69] Despite the popularity of patent race models, multiples are arguably more common in science than technology. The reason is that science is concerned with laws and facts, while technology is looking for practical ways to solve problems. Hence, while there is often only one answer to a scientific question, there usually are a variety of distinct ways of solving the practical problem.

[70] On the other hand, the common lament of interest groups that there are not enough entrants in certain races of apparent Nobel proportions (e.g., a cure for breast cancer) leads one to be cautious in making broad generalizations. It is, of course, possible that such groups are expressing the concern that victory is undervalued by the community. It is also possible that a cure is not "in the air" and applying more resources to the contest would be inefficient.

A related question is whether scientists at universities direct too much time to research as opposed to teaching. The fact that only a handful of scientists contribute the lion's share of output suggests that substantial inefficiencies arise when yeomen scientists devote long hours to research. Efficiency concerns also arise with regard to the large number of individuals working in postdoctoral positions. The work of Lazear and Rosen (1981) suggests that an efficient allocation of resources can result in a tournament model, such as that which exists in science. But while rock stars, opera singers and soccer players do not have tenure, professors do. This means that creative young scientists, despite their demonstrated ability, may find it difficult to secure a lab of their own, especially if the number of tenure-track positions does not grow.[71]

6.4. The incentive to share knowledge in a timely fashion

Despite the similarities between priority rights and proprietary rights such as patents, they differ markedly in the incentives they provide to disclose research findings in a timely fashion. On the one hand, the quest for priority requires scientists to share discoveries quickly because it is only by sharing that priority rights can be established. The quest for proprietary rights, on the other hand, discourages the rapid sharing of information, because the very purpose of proprietary rights is to provide a means for capturing the economic rents attached to a new product or technology. And, while some forms of proprietary rights require the sharing of knowledge in recognition of its public nature (e.g., the patent process), incentives to divulge the knowledge *quickly* are not present.[72]

The distinction is so crucial that Dasgupta and David (1987, p. 528) argue that the two types of property rights, and the implications they hold for appropriability and disclosure, differentiate science from technology: "If one joins the science club, one's discoveries and inventions must be completely disclosed, whereas in the technology club such findings must not be fully revealed to the rest of the membership."

This distinction between science and technology often leads to the (erroneous) conclusion that science is done by scientists at universities and public labs and results in published knowledge, while the focus of scientists working in industry is the development of proprietary technology (Nelson and Winter, 1982).[73] While location does correlate with the incentive to share knowledge in a timely fashion, the relationship is far from perfect. Some firms make the results of their research public; some academics engage in tactics that lead to the "privatization" of knowledge. In many instances

[71] Other efficiency concerns exist. One is the degree to which the process of cumulative advantage excludes talented individuals from making contributions. Another is the question of whether there is a critical point at which additional resources allocated to a scientist for research lead to marginal results. The question is of policy importance given that during the doubling of the NIH budget some extremely successful scientists went from having two R01 research grants to having three or more. Dasgupta and David (1994, pp. 506–507) discuss additional efficiency concerns.

[72] A patent application in the United States must be filed within 1 year of publication. Many other countries require that the patent application be filed prior to publication.

[73] Philippe et al. (2008) specify a model that does not rely on the public nature of research as the rationale for academic research but rather on control-rights consideration. To be more specific, they argue that "the fundamental tradeoff between academia and the private sector is one of creative control versus focus." (p. 617). In academe, scientists are free to pursue their own interests; thus academia fosters early-stage research. By way of contrast, the private sector directs scientists toward higher payoff research that is of a later stage.

agents can eat their cake and have it too, selectively publishing research findings while monopolizing other elements with the hope of realizing future returns. Eisenberg (1987) argues that such behavior is more common among academics than might initially be presumed because they can publish results and at the same time keep certain aspects of their research private by withholding data, failing to make strains available upon request, or restricting the exchange of research animals, such as mice.[74] If such were the case in 1987, one might hypothesize it to be even more so today, as academic scientists increasingly engage in patenting. But intellectual property does not appear to play a major role in restricting access to knowledge and materials used in subsequent research. Walsh et al. (2007) find that access is largely unaffected by patents, primarily because of issues related to enforceability. But access to the research materials of others, such as cell lines, reagents, and antigens, is restricted: 19% of the material requests made by their sample were denied. Competition among researchers played a major role in refusal, as did the cost of providing the material. Whether the material in question was a drug or whether the potential supplier had a history of commercial activity were also relevant factors in refusal.[75] This is not to say that instances do not exist where patents play a major restrictive role. A recent example concerns human embryonic stem cells. The University of Wisconsin, where they were discovered, has used their control, both through patents and material rights to the cell lines, to impose limits and conditions on other academics (Murray, 2007).

The ability to eat one's cake and have it too is not only facilitated by the fact that publication is not synonymous with replicability. It is also facilitated by the fact that certain kinds of knowledge, especially knowledge that relates to techniques, can often only be transferred at considerable cost, in part because their tacit nature makes it difficult, if not impossible to communicate in a written form (or codify). Tacit knowledge is thus "sticky" (Von Hippel, 1994) and requires face-to-face contact for transmission. It is one reason, as we will see, for arguing that knowledge may be geographically bounded and hence for expecting spillovers to have a geographic dimension.[76] The private aspect of technology is a major reason patents are not a necessary condition for successful research and development and underlies the willingness of industry to share knowledge through publications.

[74] Eisenberg (1987) suggests that the patent process may be more congruent with the scientific norms of disclosure and replication than the publishing process in certain areas of the life sciences. This is because patents in the biological sciences require that the material in question be placed on deposit. This is not a requirement for publication; neither are materials themselves part of the published text.

[75] There is the closely related anticommons issue of how multiple property right claims, sometimes in the hundreds, dampen research by requiring researchers to bargain across multiple players to gain access to foundational, upstream discoveries (Heller and Eisenberg, 1998). Walsh et al. (2007) ask academic respondents reasons that may have dissuaded them from moving ahead with a project. Lack of funding (62%) or being too busy (60%) were the most commonly reported reasons. Scientific competition (29%) was also an important reason given for not pursuing a project. Technology control rights related to terms demanded for access to inputs (10%) and patents (3%) were significantly less likely to be mentioned.

[76] Some aspects of technical knowledge have a strong tacit component, meaning that they cannot be completely codified and made explicit in the form of blueprints or instructions, but instead must be learned through practice. Nelson and Winter (1982) discuss tacit knowledge, particularly as it relates to skill, as does Foray (2004). Dasgupta and David (1994) use the term tacit somewhat differently to connote knowledge that, for whatever reason, is not codified and argue that the boundary between what is codified and what is tacit is not simply a question of epistemology. Rather, as suggested above, the boundary is "a matter, also of economics, for it is determined endogenously by the costs and benefits of secrecy in relation to those of codification." (p. 502).

There are other reasons why firms engage in disclosure. Foremost among these is recruitment of talent. Scientists and engineers working in industry value the ability to publish and are willing to pay for the privilege. Stern (2004) finds for a sample of postdoctoral biologists that firms which allow their employees to participate in the norms of science by publishing pay on average 25% less than firms which do not.[77] It is not only an interest in priority; the ability to publish allows scientists to maintain the option of working outside the for-profit section. The reputation of the lab, which is directly related to publication activity, also affects the ability of the company to hire scientists and engineers (Scherer, 1967); it may also affect its ability to attract government contracts (Lichtenberg, 1988). Hicks (1995) explores a number of other factors leading companies to opt for disclosure through publication. A critical element is the company's ability to screen the material that is published, thereby insuring that its proprietary interests are maintained. In the process, however, the firm must be mindful that delays can lower morale among research scientists. Hounshell and Smith (1988, p. 369) describe the loss of morale that occurred at Du Pont when research managers implemented what turned out to be a de facto moratorium on publishing.

From Table 1 we see that industry authors approximately 16,000 articles a year (measured by fractional counts) or about 7.5% of all articles with a US author. The number of articles peaked in the early 1990s and then declined for the next 10 years; output of the for-profit sector shows a modest increase in 2003. Declines were most notable in the fields of chemistry, physics, and engineering and technology (National Science Board, 2006). Coauthorship patterns, which are not shown in the table, are also of interest. During the same period, the coauthorship share (measured on a whole-count basis) of the private-for-profit sector with academe increased from 31.1% in 1988 to 47.3% in 2003 (National Science Board, 2006, Table 5-22).

7. Scientists in industry

Approximately two million researchers were employed in business enterprises in EU-15 countries, Japan and the United States as of 1999. Slightly more than 50% of these were working in the United States, where the percent of researchers working in the private sector is approximately 83%, compared to 51% in EU-15 and 67% in Japan (European Commission, 2003, Table 4.1.1 and Figure 4.1.4). Although research in industry is not restricted to those with a PhD, many PhDs do work in industry, especially in the United States where fully one-third of individuals with an S&E PhD work in the private sector. Moreover, the percent has grown in recent years, rising from around 23% in 1973 to 36% in 2003 (see Table 2).

Although the general assumption is that scientists and engineers are hired by industry to work in R&D, Stephan (2002) shows this to be an oversimplification, documenting that many scientists and engineers are employed outside the traditional R&D activities of firms. Some of this undoubtedly reflects promotion to managerial levels. But it also reflects the fact that innovation occurs in non-R&D sectors of firms, such as in sales, acquisitions and communications. Thus, studying scientists and engineers working in industry can provide another view (and measure) of innovative activity, different from such measures as R&D-expenditure data or patent-count data. Studying the employment pattern of scientists and engineers in the private sector also sheds light on sources of growth in the

[77] The finding depends on the inclusion of researcher fixed effects, leading Stern to conclude that the finding is conditional on scientific ability.

Table 2
Sector of employment of doctoral scientists in the United States, 1973–2003 (%)

Year	Total	Business/industry	Universities and 4-year colleges	Federal government	Other
1973	130,355	30,887 (23.7)	77,289 (59.3)	12,522 (9.6)	9657 (7.4)
1979	175,588	45,518 (25.9)	100,073 (57.0)	15,634 (8.9)	14,363 (8.2)
1985	218,328	64,962 (29.8)	119,365 (54.7)	16,860 (7.7)	17,141 (7.9)
1991	233,303	82,166 (35.2)	114,417 (49.0)	17,616 (7.5)	19,104 (8.2)
1997	279,430	97,300 (34.8)	133,530 (47.8)	23,670 (8.5)	24,930 (8.9)
2003	321,950	114,580 (35.6)	150,550 (46.8)	25,550 (7.9)	31,270 (9.7)

Fields included in the definition of science are: physical, mathematical, computer, environmental, and life. Self-employed are included in business and industry. "Other" includes state and local government, private not-for-profit, "other" educational institutions and "other."

Note: The dramatic changes in 1991 may in part reflect a change in survey methodology.

Source: Stephan (1996, Table 2), National Science Foundation (1999, Table 17), and National Science Foundation (2006, Table 13).

nonmanufacturing sector of the economy. For example, in recent years, PhD scientists and engineers have been increasingly employed in the service sector of the economy and have arguably contributed to the growth that the sector has experienced in recent years.

Firms engage in research for a variety of reasons. In some instances the research is a by-product of the development of a new product or process (Rosenberg, 1990). In other instances, the production of generic knowledge is, itself, the goal and is motivated by the belief that a particular new product or process innovation will result from that knowledge. Research activities (and the related publications) can also be a signal that the firm is worthy of receiving third-party funds, either in the form of research grants, such as Small Business Innovation Research (SBIR) awards, or venture capital. Allowing scientists and engineers to engage in basic research is also a recruitment mechanism, as noted above. Basic research is needed if the company is to stay abreast of developments in relevant scientific fields and more readily absorb the findings of other scientists (Cohen and Levinthal, 1989). Sometimes firms are motivated by the expectation that fundamental research will provide a scientific foundation for the company's technology. Firms have even been known to engage in basic research because of a concern that the fundamental knowledge required for the industry to advance is lacking and unlikely to be forthcoming from the academic sector. When Charles Stine made his presentation to the Executive Committee of Du Pont in 1926, for example, he argued that fundamental research was necessary because "applied research is facing a shortage of its principal raw materials" (Hounshell and Smith, 1988, p. 366).

This means that the research of some scientists and engineers working in firms is virtually indistinguishable from that of their academic counterparts.[78] This used to be especially the case for scientists employed at major industrial labs such as Bell Labs, IBM, and Du Pont. It remains the case for scientists and engineers working in certain sectors, such as pharmaceuticals, medical devices, and IT, but the

[78] A number of scientists and engineers from industry have received the top honors that their field can bestow. Bell Labs, Du Pont, IBM, Smith Kline and French, Sony, and General Electric have each been the research home to scientists who have subsequently won the Nobel Prize. In 1994, 3.8% of the then 2088 members of the US National Academy of Sciences came from industry.

demise of Bell Labs and the change in mission of IBM and Du Pont has been one of several factors contributing to the decline in "basic" research performed in industry in the United States.[79]

The reasons for industry to publish research findings, as well as the economic incentives for adopting a basic research agenda, have been noted above. This should not, however, be taken as an indication that economists (or others, for that matter) have adequately studied scientists in industry doing "science" and the role that pecuniary and nonpecuniary incentives play in innovation. Sauermann and Cohen (2007) are beginning to address this void by analyzing, at the individual level, the impact that preferences for benefits and job characteristics have on innovative effort and performance. But many questions remain unanswered and—perhaps even more fundamental—unposed.[80] For example, why do companies adopt compensation strategies that impair the productivity of scientists by frequently tying salary increases to the assumption of managerial responsibilities? Does the strategy adopted by IBM and Du Pont of creating well-paid research-fellow positions help alleviate the problem? What role do publications play in facilitating movement between the industrial and nonprofit sectors? How is basic research in industry monitored? The unpredictable nature of research, as well as the belief that creativity requires freedom of choice, suggests that success is hampered if company scientists are managed too closely. Yet firms can ill afford to fund research that has little promise of (eventually) relating to the company's objectives.[81]

8. Scientific labor markets

Science emerged from World War II with enhanced respect. Its successes were credited with shortening the war and reducing fatalities of Allied troops. There was also a growing appreciation of the important role science could play in stimulating economic growth and employment in peacetime. In a report prepared at the invitation of the White House, Bush (1945) argued that science provided an endless frontier and should be more heavily supported by the government. One response to Bush's report was the formation of the US National Science Foundation in 1950.[82]

[79] In the mid-1950s, approximately one-third of basic research performed in the United States was done by industry; in 2004, the last year for which data are available, the proportion had declined to approximately 16% (National Science Board, 2006, Table 4-8, vol. 2). Other factors contributing to the decline, in addition to the closure or refocusing of certain large industrial labs, include an increased propensity to "outsource" research to the university sector, as well as possible changes in definition and classification. At the same time that industry's share of basic research declined, their share of applied research rose from 56.3% to 61.8% (National Science Board, 2006, Table 4-12, vol. 2); the combined share of basic and applied research went from 50.1% to 40.3%.

[80] Our knowledge of scientists working in industry comes largely from a number of excellent case studies. These include Gambardella's (1995) study of the pharmaceutical industry, Hounshell and Smith's (1988) study of Du Pont, Willard Mueller's discussion of Du Pont (1962); Nelson's (1962) study of the development of the transistor, and Sobel's (1986) study of RCA. For a discussion of specific industries, see Mowery and Rosenberg (1998).

[81] Scherer (interview) reports that Bell Labs solved this problem by giving "the glassy-eyed stare" to scientists who were seen as straying too far from the Lab's purpose. Recipients knew that they had the choice of either modifying their research or being ostracized.

[82] The Bush report personifies the linear model to the extent that it argues that innovations flow out of basic research. But ironically the Bush report contributed to the nonlinearities of the innovation system by growing the scientific labor force available to work in industry. The effect was indirect but profound. Research grants awarded to PIs created a demand for doctoral students and postdocs. And the newly trained increasingly headed to industry as the academic sector proved less and less able to absorb the increased supply of newly trained PhDs.

The groundswell of support for science, heightened in the West in the 1950s by the threat of Soviet scientific and technological superiority, underscored the need to understand the workings of scientific labor markets. Stellar talent was drawn to this question. First, Blank and Stigler (1957) published a book on the demand and supply of scientific personnel; then Arrow and Capron (1959) wrote an article concerning dynamic shortages in scientific labor markets. Both studies set the stage for work to come.

8.1. A description of scientific labor markets

8.1.1. Where they train

The United States and Europe (defined here to be France, Germany, and the United Kingdom) were for many years the primary producers of PhDs in the natural sciences and engineering, as is seen in Figure 3. (The jump in the European data in 1989 is due to the inclusion of French data which prior to that date were unavailable in series form.) This pattern changed, however, in the 1990s, when the number of PhDs awarded in Asia began to rise rapidly and now surpasses the number produced in the United States. Part of the increase in the number of PhD's awarded in Asia reflects an increased proclivity for

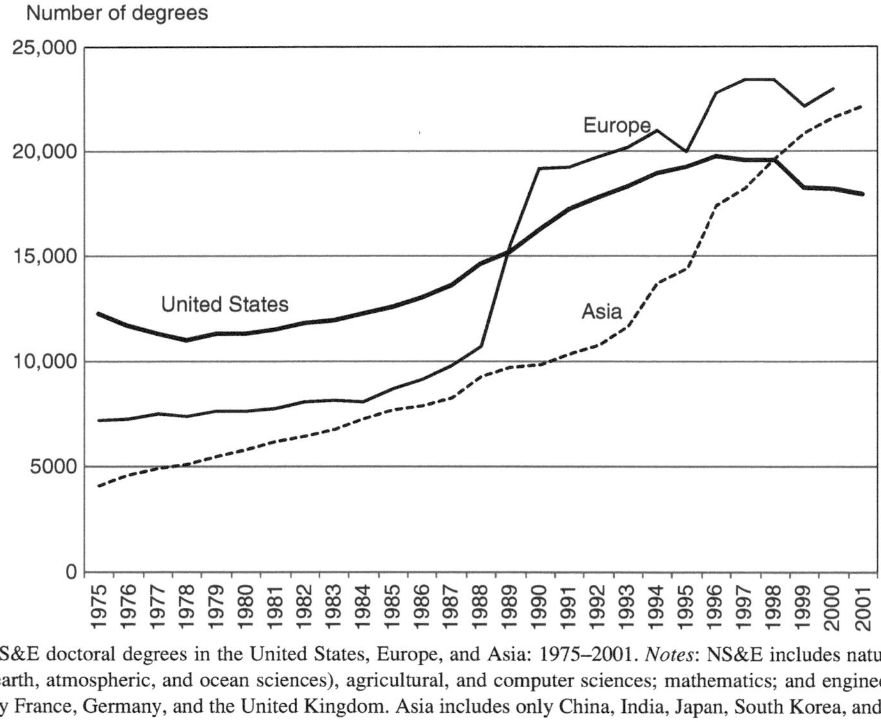

Figure 3. NS&E doctoral degrees in the United States, Europe, and Asia: 1975–2001. *Notes*: NS&E includes natural (physical, biological, earth, atmospheric, and ocean sciences), agricultural, and computer sciences; mathematics; and engineering. Europe includes only France, Germany, and the United Kingdom. Asia includes only China, India, Japan, South Korea, and Taiwan. The jump in the European data in 1989 is due to the inclusion of French data, which were unavailable in this data series before 1989. French data are estimated for 2000. Source: National Science Board (2004, Figure 2-38).

Asian students to train at home, rather than in the United States. For example, the number of PhD degrees awarded in South Korea steadily increased in the 1990s while the number of South Koreans receiving PhDs in the United States peaked in the early 1990s and declined steadily for 7 or 8 years thereafter. In recent years, however, it has risen again (Sunwoong, 2007), in part because the job prospects for Koreans receiving degrees in Korea declined after the Asian meltdown in the late 1990s.

8.1.2. Where they work

It is far harder to describe employment patterns of PhD scientists and engineers. Data simply are not readily available outside the United States and Canada to study sector of employment, although efforts are currently underway in the EU and other OECD countries to produce reliable counts of PhD employment by sector. We must thus regrettably, and for the time being, limit our discussion of sector of employment to the United States, where, since 1973, data has been routinely collected on the career outcomes of PhD scientists and engineers trained in the United States and working in the United States.

As can be seen in Table 2, the majority of doctoral scientists and engineers in the United States are employed in institutions of higher education and in business and industry. A distinct minority work at FFRDCs, the government and nonprofit institutions. Over time, the sectoral composition has shifted substantially as business and industry have employed proportionately more scientists and academe proportionately fewer. While in 1973 almost 60% of all scientists worked at universities and 4-year colleges, this had fallen to 47% by 2003.[83] At the same time, the percent working in business and industry (which includes self-employed) increased by 150%, growing from 23.7% to 35.6%.

8.1.3. International mobility patterns

Science, perhaps more than any other enterprise, is international in scope. We see this in terms of location of training, location of work and, as we have noted earlier, in coauthorship patterns. In terms of training, a very large percent of degrees, especially in Europe and the United States, are awarded to foreign students. While the percent has fluctuated over time in response to such things as changes in available funding and visa policies, overall the percent of PhDs awarded to international students in the United States has grown considerably during the past 30 years. By 2006, 36.0% of PhDs awarded in science and 58.6% of those awarded in engineering went to candidates on a temporary visa while 6.0% of science PhDs and 4.5% of engineering PhDs were awarded to noncitizens on permanent visas (National Science Foundation, 2006, Table 3).[84] A somewhat similar situation exists in Europe, especially in the United Kingdom, where in 2003 over 50% of engineering PhDs and approximately 45% of math and computer science degrees were awarded to foreign students (National Science Board, 2004, Figure 2-40). The percent of foreign PhDs awarded in France is somewhat lower, but close to 30%

[83] There has also been a structural shift away from tenure-track positions to nontenure-track positions, including staff scientists. In 1993, 78% of academic appointments were either in tenured positions or in tenure-track positions; in 2003, 62% were tenured or in tenure-track positions (National Science Foundation, 1996, Table 17; 2006, Table 21).

[84] Note that since some individuals do not respond to the question regarding citizenship, it does not follow that the remainder were all awarded to US citizens.

of PhDs in math/computer science went to foreign students in 2001; 22% of engineering PhDs were awarded to foreign students.

Science is also international in terms of work location. In 1990 (the latest date for which data is available at the time of this writing), 24.7% of all doctoral-trained scientists working in the United States were born outside the United States (Levin and Stephan, 1999). While many of these came to the United States to receive their training, a not insignificant number came to the United States after receiving either a baccalaureate degree abroad (16%) or a PhD abroad (10.7%). When newer data becomes available, we would expect to see an increase, reflecting the increased proportion of PhDs awarded to foreign-born scientists and engineers, but also the increasing number of postdoctoral fellows who come to the United States after receiving a PhD abroad, as well as the inflow of doctoral-trained foreign talent, especially from Russia, during the 1990s.

Star scientists are highly mobile, as Zucker and Darby (2007) show in recent work which tracks the most cited scientists and engineers during the period 1981–2004. They measure "home country" by address on first publication and mobility in terms of subsequent addresses appearing on publications. They find that 8.6% of the stars make a "one-way trip" from their home country to a different country; another 8.4% make a "round trip."[85]

One factor that contributes to international mobility is the wide differences that exist in funding for research and development across countries. Even among OECD countries there is considerable variability in the amount of funding for R&D, as is blatantly obvious from Figure 4, which shows R&D intensity as share of gross domestic product for eight OECD countries for the period 1981–2003. Italy (more recently, Russia) is at the bottom, and Japan and the United States are at the top. The figure also shows the wide fluctuations that occur in R&D expenditures within any country; these directly affect the market for scientists and engineers and contribute to cohort effects. The aggregate nature of the data conceals the mix of a country's R&D expenditures. In recent years this was exemplified in the United States when the NIH's budget doubled over a 5-year period, going from approximately $14 billion to $28 billion (current US), while other R&D budgets grew marginally, at best.

8.2. Studies of supply and demand for new entrants to science

A number of studies have examined the market for new entrants to science. Leslie and Oaxaca (1993) do an excellent job of surveying this literature and summarizing the major findings, as does Ehrenberg (1991, 1992).[86] Variables that are usually found to affect the supply of enrollees (or the number of graduates) in field *j* are salary in field *j*, salary in an alternative occupation such as law or business, and (for men) the draft deferment policy.[87] These variables almost always have the expected sign and are highly significant. The magnitude of the implied elasticities, however, varies considerably across studies, even when field is held constant (Ehrenberg, 1992). Another market variable often included in predicting supply is some measure of concurrent, past, or future supply. Other things being equal,

[85] The analysis is for scientists and engineers working in a top-25 science and technology country (Zucker and Darby, 2007).

[86] Most studies focus on long-run adjustments. A few, however, examine the short-run responsiveness of the market by also focusing on the movements of trained personnel between fields and sectors (Blank and Stigler, 1957).

[87] Groen and Rizzo (2007) conclude that a major reason the propensity of US men to enroll in graduate school declined in the early 1970s was the end of Vietnam War draft deferments for graduate students.

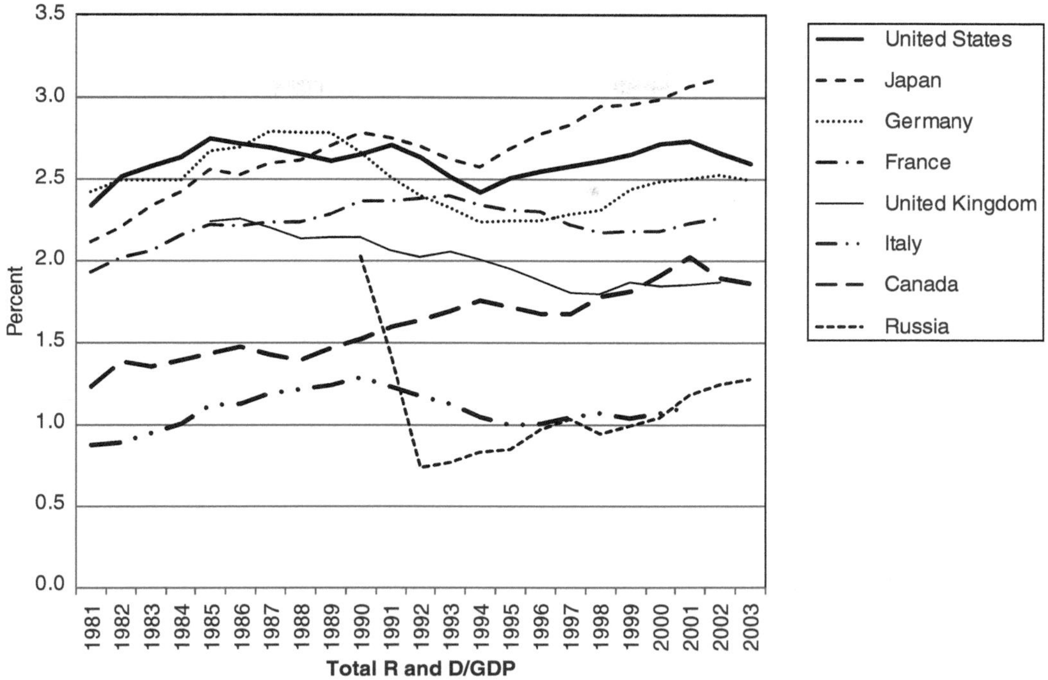

Figure 4. R&D share of gross domestic product, by selected countries: 1981–2003. Source: National Science Board (2006, Figure 4-30).

enrollments are positively associated with present cohort size. Various lag structures are used in estimating these models and it is common to assume some form of adaptive (or rational) expectations. Supply variables generally ignored by these studies (primarily because of a reliance on aggregated data) include type of support available while in school, debt level upon graduation from college, and average time to degree.

Demand equations prove more difficult to specify, partly because we know so little about the behavior of universities and governments. There is, however, convincing evidence that demand relates to R&D expenditures and that these expenditures in turn affect supply decisions. Freeman (1975) finds degrees at the B.S., M.S., and PhD level in physics for the period 1950–1972 to be significantly related to R&D expenditures. Salaries also play a role. The propensity of recent graduate students to work in industry in the United States, for example, is in part a reflection of the higher relative salaries that are available in industry (Ehrenberg, 1991). It also undoubtedly reflects the softness of the academic labor market, given that most graduate students and postdoctoral students express a strong preference for a job at a research university (Davis, 2005; Fox and Stephan, 2001).

Several factors explain the softness of the academic market in recent years in the United States. First, cutbacks in public funds and lowered endowment payouts, especially in the early 2000s, clearly affect hiring. Second, salaries of tenure-track faculty are higher than those of nontenure-track faculty and this leads to a substitution away from tenure-track positions (Ehrenberg and Zhang, 2005). Third, funding for nontenure-track positions, such as staff scientist, is available in research grants. The high cost of start-up packages also plays a role in explaining these trends. Given the costs of such packages (which range from $300,000 to well over $1 million), when universities do hire in the tenured ranks, they are tempted to recruit senior faculty away from another university, rather than hire an as yet untested junior faculty member. The financial risk is considerably lower. While the start-up packages are generally higher at the senior ranks (Ehrenberg et al., 2003), the university gets an immediate transfer of grant money, because the senior faculty generally bring existing research grants with them when they come.

It is not only in the United States that the market for scientists and engineers in the academic sector has been soft in recent years. The job prospects of young PhDs in Italy in the university sector, for example, have also been bleak. The situation was aggravated in 2003 when a "no new permanent position" policy was put into effect. This has resulted in a situation in which the share of temporary researchers at universities had reached 50% in some instances, with young people being heavily concentrated in temporary positions (Avveduto, 2005). Reflecting these problems, the average age of faculty in research positions in 2003 in Italian universities was 45; for those in associate professor positions, 51; and for those in full professor positions, 58 (Stephan, 2008).

The academic labor market also appears soft in Germany. Schulze (2008) documents that the number of professors at German universities peaked in 1993 at about 23,000 and has been, with few annual exceptions, steadily declining ever since. In 2004, the last year for which he reports data, the number stood at just slightly over 21,000. The decline is not due to a decline in the number of students. During the same period the number of high school graduates has increased significantly and the ratio of professors per 100 high school graduates "has deteriorated significantly from 11.26 in 1996 to 9.43 in 2004" (p. 23). The decline has come at the same time that the number of Habilitationen, a requirement for obtaining an appointment as a professor at most institutions and in most fields, has grown dramatically.[88] Using a back-of-the-envelope type of calculation, Schulze estimates that the ratio of new applications to job openings rose from roughly 3/2 to 5/2 during the 14-year period that he analyzes.

A similar situation exists in South Korea, where universities, particularly private universities, under pressure to reduce expenditures on teaching personnel, are relying increasingly on part-time instructors. Sunwoong (2007) estimates that the number of full-time instructors in 4-year colleges and universities in 2006 was approximately 43,000 while the number of part-time instructors in 2003 was more than 50,000. Because of the slow turnover and the sluggish expansion of new positions, the problem for newly trained PhDs is likely to exacerbate.

8.3. Forecasting scientific labor markets

Although models of scientific labor markets have been somewhat successful in providing insight into factors affecting demand and supply, reliable forecasts of scientific labor markets do not exist, partly because of the unavailability of reliable predictions of exogenous variables. While this problem is endemic to forecasting in general, the ups and downs of funding for research (see Figure 4), as well as changes in policies, make forecasts of scientific labor markets particularly unreliable.

[88] The typical academic career path in Germany involves preparing the Habilitation. After completion, and pending availability of a position, one is hired into a C3 position, which must be at an institution other than where the Habilitation was prepared.

Forecast error is common with regard to scientific labor market outcomes (Leslie and Oaxaca, 1993). In 1989, the NSF predicted an impending shortage of S&E doctorates in the United States (National Science Foundation, 1989). Others also predicted an impending shortage in the late 1980s (Atkinson, 1990; Bowen and Sosa, 1989). The underlying rationale was based on two assumptions: (1) an aging faculty, hired when higher education was expanding in the United States in the late 1950s and 1960s, would retire and be replaced; and (2) increases in the student body, as Baby Boomers' children headed to college, would increase demand for faculty. By the mid-1990s, if not before, it was clear that these forecasts had widely missed the mark, as was indicated by the dramatic increase in the proportion of new PhDs in nonpermanent jobs, the lengthening of time in postdoctoral positions and a decrease in the proportion of recent PhDs holding tenure-track positions. The reason for the forecast error related to a failure of the forecasters to predict changes in demand. These changes were brought about by the elimination of mandatory retirement, by an economic recession, and by political pressure to downsize the federal budget and the demise of the Cold War, which led to cuts or plateaus in federal funding.

In response to forecast error, a National Research Council Committee was created to examine issues involved in forecasting demand and supply. The committee was chaired by Daniel McFadden. The report, issued in 2000 (National Research Council, 2000), should be mandatory reading for anyone tempted to enter this arena. The committee concluded that forecast error could occur from: (a) misspecification of models, including variables, lag structure and error structure; (b) flawed data, or data aggregated at an inappropriate level; (c) unanticipated events. Even if model specification and lag structure are improved upon, unanticipated events continue to plague the reliability of forecasts. Both the fall of the Wall and the events of 9/11 had profound effects on scientific labor markets and would have been difficult to incorporate into any forecasting model.

Despite the report, and the well-known proclivity of forecasts to miss the mark, it is common for policy groups on both sides of the Atlantic to declare an impending shortage of scientists and engineers. A 2003 report issued by the National Science Board concluded that "Analyses of current trends ... indicate serious problems lie ahead that may threaten our long-term prosperity and national security" (p. 7). A 2004 report from the European Union concluded that "Increased investment in research will raise the demand for researchers: about 1.2 million additional research personnel, including 700,000 additional researchers, are deemed necessary to attain the objectives, on top of the expected replacement of the aging workforce in research" European Commission, 2004, (p. 11).

9. Science, productivity, and the new growth economics

The foremost reason economists have for studying science is the link between science and economic growth. That such a relationship exists has long been part of the conventional wisdom, articulated first by Smith ([1776] 1982, p. 113). That the relationship is nonlinear has more recently entered into the conventional wisdom, as the role that technology plays in shaping scientific advances has been investigated and articulated by Rosenberg and Mokyr, as well as others. The nonlinearity relates not only to the role played by equipment in scientific discovery but also to the role that technological breakthroughs have played in fostering scientific insights as well as to their role in encouraging scientists in the public sector to develop new programs and research agendas. Solid-state physics is but one of many cases in point.

It is one thing to argue that science affects economic growth or to establish that a relationship exists between R&D activity and profitability. It is another to establish the extent to which scientific knowledge spills over within and between sectors of the economy and the lags that are involved in the spillover process. To date, four distinct lines of inquiry have been followed to examine these relationships. One inquires into the relationship between published knowledge and growth. Another surveys firms, with the goal of understanding the role that public knowledge plays in innovation. A third examines how the innovative activity of firms relates to research activities of universities (and other firms) by using measures of innovation as well as paper trails provided in patents and initial public offerings. A fourth looks at the degree to which firm performance is mediated by links with public research.

Adams (1990) uses the published-knowledge line of inquiry to examine the relationship between research and growth in 18 manufacturing industries between the years 1953 and 1980. The study is ambitious; for example, Adams measures the stock of knowledge available in a field at a particular date by counting publications in the field over a long period of time, usually beginning before 1930. He creates industry "knowledge stocks" by weighting these counts by the number of scientists employed by field in each of the industries being studied. He then relates productivity growth in 18 industries over a 28-year period to stocks of "own knowledge" and stocks of knowledge that have flowed from other industries. Adams finds both knowledge stocks to be major contributors to growth of productivity. He also finds that the lags are long: in the case of own knowledge, on the order of 20 years; in the case of knowledge coming from other industries, on the order of 30 years.

A necessary step in the growth story that Adams documents is that public science "leak out" to firms. Recent work by Adams et al. (2006) estimates a measure of the lag involved in this phase by analyzing citation patterns from industry-authored papers to university-authored papers. They report an average modal lag across the six disciplines studied of 3.02 years. The lag is longest in computer science (4.12) and shortest in physics (2.06).[89]

A different way to study the relationship between public science and innovation is to survey firms with an eye to ascertaining the role that university research plays in product development. Mansfield (1991) uses such a technique. He surveys 76 firms in seven manufacturing industries to ascertain the proportion of the firm's new products and processes commercialized in the period 1975–1985 that could not have been developed (without substantial delay) in the absence of academic research carried out within 15 years of when the innovation was first introduced. He finds that 11% of the new products and 9% of the new processes introduced in these industries could not have been developed (without substantial delay) in the absence of recent academic research. Using sales data for these products and processes, he estimates a mean time lag of about 7 years. He also uses these data to estimate "social" rates of return of the magnitude of 28%.

The interaction between firms and faculty is reciprocal: Relationships with firms also enhance the productivity of faculty. Mansfield (1995) finds that academic researchers with ties to firms report that their academic research problems frequently or predominately are developed out of their industrial consulting, and that this consulting also influences the nature of work they propose for government-funded research. Agrawal and Henderson (2002), in their study of MIT patenting, find similar

[89] Tacit knowledge is most easily transmitted by face-to-face interaction. Stephan (2007) traces the placement of newly minted PhDs in industry as another means of the transmission of knowledge from the public sector to the private sector.

sentiments. An engineer whom they interview reports that "it is useful to talk to industry people with real problems because they often reveal interesting research questions . . ." (p. 58). Zucker et al. (1998a,b) find that the productivity of academic scientists is enhanced when they work with scientists in biotechnology companies.

Cohen et al. (2002) use a related approach, drawing on data from the 1994 Carnegie Mellon Survey (CMS) of industrial R&D, to determine the extent to which public knowledge is utilized by firms in their R&D activities and the means by which knowledge flows from the public sector to the private sector. They find that public research plays a major role in R&D in a few industries, particularly pharmaceuticals, and is generally more important in manufacturing than in other sectors. People and publications play a major role in transmission: firms rated publications, attendance at conferences and informal interaction as the most important channels for accessing public research. The licensing of university patents plays a substantially smaller role. Whether the licensing result would persist if the data were to be collected today remains to be seen.[90] They also find that "public research is used at least as frequently to address existing problems and needs as to suggest new research efforts." (Cohen et al., 2002, p. 2).[91]

Knowledge spillovers can also be studied by examining the relationship between some measure of innovative activity of firms and the research expenditures of universities. This production-function approach finds its roots in the work of Griliches (1979), who posited what has become known as the knowledge-production function. This line of inquiry ignores the lag structure, but focuses instead on the extent to which such spillovers exist and are geographically bounded. The rationale for expecting them to be bounded is that transmission of tacit knowledge is greatly facilitated through face-to-face communication. The approach is not restricted to examining the relationship between innovation and university research, but often includes a measure of private R&D expenditure in the geographic area to determine the extent to which spillovers occur within the private sector as well. Sometimes the measure of innovative activity is patents (Autant-Bernard, 2001; Jaffe, 1989); sometimes it is counts of innovations (Acs et al., 1992). Sometimes (Black, 2004) it is counts of SBIR grants. In any case, measured at the geographic level, innovative activity is found to relate to R&D expenditures of universities and firms in the same geographic area. There is some indication that these spillovers, particularly those coming from universities, are more important for small firms than for large firms (Acs et al., 1994).[92]

Patents provide a means of establishing a paper trail of knowledge spillovers, given the requirement that previous art be cited. Although it is the patent examiner who has the final say on which citations to include, the applicant is legally required to disclose any knowledge of prior art. Jaffe et al. (1993) use citations to other patents to analyze knowledge spillovers. They find that citing patents are in closer geographic proximity to the cited patent than they are to the sample of "control" patents that have the same temporal and technological distribution but are not linked through citation. The effect is most notable at the SMSA level but also holds, to a lesser extent, at the state and country level.

[90] Jinyoung et al. (2005) find an increasing incidence of firm patents that list one or more inventors who had previously appeared as an inventor on a patent assigned to a university.

[91] Adams (2006) surveys 220 R&D labs. He finds that state universities in the South and Midwest are more often cited as a source of knowledge by mature industries, while younger industries are more likely to look to private US universities and universities in coastal regions.

[92] Adams (2002), for a sample of 220 R&D labs, finds academic spillovers to be more localized than industrial spillovers.

Patent citations to university articles also provide a paper trail of knowledge spillovers. Here, too, there is evidence that spillovers have a geographic dimension (Hicks et al., 2001).

Information on inventors can also be used to establish a paper trail by examining the mobility of inventors over time as measured by inventor addresses recorded on the patent. Using such a paper trail for Italian patents, Breschi and Lissoni (2003) conclude that mobility of researchers between firms is the mechanism by which knowledge spills over. And, because mobility is often within the same geographic area, knowledge spillovers have a geographic dimension. Indeed, their research indicates that localization effects (as measured by citations) tend to vanish in the absence of a network relationship between inventors. Their work is consistent with that of Almeida and Kogut (1999) which analyzes the interfirm mobility of patent holders in semiconductors and finds that labor markets have strong spatial characteristics, especially in Silicon Valley, where intraregional mobility is high and interregional moves are much smaller. Zucker and Darby's work (2007) also affirms the important role that people play in the spillover process. They show that where star scientists are active plays a key role, over and above the location of universities, in determining where biotech firms develop.

Start-up firms provide another indication of knowledge spillovers. Stanford University estimates that (http://www.stanford.edu/group/wellspring/index.html) over 2400 full-time companies have been founded by members of the Stanford community during the past several decades. The BankBoston's study (1997) is widely cited to show the important role that MIT has played in creating new companies in the Boston area.

Founders and members of SABs provide still another paper trail for studying knowledge spillovers. Audretsch and Stephan (1996) examine the location of university-based scientists having such a formal relationship with a biotech firm. They find that proximity matters, but that it does not matter that much. The majority of scientists (70%) do not live in close geographic proximity to the company. They conclude that when spillovers are mediated through people, they need not be geographically bounded if firms require expertise that may not exist in the local area. This is consistent with work by Mansfield (1995) that suggests that industry, when looking for academic consultants, is likely to use local talent for applied research, but focuses on getting the "best" regardless of distance when basic research is involved.[93]

A fourth vane of studies examines the relationship between a measure of firm performance and the firm's links with open science. Zucker et al. (1998a,b) find that the more articles a biotechnology firm has coauthored with a star, the better the firm performed, whether measured by products in development, products on the market, or employment. Cockburn and Henderson (1998) find that pharmaceutical firms that coauthor with publicly funded researchers have a higher performance as measured by research productivity.

Characteristics of a firm's patent portfolio, as measured by citation patterns, also relate to the valuation of the firm. Deng et al. (1999) build a model of stock performance based on closeness to science (as measured by cited articles in the firm's patents) and the influence of the patent (as measured by cites to the patent). Using such a methodology, CHI Research, Inc. identified undervalued firms and

[93] Not addressed here, but of clear policy importance, is the degree to which university scientists are able to appropriate rents for their knowledge (Zucker et al., 1998a,b, p. 302). A related question can be asked with regard to knowledge that is transferred between firms. If it is people that provide the means by which knowledge is transferred between firms, then the resulting externalities may be fully captured or, at best, only pecuniary externalities may arise (Breschi and Lissoni, 2003).

compared them to overvalued firms. Doing retrospective analysis of the 20 most undervalued and 20 most overvalued firms, and updating the list annually, CHI found that the performance index of undervalued firms grew from 100 to approximately 2500 during the period 1990–2001, while the overvalued portfolio grew from 100 to approximately 250. In a somewhat related study, Hall et al. (2001) demonstrate that the market-to-book value of a firm is related to the number of times a firm's patent has been cited in other patent applications.

Despite the crudeness of the measures and the problems inherent in the various approaches, these studies go a long way toward demonstrating that the spillovers between scientific research and innovation are substantial, as are the lags. We cannot, however, leave the growth story here. Knowledge spillovers not only are a source of growth; they are endogenous. The story goes something like this: In an effort to seek rents, firms engage in R&D. Public aspects of this R&D then spill over to other firms, thereby creating increasing returns to scale and to long-term growth (Romer, 1994). The work of Schmookler (1966) and Scherer (1982), which demonstrates the responsiveness of R&D to demand factors, is consistent with this concept of endogenous growth. So is the work of Jaffe (1989), Acs et al. (1992), and Autant-Bernard, 2001, among others, whose work suggests that firms appropriate the R&D of other firms. Empirical work summarized above also implies that scientific research conducted in the academic sector of the economy spills over to firms.

Does this mean that research in the academic sector is an important component of the new growth economics? The answer depends upon the extent to which scientific research in the public sector is endogenous.[94] If it is not, spillovers from the public sector to firms are important, but not as a component of the new growth economics. Five aspects of science that we have developed in this chapter lead us to argue that an endogenous element of academic research exists. First, profit-seeking companies support academic research. Second, the problems that academic scientists address often come from ideas developed through consulting relationships with industry. Third, markets direct, if not completely drive, technology and technology affects science (Price, 1986; Rosenberg, 1982).[95] Fourth, government supports much of public-sector research, and the level of support available clearly relates to the overall well-being of the economy. Finally, there is evidence that relative salaries and vacancy rates affect the quantity and quality of those choosing careers in a field. "Hot fields" like biotechnology and computer science have attracted a disproportionate number of people in recent years when the rewards (at least to a few) have been extraordinary. The impact on academic research has been substantial.[96]

One could even argue (and many have) that public researchers (and the institutions where they work) have become too responsive to economic incentives for the good of science, or for the long-term good of the economy. Hundreds of patents can create thickets; competition can lead scientists to deny others access to research material; industrial sponsorship of public research can encourage secrecy and delay of publication.

[94] It goes without saying that the science performed in companies is endogenous and spills over to other companies. A portion of this chapter has been devoted to demonstrating that profit-seeking companies hire scientists, direct them to do basic research, and often allow (encourage) them to share their research findings with others.

[95] The counter thesis of "technology push" is also important. That is, in many cases the invention of a new technology leads to new demands.

[96] This is not to argue that outcome X is endogenous, but merely that the growth of public knowledge has an endogenous component. At any point in time constraints clearly exist to discovery, either through the technology that is available to address the problem or because of lack of fundamental knowledge in an area necessary to the inquiry. Many of these constraints must be viewed as being exogenously determined, at least over a specific period of time (Rosenberg, 1974).

One could also argue that public institutions have been overly successful in selling the contribution they make to local economic development. It is one thing to find that knowledge spills over; it is another to create new universities and research programs with the goal of generating significant local economic development. Yet governments are doing precisely that. The California system opened its new campus in Merced in the fall of 2005. At least part of the impetus for its construction was the California Legislature's belief that the investment could bring economic development to the San Joaquin Valley. The News from Texas in August 2006 was that the state had decided to invest $2.5 billion for science teaching and research in the University of Texas system. A primary focus of the initiative is to build up the research capacity at campuses in San Antonio, El Paso, and Arlington in an attempt to turn these cities into the next Austin, if not the next Silicon Valley. Texas and California are not unique. Across the world, governments are working to turn universities and public research institutes into engines of economic development. Such investments will undoubtedly contribute to economic growth in the long run; but the extent to which it is a rational policy for fostering local economic development is not clear.

10. Conclusion

This chapter suggests several areas of inquiry in which economists have added significantly to an understanding of science and the role that science plays in the economy. Some of our discussion draws heavily on the work of sociologists and demonstrates the continued need to approach the study of science from an interdisciplinary perspective.

First, we have begun to quantify the relationship between science and economic growth, both in terms of payoff and lag structure. We have also achieved a better understanding of how science relates to growth, as a result of two threads of research coming together. One demonstrates that firms benefit from knowledge spillovers. The other suggests that knowledge spillovers are the source of growth and that these spillovers are endogenous. Although the authors of the new growth economics focus on the role that the R&D activities of firms play in this spillover process (both as creator of spillovers and recipient of spillovers), a case can be made that research in the nonprofit sector also has endogenous elements that are set in motion by profit-seeking behavior.

Second, the priority-based reward system that has evolved in science provides incentives for scientists to behave in socially beneficial ways. In particular, the reward of priority encourages the production and sharing of knowledge and thus goes a long way toward solving the appropriability dilemma inherent in the creation of the public good knowledge.

Third, science is not only about fame; it is also about fortune. Many of the financial rewards in science are a consequence of priority: salary, for example, is positively related to both article and citation counts. Because the financial rewards often come in the form of consulting and royalty income, we will never know the full extent of the relationship until we have reliable data on nonsalary dimensions of the income of scientists. There is also evidence that reputation matters to industry. We know, for example, that some firms encourage scientists to publish. We also know that startup companies benefit from affiliations with highly cited scientists.

Fourth, economics has been brought to bear on understanding the way in which scientific labor markets function. This in turn provides insights into how various government policies, intentionally or unintentionally, affect the market for scientists and engineers. Our ability to understand and model labor markets, however, is seriously hampered in some countries by the unavailability of data.

Fifth, our understanding of the many exogenous factors affecting demand and supply has led to the conclusion that we cannot forecast market conditions for scientists and engineers with much accuracy. It has also led to expressions of caution (and skepticism) concerning forecasts (usually of shortages) that policy groups are wont to make.

Sixth, numerous studies done in the late 1990s and early 2000s have contributed considerably to our understanding of the productivity of scientists and engineers. Moreover, we have extended studies of productivity to include patents as well as publications and considerable work has been done regarding the relationship of patents to publishing and vice versa.

But much remains to be understood and modeled. Foremost is a study of labs. Economists almost always approach productivity issues by studying individual scientists rather than the labs in which the scientists work. While individuals matter, science is increasingly about teams and collaboration. Yet we continue to focus on the individual. Our bias is caused by at least three factors: (1) ease of data collection, (2) an econometric tool kit that invites analyzing individual behavior, and (3) a funding system, at least in the United States, that continues to place great emphasis on the individual scientist despite the importance of labs.

Once we shift to a study of labs, numerous questions invite exploration. For example, we need to learn more about the production function of the lab, the degree of substitutability between capital and labor and whether the capital–labor ratio has changed over time as equipment has become more sophisticated. We need to know more about how lab size is determined. To what extent do economic factors come into play? Is size determined by the tradition of giving a researcher two rooms, with eight at a bench per room? Is there an efficient lab size? Is it efficient to increase lab size, as happened with the NIH doubling?

There are other ways economists can contribute to a better understanding of the workings of science. Seven are mentioned here. First, we need a better understanding of how outcomes relate to changes in funding. By way of example, to what extent has the practice of funding departments and programs on the basis of publications and citations led to "just-in-time" hiring? To what extent has the practice changed the submission and publishing patterns of scientists, especially outside the United States, where the changes have been the most notable? How has this, in turn, affected the refereeing process? Related is the question of the degree to which networks, in which funding agencies have placed great stock, contribute to productivity.

Second, economists can contribute to a discussion of other efficiency questions: Are there too many entrants in certain scientific contests or, more generally, too many scientists? A related question concerns whether science is organized in the most efficient way, particularly in the nonprofit sector. Is the demand for graduate students as research assistants and subsequently as postdocs so strong that it masks market signals concerning the long-run availability of research positions and encourages inefficient investments in human capital? Could other kinds of personnel (e.g., permanent research scientists) substitute for graduate students and postdocs in the lab?

Third, economists have a comparative advantage in understanding and analyzing the role that risk and uncertainty play in science. We can, for example, explain why risk aversion on the part of funding agencies dissuades scientists who are by disposition willing to take risk from engaging in this kind of research. We have the tool kit required to understand choices as outcomes of games and the possibility of using experimental economics to better understand how outcomes depend on rewards and funding.

Fourth, economists can contribute to an understanding of science by extending to the study of science approaches that have proved fruitful to the study of firms. Work in industrial organization that examines the entrance and survival of new firms could provide a framework for studying careers. Another possibility is to view the production of scientists through the lens of an evolutionary model (Nelson and Winter, 1982). Diversity and selection—the heart of evolutionary economics—are clearly present in the way in which scientists are trained, promoted, and rewarded.

Fifth, economists can contribute to a better understanding of how the reward structure of science leads some scientists to behave in socially undesirable ways. Issues include the fragmentation of knowledge that a focus on article counts encourages and the temptation to engage in fraudulent behavior.

Sixth, as a discipline we need to pay considerably more attention to understanding the way scientific effort is organized, monitored, and rewarded in industry. We also need to learn more about how scientists contribute to productivity outside of the traditional industrial R&D labs. We could learn much, for example, by studying scientists and engineers working in the service sector.

Seventh, the question of how opportunities for entrepreneurial behavior affect the practice of science bears continued exploration. So, too, does the question of whether policy makers have oversubscribed to the idea that knowledge spillovers lead to local and regional economic development.

In short, economists have accomplished a reasonable amount in our study of science; but other issues await investigation. It is hoped that this chapter will encourage that process.

Acknowledgments

This chapter draws extensively from Stephan (1996). The author thanks William Amis, Cristiano Antonelli, Wesley Cohen, Richard Freeman, Bronwyn Hall, Fiona Murray, and Nathan Rosenberg for their comments. Kelly Wilkin and Erin Coffman provided research assistance. Work on this chapter was begun while the author was a Wertheim Fellow at Harvard University, spring 2007; it was completed while the author was a Fellow at the International Centre for Economic Research (ICER), Torino (Italy), Fall 2009.

References

Acs, Z., Audretsch, D., Feldman, M. (1992). "Real effects of academic research: Comment". American Economic Review 82, 363–367.

Acs, Z., Audretsch, D., Feldman, M. (1994). "R&D spillovers and recipient firm size". The Review of Economics and Statistics 76 (2), 336–340.

Adams, J. (1990). "Fundamental stocks of knowledge and productivity growth". Journal of Political Economy 98, 673–702.

Adams, J. (2002). "Comparative localization of academic and industrial spillovers". Journal of Economic Geography 2, 253–278.

Adams, J. (2006). "Learning, internal research, and spillovers". Economics of Innovation and New Technology 15 (1), 5–36.

Adams, J., Black, G., Clemmons, R., Stephan, P. (2005). "Scientific teams and institution collaborations: Evidence from U.S. universities, 1981–1999". Research Policy 34 (3), 259–285.

Adams, J., Clemmons, R., Stephan, P. (2006). How Rapidly Does Science Leak Out? NBER Working Paper 11997.

Agrawal, A., Goldfarb, A. (2008). "Restructuring Research: Communication costs and the democratization of university innovation". American Economic Review 98, 1578–1590.

Agrawal, A., Henderson, R. (2002). "Putting patents in context: Exploring knowledge transfer from MIT". Management Science 48, 44–60.

Allison, P., Stewart, J. (1974). "Productivity differences among scientists: Evidence for accumulative advantage". American Sociological Review 39, 596–606.

Almeida, P., Kogut, B. (1999). "Localization of knowledge and the mobility of engineers in regional networks". Management Science 45 (7), 905–917.

Andrews, F.M. (1979). Scientific Productivity: The Effectiveness of Research Groups in Six Countries. Cambridge University Press, Cambridge.

Another Physics Department Down. (2006). Science 314, 1363.

Arrow, K.J. (1962). Economic welfare and the allocation of resources for invention. The Rate and Direction of Inventive Activity: Economic and Social Factors. Princeton University Press, Princeton, NJ.

Arrow, K.J. (1987). "Reflections on the essays". In: Feiwel, G.R. (Ed.), Arrow and the Ascent of Modern Economic Theory. New York University Press, New York.

Arrow, K.J., Capron, W.M. (1959). "Dynamic shortages and price rises: The engineering-scientist case". Quarterly Journal of Economics 73, 292–308.

Atkinson, R.C. (1990). "Supply and demand for scientists and engineers: A national crisis in the making". Science 248, 425–432.

Audretsch, D.B., Stephan, P.E. (1996). "Company-scientist locational links: The case of biotechnology". American Economic Review 86, 641–652.

Autant-Bernard, C. (2001). "Science and knowledge flows: Evidence from the French case". Research Policy 30, 1069–1078.

Avveduto, S. (2005). Characteristics and Organization of Research in Italy: A Specific Glance on Human Resources. IRPPS-CNR, Rome.

Azoulay, P., Ding, W., Stuart, T. (2009). "The impact of academic patenting on the rate, quality, and direction of (public) research output", Journal of Industrial Economics 57, 637–676.

BankBoston. (1997). MIT: The Impact of Innovation.

Barnett, A., Ault, R.W., Kaserman, D. (1988). "The rising incidence of co-authorship in economics: Further evidence". Review of Economics and Statistics 70, 539–543.

Black, G. (2004). The Geography of Small Firm Innovation. Kluwer, New York.

Blank, D., Stigler, G.J. (1957). The Demand and Supply of Scientific Personnel. National Bureau of Economic Research, New York.

Bowen, W., Sosa, J.A. (1989). Prospects for Faculty in the Arts and Sciences: A Study of Factors Affecting Demand and Supply. Princeton University Press, Princeton, NJ.

Breschi, S., Lissoni, F. (2009). Mobility of skilled workers and co-invention networks: An anatomy of localized knowledge flows", Journal of Economic Geography 9, 439-468.

Breschi, S., Francesco, L., Montobbio, F. (2007). "The scientific productivity of academic inventors: New evidence from Italian data". Economics of Innovation and New Technology 16, 101–118.

Brodetsky, S. (1942). "Newton: Scientist and man". Nature 150, 698–699.

Bush, V. (1945). Science: The Endless Frontier. U.S. GPO, Washington, DC.

Calderini, M., Franzoni, C., Vezzulli, A. (2007). "If star scientists do not patent: The effect of productivity, basicness and impact on the decision to patent in the academic world". Research Policy 36, 303–319.

Callon, M. (1994). "Is science a public good? Fifth Mullins lecture, Virginia Polytechnic Institute". Science, Technology and Human Values 19, 395–424 23 March, 1993.

Cañibano, C., Bozeman, B. (2009). "Curriculum vitae method in science policy and research evaluation: State-of-the-art". Research Evaluation 18 (2), 86–95.

Carayol, N. (2007). "Academic incentives, research organization and patenting at a large French university". Economics of Innovation and New Technology 16 (1&2), 71–99.

Chait, B.T. (2006). "Mass spectrometry: Bottom-up or top-down". Science 314, 65–66.

Coase, R. (1937). "The nature of the firm". Economica, New Series 4 (16), 386–405.

Cockburn, I., Henderson, R.M. (1998). "Absorptive capacity, coauthoring behavior, and the organization of research in drug discovery". Journal of Industrial Economics 46, 20 (2), 157–182.

Cohen, W., Levinthal, D. (1989). "Innovation and learning: The two faces of R&D". The Economic Journal 99, 569–596.

Cohen, W., Nelson, R.R., Walsh, J.P. (2002). "Links and impacts: The influence of public research on industrial R&D". Management Science 48 (1), 1–23.
Cole, J.R., Cole, S. (1973). "Social Stratification in Science". University of Chicago Press, Chicago.
Collins, F., Morgan, M., Patrinos, A. (2003). "The human genome project". Lessons from Large-Scale Biology, Science 300 (5617), 286–290.
Couzin, J. (2006). "Scientific misconduct: Truth and consequences". Science 313, 1222–1226.
Crespi, G., Geuna, A., Nomaler, O., Verspagen, B. (2010). "University IRPs and knowledge transfer: Is university ownership more efficient?" Economics of Innovation and New Technology Forthcoming.
Dasgupta, P. (1989). "The economics of parallel research". In: Hahn, F. (Ed.), The Economics of Missing Markets, Information, and Games. Clarendon Press, Oxford.
Dasgupta, P., David, P. (1987). "Information Disclosure and the Economics of Science and Technology". In: Feiwel, G.R. (Ed.), New York University Press, New York.
Dasgupta, P., David, P. (1994). "Toward a new economics of science". Research Policy 23, 487–521.
Dasgupta, P., Maskin, E. (1987). "The simple economics of research portfolios". Economic Journal 97, 581.
Dasgupta, P., Stiglitz, J.E. (1980). "Uncertainty, industrial structure, and the speed of R&D". Bell Journal of Economics 11, 1–28.
David, P. (1993). In: Knowledge, property and the system dynamics of technological change. Proceedings of the World Bank Annual Conference on Development Economics 1992. World Bank, Washington, DC.
David, P. (1994). "Positive feedbacks and research productivity in science: Reopening another black box". In: Granstrand, O. (Ed.), The Economics of Technology. Elsevier Science, Amsterdam.
Davis, G. (2005). "Doctors without orders". American Scientist 93 (3) Supplement.
De Figueiredo, J.M., Silverman, B. (2007). "How does the government (want to) fund science". In: Stephan, P., Ehrenberg, R. (Eds.), The Economics of Science. University of Wisconsin Press, Madison, WI.
Deng, Z., Lev, B., Narin, F. (1999). "Science and technology as predictors of stock performance". Financial Analysts Journal 55 (3), 20–32.
Diamond, A.M.Jr., (1984). "An economic model of the life-cycle research productivity of scientists". Scientometrics 6, 189–196.
Diamond, A.M.Jr., (1986a). "The life-cycle research productivity of mathematicians and scientists". Journal of Gerontology 41, 520–525.
Diamond, A.M.Jr., (1986b). "What is a citation worth?" Journal of Human Resources 21, 200–215.
Ding, W., Murray, F., Stuart, T. (2006a). "Gender differences in patenting in academic life science". Science 313, 661–665.
Ding, W., Murray, F., Stuart, T. (2006b). "Commercial science: A new arena for gender stratification in scientific careers?" Paper presented at the 2006 Annual Meeting of the American Sociological Association. Montreal, Quebec, Canada.
Dreifus, C. (2007). Watching a Cell Fold, and Witnessing Disease. International Herald Tribune, April 26.
Edwards, M., Murray, F., Robert, Y. (2006). "Gold in the ivory tower: Equity rewards of outlicensing". Nature Biotechnology 24 (5), 509–515.
Ehrenberg, R.G. (1991). "Academic labor supply". In: Clotfelter, C. (Ed.), Economic Challenges in Higher Education. Part II. University of Chicago Press, Chicago, IL.
Ehrenberg, R.G. (1992). "The flow of new doctorates". Journal of Economic Literature 30, 830–875.
Ehrenberg, R.G., Zhang, L. (2005). "The changing nature of faculty employment". In: Clark, R., Ma, J. (Eds.), Recruitment, Retention and Retirement in Higher Education: Building and Managing the Faculty of the Future. Edward Elgar, Northampton, MA.
Ehrenberg, R.G., Rizzo, M., Jakubson, G. (2003). Who Bears the Growing Cost of Science at Universities? NBER Working Paper 9627.
Eisenberg, R.S. (1987). "Proprietary rights and the norms of science in biotechnology research". Yale Law Journal 97, 177–231.
Ellison, G. (2002). "Evolving standards for academic publishing: A q-r theory". Journal of Political Economy 110, 994–1034.
Enserink, M. (2006). "Unprecedented budget increase draws faint praise". Science 314, 1523–1525.
European Commission. (2003). Third European Report on Science and Technology Indicators. Directorate-General for Research, Brussels.
European Commission. (2004). Europe Needs More Scientists. Directorate-General for Research, Brussels.
Feigenbaum, S., Levy, D.M. (1993). "The market for (Ir) reproducible econometrics". Social Epistemolagy 7, 215–232.
Feist, G. (2006). The Psychology of Science and the Origins of the Scientific Mind. Yale University Press, New Haven, CT.
Feynman, R.P. (1999). The Pleasure of Finding Things Out. Perseus, Cambridge, MA.

Foray, D. (2004). The Economics of Knowledge. The MIT Press, Cambridge, MA.
Fox, M.F. (1983). "Publication productivity among scientists: A critical review". Social Studies of Science 13, 285–305.
Fox, M.F. (1991). "Gender, environmental milieu, and productivity in science". In: Zuckerman, H., Cole, J.R., Bruer, J.T. (Eds.), The Outer Circle: Women in the Scientific Community. W.W. Norton, New York.
Fox, M.F. (2005). "Gender, family characteristics, and publication productivity among scientists". Social Studies of Science 35, 131–150.
Fox, M.F., Braxton, J.M. (1994). "Misconduct and social control in science: Issues, problems, solutions". Journal of Higher Education 65, 373–383.
Fox, M.F., Stephan, P. (2001). "Careers of young scientists: Preferences, prospects and realities by gender field". Social Studies of Science 31 (1), 109–122.
Frank, R., Cook, P. (1992). Winner-Take-All Markets. Cornell University.
Franzoni, C. (2009). "Do scientists get fundamental research ideas by solving practical problems?", Industrial and Corporate Change 18, 671–699.
Freeman, R.B. (1975). "Supply and salary adjustments to the changing science manpower market: Physics, 1948–1973". American Economic Review 65, 27–39.
Fuller, S. (1994). "Comments on 'The Economics of Science'". In: Arthur, J., Diamond, M. (Eds.), University of Nebraska at Omaha, Omaha, NB.
Gambardella, A. (1995). Science and Innovation: The U.S. Pharmaceutical Industry During the 1980s. Cambridge University Press, Cambridge.
Gibbons, M., Limoges, C., Nowotny, H., Schwartzman, S., Scott, P., Throw, M. (1994). The New Production of Knowledge: The Dynamics of Science and Research in Contemporary Societies. Sage Publications, London.
Gieryn, T.F., Hirsch, R.F. (1983). "Marginality and innovation in science". Social Studies of Science 13, 87–106.
Gonzalez-Brambilia, C., Veloso, F. (2007). The Determinants of Research Productivity: A Study of Mexican Researchers. Carnegie Mellon University.
Griliches, Z. (1979). "Issues in assessing the contribution of research and development to productivity growth". Bell Journal of Economics 10, 92–116.
Groen, J.A., Rizzo, M. (2007). "The changing composition of American-citizen PhDs". In: Stephan, P., Ehrenberg, R. (Eds.), Science and the University. University of Wisconsin Press, Madison, WI.
Hagstrom, W.O. (1965). The Scientific Community. Basic Books, New York.
Hall, B., Jaffe, A., Trajtenberg, M. (2001). The NBER Patent Citation Data File: Lessons, Insights and Methodological Tools. NBER Working Paper 8498.
Hall, B., Mairesse, J., Turner, L. (2007). "Identifying age, cohort and period effects in scientific research productivity: Discussion and illustration using simulated and actual data on French physicists". Economics of Innovation and New Technology 16 (1 & 2), 159–177.
Hargens, L.L. (1988). "Scholarly consensus and journal rejection rates". American Sociological Review 53, 139–151.
Harmon, L.R. (1961). "The high school backgrounds of science doctorates". Science 133, 679–688.
Heller, M., Eisenberg, R. (1998). "Can patents deter innovation? The anticommons in biomedical research". Science 280, 698–701.
Hermanowicz, J. (2006). "What does it take to be successful?" Science Technology and Human Values 31 (2), 135–152.
Hicks, D. (1995). "Published papers, tacit competencies and corporate management of the public/private character of knowledge". Industrial and Corporate Change 4, 401–424.
Hicks, D. (2007). "Research competition affects measured U.S. academic output". In: Stephan, P., Ehrenberg, R. (Eds.), Science and the University. University of Wisconsin Press, Madison, WI.
Hicks, D., Breitzman, T., Olivastro, D., Hamilton, K. (2001). "The changing composition of innovative activity in the U.S". Research Policy 30 (4), 681–703.
Hollingsworth, J.R. (2006). "The dynamics of American science: An institutional and organizational perspective on major discoveries". In: Beckert, J., Ebbinghaus, B., Hassel, A., Manow, P. (Eds.), Transformationen Des Kapitalismus: Festschrift Für Wolfgang Streeck Zum Sechzigsten Geburstag. Campus Verlag, Frankfurt and New York.
Hounshell, D.A., Smith, J.K.Jr. (1988). Science and Corporate Strategy: Du Pont R&D, 1902–1980. Cambridge University Press, Cambridge.
Hull, D.L. (1988). Science as a Process. University of Chicago Press, Chicago, IL.

Jaffe, A.B. (1989). "Real effects of academic research". The American Economic Review 79, 957–970.

Jaffe, A.B., Lerner, J. (2004). Innovation and Its Discontents: How Our Broken Patent System is Endangering Innovation and Progress, and What to Do About It. Princeton University Press, Princeton, NJ.

Jaffe, A.B., Trajtenberg, M., Henderson, R. (1993). "Geographic localization of knowledge spillovers as evidenced by patent citations". The Quarterly Journal of Economics 108 (3), 577–598.

Jefferson, T. (1967). Foley, In J.P. (Ed.), The Jefferson Cyclopedia. Russell and Russell, New York, vol. 1.

Jinyoung, K., Sangjoon, J.L., Marschke, G. (2005). The Influence of University Research on Industrial Innovation. NBER Working Paper No. 11447.

Johnson, H.G. (1972). "Some economic aspects of science". Minerva 10, 10–18.

Jones, B. (2005a). Age and Great Inventions. NBER Working Paper 11359.

Jones, B. (2005b). The Burden of Knowledge and the Death of the Renaissance Man: Is Innovation Getting Harder?. NBER Working Paper 11360.

Kean, S. (2006). "Scientists spend nearly half their time on administrative tasks, survey finds". Chronicle of Higher Education A23.

Kelchtermans, S., Veugelers, R. (2007). Top Research Productivity and Its Persistence. Katholieke Universiteit Leuven, Leuven.

Kohn, A. (1986). False Prophets. Basil Blackwell, Oxford.

Kolbert, E. (2007). "Crash course: The world's largest particle accelerator". The New Yorker 68–78 May 14.

Lach, S., Schankerman, M. (2008). "Incentives and invention in universities". Rand Journal of Economics 39 (2), 403–433.

LaTour, B. (1987). Science in Action: How to Follow Scientists and Engineers Through Society. Harvard University Press, Cambridge, MA.

Lawani, S.M. (1986). "Some bibliometric correlates of quality in scientific research". Scientometrics 9, 13–25.

Lazear, E.P., Rosen, S. (1981). "Rank-order tournaments as optimum labor contracts". Journal of Political Economy 89, 841–864.

Leslie, L.R., Oaxaca, R.L. (1993). "Scientist and engineer supply and demand". In: Smart, J.C. (Ed.), Higher Education: Handbook of Theory and Research. Agathon Press, New York.

Levin, S., Stephan, P. (1991). "Research productivity over the life cycle: Evidence for academic scientists". American Economic Review 81, 114–132.

Levin, S., Stephan, P. (1999). "Are the foreign born a source of strength for U.S. science?" Science 285 (5431), 1213–1214.

Levin, S., Stephan, P., Winkler, A. (2006). The Diffusion of Information Technology across Institutions of Higher Education: Effects on Productivity by Type of Institution and Gender. Proposal to the Andrew W. Mellon Foundation.

Lichtenberg, F. (1988). "The private R&D investment response to federal design and technical competitions". American Economic Review 78, 550–559.

Lotka, A.J. (1926). "The frequency distribution of scientific productivity". Journal of the Washington Academy of Sciences 16, 317–323.

Malakoff, D. (2000). "The rise of the mouse, biomedicine's model mammal". Science 288, 248–253.

Mansfield, E. (1991). "Academic research and industrial innovation". Research Policy 20, 1–12.

Mansfield, E. (1995). "Academic research underlying industrial innovations: Sources, characteristics, and financing". The Review of Economic and Statistics 77, 55–65.

Markiewicz, K., Di Minin, A. (2004). Commercializing the Laboratory: The Relationship between Faculty Patenting and Publishing. University of California, Berkeley, CA.

Merton, R.K. (1957). "Priorities in scientific discoveries: A chapter in the sociology of science". American Sociological Review 22, 635–659.

Merton, R.K. (1961). "Singletons and multiples in scientific discovery: A chapter in the sociology of science". Paper read at Conference Commemorating the 400th Anniversary of the Birth of Francis Bacon. October 13, 1969.

Merton, R.K. (1968). "The Matthew effect in science". Science 159, 56–63.

Merton, R.K. (1969). "Behavior patterns of scientists". American Scientist 57, 1–23.

Merton, R.K. (1988). "The Matthew effect in science, II: Cumulative advantage and the symbolism of intellectual property". Isis 79, 606–623.

Mincer, J. (1974). Schooling, Experience, and Earnings. NBER, Columbia University Press, New York.

Mowery, D.C., Rosenberg, N. (1998). Paths of Innovation: Technological Change in 20th Century America. Cambridge University Press, Cambridge, MA.

Mueller, W.F. (1962). The origins of the basic inventions underlying DuPont's major product and process innovations, 1920 to 1950. Economic and Social Factors, NBER, Princeton University Press, Princeton, NJ, The Rate and Direction of Innovative Activity.
Murray, F. (2006). The Ocomouse that Roared: Resistance and Accommodation to Patenting in Academic Science. MIT Press, Cambridge, MA.
Murray, F. (2007). "The stem-cell market—Patents and the pursuit of scientific progress". The New England Journal of Medicine 356 (23), 2341–2343.
National Academies. (2005). Policy Implications of International Graduate Students and Postdoctoral Scholars in the United States. National Academies Press, Washington, DC.
National Research Council. (2000). Forecasting Demand and Supply of Doctoral Scientists and Engineers: Report of a Workshop on Methodology. National Academies Press, Washington, DC.
National Science Board. (2003). The Science and Engineering Workforce: Realizing America's Potential. National Science Foundation, Arlington, VA.
National Science Board. (2004). Science and Engineering Indicators. National Science Foundation, Arlington, VA.
National Science Board. (2006). Science and Engineering Indicators. National Science Foundation, Arlington, VA.
National Science Foundation. (1989). Future Scarcities of Scientists and Engineers: Problems and Solutions: Division of Policy Research and Analysis. National Science Foundation, Arlington, VA.
National Science Foundation. (1996). Characteristics of Doctoral Scientists and Engineers in the United States 1993. National Science Foundation, Arlington, VA.
National Science Foundation. (1999). Characteristics of Doctoral Scientists and Engineers in the United States 1997. National Science Foundation, Arlington, VA.
National Science Foundation. (2006). Characteristics of Doctoral Scientists and Engineers in the United States 2003. National Science Foundation, Arlington, VA.
National Science Foundation. (2007). Changing U.S. Output of Scientific Articles: 1988–2003. National Science Foundation, Arlington, VA.
Nelson, R.R. (1959). "The simple economics of basic scientific research". The Journal of Political Economy 67, 297–306.
Nelson, R.R. (1962). "The link between science and invention: The case of the transistor". In: The Rate and Direction of Inventive Activity: Economic and Social Factors. Princeton University Press, Princeton, NJ.
Nelson, R.R., Winter, S. (1982). An Evolutionary Theory of Economic Change. Belknap Press of Harvard University Press, Cambridge, MA.
Overbye, D. (2007). "A Giant Takes on Physics' Biggest Questions". New York Times D1, D4–D5 May 15, CLVI.
Oyer, P. (2006). "Initial labor market conditions and long-term outcomes for economists". Journal of Economic Perspectives 20 (3), 143–160.
Pavitt, K. (2000). "Why European Union funding of academic research should be increased: A radical proposal". Science and Public Policy 27 (6), 455–460.
Philippe, A., Dewatripont, M., Stein, J.C. (2008). "Academic freedom, private-sector focus, and the process of innovation". Rand Journal of Economics 39 (3), 617–635.
Polanyi, M. (1962). "The republic of science". Minerva 1, 54–74.
Pollak, R., Wachter, M. (1975). "The relevance of the household production function and its implications for the allocation of time". Journal of Political Economy 83, 255–277.
Price, D.J.D., Solla, D.E. (1986). Little Science, Big Science... and Beyond. Columbia University Press, New York.
Roe, A. (1953). The Making of a Scientist. Dodd, Mead, New York.
Romer, P. (1994). "The origins of endogenous growth". Journal of Economic Perspectives 8, 3–22.
Rosenberg, N. (1974). "Science, invention and economic growth". Economic Journal 84, 90–108.
Rosenberg, N. (1982). Inside the Black Box: Technology and Economics. Cambridge University Press, Cambridge.
Rosenberg, N. (1990). "Why do firms do basic research (with their own money)?" Research Policy 19, 165–174.
Rosovsky, H. (1990). The University: An Owner's Manual. Norton, New York.
Sauermann, H., Cohen, W. (2007). What Makes Them Tick? Employee Motives and Industrial Innovation. Fuqua School of Business.
Scherer, F.M. (1966). "Time-cost tradeoffs in uncertain empirical research projects". Naval Research Logistics Quarterly 13, 71–82.

Scherer, F.M. (1967). "Review of technology, economic growth and public policy, by Richard R. Nelson, M.J. Peck, and E.D. Kalachek". Journal Finance 22, 703–704.
Scherer, F.M. (1982). "Demand-pull and technological innovation: Schmookler revisited". Journal of Industrial Economics 30, 225–238.
Schmookler, J. (1966). Invention and Economic Growth. Harvard University Press, Cambridge, MA.
Schulze, G.G. (2008). "Tertiary Education in a Federal System—the Case of Germany". In: Albert, M., Schmidtchen, D., Voigt, S. (Eds.), Scientific Competition. Mohr Siebeck, Tübingen.
Smith, A. (1982). The Wealth of Nations. Penguin, Harmondsworth, UK.
Sobel, R. (1986). RCA. Stein and Day, New York.
Stephan, P. (1996). "The economics of science". Journal of Economic Literature 34, 1199–1235.
Stephan, P. (2002). "Using human resource data to illuminate innovation and research utilization". In: Merrill, S., McGeary, M. (Eds.), Using Human Resource Data to Track Innovation. National Academy Press, Washington, DC.
Stephan, P. (2004). "Robert K. Merton's perspective on priority and the provision of the public good knowledge". Scientometrics 60, 81–87.
Stephan, P. (2007). "Wrapping it up in a person: The location decision of new phds going to industry". In: Jaffe, A., Lerner, J., Stern, S. (Eds.), Innovation Policy and the Economy. MIT Press, Cambridge, MA.
Stephan, P. (2008). "Job market effects on scientific productivity". In: Albert, M., Schmidtchen, D., Voigt, S. (Eds.), Scientific Competition. Mohr Siebeck, Tübingen.
Stephan, P., El-Ganainy, A. (2007). "The entrepreneurial puzzle: Explaining the gender gap". Journal of Technology Transfer 32, 475–487.
Stephan, P., Levin, S. (1992). Striking the Mother Lode in Science: The Importance of Age, Place, and Time. Oxford University Press, New York.
Stephan, P., Levin, S. (1993). "Age and the Nobel prize revisited". Scientometrics 28 (3), 387–399.
Stephan, P.E., Gurmu, S., Sumell, A., Black, G. (2007a). "Who's patenting in the university? Evidence from the survey of doctorate recipients". Economics of Innovation and New Technology 16 (2), 71–99.
Stephan, P.E., Black, G., Chaing, T. (2007b). "The small size of the small-scale market: The early-stage labor market for highly skilled nanotechnology workers". Research Policy 36 (6), 887–892.
Stern, S. (2004). "Do scientists pay to be scientists?" Management Science 50 (6), 835–853.
Stigler, G.J. (1980). Merton on multiples, denied and affirmed. In: Science and Social Structure: A Festschrift for Robert K. Merton. Transactions of the New York Academy of Science.
Stigler, G.J. (1983). "Nobel lecture: The process and progress of economics". Journal of Political Economy 91, 529–545.
Stokes, D. (1997). Pasteur's Quadrant. Brookings Institution Press, Washington, DC.
Sunwoong, K. (2007). Brain Drain and/or Brain Gain: Education and International Migration of Highly Educated Koreans. University of Wisconsin, Milwaukee, WI.
Terviö, M. (2006). Network Analysis of Three Academic Labor Markets. Hass School of Business.
Thursby, J., Thursby, M. (2007). "Patterns of research and licensing activity of science and engineering faculty". In: Stephan, P., Ehrenberg, R. (Eds.), Science and the University. University of Wisconsin Press, Madison, WI.
Thursby, J., Thursby, M., Mukherjee, S. (2005). Are There Real Effects of Licensing on Academic Research: A Life-Cycle View. Georgia Institute of Technology, Atlanta, GA.
Thursby, J., Thursby, M., Fuller, A. (2009). "U.S. patenting activity: Inside and outside the university". Research Policy 38, 14–25.
Tuckman, H., Leahey, J. (1975). "What is an article worth?" Journal of Political Economy 83, 951–968.
Turner, S. (1994). "Comments on 'The Economics of Science'". In: Arthur, J., Diamond, M. (Eds.), University of Nebraska at Omaha, Omaha, NE.
Turner, L., Mairesse, J. (2003). "Individual productivity differences in scientific research: An econometric study of publishing of French physicists". Paper presented at the Zvi Griliches Memorial Conference Paris, France.
Vogel, G. (2000). "The mouse house as a recruiting tool". Science 288, 2254–2255.
Vogel, G. (2006). "Basic science agency gets a tag-team leadership". Science 313, 1371.
Von Hippel, E. (1994). "Sticky information and the locus of problem solving: Implications for innovation". Management Science 40 (4), 429–439.
Walsh, J.P., Cho, C., Cohen, W. (2005). "View from the bench: Patents and material transfers". Science 309, 2002–2003.

Walsh, J., Cho, C., Cohen, W. (2007). "Where excludability matters: Material vs. intellectual property in academic biomedical research". Research Policy 36, 1184–1203.
Weiss, Y., Lillard, L. (1982). "Output variability, academic labor contracts, and waiting times for promotion". In: Ehrenberg, R.G. (Ed.), Research in Labor Economics, vol. 5, pp. 157–188.
Wilson, R. (2000). "They may not wear armani to class, but some professors are filthy rich". Chronicle of Higher Education 46 (26), A16–A18 March 3.
Wolpert, L., Richards, A. (1988). A Passion for Science. Oxford University Press, New York.
Wuchty, S., Jones, B., Uzzi, B. (2007). "The increasing dominance of teams in the production of knowledge". Science 316, 1030–1036.
Xie, Y., Shauman, K. (1998). "Sex differences in research productivity: New evidence about an old puzzle". American Sociological Review 63, 847–870.
Xie, Y., Shauman, K. (2003). Women in Science: Career Processes and Outcomes. Harvard University Press, Cambridge, MA.
Xin, H. (2006). "Scandals shake Chinese science". Science 312, 1464–1466.
Ziman, J. (1968). Public Knowledge. Cambridge Press, Cambridge, UK.
Zucker, L., Darby, M. (2007). "Star scientists, innovation and regional and national immigration". Paper read at Kauffman-Max Planck Research Conference on Entrepreneurship July 19–21, Dana Point.
Zucker, L., Darby, M., Brewer, M. (1998a). "Intellectual human capital and the birth of U.S. biotechnology enterprise". The American Economic Review 88 (1), 290–306.
Zucker, L.G., Darby, M., Armstrong, J. (1998b). "Geographically localized knowledge: Spillovers or markets". Economic Inquiry 36 (1), 65–86.
Zuckerman, H.A. (1992). "The proliferation of prizes: Nobel complements and Nobel surrogates in the reward system of science". Theoretical Medicine 13, 217–231.

Chapter 6

UNIVERSITY RESEARCH AND PUBLIC–PRIVATE INTERACTION

DOMINIQUE FORAY* *AND* FRANCESCO LISSONI[†,‡]

**Chaire en Economie et Management de l'Innovation*

Ecole Polytechnique Fédérale de Lausanne

Lausanne, Switzerland

[†]DIMI, Università degli Studi di Brescia

Brescia, Italy

[‡]KITeS, Università "L. Bocconi"

Milano, Italy

Contents

Handbooks in Economics, Volume 01

DOI: 10.1016/S0169-7218(10)01006-3

Abstract

Universities' centrality within the public research systems has been increasing over time, as it has their interactions with industry. Such interaction poses two dilemmas. One concerns individual scientists and the potential trade-off between basic research activities and those activities required to successfully develop and commercialize academic inventions. The second dilemma occurs at the system level, and it has to do with the tension between the industry's need to rely upon clear and solid intellectual property rights (IPRs), and the cumulativeness of the scientific enterprise, which requires the results of academic research to be freely accessible. The empirical literature suggests that the first dilemma may not be as dramatic as expected by many. On the contrary, some evidence exists on the relevance of the second dilemma: commercial interests may exacerbate common threats to the commonality of research efforts; and the existence of IPRs over academic research results may discourage some scientists to build upon those results in order to advance knowledge. Existing bridging institutions, both internal and external to universities, seem to give only marginal contributions to the solution of both dilemmas.

Keywords

academic entrepreneurship, intellectual property, research, technology transfer, university

JEL classification: I23, O31, O34

1. Introduction

The university is among the oldest institutions active today in all the developed countries. Over the course of its long history it has managed not only to adapt to many external shocks, but also to expand considerably, both in size and diversity of activities (Ben-David, 1977). In present times, its role is to couple basic research and teaching—two activities of wide relevance in the economy to the extent that they provide for the generation of externalities in the form of human capital and basic knowledge, both of which have the characteristics of quasi-public goods (Clark, 1993).[1] As countries progressively shift towards knowledge-based economies, there is a positive supply response on the part of universities to the increasing demand for basic knowledge and highly skilled people. In this respect, universities play a critical, but indirect role in the productivity growth and expansion of industry and services.

Universities also contribute directly to innovation, by providing industry and services with technical solutions or devices, or by getting involved in applied research activities. Such a role is in accordance with a view of the university as a "permeable institution" (Lécuyer, 1998), which allocates efforts and attention to problem-solving activities that have immediate relevance for business firms (most often the national or local ones). Such a view is not at all new, as it dates back at least to the nineteenth century, sometimes in coexistence, sometimes in competition, with the emphasis on basic research and teaching (Rothblatt and Wittrock, 1993). More recently, however, governments and large sections of the public opinion have placed more emphasis on demands that universities fulfil this type of task by commercializing their own academic inventions. This requires them to get involved into the creation and management of intellectual property rights (IPRs), and even into entrepreneurial activities such as the foundation of new firms (Martin, 2003; Slaughter and Leslie, 1997; Yusuf and Nabeshima, 2007). A major witness of this change is the wave of legislation aimed at encouraging universities to take patents and license them under profitable conditions, started in the United States with the Bayh-Dole Act of 1980 and continued elsewhere with many imitations of this Act and, in several European countries, with the abolition of the "professor's privilege" typical of the German academic model.[2] The increase of direct government funding of research projects (as opposed to general university funds or "block grants"), many of which are explicitly targeted at technology areas, can also be interpreted as the result of this new attitude.[3]

[1] The human capital embodied in graduates is not highly specific, and is in fact general enough that it constitutes a public good, or at least without labor contracts that are tantamount to indentured service, a nonappropriable good.

[2] The Patent and Trademark Laws Amendment, better known as Bayh-Dole Act, was issued in 1980, following a decade-long debate on the US research system's apparent failure to turn scientific achievements into innovations. It entitled universities and other not-for-profit research organizations with the intellectual property rights over the results of research funded by the federal government and with the possibility to issue exclusive licenses. It followed the Stevenson-Wydler Technology Innovation Act, which issued similar provisions for federal laboratories (Jaffe, 2000). On the wave of Bayh-Dole-like European legislation in the 1990s, see OECD (2003) and Mowery and Sampat (2005). On the "professor's privilege," see Section 4.2.1, and references therein.

[3] Vincent-Lancrin (2006) shows that the average OECD share of direct funding over total government funding of academic research has grown from 27% in 1981 to 39% in 2003, while the share of general funding has declined from 78% to 65%. A related change in funding policies is the diffusion of performance-based funding, with research performance measured mainly in terms of publications' quantity and quality, but also in terms of patents and technology transfer activities (Geuna and Martin, 2003).

This change of perspective has gone hand in hand with the increasing attention paid by industry to universities' research, as part of a general strategy to move away from a "vertical" model of R&D to a "network strategy" of innovation, based upon the exploitation of external knowledge resources.[4] Since the 1980s, industrial funding of academic science in OECD countries has grown considerably both in real terms and as a percentage of GDP. Public funding has also grown in real terms, but it has not kept up with the growth both of GDP and of industrial funding, so that in 2003 the share of government-funded academic research was down to 72%, from over 80% in 1981. In the meantime, the share of industrial funding had doubled, from 3% to 6%; and universities' self-financing share has gone up from 13% to 16%, thanks largely to the expansion of new entrepreneurial activities both in the field of education and in technology commercialization (Vincent-Lancrin, 2006). Although governments are still eager to pay most of the bill for academic science, these are further signals that those same governments increasingly expect universities to look elsewhere for resources, and in particular to research partnerships with industry and to markets for technologies.

At the present time, the most research-oriented of modern universities look quite like the "multiversity" envisaged by Clark Kerr, the prescient president of the University of California of the 1960s: a "knowledge factory ... to which policy wonks turn for expertise, industrialists turn for research, government agencies turn for funding proposals, and donors turn for leveraging their philanthropy into the greatest impact" (Wagner, 2007); and, one may wish to add, university administrators turn for self-financing.[5]

All of these stakeholders combine to mould the fundamental incentive structures of academic scientists, setting the balance between the marginal returns, respectively associated to basic research, education, and involvement in commercialization. This evolution both generates opportunities and entails the risk to damage to the overall universities' contribution to scientific advancement and human well being.

In particular, fears have been expressed that universities will be forced to limit their production of basic research and teaching, the quasi-public goods that market-oriented organizations often fail to provide. Such a risk appears paradoxical, at a time when the provision of such public goods is of strategic importance as countries progressively shift towards knowledge-based economies.

In short, two types of interaction between universities and industry seem to coexist, both of which aim at realizing effectively the potential for complementarities between the two in the domain of innovation. Interactions can be of the traditional type, covering networks of people, collaborative funding of research programs, and informal contacts. The recruitment of graduates in the business sector is part of this concept and is often the strongest channel of interaction between the two worlds. The other type of interaction is that from universities better exploiting their inventions—through professional management of intellectual property, opening technology licensing offices, and launching their own spin-offs and start ups.

It is clearly difficult to know whether this second, emerging model of university–industry interaction will contribute to scientific advancement and long-term economic growth more or less than those that preceded it. It is also hard to tell how generalized and effective has been the transition to the new model

[4] On this change of strategy see Powell, this volume. On open innovation, see also von Hippel, this volume.

[5] The term "multiversity" was coined by Clark Kerr in his Godkin Lecture at Harvard University in 1963, now republished along with many related essays in Kerr (2001).

in countries other than the United States, which have been the most important institutional laboratory for academic life since World War II, and where the new model of the university has made the most inroads. However, both economic theory and applied studies have already produced enough material for a first assessment, as well as significant guidelines for future research directions.

In what follows, we place the role of universities in context, and show that their centrality within the public research systems has been increasing over time, even in countries which traditionally entrusted public research to different institutions (Section 2). We then develop a general formulation of the opportunities and problems generated by the interaction between university and industry (Section 3). In Section 4 we examine the main issues explored by the growing empirical literature on the economics of university–industry technology transfer. Finally, in Section 5 we discuss policy implications and directions for future research.

2. Government laboratories and research universities: Two different public research organizations

Knowledge—defined as a quasi-public good—requires special socioeconomic institutions upon which society can rely to produce and allocate it in an efficient manner. Private markets (involving IPRs as well as other mechanisms to help private agents to capture economic rents) and the public sector form the two main institutions which we need to study in order to design an empirically and analytically informed knowledge policy. This chapter focuses on their interaction, but first we discuss the public sector institutions in a bit more detail.

In the public sector, there are clearly (at least at the conceptual level) two different types of institutions (Dasgupta, 1988): the first consists in the government engaging itself directly in the production of knowledge; the second consists in private agents undertaking the research, who in turn are subsidized for their effort by the public purse. While the first arrangement characterizes the so-called government research laboratories (GRLs), the second one characterizes research universities (RUs).[6] The RU solution is a decentralized mechanism, in which production decisions are independently taken by members of a self-regulating profession (scientists), and whose work is subsidized by the government, while the GRL arrangement is closer to a kind of "command mode of planning," such that the decision of what to produce and how much to produce is made by the government. GRLs comprise both the large institutes dedicated to fundamental research activities (such as Max Planck in Germany or CNRS in France) and a number of mission-oriented organizations dedicated to the advancement of specific scientific fields and technologies, often under direct ministerial supervision (such as national space agencies, institutes of health, or atomic energy organizations). Networks of laboratories for applied research and development, most often in support of small and medium enterprises (SMEs)

[6] The term "research universities" is mainly used in the United States to distinguish doctoral-granting higher education institutions from master's colleges and universities, as well as from other colleges with no research activity (doctoral activity being a proxy for research orientation). It was systematized and diffused by the first report of the Carnegie Commission on Higher Education, published in 1967, whose updates have contributed to refine it (Carnegie, 2009). The latest Carnegie report identifies around 200 RUs, both private and public. Academic jargon in Europe and Asia also refer increasingly to RUs in order to identify those institutions whose international standing in the research arena is comparable to their US counterparts (at least in their administrators' intentions).

can also be regarded as GRLs, a classical example being the Fraunhofer Gesellschaft in Germany (Beise and Stahl, 1999). While several GRLs host laboratories that often operate according to a logic and under provisions which are closer to that of RUs (so that their scientists regard themselves as part of the academic community), most of them pursue more strictly defined objectives, even when they rely on academic scientists' services (such as contract research or consultancy).[7]

GRLs and RUs form what is commonly known as public sector research, and are related by exchanges of knowledge, personnel, and finances (large GRLs are often in charge of administering public funds directed also at universities, and recruit scientists in the same labor market of RUs). Yet it is important to maintain the distinction between the two forms of public research because the economic incentives and resources allocation mechanisms are fundamentally different. In the RU system, individuals are free to pursue research targets of their own choice (although the system of grants often selects a few main research areas). In return for financing, individuals and institutions must provide educational services, such as teaching and supervision of qualification into professional associations (such as those of medical doctors, lawyers, and engineers). Modern scientists receive a fixed salary for their teaching and examination tasks, in addition to other rewards (e.g., promotions and increased reputation) for successful research.[8] By contrast, in the GRL system research is organized by the state in relation to targeted objectives. Individuals are not as "free" as in RUs, due to commitments to follow certain research directions. It follows that they do not have to provide as many other services, such as lecturing, in order to create a fair balance of advantages and constraints.

Both GRLs and RUs have significant shortcomings as methods of resource allocation. In the RU system, mechanisms for the allocation of research grants to individuals and teams exhibit hysteresis effects (reputation increases the probability of receiving a new grant which, in turn, has the effect of increasing reputation even more). This may weaken the system's capacity to identify and maintain the "best" researchers. RU systems face tremendous difficulties in generating (in a decentralized way) new disciplines or research activities at the interstices of existing fields. In the GRL system, problems of asymmetric information make it difficult for research administrators to manage the scientists' activity. Government failures (instead of market failures) may occur. In addition, large basic-science- and mission-oriented GRLs projects are high risk ventures, with a few large bets are placed on a small number of races. These ventures may also create distortions in competition, to the extent that they favor selected industries and the "national champions" therein.

These two arrangements have specific functionalities and are therefore complementary. These differences are reflected in the way knowledge flows to industry and society are managed in the two systems. While maximizing knowledge externalities is the *raison d'être* of the RU system, this is not the case in the GRL system. Spillovers from the latter can be either massive or very weak, depending on the administrators' intentions; in any case, they cannot be considered the key rationale for the public funding of GRLs.

Historically, most countries that are now at the technological frontier have experienced a slow shift from a system involving government laboratories and *teaching* universities as the main "knowledge

[7] For a taxonomy of GRLs, see Nelson (1993; especially Chapters 1, 2, 4, and 6) and Ergas (1987). Good examples of SME-oriented GRLs are provided by Semlinger (1993), Kelley and Arora (1996), Feller (1997), and Beise and Stahl (1999).

[8] See also Paula Stephan's chapter in this book for an in depth discussion of incentives and organizational structures in scientific and research activities.

institutions" to a system characterized by the research centrality of RUs. There are of course variations across countries (e.g., in France the GRL role as an R&D performer has been maintained at high level), but the direction of the trend is clear across most OECD countries (Figure 1).[9]

Heavy reliance on GRLs can be seen as a legacy of the past: it was appropriate at a certain stage of economic development, when the main challenge for Western countries was to build a science and technology infrastructure, and the fastest way to do so was to create these "mission-oriented" institutions. However, as those countries approach the technological frontier (i.e., are no longer catching up and imitative but rather are leading the international innovation process[10]), the need for more resources in RUs is obvious. RUs can generate externalities in the form of both human capital and basic research that have the status of "joint products" (giving rise therefore to economies of scope and internal

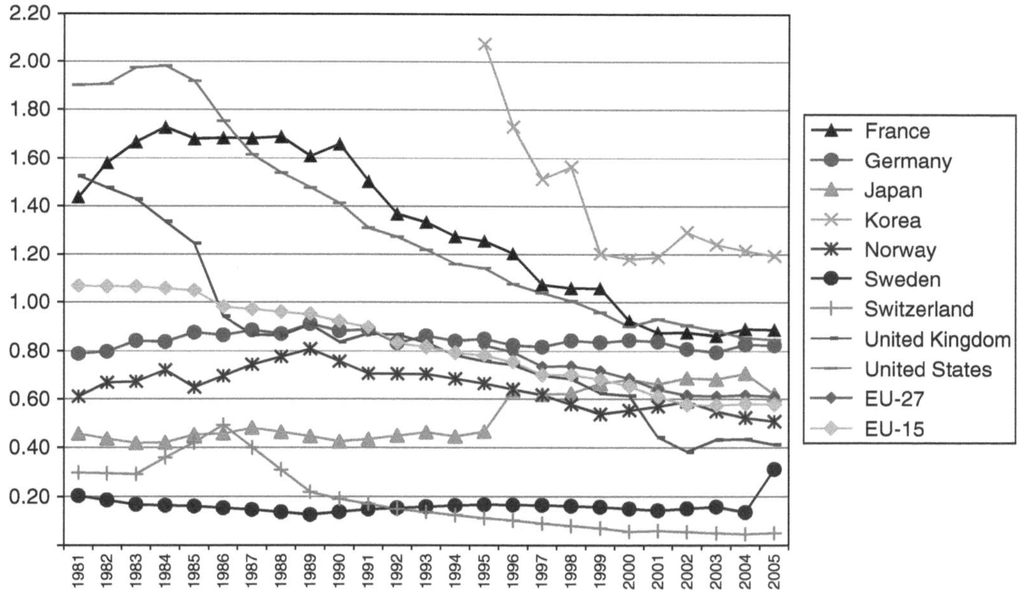

Figure 1. Relative weight of R&D performed by governments (GL) versus performed by universities (RU). Source: OECD (2007), calculations made by S. Lhuillery.

[9] Many reasons are behind such a shift and they differ across countries. Among the main factors, the reduction of spending for noncivil R&D and nuclear energy (both being performed mainly outside the RU system) as well as the privatization of GRLs in some countries (UK) should be highlighted. But whatever reasons, the consequence is the increasing dominance of research universities as R&D performers across OECD countries.

[10] A formal definition of "proximity to the technological frontier" for an economy at a given time is the ratio of total factor productivity in that economy at time t and the highest TFP at time t among all countries. Proximity varies from 0 (for very inefficient economies) to 1 (for the most efficient).

spillovers) while GRLs break the intimate relations between research and high education and only provide a small fraction of the total amount of positive externalities that RUs are able to provide. As explained by Zucker and Darby (1998, p. 62):

> *"the idea of research institutes sounds very attractive, particularly in a small country that sees them as a vehicle to achieve a critical mass by concentrating the nation's best scientists in one place. In fact, we ourselves would like to have our research well funded until retirement and the opportunity to build a more permanent research group without the need to educate and train successive generations of graduate students and post doctoral fellows. Despite the personal attractions, we can also see how that situation might cool the entrepreneurial spirit as well as our impact on the most important objective of any knowledge institution: the generation of high quality human capital."*

The focus of the rest of this chapter is on RUs, for two reasons. First and most importantly, RUs have become more central to the knowledge economy and innovation systems than government laboratories. Second, the literature on the organization and impact of government laboratories is more limited and sparse than that on RUs.[11]

3. A conceptual approach to the problem of managing complementarities between universities' and industry's research

A report written by David and Metcalfe (2008) for the expert group "Knowledge for Growth" of the European Commission makes a strong argument that there is much more to the process of innovation than R&D. Achievement of innovation requires accessing and combining many more types of knowledge and capabilities than is summed up by the phrase "science and technology," such as knowledge of markets and organizations, as well as of the availability and quality of inputs. Production of these knowledge assets is a key aspect of the innovation process, but it does not take place in universities or other public research organizations. Universities are not organized and governed to be producers of innovations in their own right—they are first and foremost designed to achieve a new understanding of natural phenomena and technologies: in this task they are naturally inventive. Conversely, in modern free market economies, it is firms that have the incentives and governance structures to make innovation their central goal, and are expected to be the almost exclusive sources of innovation. In the realm of innovation, a public research organization will never be more than a second rank institution.

So it seems wise to acknowledge the virtues of the division of labor between universities and business firms regarding the knowledge production function and to allocate the innovation function to the business sector. However, as with any division of labor, the increased efficiency of the various tasks (invention on one side and innovation on the other side) comes at the price of introducing problems of connection between the two worlds: boundary issues may impede interactions between the various organizations.

[11] Noticeable exceptions are Jaffe and Lerner (2001) and Jaffe et al. (1998).

3.1. Economic opportunities

A large number of economic opportunities exist for exploiting potential transfers from academic research to industry. When the two systems are institutionalized in specialized, dedicated organizations that permit their respective advantages to be exploited most fully, their interactions are complementary and, historically, have proved to be highly conducive to sustaining long-term economic growth and improvements in human welfare and well-being. Three sources of interactions are typically identified (David, 1993). One source is found at the macroeconomic level and consists in the "externalities" that advancements in fundamental scientific knowledge provide to applied researchers (see David et al., 1992).

A second and no less important economic opportunity lies in the connection between the effective training of researchers and research managers and the profitability of corporate R&D programs. The coupling of open science research activities with graduate training of scientists and engineers has turned out to be particularly effective not only for the quality of human capital created, but also in providing industrial employers with an efficient and very inexpensive process for screening talent.

A third opportunity channel is the open access provided by universities to new information about research methods and findings, which greatly facilitates the ability of research intensive enterprises to monitor scientific advances that are likely to transform technologies and markets.

All of these three channels represent complementarities that university research generates in favor of industrial R&D. There are also complementarities that run in the opposite direction such as industrial research playing a role in "equipping" university scientists with new and powerful tools and instruments (Rosenberg, 2004). Industrial research also provides *antidotes* for academic researchers' conservatism, such as challenging questions, experimental evidence, and support for the expansion of new disciplines. In other words it contributes to set the research agenda of universities, without necessarily constraining it towards immediate or menial objectives.

The main effects of these complementarities are to raise the expected rates of return, and to reduce the risk of investing in applied R&D. A central policy concern, therefore, is to ensure that these complementarities are properly managed to achieve those purposes, but also that concern with the immediate exploitation of those spillovers to the private sector from the activities of public research organizations will not be detrimental to the long-term vitality of the latter, and hence to the regeneration of opportunities for profitable R&D investments.

3.2. Institutional obstacles

Direct transfers of knowledge between academic science communities and the proprietary R&D organizations of the private business sector are especially problematic to institutionalize, since the coexistence of two reward systems typical of each system makes the participants' behavior difficult to anticipate, and tends to undermine the establishment of coherent cultural norms for the promotion of cooperation among team members (David et al., 1999). Clearly the difficulties of technology transfer are not caused in the first instance by inappropriate or ill-adapted institutional frameworks, legal systems or cultural norms. Rather the difficulties are inherently associated with the process itself, and all countries are facing the same problem, which consists in managing a trade-off between two good things: getting more academic knowledge used by the economy versus maintaining the fundamental missions (long-term research and education) of universities.

One feature of knowledge creation and transfer which makes the problem greater is the importance of the postinvention process, which starts with the invention in universities and finishes with its commercial exploitation. Cases of university inventions that with slight modification can be commercialized or incorporated into a process by a private firm are relatively rare. Most university inventions require much more substantial modification and additional development for commercial introduction (see Section 4.2.1).

We cannot discuss in details all the issues that are involved in the process of transferability and operation of new knowledge, as produced in academic institutions. In what follows, we will restrict our discussion to a few points that are relevant for further study by economists.

3.2.1. On rewards, spillovers, and the distribution of IP

University and industry follow very different economic logic with respect to the relative importance of "appropriability" versus the benefits of full and costless knowledge diffusion.[12] The private industry model of innovation assumes economic returns resulting from private goods and the ability to control the exclusive use of the new knowledge. In this model any freely revealed or uncompensated dissemination of proprietary technologies will reduce the innovator's profits from his investments. Within the academic research community, on the contrary, a different reward system has evolved over time, which is based on the individual scientist's rapid publication and dissemination in order to achieve a prior claim as author, either of a discovery or an invention.[13] Such a system is so different from the common practices according to which most industrial firms operate, that it is not surprising to observe tensions arising in settings where the conventions of one world come up against the conventions of another (Hall, 2004).

The fundamental tension related to the ownership and control of technologies arises from the problem of making compatible the granting of exclusive rights to one sphere (private industry) and the granting of freedom to operate and publish to the other (public R&D and university science).

This tension is exacerbated when the academic invention requires heavy postinvention investments to go to the market. In such a case, there is a need to provide a secure economic environment for the investment that converts ideas into reality: firms would be unwilling to support these costs without some assurance of protection from competition. This was one of the rationales behind the Bayh-Dole Act of 1980 in the United States, and of similar provisions both in Europe and Japan throughout the 1990s, which have led to a higher level of direct involvement by universities in the management of patenting and licensing activities during the recent period (Mowery and Sampat, 2005).[14] Such involvement creates a potential risk of distortion of the whole incentive structure which traditionally underlies the activity of knowledge transfer: by focusing on exclusive licensing, these laws are based on a narrow view of the channels through which public research interacts with industry. In reality these channels are multiple and all contribute to the transfer of knowledge, while the incentives created by such laws promote only one channel (patenting and licenses) with the risk of blocking the others (Mowery et al., 2004). This also increases the likelihood of an institutional clash between the objective of maintaining a space for

[12] See also Katherine Rockett, this volume.
[13] See also Paula Stephan, this volume.
[14] See the "Introduction" and footnote 1 therein.

social sharing and distribution of scientific knowledge, tools and information and the objective of securing the private investments made to develop that knowledge (Cockburn and Henderson, 1998).

The institutional clash may be particularly acute when it comes to patent protection and exclusive licensing of research tools, such as scientific instruments, data, and (above all) genetic material for biological research. Exclusive licenses on research tools, or too high prices for universal licenses, may contribute to create barriers to entry in the field of scientific research, with perverse effects such as narrowing the base of researchers and thus limiting the advancement of science. Heller and Eisenberg (1998) suggest that from the end of the 1980s patenting practices have made inroads into the early stages of biomedical research, so that the research tools needed by individual researchers risk being both proprietary and sold for a profit, instead of being accessible at no or low cost to all researchers. To make things worse, the IPRs covering all tools necessary for research may be fragmented into numerous patents controlled by several different owners.[15] Following Heller and Eisenberg, this phenomenon is now referred to as "tragedy of the anticommons," because it requires the researcher to ask too many licenses to access the common pool of scientific discoveries and continue in her research. The negative effects of the phenomenon is exacerbated by the possibility of so-called reach-through license agreements, for which the owner of a patented invention derived from basic research preserves the rights to downstream discoveries.[16]

3.2.2. On cognitive focus, mental mobilization, and time horizon

Another example of institutional difficulties involves the divergence of opinion concerning what may be identified as "the optimal quality of invention." On both the academic and industry sides, "optimal quality" is sought. However, the optima are not the same. From the point of view of academic research, optimal quality will entail the novelty gap or inventive step, elegance of the solution, or importance and generality of the new knowledge (able to generate cumulative effects across different fields). From the industry point of view, optimal quality entails cost effectiveness, reliability of the new system, time to market, and economic availability of the various inputs of the new production function. This is a major tension: academic researchers are looking for hyperinnovative solutions which can fuel interesting and challenging discussions among colleagues while industry engineers are focusing on reliability and

[15] In the United States, extension of patents to genetically modified organisms (and more generally to the biotech products) followed and built upon an important a Supreme Court decision of 1980 (Diamond vs. Chakrabarty), which established that a "living, man-made microorganism is patentable subject matter as a "manufacture" or "composition of matter." European and Asian countries have followed the US example mainly through legislative action.

[16] The case of Madey versus Duke University (307 F3d 1351—Fed Cir 2002) is frequently recalled to illustrate the anticommons effects of research tool patenting. It refers to a US Federal Court's decision that established the violation, on the part of Duke University, to certain patents of Prof. John Madey on the use of free electron lasers (FELs). These patents were assigned to Prof. Madey before his appointment as professor and director of the FEL laboratory at Duke University. Following the removal from the post of Director, and his subsequent resignation as a teacher, Prof. Madey denounced the Duke University for continuing to use equipment and methods covered by its patents. Duke University's defense relied on the experimental exception, but in the process of appeal the exception was found to be valid allowed only for attempts to "fun, the satisfaction of mere curiosity or for philosophical investigations closely," and that the exception could not apply when research had "a defined, recognizable and substantial economic purpose," as in universities. The nonprofit status of the university has been judged irrelevant. Prof. Madey's success increased concern within the academic community, that the holders of patents (e.g., DNA sequences or structures of proteins) can prosecute academic scientists who use such material in their research (see Argyres and Liebeskind, 1998).

cost-effectiveness. Thus, the "mental mobilization" and "the cognitive focus" address different aspects of a problem. In the worst case, the gap will never be filled. In the best cases, people work together and at certain points the parameters of optimal quality gradually change, shifting away from "curiosity-driven" research towards practical application (Argyres and Liebeskind, 1998).

3.2.3. On the allocation of resource: Faculty's time and effort

Invention disclosure is a good example of these difficulties. The academic career system provides no natural incentive for scientists to patent new research methods or instruments that have been created for internal use. To the extent that her career will be based on reputation, the scientist's goal is to solve problems through some kind of "do-it-yourself" practices. Profit objectives are, therefore, not present at this stage. Dedicated incentives or regulatory structures (such as technology audits, compulsory notification of invention, etc.) are then necessary in order to avoid the danger that many innovation opportunities will be missed because the scientist will simply not disclose them. Such incentives and regulations are even more necessary when it comes to the postinvention process (development), where and when it is needed.

First, faculty plays a crucial role in helping firms to identify relevant inventions. The most important mechanism here is the one-on-one approach based on personal contact (between industry and faculty), followed by private sector firms surveying the publicly available information, while the direct marketing effort of TTOs is possibly the least important (Thursby and Thursby, 2002, 2003).

Second, the involvement of faculty in further development of the technology is a key factor for successful technology transfer. This is particularly important when the technology is still at an early stage of development at the time the agreement is negotiated. The main reason is that faculty has specialized knowledge about the technology—which is hardly transferable in a codified form through the licensing agreement (Zucker and Darby, 1996; Zucker et al., 1998b). However, the existence of different norms and cultures makes this mobility difficult to realize.

To summarize, the role of faculty may be critical in successful technology transfer. This role is unquestionably important at the invention disclosure stage but obviously extends beyond it. However, the involvement of faculty in the postinvention process of development and transfer implies the mastery of a difficult art: the art of combining academic missions and transfer (and perhaps commercialization) activities.

3.2.4. Dilemmas

To conclude, most arguments above indicate two *dilemmas* (Thursby and Thursby, 2003):

- Successful relations with industry require faculty efforts in the management of those relations (invention disclosure, identification of partners, contribution to the development of the technology), but that effort potentially diverts faculty from its role in academic research.
- The willingness of firms to engage resources in postinvention activities is conditional to the creation of a secure economic environment for their investments. The "ideal" mechanism for them is based on exclusive licensing. However, this "security" has the potential to adversely affect the whole system by weakening the social norm of knowledge openness and sharing through various feedbacks and influences, one of which is the fragmentation of IPRs, as described by proponents of the anticommons metaphor.

3.3. Structural factors for managing complementarities

We turn now to the analysis of some structural factors that can be viewed as particularly effective in minimizing the tensions and conflicts described above and in improving the management of complementarities.

3.3.1. The role of engineering sciences

The institutionalization and development of the so-called "transfer sciences" or engineering constitutes a good case in point. A pivotal element in the "chain of events" occurring between the two spheres (abstract research and concrete applications) is a powerful engineering discipline (computer-, chemical-, aeronautical-, electrical engineering). Engineering sciences support the gradual transformation of knowledge from ideas into operational concepts and from one codified form (adapted at a high level of abstraction) to another codified form (that is adapted to application). The tensions described above are therefore expected to be weaker than in the context of pure fundamental research activities. According to Nelson and Rosenberg (1994), the early recognition of engineering as an academic discipline within US universities explains much of the latter's success in transferring knowledge to industry. Such recognition laid the foundations for the profitability of scientific research, because it allowed the creation of learning programs aimed explicitly at putting engineers in the condition to improve products and processes on the basis of scientific notions. Being engineering schools more "permeable" to industry needs than the colleges of arts and sciences, they could also be charged with research missions distinctive from those of either traditional academic science or profit-oriented R&D laboratories, and quite effective in facilitating technology transfer (Lécuyer, 1998).

3.3.2. New managerial practices in industry

With the increasing importance of "science-driven discoveries and innovation," there is a strong need for changing managerial practices at firm levels as a means of improving absorptive capacities (or to take a Marshallian reference the "external organization" of companies). Science-oriented discovery and innovation is both a technology for discovering and developing new products and a set of managerial practices for organizing and motivating research workers in companies (Cockburn et al., 2000). Thus, science-driven R&D requires that firms should become participants in science rather than mere users of scientific knowledge. This means that the design and adoption of new human resources management practices in firms are part of the solution (for improving structural conditions for effective knowledge transfer), although they are rather neglected in policy discussion and indicator building.

3.3.3. Bridging institutions

As David and Metcalfe (2008) remind us, effective management of complementarities requires the explicit creation of organizations that can bridge with business. To the extent that universities contribute indirectly to technical progress (through basic research and training) the bridging task falls generally on organizations that are *external* to universities themselves, such as public agencies for local or sectoral development or for the support of specific categories of firms (typically, the SMEs). This category includes large scale programs for technology transfer (such as those run once by ANVAR, *Agence Nationale de Valorisation de la Recherche* in France; or by BTG, the *British Technology Group*, now privatized) as well as networks of universities' laboratories (such as the *Steinbeis Institute* in Germany, or the academic partners of MEP and ATP, the *Manufacturing Extension* and *Advanced Technology* Programs run by the US National Institute of Standards and Technology).[17]

In more recent years, universities' direct involvement in the commercialization of their faculty's inventions and expertise has led to an increasing *internalization* of many bridging activities, through the creation of technology transfer and industry liaison offices. Earlier experiences of universities' "direct marketing" of their knowledge (and infrastructure) were the science parks that boomed throughout the 1980s and 1990s.

In principle, bridging institutions, either external or internal, ought to facilitate knowledge transfer from university to industry. It is often the case that policy makers, when assisting or promoting their creation, place emphasis on their value to SMEs, which need more assistance than large companies in approaching academic science. Two difficulties stand in the way of this mission.

First, bridging institutions face a problem of legitimacy in the eyes of all parts they are supposed to serve. This difficulty is well illustrated by (but by no means restricted to) the case of universities' technology transfer offices (TTOs), whose effectiveness is often hampered by a series of principal-agent problems. As organizations, TTOs need to legitimize their existence by achieving some objectives (number of patents, licenses or spin-off, and revenues from the latter) which may contrast with the objectives of both the parties they are supposed to serve, namely the academic scientists and the industry representatives. Such parties may have consultancy or collaboration agreements that pre-date the intervention or the creation of the TTO. If it is so, they will see the latter more as an agent of the university administration whose aim is to alter the existing arrangements in the latter's interest, rather than as a facilitator. Indeed, this is the case whenever the university administrators' or the policy makers' nurture great expectations of financial returns from the TTO's activity, especially through patent royalties. Such expectations may clash against the scientists' preference for payments in the form of research sponsorship, and the industry's resentment for what is perceived as the university administration's greed. A theoretical treatment of this problem is provided by Jensen et al. (2003), while evidence related to this treatment can be found in Siegel et al. (2004) (we come back to this in Section 4.4).

Second, bridging institutions may fail to act as two-way channels of communications between university and industry, that is to facilitate not only the transfer of knowledge from university to industry, but also the reverse flow of data, access to instruments, and interesting research questions (Meyer-Krahmer and Schmoch, 1998). If this is the case, bridging institutions will not be able to elicit the academic scientists' interest in their activities, and ultimately stifle the latter's development.

More generally, the challenge faced by all bridging institutions is to build and maintain their links with both parties (academic scientists and industry) while at the same time acting in the interest of a third party, whether the university administration or the local government. In order to understand these

[17] On the experience of ANVAR, which is now part of OSEO, a larger organization for the support of SMEs, see Laredo and Mustar (2002). On Steinbeis, MEP, and ATP, see respectively Hassink (1996), Feller et al. (1996), and Hall et al. (2003). On BTG see Section 4.2.1 in this chapter.

difficulties, one needs to frame the study and planning of bridging institutions within the more general context of the existing social ties between scientists in university and industry, and ground the analysis on a clear understanding of both parties' incentives to collaborate.

4. The empirical literature: Issues and results

The empirical literature on the research relationships between university and industry has been growing continuously over the past 20 years. We do not aim here to cover it entirely. On one hand, we will be highly selective and focus as much as possible on issues related to the explicit interaction between universities and industry, as opposed to more general and mediated forms of knowledge exchange (such as education or long-term impact of science on productivity). On the other hand, we will survey many forms of interaction, ranging from informal exchanges to universities' involvement in commercial R&D and patenting. The reader may wish to integrate this chapter with other surveys, either more general or more specific than ours, of which five stand out as particularly useful: Mowery and Sampat (2005), who discuss the role of universities in national innovation systems, placing more emphasis than we do on policy issues; Agrawal (2001), who delves into a number of methodological details and places special emphasis on the characteristics of firms that choose to interact with universities; Verspagen (2006), whose survey is entirely dedicated to the emerging phenomenon of university patenting; Rothaermel et al. (2007), whose review of the literature on university entrepreneurship is by far the most complete we are aware of; and Link and Scott (2007), who survey the more recent literature on university science parks.

In what follows we first discuss two classic lines of enquiry on the contribution of academic research to industrial innovation, the first approach being based on questionnaire data, the other on patent and innovation counts (Section 4.1). We then move on to examine the more recent literature on universities' direct involvement in commercial innovation activities, either through patenting or firm creation (4.2). In Section 4.3, we discuss the very first quantitative studies that try to assess how academic research can either benefit or suffer from interaction with industry. Finally, in Section 4.4, we provide a synthetic survey of the empirical literature on the effectiveness of two types of bridging institutions, namely TTOs and science parks.

4.1. From university to industry: The quest for "relevant knowledge"

Understanding the way academic science impacts technological change has been a longstanding objective of empirical research concerning the relationship between university and industry. Three sets of questions have been addressed:

I. How relevant to firms is academic research as a source of innovation, compared to other sources such as internal R&D, users (customers) and suppliers?
II. Does the relevance of academic research vary by industry or firm size? Do large firms in R&D intensive sectors benefit of academic research results more than other companies? Or is it the case that SMEs, facing too high fixed costs for setting up internal R&D facilities, have more incentives to keep in touch with academia?

III. Does access to academic knowledge vary with geographical distance? Is location in the proximity of a leading RU a source of competitive advantage? Can bridging institutions help fostering technology transfer to local industry, or change the latter's specialization by attracting or creating companies active in hi-tech sectors?

More recently, a fourth question has resonated in policy-led empirical research:

IV. Which property regime for the results of academic research is more effective in supporting the diffusion of those results? Do firms access those results as a public good, so that universities can be seen as producers of a positive externality, or do they engage in contractual relationships with universities (such as when they license their inventions or put out some contract research)?

Answers to these questions have been produced on the basis of a number of data sources. Among them are innovation surveys, which are also discussed in Chapter 29 of this handbook (by Jacques Mairesse and Pierre Mohnen).

Another stream of relevant empirical research has made use of patent data and innovation counts. This tradition has often relied upon the modeling tool of the "knowledge production function" and the related concept of "knowledge spillover."

Both research traditions have addressed all of the four research questions listed above. In this section, we examine the first three of them, and postpone the treatment of the fourth to Sections 4.2– 4.4.

4.1.1. *Evidence from innovation surveys*

In the relatively short history of quantitative research on universities' contribution to innovation, four surveys stand out for having provided economists and business students with most of the data on the issue: the Yale survey, conducted on a sample of medium–large R&D-performing companies in the United States; the Carnegie Mellon survey, which can be regarded as a follow-up of the Yale survey, and was conducted in the early 1990s; the PACE survey, also conducted in the early 1990s and conceived as the European equivalent of the Yale survey; and the four editions of the Community Innovation Survey (from 1991 to 2004), also modeled upon the Yale survey, but gradually extended to firms of all size (except those with fewer than 10 employees), R&D intensity, EU countries and sectors. To these large, general-purpose surveys one may wish to add the three smaller, on-purpose surveys run by Edwin Mansfield in the 1990s, whose results still nowadays provide us with outstanding evidence and challenging questions.

Data on the role of university research produced by the Yale survey were limited. The surveyed firms were merely asked (among many other things) to rank the direct contribution of research conducted by scientific institutions to their innovation activities, as opposed to the contribution of internal R&D and information or artefacts from suppliers, customers, and rivals. Other questions related to the importance of science in general as a useful stock of knowledge. While universities (as part of the broader category of scientific institutions) were found to contribute less than other actors to the respondents' innovation activities, science as such was found to be quite important. Nelson (1986) and Klevorick et al. (1995) interpret this evidence as supportive of Nelson's (1959) original theory of the economics of basic research, namely that the latter hardly meets industrial needs in the short run, but turn out to be most useful over the long run, as a stock of knowledge which all firms can access when looking for technical solutions to unforeseen problems or market opportunities. Most notable exceptions to this pattern are provided by a few industries where scientific novelties may turn out to be of immediate practical relevance, such as pharmaceuticals and chemicals, as well as some areas of electronics.

More recently, Cohen et al. (2002) have reported evidence from the Carnegie Mellon survey, whose questionnaire addressed more directly issues of university–industry interaction than its predecessor. In particular, questions were asked to managers on what products of academic research were of most interest for industry, what disciplines were most relevant, and which information channels were most often used.

Among academic research outputs, the respondents assigned greater relevance to new discoveries and scientific instruments, as opposed to prototypes, a result which, according to Cohen et al. (2002), goes against the rationale for encouraging universities to take patents.[18] As for disciplines, pure scientific ones, with the exception of Chemistry, are found to be less relevant than engineering. Finally, scientific publications are the most highly rated channel of communication from university to industry, followed by two other "open science" channels such as attendance of meetings/conferences and informal interaction. These last two, however, are closely trailed from behind by consulting, a "private channel" which is found to be most often used in conjunction with the open science ones. Most strikingly, patents and licenses are poorly rated and hardly used in conjunction with other information channels, a piece of evidence which Cohen et al. (2002) once more level against theories and policies that emphasize the importance of IPRs for technology transfer.[19] All of these results vary greatly by industry and firm size, very much like what was previously found by the Yale survey.

Mansfield's (1991a,b, 1995, 1998) evidence for three samples of over 50 large R&D-intensive firms, confirms that the direct impact of academic research on industry is quite limited, relatively to other sources of innovation inputs, and that it varies across sectors. However, for the period 1986–1994, Mansfield observed some 9% of new products and 3.5% of new processes whose development either required or greatly benefited from academic inputs, for an overall value of over 100 billion dollars of the time. Mansfield also found that, measured in such way, the contribution of academic research had been increasing from his previous assessment for the 1975–1985 period. Finally, when asked to name the most influential academic researchers they had been in touch with, the survey respondents pointed at scholars of quite high standing, who entertained continuing consulting relationships with industry. When interviewed, these scholars were found to be recipients of governmental support, which made them not at all dependent from contracts with industry; even more interestingly, they declared their scientific work to be influenced by the research questions posed by industry, a result in line with those from the recent quantitative assessments we survey in Section 4.3. Mansfield (1995) also explored the role of geographical proximity in fostering university–industry contacts, and found that a positive effect could be detected only for applied research; on the contrary, when it comes to accessing fundamental research, firms are ready to travel any distance.

[18] See Section 4.2.1.

[19] Cohen et al. (2002), however, do not test wheter exclusive licenses over university patents may be necessary to provide industry with the proper incentives to develop the inventions covered by such patents. Their analysis is limited to the information value of the latter. For empirical evidence on the incentive problem, see Section 4.2.2 and Arora and Gambardella, in this volume.

Arundel and Geuna (2004) also explore the geographic dimension of university–industry knowledge flows through the PACE survey. They find that public science is the source of innovation which most require proximity for being accessed, as opposed to inputs from suppliers and customers. However, the two authors do not have a ready explanation for this result. Contrary to their expectations (and many theories) they find that firms that access public science through informal contacts with individual researchers are also those that rely less upon domestic scientific institutions; this goes against the intuition that informal contacts convey tacit knowledge, which requires frequent personal exchanges and cannot be transmitted over long distances.

CIS data for the United Kingdom have been exploited by Laursen and Salter (2004) to examine whether the relevance of academic research for industrial innovation depends not only on structural variables such as industry and firm size, but also on the firms' strategic profile. In that respect, they find that firms with an "open" approach to innovative search, that is firms that rate highly all external sources of innovation inputs, are also those that attach the greatest importance to academic research. Similar results are found by Veugelers and Cassiman (2005) for Belgium. In this case, CIS data suggest that engaging in cooperative agreements with universities is part of a broader strategy of exploitation of public information sources and, possibly, of cooperation with suppliers and customers. In general, it seems that studies based upon CIS data find a greater role for formal university–industry collaboration than other surveys (see also, for France, Monjon and Waelbroeck, 2003). However, Mohnen and Hoareau (2003) find that collaborations are typical only of firms which are large and/or patent-intensive, and that government financing seems to play a role in making collaborations possible. Firms that are both R&D intensive and dedicated to radical innovation are found to make use of academic research results, but not necessarily to engage in formal collaboration.

A recurrent finding of studies based upon CIS data is that, compared to the previous surveys, many fewer respondents assign some relevance to academic inputs to the innovation process. However, this is largely explained by the fact that while those previous surveys addressed only large firms with both an innovative record and internal R&D facilities, the CIS samples include firms of any size, many of which have no record of innovation or have only undertaken incremental innovations, and no R&D activity (Mohnen and Hoareau, 2003). Arundel and Geuna (2004) point out that, when resampling CIS data in order to make them comparable to PACE ones, most differences between the two surveys disappear.

4.1.2. Patent data and innovation counts: The "knowledge spillover" approach[20]

Empirical studies in the economics of innovation have for long relied upon patent data or, to a lesser extent, innovation counts. In particular, patents and innovation counts have been used as output measures in studies based upon the modeling tool of the knowledge production function (with R&D as an input; see Griliches, 1979).

These studies have traditionally assigned great importance to the concept of "knowledge spillover" or "externality." Within the framework of the knowledge production function, in fact, one has to provide some explanation for the common finding that a firm's patent or innovation output does not depend entirely from internal R&D; and that other firms' R&D activities or public research efforts also bear

[20] This section draws in part from Breschi et al. (2005a).

some positive influence (Griliches, 1992). Considering academic research as a public good is therefore a natural complement of the knowledge production function approach.

Starting with the 1990s, most econometric attempts to measure the extent of knowledge spillovers from academic research have been coupled with exercises aimed at measuring the geographical scope of those spillovers.[21]

Jaffe (1989) is generally acknowledged as the pioneering paper in this field. Aiming to assess the *Real effects of academic research*, Jaffe estimated a "modified knowledge production function" in which the dependent variable is given by the number of private corporate patents produced in a given technology by each state of the United States, and the explanatory variables include, among others, the research expenditures of universities and a measure of within-state geographic coincidence of corporate R&D labs and university research.

Jaffe's results show that the number of corporate patents is positively affected by the R&D performed by local universities, after controlling for both private R&D inputs and the state size, as measured by population.

Many authors have replicated Jaffe's exercise. Using innovation counts from the Small Business Innovation Data Base (SBDIB), Audretsch and Feldman (1996) and Feldman and Audretsch (1999) show that, even after controlling for the geographic concentration of production, innovative activities present a greater propensity to cluster spatially in those industries in which industry R&D, university research and skilled labor are important inputs. Acs et al. (1994) also find that the elasticity of innovation output with respect to university R&D is greater for small firms than for large ones. This is interpreted as evidence that small firms, while lacking internal knowledge inputs, have a comparative advantage at exploiting spillovers from university laboratories. Along similar lines, Anselin et al. (1997) refine Jaffe's original methodology to take into account cross-border effects, and show that university research has a positive impact on regional rates of innovation.[22]

In recent years, a debate has arisen over the proper interpretation one should give to these findings. Originally, the most common explanation was that knowledge is indeed a public good, but one which contains tacit elements, so that its transmission through written publications is not complete, and requires to be supplemented by fact-to-face contacts (which are much easier to arrange or more likely to occur accidentally at short physical distances).

This explanation, however, hides a contradiction. Knowledge tacitness, in fact, is a powerful *exclusionary* means. Lack of codification, which may occur because of the novelty of the knowledge produced, or as a result of an explicit strategy of the knowledge producers, may be used to prevent other actors from fully understanding the contents of scientific and technical messages (Foray, 2004). Local flows of knowledge, far from being pure externalities may turn out to be, at a more careful scrutiny, knowledge exchanges entirely mediated by market mechanisms (Geroski, 1995). These observations, of course, concern not only academic knowledge, but scientific and technical knowledge at large.

A few recent papers provide evidence in this direction. Building upon his own previous work on spatial econometrics, Varga (2000) estimates the innovation elasticity with respect to academic R&D

[21] See Feldman and Kogler, in this volume.

[22] A more comprehensive review of the econometric literature on localized knowledge spillovers can be found in Breschi and Lissoni (2001a,b).

for a number of US metropolitan areas characterized by markets for business services of different size, and a different degree of specialization in high-tech industries. He finds that academic R&D expenditures impact significantly on innovation only within areas where business services and the high-tech industries have achieved a substantial critical mass.

Agrawal and Cockburn (2003) propose a set of cross-section regressions of the number of patents over the number of university publications in over 200 US metropolitan areas, for three science-based technological fields. After controlling for the size and specialization of the areas, they find that the patent–papers association is the strongest for those areas hosting at least one "anchor tenant," namely a large, patent-intensive firm, with some absorptive capacity in the relevant technology. The authors suggest that *vertical spillovers* may exist (from universities to the local companies), but they require a mediation of a large, R&D-intensive firm.

Results like these call to mind the findings of Mansfield (1995) we reported above. They point to the necessity of setting aside any presumption that academic knowledge is by definition a public good, and force us to look at the place of universities and academic scientists within markets for technologies, especially if we are interested in the impact of academic research on local development.

4.2. Universities in the market place

Universities participate in market or market-like activities both in the field of education and in that of research. Such participation has increased over the past 20 years or so, both as a result of strategic choices by universities and as a consequence of changes in the way governments allocate funds, which have been increasingly inspired to criteria of competition and market-like mechanisms (Bok, 2003; Clark, 1998).

Here we concentrate on the empirical literature dealing with two aspects of universities' involvement in the market place, namely the extent of university patenting and of academic entrepreneurship, and the relevance of both kinds of commercial activities for university–industry technology transfer.

4.2.1. Academic patenting

Over the past 20 years, the issue of university patenting has moved to the forefront of economic analysis, due to the impressive growth registered in the number of patent applications by US universities after the introduction of the Bayh-Dole Act in 1980 (see Section 3.2.1 above).

In particular, USPTO patent applications by universities have increased at a much faster rate than those by business companies and individuals. The number of academic institutions entering for the first time in the patent system has also increased, from 30 in 1965 to 150 in 1991 (Henderson et al., 1998). Most patents, however, remain concentrated in the hands of the major RUs: in 1991, the top 20 universities held 70% of patents. Biotechnology, and later on software, have been the fields where university patenting has thrived most.[23]

[23] Mowery et al. (2004) reach similar conclusions.

University patenting was common in the US academia well before the introduction of the Bayh-Dole Act, but for long it was hardly associated with a profit motive, at least on the part of the university. Mowery and Sampat (2001) remind us of the historical role of Frederick Cottrell, professor at UC Berkeley, who in 1912 founded the Research Corporation, a no-profit company he endowed with his own patents and later on became a key broker of academic inventions. Apple (1989) and George (2005) offer a similar story for Wisconsin's professor Harry Steenbock, who in 1925 founded the Wisconsin Alumni Research Foundation (WARF).

The Research Corporation and WARF were instrumental in diffusing IPR management expertise in the US academic system. In Europe, only Britain had a similar experience with the BTG, which was founded in 1948 (originally with the name of National Research Development Corporation) with the specific aims of commercializing the results of British public research and of reinvesting the proceedings in the university system (Clarke, 1985; Gee, 1991).[24]

Assessing the impact of the Bayh-Dole Act has been a major line of research in the United States. Among the many questions investigated, two are of particular interest here: Did the Act really increase the number of university patents, or is it the case that progress in biotechnology and software (and the concurrent strengthening of IPR laws) would have led anyway to the observed growth? Did the Act change the economic incentives attached to patenting and alter the research pattern of universities, either by increasing the overall research effort, or by addressing it towards more applied fields?

Research on these questions has first investigated the kind of inventions patented by universities. The Bayh-Dole Act aimed at creating a marketplace for proofs of concepts and prototypes, to be acquired, developed, and finally placed on the market. Granting IPRs to universities was seen as necessary to overcome any potential market failure. Case studies by Zucker et al. (1998a) suggest that prominent US biotech scientists whose patents are licensed either to established or new companies, play a prominent role after licensing for the precise reason that their expertise and skills are needed to further develop the inventions. In this vein, Thursby and Thursby (2002) suggest, on the basis of survey data, that growth in university patenting and licensing may be explained by "universities becoming more entrepreneurial" at all levels, after the Act: scientists became more willing to disclose their inventions, while the university administrations increased the patenting rate of disclosed inventions; academic research did not shift from basic to applied, but commercialization efforts became so aggressive that also inventions of minor importance have been patented. These surveys results may also explain early findings by Henderson et al. (1998) on the decline in quality of university patents (as measured by citations received) after the Bayh-Dole Act.

These results have not gone unchallenged. Studying the cases of Stanford, University of California, and Columbia University, Mowery et al. (2001) reach the conclusion that the influence of the Bayh Dole Act on recent historical developments has been overstated. Broader legislative changes in the direction of strengthening the overall IPR regime in the United States may have exercised a greater influence. In particular, the increasing freedom to patent the results of biomedical research has meant a lot for the academic world. Mowery et al. (2001) also conclude, on the basis of patent data, that academic research has not been diverted from basic targets, and criticize the methodology of previous studies. They also

[24] BTG lost its monopoly rights over academic inventions in 1985, and in 1992 it was privatized. However, it still retains a large portfolio of university patents.

suggest that that the main effect of the Bayh-Dole Act has been that of pushing a few large, private universities into the patenting arena, from which they had abstained for ethical reasons until its approval.

Colyvas et al. (2002) examine 11 blockbuster patents from Columbia and Stanford, and find that they did not originate from applied research, but rather from basic research aimed at the solution of practical problems. In contrast to the "proof-of-concept and prototype" view of academic inventions, these patents were of immediate use to industry, which either sponsored or closely monitored the related scientific advancements.

More recently, a number of empirical contributions have tested the hypothesis of a trade-off between commitment to scientific research and patenting at the individual scientist level; we review them below, in Section 4.3.

Compared to the United States, European research on academic patenting is much more recent. The largest part of it has dealt with the institutional differences between the European and the US academic systems. Discussion of these differences has served two different purposes: first, as a possible explanation of size differences between the patent portfolios of European versus US universities; second, as a justification for adopting different methodologies for measuring academic patenting activity in the two systems.

Among these institutional differences, two are of particular interest here:

(a) The legal ownership of IPRs over academic research, epitomized by the so-called "professor's privilege," which exempts academic personnel from attributing the rights over their inventions to their employers.
(b) The comparatively little autonomy and competencies of the European university administrations in matters of IPRs.

The professor's privilege used to be a typical institution of the German patent law, which reflected the power achieved by academic scientists in the late 1800s. Over the twentieth century, it was also adopted by many of the countries which imitated the German academic system and science policies. Policy concerns over the infrequent use of this privilege by professors has recently led to its abolition by Germany, Austria, and Denmark, while Sweden is also considering abandoning it (OECD, 2003; PVA-MV, 2003).[25]

More generally, no matter whether the national legislation imposed the academic privilege, most European universities have for long lacked the autonomy and administrative skills required in order to take advantage from their professors' patenting activities. They traditionally resisted being involved in such activities, and took the shortcut of allowing scientists engaged in cooperative or contract research with business companies and GRLs to sign blanket agreements that left all IPRs in their partners' hands.

This suggests that a large part of academic patents in Europe may simply escape the most commonly available statistics, which classify the origin of the patent according to the identity of the grantees or applicants, instead of the inventors.

Following this clue, Meyer (2003), Balconi et al. (2004), Iversen et al. (2007), and Lissoni et al. (2008) have reclassified patents by inventor, and matched the inventor's names with available datasets on university faculties, thus producing the first estimates of academic patenting in Finland, Italy, Norway, France, and Sweden, respectively. In all of these countries a significant percentage (from 3% to 8%) of the business companies' patents is found to cover inventions of academic scientists.

[25] Italy is the main exception to this trend, having introduced the academic privilege in 2001.

CNRS, CNR, and VTT (the three most prominent GRLs of France, Italy, and Finland, respectively) also hold many patents signed by academic inventors; the same applies to individual professors in Sweden (where the professor's privilege rules).[26]

That the US case may be an exception, when it comes to academic patenting, seems to be confirmed also by Walsh and Nagaoka (2009), who find that Japanese universities (very much like European ones) own a minority share of their scientists' patents (around 18%). The latter are by large owned by business companies and rarely used as the basis for an academic start-up.

Sample data collected by Thursby et al. (2007) suggest that in the United States the percentage of academic patents held by business companies rather than universities is much lower than in Europe, This implies that the gap between US and European universities in terms of contribution to technology transfer via patented inventions is not as big as it seems when looking only at universities' patent portfolios. Ongoing research is therefore focussing on whether the different property regime of academic patents affects their commercial value and exploitation possibilities (Crespi et al., 2006).[27]

A final line of enquiry in the field of academic patenting has explored individual incentives. Lach and Schankerman (2004) show that the design of monetary rewards can have real effects on academic scientists' eagerness to disclose their inventions to their universities' TTOs. The two authors observe cross-university variations in the share of licensing royalties received by academic scientists and estimate a positive impact of such monetary incentives on disclosure rates. Their study is quite unique in that it focuses on disclosure, that is on a stage of commercialization that comes before patenting and patent exploitation. Most of the literature on academic scientists' incentives, on the contrary, makes use of data from the opposite end of the disclosure-exploitation spectrum, namely data on licensing or commercialization *via* academic spin-offs. It is to this literature that we turn now.

4.2.2. Academic entrepreneurship[28]

Empirical research on academic entrepreneurship was originally focussed on academic start-ups as an alternative to the licensing of academic patents to established business companies. When the academic invention is disclosed at a proof-of-concept stage, it may be hard to convince a firm to take on the long and risky development work needed to bring the new product to the market. This development work cannot be done effectively by an external firm alone, because the tacit and know-how dimension of the knowledge involved is too high (Audretsch, 1995; Audretsch and Stephan, 1999; Jensen et al., 2003;

[26] Attempts to measure the number of academic patents in Germany have relied on a thinner tactic, namely that of looking for the academic title "Professor" in the inventor's field of patent applications, given that the title, in Germany, is awarded only to academics with tenured positions. Schmiemann and Durvy (2003) suggest that, according to this kind of calculation, 5% of German patents at the European Patent Office can be attributed to universities. Gering's and Schmoch's (2003) calculations suggest that academic inventors' patents at the German patent office have grown from about 200 to almost 1800 between 1970 and 2000. Relying on the same approach, Czarnitzki et al. (2007, 2008) have also assembled a large set of German academic patents, whose characteristics they examine either in contrast to nonacademic patents and as a function of ownership (in particular, they compare academic patents owned by universities and individual scientists to those owned by business companies).

[27] Czarnitzki et al. (2009a,b) find that German academic patents are more highly cited than nonacademic ones (an indicator of quality), and less prone to be opposed during the granting phase (an indicator of basicness). These characteristics are less marked when it comes to more recent patents or to (academic) patents owned by business companies, rather than universities.

[28] This section draws in part from Franzoni and Lissoni (2009).

Thursby et al., 2001). Whenever knowledge is characterized by natural excludability, the creation of a company dedicated to exploiting the scientist's idiosyncratic knowledge may become the only viable transfer option (Shane, 2004).

Many technology managers still see academic spin-offs as a sort of advanced solution to technology transfer, which helps finding viable commercialization strategies to growing patent portfolios (Franklin et al., 2001).

Some empirical evidence in support of this thesis has been provided both by case studies and quantitative analyses. Shane (2001b, 2002) finds that the probability of an invention to result in the establishment of a firm is higher in technologies characterized by a strong appropriability regimes. In a related study he also finds that the spin-off foundation rate increases with the novelty of the technology behind it (Shane, 2001a). In a study of the technology transfer activities at University of California, Lowe (2006) finds that patents characterized by a stronger scientific base and a higher degree of tacitness are significantly more likely to be licensed to their original inventors, thus supporting the idea that spin-off creation is necessary when the scientist's knowledge is highly uncodified and idiosyncratic.[29]

A related hypothesis to be tested is whether academic start-ups enjoy a comparative advantage over rival high-tech companies that cannot count upon the direct involvement of academic inventors. Some evidence in this direction was first produced for the bio-tech industries.[30] Zucker and Darby (1996) suggest that the commercial success of biotech companies is positively associated to the scientific eminence of academic researchers participating in the scientific board and holding equity stakes. The same authors show that copublications by academics and companies' researchers help predicting the citation rate of the companies' patents, which suggests that a stronger academic base would boost the quality of inventive activity (Zucker et al., 1998b). Mustar (1997) reports that the R&D intensity of French academic spin-offs is higher than that of other new-technology-based start-ups. Similar results are found for samples of UK firms (Shane, 2004).

Shane and Stuart (2002) study the probability of success of 134 new ventures exploiting MIT inventions, and find that both the academic rank of the inventor and the number of MIT patents in the company's portfolio were likely to increase the probability of an IPO and decrease the failure rate.[31] However, this evidence is far from undisputed. For instance, Nerkar and Shane (2003) find that the technological level of MIT start-ups reduces failure rates only in low-concentration industries. Field studies and extensive interviews to technology managers portray scientists involved in such firms as individuals with a good taste for science, but with relatively naive ideas about the pursuit of market goals (Thursby and Thursby, 2003).

[29] Feldman et al. (2002) report that the willingness of US universities to take an equity in a new venture is generally higher among longer experienced technology offices, which suggests that equity positions of university-administrations may offer a second-best solution to the problem of achieving higher transfer of knowledge to the market, one that perhaps involves a lower risk to divert good scientists from their original tasks.

[30] See also Darby and Zucker, in this volume.

[31] In highly incomplete informational contexts, the scientific reputation of the academic entrepreneur, or the rank of the related institution, may serve as a signal on the perspective value of the venture (Shane and Khurana 2003; Stuart and Ding, 2006). In a study of biotechnology initial public offerings, Stephan and Everhart (1998) find that the amount of funds raised and the initial stock evaluation of firms are positively associated to the reputation of the university-based scientist associated to the firm. Similarly, Di Gregorio and Shane (2003) find that spin-off companies from top universities are more likely to attract venture capitals than less prestigious ones.

More generally, it has been found that many academic scientists engage in entrepreneurial activities not so much because they expect to profit from the new venture, but because they see such ventures as a way to increase the availability of funds for their own scientific projects (Shinn and Lamy, 2006). Therefore, the opportunity costs faced by potential academic entrepreneur do not just depend on exogenous preferences and personal interests, but also on the availability of research funds.

Life-cycle effects may also matter. Older scientists may be more willing to cash-in the market gains of their knowledge assets than their younger colleagues, who need to invest more intensively in increasing their scientific reputation within the academy (Audretsch and Stephan, 1996). This can be especially true when the academic context discourages for-profit activities, in accordance with social norms that only senior and highly reputed scientists dare to challenge (Stuart and Ding, 2006). However, other studies suggest that founding of a new company may be an appealing strategy for younger scientists, such as fresh PhD graduates and research assistants, whose career perspectives are limited but wish to continue to do research in close contact with their university (see Franklin et al., 2001; Roberts, 1991; see also the history of Varian Associates by Lenoir, 1997).

Finally, cohort effects may also be detected, to the extent that younger generations of scientists may enter the academic career with a different perception of the cost and benefits of commercialization and interaction with industry, in particular a more positive one. Although no quantitative evidence has been produced yet on this point, some qualitative results have come from Owen-Smith and Powell (2001).

4.3. From industry to university: Individual and system level interactions

So far we have examined empirical studies concerned with the knowledge flow from university to industry. A number of contributions to the history of technology and to the sociology of science, however, suggest that industry contributes to the advancement of academic science in a number of ways.

As discussed in Chapter 3 by Nathan Rosenberg and Scott Stern, academic scientists have traditionally entertained close contacts with industry in order to get not only funds, but also cognitive inputs such access to data, scientific instruments, and, above all, interesting research questions. Hints in these direction can be traced also in some of studies of academic entrepreneurs' incentives mentioned in Section 4.2 above.

Throughout contemporary history, industry has also provided emerging disciplines with the legitimization and consensus they could not originally gain within the academia, whose conservative tendencies may often stifle disciplinary innovations. Lenoir (1997) and Murmann (2003) provide historical accounts of the importance of links to industry for German "discipline-building" scientists of the nineteenth century, in the medical and chemical sciences. Latour (1988) describes Louis Pasteur's debt to French business sector in a similar fashion. Even a much more recent discipline such as molecular biology had to overcome resistance from within universities, and found in industry a useful ally (Jong, 2006).

For academic science to benefit from ties to industry, however, the former has to be able to resist pressure from the latter in order to deliver immediate results and to limit the codification and diffusion of such results. Philanthropic and public funding of academic science have always been crucial in ensuring the scientists' independence from business funds, from which a stronger bargaining position with industry follows, one that enables resistance to short-termism and secrecy pressures.

The recent explosion of commercial interests in academic research we described in Section 4.2 has been perceived by many economists, social scientists, and practitioners as threatening the public good nature of scientific knowledge.

As a consequence, quantifying the net effects of scientists' involvement with industry has become a priority of empirical research. A large number of survey data analyses and econometric exercises on patent, publication, and citation data have been recently produced, which investigate the extent of two different, but related phenomena. One is the possibility that short-termism and loss of scientific productivity will occur at the individual level, due to the existence of trade-offs between fundamental scientific research and applied research for commercial purposes. The other is the anticommons hypothesis we described in Section 3.2.1 above. We examine them in turn.

4.3.1. Scientific productivity of academic inventors and industry-sponsored researchers

In the last 5 years, the increasing availability of electronic data both for patents and for publications has been exploited to test whether commercial interests impact negatively or positively on a scientist's publication activity, either quantitatively or qualitatively.[32]

Academic inventors are invariably found to be highly productive scientists, indeed more productive than their "noninventing" colleagues. However, it is not clear whether this is due to their individual characteristics (highly productive scientists are expected to produce both more patents and more publications than less productive ones) or to some beneficial feedback from patenting to publishing (such as when a scientist sells her IPRs to industry, from which obtains both cognitive and financial resources for further research).[33]

In order to deal with endogeneity problems, all studies rely upon panel data on the publication activity of large samples of academic scientists and deal with patenting as a *treatment effect*: they test whether the productivity advantage of academic inventors (the treated group) over their colleagues (the control group) increases after signing a patent. So far, all studies have not been able to reject this hypothesis. However, patents are an endogenous *treatment effect*, because it is only highly productive scientists who may hope to turn into inventors. Attempts to solve this second element of endogeneity have been made by Azoulay et al. (2007) and Breschi et al. (2005b), but a consensus has not yet been reached on their validity.

Another finding is that patenting does not seem to affect the quality and direction of research: academic inventors' publications are found to be more highly cited and to address more fundamental issues than those of the control groups. This result is reminiscent of Mansfield's (1995) evidence on academic consultants of large R&D-intensive US firms.

[32] A tentative list of these studies include Agrawal and Henderson (2002), Azoulay et al. (2006), Breschi et al. (2007), Calderini et al. (2007), Fabrizio and Di Minin (2004), Meyer (2006), and Thursby et al. (2005). An ancillary line of enquiry explores the opposite causal links, in order to assess to what extent patents and commercial initiatives are more likely to come from highly productive scientists (Azoulay et al., 2006; Breschi et al., 2005b; Stephan et al., 2007). See also Czarnitki et al. (2007) on German academic inventors.

[33] A clear hint in the direction of the importance of individual characteristics is the fact that academic inventors are found to enjoy a productivity advantage even *before* signing any patent. Lee (2000) provides an indirect confirmation that academic patenting, which often stems from collaboration with industry, may be connected to more resources (financial and cognitive) for fundamental research: his large survey of faculty members with collaboration experience confirms that the main expected benefits consisted in funding for graduate students and useful research insights.

Recent case studies (Callaert et al., 2008) highlight the importance of two conditions under which academic research can be reconciled with an emphasis on commercialization:

- a high degree of topic overlap, which makes the application and commercial development a joint product of basic research and creates a potential for economies of scope;
- the alignment of the size and composition of the research team to the multitask agenda.

Finally, it is worth mentioning a study by Beherens and Gray (2001) on a sample of young graduate students from six US universities, some of whom received sponsorship from industry. Compared to students with no sponsorship, or with a public sponsor, industry-sponsored students are found to publish more papers and to aim at longer term research objectives.

4.3.2. The anticommons hypothesis

Although studies on scientific productivity seem to dispel many fears about the possibility that commercial interests impact negatively on scientific progress at the individual level, this does not exclude the possibility of negative effects at the system level.

To the extent that science is a cumulative enterprise, it is important that all scientists may access their colleagues' research results, data, and tools, in order to avoid the anticommons effects we described in Section 3.2.1. More generally, excessive reliance on industry's resources may expose the scientist to the business partners' pressures in order not to share their data or not to publish inconvenient results. These circumstances are particularly relevant in medical research, which may explain the great number of surveys on data retention and selective publication choices published by the leading journals in the field.[34]

Blumenthal et al. (1996) analyze the impact of industrial funding on scientists' openness and ethical conduct, by means of a questionnaire distributed to over 2000 medical researchers from 50 US universities. The interviewees who declared to be recipients of industrial funds were on average more productive scientists than the nonrecipients. However, the most productive scientists among them are also those who rely the least on industrial funding. One tenth or so of the recipients declared to have denied other scientists access to their research results, and a slightly higher percentage admitted to have complied with requests from their business sponsor to maintain their results secret.

Campbell et al. (2002) build upon these results by investigating the behavior of genetists in over 100 US universities. Almost half of the interviewees signaled to have been denied access to data from a colleague at least once in their career, but only a tenth of them admitted to have behaved in the same way when faced with access requests. One of the reasons for denying access was the need to protect the economic value of the research results; however, the number of scientists who put forward this justification was dwarfed by those providing reasons entirely within the logic of scientific competition, such as the wish to preserve intact one's own chances to be the first to publish the next article on the topic.

It is interesting to notice, however, that Campbell et al. find that data access denial is a much more common phenomenon among genetists than other medical scientists, a result they explain with the

[34] For a comprehensive survey of the literature concerning the effects of industry's involvement in medical research, see Bekelman et al. (2003).

higher economic value of genetic discoveries compared to other medical advancements. They also report an increase of data access denial over the 1990s.

In a follow up of this research, Blumenthal et al. (2006) find that participation in relationships with industry positively affects data withholding by young geneticists, but that gender, mentors' advice or formal instruction, and negative past experiences in the publication race also play an important role.

Overall, the evidence produced by these surveys is rather inconclusive: commercial interests encourage scientific misconduct, but it is hard to tell this influence apart from that of intense scientific competition (Stossel, 2005).

More recently, economists and other social scientists have produced their own survey enquiries. In particular, Walsh et al. (2005) find some evidence that data retention is more likely to occur when scientists receive industry sponsorship. However, the extent of the phenomenon is quite limited and does not seem to be influenced by the scientists' patenting activity, if any.

One possible explanation for the lack of links between patenting and data retention is that scientists act on the basis of "double standards": although willing to take patents and comply with their implications when dealing with industry, they need to maintain smooth relationship with their colleagues, to whom they do not even deny access to patented research tools. Murray (2005) provides an historical account of the "double standards" applied by the scientific community dealing with the "oncomouse patent," signed by Phil Leder, from Harvard University, and granted to DuPont in 1984. Cassier and Foray (2002) have documented an abundant production of rules and institutional innovations in the area of managing and negotiating the attribution of IPRs while preserving some information commons.

Academic researchers seem capable to learn how to negotiate their industrial contracts in order to preserve areas of public knowledge and to maintain a clear distinction between the generic knowledge—that should be maintained under a public good regime—and the knowledge which is developed within the public–private partnership and that may be subject of private appropriation. At the same time, firms are often aware of the advantages of not completely undermining open and independent academic research (a shared collection of basic knowledge being always needed to provide the building blocks for new inventions). As a consequence, they try to establish good practices to allow universities to work *with* and not *for* industry. Whatever motivations they have, the fact remains that some firms are pursuing "a strategy of the commons" (Agrawal and Garlappi, 2007).

However, it may still be the case that the presence of patents in a given research area may discourage scientists to move into that area, for fear of infringing some property rights or of being force to sustain too high licensing expenditures.

In order to test this hypothesis, Murray and Stern (2007) have produced a "natural experiment," based upon a number of "patent–publication" pairs, that is a number of scientific discoveries (genetic sequences) which have been both patented and described in academic publications by the same scientists. The authors test whether citations to the relevant publication decline after the scientific community discover the existence of the related patents, compared to publications unrelated to any patent (citations are taken as an indicator of ongoing cumulative research on the subject of the cited publication). Formulated in this way, the anticommons hypothesis is found not be rejected, although the negative impact of patents is rather limited. Similar results are found by Sampat (2004), who apply the same methodology. Fabrizio (2007) observes that citations to academic patents have declined over time, along with the growth of university patenting and the related phenomena of reduced diffusion, restricted use, or more costly negotiated access to academic science. A very recent paper by Murray et al. (2009)

also finds evidence that restricted access to genetically modified mice for laboratory testing hampered scientific progress by limiting the diversity of experimental approaches.

4.4. Bridging institutions

The empirical literature on bridging institutions is quite sparse and hardly coherent. Two reasons for this characteristic is the heterogeneity of the organizations that qualify as "bridges" of some kind, and the origin of the many (small) datasets and case studies from contingent policy evaluation efforts, rather than from systematic enquiries driven by deeper theoretical questions. We focus here only on the "internal" institutions, such as TTOs and science parks, which are most often under the direct control of universities.[35]

The largest collection of studies on TTOs' functions and performance is contained in two special issues of the *Journal of Technology Transfer* (Siegel et al., 2001) and a number of related papers published in later years. Bercovitz et al. (2001) compare the organization of technology transfer activities of three US universities, and evaluate their performance in terms of:

- coordination between licensing and sponsored research, and between different units charged with technology transfer duties;
- information processing capacity (number of disclosures, licenses, sponsored research agreements, and other technology transfer transactions);
- incentive alignment between different transfer mechanisms, such as licenses and research agreements.

Technology transfer in the examined universities is organized quite differently, according to models that roughly correspond either to Chandler's M-form or Williamson's H-form, or a matrix structure. Each model is found to have distinctive advantages or disadvantages along the three performance dimensions. Changing the organizational model of technology transfer, however, is not just a matter or reorganizing transfer activities, since each model is also the result of the long and complex history of each university in terms of relative weight of disciplines, autonomy of schools and faculties, and mission with respect of the local economy. A comparable case study for Europe is the one produced by Debackere and Veugelers (2005) on the Catholic University of Leuven, the flagship institution of the Flemish higher education system (see also Clark, 1998). Since 1972, the university has trusted the coordination of all its technology transfer activities (including the management of its science park) to a separate organization, the KU-Leuven Research and Development (LRD). Debackere and Veugelers identify a number of original features of LRD, which may explain its success: its long historical record, which legitimizes it as an integral part of the academic institution; its autonomy with respect of budget and human resource management issues; its reliance on "research divisions," voluntary associations of researchers from different departments who LRD assists and helps meeting their commercialization targets; an incentive system that allows scientists to appropriate a large part of the proceedings of their transfer activities as financial resources for further research. Another interesting example of support comes from Jain and George (2007), who describe how the activities of WARF (the WARF; see Section 4.2.1) have

[35] For a general discussion of all sorts of bridging institutions, see Martin and Scott (2000).

contributed decisively to mobilize resources (both political and financial) for the development of human embryonic stem cell technologies. In this case, technology transfer was not limited to the commercialization of one or more inventions, but extended to building the institutional framework that makes research acceptable for society at large.

Extensive surveys of TTOs' practices are less common, possibly because of the difficulty of administering comparable questionnaires to heterogeneous entities such as TTOs. The available results, however, tend to spot more problems than successes. A good example is provided by Siegel et al. (2004), who examine 55 US TTOs and find them organized according to a linear view of technology transfer, which contrasts with the complexity of incentives and university–industry ties we described in Section 3. Interviews with all types of stakeholders (scientists, entrepreneurs, and technology transfer officers) reveal a misalignment of perceptions regarding the expected output, the barriers, and the most relevant type of relationships involved in the transfer process. More alarmingly, TTOs are found to be more at odds with both scientists and entrepreneurs, than the latter are between themselves. It does not come out as surprise, then, that TTOs also lack legitimacy and are often circumvented by the other stakeholders, rather than involved in their transactions. Based on a review of technology transfer activities at US universities and government laboratories, Bozeman (2000) identifies several performance evaluation criteria that have been employed by the various organizations. Some of these criteria, especially those that place more emphasis on quantitative measures, are found to be rather ineffective in shaping the TTOs' actions, although their popularity may be explained by policy-makers' and administrators' appetite for synthetic evaluation exercises.

The literature on science parks is of limited help when it comes to getting a better understanding of university–industry relationship. By and large, in fact, it is a chronicle of repeated failures, and of a sequence of evaluation attempts aimed at elusive targets. As pointed out by Link and Scott (2003, 2007), there is no generally accepted definition of science park, a term which is prevalent in Europe but not as popular in the United States (where "research park" or "university research park" are more common) and in Asia (where "technology park" is more diffused). In general, science parks (and their synonyms) are intended to be real estate developments aimed at hosting hi-tech or science-based firms, which provide for some technology transfer activities and involve a local university, some level of government, and possibly the private sector. Link and Scott (2003) trace their proliferation in the United States back to the 1980s, a decade when also the UK local governments set up many of them, soon to be followed by many other European and Asian countries (Bakouros et al., 2002; Lee and Yang, 2001; Phillimore, 1999; Vedovello, 1997). Founders of new science parks, inevitably invoked the Stanford Science Park as the model to imitate, but proved to have little knowledge of the unique circumstances that surrounded its creation and made its replication very hard to achieve.[36] Early criticism of the UK experience, especially of the idea that science parks could be useful tools both to revitalize deindustrialized areas and support local universities, did not deter subsequent imitation (MacDonald, 1987; Massey et al., 1992).

Most of the quantitative evaluation attempts of science parks' effectiveness focus on firms' performance. Typically, they compare on-park companies with a control sample of off-park ones, either in terms of R&D intensity, growth, or survival chances (Löfsten and Lindelöf, 2002; Phan et al., 2005;

[36] See Leslie's and Kargon's (1996) chronicle of Frederick Terman's failed attempts to replicate the success he had met when, as president of Stanford, he oversaw the creation of the Park. See also Saxenian (1985).

Siegel et al., 2003; Westhead and Storey, 1995). In many cases no advantage for on-park firms is found, and even positive results have to be considered with extreme caution, since they are based on cross-section analyses that hardly control for endogeneity and self-selection. Case studies such as Hansson et al.'s (2005) do not find much evidence of a privileged access of on-park firms to academic knowledge.

5. Policy issues and open questions

5.1. Academics in the market place: Overcoming the dilemmas

In Sections 2 and 3 we have proposed a conceptual approach to university–industry interaction that highlights two dilemmas. The first dilemma concerns individual scientists, and it originates from the potential trade-off between basic research activities and those activities required to successfully develop and commercialize academic inventions. The second dilemma occurs at the system level, and it has to do with the tension between the need of firms involved in the commercialization of academic research to rely upon clear and solid IPRs, and the cumulativeness of the scientific enterprise, which requires the results of academic research to be freely accessible.

The empirical literature we surveyed in Section 4 suggests that the first dilemma may not be dramatic: individual scientists who engage in patenting do not seem to suffer a decline of scientific productivity, nor firms seem to force them to give up the pursuit of fundamental research objective. On the contrary, some evidence exists on the relevance of the second dilemma: commercial interests may exacerbate common threats to the commonality of research efforts; and the existence of IPRs over academic research results may discourage some scientists to build upon those results in order to advance knowledge.

Contrasts may then arise between those faculty members who seek active involvement in commercial exploitation of their research findings, and those whose do not. This is a "system balance" problem both for the individual institution and for the assembly of institutions. It is here that the central administration's attitude can be critical. Do they encourage the movement towards technological commercialization as a legitimate, indeed, institutionally rewarded activity for faculty? Is the administration simply permissive of a drift in that direction, accommodating the requirements of industry in licensing arrangements that permit suppression of research findings from research publications? Or does it seek to create a reward structure that is "neutral" in so far as it does not allow the earnings of those who choose not to get directly involved in commercialization to lag behind those of their entrepreneurial colleagues?

The dilemma between the granting of exclusive rights to maintain firm's incentives and the granting of freedom to operate through the preservation of some sort of "IP-free zone" may also be overcome through the invention of practices and rules dealing with the issues of attributing property rights on clear and well defined portions of knowledge and of protecting free access to some other parts of knowledge, information and tools. These practices and rules are most often produced by researchers, as private arrangements between actors, organized under the principles of "self-discipline of a professional partnership."

5.2. Manipulating incentives: From a "by-product economy" to a "joint product economy"

The literature on academic patenting and entrepreneurship we surveyed in Section 4 provides evidence on the responsiveness of academic scientists to economic incentives. The faculty decision regarding a potential involvement in activities dealing with knowledge transfer and development in industries is obviously based on comparing the various costs and benefits of this activity with the costs and benefits of other more traditional academic tasks.

In the absence of policies and organizational practices aimed at inducing commercialization, the dominant incentive structures for faculties creates a strong imbalance in favor of traditional academic missions: fundamental research and education. These two missions are the ones that potentially generate the two fundamental kinds of spillovers that benefit industry. In this incentives regime, however, all activities related to development, industrial problem-solving and commercialization end up having the status of some sort of *by-product*. In this by-product regime, compromises and trade-offs are easier to achieve since traditional academic missions and priorities are maintained. However, one can also expect a lot of lost opportunities: some of the best inventions may not be disclosed; the most productive faculties are less ready to take time away from new projects in order to disclose inventions, and even less so to work on further development.

The challenge should be then to shift university research from a situation in which technology transfer and commercialization are seen as *by-products* to a situation in which these functions acquire a new higher status: that of *joint product*. We derive the definition of these concepts from accounting: *joint products are two products that are simultaneously yielded from one shared cost and they have comparably high (sales) value*. By-products on the contrary are produced along with a main product. The latter constitutes the major portion of the total (sales) value. By-products have a considerably lower (sales) value than these main products. We can apply these terms to think about basic research and technological applications, substituting "perceived value to the academic professor" for sales value. Such a shift involves increasing the "perceived value to the academic professor" of development and commercialization, and this requires creating a new balance in the incentive structure.

Increasing monetary incentives to encourage faculty toward more disclosure (and more involvement in further development) may have an effect on faculty's motivations to be involved in technology transfer. However, this strategy also entails risk. As already mentioned we know from multitask problems in principal-agent theory that when output is generated by workers exerting efforts on two or more different tasks, there is need to optimally balance incentives across these tasks. Otherwise, people will inefficiently devote too much effort to those tasks that provide them with the highest marginal return (Cockburn and Henderson, 1998).

Since the long-term level of research productivity depends on the level of effort devoted to basic research, it is important to avoid any incentive bias. An important issue is, for example, that any change in incentive structures (to increase effort toward disclosure and commercialization) has to be designed in an integrative and concerted way with the bodies in charge of academic incentives.

5.3. Directions for future research

The literature we have surveyed in this chapter has many limitations and gaps, which future research ought to overcome and fill.

From the theoretical viewpoint, there is still little integration between the economics of science and the economics of technology transfer. Some of the empirical evidence we surveyed in Section 4 explores the complementarities and trade-offs, at the individual level, between fundamental research and cooperation with industry, and between publishing and patenting. However, interpretations of these results rely on little more than intuitive *ex post* explanations; nor any connection has been traced with the systemic effects of increases in cooperation and commercialization efforts. Answering these research questions would require putting technology transfer and commercialization at center-stage of any theory of academic careers and scientific productivity, alongside with fundamental research and publishing. Such representation of academic scientists' activities would be both more accurate and up-to-date than those derived from the classic sociology of science, and possibly more fruitful in terms of suggestions for empirical research.

Another important limitation of the empirical literature is its US-centric bias. This is both a theoretical and an empirical deficiency. The few theoretical propositions on the relationship between scientists, TTOs and university administration have been openly inspired by fieldwork on US research universities. More generally, a number of implicit assumptions can be found in the literature on the mechanisms of academic career, the mobility of scientists, and the relative importance of publications and transfer activities, which are clearly inspired by the US university system. Unfortunately, such system is quite unique, and very different from that of other countries. US universities exercise a degree of control over their academic staff which is uncommon in most countries, where university scientists are or regard themselves as civil servants rather than university employees. US public universities also enjoy an unrivaled degree of autonomy from central government, while the size and number of US private RUs also constitute a worldwide exception. Finally, the US industry's appetite for new technologies and for PhD laureates has no rivals in the world, and allows for very large markets for ideas and for scientists and engineers. Within such markets, mobility between university and industry, and opportunities for hi-tech consultancy, are conspicuous phenomena, which cannot be said of most other countries in the world, including many advanced ones. How do scientists react to opportunities for technology transfer in academic systems with little mobility across universities, and between university and industry? What incentives do academic scientists have to commercialize their inventions or collaborate with business companies, when universities have no means to reward successful technology transfer, and possibly not even scientific excellence, but at the same time exercise little control on their employees' activities?

In the absence of answers to these fundamental questions, further empirical research on bridging institutions such as TTOs or science parks will also be of little interest, being inspired, as it has been for a long time, by a very abstract and normative portrait of university–industry relationship, and not by a vision rooted in the institutional characteristics of most countries.

As for opportunities for empirical research, much work has yet to be done on exploring the importance of interaction with industry for academic careers and scientific productivity, both in the United States and in other countries, especially in discipline different from biotechnology-related ones. Here the challenge is to produce quantitative evidence both on the importance of scientific advancements for technological progress, and of access to industry's knowledge and financial resources for scientific progress. Research in this direction will require to deepen the exploitation of existing indicators, and to produce new ones. As for existing indicators, such as publications and patents, efforts are already under way to reclassify them by authors and inventors, in order to unveil patterns of collaborations and mobility at the individual level, as well as the resulting social and professional networks. New indicators will also need to produce information at the level of individuals, in order to assess the importance of labor markers for scientists and engineers for technology transfer: while much is said of its importance in interview-based studies, little evidence has been produced so far of its scale and scope. This is possibly the most important empirical challenge for the years to come.

References

Acs, Z.J., Audretsch, D.B., Feldman, M.P. (1994). “R&D spillovers and recipient firm size”. Review of Economics and Statistics 76 (2), 336–340.

Agrawal, A. (2001). “University-to-industry knowledge transfer: Literature review and unanswered questions”. International Journal of Management Reviews 3 (4), 285–302.

Agrawal, A., Cockburn, I. (2003). “The anchor tenant hypothesis: Exploring the role of large, local, R&D-intensive firms in regional innovation systems”. International Journal of Industrial Organization 21 (9), 1227–1253.

Agrawal, A., Garlappi, L. (2007). “Public sector science and the strategy of the commons”. Economics of Innovation and New Technology 16 (7), 517–539.

Agrawal, A., Henderson, R. (2002). “Putting patents in context: Exploring knowledge transfer from MIT”. Management Science 48 (1), 44–60.

Anselin, L., Varga, A., Acs, Z.J. (1997). “Geographic spillovers and university research: A spatial econometric approach”. Growth and Change 31, 501–515.

Apple, R.D. (1989). “Patenting university research. Harry Steenbock and the Wisconsin Alumni Research Foundation”. Isis 80, 375–394.

Argyres, N., Liebeskind, J.P. (1998). “Privatizing the intellectual commons: Universities and the commercialization of biotechnology”. Journal of Economic Behavior and Organization 35, 427–454.

Arundel, A., Geuna, A. (2004). “Proximity and the use of public science by innovative European firms”. Economics of Innovation and New Technology 13 (6), 559–580.

Audretsch, D.B. (1995). Innovation and Industry Evolution. MIT Press, Cambridge, MA.

Audretsch, D.B., Feldman, M.P. (1996). “R&D spillovers and the geography of innovation and production”. American Economic Review 86 (3), 630–640.

Audretsch, D.B., Stephan, P.E. (1996). “Company-scientists locational links: The case of biotechnology”. American Economic Review 86 (3), 641–652.

Audretsch, D.B., Stephan, P.E. (1999). “Knowledge spillovers in biotechnology: Sources and incentives”. Journal of Evolutionary Economics 9, 97–107.

Azoulay, P., Ding, W., Stuart, T. (2006). “The Effect of Academic Patenting on (Public) Research Output”, http://www.nber.org/papers/w11917, NBER Working Paper No. 11917.

Azoulay, P., Ding, W., Stuart, T. (2007). “The determinants of faculty patenting behaviour: Demographics or opportunities?” Journal of Economic Behaviour and Organization 63 (4), 599–623.

Bakouros, L., Mardas, D., Varsakelis, C. (2002). “Science Park an hi tech fantasy? An analysis of the science parks of Greece”. Technovation 22, 123–128.

Balconi, M., Breschi, S., Lissoni, F. (2004). “Networks of inventors and the role of academia: An exploration of Italian patent data”. Research Policy 33, 127–145.

Beherens, T.R., Gray, D.O. (2001). “Unintended consequences of cooperative research: Impact of industry sponsorship on climate for academic freedom and other graduate student outcome”. Research Policy 30, 179–199.

Beise, M., Stahl, H. (1999). “Public research and industrial innovation in Germany”. Research Policy 28, 397–422.

Bekelman, J.D., Li, Y., Gross, C.P. (2003). “Scope and impact of financial conflicts of interest in biomedical research. A systematic review”. Journal of American Medical Association 289, 454–465.

Ben-David, J. (1977). Centres of Learning: Britain, France, Germany, United States. McGraw-Hill, New York (repr. 1992, Transaction Publishers).

Bercovitz, J., Feldman, M., Feller, I., Burton, R. (2001). "Organizational structure as a determinant of academic patent and licensing behavior: An exploratory study of Duke, Johns Hopkins, and Pennsylvania State Universities". Journal of Technology Transfer 26 (1–2), 21–35.

Blumenthal, D., Campbell, E.G., Causino, N., Louis, K.S. (1996). "Participation of life-science faculty in research relationship with industry". New England Journal of Medicine 335, 1734–1739.

Blumenthal, D., Campbell, E.G., Gokhale, M., Yucel, R., Clarridge, B., Hilgartner, S., Holtzman, N.A. (2006). "Data withholding in genetics and the other life sciences: Prevalences and predictors". Academic Medicine 81 (2), 137–145.

Bok, D.C. (2003). Universities in the Marketplace. Princeton University Press, Princeton, NJ.

Bozeman, B. (2000). "Technology transfer and public policy: A review of research and theory". Research Policy 29 (4), 627–655.

Breschi, S., Lissoni, F. (2001a). "Knowledge spillovers and local innovation systems: A critical survey". Industrial and Corporate Change 10 (4), 975–1005.

Breschi, S., Lissoni, F. (2001b). "Localised knowledge spillovers vs. innovative milieux: Knowledge 'tacitness' reconsidered". Papers in Regional Science 80 (3).

Breschi, S., Lissoni, F., Montobbio, F. (2005a). "The geography of knowledge spillovers: Conceptual issues and measurement problems". In: Breschi, S., Malerba, F. (Eds.), Clusters, Networks and Innovation. Oxford University Press.

Breschi, S., Lissoni, F., Montobbio, F. (2005b). "From publishing to patenting: Do productive scientists turn into academic inventors?" Revue d'Economie Industrielle 110 (2), 75–102.

Breschi, S., Lissoni, F., Montobbio, F. (2007). "The scientific productivity of academic inventors: New evidence from Italian data". Economics of Innovation and New Technology 16 (2), 101–118.

Calderini, M., Franzoni, C., Vezzulli, A. (2007). "If star scientists do not patent. The effect of productivity, basicness and impact on the decision to patent in the academic world". Research Policy 36 (3), 303–319.

Callaert, J., Van Loy, B., Foray, D., Debackere, K. (2008). "Combining the production and the valorization of academic research: A qualitative investigation of enacted mechanisms". In: Mazza, C., Quattrone, P., Riccaboni, A. (Eds.), European Universities in Transition. Edward Elgar, Cheltenham, UK.

Campbell, E.G., Clarridge, B.R., Gokhale, M., Birenbaum, L. (2002). "Data withholding in academic genetics: Evidence from a national survey". Journal of the American Medical Association 287, 473–480.

Carnegie, (2009). Carnegie Classification of U.S. Institutions of Higher Education. Carnegie Foundation for the Advancement of Teaching. http://www.carnegiefoundation.org/classifications/last visited: March 2009.

Cassier, M., Foray, D. (2002). "Public knowledge, private property and the economics of high-tech consortia". Economics of Innovation and New Technology 11 (2), 123–132.

Clark, B. (1993). The Research Foundations of Graduate Education: Germany, Britain, France, United States, Japan. University of California Press, California.

Clark, B.R. (1998). Creating Entrepreneurial Universities: Organizational Pathways of Transformation. Pergamon-Elsevier Science.

Clarke, M. (1985). "British technology group—UK technology transfer grows". Nature 316 (6027), 385.

Cockburn, I.M., Henderson, R.M. (1998). "Absorptive capacity, coauthoring behavior, and the organization of research in drug discovery". Journal of Industrial Economics XLVI, 157–182.

Cockburn, I.M., Henderson, R.M., Stern, S. (2000). "Untangling the origins of competitive advantage". Strategic Management Journal 21, 1123–1145.

Cohen, W.M., Nelson, R.R., Walsh, J.P. (2002). "Links and impacts: The influence of public research on industrial R&D". Management Science 48 (1), 1–23.

Colyvas, J., Crow, M., Gelijns, A., Mazzoleni, R., Nelson, R.R., Rosenberg, N., Sampat, B.N. (2002). "How do university inventions get into practice?" Management Science 48 (1), 61–72.

Crespi, G., Geuna, A., Verspagen, B. (2006). "University IPRs and Knowledge Transfer. Is the IPR Ownership Model More Efficient?" http://www.sussex.ac.uk/spru/documents/sewp154.pdf SPRU Electronic Working Paper 154, University of Sussex.

Czarnitzki, D., Glänzel, W., Hussinger, K. (2007). "Patent and publication activities of German professors: An empirical assessment of their co-activity". Research Evaluation 16 (4), 311–319.

Czarnitzki, D., Hussinger, K., Schneider, C. (2008). "Commercializing Academic Research: The Quality of Faculty Patenting". ZEW Discussion Paper No. 08-069, Mannheim.

Czarnitzki, D., Hussinger, K., Schneider, C. (2009a). “The Nexus Between Science and Industry: Evidence from Faculty Inventions”. ZEW Discussion Paper No. 09-028, Mannheim.

Czarnitzki, D., Hussinger, K., Schneider, C. (2009b). “Why Challenge the Ivory Tower? New Evidence on the Basicness of Academic Patents”. ZEW Discussion Paper No. 02-029, Mannheim.

Dasgupta, P. (1988). “The welfare economics of knowledge production”. Oxford Review of Economic Policy 4, 1–12.

David, P.A. (1993). Knowledge, property and the system dynamics of technological change. Proceedings of the World Bank Annual Conference on Development Economics. The World Bank, Washington, DC.

David, P.A., Metcalfe, S. (2008). “Universities and public research organisations in the ERA; fulfilling universities’ critical societal roles in the advancement of knowledge and the support of sustained innovation-driven economic growth in Europe”, http://ec.europa.eu/invest-in-research/monitoring/knowledge_en.htm. Knowledge for Growth Expert Group, Director General for Research, Directorate C, European Commission.

David, P.A., Mowery, D.C., Steinmueller, W.E. (1992). “Analyzing the economic payoffs from basic research”. Economics of Innovation and New Technology 2, 73–90.

David, P.A., Foray, D., Steinmueller, W.E. (1999). “The research network and the new economics of science: From metaphors to organizational behaviors”. In: Gambardella, A., Malerba, F. (Eds.), The Organization of Inventive Activity in Europe. Cambridge University Press, Cambridge, pp. 303–342.

Debackere, K., Veugelers, R. (2005). “The role of academic technology transfer organizations in improving industry science links”. Research Policy 34 (3), 321–342.

Di Gregorio, D., Shane, S. (2003). “Why do some universities generate more start-ups than others?” Research Policy 32, 209–227.

Ergas, H. (1987). “Does technology policy matter?” In: Guile, B., Brooks, H. (Eds.), Technology and Global Industry, Company and Nations in the World Economy. National Academy Press, Washington, DC.

Fabrizio, K.R. (2007). “University patenting and the pace of industrial innovation”. Industrial and Corporate Change 16 (4), 505–534.

Fabrizio, K.R., Di Minin, A. (2004). “Commercializing the laboratory: The relationship between patenting and publishing”, http://mgt.gatech.edu/news_room/news/2005/reer/index.html. Paper Presented at Fifth Annual Roundtable for Engineering Entrepreneurship Research (REER), Atlanta GA, December 4–5.

Feldman, M.P., Audretsch, D.B. (1999). “Innovation in cities: Science-based diversity, specialisation and localised competition”. European Economic Review 43, 409–429.

Feldman, M., Feller, I., Bercovitz, J., Burton, R. (2002). “Equity and the technology transfer strategies of American Research Universities”. Management Science 48 (1), 105–121.

Feller, I. (1997). “Manufacturing technology centers as components of regional technology infrastructures”. Regional Science and Urban Economics 27, 181–197.

Feller, I., Glasmeier, A., Mark, M. (1996). “Issues and perspectives on evaluating manufacturing modernization programs”. Research Policy 25, 309–319.

Foray, D. (2004). Economics of Knowledge. MIT Press, Cambridge, MA.

Franklin, S.J., Wright, M., Lockett, A. (2001). “Academic and surrogate entrepreneurs in university spin-out companies”. Journal of Technology Transfer 26, 127–141.

Franzoni, C., Lissoni, F. (2009). “Academic entrepreneurs: Critical issues and lessons for Europe”. In: Varga, A. (Ed.), Universities and Regional Economic Development. Edward Elgar, Cheltenham, UK.

Gee, H. (1991). “Privatization ahead for BTG”. Nature 349 (6307), 272.

George, G. (2005). “Learning to be capable: Patenting and licensing at the Wisconsin Alumni Research Foundation 1925–2002”. Industrial and Corporate Change 14 (1), 119–151.

Gering, T., Schmoch, U. (2003). Management of Intellectual Assets by German Public Research Organisations. OECD.

Geroski, P. (1995). “Markets for technology: Knowledge, innovation and appropriability”. In: Stoneman, P. (Ed.), Handbook of the Economics of Innovation and Technological Change. Blackwell, Oxford, pp. 90–131.

Geuna, A., Martin, B.R. (2003). “University research evaluation and funding: An international comparison”. Minerva 41, 277–304.

Griliches, Z. (1979). “Issues in assessing the contribution of R&D to productivity growth”. Bell Journal of Economics 10, 92–116.

Griliches, Z. (1992). “The search for R&D spillovers”. Scandinavian Journal of Economics 94 (Suppl.), 29–47.

Hall, B.H. (2004). “On copyright and patent protection for software and databases: A tale of two worlds”. In: Granstrand, O. (Ed.), Economics, Law, and Intellectual Property. Kluwer Publishing Co., Amsterdam/Dordecht.

Hall, B., Link, A., Scott, J.T. (2003). "Universities as research partners". Review of Economics and Statistics 85, 485–491.

Hansson, F., Husted, K., Vestergaard, J. (2005). "Second generation science parks: From structural holes jockeys to social capital catalysts of the knowledge society". Technovation 25 (9), 1039–1049.

Hassink, R. (1996). "Regional technology policies in the old and new Länder of Germany case-studies from Baden-Württemberg and Thuringia". European Urban and Regional Studies 3 (4), 287–303.

Heller, M.A., Eisenberg, R.S. (1998). "Can patents deter innovation? The anticommons in biomedical research". Science 280, 698–701.

Henderson, R., Jaffe, A., Trajtenberg, M. (1998). "Universities as a source of commercial technology: A detailed analysis of university patenting 1965–1988". Review of Economics and Statistics 80, 119–127.

Iversen, E.J., Gulbrandsen, M., Klitkou, A. (2007). "A baseline for the impact of academic patenting legislation in Norway". Scientometrics 70, 393–414.

Jaffe, A.B. (1989). "Real effects of academic research". American Economic Review 79 (5), 957–970.

Jaffe, A.B. (2000). "The U.S. patent system in transition: Policy innovation and the innovation process". Research Policy 29 (4–5), 531–557.

Jaffe, A.B., Lerner, J. (2001). "Reinventing public R&D: Patent policy and the commercialization of national laboratory technologies". RAND Journal of Economics 32, 167–198.

Jaffe, A.B., Fogarty, M.S., Banks, B.A. (1998). "Evidence from patents and patent citations on the impact of NASA and other federal labs on commercial innovation". Journal of Industrial Economics 46, 183–205.

Jain, S., George, G. (2007). "Technology transfer offices as institutional entrepreneurs: The case of Wisconsin Alumni Research Foundation and human embryonic stem cells". Industrial and Corporate Change 16, 535–567.

Jensen, R., Thursby, J.G., Thursby, M.C. (2003). "Disclosure and licensing of university inventions: 'The best we can do with the s**t we get to work with'". International Journal of Industrial Organization 21, 1271–1300.

Jong, S. (2006). "How organizational structures in science shape spin-off firms: The biochemistry departments of Berkeley, Stanford, and UCSF and the birth of the biotech industry". Industrial and Corporate Change 15 (2), 251–283.

Kelley, M., Arora, A. (1996). "The role of institution-building in US industrial modernization programs". Research Policy 25, 265–280.

Kerr, C. (2001). The Uses of the University. Harvard University Press.

Klevorick, A., Levin, R., Nelson, R., Winter, S. (1995). "On the sources and significance of interindustry differences in technological opportunities". Research Policy 24 (2), 195–205.

Lach, S., Schankerman, M. (2004). "Royalty sharing and technology licensing in universities". Journal of the European Economic Association 2, 252–264.

Laredo, P., Mustar, P. (2002). "Innovation and research policy in France (1980–2000) or the disappearance of the Colbertist state". Research Policy 31, 55–72.

Latour, B. (1988). The Pasteurization of France. Harvard University Press.

Laursen, K., Salter, A. (2004). "Searching high and low: What types of firms use universities as a source of innovation?" Research Policy 33 (8), 1201–1215.

Lécuyer, C. (1998). "Academic science and technology in the service of industry: MIT creates a 'permeable' engineering school". American Economic Review 88 (2), 28–33.

Lee, Y.S. (2000). "The sustainability of university–industry research collaboration". Journal of Technology Transfer 25 (2), 111–133.

Lee, W., Yang, W. (2001). "The cradle of Taiwan high technology industry development, Hsinchu Science Parks (HSP)". Technovation 20, 55–59.

Lenoir, T. (1997). Instituting Science. The Cultural Productivity of Scientific Disciplines. Stanford University Press, California.

Leslie, S.W., Kargon, R.H. (1996). "Selling Silicon Valley: Frederick Terman's model for regional advantage". Business History Review 70, 35–72.

Link, A.N., Scott, J.T. (2003). "U.S. science parks: The diffusion of an innovation and its effects on the academic missions of universities". International Journal of Industrial Organization 21 (9), 1323–1356.

Link, A.N., Scott, J.T. (2007). "The economics of university research parks". Oxford Review of Economic Policy 23, 661–674.

Lissoni, F., Llerena, P., McKelvey, M., Sanditov, B. (2008). "Academic patenting in Europe: New evidence from the KEINS database". Research Evaluation 16, 87–102.

Löfsten, H., Lindelöf, P. (2002). "Science parks and the growth of new technology-based firms-academic-industry links, innovation and markets". Research Policy 31 (6), 859–876.
Lowe, R.A. (2006). "Who develops a university invention? The impact of tacit knowledge and licensing policies". Journal of Technology Transfer 31, 415–429.
Macdonald, S. (1987). "British science parks: Reflections on the politics of high technology". R&D Management 17 (1), 25–37.
Mansfield, E. (1991a). "Academic research and industrial innovation". Research Policy 20, 1–12.
Mansfield, E. (1991b). "Academic research and industrial innovation: A further note". Research Policy 21, 295–296.
Mansfield, E. (1995). "Academic research underlying industrial innovations: Sources, characteristics, and financing". Review of Economics and Statistics 77, 55–65.
Mansfield, E. (1998). "Academic research and industrial innovation: An update of empirical findings". Research Policy 26, 773–776.
Martin, B.R. (2003). "The changing social contract for science and the evolution of the university". In: Geuna, A., Salter, A.M., Steinmueller, E.D. (Eds.), Science and Innovation: Rethinking the Rationales for Funding and Governance. Edward Elgar, Cheltenham, UK.
Martin, S., Scott, J.T. (2000). "The nature of innovation market failure and the design of public support for private innovation". Research Policy 29, 437–444.
Massey, D.B., Quintas, P., Wield, D. (1992). High-Tech Fantasies: Science Parks in Society, Science and Space. Routledge, London.
Meyer, M. (2003). "Academic patents as an indicator of useful research? A new approach to measure academic inventiveness". Research Evaluation 12, 17–27.
Meyer, M. (2006). "Knowledge integrators or weak links? An exploratory comparison of patenting researchers with their non-inventing peers in nano-science and technology". Scientometrics 68 (3), 545–560.
Meyer-Krahmer, F., Schmoch, U. (1998). "Science-based technologies: University–industry interactions in four fields". Research Policy 27, 835–851.
Mohnen, P., Hoareau, C. (2003). "What type of enterprise forges close links with universities and government labs? Evidence from CIS 2". Managerial and Decision Economics 24 (2–3), 133–145.
Monjon, S., Waelbroeck, P. (2003). "Assessing spillovers from universities to firms: Evidence from French firm-level data". International Journal of Industrial Organization 21 (9), 1255–1270.
Mowery, D.C., Sampat, B.N. (2001). "Patenting and licensing university inventions: Lessons from the history of the research corporations". Industrial and Corporate Change 10 (2), 317–355.
Mowery, D.C., Sampat, B.N. (2005). "The Bayh-Dole act of 1980 and university–industry technology transfer: A model for other OECD governments?" Journal of Technology Transfer 30, 115–127.
Mowery, D.C., Nelson, R.R., Sampat, B.N., Ziedonis, A.A. (2001). "The growth of patenting and licensing by US universities: An assessment of the effects of the Bayh-Dole act of 1980". Research Policy 30 (2), 99–119.
Mowery, D.C., Nelson, R.R., Sampat, B., Ziedonis, A.A. (2004). Ivory Tower and Industrial Innovation: University–Industry Technology Transfer Before and After the Bayh-Dole Act in the United States. Stanford Business Books.
Murmann, J.P. (2003). Knowledge and Competitive Advantage. The Coevolution of Firms, Technology, and National Institutions. Cambridge University Press, Cambridge.
Murray, F. (2005). "The oncomouse that roared: Resistance and accommodation to patenting in academic science", http://web.mit.edu/fmurray/www/papers/ Unpublished manuscript.
Murray, F., Stern, S. (2007). "Do formal intellectual property rights hinder the free flow of scientific knowledge? An empirical test of the anti-commons hypothesis". Journal of Economic Behaviour and Organization 63 (4), 648–687.
Murray, F., Aghion, P., Dewatripont, M., Kolev, J., Stern, S. (2009). Of Mice and Academics: Examining the Effect of Openness on Innovation. National Bureau of Economic Research, Cambridge, MA NBER Working Paper No. 14819.
Mustar, P. (1997). "Spin-off enterprises. How French academics creates high tech companies: Conditions for success or failure". Science and Public Policy 24 (1), 37–43.
Nelson, R. (1959). "The simple economics of basic scientific research". Journal of Political Economy 67, 297–306.
Nelson, R. (1986). "Innovation and public policy, institutions supporting technical advance in industry". American Economic Review 76 (2), 186–189.
Nelson, R.R. (1993). National Innovation Systems. Oxford University Press, Oxford.

Nelson, R.R., Rosenberg, N. (1994). American universities and technical advances in industry. Stanford Institute for Economic Policy Research, Stanford University CEPR Publication 342.

Nerkar, A., Shane, S. (2003). "When do start-ups that exploit patented academic knowledge survive?" International Journal of Industrial Organization, 21 (9), 1391–1410.

OECD. (2003). Turning Science into Business. Patenting and Licensing at Public Research Organizations. Organisation for Economic Co-operation and Development, Paris.

OECD. (2007). Science, Technology and Industry Scoreboard. Organisation for Economic Co-operation and Development, Paris.

Owen-Smith, J., Powell, W.W. (2001). "Careers and contradictions: Faculty responses to the transformation of knowledge and its uses in the life sciences". Research in the Sociology of Work 10, 109–140.

Phan, P.H., Siegel, D.S., Wright, M. (2005). "Science parks and incubators: Observations, synthesis and future research". Journal of Business Venturing 20 (2), 165–182.

Phillimore, J. (1999). "Beyond the linear view of innovation in science park evaluation, an analysis of Western Australian Technology Park". Technovation 19, 673–680.

PVA-MV. (2003). Report on the Abolition of the German Professors Privilege: Overview of Changes and Challenges. Vinnova—Swedish Agency for Innovation Systems, Stockholm.

Roberts, E.B. (1991). Entrepreneurs in High Technology. Oxford University Press.

Rosenberg, N. (2004). "Science and technology: Which way does the causation run?" http://www.crei.cat/activities/sc_conferences/23/papers/rosenberg.pdf. Center for Interdisciplinary Studies in Science and Technology, Stanford University.

Rothaermel, F.T., Agung, S., Jiang, L. (2007). "University entrepreneurship: A taxonomy of the literature". Industrial and Corporate Change 16, forthcoming.

Rothblatt, S., Wittrock, B. (1993). The European and American University Since 1800. Historical and Sociological Essays. Cambridge University Press, Cambridge.

Sampat, B.N. (2004). "Genomic patenting by academic researchers: Bad for science?" http://mgt.gatech.edu/news_room/news/2004/reer/index.html. Paper Presented at Fourth Annual Roundtable for Engineering Entrepreneurship Research (REER), Atlanta, GA, December 3–5.

Saxenian, A. (1985). "The Genesis of Silicon Valley". In: Hall, P.G., Markusen, A.R. (Eds.), Silicon Landscapes. Taylor and Francis.

Schmiemann, M., Durvy, J.-N. (2003). "New approaches to technology transfer from publicly funded research". Journal of Technology Transfer 28, 9–15.

Semlinger, K. (1993). "Economic development and industrial policy in Baden-Württemberg: Small firms in a benevolent environment". European Planning Studies 1, 435–464.

Shane, S. (2001a). "Technological opportunities and new firm formation". Management Science 47 (2), 205–220.

Shane, S. (2001b). "Technological regimes and new firm formation". Management Science 47 (2), 1173–1190.

Shane, S. (2002). "Selling university technology: Patterns from MIT". Management Science 48 (1), 122–137.

Shane, S. (2004). Academic Entrepreneurship. University Spin-Offs and Wealth Creation. Edward Elgar, Cheltenham, UK.

Shane, S., Khurana, R. (2003). "Bringing individuals back in: The effect of career experience on new firm founding". Industrial and Corporate Change 12 (3), 519–543.

Shane, S., Stuart, T. (2002). "Organizational endowments and the performance of university start-ups". Management Science 48 (1), 154–170.

Shinn, T., Lamy, E. (2006). "Paths of commercial knowledge: Forms and consequences of university–enterprise synergy in scientist-sponsored firms". Research Policy 35, 1465–1476.

Siegel, D.S., Thursby, J.G., Thursby, M.C., Ziedonis, A.A. (2001). "Organizational issues in university–industry technology transfer: An overview of the symposium issue". Journal of Technology Transfer 26 (1–2), 5–11.

Siegel, D.S., Westhead, P., Wright, M. (2003). "Assessing the impact of university science parks on research productivity: Exploratory firm-level evidence from the UK". International Journal of Industrial Organization 21 (9), 1357–1369.

Siegel, D.S., Waldman, D.A., Atwater, L.E., Link, A.N. (2004). "Toward a model of the effective transfer of scientific knowledge from academicians to practitioners: Qualitative evidence from the commercialization of university technologies". Journal of Engineering and Technology Management 21 (1), 115–142.

Slaughter, S., Leslie, L. (1997). Academic Capitalism: Politics, Policies, and the Entrepreneurial University. Johns Hopkins University Press.

Stephan, P.E., Everhart, S.S. (1998). "The changing rewards to science: The case of biotechnology". Small Business Economics 10, 141–151.
Stephan, P.E., Gurmu, S., Sumell, A.J., Black, G. (2007). "Who's patenting in the university? Evidence from a survey of doctorate recipients". Economics of Innovation and New Technology 16 (2), 71–99.
Stossel, T.P. (2005). "Regulating academic–industrial research relationship: Solving problems or stifling progress?" New England Journal of Medicine 353 (10), 1060–1065.
Stuart, T.E., Ding, W.W. (2006). "When do scientists become entrepreneurs? The social structural antecedents of commercial activity in the academic life sciences". American Journal of Sociology 112, 97–144.
Thursby, J.G., Thursby, M.C. (2002). "Who is selling the ivory tower? Sources of growth in university licensing". Management Science 48 (1), 90–104.
Thursby, J.G., Thursby, M.C. (2003). "Industry/university licensing: Characteristics, concerns and issues from the perspective of the buyer". Journal of Technology Transfer 28, 207–213.
Thursby, J.G., Jensen, R., Thursby, M.C. (2001). "Objectives, characteristics and outcomes a survey of major U.S. universities". Journal of Technology Transfer 26, 59–72.
Thursby, M.C., Thursby, J.G., Mukherjee, S. (2005). "Are there real effects of licensing on academic research? A life cycle view". http://www.nber.org/papers/w11497. NBER Working Paper No. 11497.
Thursby, J.G., Fuller, A., Thursby, M. (2007). "US faculty patenting: Inside and outside the university", http://www.nber.org/papers/w13256. NBER Working Paper No. 13256.
Varga, A. (2000). "Local academic knowledge transfers and the concentration of economic activity". Journal of Regional Science 40 (2), 289–309.
Vedovello, C. (1997). "Science parks and university–industry interaction: Geographical proximity between the agents as a driving force". Technovation 17, 491–502.
Verspagen, B. (2006). "University research, intellectual property rights and European innovation systems". Journal of Economic Surveys 20 (4), 607–632.
Veugelers, R., Cassiman, B. (2005). "R&D cooperation between firms and universities. Some empirical evidence from Belgian manufacturing". International Journal of Industrial Organization 23 (5–6), 355–379.
Vincent-Lancrin, S. (2006). "What is changing in academic research? Trends and futures scenarios". European Journal of Education 41 (2), 169–202.
Wagner, J.W. (2007). "Multiversity or University? Pursuing competing goods simultaneously", http://www.emory.edu/ACAD_EXCHANGE/ Academic Exchange 9/4.
Walsh, J.P., Nagaoka, S. (2009). "The R&D process in the US and Japan: Major findings from the RIETI-Georgia Tech Inventor Survey". http://hdl.handle.net/1853/27804 School of Public Policy Working Paper 48, Georgia Institute of Technology.
Walsh, J.P., Cho, C., Cohen, W.M. (2005). "View from the bench: Patents and material transfers". Science 309 (5743), 2002–2003.
Westhead, P., Storey, D.J. (1995). "Links between higher education institutions and high technology firms". Omega 4, 345–360.
Yusuf, S., Nabeshima, K. (2007). How Universities Promote Economic Growth. The World Bank, Directions in Development, Washington, DC.
Zucker, L.G., Darby, M.R. (1996). "Star scientists and institutional transformation: Patterns of invention and innovation in the formation of biotechnology industry". Proceedings of the National Academy of Sciences of the United States of America 93, 12709–12716. Colloquium Paper 93.
Zucker, L., Darby, M. (1998). "The economists' case for biomedical research". In: Barfield, C., Smith, B. (Eds.), The Future of Biomedical Research. AEI Press, Washington, DC.
Zucker, L.G., Darby, M.R., Armstrong, J. (1998a). "Geographically localised knowledge: Spill-overs or markets?" Economic Inquiry 36, 65–86.
Zucker, L.G., Darby, M.R., Brewer, M.B. (1998b). "Intellectual human capital and the birth of US biotechnology enterprises". American Economic Review 88, 290–306.

Chapter 7

PROPERTY RIGHTS AND INVENTION

KATHARINE ROCKETT*,†,‡

**Department of Economics*

University of Essex, Colchester

Essex, United Kingdom

†University of Tromsø

Tromsø, Norway

‡CEPR, London, United Kingdom

Contents

Handbooks in Economics, Volume 01

DOI: 10.1016/S0169-7218(10)01007-5

Abstract

We present a selective survey of the economic theory of intellectual property rights. After a brief description of the institutional framework, we discuss policy objectives and some basic welfare tradeoffs in intellectual property design. We consider the extent to which social objectives can be attained without intellectual property protection before passing on to intellectual property right design. We derive conclusions in the simplest, one-time innovation, case then investigate how these conclusions change when innovations build on each other or fit together as complements. Modifications of existing protection and optimal procurement of innovation are considered. Finally, we sketch enforcement and competition policy issues.

Keywords

incentives, innovation, intellectual property rights, patents, procurement

JEL classification: O3, D23, H41, K11, L51

1. Introduction

This chapter presents a selective survey of the literature on the theory of intellectual property and innovation incentives. We will outline the broad effects that have been identified with an eye to presenting a good range of modeling styles and issues. With regret, length considerations mean that we only cover patents.

We start by sketching the patent right and the associated institutional background that we will refer back to later when discussing the models. Next, we pass to some broad modeling issues, including the goals of the patent system and how they have been translated into objectives for policymakers. We outline basic theories of intellectual property design and the welfare tradeoffs they suggest. As a benchmark, we briefly consider how far we can get toward our social goals without intellectual property protection or, alternatively, how the incentive to innovate can be conceptualized when actors opt for secrecy as a means of protecting intellectual property. Next, we will look at the issues in patent design that have been investigated over the last 40 years or so. We start with the simplest case of a single, one-time innovation and then look at how our conclusions on design change when we consider innovations that either build on each other or fit together as complements. We start from models that consider a system of intellectual property protection that is quite similar to existing institutions, then move on to mechanisms that start from something closer to a "blank slate" to optimally procure innovation. We consider issues of enforcement and the interplay between competition policy and intellectual property policy briefly at the end of the survey.

2. Brief sketch of the patent right

To set the stage, we elaborate here some of the salient features of the patent right that underlie the models that follow.[1] The United States and Europe will be our focus. Indeed, the differences between the United States and Europe illustrate the range of policy tools that can be brought to bear on patent design. Further, the differences add up to a somewhat different patent right on the two sides of the Atlantic, with a tighter, more expensive, and more industrially oriented version in Europe.

2.1. What is a patent?

A patent[2] refers to a temporary property right on an invention. The patent provides a right—but not a guarantee—to exclude others from making, using, or selling the patented property. Indeed, the patent holder generally has no obligation or necessarily even the right to practice the innovation. For example, if inventor A is granted a patent, where the exercise of that patent would infringe the patent rights of inventor B, inventor A has no automatic authorization to exercise her patent. Her right is dependent on

[1] A more complete description of the patent right, including its history, can be found in Scotchmer (2004), Guellec and van Pottelsberghe (2007), and Jaffe and Lerner (2006).

[2] See Guellec and van Pottelsberghe (2007) and Jaffe and Lerner (2006) for more detailed discussion of the philosophical differences between European and US views of patents. An argument for a constitutional underpinning of patent rights is contained in Nard and Morriss (2006).

the patent rights of B.[3] Furthermore, the exercise of patent rights must fall within other laws, such as antitrust laws. In the same way as an individual may own a gun but not be allowed to shoot people with it, an inventor may be granted a patent but not be allowed to use it entirely as she pleases: she must do so within the scope of the general body of law.

In exchange for these exclusionary rights, the patent holder must disclose the invention as part of a publicly available patent document. In the United States, this public disclosure must be such that a person with sufficient background knowledge ("skilled in the art") could make or use the innovation in its "best mode" at the time of filing. Of course, for patents filed relatively early in the development cycle the "best mode" may be quite rudimentary, lacking many of the improvements that make the invention economical to exploit. While there are some differences in how it is interpreted between Europe and the United States,[4] the disclosure should be viewed as broadly helpful to third parties wishing to understand the nature of the innovation. While the embodiment of the innovation is protected by the patent, the underlying idea is not. Furthermore, the idea should be—and generally is—relatively transparent in the disclosure.

The features of the innovation are described in a set of claims, which define the metes and bounds of the patent. Features not claimed are not covered by patent protection. While claims generally are interpreted as real and proven features of the innovation, the distinction between real and suspected features can be difficult to establish.[5] The *ex post* interpretation of claims in rapidly developing fields, where changes in the dominant approach affects the interpretation of claims, may be challenging despite efforts made to write them clearly at the time of filing.

Patentable subject matter is varied. Examples could include a process or product, a composition of matter (such as a chemical composition) or machine, or a new and useful improvement on any of these. Indeed, as a result of a series of court decisions[6] patentable subject matter in the United States has broadened over the past 30 years to include the products of genetic manipulation, software and business methods. Indeed, "anything under the sun made by man" could potentially be patentable according to one decision.[7] Patentable subject matter in the United States remains relatively broad compared to other countries, despite extensions that have occurred elsewhere. This partly stems from some differences in the general philosophy of patentable subject matter, tending more toward technicality and industrial applicability in Europe than in the United States. These differences have been cited as resulting in the slower movement in Europe toward allowing patents in areas such as business methods, genetic material and surgical methods.[8]

[3] In other words, an innovation can be patentable but still infringe another patent.

[4] See Guellec and van Pottelsberghe (2007), especially pp. 39–41.

[5] See Guellec and van Pottelsberghe (2007, p. 139) and Bidgoli (2010).

[6] See, for example, *Diamond v. Chakrabarty*, 447 U.S. 303 (1980), *Diamond v. Diehr*, 450 U.S. 175 (1981) and *State Street Bank & Trust Co. v. Signature Financial Group, Inc. 149 F.3d 1368 (Fed. Cir. 1998)*. Commenting on and restricting the *State Street* criteria, *In re Bernard L. Bilski and Rand A. Warsaw 545 F.3d 943, 88 U.S.P.Q.2d 1385 (2008)* specifies that a "claimed process is surely patent-eligible under Section 101 if (1) it is tied to a particular machine or apparatus, or (2) it transforms a particular article into a different state or thing."

[7] Contained in *Diamond v. Chakrabarty*, 447 U.S. 303 (1980).

[8] For a discussion of business method patents in practice in Europe, see Harhoff and Wagner (2006). For a more general discussion of trends in patentable subject matter in Europe, see Guellec and van Pottelsberghe (2007), especially pp. 119–132.

In most patent systems, a patentable innovation must represent a significant innovative step. Indeed, one of the distinctive features of a patent compared to other forms of intellectual property protection is that patent systems have often been thought of as applying to relatively large advances. In other words, the strong exclusionary patent right is granted only in exchange for the disclosure of "valuable" information. In point of fact, the requirement of significance differs across countries and has differed over time. In the United States, a patentable innovation is required to be nonobvious as well as new. Judging whether an innovation indeed satisfies the conditions of patentability is a main role of the US Patent and Trademark Office (USPTO). In contrast, some patent systems have traditionally been mere "registration systems," where little or no screening is done to weed out little from big steps. Jaffe and Lerner (2006) document a recent trend in the United States toward weaker requirements of "significance" due to workload, incentive system, and other pressures at the USPTO. Hence, the concept of significance cannot be considered static but instead responds to intentional or unintentional changes in patent approval practice. The European Patent Office (EPO) has attempted to unify for European states the inventive step that is required to qualify for a patent. This requirement has varied across states from a "scintilla of invention" to a relatively high standard of novelty and nonobviousness. In both the United States and Europe, the evaluation of significance is generally thought of as an evaluation of how big a step the innovation makes in a scientific or technical sense. Commercial success can be used *ex post* only in a limited way as an argument that an innovation was significant, with the United States viewing this sort of evidence somewhat more favorably than Europe.[9]

Once granted, a patent may be exercised, traded (sold or "rented" via a licensing contract, or otherwise transferred), or abandoned, like other forms of property. Indeed, contracts of trade are very common, amounting to somewhere between 10% and 20% of the patents issued.[10] While it is common for licenses to be agreed *ex post*—after a patent has issued—this is not the only timing that is observed. Licenses can, in principle, be agreed before discovery or even before investment in a research path has begun. These sorts of prospective or *ex ante* agreements specify sharing arrangements for any patents that *might* issue as a result of a research program. A wide variety of pricing arrangements from no price at all to upfront fixed fees, per-unit or revenue-based royalties, profit shares, reciprocal trades of other patents, or other in kind payments can be observed in licensing agreements.[11] Even if quite standard pricing schemes, such as simple royalties, are specified, contracts can vary as to how these payments are spread over time.

Patents differ from many other forms of property in the sense that they are temporary and not permanent rights. Patent protection lasts a statutory maximum of 20 years from the date of filing. In the United States, this represents a change from the 17 years that were available from the date of grant prior to the 1994 agreement on Trade-Related Aspects of Intellectual Property Rights (TRIPS). Indeed, the patent term has varied over time in a wide number of countries.[12] Statutory protection need not last this long, however, as periodic renewal payments often are required to maintain the right up to its statutory maximum term. Patents frequently are allowed to lapse. Only about 8% of all patents go to full

[9] See Guellec and van Pottelsberghe (2007, p. 137).
[10] See Guellec and van Pottelsberghe (2007, p. 92).
[11] See Anand and Khanna (2000) for observed license contract structure.
[12] See Jaffe and Lerner (2006), especially pp. 82–94.

term in Europe, although considerably more reach term in the United States.[13] This discrepancy is likely due in part to the differences in the maintenance expenses involved. When translation, maintenance, processing and external (including legal) costs are included, a European patent valid in all member states could cost ten times more than a US patent for a 20-year term.[14] Extension of the patent term beyond 20 years is clearly much more difficult, and can only be attempted by indirect strategies.[15] A "continuation," whereby modifications of an original patent application can be filed over time, can be used to attempt something like a term extension in the United States. Continuations constitute a significant amount (about a third) of US examiner workload at the moment, so they are not rare occurrences.[16] Indeed, abuse of this system has been discussed recently by Jaffe and Lerner (2006).[17] In Europe, the closest proxy of term extensions is probably to exploit so-called "dependent" patents. Generally, however, the scope for this sort of behavior in Europe appears more limited than in the United States.[18]

2.2. *Patent agreements and administration*

The basic system of patents that is used in the United States is set forth in Article 1, Section 8 of the US Constitution, where Congress is granted the power "To Promote the Progress of Science and Useful Arts by securing for limited Times to Authors and Inventors the exclusive Right to their respective Writings and Discoveries." Under this power, Congress has enacted various patent laws, the first in 1790, with various revisions added over time. These are codified in title 35 of the United States Code. Congress has also created the USPTO to administer these laws and perform other duties related to patent protection. More recently, patent laws in a variety of countries, including the United States, have been aligned under the TRIPS agreements. Other key treaties have included the Paris Convention (which specifies that a first patent filing date in any member state can serve as the patent application date for any subsequent member state filing), the European Patent Convention (which establishes the EPO as a means of coordinating patent grants within Europe), and the Patent Cooperation Treaty (which establishes a uniform procedure for filing patent applications in the member states).

In Europe, patents may issue from either the patent office of individual countries or from the EPO, or both. In fact, a common practice is to file at the EPO only after having filed at the patent office of a

[13] See Guellec and van Pottelsberghe (2007, p. 148) for European figures. For the United States, the percentage reaching term appears to be about a third, see Lemley (2001).

[14] See Guellec and van Pottelsberghe (2007, Chapter 7) for an extensive comparison of operations at the USPTO, EPO, and Japanese Patent Office (JPO). Under the London Agreement, which came into force in May 2008, translation fees are anticipated to fall considerably.

[15] A brief scan of the web leads to hits on a variety of ideas to "game the system" to obtain effective extensions. http://www.mewburn.com/Patent/US_Patents:_Term_extensions.htm, accessed July 29, 2008 for one such example.

[16] See http://www.uspto.gov for details. For a more complete discussion of strategic uses of continuations, see Lemley and Shapiro (2005). Hedge et al. (2007) document empirically that, while certain patent procedures are used disproportionately by certain types of firms and values of patents, the use of continuations has fallen overall in more recent years. Empirical analysis of European patent process strategic issues is documented in Graham and Harhoff (2006).

[17] More generally, strategic (ab)use of filing procedures has been documented in Van Zeebroeck and Van Pottelsberghe (2008) and references therein.

[18] See Guellec and van Pottelsberghe (2007, p. 145).

specific European country or the United States.[19] The administrative body responsible for implementing patent grants, such as the USPTO or the EPO, reviews patent applications to determine if the candidate invention satisfies the minimum standards for patentability: novelty and nonobviousness being the most salient characteristics. If the candidate technology is determined not to satisfy this minimum standard, the patent can be denied. A patent can issue after the review process has concluded.[20] This often takes years, with about one out of three patent applications being finally rejected in the United States.[21] While the period of exclusivity starts at the date of issuance or grant, of the patent, the patent disclosure—the information on the nature of the innovation—is now published 18 months after the initial filing with the patent office (although this is not required if the applicant declares the intent not to request coverage outside the United States). This represents a change in the United States compared to the older system of being published only at the time of patent grant, which was in force prior to the implementation of the American Inventors Protection Act of 1999.

2.3. Enforcement

Patents are only as strong as their enforcement. Enforcement is handled privately in the United States (for the most part in civil suits) through the court system. For example, if a patent holder detects infringement (that is, unauthorised manufacture or use of patented material) within the jurisdiction of the patent, the patent holder can sue the violator in court. Infringement suits tend to be extremely expensive[22] and, indeed, can constitute a substantial proportion of total research expenditure.[23] Further, the cost can include a substantial joint loss of wealth rather than a simple transfer from infringer to patent holder (Bessen and Meurer, 2008b). If the court finds for the patent holder it may impose an injunction on the violator prohibiting sales of the infringing item and may impose monetary compensation of another type, such as damages. Indeed, a temporary injunction may be imposed even before this. Of course, the defendant can countersue as well. A common response to an infringement suit is a counterclaim that the infringed patent was not valid in the first place. Overall, few patents—on average 1.5% of all patents granted—are ever litigated, and fewer still—on the order of 0.1%—ever go to trial.[24] Those that are litigated appear to be the

[19] This may be done in order to obtain several chances at obtaining a valid national patent. As the EPO tends to produce feedback on the patentability of the innovation slowly compared to national offices or the United States, this strategy also has the advantage of providing valuable early feedback to the applicant as to whether the application is worthwhile to pursue. Furthermore, as the USPTO allows more substantial changes to be introduced to the patent document after filing, it may make sense to file at the USPTO early and then file a better-crafted document at the EPO later. See Guellec and van Pottelsberghe (2007, pp. 155–159) for patent filing routes.

[20] Other possible outcomes, which differ between the United States and Europe, include withdrawal or revision of the document. See Guellec and van Pottelsberghe (2007) or Jaffe and Lerner (2006) for details.

[21] See http://www.uspto.gov/web/offices/com/speeches/07-46.htm, for example. From this press release, it is clear that there is considerable variance across years included in this figure. See also Harhoff and Wagner (2005). Ebert (2004), commenting on the role of patent continuations in the US system, obtains a somewhat lower adjusted rejection rate of 1/4.

[22] See Bessen and Meurer (2008a) for estimates.

[23] Lerner (1995) estimates that cases begun in 1991 in the United States would eventually incur legal expenditures, in 1991 dollars, of about 27% of all of the basic research expenditure in the United States in 1991.

[24] See Lanjouw and Schankerman (2001) and Lemley (2001). In other words, about 95% of litigated patents settle out of court.

high value patents and those drawn from a subset of particularly litigious technology areas.[25] The success rate for patent holders in trials has risen over time,[26] a change often attributed to the creation of the unified Court of Appeals of the Federal Circuit in 1982.

The European approach relies much more on an oppositions system to weed out "bad" patent grants relatively promptly. Third parties may submit opinions during the patent examination—to a limited extent—on whether a patent grant should be made and to a greater extent during a centralized postgrant oppositions procedure.[27] This is a less expensive route to challenge patents in terms of both time and money than full litigation in a court of law and is accessed somewhat more frequently than litigation in the United States: somewhat over 6% of all issued patents are challenged in this way.[28] Even though few granted patents go through oppositions, the system and the potential to go through oppositions is cited as a pillar of quality control at the EPO, allowing bad patents—those that do not "in truth" satisfy patenting criteria—to be screened out shortly after issue.[29]

Aside from the oppositions procedure, a second difference in enforcement between Europe and the United States is that an EPO patent is, in fact, a bundle of national patents (in the countries designated by the applicant). National laws apply to these patents, with any legal challenges being pursued at the national level. In other words, there is no unified court to deal with patent disputes for the whole of Europe. This can make for multiple litigations in many European jurisdictions with potentially contradictory outcomes.[30]

A final area of difference across the Atlantic is what constitutes infringement. A "doctrine of equivalents" may be used to judge whether infringement has occurred even if the infringing item is not a perfect replica of the patented invention. In other words, the looser the interpretation of what constitutes the invention, the greater the "breadth" of the patent protection. Further, an "experimental exemption" allows use of the patented technology for research purposes. Europe has had a more restrictive interpretation than the United States on the doctrine of equivalents, but has tended to take a broader view of the research exemption when it comes to university research. This helps explain why infringement decisions for similar types of patents sometimes differ in Europe and the United States.[31]

[25] See Lanjouw and Schankerman (2001, 2004) and Lerner (1995) for statistics and discussion. While some industries have litigation rates as high as 6%, Lanjouw and Schankerman find, in fact, a much lower average rate of litigation at less than 2%. See also Scotchmer (2004, Chapter 7) for an overview.

[26] See Jaffe (2000). Allison and Lemley (1998) find that, of the patents litigated to a final determination in their dataset, 46% are held invalid.

[27] Hall et al. (2003) and Harhoff and Reitzig (2004) document the use of this system. See also Guellec and van Pottelsberghe (2007, p. 176).

[28] See Harhoff and Reitzig (2004) for figures and detailed discussion of EPO oppositions. The frequency of opposition seems to have fallen modestly over time.

[29] See Guellec and van Pottelsberghe (2007, p. 178).

[30] The draft European Patent Litigation Agreement aims to centralize litigation, but has not yet been adopted. For more information see http://www.epo.org/patents/law/legislative-initiatives/epla.html.

[31] An important recent decision involving a change in the interpretation of the doctrine of equivalents in the United States is *Festo Corp. v. Shoketsu Kinzoku Kogyo Kabushiki Co.*, 535 U.S. 722 (2002). See also Weston (1998) for a comparative treatment of the doctrine of equivalents in Europe and the United States. EU Directives 2001/82/EC Articles 13(1) to 13(7) and 2001/83/EC Articles 10(1) to 10(5) attempt to harmonize European approaches to the experimental exemption. For discussion of cross-Atlantic differences in the treatment of university research, see also Guellec and van Pottelsberghe (2007, pp. 190–191). A recent and significant case on the experimental exemption is *Madey v. Duke University, 307 F.3d 1351 (Fed. Cir. 2002)*, summarized in Janis (2003).

2.4. Summary

This brief sketch of the patent right suggests a number of policy levers that can be exercised in order to affect the innovation incentives that actually derive from the patent system. Clearly, one set of policy levers applies to the design of the patent right itself: notably the statutory length of protection, how broadly we interpret the exclusive right, what is disclosed and when this occurs, the size of the inventive step required to earn protection, and what constitutes patentable subject matter. Intervention exercising these levers has occurred in both the United States and Europe over time, resulting in a process of evolution of the patent right.[32] A second set of levers applies to the administration of the patent, including how procedures implement the general aims of patent protection. These include the structure of the review process, the structure of fees, the incentives of patent examiners, and other administrative features.[33] Again, all these features have evolved over time. A third set of levers applies to the enforcement of the patent in court and includes success rates for patent holders in infringement suits, the types and sizes of remedies imposed, and the fora in which patent defense and attack can occur.[34] A final set of levers applies to the freedom with which patent holders can exercise their rights under other bodies of law. A main case in point is the effect of competition policy on the amount of profit that can be extracted from the patent right.[35] This could vary from simple limits on excessive pricing to limits on the ability or necessity to license as well as the licensing contract structure. The literature has examined interventions at all these levels and, in some cases, interactions between different levers. A case in point is the interaction between enforcement methods and patent quality, which will be discussed later. Translating the available administrative and legal policy levers into features of economic models that accurately reflect these levers is, of course, a major challenge. We now turn to how this challenge has been addressed.

3. Economic interpretation of the patent process

We will now use the US system to outline issues in the interpretation of the patent process that determine the basic building blocks on which the rest of the modeling rests. The issues we deal with are the objective function of policymakers, the source of value of the patent, and the basic functions of the patent that determine its private and social welfare effects. By and large, these basic features are shared with non-US patent systems, so the focus on the United States is for expositional convenience only.

[32] For example, the recent patent term extensions in the United States in accord with the TRIPS agreements and the somewhat slower evolution of patentable subject matter in Europe versus the United States.

[33] See Jaffe and Lerner (2006) and Guellec and van Pottelsberghe (2007) for a detailed discussion of change in administrative procedures in Europe and the United States.

[34] Allison and Lemley (1998) and Jaffe and Lerner (2006) discuss how the move to a unified court of appeals in the United States has changed success rates of litigation and litigation strategy.

[35] A simple example to keep in mind for later parts of this chapter could be strict enforcement of EU Article 82(a), which forbids "excessive pricing" by a dominant firm.

3.1. Interpreting the goal(s) of patent policy

Article 1, Section 8 of the US Constitution is quite explicit that the objective of the intellectual property rights system is the progress of "Science and the Useful Arts."[36] If one were to take this at its word, one would not necessarily want to use social welfare—or even economic growth—as the standard of optimality in a model of the intellectual property rights system. Instead, one might wish to use the rate of innovation or, less directly, the rate of research and development spending: the more the better. The interpretation one takes is important to the conclusion one reaches about the optimality of any intellectual property protection system. For example, Horowitz and Lai (1996) compare the optimal design of patents when the objective is to maximize the rate of innovation to the optimal design when the objective is to maximize discounted consumers' surplus. A system that aims to maximize consumers' surplus places more value on frequent innovation than a system that maximizes the rate of innovation since intermediate steps generate surplus gains as each quality step enters consumption. Despite the ambiguity in how one should interpret the goal of establishing a system of intellectual property rights in the first place, however, the bulk of the economics literature has taken social welfare to be the appropriate objective that is maximized by policymakers.

A second issue of interpretation of Article 1, Section 8 is how we understand "progress" in "Science and the Useful Arts." Most models capture the significance of patented innovation by some value which is created for society. In some models this value is interpreted as a private market value.[37] However, the actual patent approval process does not make such a direct link between commercial and scientific or technical value. Indeed, the patent document and the review process explicitly identify a technical value and the source of that value, as well as "usefulness," but do not go farther to determine any kind of monetary value. Therefore, the patent office makes no direct judgment at the time of grant on market or any other private value to the inventor, has no particular expertise in this area, and does not make market value an explicit criterion for patentability.[38] Protection is not tailored *ex ante* to such a notion of market value.

While the link between value that could be captured by a profit-maximizing firm and value in terms of technical progress could, in principle, be quite tenuous, as a practical matter they are more closely linked. Since patents are expensive to obtain inventors who are concerned with their own profits would not apply for patents if they anticipated no resulting private commercial value. This leads to the interpretation, taken by Scotchmer (1999) and Cornelli and Schankerman (1999), that inventors have—and seek—information about whether or not a patent will generate private market value even if the patent office has—and seeks—little information on this count. The inventors' information is revealed by their patenting behavior. In particular, inventions with higher private value may precisely be those that are observed to be patented and where that patent is observed to be renewed despite

[36] Not all patent systems have a constitutional basis, so in this sense a focus on the US system is somewhat special. While we will not pursue the consequences of this institutional feature of the US patent system, Nard and Morriss (2006) argue that constitutional patent law strengthens the bargain between the state and inventor compared to systems such as patronage. See Jaffe and Lerner (2006) and Scotchmer (2004) for histories of the patent systems and their legal bases.

[37] For example, Gilbert and Shapiro (1990) interpret this value as a flow rate of profits.

[38] As was pointed out earlier, however, the commercial value of an innovation can be brought in a limited way in some jurisdictions as evidence *ex post* that an innovation was a "nonobvious" or "useful" step.

renewal fees. Hence, (continued) patent protection is correlated with innovations that have higher private value in the eyes of their inventors.[39]

Finally, regardless of the *magnitude* of the invention's contribution, there is a question of *when* the value is realized for society or for the inventor compared to when the patent right is awarded. La Manna (1994) points out that if the patent right is awarded early, before much of the expenditure to develop the innovation has been incurred, then the exclusionary right assures the patent holder that she can reap the entire reward to its expenditure before that expenditure is incurred. If the right is granted late, and many firms may compete for that right, then the potential patent holder only faces an expected benefit at the time of her research investment. The difference between these two scenarios can affect the incentives to invest, as the patent holder is in a "race" for the right to the fruits of its investment in the latter case but is not in the former case.

More generally, the issue of "when" a stream of social benefits is created also turns on "how" the benefits are created. The social and private value of a patent need not flow directly from the technology that is patented, but may be largely derivative. Value may flow primarily from the innovations a patented advance inspires ("follow-on innovations") or from companion innovations that are used together to create a valuable product ("complementary innovations"). In both these cases, a single patent in isolation may have no private value at all. Indeed, a case we will examine later is that of "pure research tools": innovations that have technical value but no monetary value in isolation. In such cases, a main function of the patent right is to facilitate the transfer of value via licensing contracts from the follow-on innovations or the complementary innovations back to the holder of a "key" patent. Indeed, Hall (2007) presents evidence that patents do actually facilitate such trade in intellectual property.

3.2. The reward and contract theories of patents

Article 1, Section 8 goes on to specify a method that should be used to achieve its stated goal of the "Progress of Science and the Useful Arts." Specifically, a system of exclusive rights *to make*, *sell*, and *use* the innovation is granted for a limited time. There are two main ways one can imagine that the patent right could promote the progress of science.

The first mechanism is the establishment of a private reward for innovation. This is sometimes called the "reward theory" of patent protection and is presented in the classic work of Nordhaus (1969). The argument is that by generating potential monopoly power—and thus patent monopoly rents—exclusivity provides remuneration for successful innovators. If the cost to generate an innovation is privately borne, then the anticipation of such private compensation is a necessary "reward" to induce innovation in a market setting with profit-maximizing agents. If exclusive rights were not available to the innovator and if the underlying knowledge were a pure public good, any party could use this information to duplicate the invention and compete with the patent holder to provide it to purchasers. This kind of competition could reduce the rewards to innovation to the point where it would not be worthwhile to conduct the activity in the first place. Hence, the patent system promotes innovation that would otherwise be underprovided by the market due to a positive informational externality.

[39] Many papers have examined patent value and its correlates, see Bessen (2006) and references therein.

Consider first the "classic case" where a single inventor has exclusive rights to supply an invention that is deemed useful. This inventor is, then, a monopolist over some (residual) demand curve. If the inventor sets a single price, as a monopolist, it can earn profits labeled π in Figure 1. This is the private "reward" for the inventive effort. Of course, these monopoly profits come hand in hand with consumer's surplus, s, but also a deadweight loss, d, created by the monopoly pricing. Hence, there is a social cost to ensuring the reward to innovation. The private value captured by the inventor is less than the social value created by the innovation: only by awarding the entire social surplus, the triangle $W = (\pi + s + d)$, could firms' incentives be brought in line with society. Hence, the incentive to generate scientific progress, while positive, is socially too low in such a system, creating a dynamic welfare loss.

Despite this argument, one cannot conclude, in fact, that a system of temporary exclusion rights necessarily creates socially insufficient incentives to innovate. Exclusivity also generates forces that can create socially excessive incentives to innovate. The patent right designates no single party that is allowed to *attempt* the innovation: anyone can potentially compete for the intellectual property right and the benefits derived from it. In fact, if there are several potential innovators who can compete for the right to earn the "reward" to innovation, there may be socially too strong incentives to invest in innovation. Each potential innovator has the incentive to win the prize to "steal business" from its rival. In other words, the competitors for this prize do not take into account a negative externality that each exerts on others when making an effort to win, leaving losers with nothing to show for their—privately and socially costly—efforts. Hence, even in the context of a single innovation if there is more than one potential innovator, the market may generate socially excessive incentives to innovate.

More formally, and following Scotchmer (2004), consider a case where two firms are potential researchers. Successful innovation by either firm generates social value W, but there is no additional social value generated by duplicative innovation. Let the prize a firm wins for being the sole innovator be π. Suppose that innovation is probabilistic, so investment generates a probability, p, of successful innovation for the investor and a probability $1 - p$ of failure. Of course, it could be possible that both would simultaneously be successful. In this case, the prize is split evenly between the two firms. In this framework, the contribution of each additional researcher to social surplus, when the success probabilities are independent and uncorrelated, is

$$p(1-p)W.$$

Having a second researcher is only valuable if the first researcher fails but the second is successful. If this exceeds the incremental cost of research of an additional researcher, then *society* would benefit from the effort of a second researcher. The private contribution, which determines whether a single researcher actually enters or not, is the reward to being the sole winner in the event of failure by the other researcher plus the reward in the event of joint success:

$$p(1-p)\pi + p^2(1/2)\pi = (p - p^2)\pi + p^2(\pi/2).$$

If this exceeds the incremental cost of research, then a private firm would benefit from entry.

There are two differences between these two expressions. First, the value is W in the former and only π in the latter. In this sense, there is an underincentive to conduct research. Now, refer back to Figure 1 and set $W = \pi$, so that π reflects the entire social surplus of the innovation. The first term in the private incentive is now the same as the social benefit; however, we also have a second term in the private incentive reflecting the gain to any one inventor when both firms invest and are "lucky." This raises the

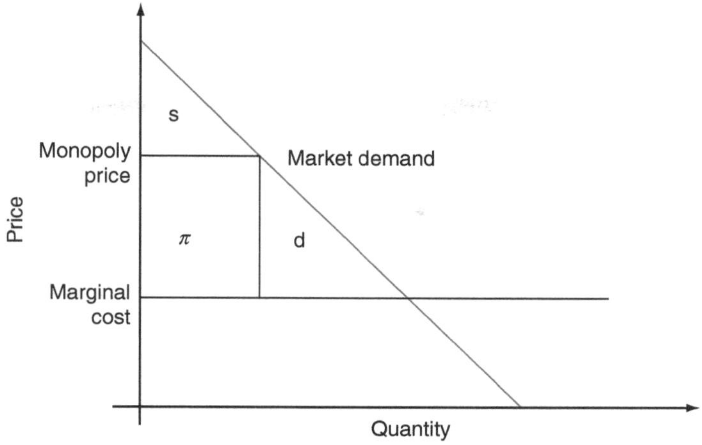

Figure 1. Profit, consumer surplus, and deadweight loss from monopoly.

private incentive for the inventor. An analogous term does not enter into the social benefit because society does not obtain any more surplus when both firms succeed than when one succeeds: society only cares that someone—anyone—makes the discovery. In contrast, individual inventors care very much who succeeds: the winner obtains a prize whereas the loser obtains nothing. Hence, the private incentive exceeds the social incentive due to this second term. In fact, if the research strategies are perfectly positively correlated (so that success by one firm always accompanies success by the other firm) there is no benefit at all to society of adding a second researcher. On the other hand, there is a positive private benefit for each individual firm since each stands an even chance of receiving the prize. Hence, the patent reward for innovation can create incentives to invent, but these incentives can—in general—either exceed or fall short of the social optimum.

A conclusion from this line of reasoning is that the basic assumptions we make about entry conditions into new research trajectories are likely to affect the conclusions we reach about whether incentives to invent are likely to need to be increased or decreased. If the trajectories are publicly known and the resources to pursue them are freely available, we may be rightly concerned about excessive incentives to innovate. If ideas are revealed to innovators in a unique way, then the concern about socially excessive entry into a "common pool" may be irrelevant. We may need to increase the incentives to invent in this "private information" case.

There is a second mechanism by which the exclusionary rights of the patent can create benefits from invention and "promote the progress of science." When innovations are created, so is information. This can be specific information on the nature of the innovation or it can be in the form of showing that a particular approach to a problem is possible and fruitful. Creating this information is privately costly, but socially useful since it can facilitate innovation by others. The usefulness may be in entirely different fields or markets, so the reward to private innovation need not fall below the investment cost. Still, there is a positive externality exerted by the creators of information. This drives a wedge between the private and social incentives to exert inventive effort, suggesting that information provision needs to be encouraged as it will tend to be underprovided in the market.[40] In particular, even if adequate rewards are available to induce the creation of information, it may be held secret.

The diffusion of information is an explicit goal of the patent system. The "contract theory" of patenting views patents as "contracts" between inventors and society where the patent right is granted in exchange for diffusion of the information that is useful to imitators, followers, or others who exploit the information to generate further social gains.[41] The patent documents create a—freely accessible—library of such information. Even if the rewards to patents and secrecy were the same for the original innovators (so that the same incentive to create information exists) disclosure considerations could make the patent system superior due to the benefits to others of the "library". Of course, this argument relies on an assumption that systems that *allow* secrecy actually *generate* secrecy and that the patent library generates accessible and interpretable information. We will consider these issues below.

Summarizing, the "reward" and the "contract" theories of patents form the underpinning of how patents generate scientific progress. Reward and information benefits can be seen as underlying both single-innovation models of patent design and models of multiple innovations, which we review below but their weighting determines the focus and design of the model. Before discussing these, however, we look at a benchmark where we assume that no intellectual property protection is available. In this framework, we will see under what conditions we can, indeed, generate innovation incentives despite the lack of protection.

4. Incentives to innovate in the absence of intellectual property protection

Is a system granting exclusive rights to innovators necessary to generate a reward or disclosure? Let no intellectual property right exist. Further, as soon as an innovative product is sold or used let a variety of individuals become familiar with the invention, creating the seeds for imitation. If the innovation generates profits, potential imitators are attracted to the innovation to produce their own versions of it. This process creates a variety of suppliers of the innovative product or process, driving down its price and so the profits of the original innovator. If this process is quick or very cheap, then very little surplus is captured by the initial innovator. Indeed, if the cost of developing the innovation in the first place was privately borne, the rapid imitation can reduce the benefits from innovating below the original cost. Any innovator anticipating this process will not invest in the innovation in the first place. In essence, the innovator contributes to a common pool of knowledge when she creates and practices an innovation. This positive externality, if it is not captured by the inventor, generates a private underincentive to innovate. The patent resolves this problem by making the embodiment of the innovation—in other words, the "object" that is actually traded in the marketplace—a private good even though the underlying knowledge remains a public good.

Many objections have been raised to this argument. These focus on its underlying assumptions that monetary gain is the motivator for invention, and that the imitation process is quick, costless, and purely

[40] Before, $W > \pi$ due to monopoly pricing that caused a deadweight loss. Here, $W > \pi$ even if pricing is not an issue.

[41] See Denicolo and Franzoni (2004) for discussion. Also see Eisenberg (1989) and Miller and Davis (1983) for more complete presentations of legal theories of patent rights linked to diffusion and the "natural rights" doctrine. In an early paper, Arrow (1962) pointed to patents as a way of encouraging information disclosure.

duplicative. A related point is that the patent system may be the wrong solution to the problem, to the extent that a problem exists at all. We will address only the former points here, leaving optimal design issues to later in this chapter.

A common concern is that many inventors do not care about monetary gain and are instead altruistic givers to society who place utility value on the creation of this externality or simply place utility value on the process of creation.[42] Of course, since the patent system is voluntary, the fact that rewards are not necessary to induce the creation and sharing of innovation is not really an objection to the system: inventors are always free to donate their innovations to society. Furthermore, while the pleasure of creation is enough to generate prototypes, it may not be enough to generate the much larger development and commercialization expenditure necessary to bring the innovation to consumers.

Alternatively, private incentives to invent for individual inventors may derive mainly from signaling to the job market. While some means of attributing innovation to its creator is necessary for this mechanism to work (e.g., there may be slow spread of information about the innovation to other job market candidates who could claim authorship), patent protection is only indirectly relevant to innovation incentives in this case. Indeed, depending on the competitive structure of the job market, one could generate either socially insufficient or socially excessive incentives to innovate as individuals compete to develop and showcase their talent. Lerner and Tirole (2002) and Lakhani and Wolf (2005) suggest that open-source code developers often invent to develop their own skills.[43] If this private benefit exceeds the cost of innovation then the innovation will be provided despite the external benefits it might confer on others—and regardless of intellectual property protection—even though social and private incentives might remain not fully aligned.

While some argue that the patent system is not necessary, others argue that even if it is necessary it is not very effective. Survey evidence of Cohen et al. (2000) has indicated that managers do not view patents as very effective at generating direct rewards to innovation. While certain sectors, such as biotechnology and pharmaceuticals, appear to get great benefit from patents, various first mover advantages (such as learning by doing) are credited with generating greater reward to innovation than intellectual property rights. If firms rely on other "frictions" such as barriers to entry to generate profits from invention, patents may at best be redundant. On the other hand, Farrell (1995) argues that the "honeymoon" period of patent protection may allow these other potentially long lasting first mover advantages to get going. Hence, patents may contribute more to profits than is acknowledged in the survey results. Still, if frictions and not patents are generating the rewards, then perhaps we should consider weakening or eliminating patents, since the patent system is costly to maintain and may generate few benefits.

An early consideration of the benefits of weakening intellectual property rights in the face of frictions is Cohen and Levinthal (1989), who postulate that imitation is not "free and immediate." Rather,

[42] See Middendorf (1981) and Maurer and Scotchmer (2006b). Giuri et al. (2006) provide extensive and recent information about inventor behavior and motivation based on the PATVAL-EU survey.

[43] For a discussion of open-source institutions and the possible consequences of patenting for the open-source community, see Lerner and Tirole (2005).

imitation is a skill requiring investment and costly effort to "absorb" knowledge from the "common pool."[44] If *absorption* as a function is costly and can be separated from the function of *contributing* to a common pool, weaker patent protection can have a benefit. A weaker patent makes the common element of knowledge greater and so allows a larger pool. As absorption from the pool is costly, there is a friction—or barrier—that can allow profits drawn off the larger pool despite weak patent protection. Henkel (2004), in a related model, makes an analogy to a juke box: users each individually and privately contribute to the musical enjoyment of all despite heterogeneous musical tastes and the positive externality they create. In a similar vein, Harhoff et al. (2003) suggest that a pool of innovations can contribute as an input to a process of improvement that is fostered by the market but also cannot be fully appropriated by competitors due to "idiosyncrasies." Hence, while each innovation contributes to a public pool, all contributors may be able to draw off benefits that exceed their private contributions. *Weaker* intellectual property rights can spur innovation precisely because of the existence of a common pool and the link between weakness and pool size. In addition, the free riding on research by others that is possible in the common pool also has the benefit of eliminating duplicative research spending.

Bessen and Maskin (2009) systematize the friction-based argument and show that even very small frictions can result in the dominance of a system where all information is available to some (exogenous) degree when compared to—a particular variant of—the patent system. Bessen and Maskin's result depends crucially on three elements. First, even if they are small, there are frictions that mean that an inventor's current profit is not reduced to zero immediately by imitation: imitation is costly, time consuming, or both. Second, each pool member is both a contributor to and a benefactor of the pool. In other words, the externality from the innovation runs both ways, from and to an inventor.[45] Third, each firm's private share of a common pool of benefits expands with membership in the pool so that "the more the better." This could be due to the presence of complementarities or network externalities or both. Relatedly, there is a benefit to fragmenting research spending across a large number of firms in that the date of discovery is brought forward under increased fragmentation.[46]

Clearly, the assumptions underpinning this model apply much more readily to some industries (perhaps information and communication technology) than others (perhaps pharmaceuticals). Further, the benefit of the "open" system is not compared to an optimized patent system. Indeed, Maurer and Scotchmer (2006b) note that the assumptions on the efficiency of licensing determine the relative desirability of the patent or the open system.[47]

However, costly or slow adoption need not imply the dominance of open systems. Some have modeled the adoption process more fully so as to diagnose where precisely the benefits of a patent occur compared to a system of free access. One such benefit to patents could be coordination of the

[44] See Cassiman and Veugelers (2002) and Bloom et al. (2007) for discussion and references to the vast spillover literature. Contributions to the common pool are termed "outgoing spillovers" of knowledge and absorption is termed "incoming spillovers."

[45] Belenzon (2006) has presented some evidence that spillovers are "reabsorbed" by the initial inventor in the context of sequential or cumulative innovation in a panel of US firms.

[46] In later models of cumulative innovation, we will see that in some models ideas are assumed to be "scarce." The probability of discovery of an improvement to an innovation may be zero if only the original innovator is present because the original innovator may not have the "idea" for the improvement. Participation of an independent entity is necessary to generate further advance after a first step.

[47] For a further model of the benefits of weak or nonexistent patents that relies on frictions, see Boldrin and Levine (2003).

adoption process. A paper that presents a simple version of this is Glachant and Meniere (2008). Suppose that demonstration effects facilitate future adoptions. Hence, early adopters exert a positive externality on later adopters that need not be reflected in early adopter behavior. On the other hand, an upstream patent holder—as a monopoly holder of the technology—will internalize the externality. Furthermore, an upstream patent holder has the tools to control the incentives for adoption by means of intertemporal price discrimination. Learning spillovers create two types of inefficiency in their model: first, even if an early adoption is socially desirable, it may not be privately profitable. Second, there is an incentive to delay adoption so as to benefit from the fall in cost. It may be necessary to both "kick off" the process and ensure that excess delay does not occur. Indeed, to the extent that imitation constrains the ability of a patent monopolist to price discriminate over time, imitation can be socially undesirable in this model as it can generate suboptimal patterns of adoption. Of course, while patents are a way to solve this problem of adoption externalities, other government instruments could well dominate them. A monopolist would not necessarily coordinate the market in the same way as a social planner would. The point is, however, that free access may need to be supplemented with some form of intervention in cases where positive externalities to adoption exist. Earlier papers (Katz and Shapiro, 1986, 1987, among others) incorporate these same effects into more complex models that include strategic effects, as well as extensive modeling of the source of the adoption externality (such as a network externality on the demand side).

The papers we have examined so far compare a patenting regime to an open system without rights. Anton and Yao (1994) take the polar opposite case of examining how well an alternative system of *secrecy* can reward innovators. In other words, there are no patents, but it is possible—via an unmodeled legal framework—to keep "ironclad" secrets. Is a system of secrecy enough to generate private rewards to innovation? Consider Ms. A, who has an innovation that can be kept secret despite its exploitation (e.g., the innovation is a process). Ms. A knows that this is a valuable innovation, but potential buyer Mr. B does not have this information. There is a potential market failure in trading this innovation due to asymmetric information: B does not want to pay for an innovation of little value. If the profit potential of the innovation relies on sale, we have an underincentive to innovate. Ms. A can attempt to resolve this failure by revealing the innovation to Mr. B. Absent patent protection, however, revealing the information to Mr. B gives him all the knowledge he would need to exploit the innovation without payment to Ms. A. How can Ms. A possibly get any reward for her innovation in this case?

If Ms. A has a secret and reveals it to a purchaser to exploit, she still may be able to gather full profits from this secret because she can threaten—credibly—to expose the secret to a third party, Ms. C. Hence, revelation of the value of the innovation also reveals the magnitude of loss under "punishment": the more valuable the innovation, the more value could be lost if the secret were revealed to Ms. C. More precisely, since the innovation is a secret that is not patented, the threat to reveal comes along with it the *credible* threat not to attach strings to the innovation such that Ms. C. would be a restrained competitor. It is precisely this "weakness" of secrecy that makes it a strong negotiating tool. Further, not only does a threat exist (because Ms. C is available outside the private bargaining between Ms. A and Mr. B) but also the strength of the threat is linked to the value of the technology. Now, Ms. A would always approach Ms. C if the payments were not contingent on profits actually earned, but if the license payments are contingent on the gross profits actually earned by the licensees, which are assumed to be observable and verifiable, then Ms. A could prefer to maintain a monopoly structure as long as she gets a share of the gains. Under the relatively mild assumption that the gains from maintaining a monopoly structure rather than duopoly are large while the profit gains for Ms. C from accessing the same technology as Mr. B are small, secrecy can be associated with trade and also a reward to innovators.[48]

This argument relies on the parties being able to keep a secret so that imitation cannot simply destroy the innovation's value out of the control of the parties to the transaction. Either one can interpret this as an implicit friction in the model that allows for profits to be reaped despite the absence of strong property rights or one can interpret this as a very robust legal framework enforcing an—alternative—trade secrecy system. Clearly, the situation analyzed is quite special in that it relies on the innovator's not being able to exploit the innovation herself after revelation, as this could affect the negotiations. Nor can the innovation be reverse engineered based on *ex post* observation of market products. Hence, secrecy comes with a great deal of control in this model, in contrast to the story we told at the beginning of this section where secrecy really was not an option. Their paper suggests, however, that in situations where information can be controlled extensively sufficient rewards to induce innovation can be created even in the absence of any *patenting*. Furthermore, a complete lack of *patent* rights is consistent with some disclosure when this degree of control exists. When we compare a patent system to an alternative of secrecy, we should not automatically assume that all information is disclosed under patenting and none under secrecy.

Hence, one can make a theoretical argument against patent rights if the circumstances are right. If frictions make patents redundant, or if secrecy is a very effective tool, then they may not be worthwhile. Do the conditions exist for weaker protection systems to generate more innovation? Empirically, this is not yet a settled question. Hall (2007) suggests that strengthening a patent system (in terms of lengthening the patent term, broadening subject matter coverage, or improving enforcement), while associated empirically with more use of the patent system, has less clear effect on aggregate innovative activity. Indeed, Sakakibara and Branstetter (2001) found that the effect of strengthening patents in Japan had only a very small effect on R&D activity. Empirical investigation raises the issue of simultaneity between research intensity and intellectual property protection. Indeed, Qian (2007) controls for this link, and finds relatively little evidence of a relation between strength of protection and investment activity. In a cross-country study using a broad measure of patent strength, Lerner (2005) finds that strengthening patents has a positive effect on innovation if patents are very weak, but a negative effect if patent protection is quite strong, so that intermediate levels of protection seem to work best at inducing innovation. The test is a step away from the specifics of the theories discussed here, however, so it is difficult to tell what mechanism is at play. To the extent that there is a linkage between innovation and patent strength, the main effect appears centered on a few industries where patents tend to be viewed as quite effective, notably pharmaceuticals, medical devices, and biotechnology (see Arora et al., 2003). In Section 6, we will consider optimal alternative systems of rewarding innovative effort that, combined with free access to innovations, might dominate patent systems.

[48] Formally, for Mr. B to accept the license it must be the case that the (contingent) fee not exceed his gains from moving from symmetric duopoly to monopoly, $\pi_M - \pi_D$. For Ms. A not to approach Ms. C, it must be the case that the fee from B exceeds what she could earn from Ms. C ($\pi_D - \pi_L$), where π_L is the profit Ms. C would earn if excluded from the innovation, plus the fee that B would pay knowing that the secret would be disclosed afterward to Ms. C. If $\pi_M + \pi_L > 2\pi_D$, as is true for many standard industry structures, then even with a positive fee in the case of disclosure to all parties, a parameter range exists where disclosure of a valuable innovation to only Mr. B is the equilibrium.

5. Optimal patent design: Length and breadth of protection

We have argued that patents can be seen as generating some kind of social benefits by rewarding technical progress, but it may be hard to argue that they generate progress optimally due to the deadweight loss they potentially create. We have also said that there are reasons to believe that the reward of a patent could generate either socially insufficient or excessive incentives to innovate—in addition to generating the deadweight loss. However, the argument for moving to no intellectual property protection at all or to a system of pure trade secrecy relies on relatively specific assumptions. An alternative would be, then, to retain the basic features of the current system but rebalance its parameters to generate improved performance. There is a long tradition of papers within this approach. We will consider first models where there is only a single innovation, followed by models of multiple innovations. These models will be primarily in the "reward theory" style. We will look at models that focus on disclosure issues in the final section.

5.1. Single-innovation models

A first set of papers examines a single innovation. Nordhaus (1969) set the stage for this work by suggesting that the length of patent protection should balance off two forces: first, for an innovation that will potentially yield benefits to society forever, the length of protection should be long (potentially infinite). Since protection is based on exclusive ownership, however, this creates a potential deadweight loss due to monopoly pricing. Minimizing this deadweight loss argues for short protection. The optimal length of protection needs to balance these concerns: the longer the protection the more innovation is induced, but the worse is the deadweight loss problem. Suppose that the innovation generates a notional maximum discounted social value $\bar{W}$ that could be earned if it were available for free immediately but a deadweight loss, d, is incurred during each period of protection.[49] Let the flow profits for each period of protection be π for the innovator. Profits fall to a baseline level of zero after protection expires. If we define $X(T) = (1 - e^{-rT})/r$ when there are T periods of patent protection, the innovation generates net benefit $\bar{W} - dX(T)$. This expression is decreasing in T. We can either think of maximizing this expression with respect to T or minimizing $dX(T)$ with respect to T, subject to the constraint that the discounted benefits generated from innovation, $X(T)\pi$, meet a value, c, required to induce innovation. Noting that $X(T)\pi$ is increasing in T, the solution to this problem is the minimum T that allows the constraint to be met.

Of course, any tool affecting monopoly pricing, including competition policy, could have a similar effect. For example, one could invoke a limit on prices (via, e.g., Article 82(a) in Europe) in each period but allow protection to persist for a very long time so that the small return in each period cumulates to the desired reward. Hence, the recommendation is that patents should last a long time, but should be combined with a strict limit on pricing. This argument was developed by Tandon (1982), where the limit was imposed via compulsory licensing guidelines. Formally, society's problem is to maximize total discounted social welfare from innovating (or minimize the total discounted social costs), where welfare

[49] In the earlier notation, if welfare equals $d + s + \pi$ per period, as before, society obtains only $s + \pi$ during the period of protection, but the entire welfare triangle after the patent expires.

in each period decreases with the price premium over marginal cost and the patent expires at time T. If we take π to reflect the price–cost margin, we have the following social planner's problem:

$$\max_{\pi,T} \bar{W} - X(T)d(\pi) \quad \text{or,} \quad \text{equivalently,} \quad \min_{\pi,T} X(T)d(\pi)$$
$$\text{subject to} \quad c \leq X(T)\pi,$$

where c is the value that must be covered to induce innovation. The difference between this formulation and the one above is that there are now two instruments of control (the price–cost margin and statutory length) and one, the price–cost margin, is an argument in deadweight loss so that we now have function $d(\pi)$. Tandon shows for the case of linear demand that while the minimand varies proportionately with the discount factor $X(T)$, it is proportionate to the square of the royalty rate (which determines the price premium) via deadweight loss. This makes the length of the patent, T, a relatively efficient instrument to compensate innovators relative to the price–cost margin. Commenting on this problem, Ayres and Klemperer (1999) point out that lengthening the patent life so as to hold the expected profit constant while restricting the monopoly distortion is a form of Ramsey pricing: price is set to minimize its distortionary effect while still generating a target amount of revenue.[50]

In this model, the patent designer wields a great deal of control: both the price–cost margin and the length of time during which that margin can be charged are direct instruments. No imitation limits the time during which rents can be earned (so that the effective patent length is the statutory patent length) and no imitation or other competitive concerns limit the price–cost margins that can be charged. We shall see later that imitation can modify the model results and the policy implications.

Gilbert and Shapiro (1990) extend this line of research in two directions. First, they associate the freedom to charge a large price–cost margin with "patent breadth." Patent breadth can be thought of as a strength index for the patent where a stronger patent is associated with higher flow profits. One interpretation of this would be that broader claims could be approved by the patent office, resulting in a larger "exclusion zone" around an innovation in product space. This could translate into higher monopoly profits if close substitutes are not permissible. Hence, patent policy now consists of two policy instruments: patent length and breadth, where the policy behind breadth is interpreted more generally than in Tandon's work.

Second, Gilbert and Shapiro show that either long, narrow patents or short, broad patents can minimize the deadweight loss cost of patent policy, subject to the constraint that the innovator earns a reward that induces some desired level of investment. Which of these designs is better depends on how welfare is related to profits: the relation is assumed negative, but the second derivative can realistically take either sign. It is this second derivative that determines optimal policy. The welfare maximization problem they consider is the same as above. Their approach is to define the "required profit," $\pi(T)$, as the value of flow profit, π, that satisfies the constraint for some given level of c. Total welfare, $W(T, \pi(T))$ can then be obtained solely as a function of T. They analyze the optimal policy by considering the shape of this function, W, as T changes. This shape is determined by both direct and indirect effects:

[50] Allowing a supracompetitive price is like allowing the patent holder to impose a tax. It could be even more efficient to spread a centrally collected tax over all *goods* as well as all time periods, but this would require intervention by a government body that knows the appropriate target level of reward. As we discuss below, knowing this target is an unrealistic informational assumption in many cases. See Ayres and Klemperer (1999) for more discussion.

$$\frac{dW}{dT} = \frac{\partial W}{\partial T} + \frac{\partial W}{\partial \pi}\frac{d\pi}{dT}.$$

The first term on the right-hand side, the direct effect of lengthening the patent, is clearly negative due to the deadweight loss per period of protection. The second partial derivative on the right-hand side is the direct effect of increasing profits on welfare. This is also negative via the deadweight loss. However, the third term, the effect of lengthening the protection on the "required" profit level, is also negative. As a result, the sign of the right-hand side depends on the weighting of the effects, which is determined by the second derivative of W. In order for an infinitely lived patent to be the optimal design, we need the entire expression on the right-hand side to be positive. This occurs if welfare is concave in the patent holder's profits, $\partial^2 W/\partial\pi^2 < 0$, since the patent reward becomes increasingly costly to welfare via the deadweight loss in this case. The reward should be constructed of flow profits that are very small, but cumulate to the target reward level that we wish to achieve in order to maximize dynamic benefits. Hence, the welfare-maximizing policy is narrow, long patents. This is essentially the situation analyzed by Tandon. Gilbert and Shapiro show that there is, however, a second case where deadweight loss is decreasingly costly as breadth rises (the second derivative of welfare takes the opposite sign). Here, short, broad patents are optimal. The optimal patent design is no longer so clear.

It may not, however, be appropriate to consider patent breadth to be an absolute exclusion zone. After all, patent protection does not prevent competition from other firms if they come up with noninfringing substitutes. Gallini (1992) suggests that a broader patent should instead be thought of as one that is more costly—not impossible—to invent around, noting that neither the Gilbert and Shapiro nor the Tandon approach really accounted for imitation possibilities. In this optic, patent policy consists of length and breadth—the latter now defined as the cost of imitation. The constraint we have included before[51] is now accompanied by a second consideration that entry will occur freely up to the point where the entry cost, E, is just offset by the gains from competing as an oligopolist in the industry. The policymaker must take into account that the increased reward earned by patent holders makes entry more attractive. Specifically, for a given cost of imitation, lengthening the period of patent protection now makes imitation more attractive. The entry costs incurred are, of course, also social costs. Hence, the social planner's problem is to create a reward that compensates innovators, minimizes deadweight loss, and minimizes duplicative spending. More formally, let the profits earned by any firm actually entering the industry, π, be a continuous and decreasing function of the number of entrants, m. We have entry determined by the condition: $E = X(T)\pi(m)$. If we take $m(\pi)$ to be the inverse of this profit function, we have the social planner's problem:

$$\max_{\pi,T} \bar{W} - X(T)d(\pi) - m(\pi)E$$
$$\text{subject to} \quad E = X(T)\pi \quad \text{and} \quad c \le X(T)\pi.$$

[51] The "innovation constraint" in Gallini's model has a different interpretation. It is a constraint that the innovator uses the patent system rather than secrecy. Hence, while written the same way in our notation, the interpretation of c is "the value that could be earned by not patenting" and the constraint is now that patenting as a protection option dominates not patenting and using secrecy. This interpretation is in keeping with the emphasis on imitation in the model.

In other words, we now maximize the surplus net of deadweight loss and entry costs, subject to a free entry constraint, and generating rewards sufficient to induce patenting. The patent length that maximizes social surplus when E is small enough that the threat of imitation is real will be the one that just discourages imitation—under the condition that the output elasticity with respect to the total number of firms, $(m + 1)$, does not exceed 1. The elasticity condition means that the relative cost of excess entry is large compared to the deadweight loss benefit of an additional entrant. If imitation is just discouraged, there is no resource cost to imitation for any given level of industry profits. Also, any given level of industry profits can be channeled entirely to the innovator if no imitation occurs. Hence, the optimal length for any given E is capped such that the policy just discourages all entry ($m = 0$).[52] If both the length of protection, T, and the breadth of protection, E, are policy levers of the government, then the best policy (when the same output elasticity condition holds) is to set E—the direct instrument to control imitation—large enough to discourage all imitation and the length, T, to generate the desired reward for innovation. Hence, we solve for length from the second constraint (assuming no entry will occur), then set the entry cost level such that the first constraint generates no entry. Entry cost, as it both loosens the cap on profits earned by the patent holder and also discourages imitators so that all of those profits are channeled to the innovator, is a very efficient instrument in this framework, so there is an argument for patents that are optimally broad and short.

Klemperer (1990) conceives of patent breadth as a zone of exclusion in product space around any given invention: the best current design is protected by the patent so imitators only offer inferior products. Hence, increased breadth has the cost of redirecting "imitation" away from desirable designs and into less desirable products, inducing a welfare loss for any consumer who buys a "knock-off" rather than a more desirable product. In this sense, the formulation extends imitation considerations to envisage the possibility of (inferior) imitation and a zone of exclusion. Breadth is interpreted as a portion of a product spectrum reserved for a patent holder[53] with imitation supplied competitively (so that the best noninfringing design is supplied "at cost"). Now, the social cost includes not only the deadweight loss due to a price distortion on the patented product, but also the "transportation costs" for consumers who "travel" to the patent boundary to purchase there, and any additional reduction in demand for traveling customers due to the cost of transport. Given that imitation is redirected so as to occur outside the zone of exclusion, the pattern of consumption induced by breadth as a tool may be undesirable for society as a whole since it may result in consumption switched to the "wrong" product variety. Call losses from "travel" $\tau(z)$. We now think of the social planner's problem as minimizing total discounted social costs:

$$\min_{z,T} X(T)[d(z) + \tau(z)]$$
$$\text{subject to} \quad c = X(T)\pi(z),$$

[52] This can be obtained from the entry constraint.

[53] As patents are specified in *technical* terms, there is no precise legal equivalent of such an exclusion zone in *product* space for many patents. Still, one can think of some patents that are quite close to this specification. For example, a patented shoe (US patent number 5255452) used in stage illusions includes claims on both lace-up and strap-on variants. To the extent that a "knock-off" would have to use an attachment mechanism other than the standard ways a shoe attaches to a foot, this could create the sort of inferior knock-off in product space that this model envisages.

where z is the width of the exclusion zone of the patent, and both deadweight loss and profits, π, vary with this width. Combining this constraint into the minimand, we can reformulate this problem as minimizing the ratio of patenting's social cost per unit of money spent, with the patent's lifetime set to be the minimum that satisfies the innovation constraint. That is, choose z and T to satisfy the following:

$$\min_z \frac{c[d(z) + \tau(z)]}{\pi(z)} \quad \text{and} \quad X(T) = \frac{c}{\pi(z)}.$$

For very wide patent breadths, the social cost is primarily a simple monopoly distortion since little substitution actually occurs. For very narrow patent breadths, however, there is high competitive pressure from close substitutes so that deadweight loss is low. On the other hand, substitution occurs readily for narrow breadth so that the cost of travel contributes significantly to social cost. Optimal patent design is, therefore, very sensitive to the pattern of consumer preferences. If all consumers have identical transport costs, the patent holder *must* set prices low enough that no consumers switch. If prices are set such that no switching occurs, however, the only social cost that is relevant to the social planner's decision is the standard deadweight loss. As we have just seen, this is minimized by setting the breadth as narrowly as possible. In contrast, if all consumers have identical reservation prices for their most preferred variety (so demand is inelastic), then the patent holder optimally charges consumers this (common) reservation price. All consumers purchase, but there is zero deadweight loss. If the patent breadth is set as wide as possible, then all industry profits accrue to the patent holder and travel costs are minimized. Hence, wide breadth minimizes the social cost per unit of monetary incentive for the patent holder. In all cases, the patent lifetime is set so as to just satisfy the constraint that the patent holder have the incentive to create the innovation in the first place: narrow patents must be long; broad patents must be short.[54]

These three papers, Gilbert and Shapiro (1990), Klemperer (1990), and Gallini (1992), taken jointly suggest that there is no clear-cut answer to whether larger or smaller breadth is better for social welfare. Gilbert–Shapiro's result that narrow, long patents are associated with a decreasing and concave flow social welfare function is intriguing, but they show that this is not the only case. Indeed, Gallini and Klemperer provide examples where social welfare can be convex or simply increasing in patent breadth. These different shapes give rise to drastically different optimal policies.

Two general comments have been made about this type of patent design story. First, most of these models take the identity of the original innovator as given. Denicolo (1996) notes that reduced breadth may be accompanied by more entry by researchers. If more entry into the research stage is accompanied by the presence of inefficient producers, insufficient product variety, or duplicative research costs, then clearly whether or not increased breadth is desirable on balance will depend on how these costs and benefits weigh up in social welfare. His paper generalizes the reasoning of the previous models to obtain

[54] Waterson (1990) also takes a spatial exclusion zone interpretation of patent breadth. His result is that flow welfare can increase in patent breadth due to better locational decisions by firms. In other words, if an imitative entrant unconstrained by patent breadth would choose to locate socially too close to the incumbent firm for business stealing reasons, an exclusion zone around the first entrant can benefit both innovators and society. For the innovator, an exclusion zone ensures a greater degree of product differentiation, hence attenuating competition, and for society it ensures greater variety, reducing "transport" cost losses.

sufficient conditions under which maximum or minimum patent breadth is optimal when the identity of the original innovator is determined by means of a patent race. In his formulation, then, we have

$$\max_{k,T} \bar{W} - X(T)d(k)$$
$$\text{subject to} \quad c \leq X(T)I(k),$$

where I, the "incentive" to win, is now measured by the profits of a losing racer weighted by the probability of losing, and the profit gain from patent protection, weighted by the probability of winning. The term I is a function of the flow profits due to the patent holder. Denicolo postulates that narrower breadth is associated with a smaller profit for winners compared to losers. Hence, denoting a patent breadth index by k, we have the incentive to win and deadweight loss both as positive functions of breadth. Narrower breadth tends to bring about more competition, but what this means in terms of social welfare is ambiguous, since social welfare could decrease (due to duplicative research expenditure in the R&D race or inefficient production) or increase (due to reduced deadweight loss) and could have almost any second derivative with respect to breadth. The paper takes a reduced form approach, showing that the optimal patent policy of long, narrow patents or short, broad patents depends on the second derivative of flow welfare with respect to breadth. This chapter shows, then, that the general reasoning of the earlier models is not affected by the incorporation of racing concerns even if the precise breadth and length cutoffs might be.

Second, Gallini (1984) has pointed out in earlier work that licensing activity has the significant benefit of allowing market participants to economize on duplicative research expenditure. Surely, an incumbent facing a threat of entry by imitative spending should have an incentive to contract *ex ante* with the potential entrant to save on the imitative expenditure. Both parties are at least as well off under this scenario. This sort of insight should suggest that introducing licensing along with imitation concerns could dramatically change the conclusions we draw on the length–breadth tradeoff. The precise effect depends on the modeling. Consider the Klemperer (1990) and the Gallini (1992) formulations. In the Klemperer model the entry cost is zero, so licensing activity would not alter the basic conclusions: the patent holder would potentially have to license an infinity of firms in order to avoid imitation, and this it cannot do. On the other hand, in the Gallini paper, the entry fee is both positive and central to the analysis. *Ex ante* licensing by the patent holder allows the entry fee of imitators to count as income for the innovator. Hence, it becomes an innovation incentive for the patent holder, but does not enter as a welfare loss because it becomes a pure transfer between market participants.[55] This alters the problem considerably, so that the setup really becomes one of minimizing deadweight loss subject to an innovation constraint that now includes licensing revenue from would-be imitators. This can weaken the argument for broad protection. Hence, when we include licensing considerations, there is a stronger argument in this model for long, low-breadth patents.

[55] Indeed, in the context of arguing for an independent innovator defense, Maurer and Scotchmer (2002) argue that imitation costs should not determine patent design, as their impact should be minimized by privately organized licensing activity. We will see damages used to (optimally) compensate innovators for infringing in later models in this survey.

5.2. Cumulative innovation models

All the above papers were set in the context of a single innovation, albeit with possible imitators or knock-offs. This is a simple and instructive case, but not necessarily a commonly observed one. Two cases of multiple innovation streams which raise issues quite distinct from those discussed above have been analyzed in the literature in response to this concern. The first of these cases is discussed in this section. It is the case of cumulative innovation: where innovations build on previous advances. This case raises new challenges to patent design in the following sense. Suppose that, without a first innovation, the idea for an improvement cannot exist. The fact that the first innovation creates the seeds for its own improvement means that there is a positive externality running from the first innovation to the second. This externality need not be internalized if the follow-on innovator is distinct from the first innovator. How to best divide a single profit stream so as to both reward the first innovator for this externality and induce follow-on innovation is the focus of this literature.

More precisely, following Scotchmer (1991, 2004) and using our earlier notation, suppose that one firm has generated an innovation that could give rise to further innovations: without the first innovation, the follow-on would not be possible. For example, think of a basic innovation that opens an entirely new field of research that had heretofore not been contemplated. The first innovation generates a positive externality by its revelation, as it identifies the new field. From society's perspective, the full benefit of the first innovation includes creating the possibility of a stream of innovations that cumulate to produce benefits, ultimately for consumers. If these innovations are separately held by independent inventors, however, we face the challenge of simultaneously generating full incentives for the first innovator to "kick off" the innovation path and also generating full incentives for any subsequent improver to produce follow-on innovations. Suppose, for example, that the second innovation generates a positive total discounted social value of W_2 on its own. If we award W_2 entirely to the second innovator, then we create full incentives to invest in the follow-on. However, we still face the difficulty that we should attribute both the direct value of the first innovation, W_1, as well as the value of the second innovation (which would not have existed without the creation of the first innovation), W_2, to the *first* innovation. Hence, to maintain full incentives to create the first innovation we need to allocate W_2 twice. Otherwise, innovation incentives will be socially too low.

Establishing exclusive rights can partially address this "double-allocation" problem. If a single innovator has control over the rights to an entire stream of innovations there is no need to allocate W_2 twice. A social planner or any other single inventor would internalize the externality and so we would not have any trouble achieving efficiency. This solution is straightforward if the same innovator is able to efficiently obtain both the initial innovation and its follow-ons. However, if a single entity does not have the ability to create all inventions that stem from the information revealed by a single invention, property rights may be used to allocate the benefits of the externality so as to achieve the desired technical progress despite the participation of multiple parties. This is where the possibility of licensing matters. Licensing makes it possible for a first innovator who has exclusionary rights to follow-on innovations but not the ability to develop them to trade access to those rights for a benefit flow from the second innovation. As long as enough benefit is left to the second innovator to cover the costs of creating the second invention, it is in the interests of both innovators to agree access and also to conduct research to generate the second invention. In this way, the presence of exclusive property rights does nothing to impede the pace of innovation. To the contrary, and recalling the Coase Theorem, property rights facilitate net benefit transfers from future innovations to the first innovator, improving the incentive to develop innovations in the first place.

Using the terminology of O'Donoghue et al. (1998), we must now distinguish in our patent design problem between patent breadth as protection from pure imitation ("lagging breadth"), and patent breadth as protection from different—and perhaps better quality—follow-on innovations ("leading breadth"). In other words, a literal copy (such as a drug that uses the same molecule as a patented drug and with the same delivery) could violate the lagging breadth of coverage while a small improvement (such as a modification that slightly improves the delivery of a patented drug) could violate the leading breadth. In the single-innovation case, only the former was relevant. With cumulative innovation, however, the leading breadth granted to the first innovator determines whether a follow-on innovation infringes the original patent and therefore can be barred from sale by the first innovator. A patentable innovation outside the scope of (leading) protection is noninfringing, while one inside this scope infringes. A broad patent on the first innovation implies, then, that a follow-on innovator would need the express agreement of the first innovator to exploit the follow-on. Hence, the first innovator can use the agreement to allocate the externality back to the first in the chain of innovations.

Green and Scotchmer (1995) enunciate and formalize this basic insight. Consider the case of a "research tool" where the value of the first innovation on its own is nil ($v_1 = 0$).[56] If there is only one potential innovator, then the fact that there is an externality running from innovation 1 to innovation 2 creates no inefficiency in the decision to innovate as long as the patent runs long enough that the total profit of the innovator exceeds the total development cost. While the innovation constraint becomes $c_1 + c_2 \leq X(T)\pi_2$, the problem is essentially unchanged from the single-innovation case. The innovation stream should be undertaken whenever $W_2 - c_1 - c_2 > 0$ and policy must set patent length to allow the monopoly profits to cover the full cost of investment, $c_1 + c_2$. Breadth has no new role in this story. If the inventor of the second innovation is different from the first inventor, however, then we must be concerned not only with the total profit but also its division, as both inventors must obtain a large enough percent of the earnings to cover the development cost. Now we have two innovation constraints: the first innovator's earnings over the patent period must cover investment cost c_1, and the second innovator's earnings must cover investment cost c_2. Because the earnings may now include transfers between the parties, bargaining is now the focus of the analysis.

Leading patent breadth can both affect the bargaining positions of the parties to the technology access agreements (licenses) and the need of each party to "come to the table" in the first place. As the first innovation has zero value on its own, the first innovator would never invest unless some of the second generation profit was transferred to it. Such a transfer can occur via a licensing contract, but the timing of this agreement matters: whether this contract is executed before (*ex ante*) or after (*ex post*) the second innovator invests c_2 can determine what terms will actually result from bargaining between the two parties.[57] This is because *ex post* agreement allows the second innovator to be "held up" for (sunk)

[56] There is some evidence that research tool patents have increased over the last twenty years, in particular in the area of biotechnology, making this a pertinent example. See Walsh et al. (2003).

[57] One could also think of very early agreements, before innovation 1 is created. This would be closer to a research joint venture, on which there is a considerable literature. See Tao and Wu (1997) or references included in Miyagiwa (2007).

cost c_2. Hence, the bargaining equilibrium potentially depends both on the breadth of the patent and on the licensing timing regime that is permitted.

Suppose that only *ex post* licensing is permitted and while both innovations are patentable, the patent breadth is such that the second innovation infringes the first. In this case, either innovator could potentially prevent the second innovation from coming to market. If the firms fail to agree a license, no transfer is possible and the first innovator cannot benefit from the second innovation. Further, the second innovator stands to lose its development cost but obtain no return for it if the innovation is blocked by the first patent holder. On the other hand, if the firms agree to split the surplus evenly, the first innovator potentially earns half the profits from the follow-on innovation, as does the second innovator. If π_2 is a per-period reward, the cumulative reward for a patent of length T is $\Pi_2 = X(T)\pi_2$.[58] Hence, *ex post* licensing results in profits $(\frac{1}{2}\Pi_2 - c_1, \frac{1}{2}\Pi_2 - c_2)$ for the first and second innovator, respectively. Because the second innovator earns only half the profits from innovation 2, innovation may be deterred when $[\frac{1}{2}\Pi_2 < c_2 < \Pi_2]$—due to "holdup." In other words, the innovation is deterred even though its profits would justify its costs. Hence, *ex post* licensing does not fully resolve the reward problem at both the levels of innovation one and innovation two. Furthermore, if we narrowed patent breadth so that the second innovation no longer infringed the first, innovation 1 would never be undertaken at all: the first innovator would have no basis on which to capture value in exchange for its cost of investment, c_1.

An *ex ante* agreement can resolve this problem by allowing the first innovator to commit to a lower licensing fee by means of negotiating at a time when the second innovator has yet to sink development cost c_2. The second firm can ensure itself a payoff that covers its investment cost. Further, both innovations will obtain enough surplus to be innovated as long as profits cover the entire costs, $\Pi_2 > c_1 + c_2$. Hence, the combination of *ex ante* licensing and large patent breadth for the first innovation generates desirable investment incentives. If the second innovator knows Π_2 and c_2 before investing, the optimal policy is infinite breadth, in fact, so that *all* follow-on products infringe the basic innovation. This minimizes the second innovator's profit in an *ex ante* agreement. In other words, the situation where the second innovator's product infringes puts it in the weakest bargaining position, allowing the first innovator to give it only just enough to induce it to innovate. This "outsourcing" in turn ensures that profits are channeled to the first innovation as a reward for the externality it generates.

The authors comment that the legal status of *ex ante* licensing agreements such as the one we have just discussed is questionable under competition policy since one could claim that such agreements could amount to *ex ante* collusion. On the other hand, if one restricted all licensing to be *ex post*, one would have to recognize that this could restrict the cases where the follow-on innovations are developed or could require that patents be lengthened in order to increase the reward of the patent holders sufficiently to satisfy their innovation constraints. In other words, we would need to "scale up" the term Π_2. Whether or not this is desirable depends on the deadweight loss associated with the patent period. Notice that, since the effect of the stringent infringement standard and *ex ante* licensing is to obtain a better distribution of licensing revenues—which are a pure transfer—the change in "breadth" has no direct effect on deadweight loss. Any deadweight loss is via the patent term, and this can be minimized while inducing both innovations to occur when the first patent holder is given broad control and *ex ante* licensing ability.

58 The underlying story could be that Nash bargaining determines an even split in the licensing negotiations.

The emphasis of this argument is on the importance of infringement as the salient aspect of "breadth" in a cumulative setting. This, however, is not the only policy tool that could be relevant to the allocation of surplus between initial and follow-on innovators. Scotchmer (1996) investigates how the division of profit is affected by the *patentability* of the second product. This plays a role when the identity of the second innovator is not known *ex ante*.

Again, consider the case when the first innovation has no stand-alone value and assume that the follow-on innovator need not be the same as the first innovator. Let the second innovation infringe the first. The first innovator would potentially issue an exclusive license (before research into the follow-on has occurred) to a single agent, hoping to collect profits so as to provide a payoff to its own basic innovation. If the second innovation is separately patentable, however, then *any* independent second innovator (regardless of whether it was a licensee) can block the follow-on's sale if she obtains a patent. Suppose, then, that two independent innovators potentially could invest c_2, each innovator potentially patenting the follow-on with probability 1/2. The first and second innovator must then bargain *ex post* over surplus Π_2 regardless of whether the second innovator had previously received a license. Assume the any bargaining parties split this surplus evenly so that each receives $\Pi_2/2$. A potential second innovator who obtains a license faces a probability of 1/2 of obtaining the follow-on and earning Π_2 and a chance of 1/2 of losing. In the latter case, the license to the first innovation is useless without a license to the second. Hence, the licensee ends up with expected profit $\frac{1}{2}\Pi_2 + \frac{1}{2}\Pi_2/2 - c_2$. The entire stream of innovation is now expected to generate $3\Pi_2/4 - c_2$ for the licensee. The firm that was a nonlicensee earns nothing if she loses the race for the follow-on, but can bargain for $\Pi_2/2$ if she wins, which she does with probability 1/2. Hence, a nonlicensee expects to earn $\Pi_2/4 - c_2$. The winning bid for the exclusive license, the difference between these, is only $\Pi_2/2$ which is less than $\Pi_2 - c_2$ whenever the losing bid would be positive. If, on the other hand, the second innovation is not separately patentable, an independent follow-on innovator has no exclusionary rights. As a result, the first innovator never bargains with a nonlicensee, and nonlicensees never invest. The first innovator can earn the entire net stream of returns in this case, $\Pi_2 - c_2$, as a payment for an exclusive (*ex ante*) license.

We see, then, that the first innovator receives a lower payoff when the second generation innovation is patentable than when it is not. Patentability of the second generation product has two drawbacks in this story: it potentially encourages duplicative R&D costs for the follow-on product—reducing the surplus available to the bargaining parties—and also it transfers some of the profit stream to the follow-on inventor. We have an argument based on these two papers for very strong rights to seminal innovations but relatively weak protection for any follow-ons.

Of course, policymakers would generally have more instruments than patentability to work with. As before, we could consider the value generated by the patent, Π, to be an increasing function of patent term, T. The length of the patent serves to scale the reward. In this case, we could examine how patentability and patent term could work together to create rewards for the innovators. If there is no deadweight loss to the patent, then this yields an answer that infinite protection is optimal. More generally, let there be a deadweight loss to protection that we wish to minimize, subject to an innovation constraint. Then to cover the first innovator's cost c_1, patent life could now be adjusted upward when the second generation is patentable in order to induce the first innovator to invest in the first place. Combining this possibility with our previous observations on patentability, *ex ante* licensing tends to allow for shorter patent lives, as the rewards can be adjusted in the licensing contract to internalize the externality before any costs are sunk. Even in the case of efficient contracting, however, patentability of the second innovation tends to require longer patent lives to ensure that innovation incentives are maintained. If a longer patent period is undesirable because of the deadweight loss, this structure of the patent can be dominated by a structure with strong novelty and nonobviousness requirements for cumulative innovations. Under this policy, relatively small steps (clear follow-ons) would often not be patentable.[59]

One way to think of cumulative innovation is to think that innovations now move up a quality ladder so that improvement innovations can make the earlier innovations obsolete. A process of Schumpeterian "creative destruction" occurs as we move up this ladder. This process, however, can render the statutory patent life irrelevant since innovations are eclipsed before the statutory length of protection is reached. It is not clear, then, that statutory length can have the same "scaling" function that we have attributed to it. In fact, it is no longer clear that we can make the strict separation between length and breadth of protection as independent policy tools that we did before. Now, small leading breadth is no longer consistent with an infinite stream of monopoly rents. While the notion of lagging breadth is well defined, leading breadth and the statutory length of protection combine to determine an expected effective length of protection. Another way of seeing this is to say that the statutory length may scale up profits if improvements are slow to emerge, but cannot necessarily be relied upon to create such scaling if improvements come quickly. One might then need to rely on other tools, such as leading breadth, to do this.

A second comment on the cumulative innovation models we have reviewed is that the previous models assumed that the "roles" of first and second innovator were clearly assigned. In point of fact, the same firm may sometimes function as a follow-on innovator and may sometimes be the first innovator. The distinction between first and second *inventors* then becomes blurred even if the distinction between first and second *innovations* is clear. Despite the prominence of bargaining, the role of patent design may not, then, be to transfer profits from one "type" of innovator to another: all firms are potentially of all types. Instead, the aim is to balance *total* profits to innovation for each innovator against deadweight loss. In this sense, we move back toward the tradeoffs found in the single-innovation literature.

O'Donoghue et al. (1998) examine such a quality ladder setting where each firm can take on both leader and follower roles and where statutory length and leading breadth interact to jointly determine patent rewards. In the phrasing of Hopenhayn and Mitchell (2001), the patent system in a quality ladder framework establishes a clock that is running on monopoly rights for lower quality firms as well as a promise of rights for the firm currently holding the high quality innovation. Suppose that the magnitude of each innovation's quality improvement over the previous frontier technology is v, and this also indexes the profitability of the improvement until it is supplanted by the next innovation or the patent expires, whichever comes first. Then each improvement generates a potential value to society of $v/r - c$, where c is the development cost of the improvement and the improvement generates value forever discounted at rate r. On the other hand, if creative destruction occurs each period, the innovator only earns v for one period, after which the product becomes "obsolete." This private reward may be insufficient to cover development cost c. We have insufficient innovation incentives since each innovation creates value that benefits society forever, but the innovator collects this value only for a

[59] There are limits to this argument, as delays in research plus a short patent on the first innovation mean that the second innovation does not infringe for at least some of its life. Furthermore, if it is not clear that follow-ons depend heavily on the first innovation, the externality argument gets weaker. In both these cases, the argument for patentability of the second innovation gets stronger. See Scotchmer (1996) for more discussion.

single period. Indeed, even if statutory patent life is infinite, an innovator will never have full incentives to innovate as long as creative destruction occurs at some rate.

Define leading breadth in this framework as a quality margin, k, such that if an improvement possesses a quality margin less than this it infringes the patent. Now consider alternative patent protection designs in this framework. Suppose first that for the duration of a patent, all improvements infringe (so that leading breadth is infinite). Ideas for improvements arrive at some rate to independent agents. Since *ex ante* licensing is permitted, all improvements where the net value is positive will be made but will incur a licensing fee that splits the surplus between the improver and the holder of the infringed patent. If patents last a number of periods, T, each innovation v would then earn "direct" discounted profits $X(T)v$, but would also earn licensing revenues from improvers—and would result in licensing *payments* to earlier infringed patents—during its life. Call net licensing revenues $L(T, h)$, where h describes the history of previous quality improvements. Hence, the innovation constraint now becomes $c \leq X(T)v + L(T, h)$. We can define $v(T)$ as the schedule of quality steps that satisfy this constraint with equality for different patent terms: this set of steps defines the marginal innovations that will be invented.

Consider now the alternative design where all patents have infinite life but limited breadth so that only creative destruction causes them to "expire." Any improvement within margin k creates profits until it is supplanted by a noninfringing improvement: that is, an improvement lying outside margin k. Infringing innovations will be created as long as the profit surplus they create is nonnegative due to *ex ante* licensing. Hence, during the period of protection, the patent holder earns revenue composed of direct returns plus licensing fees and payments. Until a noninfringing innovation is discovered, this innovator will remain the market incumbent, earning the revenue stream. Formally, if the discounted profits of a patent lasts some set of periods, t, before being replaced by creative destruction, but this duration t is distributed according to a Poisson process with arrival rate Γ (reflecting an uncertain research process), the authors assert that the expected net discounted profit from any improvement, v, is $\{v/[r+\Gamma(k)]\} - c$. Define $v(k)$ as the quality step that just sets this expected net discounted profit equal to zero for patent breadth k. All innovations at least as large as $v(k)$ will be created, even if they infringe and so require a license, so that $v(k)$ is the marginal innovation under the alternative regime.

The marginal innovations are not necessarily the same under the two policies, giving rise to differences in the rate of innovation and the research expenditure under the two protection regimes. As protection increases toward infinity on both dimensions —breadth and length— the rate of innovation approaches the social optimum. Hence, the flavor of the result is similar to earlier cumulative innovation models where there is a tendency for very strong protection to be optimal.

As we have noted, however, the two policies we have just considered are not equivalent. To induce the same rate of innovation the first policy is associated with a shorter effective patent life. In the first policy, the binding dimension of patent protection is its statutory length so that a patent holder has claims on future innovations. The same level of investment incentive can be created with relatively short protection in this case. Rewards to the innovation are high, so statutory protection can be short because the total reward quickly surpasses the cost of innovation. On the other hand, broad short patents potentially create deadweight loss by concentrating the rights to use innovations in a few hands with little "close" quality competition, recalling the single-period models. In the second policy, the binding dimension is patent breadth, so that follow-ons tend not to infringe, and effective life must be determined as a consequence of breadth. In other words, the claims on future generations of innovations are quite limited in this case. To achieve the same initial investment incentive, the effective patent life must be adjusted to be longer for narrower breadth and shorter for larger breadth.[60] When demand is inelastic, the lower R&D costs that come with the latter policy make it preferable since the longer patent period does not create deadweight loss. If there is a deadweight loss associated with the period of patent protection, then the first policy can be better as patent protection is shorter.

Translating the policy in this chapter into patent statutes is, as usual, tricky. A way to think of the policy of narrow leading breadth but long statutory length patents is perhaps by applying a strict interpretation of the doctrine of equivalents (where equivalence is in terms of quality step), which can have the effect of granting very limited scope to patent claims beyond what is actually enunciated in the patent claims themselves.[61]

As O'Donoghue (1998) points out, the above result on leading breadth relies on a well-functioning licensing market. When efficient licensing is not possible, (perhaps because it is difficult to identify subsequent innovators as a practical matter or because transactions costs are high) we obtain a stronger argument for the importance of a *patentability* requirement to obtain optimal innovative behavior in a quality ladder framework.[62] Note that we considered infringement standards in the O'Donoghue et al. framework, but we did not consider whether follow-on innovations should be separately patentable. O'Donoghue reasons that, if it is assumed that an unpatentable innovation earns no profits, there is no point in targeting unpatentable innovations in research. As a result, a larger patentability requirement can induce firms to target larger innovations—since these are the profitable ones. If these big steps take longer to accomplish, then this policy comes hand in hand with increased rewards to innovation since larger steps tend to prolong the effective period of incumbency. As a result, the reward to research can be increased by a tough patentability requirement. In other words, patents promote, but also *retard* research by effectively discouraging innovations inside a quality threshold, and so a patentability requirement can modify the chosen step size on a quality ladder. Imposing a patentability requirement so that firms target innovations larger than the social optimum can, in fact, improve dynamic efficiency. This is the case because firms tend to invest too little when they can be eclipsed by followers. The patentability requirement tends to increase R&D incentives, which has a first order effect on welfare, while the adjustment to the innovation "step" has a second order effect when that step is close to the social optimum. The point is, then, that patentability requirements have "bite" if licensing functions poorly. As an empirical matter, Moser (2005) notes

[60] Horowitz and Lai (1996) anticipate this point using patent length and the frequency of "creative destruction" to determine the incentives to produce a "big" innovative step. Their work interprets length as statutory length, while O'Donoghue et al. (1998) make it clear that effective length can be determined by either statutory length or leading breadth.

[61] For a recent legal decision in this area, see *Festo Corp. v. Shoketsu Kinzoku Kogyo Kabushiki Co.*, 535 U.S. 722 (2002).

[62] See Gallini (2002), Merges and Nelson (1990), and Heller and Eisenberg (1998) among others, for a discussion of impediments to licensing. Comino et al. (2007) show that both early innovators and followers may benefit from decreased breadth when licensing is inefficient because the first innovator cannot observe whether follow-on innovators have already undertaken R&D activity, which decreases licensing's effectiveness as a tool.

that there is evidence that firms direct their research toward patentable rather than unpatentable subject matter so the underlying assumptions of the model seem to receive some support.[63]

In the cumulative innovation papers considered so far, it would be best if a single firm had the ability and resources to carry out the entire stream of innovation itself. If a single agent were responsible for the entire stream of innovations, the externality would be internalized. This is the root of the tendency for these models to favor very strong protection for seminal contributions. However, this single firm benchmark neglects the potential benefits and costs of having several potential innovators "race" for the rights to a given "idea." If "ideas" are not public knowledge, then these potential benefits are irrelevant: each innovator pursues his or her own "ideas" without the fear of being beaten to the punch by a rival. If, however, research ideas have a significant public knowledge dimension, then the potential benefits and costs from "racing" cannot be neglected in patent design. This issue is addressed by Denicolo (2000) who points out that, because of potential duplication of efforts and the incentives to pre-empt, the private market may over-provide innovation. It may be better, then, to reduce the reward to innovation. This effect can, then, dampen the optimality of heavily rewarding the firms that create seminal innovations. In fact, once we introduce the possibility that firms race for innovations, aligning the private and the social reward to innovation without considering duplication may be the wrong policy as the losses from duplication may be very large.

Maurer and Scotchmer (2002) address the issue of racing by arguing for an independent invention defense to patent infringement (currently available for copyrighted material). In other words, they suggest that racing concerns could be addressed by another policy tool: that of allowing a firm that has conducted duplicative but nonimitative effort to commercialize its invention. This has the benefit of reducing deadweight loss by introducing more competition into the final market. It also puts a "cap" on earnings and so reduces entry into the race, which dampens duplicative expenditure. If the social benefit of reducing this deadweight loss exceeds the negative impact on the innovator's incentives to invent, then it can be socially beneficial to allow independent inventors to coexist in the market. Of course, determining whether invention was truly independent or simply imitative could be a daunting task.

Summarizing the papers examining patent length and breadth in the cumulative innovation case, one can suggest the following conclusions. First, there appears to be a relatively strong argument for protection from literal imitation (large lagging breadth). Leading breadth has more qualified support: its benefits rely on the assumptions one makes about the scope for licensing. If licensing is fully flexible and efficient, then a strong argument for leading breadth exists. If licensing possibilities are restricted, then a much more limited case for leading breadth can be made. Strong patentability requirements receive some support when licensing does not function well. When duplicative investment as a result of racing is taken into account, there is an argument to be made against very large rewards for any invention. Indeed, racing considerations generally limit the argument for strong patent rights.

[63] More precisely, when no patents are available, firms' investments focus in areas where other appropriability mechanisms are present, while when patents are present, investments are more diversified. See also Lerner (1995) for related work. In a model similar to O'Donoghue (1998), Hunt (2004) finds that the inventive step requirement for patentability that maximizes the rate of innovation is at an intermediate level. While increasing the inventive step requirement makes the marginal discovery unpatentable (so that R&D expenditure is "wasted"), it prolongs the reward to patentable steps since discoveries exceed the patentable threshold less frequently. This increases the incumbency period. If exogenous parameters are such than an industry tends to invent frequently, increasing the inventive step has a large marginal effect on rents as they are discounted little. Hence, "high tech" industries—with frequent innovation—optimally require a more stringent inventive step than those with infrequent innovation.

5.3. Complementary innovation

Lemley and Shapiro (2007) suggest that it is not just the cumulativeness of innovation that creates a difficulty in allocating an externality. If patents are complementary, with synergistic benefits such that the sum of the patents adds up to more than the separate parts (e.g., if a product is made possible only by the combination of the patents) then we can also get socially incorrect levels of innovation investment. The reason is that there is both an externality and an investment coordination problem that did not exist before. In the case of cumulative innovation, there was a (positive) externality that ran only one way, from the first innovation toward the follow-on. Now, the externality runs two ways, as each innovation is a necessary "piece of the puzzle" in the final composite good. Further, one innovation does not necessarily completely precede the other in time. That is, when innovations cumulate, the follow-on investment does not begin until the first innovation exists. In the case of complementary innovation, however, all investments could potentially occur simultaneously. It could, then, be possible that multiple innovation equilibria exist: it could be an equilibrium for all innovations to be created or for none of them to be. Hence, as a result of the two-way externality and this difference in timing, there is a pure coordination problem in investment to be solved that was not present before.

If pooling a variety of patent rights is necessary to create a final product and the licensing transaction is costly, Heller and Eisenberg (1998) make a general point that when multiple, separate, rights holders must be brought on board to create social value, innovation may be underprovided due to transaction cost considerations. They identify this as a "tragedy of the anticommons," in contrast to the more classic tragedy of the commons. While this issue existed in the cumulative innovation case, it may be more severe in the case of products that read on a wide number of patents in a variety of fields simply because the relevant patents may not be filed over time but may instead be simultaneous and so in force for a long time. If licenses are not negotiated, then there is a potential for an innovative good never to make it to market in the first place, resulting in social loss.[64]

Shapiro (2001) examines formally the case where multiple rights owners contribute to a new product or process, creating a "patent thicket" that a new product could potentially infringe. Shapiro draws an analogy to the "Cournot complements" problem where a manufacturer must purchase n essential inputs from n distinct monopolists. Suppose that each $i = 1,\ldots,n$ separate firms owns a patent that is essential to the production of a final product to be sold on a competitive market. Each firm sets a per-unit royalty, r_i, for its patented "input" and each patented "input" is produced at marginal cost o_i. The final good price, p, will be composed of some manufacturing cost for the assembler plus the sum of all the royalty payments charged for access:

$$p = c + \sum_{i=1}^{n} r_i.$$

If each of the royalties is set independently and noncooperatively, then for price elasticity of final demand ε the markup of the price over the marginal cost of "input" production will be

[64] Walsh et al. (2003, 2005) suggest that at least in the case of research tools in the biomedical industry, the "anticommons" problem may not be very severe empirically.

$$\frac{p-(c+\sum_{i=1}^{n}o_i)}{p}=\frac{n}{\varepsilon},$$

which is n times the standard monopoly markup. The final price of the manufactured good is higher under this vertically separated structure than it would be if a single vertically integrated firm provided (all) the inputs and output. It is also higher than the price that would be charged by a competitively organized final product market, which purchases from a single, monopolistic supplier of all essential inputs. The profitability of the innovation as a whole falls because individual firms fail to internalize a (negative) pricing externality. As a result, there is a socially undesirable reduction in the research incentive.

Since the Cournot complements problem penalizes members of the industry as well as consumers, one would expect institutions to have arisen to limit this behavior. Where high technology products rely on technological standards that are composed of multiple essential patents owned by different parties, the patents are often required to be licensed at "reasonable and nondiscriminatory" (RAND) royalties. While this can be seen as a way to limit royalty overcharges, Schmidt (2008) comments that it would be very difficult to implement vague words like "reasonable" in any systematic way and, indeed, quotes Swanson and Baumol (2005) who state that "It is widely acknowledged that, in fact, there are no generally agreed tests to determine whether a particular license does or does not satisfy a RAND commitment."

The Cournot complements line of reasoning we have just developed clearly yields social efficiency arguments for various policies. Schmidt (2008) examines horizontal merger among patent holders as a remedy. He also allows for market power downstream so that a double-marginalization problem exists on top of the complementarity problem. Under sufficiently flexible licensing contracts (such as two-part tariffs), merger solves both inefficiencies. He notes, however, that Layne-Farrar and Lerner (2008) find that all the patent pools they investigate used linear royalties. When he restricts contracts in this way, horizontal merger among inventors continues to perform well, but vertical integration does not. The reason is that each vertically integrated entity does not internalize the externality it exerts via its royalty rate charged to other (vertically integrated) entities. To the contrary, by raising the royalty, each entity can raise rivals' costs. Further, each entity suffers from some double marginalization for the patents it must buy in.

If one assumes that the set of patents in the "thicket" is not "fixed," but is accumulated over time due to continuing research, Noel and Schankerman (2006) hypothesize that a reasonable reaction to the Cournot complements problem could be for firms to accumulate large patent portfolios. Indeed, they find some empirical evidence for excessive incentives to patent in order to "hoard" in the software industry.[65] Related work by Arora et al. (2001), Hall and Ziedonis (2001a), and von Graevenitz et al. (2008) finds that the recent growth in patent applications can be attributed to defensive use[66] of patents in "complex" industries—those where patent thickets are present.

Alternatively, allowing complementary patents to be traded as a "package" for a single price rather than traded separately could yield gains. Hence, we might wish to treat patent pool agreements—agreements among multiple patent holders to aggregate a set of patents among pool members or license

[65] They hypothesize that a larger patent "arsenal" also strengthens the bargaining position of an inventor and reduces transaction costs as the number of potential negotiations fall. Dewatripont and Legros (2008), in an analysis of patents' contributions to standards, appeal to a version of a Shapley value to justify the relation of bargaining strength to the proportion of patents owned.

[66] This defensive use can include litigation concerns, which will be discussed below.

as a package to nonmembers—leniently when they involve complementary patent rights.[67] Cross-licenses could serve the same purpose.[68]

Not all industries are equally susceptible to complementarity problems. Cohen et al. (2000) classifies industries according to whether they are "complex"—so that value is derived from complementary components—or "discrete." If this is the case, targeted industrial policy toward patent pools or merger could address the complementarity problem. Alternatively, one could think of the complexity of an industry as the result of patent design: if patents are granted very narrowly, then many complementary "bits" would necessarily contribute to almost any product. The appropriate policy response in this optic is to make patents broader to reduce the cases where complementarity issues arise.

However, both of these solutions could be hasty. The reason is that both assume that the degree of complementarity is not a choice variable for the producers. Lerner and Tirole (2004) take the opposite tack suggesting that, while the Cournot analogy provides a good starting point, it is often difficult in practice to pinpoint whether a patent is a complement or a substitute for another. The complementarity of patents may be less an unchanging "objective" characteristic of the patents than a characteristic of how a particular manufacturer optimally decides to combine technologies into a final product. Worse, these characteristics may change over time as technology and its applications progress. This could call our policy responses into question.

To sketch Lerner and Tirole's argument, consider a case where users can purchase patent "inputs," supplied by n upstream owners, each of whom owns one patent. Users combine these patents in various ways to create a valuable product. Users will do this in a surplus-maximizing way, which necessarily takes into account the input cost (i.e., the price at which the patent is licensed). Users may create surplus either by using a subset, m, of the n available patents or the entire available set. More precisely, a user can combine patents to create value $\theta + V(m)$, where V is an increasing function of the number of patents actually used, $m \leq n$, and the values θ are distributed according to some cumulative distribution function, G, over the user population. Hence, just using a single patent creates value, but the set of all patents creates even more value and further, users are heterogeneous in how much value they derive from the final "product."

A patent owner can license these innovations to users at a price, P. User demand will be determined by the value extracted from the patents, $\theta + V(m)$, net of the price at which they are sold, P. For some P, it may very well be the case that the net value extracted from the patents is highest when all n patents are bought. This would be the case when the price for each patent, p_i, is less than the value added of the mth patent, $V(m) - V(m-1)$, for all m. Note that, when all patents will be used, a rise in the price of one patent will tend to decrease the attractiveness of the final good as a whole because its price will rise. Hence, a price rise for patent i decreases the demand for other patents. This means that the patents are demand complements at low prices: the rise in the price of one causes a fall in the demand for another. On the other hand, it is possible for the prices of a patent to exceed its marginal contribution over some

[67] See Schmidt (2008) for a summary and comments on recent US policy moves toward a "rule of reason" approach to whether patent pools must contain only complementary patents or whether substitutes can be included as well. Layne-Farrar and Lerner (2008) give a history of patent pool policy in the United States.

[68] For examples of patent pools and their diversity (from mega pools comprising a broad-based governance structure for huge numbers of patents to small pools that amount to no more than a few multilateral contracts establishing a way to consolidate patent rights and a rule to divide up licensing revenues) see Merges (1999).

base number of patents. In this case, only a subset of patents will be purchased and combined to create end value. For example, if $n = 2$ and each patent is priced above the contribution of a single patent, $\theta + V(1)$, but below the marginal contribution of the second, $V(2) - V(1)$, then only one of the patents will be purchased and used. In this case, a rise in the price for one patent causes the demand for the other patent to rise because use switches to the cheaper of the two. Hence, the patents are substitutes.

In setting a license fee, then, a patent owner needs to take into account two effects. First, she needs to think about whether the patent will be retained in the "basket" of patents that are purchased. Second, she needs to take into account the effect of her own fee on the final price of the good that the patents are used to create (and hence the final demand for the "basket" of patents). If the second effect is dominant, then under noncoordinated pricing, each patent holder exerts a positive externality on other patent holders when she lowers her price since she raises demand for the entire "basket" of patents. The price for the basket will fall when this externality is internalized, so that coordinated "pool" pricing reduces user price and raises welfare in general. This argument recalls the Cournot reasoning, above. On the other hand, if the dominant effect is the first, noncoordinated pricing may induce an incentive to lower each patent price so as to "steal business" from other patent holders. This can create welfare gains to noncoordinated pricing over pool pricing if patents are sufficiently substitutable. Hence, the recommendation for public policy towards patent pools is nuanced: we only want to be lenient and allow pooling when an endogenous "complementarity effect" (and not some exogenously determined "complementarity characteristic") is dominant.

5.4. Disclosure issues

As we said earlier in this chapter, a major function of the patent system is to disseminate information. One could think of this as transforming private ideas and their embodiment into public knowledge by means of the patent disclosure requirement. Hence, the degree to which ideas are "private" is a policy instrument of the patent system as well as a choice variable for firms that can select between patenting their innovations and exploiting them as trade secrets. The degree to which information will actually be revealed in a patent system and the degree to which it will be withheld in a secrecy system is, however, debatable as we will see below.

Maurer and Scotchmer (2006a) emphasize the coordination role of the disclosures, arguing that one of the benefits of disclosure comes from informing the inventing community generally of who is working on what, which results have been obtained and which have not in the same way that publishing serves the academic community. In a sense, the disclosure requirement of patent law creates a "public repository of knowledge."[69] If a first inventor could easily identify the best qualified "next" inventor, she could disclose any relevant information to subsequent inventors privately for a fee, thereby profiting from the increased efficiency of the research path. Without this information—with unanticipated applications of technologies coming from unlikely sources—a public repository of knowledge may be the most efficient way to allow those with the skill and creativity to make the next step to actually contribute. To the extent that licensing actually occurs, allowing technologies to work together, the

[69] Aoki and Spiegel (1998) suggest that the recent move in the United States toward earlier disclosure may have significantly sped up the development of technology by improving the available research base.

"public repository" should also allow inventors to specialize in their area of technical competence. This coordination benefit becomes more significant the more efficient the licensing market is.

Denicolo and Franzoni (2004) add to the argument for the coordination benefits of the patent library by suggesting that the patent disclosure can reduce—perhaps unintentional—duplicative research effort. Whether the disclosure is a tool that actually publicizes this information in a form that can be interpreted and accessed readily is, perhaps, more debatable. Bessen and Meurer (2008a) have argued that, in fact, many infringement cases are inadvertent. This could suggest that potential the coordination benefits of a library are not being realized fully.

Even if competitors are already aware that a potentially profitable investment opportunity exists, the information in the disclosure can affect the nature of the race toward discovery within this general area. Disclosures make the information structure in an R&D race a choice variable for the participants. For example, suppose that Ms. A possesses an innovation that is secret and that gives her a hint about how a future innovation could be designed. Ms. A knows that Mr. B. is working on the same problem, but has not yet obtained such an intermediate result. If Ms. A discloses this information via obtaining a patent, she gives up her informational lead in the R&D race. This can create a large disadvantage to the disclosure system for users. On the other hand, Ms. A. can commercialize her innovation without fear of imitation due to the protection the patent affords on her intermediate step. In a system where taking a patent is voluntary and secrecy is always an alternative means of protecting the gains to innovative effort, the tradeoff faced by Ms. A. suggests that not all innovations will be patented and disclosed. Only those innovations will be patented for which the tradeoff goes in the direction of large gains to commercializing under patent protection and little loss in terms of an R&D race. Hence, a patent system with disclosure only ensures that some innovations may be disclosed, not that all innovations are disclosed. This, too, can hamper the coordination role of the patent system.

If disclosure has a benefit, perhaps society would be better off in a system where secrecy is not an option. Aside from the difficulty of enforcing such a policy, there are reasons why allowing firms the option of patenting could be beneficial. If firms have an observable choice between secrecy and patenting (so that it is possible to observe that a firm is keeping a secret, but it is not possible to know what the precise nature of the secret is), then the act of patenting can have signaling value. An early contribution by Horstmann et al. (1985) takes the view that the simple act of patenting signals information accrued by the inventor during an R&D stage. If the information thus revealed makes imitation around the patent more profitable for a competitor, the propensity to patent falls. Forcing full revelation is not necessarily welfare improving due to welfare losses from increased imitative R&D expenditure.[70]

Further, we have already seen that Anton and Yao (1994) show that under certain conditions a limited amount of revelation will occur under secrecy, as inventors reveal their innovations privately to a limited set of licensees. Hence, a system of secrecy will be associated with some disclosure. Anton and Yao (2004) examine which *types* of innovation may tend to be patented rather than kept secret in a signaling framework. Their model shows that it is the smaller innovations that will tend to be patented (and disclosed), rather than the larger ones. This could potentially lower the benefit of the patent disclosure since only small steps will appear in the "repository of knowledge" that the patents create. More precisely, the authors assume that the enabling information in the patent need not, in fact, allow

[70] More recently, Langinier (2005) develops this line modeling.

rivals to completely duplicate an innovation. Innovators may choose to disclose a lot of information in the patent document, thereby convincing rivals that innovations are quite significant. Such disclosure triggers imitative behavior, of course, and may result in damage payments from the imitation. Innovators could also opt for trade secrets that disclose very little but also give no rights to damages in the case of imitation. A separating equilibrium exists where small innovations are patented, fully disclosed in the patent document, and are not imitated; large innovations are kept secret and are not imitated because no information was disclosed. This model assumes a weak enablement requirement so that partial disclosure is possible. The informational requirements apart from the specific information in the enabling requirement are quite large: it must also be possible to know what *proportion* of a total amount of information was disclosed in order to derive the equilibrium in the first place. The basic point they are making, however, is that as long as secrecy comes along with sufficient control, it can generate selective disclosure. Further, as long as patenting is a choice and not an obligation, only certain types of innovation will be disclosed via patents. If the patented innovations are not the most socially valuable types to disclose, then the patent disclosure does not function optimally.[71]

The enablement requirement, while clearly linked to how much information is disclosed in the patent system, is not the only tool that affects the disclosure function. How should other aspects of the patent be designed to obtain the most out of disclosure? Scotchmer and Green (1990) examine the novelty requirement for patentability and how this can be managed to promote disclosure. Disclosure may be well served by a weak novelty requirement for patentability where even small improvements can be patented. If the uptake of patents on these intermediate steps is good, scientific progress building on known art can be rapid. This advantage is undermined if firms do not choose to patent the interim innovations in order to avoid giving away valuable information; however, a strong novelty requirement does nothing to help resolve this problem, as no further disclosure will occur under this regime. However, the novelty requirement also affects the incentives in the patent race. A weak novelty requirement could have the effect of ensuring that the market is populated with products that are relatively close substitutes and, hence, are not very profitable. While a strong novelty requirement could lead to slower discovery by any one researcher, the larger reward that it promises to those who remain in the race could lead to increased entry into innovative activity. Entry could ultimately speed up final discovery. Patenting is voluntary, however, and the fact that the weak novelty requirement opens the *possibility* that close substitutes would be provided does not *ensure* that patenting occurs. Scotchmer and Green show, to the contrary, that firms choose to suppress the interim discovery precisely when profits would be eroded. Hence, the weak novelty requirement does not necessarily lead to low rewards. One can, however, make a signaling argument for a strong novelty requirement: if the novelty requirement is weak, a firm can infer something from the very suppression of an invention when patenting was a viable option. The inference that an invention has been discovered but suppressed can

[71] Anton and Yao (2002) examine the case where ideas can be partially disclosed so that some information can be revealed, but other information can be traded privately as unpatented "know how." In their 1994 paper, where partial disclosure was not possible, the licensing contract offered by the purchaser to the inventor only had to eliminate the incentive to disclosure the information to a third party. In Anton and Yao (2002), disclosure can also be used to signal the extent of other—undisclosed—knowledge that the seller possesses. The undisclosed knowledge will be bid for by competing buyers and not all knowledge may be disclosed in the equilibrium of interest.

discourage innovation investment by a rival who thinks she has fallen behind in the race.[72] If the novelty requirement is strong, there is no option to patent so there is no signaling value in the observation of no patenting. Of course, a weaker enablement requirement—so that information in the patent disclosure need not be very complete—could undermine this argument as the overall disclosure benefit as well as the signaling value of patenting would fall.[73]

In related work with a somewhat different legal interpretation, Baker and Mezzetti (2005) and Bar (2006) develop the idea that the disclosure of intermediate steps in a patent race affects prior art. This means that disclosure of intermediate steps can affect the patentability of subsequent innovations because the subsequent innovations must be novel when held up against this prior art. Disclosures may be optimal in their framework even if they are not accompanied by patent protection. As prior art is built up of any public information (patented or not), the leading firm must make a greater improvement to obtain an innovation viewed as sufficiently novel to patent. Hence, laggard firms may wish to disclose in order to prolong the race toward a prize that gets ever farther away: the disclosure buys them needed time to attempt to pull ahead in a stochastic R&D race framework. The decision to patent, then, comes hand in hand with a decision to make the R&D stage a more complete information race. The exact interaction between the information, the exclusive rights, and the patentability criteria determines whether firms race more or less intensely and so whether discovery comes sooner or later.

Matutes et al. (1996) focus on the distinction between a patent's *disclosure* of both the embodiment of an innovation and the idea underlying it but the patent's *protection* applying only to the embodiment. Further, in the other models we have reviewed, it is supposed that all parties realize that a secret is being kept and all have a rough idea of its nature so that there was a lot that could be inferred from "no patent" or "no information." Matutes et al. take the opposite tack of assuming that until the seminal information is revealed, competitors have no clue of its existence. Hence, "no information" does not act as a signal. Instead, they focus on when to disclose when disclosure both allows the innovation to be commercialized under patent and also initiates a race for the remaining unpatented applications of the underlying idea. Sketching their "waiting game," let it take one unit of time to develop each profitable application (claim) of the basic insight. Then an innovator has an incentive to keep the insight secret by waiting a period of time before introducing any of its applications, as this postpones the time others realize that a fertile insight is available to be built upon, and so start developing applications of their own. In other words, once the "cat is let out of the bag," m potential entrants will start to develop any applications that have not already been developed and protected by patent by the first innovator. Hence, there is a positive externality of the disclosure that is not internalized by the first innovator: the first innovator will use trade secrecy to postpone this race to grab applications, even though waiting is socially harmful because it delays commercialization of the applications. From the initial inventor's point of view, the impatience to commercialize the applications that it has already developed creates an incentive to patent early to weigh against the incentive to prolong the period of development "in secret." More precisely, if

[72] See Gill (2008) for more discussion of strategic disclosure as a tool to make a competitor drop out of an R&D race.

[73] In a related point, Aoki and Spiegel (1998) have suggested that the recent move in the United States toward earlier disclosure can significantly speed up the development of technology by improving the available research base. Shapiro (2004) suggests that early disclosure may have been associated empirically with less strategic activity that could be undesirable, such as so-called "submarine" patents.

A potential applications exist and the initial innovator waits λ units of time before disclosing, then when each application earns the same fixed sum, the innovator would choose to wait according to

$$\max_{\lambda}\left(\lambda e^{-\lambda r}+\int_{\lambda}^{[A+(m-1)\lambda]/m} e^{-rt}\mathrm{d}t\right).$$

They investigate the extent to which limiting this kind of waiting to access the patent system in the first place can be addressed legally by granting claims on applications that are not yet fully worked out. This has been termed granting a "license to hunt" by means of a patent. Granting a limited license to hunt increases the incentive to disclose the first step because if creates an entry barrier—a zone of exclusivity—into a set of potentially profitable innovations.[74] Indeed, under this lens, leniency on large patents, with many (nonoverlapping) claims, submitted at an early stage has a welfare benefit of speeding up useful disclosure.[75] Of course, the disclosure is assumed to be useful here: potential researchers face no difficulty in wading through and digesting reams of claims.[76] This chapter suggests that delay that facilitates "hoarding" may occur in the case where accessories could follow on to an original "platform."[77]

Hence, both the enabling disclosure in the patent and the act of patenting *per se* carry information that affects imitative and innovative behavior. While the positive externality conferred by the enabling information has social benefit, private parties may be expected to disclose less than the socially optimal amount. This limits the value of patents as a "repository of knowledge." Secrecy does promote some limited disclosure, although the disclosure may be private to only licensing partners. On the other hand, the standards of enablement, the novelty requirement, the role of previous patents in the definition of prior art, and the leniency in granting claims on prospective applications are all available tools to affect the amount of disclosure actually obtained by the patent system. More complex strategic reasons to decide to patent or not also exist and can be used to manipulate the behavior of rivals in a model of disclosure. This latter function derives from the optional nature of patenting, and the existence—in some cases—of commercially viable alternatives.

[74] The claims in new and quickly developing fields can be rather speculative, in particular in some pharmaceutical areas according to Bidgoli (2010, Chapter 216 "Innovation and Intellectual Property").

[75] The idea of giving prospective protection on claims that are not fully developed recalls the work of La Manna (1994), discussed above, where there is a benefit to reserving territory to a single patent-holder before investment occurs, although diffusion benefits are not a concern in that model. Kitch (1977) identifies a "prospect function" for patents when they are granted early in development. See Merges and Nelson (1990) for a discussion of this view.

[76] Chen and Iyigun (2006) incorporate concerns about delay in patenting and disclosure into a model of economic growth. Duplicative research expenditure is less of a concern in their framework, so imitation will be observed in their optimal patent design, in contrast with most of the papers reviewed here.

[77] Noel and Schankerman (2006) finds evidence for hoarding in complex industries. No complementarity is required to obtain hoarding in the Matutes et al. model, as the accessories are assumed to be independent "pots of gold" of equal size for whoever innovates them.

5.5. Alternatives to a patent system: Optimal procurement of innovation

The discussion of the preceding section took for granted that some system rather like the current patent system would be used to generate innovation incentives. We considered modest modifications of this system, using existing tools such as statutory patent length, infringement standards, a patentability requirement, the enablement requirement, and the interpretation of claims to achieve social goals in the most efficient way. If we were to start from a blank slate, it is not clear, however, that patents as we know them would be our chosen optimal scheme to "promote the progress of science and the useful arts." Indeed, many other schemes have been and are still used to achieve this goal.[78]

Wright (1983) develops the point that a patent-like system might not be the best mechanism in a fully optimized model of innovation incentives. Instead of using the monopoly mechanism to create a reward for innovation, the state could instead simply award compensation in the same amount directly to the innovator and obtain a welfare gain. In other words, whatever the reward available through the patent system, transferring that reward to the innovator as a lump sum rather than as a result of a market distortion achieves the same innovation incentives with less deadweight loss. He refers to this sort of payment made by an authority and conditional on the delivery of a completed advance as a "prize." Hence, if the value and cost of the innovation are publicly observable, and if the funding of the prize is relatively nondistortionary, then the prize system will dominate the patent as an incentive mechanism. Wright goes on to suggest that subsidies could dominate patents as well. Competitive bidding can be used to contract out the research before its completion to ensure that only the most efficient researchers are used. While this removes the normal racing incentive, this system can achieve higher welfare than a patent if timely innovation can be induced via performance requirements.

Information is rarely this good. In particular, it may be more realistic to assume that the authority awarding the prize has less information about the value of any candidate innovation, let alone its cost, than the innovators themselves. In the presence of asymmetric information, Wright shows that any one of the three mechanisms—patents, prizes, or contract research—could be the best mechanism. Patents have the advantage that they delegate the decision of which investments to put forward to the "informed party," the inventor. If the inventor is the one who knows which investment will generate the most value, but the sponsor does not (without incurring a cost), this represents an improvement over a prize system that would require the government to "pick winners."[79] This gain can outweigh the deadweight loss associated with the patent system. Further, a prize has a drawback of its own in the case where innovative activity can be conducted by many parties since the prize does not limit entry. Hence, it suffers from generating excessive research expenditure due to the "common pool problem" that was discussed above as part of the reward theory. One alternative to reduce this incentive for duplicative expenditure is to reduce the size of the prize. If the funding body has poor information on the value or cost of the research, however, this system might only elicit low value ideas. Contracting out research

[78] See Scotchmer (2004) for an in-depth review and history of other schemes to promote innovation, including prizes, subsidies and direct procurement.

[79] Shavell and Ypersele (2001) show that a combined reward *cum* patent system always dominates a pure patent, taking as an example a case where innovations can be ranked by their value and the prize is set so as to induce the lowest innovation "type" (value) to switch from patent protection to the prize. This reduces deadweight loss and does not suffer from a problem of picking winners, although it does require enough knowledge about possible innovation types to be able to set this minimum reward.

potentially avoids excessive research expenditure, but it may not result in sufficient incentives to create value in the first place, precisely because it eliminates research competition. In other words, while the incentive to "race" for the patent may be too large, the incentive to invent at all for a single, designated contract researcher may be too small. The contract would need to be designed to ensure that the best idea was selected for funding and that invention incentives were maintained. The possibility of designing such a contract depends on what is observable to the funding body and on the credibility of the promise to pay for deliverables.[80]

When information is asymmetric, Gallini and Scotchmer (2002) point out that designing an incentive mechanism for innovation can be broken down into three "steps." First, there is the decision problem of whether a project should be undertaken. Second, there is the delegation problem of which firms should undertake the investment and at what rate. Finally, there is the funding problem of how to reward the investments. A set of papers applying mechanism design machinery to innovation incentives have begun to address these three points.

One approach has been to incorporate a modified "prize" into the patenting system by means of patent buyouts. Kremer (1998) suggests a system that effectively awards prizes but does not rely on the planners identifying the value of potential innovations beforehand. Let a period of time elapse during which firms hold their patents in the normal way. After this elapsed time other firms, besides the one that patents, are likely to have an idea of the private value of the innovation. This information can then be marshaled by the planner to create a reward for the innovation using a second price sealed bid auction to elicit payments for the right to the innovation from the private parties. In order to maintain the incentive to bid, with small probability the patent will be transferred to the highest bidder in exchange for the winning bid. Otherwise, the innovation is placed in the public domain. In either case, the price determined by the auction would be paid as a prize to the original patent holder by the government out of general tax funds (or some other budget associated with the office responsible for the buyout mechanism). In fact, to reflect the difference between the private and social value of the patent, the government could apply a positive markup to the winning bid when making its payment. Whether to put a patent up for auction (or "buyout") would be at the discretion of the patent holder. If the bids are relatively low, the patent holder can refuse to sell. Clearly, this proposal is meant to complement the patent system,[81] not replace it: the initial award of a patent is essential to the mechanism. The proposed approach will dominate a pure patent system as long as the administrative costs of the buyout and the cost of public funds are not too high.

Hopenhayn et al. (2006) investigate buyouts in a quality ladder framework. Formally, the authors first consider the case when an innovator's "type"—their innovative ability—is observable. They show that the optimal patent system can take an "exclusive" form where a quality leader above some threshold

[80] See Scotchmer (2004) for a discussion of contests for sponsored research. The more knowledge the planner has about what the target of research should be, its value, or its cost the wider the set of alternative instruments to patents. Maurer and Scotchmer (2004) discuss a variety of procurement mechanisms, including auctions, prototype contests, grants, and matching funds. Trajtenberg (2002) provides extensive discussion of how these alternative government supports have been used to advantage in Israel.

[81] Brunt et al. (2008) observe that prize awards, even nonpecuniary ones, can have a large inducement effect for innovation when used in conjunction with a patent system; however, the prizes studied are not buyouts. Rather they are awards to scientists to undertake research that can lead to patents.

ability receives patent rights while all other previous innovators' rights terminate. More precisely, suppose that a social planner can make promises of durations, k, in the form of a set of time periods during which no other innovation may be implemented. In each period a new innovator arises. In such a case, allocating monopoly power to the current innovator potentially curtails the planner's ability to allocate monopoly power to future innovators. Furthermore, allocating a period of exclusionary rights today may postpone the benefits one could earn from a superior innovator in the future. The planner's problem needs to take into account several constraints. First, innovators will choose the quality increment v that they target as a function both of their basic ability, θ, (their "type") and the duration of their monopoly power, k, so that we have $v(k, \theta)$ when they are assumed only to earn profits during the k periods that they are allocated. Second, the planner must keep her promises in the sense that the cumulative duration already allocated, K, must equal the duration promised to all innovators who have already implemented their innovations. Hence, if $k_{\mathrm{p}}(\theta)$ is allocated to previous innovators in each period, we have $K = \int k_{\mathrm{p}}(\theta) g(\theta) \mathrm{d}\theta$, where the distribution of types is described by density function $g(\theta)$. Third, the total duration allocated in each period to previous and current innovators, $k_{\mathrm{p}}(\theta) + k_{\mathrm{c}}(\theta)$ cannot exceed the total discounted time horizon $(1 - \beta)^{-1}$ when discount factor β is used. Finally, the rule by which K changes in each period is that, during a single period $k_{\mathrm{p}}(\theta) + k_{\mathrm{c}}(\theta)$ is allocated, but at the same time a single period elapses so that $\tilde{K}(\theta) = (1/\beta)[k_{\mathrm{p}}(\theta) + k_{\mathrm{c}}(\theta) - 1]$ is the balance of duration that remains next period. Under all these constraints, the social planner grants $k_{\mathrm{p}}(\theta)$ and $k_{\mathrm{c}}(\theta)$ so as to maximize the expected present value of all future innovations, W, given that K units of time have already been allocated. This value is composed of the contribution of the innovator to social welfare, which is the contribution of the quality improvement over its development cost, chosen optimally as a result of the policy set by the planner, times the entire future duration over which the innovation will not be excluded, $(1 - \beta)^{-1} - k_{\mathrm{p}}(\theta)$. Summarizing, then, we have the following expression for W as a function of K:

$$W(K) = \max_{k_{\mathrm{p}}(\theta), k_{\mathrm{c}}(\theta)} \int \left[\left[\frac{1}{1-\beta} - k_{\mathrm{p}}(\theta) \right] v(k_{\mathrm{c}}(\theta), \theta) - c(v(k_{\mathrm{c}}(\theta), \theta)) + \beta W(\tilde{K}(\theta)) \right] g(\theta) \mathrm{d}\theta$$

subject to the constraints discussed above. Given the appropriate sorting assumptions, they establish sufficient conditions under which the optimal patent system is of a form whereby innovators who report type above some threshold θ obtain exclusionary rights, which begin immediately upon grant without delay. In other words, when $k_{\mathrm{c}}(\theta)$ is positive, $k_{\mathrm{p}}(\theta)$ is set to zero. Conditional on not being replaced, a current rights holder retains unchanging protection.

Notice that the proposed mechanism addresses the problem of whether investment should occur at all as well as the delegation problem identified by Gallini and Scotchmer (2002) in the sense that only firms with types above a threshold ability find it profitable to undertake investment. It also designs a system of rewards consistent with this. The mechanism is similar to O'Donoghue et al.'s (1998) result of limiting (and constant) breadth and infinite statutory length. The framework is much less tied to legal institutions of the existing patent system, however, so it is difficult to translate the assumptions of the model into precise legal principles that are currently observed.

We have not discussed yet how buyouts enter the Hopenhayn et al. model. When it is not possible to observe type θ, the "exclusive patent system" we have described can be decentralized into a *mandatory* buyout system. The buyout takes the form of a payment to the current market leader to displace her, as well as a specified buyout amount that the new leader would accept to be displaced by another. The buyout also involves a transfer fee, paid to the granting authority, and potentially varying by the innovator's type. To find the buyouts that result in the same duration, $k_c(\theta)$ as in the optimal patent system described above, the authors derive a revelation mechanism such that an innovator of type θ will report his true type. The payment that implements the optimal policy can be shown to be separable into two parts. The first, $\sigma(\theta)$, is a function only of the innovator's type and the second, $\gamma(K)$, is a function only of the cumulative protection, K. The authors propose that a new innovator must pay buyout $\gamma(K)$ to the existing (exclusive) patent holder and a fee $f(\theta)$ to the planner that entitles him to a buyout $\gamma(k_c(\theta))$ in the future. Since the fee entitles the patent holder to a buyout, the fee includes a component dependent on the innovator's type ($\sigma(\theta)$) and the future buyout fee. In this way, the innovator buys out the current innovator, while the fee he pays to the planner incorporates a payment that will eventually be "refunded" by a future innovator. This buyout scheme is simple in the sense that it involves only a list of fees and buyouts. The authors point out, however, that for such a system to be derived, the planner must know a great deal about the structure of the innovation system, including the cost of development and the distribution of the types. Furthermore, the model analyzes only a single and definable "ladder." In point of fact, it may not be at all clear which ladder(s) a particular innovation is on. In terms of whether the patent system could be amended to look more like the proposed mechanism, the authors note that optimal prespecified and efficient licensing payments can potentially serve some of the same functions as a buyout organized by sponsors and could effectively implement the optimal system. This would put us back much closer to the earlier literature on optimal patent design with efficient licensing.

Kremer (1998), in his discussion of his own proposals, documents many practical problems with buyout schemes. A few examples will suffice to give an idea of the difficulties of moving toward such a system. First, as emphasized by Wright (1983), the "true private valuations" revealed by this system should include business stealing effects. As we discussed above, the private value may exceed the social value when business stealing effects are present. Kremer's "markup" reflecting the gap between social and private values could sometimes be negative if business stealing was present. More generally, this markup would depend on the innovative industry's structure as well as the nature of the patented information. Referring to his own "voluntary buy out system," Kremer suggests that there is a lemons problem in the sense that firms that know that a new innovation in their own research pipeline will eclipse an existing innovation might tend to be the ones putting their innovations up for sale in order to exploit this. Hence, there is an underlying signaling problem that could affect the performance of the system and would affect its design. On the other hand, a mandatory system, such as Hopenhayn et al.'s formulation, may avoid this by dint of being mandatory. Third, the optimal system depends on how patents interact: whether patents are complements, substitutes, or on the same—independent—quality ladders. We have already argued that the substitutability or general interaction between patents is not clear cut in many cases, following Lerner and Tirole's analysis. Fourth, auction-based buyout mechanisms can only be as good as the auction mechanism on which they rely. No auction mechanism is perfect. For example, the second price sealed bid auction is vulnerable to collusion amongst participants. Finally, there are knotty political economy issues associated with making a buyout system widely available to all patents, even if one takes for granted that the economic issues can be solved. For example, if it were widely publicized that frivolous patents—of which there appear to be many[82]—were receiving payouts from the government, those paying into the general tax funds might not react well.[83]

Scotchmer (1999) and Cornelli and Schankerman (1999) suggest pairing the patent system with other incentive tools, focusing on subsidies. Both start from the idea that one wishes to preserve the desirable self-selection characteristics of the patent system, while not sacrificing the advantage that subsidies have of not creating deadweight loss. Weighed against this advantage is the disadvantage that subsidies potentially encourage those applicants whose inventions have little social value to come in search of a handout.

Using Cornelli and Schankerman's presentation of the mechanism design problem, suppose that firms may be of a variety of types, where the government wishes to shift the distribution of R&D effort toward the types that are highly productive in order to minimize the social cost of producing innovations. The optimal patent policy, then, is a time of protection, T, which is a function of the announced type of the innovator, θ. In this case, we can think of θ as indexing the skill of the researcher in terms of producing an innovation (as in the Hopenhayn et al. paper, discussed above) or the value of the innovation for both society and the researcher. For example, we could think of the profit from innovation as $\pi = \theta e$, where e is effort. Hence, given private information θ the firm chooses the "size" of innovation, π by setting e. The government only knows the distribution of the θ. The cost of effort is some nonconcave function.

Suppose that the researcher announces $\hat{\theta}$ and the planner determines a patent length, T, and fee, f, (to be paid into the system by a patent applicant) according to the announced value, $\{T(\hat{\theta}),f(\hat{\theta})\}$. The researcher responds by choosing effort, e^*, as a function of this schedule. The welfare maximization problem, where w denotes the flow welfare gain from the patent (profits plus consumer surplus that are created by the innovation and contribute indefinitely to welfare) and d denotes the flow deadweight loss due to patent protection (available as a gain once the patent expires at time T) becomes

$$\max_{T,f} \int_0^{\bar{\theta}} \left[\frac{w(\pi(\theta, e^*))}{r} + \frac{d(\pi(\theta, e^*))}{r} \mathrm{e}^{-rT(\theta)} - c(e) \right] \mathrm{d}G(\theta)$$

$$\begin{aligned} &\text{subject to:} \\ &U(\theta,\theta) \geq 0 \quad \text{(individual rationality)}, \\ &\theta = \arg\max_{\theta} U(\theta,\hat{\theta}) \quad \text{(incentive compatibility)}, \end{aligned}$$

where

$$U(\theta,\hat{\theta}) = \int_0^{T(\hat{\theta})} [\pi(\theta, e^*) - c(e^*) - f(\hat{\theta})]\mathrm{d}G(\theta).$$

[82] Jaffe and Lerner (2006) discuss the quality of patents.

[83] If the buyout system is not self-financing, we need to take account of the effect of financing on the general economy. An advantage of patents is that they "tax" those who participate in the market for the innovation. While general tax payers under a prize/buyout system arguably gain from the elimination of deadweight loss in a system where transfers occur, the transfers would need to be established. Further, these deadweight loss benefits are more hidden than the explicit payments made to innovators.

The solution of this could generally include negative fees for low type θ firms: in other words payments—subsidies—from the government to particular types of researchers. The authors point out that this would require both monitoring schemes to ensure that low types actually innovate and public funds to provide the subsidy. Not surprisingly, the length of protection increases with type θ, in order to satisfy incentive compatibility. Hence, the length of protection, $T^*(\theta)$, (strictly) increases in θ so that heterogeneous types of researchers would generally be associated with heterogeneous protection regimes. Indeed, this optimal direct mechanism can be implemented by using either an upfront menu of patent lengths and fees or a renewal fee scheme.[84]

In a more general framework, Scotchmer (1999) shows that a system that does look like a patent, a mechanism like the Cornelli and Schankerman system, is what an incentive compatible mechanism *must* look like when the economy has a single firm innovating once and where the cost and value of the innovation are not observed by the social planner (but are known to the innovator). In this system the payoff to reporting a value and cost pair such that the patent authority would ask the firm to conduct the research in the first place must at least equal the payoff to saying that the research is not socially worthwhile (individual rationality). Further, low value innovations must get a subsidy but little patent protection while high value innovations must pay a fee and get high patent protection in a way that achieves incentive compatibility. Low value innovations do not mimic high value ones as they would have to *pay* a fee instead of *receiving* a subsidy; furthermore, the inventor of a low value innovation would get little value from the stronger patent protection. The subsidy instrument is set to optimally tax firms once incentive compatibility is achieved.

The interesting result of these two papers is that the optimal system does not look "far off" from the patent *cum* renewal fee system that is actually observed. Specifically, the best system in these papers is a renewal system that specifies a menu of fee payments in exchange for extensions of the patent life. The patent holder purchases these extensions to patent life more readily for high value innovations, so only these patents are long-lived.[85] Further, the incentive to develop high value innovations is stronger since precisely these innovations receive longer protection. Cornelli and Schankerman show, however, that there are significant differences between the shape of the renewal fee schedule predicted by their theoretical framework and that observed in practice. First, the scheme derived from the optimal mechanism suggests subsidies for small innovations, which are not observed in practice (at the patent office, at least). Second, fees should be rising sharply over time in the optimal structure, whereas in their sample of European countries, fees actually fall. They comment that these differences should not be surprising in light of the fact that patent renewal fees tend to be set so as to finance patent offices rather than with any sort of optimal mechanism for eliciting innovation in mind. Still, these papers suggest that the "tools" exist and could be adjusted to implement a socially desirable system.

[84] Since $T^*(\theta)$ is strictly increasing, it can be inverted to obtain the type associated with each patent length. This can be substituted into the optimal fee schedule, $f^*(\theta)$, that is derived from the constraints on the problem, above, at the optimal schedule of patent lengths. Hence, we obtain $F(T) \equiv f^*(\theta(T))$ as the fees associated with each patent length. Alternatively, an annual renewal fee such that the sum of the renewals during the lifetime, $T^*(\theta)$, adds up to $f^*(\theta)$ would implement this solution.

[85] Pakes (1986) models patents as options in the sense that they are applied for at an exploratory stage, and information about their value is "revealed" over time. Early renewal decisions are based on both current (known) value and an option value to renew in the future. Higher value patents are those that are renewed longer, but this value is revealed over time, including an ever smaller option component and a greater certainty component. This concept of value is consistent with the models discussed here.

As was mentioned above, a further characteristic of these models is that the effective patent protection of different types of innovations differs: the uniform patent protection that is generally observed in practice is not optimal. Cornelli and Schankerman conduct simulations to illustrate the welfare loss of moving to a uniform level of protection from a heterogeneous system, finding that the optimal mechanism generally raises welfare 2–7% above uniform protection.

Indeed, Hopenhayn and Mitchell (2001) explore heterogeneous protection more fully in an optimal mechanism that allows patent authorities to choose the length and (leading) breadth of protection as well as the renewal fee schedule. In other words, rather than consider just fees and the length of protection as instruments, Hopenhayn and Mitchell (2001) reintroduce breadth (which can be thought of most straightforwardly in terms of quality increments in a quality ladder, but could also be an exclusion zone in product variety, similar to Klemperer, 1990 in their general framework) as an instrument to determine how all three of these tools can be combined optimally by social planners. Their paper brings the patent design literature "full circle" in the sense of bringing the tools explored in the earlier literature into a mechanism design framework. The paper also derives the result that in this setting, the optimal fee is zero. This latter result questions, then, the conclusions we drew on optimal fee schedules based on the papers reviewed above where breadth was omitted as an instrument.

Hopenhayn and Mitchell (2001) suppose that ideas of type $\theta \in \Theta$ arrive to innovators with probability $g(\theta)$. This type is not observable by the planner. The development cost of the innovation is c. Innovators can make profits only if they obtain both length, T, and breadth, k, of protection so that profits are a function $\pi(k, T, \theta)$. The planner must maximize the social benefits of this protection regime, W, subject to the usual individual rationality and incentive compatibility constraints when it can set breadth, length of protection and also impose a fee. Hence, the problem is straightforward:

$$\max_{k(\theta),T(\theta),f(\theta)} \sum_{\Theta} W[k(\theta), T(\theta), \theta]g(\theta)$$

such that:

$$\pi(k(\theta), T(\theta), \theta) - c - f(\theta) \geq 0 \text{ (individual rationality)},$$

$$\pi(k(\theta), T(\theta), \theta) - c - f(\theta) \geq \pi(k(\hat{\theta}), t(\hat{\theta}), \theta) - c - f(\hat{\theta}) \ \forall\hat{\theta} \text{ (incentive compatibility)},$$

and

$$f(\theta) \geq 0.$$

Notice that fees, as a pure transfer, do not enter directly in the objective function. In this setting, the fees are set optimally to zero under the appropriate sorting conditions. The reason is that the policy design problem optimizes social welfare under constraints including covering research costs. Since fees raise these costs, they also tighten the constraint. This makes fees a relatively inefficient instrument compared to breadth or length of protection, which operate instead by raising value.

If innovations can be ranked according to how efficient length of protection is at generating surplus for the innovator, then the optimal contract involves a menu of length and breadth offered to different types of innovation where those innovations that get little value out of length get primarily breadth protection (large breadth and small length) while those that get large value out of length get primarily length protection (small breadth, large length). The justification is that some innovations are "fertile": they will generate follow-ons—developed by other firms—that replace the original innovation in a short time span. As we have seen elsewhere, statutory length protection is of little value in this case since "effective length" will be much shorter. For example, suppose that the probability of arrival of a follow-on innovation in the first t periods after patenting for an innovation of fertility type θ is $p(t, \theta)$. The patent holder makes profits π per period and the basic innovation costs c to develop. Imagine that for some low fertility innovation type, θ_1, it is the case that an innovation with protection length T_1 and no breadth at all (so that all improvements are free to infringe) is just expected to generate enough discounted profits to cover the development cost. This may be the case because it is unlikely that improvements will be found during time span T_1. On the other hand, another innovation might have high fertility type, θ_2. Even with an infinitely lived patent, it could be that this type might not be anticipated to cover its development cost with a zero breadth of protection. Instead, since improvements arrive very quickly, such an innovation requires positive breadth of protection, k: all improvements would be barred from the market for the duration of the patent. Hence, a patent system could involve protection levels $(0, T_1)$ and (k, T_2), where $T_1 > T_2$. The low fertility type would strictly prefer the first type of protection and the high fertility type would strictly prefer the latter type. The patent authority can then screen innovations when it offers such a menu of protections.

In this framework, adding (leading) breadth is more effective at generating profits for high fertility innovations since this "slows down" the time at which the innovation will be replaced in a way that recalls O'Donoghue's (1998) work. Notice that there is no licensing allowed here and that patentability and infringement are tightly linked. Further, length is adjusted in each case so that the "minimum" level of profits to induce innovation is always sent to the innovator in the optimal scheme. This begins to look like a very complex set of requirements to explain to the inventing public, let alone the public at large, and to implement at reasonable cost with realistic levers. That being said, some new proposals from industry have suggested having "deluxe" patents and "run of the mill" patents in a menu that would be at the disposal of patent applicants.[86] On the other hand, the point remains that while the Cornelli and Schankerman (1999) model appears practically implementable with existing tools, this mechanism—especially its conception and manipulation of breadth—would require considerable development to imagine in practice.

6. How is the right secured? Enforcement

Intellectual property rights are only as good as their enforcement, and enforcement largely occurs by means of private suits brought by individual inventors or groups of inventors against other inventors. Indeed, Crampes and Langinier (2002) note that a patent merely grants the right to sue intruders that have been identified. Identification must be done by the patent holder at some monitoring cost, and even if infringers have been identified, the patent holder has the choice of how to react: by defending the patent in a court suit, settling out of court for some negotiated value or simply accommodating the entry. If the defense involves a countersuit questioning the validity of the patent, the patent holder could find that the upshot of litigation is to lose all rights to the intellectual property. Lemley and Shapiro (2005), developing an idea also put forward by Ayres and Klemperer (1999), propose modeling patents as

[86] IBM has proposed this two tier Community patent, see http://www.epip.eu/conferences/epip02/lectures/European%Inter-operabily%20Patent%201.1.pdf.

"probabilistic" to capture the idea that patents only give a possibility—and not a guarantee—of a reward. The entry decision of a potential imitator depends on how aggressive the response to entry will be but also on the prior belief that the parties have about the likely strength of the patent, should it be challenged in court. The response to entry will, in turn, depend on the underlying patent strength, the characteristics of the firms involved, the market, and the cost of the various alternative strategies.

Lanjouw and Schankerman (2001) note that while the average litigation rate of patents is low—on the order of 1% of all patents—the probability of litigation of more valuable patents can be above 10% in some fields and more than 25% in pharmaceuticals. Infringement suits, such as those modeled by Crampes and Langinier, are more common than invalidity suits. Aside from the inherent reasons why the number of invalidity suits should be lower—an invalidity would generally arise in response to an infringement accusation—Lanjouw and Schankerman note that due to the positive externality that a litigant creates by bringing an invalidity suit, their seeming paucity may not be surprising. In other words, a single litigant carries the entire burden of the trial costs, but if a patent is found to be invalid all potential users of the technology could benefit. Parties bringing private suits must weigh the high cost of the suits against the likelihood of covering these costs by extracting a high value in settlement. In terms of costs for those accused of infringement, Lanjouw and Lerner (2001) find that preliminary injunctions—bars on the allegedly infringing activity—are used quite often (almost 20% of the cases they studied) as a remedy to infringing behavior. An injunction can potentially "shut down" a business, which can run up the costs of being accused—rightly or wrongly to high levels.

The probabilistic nature of patents creates several effects. The first, following Ayres and Klemperer (1999), is that the probabilistic nature of patents induces a certain amount of infringement because infringement may pay off: the patent may not be upheld as valid (or may simply not be enforced).[87] This compensates infringers for the risk they run of being held in violation of a valid patent. Their level of risk depends on the damages they must pay in the case of a finding of misconduct, and also the delay in the finding of infringement. A regime where patents are probabilistic but where disputes resolve slowly may benefit consumers without compromising innovation incentives[88] in the same way as compulsory licensing helped consumers in Tandon (1982) or restrictive competition policy helped in Gilbert and Shapiro (1990). Ayres and Klemperer suggest that an optimal regime would allow patent holders to choose from a menu of lengths of protection and probabilities of enforcement. Such a menu allows efficiency gains via lower deadweight loss when there is limited infringement, at the same time returning a target reward level to innovators, and harnessing the private information of patent holders. In this sense, their ideas extend some of the insights of the Cornelli and Schankerman (1999) and Scotchmer (1999) frameworks to the case of probabilistic patents.

[87] Hall and Ziedonis (2001a) note, based on interviews with industry participants that, "until the Kodak–Polaroid case, infringing firms generally expected to pay royalties on past use of the property covered by the infringed patent (a reasonable risk and slightly less expensive in an expected value sense than paying royalties from the beginning); in contrast, after the Kodak–Polaroid ruling, firms perceived that they could be shut down in an injunction rather than simply paying the infringed firm."

[88] It is important to have both uncertainty and delay to get full benefits. If disputes are resolved immediately, then with some probability the patentee sets the unconstrained monopoly price. If there is no uncertainty, but damages are set high enough to fully reimburse the patentee, then infringers cannot break even and so do not enter in the case where the patent is known to be "ironclad."

Ayres and Klemperer (1999) list a long set of caveats to their generally positive take on uncertain patent rights. This more negative side of the uncertainty coin has been developed by others in a series of papers.[89] One such negative is that the probabilistic nature of patents can create bias in the type of research that patents induce. The argument, put forward by Farrell and Shapiro (2007) is as follows. A standard response of an infringer is to challenge the validity of the patent at issue in court. One could think of an index of patent strength, S, where this index reflects the probability that the patent would withstand a test of its own validity (e.g., showing that an invention fails to meet novelty, nonobviousness or usefulness criteria). Patents that are weak—have a low value of S—can "punch above their weight" in terms of licensing revenues if private suits are not always brought.

As an example, consider an upstream lab that relies on licensing revenues from downstream industry for its income. A licensing scheme using per-unit royalties will be optimal for this patent holder as it allows the effective marginal cost of the licensees to be raised thereby restraining downstream competition. Indeed, the monopoly outcome can be mimicked with the right choice of royalty. If there is also a fixed fee component to the licenses, then the fee can be used to distribute the profits from this newly "collusive" industry among the participants in this scheme, including the lab. Even a weak patent can be put to such a cynical purpose.

The antitrust status of the licensing agreement we have just outlined would be tenuous, of course. Even if we do not allow this type of collusive scheme, however, a weak patent can generate surprisingly large revenues. As there is a positive externality for any single litigant to bringing a suit that reveals the true weakness of a patent, if a license for the patent is held by many firms, there is an underincentive for anyone to invest in the privately borne cost of litigation. For example, suppose a process innovation reduces marginal cost from c to $c - \varepsilon$. Let a potential licensee face the choice of joining up to a licensing agreement (now) or being excluded (forever) if she brings invalidity litigation. Additional litigants do not improve the chances that the patent will be found invalid in court. A potential licensee will accept the royalty rate proposed by the patent holder as long as it exceeds the payoff to litigation:

$$\pi(c - \varepsilon + r, c - \varepsilon + r) \geq S\pi(c, c - \varepsilon + r) + (1 - S)\pi(c - \varepsilon, c - \varepsilon),$$

where profit, π, has two arguments: the effective marginal cost of the licensee and that of its rivals. The left-hand side of the inequality represents the profit earned if all firms accept the proposed royalty rate, r, on top of the lowered marginal cost of production that the innovation generates, $c - \varepsilon$. The first term on the right of the inequality is the payoff to being the one "excluded" firm that brings suit—and loses—weighted by the probability that this will occur. The last right-hand term is the payoff to free access to the technology in the case that the suit is successful. Only one firm needs to bring the suit in order to generate this gain to all, so only a single firm—at most—will decide to be the "excluded" party. All the others would do better accepting the proposed royalty and awaiting the outcome of litigation. The maximum royalty rate that all firms will accept will depend on the importance to a firm of small changes in its own cost compared to small changes in the cost of the entire industry, as this determines how a change in royalty rate affects the right and left sides of this equation. In particular, if one thinks that the "fair" royalty rate for a patent of strength S that reduces marginal cost by amount ε is $S\varepsilon$ there are many cases where the royalty will exceed this amount.

[89] Farrell and Shapiro (2007), Lemley and Shapiro (2005, 2007), and Shapiro (2008).

Hence, weak patents may be "overcompensated" relative to their true strength (as measured by benchmark $S\varepsilon$). Not only do we suffer deadweight loss because royalties are relatively high, but also research incentives are biased toward paths that result in "little steps" that would normally fall below patenting criteria because precisely these steps are overcompensated in the market.

The policy implication of Farrell and Shapiro's model is that there is a benefit of patent review by the patent office to "weed out" weak patents before this licensing game ever occurs. The review would need to be directed at novelty and nonobviousness to implement the concern underlying the Farrell and Shapiro framework: if instead patents lack strength due to incorporating subject matter that has not heretofore been patented, it may be precisely those more creative and big steps that result in "weak patents." Hence, the policy recommendation for the review must place a specific interpretation on "weakness." Encoua and Lefouili (2008) note that the argument we have made applies mainly to "small enough" innovations. If one allows larger—drastic—innovations in a similar framework, they can generate the result that weak patents punch *below* their weight under the right assumptions.[90] Despite this, the point made by Farrell and Shapiro is that correcting "errors" in awarding patents via the courts may create distortions that could be avoided by more careful patent office review. While totally eliminating those errors is probably too costly, as Lemley (2001) suggests, simply relying on private suits to sort out errors also carries significant costs that could be avoided by patent office review.[91] A compromise that eliminates more errors would probably be desirable.

Hence, one "take" on modeling enforcement issues is to model patents as generating a probability of benefits, but not a guarantee. The outcome of the probabilistic models depends, of course, on what remedies are allowed to the patent holder in the case of a successful outcome in court. This leads to modeling that compares damage systems to injunction systems as means of upholding the patent right. Indeed, a fundamental legal issue is whether market conduct should be managed by means of completely suppressing certain practices, for example by enjoining behavior, or by allowing these practices subject to the payment of damages in the case of harm. Hylton (2006), commenting on an argument made earlier by Calabresi and Melamed (1972), notes that property rules—such as the intellectual property right we have analyzed so far—prohibit others from infringing the property right without first gaining consent. In contrast, liability rules do not require consent, but rather simply require the payment of damages when loss has occurred. Whereas liability may simply make an activity unprofitable, an injunction can directly prevent conduct that infringes that right.[92] Formalizing this idea, Anton and Yao (2004), discussed earlier, explicitly include the possibility of damages as a "reward" that inventors can collect in the case of imitation that is triggered by the patent disclosure. In such a system, triggering imitation can be good for the patent holder, if we assume that damages are not too expensive to collect and are awarded based on the true value of the innovation. Indeed, in that model damages and licenses would serve some of the same functions. Hylton focuses more narrowly on the cost side, examining the role of transaction costs and the distribution of valuations of the property to

[90] See Encoua and Lefouili (2008) for derivations.

[91] Uncertainty about which of two competing innovators' patent claims might be valid might result in spreading the expected rewards to the invention. This, in turn, could induce entry as "joint ownership" of the right to produce could result across parties. See Footnote 1 and references contained in Ayres and Klemperer (1999).

[92] Ayres and Klemperer draw an analogy to type I and type II errors: Let the null hypothesis be that the patent is valid. A damage system corresponds to a system where a valid patent might not be enforced, creating a type I error. An injunctions system corresponds to one where an invalid patent might be enforced, creating a type II error.

determine whether property rules or liability rules are more desirable. For example, if bargaining is not possible, then a high value user might effectively be barred from creating value under property rules. This case would be equivalent to the case of no licensing in the earlier papers we reviewed on patent design. Damages also have their own problems: asymmetric information on the magnitude of the damages could create a wedge between the actual damage award granted by the court and the damage truly incurred by the property holder. Hence, the nature of market failures in the transfer of intellectual property can induce a ranking of liability and property rule systems.

Ayres and Klemperer (1999) focus largely on damages, arguing that these will promote limited infringement as imitators "try their luck" against a patent holder who may not (successfully) enforce her patent. Limited infringement may be just what an efficient system should aim for, as limited infringement effectively puts a cap on deadweight loss, but still allows some per-period rewards to research activity to accumulate. In this sense, damages have some of the benefits that a royalty cap did in Tandon's compulsory licensing scheme, which we discussed in the single-innovation model section. Injunctions do not have this benefit, although to be fair the use of triple damages in the United States mitigates this difference. Injunctions amount to blanket prohibitions and so cannot achieve the same efficiency gains of limited infringement: either infringement is unlimited or it does not happen at all.[93]

Schankerman and Scotchmer (2001) make the bargaining role of damages and injunctions more precise to compare their effects. They show that different rules allow for different credible threats in the case that negotiations for access to a patent break down. These different levels of credible threats can affect the division of profits between a violator and a victim. For example, suppose that an independent firm has developed a profitable application of a patented innovation that infringes the original patent. If the first patent holder has the ability to enjoin an infringing firm, then a violator has a "credible threat" *not* to infringe. If a violator credibly "refuses" to create value, the threat of refusal can be used to extract value from the original patent holder in an access agreement. Instead, suppose that an inventor facing infringement by a high-value violator could use damages to collect this value *ex post*. Infringement is good for the inventor if the infringer can use the invention to create more value than the inventor could. This value can simply be extracted *ex post* through a damage settlement. Hence, the ability of damages to allow for value creating activity while distributing the gains back to the original patent holder can be more valuable to the inventor than the ability to enjoin an infringing firm. This sort of role for damages is very much in the spirit of the socially valuable role that licensing performed in Scotchmer's earlier work on cumulative innovation, summarized above. Hylton's addition to this is to caution that licensing markets may not work well, and courts may also not have the information they need to award appropriate damages.[94]

A final set of papers on enforcement pairs litigation issues with complementary innovation ("patent thickets"). One of the basic forces in models with both these considerations is illustrated by Lemley and Shapiro (2007). These authors investigate the effect of injunctions on bargaining for (licensing) access

[93] Maurer and Scotchmer (1999) use a similar argument for the benefits of limited entry to suggest that an independent invention defense should be allowed in cases of infringement.

[94] Boyce and Hollis (2007) make an even stronger point against injunctions, suggesting that the way they are implemented creates the possibility that they can be used as a "court-ordered collusive scheme" since they protect the patent monopoly without compensating consumers. Furthermore, they claim that US patent law can allow patent holders to gain more from the injunction process if the imitator is found not to have infringed. This creates an incentive to seek a preliminary injunction when patents are weakest.

to patents. Similar to the benchmarking exercise conducted by Farrell and Shapiro, above, Lemley and Shapiro propose a benchmark "fair" level of compensation for any one patent that reflects the value contributed by that patent to the product, a measure of bargaining skill in negotiations, and the strength of the patent. It then compares the actual compensation received in a bargaining game to this benchmark level. Hence, the exercise is similar to Farrell and Shapiro (2007), but is in the context of "complex" products composed of a large number of patented elements. Here, the benchmark is more complex as well.

Lemley and Shapiro find that, where a product reads on a large number of patents, the negotiated royalty rate can exceed this benchmark level. The reason is based on holdup. Suppose that the cost of developing the product is largely sunk at the point when the inventor and the producer attempt to negotiate a royalty. In this case, the holder of a single patent can "hold up" the producer for the entire value of the product if royalty negotiations break down when the alternative to a license is an injunction against the product's sale. This potentially gives the holder of a single patent the power to extract much more than her own contribution to the final value of the product. Indeed, if a product reads on many small patents, each individual patent holder may be very keen to exploit this form of holdup to raise her licensing revenues. Hall and Ziedonis (2001b) document such holdup in the face of large sunk investments.[95]

This type of holdup can have several undesirable effects. First, if patent holders can extract very large royalties by negotiating independently with the producer, it can result in the patent holders' not wishing to join standard setting groups that might license patents in a package. While the patent holders could potentially increase the total surplus to be shared by pooling together their patents, a coordinated solution is unlikely to arise given that each patent holder wishes to exploit holdup to her own advantage. Further, negotiating licenses individually for a large number of patents could be an extremely costly process, which could create direct social welfare losses. Secondly, producers might avoid entering an industry where this form of holdup might arise, causing underprovision of certain types of product and further attendant welfare losses. Of course, designing around such patents may be the best way to avoid these overcharges, but if redesign is costly it also causes losses in its turn. The authors conclude that various policy solutions, including limiting the use of injunctions, and some imposition of reasonable royalty calculations, could go a long way to resolving these problems. While this approach hammers home a point that Scotchmer brought up earlier to wit, *ex ante* licensing works better than *ex post* licensing to efficiently coordinate technology sharing in the case of externalities, the presumption of this work is that *ex post* licensing is the more relevant case. After all, it might be in the interests of the "troll" to *wait* until investments have been sunk to make his presence known to potential victims. In other words, the patent troll "hides under the bridge" and then emerges to extract large fees. In terms of the empirical relevance of this potential behavior, Siebert and von Graevenitz (2005) find evidence for increased *ex ante* licensing activity in the semiconductor industry, where patent thickets are generally thought to be present.[96]

[95] For related work, see also Gilbert and Katz (2006, 2007). The term "patent troll" has been coined to refer to entities that aggressively enforce patents in order to extract exorbitant licensing fees by exploiting holdup. A series of actions involving NTP, Inc. and Research in Motion, maker of the BlackBerry, has been held up as a canonical example of patent trolling. In that case, it was claimed that the "troll" was in hiding until the value of the final product was created.

[96] Geradin et al. (2007) evaluate proposals to implement a test for whether licensing terms are "reasonable and nondiscriminatory".

These arguments rely on an assumption of complementarity and, as Lerner and Tirole have pointed out, it may be quite difficult to determine how complementary or substitutable a set of patents are and further, the degree of substitutability may change.[97] Indeed, Galasso and Schankerman (2008) analyze the effect on legal disputes of fragmentation of patent claims to a valuable "reward" when patents may be less than perfect complements. Imagine that a single product that generates market value reads on a large number of patents. If greater fragmentation reduces the contribution of any single patent to the final product—that is, if patents are not perfect complements—the value of litigating (or continuing to litigate) any single patent falls because the expected damages fall.[98] They show empirically that greater fragmentation is associated with *less* delay *per dispute* in settling patent litigation, which suggests that each litigated patent must have smaller value. If we were in a world where fragmentation involved strictly complementary patents, this should not be the case since each patent is vital to the final product and so has equal "value." Hence, they reason that the sort of holdup effects we saw in the Lemley and Shapiro work are being dominated empirically by the effect of decreased significance of any single patent.

In a separate point, Galasso and Schankerman (2008) note that, to the extent that the Unified Court of Appeals has introduced less uncertainty about the outcome of court disputes, this decreases the role for information asymmetries in the bargaining process that spawns litigation. In other words, whether the court is or is not biased in its judgments is less relevant to the speed of settlement than the fact that the outcome may be more *certain* with a court that always rules a particular way. As a result, patent litigation should be observed to progress more quickly to settlement under the Unified Court of Appeals. This is, indeed, what they find, but the finding runs counter to some of the received wisdom on the events of recent years that litigation has been "excessive" in areas where rights are fragmented. They reconcile these views by suggesting that even though they find that delay has fallen per dispute, the total settlement time could have increased because the total number of disputes would be expected to rise with the degree of fragmentation of rights. In other words, if technologies are more "complex" nowadays there are more points of conflict, so the benchmark for the "expected" amount of litigation should rise with fragmentation.

Summarizing, one way of modeling the role of enforcement is to say that patents are not "ironclad" rights, but instead are probabilistic. Here, a main finding is that even patents that are unlikely to withstand a court challenge could generate very large licensing revenues because the incentive for any one user to challenge the patent could be socially too small. This could argue—in the United States—for instating postpatent review along the lines of the European system so as to reduce the burden of weak patents. A second tack to take is to examine the tools at the disposal of a patent holder in the case of infringement and see how varying this set of tools affects efficiency. A role of the tool is to set the threat that can be held out in case negotiations for access to patent rights break down. Liability rules that simply assign damages in the case of harm from infringement can act to compensate innovators for use of their innovations while allocating the innovation to the high value users. Some have argued for injunctions on the basis of courts' inability to award damages that reflect value to innovations' creators. However, injunctions combined with holdup can allow owners of individual patents to extract such high

[97] Other work involving litigation and patent pools includes Choi (2003), who incorporates litigation into a patent pool model, finding that pool members with weak patents tend not to bring suit against each other because suits provoke countersuits.
[98] See also Lichtman (2006) on this point.

royalty rates that producers who must combine patents to create valuable end products could be put off entering in the first place. This argues for damages to be used instead as a remedy for infringement.

7. Patent rights and competition policy

Competition policy and intellectual property rights have developed as two separate areas of law, each with its own goals and methods. While not a focus of the review, it should be clear from the preceding sections that competition policy and patent policy interact jointly to determine innovation incentives. Recalling Gilbert and Shapiro's (1990) perspective, competition policy can determine the flow rewards to patenting for any given patent right policy. Competition policy serves this function by determining the parameters for use, while patent rights determine the parameters for exclusion.

While one might think that a limit on flow rewards might necessarily contradict the goals of patent policy, Ayres and Klemperer (1999) put forward the argument that competition policy and reward to innovators need not be "at odds." As we saw in the review of single-innovation models, small restrictions on the patent holder's pricing from the unconstrained monopoly level will have a second-order effect on the monopolist's profits (since profits were maximized at the unconstrained price) but a first order effect on deadweight loss (since social welfare was not).[99] By rebalancing the patent right and the parameters of use to yield a longer but lower level of market power, one could create efficiency gains that would at once benefit consumers and not hurt inventors. To implement this idea some have used Kaplow's (1984) ratio test: the ratio of the patentee's incremental reward to the incremental social loss of a given practice. This ratio is a litmus test for allowing a practice: the higher the ratio, the more desirable the practice. Ayres and Klemperer's point is that under such a "ratio test," allowing unconstrained monopoly power probably looks like a bad idea: the ratio is zero at the unconstrained monopoly point.

Evans and Schmalensee (2002) point out that innovators often compete *for* markets, rather than compete on the margin *in* a market, as in Ayres and Klemperer's treatment. Competition policy issues for innovative industries may center less on controlling per-period reward for a given market than on practices that tend to exclude other competitors and their innovations from emerging markets. This gains particular force in network industries, of course. A third area where competition policy can interact with innovation incentives is that it can define when and how coordination can occur amongst independent players in an industry. As such, it affects the scope for licensing agreements, which we have seen plays a crucial role in patent design for cumulative and complementary innovation.

A further difficulty is that, if we think of competition policy as regulating structural characteristics of markets, it is not at all clear what sort of market structure promotes innovation best. In general, a firm that already is earning high profits in an industry will have little incentive to "replace" itself by innovating more. On the other hand, this sort of firm might wish to innovate in order to prevent another firm from taking over the market from outside. The effect of product market concentration on innovation is unclear, then. Indeed, which product market structure is socially best in terms of innovation incentives depends on the interplay of various factors that must be balanced quite specifically to

[99] For demand that is not too convex, this will hold for larger reductions in price as well.

generate any clear guidance.[100] Further, the structure of the innovative process itself determines whether there is socially too much or too little incentive for innovation in the economy in the first place.[101] As Shapiro (2008) points out, it is not clear whether we would align social and private incentives to innovate by increasing or decreasing innovation incentives. If we do not know what market structure to aim for or whether we are even above or below "target" as a starting point for analyzing policy, attempting to give any general guidance on structure is daunting.

Regibeau and Rockett (2007) suggest some overall rules for the interface between competition policy and intellectual property.[102] Their starting point is that the exclusionary rights granted by intellectual property law do not necessarily confer (monopoly) rents, but they can only be effective at stimulating innovation if they sometimes do. Indeed, as many of the models we have reviewed have shown, any single innovation may either fail to find a profitable market or may be pre-empted so rapidly by further advances that it generates very little value for the innovator. Still, it is the expectation of rents *ex ante* that generates innovation in the first place. Hence, some level of rents must potentially be made available. In point of fact, competition policy only intervenes selectively after an asset—of any type—becomes the basis for significant monopoly rents. A consequence of this difference in timing is that—even if the goals and skills of intellectual rights policy and competition policy authorities were the same—the information available to each of these authorities can differ. Indeed, to the extent that competition authorities tend to have access to more detailed information than was available at the time property rights were granted, there is a great temptation to revisit the tradeoff between innovation incentives and the deadweight loss resulting from intellectual property rights. Another consequence is that the intervention does not fall evenly across all property, which can feed back on the level of expected reward.

If one takes the perspective that intellectual property is there to generate a certain reward for innovators then, as Ayres and Klemperer (1999) and Maurer and Scotchmer (2002) have emphasized, that reward can be generated even if competition law limits the extent to which a rights holder can exercise the property right. In this sense, once competition law is "fixed" property rights design can be adjusted to accommodate it and still achieve its basic goals. The only requirement is that competition policy does not completely expropriate those gains. Second, as the *expectation* of gains is what counts, intellectual property law need not adjust to *individual* competition law decisions except to the extent that they represent a change in competition law policy. On the other side of the balance, any systematic attempt to revisit the tradeoff between static and dynamic efficiency by competition law could be undesirable. These tradeoffs—evaluated from an *ex ante* viewpoint—are what determine intellectual property right design in the first place, so revisiting these tradeoffs later—from an *ex post* viewpoint—are unlikely to improve innovation incentives (which are determined by *ex ante* evaluations). Indeed, there is a real risk of regulatory opportunism at later stages. Given that *ex post* the socially optimal policy is free access to intellectual property, the incentives of authorities intervening

[100] See seminal work by Arrow (1962), followed by a series of other papers including Gilbert and Newbery (1982), Vickers (1986), and more recently Aghion et al. (2005).

[101] See Gilbert (1995) for a discussion of "innovation markets," how they are distinguished from product markets and how their structure is taken into account in antitrust policy.

[102] See also Anderson and Gallini (1998) for a full volume covering many aspects of intellectual property law and competition policy.

selectively *after* the property has been created are not aligned with those of an authority that wishes to optimize *ex ante* innovation incentives. This is particularly salient at the level of an individual case: to the extent that each case affects expectations of reward very little, it is unclear what there is to "lose" by expropriating rights in a single case. Hence, there is an argument for commitment, whatever policy is chosen.

Another way of thinking of this interaction is to advocate policy made in a coordinated way, rather than delegated to individual court decisions, a point stressed by Scotchmer (2004). As we have argued, individual courts may not have the incentives to take decisions in a way that preserves *ex ante* innovation incentives. In this sense, it is well to make policy at the level of the legislature rather than in the courts.

A final general issue is clarity. If antitrust policy is confusing or contradictory in its treatment of certain practices, it can be difficult for inventors to calculate any consistent reward for investment.

This set of principles does not give much guidance on a more specific level for the desirability of particular competition laws and particular elements of intellectual property policy. While the argument in the previous paragraphs applies quite generally—even beyond intellectual property to other types of property that can be the basis of monopoly power—there are interactions between these two bodies of law that are quite specific to intellectual property and must be modeled specifically. Licensing policy is an area that has received a lot of attention, where the concern is to preserve the efficiency enhancing aspects of licenses while preventing such undesirable elements as overpricing. Maurer and Scotchmer (2006a) recommend that a reasonable policy objective should be that a licensor should be able to earn as much profit by licensing as by producing the product herself in an equivalent production environment. The idea here is not to penalize licensing activity that may, for example, be necessary because of financial constraints faced by the inventor. This principle of "profit neutrality" can give a useful benchmark to authorities generating specific rules toward licensing activity and fees.

Shapiro (2003) examines another type of "neutrality" to address antitrust treatment of proposed settlements of patent disputes. He suggests that these should generate at least as much consumers' surplus as they would have accrued if litigation had ensued. The exact nature of the settlement does not need to be specified: it could be merger, joint venture, licensing, or other agreements. The point is that while the invention incentives need to be preserved, efficiency is a concern as well. We need a precise rule to balance the two. Formally, the problem of settlement design is the "dual" of a Ramsey pricing problem where we solve

$$\max_{x} \pi(x) \quad \text{subject to} \quad s(x) \geq s^*,$$

where x is a set of actions that the two parties can specify in a contract *ex ante*, π is industry profits, s is consumers' surplus, and s^* is the surplus that would result from litigation. Given that litigation remains an alternative, the parties can always do better to settle.

What remains is to translate this general rule into settlement policies. For example, in the case of licensing, we could propose capping the royalty rate so as to guarantee a minimum level of surplus to consumers. In the case of mergers, we could apply a factor—taking into account that a merger involves intellectual property issues—to the normal offsetting efficiencies we might demand to justify a merger's anticompetitive harm. For example, define a ratio[103] of the harm caused to consumers by the merger with patent issues over the harm that would be caused by a merger with no patent issues. If the standard merger would require offsetting efficiencies, E, to benefit consumers, a merger involving patenting parties would require E times that ratio to be justified. Hence, we have a "scaling factor" to apply to standard merger guidelines and procedures, where the scaling depends on patent strength.[104]

More specific issues have been the subject of a range of models incorporating both competition and intellectual property policy. The Lemley and Shapiro (2007) benchmark licensing pricing policy in the context of multiple (complementary) innovations can be used to evaluate whether a license is being issued at fair and reasonable terms.[105] Licensing restrictions such as grantbacks (Choi, 2002; Van Dijk, 2000), cross-licensing (Choi, 2003; Eswaran, 1994; Fershtman and Kamien, 1992), licensing for standard setting (Farrell et al., 2007), as well as joint ventures (Scotchmer, 1998) and tying (Whinston, 1990, 2001) have been investigated.[106] An area where many of the conflicts have emerged has been in industries with network effects, so special attention has been paid to modeling in this area (see, e.g., Farrell and Katz, 1998, 2000). Most of these models rely on specific structural features to generate their results or require judgments that may be difficult in practice (such whether patents are substitutable or complementary). Hence, it is difficult to make watertight rules based on them that policymakers can easily implement in a wide variety of cases. Developing rules based on a consensus in the literature may need to wait until more modeling is done so that a considered consensus can emerge. Alternatively, advocating a rule of reason approach under some broad guidelines may be about as far as theory can go at this point.

Finally, in contrast to most of the papers outlined here, the interaction between competition policy and innovation has been treated in the growth context to give slightly different conclusions. In a recent example of this, Segal and Whinston (2007) note that competition policies that protect new entrants from exclusionary behavior by incumbents can raise entrant profits, thereby encouraging innovation by entrants. On the other hand these same policies will come back to haunt the entrants in the future, should they be successful. This future lower profit to incumbency can eventually slow the rate of innovation. Hence, an antitrust policy that favors new entrants affects the time distribution of benefits from innovation, "front loading" those benefits. This sort of timing issue is something that a growth model

[103] Formally, the ratio equals $(\bar{s} - s_M)/(s_D - s_M)$, where s_D is consumers' surplus in the case of duopoly and s_M is surplus in the case of monopoly and $\bar{s}$ is a maximal value of surplus. To see how this works in practice, take two polar cases of patent strength. If a patent is ironclad and if the resolution of litigation would be immediate the "reservation" level of consumers' surplus is just the monopoly level and merger creates no further harm to consumers when patents are present than without. Indeed, the numerator of the ratio is zero, so no efficiency gains are necessary to justify the merger. If the patent has no chance of surviving the court challenge, on the other hand, the outcome of litigation is the duopoly payoff. In this case, the ratio is one since both the numerator and denominator are the difference between duopoly and monopoly surplus. Here, the efficiency gains that would be required neglecting patent issues would be the same as those required to justify the merger in the presence of patents. See Shapiro (2003) for a full discussion.

[104] As a practical issue, both the solution to the maximization problem and the level of the ratio depend on the strength of the patent, as this determines the outcome to litigation. Shapiro notes that the strength could be difficult for any of the parties, let alone an outside authority, to evaluate accurately. Katz and Shelanski (2006, 2007) and Carlton and Gerntner (2003) weigh other factors, such as the reduction in duplicative effort, the benefits of specialization, and various spillover benefits against the reduction in competition from merger when comparing merger in innovative industries to "standard" cases.

[105] Also see Gilbert (1995) and Gilbert and Sunshine (1995) on the Department of Justice 1995 guidelines and provisions involving licensing of complementary factors.

[106] See also Gilbert and Shapiro (1997), who map out a large number of issues that are developed in later work by both authors.

can highlight particularly well. Segal and Whinston analyze the tension between rewarding entrants and incumbents in the light of growth goals and several specific antitrust policies, finding that in some cases, policies benefit both entrants and incumbents while in others there is a conflict of interest between the two parties.

8. Conclusions

This is far from a settled literature. Many basic points remain unresolved, such as what style of protection should be used to promote innovation, and the level of incentives compared to the social optimum. Economic modeling of the complex features of the administrative process of getting a patent, the litigation process, and the interaction between specific areas of intellectual property and competition law are ample, but still do not cover the range of issues actually faced by the wide range of actors involved in getting an innovation out to market. Some topics are areas where feeling runs high. A quick search of the web, for example, indicates that "patent trolls" can be viewed as heroes or as demons, with a wide number of arguments on either side that have only been partially modeled in the academic literature. Finally, innovation policy is an area where there is intense interest on the part of policy-makers. If one wishes to make a contribution to a debate that is viewed as pressing, this is a good area to enter.

There are many issues that have not been treated much in this chapter—since they have been developed relatively less in existing work—but which are nonetheless very important. One such issue is how to compensate and motivate scientists to conduct innovative activity. Aghion and Tirole (1994) consider the question of whether research units (such as the scientists themselves) or customers (such as the manufacturing firm that might use the patent to produce value) should own patents: the issue is not so much the design of the property right as the allocation of it across interested parties when financing, creating value, and rewards potentially result from the effort of a variety of differing agents. These authors note that the magnitude of research activity as well as its nature (such as the size of innovation) can change with different organizational structures. While some work has followed on scientists, their contracts and incentives, and the organizations in which they work,[107] a lot still remains to be done.

This survey has been quite narrow: it has not covered all of intellectual property. Copyright and trademark law have been excluded—with regret. Many of the issues are similar between copyright and patent, but not all. Indeed, technology is forging ahead so rapidly in the area of digital media that copyright issues are a fertile area of work. For example, an issue that comes up with copyrighted digital media but less with patented material is whether to protect the intellectual property with legal rights or with technical boundaries (such as encryption).[108] The economics of trademarks is rather distinct from that of patents and copyright, involving reputation mechanisms, consumer search and repeat purchase. Still, to the extent that many products employ many methods of protection all at once—patenting some

[107] Kim and Marschke (2005) study labor mobility issues and Sena (2004) reviews spillovers and cooperative agreements. Sorenson and Fleming (2004) relate norms and institutions of science and their effect on patenting activity. Severinov (2001) studies confidentiality agreements.

[108] See, for example, Menell and Scotchmer (2007) for a discussion of the law and economics of trademark and copyright. Varian (2005) provides a recent review and opinion on copyright.

aspects, employing secrecy for others, copyrighting design aspects and trademarking the product as a whole—investigating these rights as they are used together could be a fruitful area to pursue as well. Indeed, using these three tools together to create the sort of exclusion zone around a product that is assumed in some of the single-innovation models is a goal of this sort of strategy but little work outside of the interaction of trade secrecy and patents has been done.[109]

A third area where there is relatively little work is the exploration of alternative instruments and institutions that are currently available to influence innovative activity. For example, Acharya and Subramanian (2007) examine the effect of bankruptcy codes on innovation, where innovation is a relatively "risky" investment. Financial structure affects innovation, so that the incentive to innovate depends on how creditor-friendly or how debtor-friendly bankruptcy codes are. Given that there appears to be an empirical link between financial structure and innovation (see Hall, 2002 and references therein), a more systematic investigation of financial regulations' effect on innovation could be useful. Related to this, Hall and Lerner (2010), included in this volume, overviews work on the effects of methods of financing R&D on innovation.

Finally, some recent work has examined the effect of local versus global diffusion of information on behavior (e.g., see Boncinelli, 2008). While one could imagine that information about innovation might travel easily amongst members of a focused set of researchers by word of mouth or on the conference circuit, it is possible that the function of patents as repositories of information might help to spread information to those who would not normally receive it. This could affect the trajectory of future developments as well as their speed. Developments in the field of social network theory could be helpful to diagnose the effectiveness of patent repositories versus the use of other tools targeted more at "key" individuals.

References

Acharya, V., Subramanian, K. (2007). "Bankruptcy codes and innovation". Centre for Economic Policy Research Discussion Paper 6307.

Aghion, P., Tirole, J. (1994). "The management of innovation". Quarterly Journal of Economics 109 (4), 1185–1209.

Aghion, P., Bloom, N., Blundell, R., Griffith, R., Howitt, P. (2005). "Competition and innovation: An inverted-U relationship". Quarterly Journal of Economics 120 (2), 701–728.

Allison, J., Lemley, M. (1998). "Empirical analysis of the validity of litigated patents". American Intellectual Property Law Association Quarterly Journal 26 (3), 185–275.

Anand, B., Khanna, T. (2000). "The structure of licensing contracts". Journal of Industrial Economics 48 (1), 103–135.

Anderson, R., Gallini, N. (Eds.), (1998). Competition policy and intellectual property rights in the knowledge-based economy. University of Calgary Press, Calgary.

Anton, J., Yao, D. (1994). "Expropriation and inventions: Appropriable rents in the absence of property rights". American Economic Review 84 (1), 190–209.

Anton, J., Yao, D. (2002). "The sale of ideas: Strategic disclosure, property rights and contracting". Review of Economics Studies 69 (3), 513–531.

Anton, J., Yao, D. (2004). "Little patents and big secrets: Managing intellectual property". RAND Journal of Economics 35 (1), 1–122.

Aoki, R., Spiegel, Y. (1998). "Public disclosure of patent applications, R&D and welfare". Foerder Institute Working Paper 30–98.

[109] For a discussion from the legal perspective, see Merges et al. (2006). The interaction trademarks and patents are studied in Manelli et al. (1994). Recent empirical work on trademarks includes Von Graevenitz (2007) and Greenhalgh and Rogers (2007a,b).

Arora, A., Fosfuri, A., Gambardella, A. (2001). Markets for Technology: The Economics of Innovation and Corporate Strategy. MIT Press, Cambridge, MA.

Arora, A., Ceccagnoli, M., Cohen, W. (2003). "R&D and the patent premium". National Bureau of Economic Research Working Paper 9431.

Arrow, K. (1962). "Economic welfare and the allocation of resources to invention". In: Nelson, R. (Ed.), The Rate and Direction of Inventive Activity: Economic and Social Factors. National Bureau of Economic Research, Cambridge, MA, pp. 609–626.

Ayres, I., Klemperer, P. (1999). "Limiting patentees' market power without reducing innovation incentives: The perverse benefits of uncertainty and non-injunctive remedies". Michigan Law Review 97 (2), 985–1033.

Baker, S., Mezzetti, C. (2005). "Disclosure as a strategy in the patent race". Journal of Law and Economics 48 (1), 173–194.

Bar, T. (2006). "Defensive publications in an R&D race". Journal of Economics & Management Strategy 15 (1), 229–254.

Belenzon, S. (2006). "Knowledge flow and sequential innovation: Implications for technology diffusion R&D and market value". Oxford University Working Paper 259.

Bessen, J. (2006). "The value of U.S. patents by owner and patent characteristics". Boston University School of Law Working Paper 06-46.

Bessen, J., Maskin, E. (2009). "Sequential innovation, patents and imitation". RAND Journal of Economics 40 (4), 611–635.

Bessen, J., Meurer, M. (2008a). "The private costs of patent litigation". Boston University School of Law Working Paper 07-08.

Bessen, J., Meurer, M. (2008b). "The value of U.S. patents by owner and patent characteristics". Research Policy 37 (5), 932–945.

Bidgoli, H. (2010). The Handbook of Technology Management, vol. 3. Management Support Systems, Electronic Commerce, Legal and Security Considerations. Wiley, New York.

Bloom, N., Schankerman, M., Van Reenen, J. (2007). "Identifying technology spillovers and product market rivalry". National Bureau of Economic Research Working Paper 13060 (April).

Boldrin, M., Levine, D. (2003). "Rent-seeking and innovation". Journal of Monetary Economics 51 (1), 127–160.

Boncinelli, L. (2008). "Global vs. local information in (anti-)coordination problems with imitators". The B.E. Journal of Theoretical Economics 8 (1), (Article 17).

Boyce, J., Hollis, A. (2007). "Preliminary injunctions and damage rules in patent law". Journal of Economics & Management Strategy 16 (2), 385–405.

Brunt, L., Lerner, J., Nicholas, T. (2008). "Inducement prizes and innovation". Centre for Economic Policy research Discussion Paper 6917.

Calabresi, G., Melamed, D. (1972). "Property rules, liability rules and inalienability: One view of the cathedral". Harvard Law Review 85 (6), 1089–1128.

Carlton, D., Gertner, R. (2003). "Intellectual property, antitrust, and strategic behavior". In: Jaffe, A., Lerner, J., Stern, S. (Eds.), Innovation Policy and the Economy, vol. 1. MIT Press, Cambridge, MA, pp. 29–59.

Cassiman, B., Veugelers, R. (2002). "R&D cooperation and spillovers: Some empirical evidence from Belgium". American Economic Review 92 (4), 1169–1184.

Chen, M., Iyigun, M. (2006). "Patent protection and strategic delays in technology development: Implications for economic growth". University of Colorado Working Paper 04-09 (Revised 2007).

Choi, J. (2002). "A dynamic analysis of licensing: The 'boomerang' effect and grant-back clauses". International Economic Review 43 (3), 803–829.

Choi, J. (2003). "Patent pools and cross-licensing in the shadow of patent litigation". CESifo Working Paper 1070.

Cohen, W., Levinthal, D. (1989). "Innovation and learning: The two faces of R&D". The Economic Journal 99 (397), 569–596.

Cohen, W., Nelson, R., Walsh, J. (2000). "Protecting their intellectual assets: Appropriability conditions and why U.S. manufacturing firms patent (or not)". National Bureau of Economic Research Working Paper 7552.

Comino, S., Manenti, F., Nicolo, A. (2007). "Sequential innovations with unobservable follow-on investments". Universita Degli Studi Di Padova "Marco Fanno" Working Paper 41.

Cornelli, F., Schankerman, M. (1999). "Patent renewals and R&D incentives". RAND Journal of Economics 30 (2), 197–213.

Crampes, C., Langinier, C. (2002). "Litigation and settlement in patent infringement cases". RAND Journal of Economics 33 (4), 258–274.

Denicolo, V. (1996). "Patent races and optimal patent breadth and length". The Journal of Industrial Economics 44 (3), 249–265.

Denicolo, V. (2000). "Two-stage patent races and patent policy". RAND Journal of Economics 31 (3), 488–501.

Denicolo, V., Franzoni, L. (2004). "The contract theory of patents". International Review of Law and Economics 23 (4), 365–380.

Dewatripont, M., Legros, P. (2008). "'Essential' patents, FRAND royalties and technological standards". Centre for Economic Policy Research Discussion Paper 6925.

Ebert, L. (2004). "How high are the grant rates at the USPTO?" Journal of the Patent and Trademark Office Society 86 (7), 568–576.

Eisenberg, R. (1989). "Patents and the progress of science: Exclusive rights and experimental use". The University of Chicago Law Review 56, 1017–1086.

Encoua, D., Lefouili, Y. (2008). Licensing Uncertain Patents: Per-Unit Royalty vs. Up-Front Fee. Paris School of Economics, University of Paris-I Pantheon Sorbonne, Paris.

Eswaran, M. (1994). "Cross-licensing of competing patents and a facilitating device". Canadian Journal of Economics 28 (3), 698–708.

Evans, D., Schmalensee, R. (2002). "Some economic aspects of antitrust analysis in dynamically competitive industries". In: Jaffe, A., Lerner, J., Stern, S. (Eds.), Innovation Policy and the Economy, vol. 2. MIT Press, Cambridge, MA, pp. 1–49.

Farrell, J. (1995). "Some arguments for weaker intellectual property protection in network industries". StandardView 3 (2), 46–49.

Farrell, J., Katz, M. (1998). "The effects of antitrust and intellectual property law on compatibility and innovation". Antitrust Law Bulletin 43 (3), 609–650.

Farrell, J., Katz, M. (2000). "Innovation, rent extraction, and integration in system markets". Journal of Industrial Economics 48 (4), 413–432.

Farrell, J., Shapiro, C. (2007). "How strong are weak patents?" Competition Policy Center Paper CPC05-54.

Farrell, J., Hayes, J., Shapiro, C., Sullivan, T. (2007). Standard Setting, Patents, and Hold-Up. University of California at Berkeley, Berkeley.

Fershtman, C., Kamien, M. (1992). "Cross-licensing of complementary technologies". International Journal of Industrial Organization 10 (3), 329–348.

Galasso, A., Schankerman, M. (2008). "Patent thickets and the market for innovation: Evidence from settlement of patent disputes". Centre for Economics Policy Research Discussion Paper 6946.

Gallini, N. (1984). "Deterrence by market sharing: A strategic incentive for licensing". American Economic Review 74 (5), 931–941.

Gallini, N. (1992). "Patent policy and costly imitation". RAND Journal of Economics 23 (1), 52–63.

Gallini, N. (2002). "The economics of patents: Lessons from recent U.S. patent reform". Journal of Economic Perspectives 16 (2), 131–154.

Gallini, N., Scotchmer, S. (2002). "Intellectual property: When is it the best incentive system?" In: Jaffe, A., Lerner, J., Stern, S. (Eds.), Innovation Policy and the Economy, vol. 2. MIT Press, Cambridge, MA, pp. 51–77.

Geradin, D., Layne-Farrar, A., Padilla, A.J. (2007). "The ex ante auction model for the control of market power in standard setting organisations". Centre for Economic Policy Research Discussion Paper 6304.

Gilbert, R. (1995). The 1995 Antitrust Guidelines for the Licensing of Intellectual Property: New Signposts for the Intersection of Intellectual Property and the Antitrust Law. Department of Justice, Washington, DC.

Gilbert, R., Katz, M. (2006). "Should good patents come in small packages? A welfare analysis of intellectual property bundling". International Journal of Industrial Organization 24 (5), 931–952.

Gilbert, R., Katz, M. (2007). "Efficient division of profits from complementary innovations". Competition Policy Center Paper CPC07-076.

Gilbert, R., Newbery, D. (1982). "Preemptive patenting and the persistence of monopoly". American Economic Review 72 (3), 514–526.

Gilbert, R., Shapiro, C. (1990). "Optimal patent length and breadth". RAND Journal of Economics 21 (1), 106–112.

Gilbert, R., Shapiro, C. (1997). "Antitrust issues in the licensing of intellectual property: The nine no-no's meet the nineties". In: Winston, M.N., Baily, M.N., Reiss, P.C. (Eds.), Brookings Papers on Economic Activity, Microeconomics 1997. Brookings Institution Press, Washington, DC, pp. 283–349.

Gilbert, R., Sunshine, S. (1995). "Incorporating dynamic efficiency concerns in merger analysis: The use of innovation markets". Antitrust Law Journal 63, 569–602.

Gill, D. (2008). "Strategic disclosure of intermediate research results". Journal of Economics & Management Strategy 17 (3), 733–758.

Giuri, P., Mariani, M., Brusoni, S., Crespi, G., Francoz, D., Gambardella, A., Garcia-Fontes, W., Geuna, A., Gonzales, R., Harhoff, K., Hoisl, K., Lebas, C., et al. (2006). "Everything you always wanted to know about inventor (but never asked): evidence from the PATVAL-EU survey". Centre for Economic Policy Research Discussion Paper 5752.

Glachant, M., Meniere, Y. (2008). Technology Diffusion with Learning Spillovers: Patent versus Free Access. Ecole des Mines, Paris.

Graham, S., Harhoff, D. (2006). "Can post-grant reviews improve patent system design? A twin study of US and European patents". Centre for Economic Policy Research Discussion Paper 5680 (May).

Green, J., Scotchmer, S. (1995). "On the division of profit in sequential innovation". RAND Journal of Economics 26 (1), 20–33.

Greenhalgh, C., Rogers, M. (2007a). Trade Marks and Performance in UK Firms: Evidence of Schumpeterian Competition through Innovation. Oxford Intellectual Property Research Centre, Oxford University, Oxford.

Greenhalgh, C., Rogers, M. (2007b). The Value of Intellectual Property Rights to Firms. Oxford Intellectual Property Research Centre, Oxford University, Oxford.

Guellec, D., van Pottelsberghe de la Potterie, B. (2007). The Economics of the European Patent System: IP Policy for Competition and Innovation. Oxford University Press, Oxford.

Hall, B. (2002). "The financing of research and development". Oxford Review of Economic Policy 18 (1), 35–51.

Hall, B. (2007). "Patents and patent policy". Oxford Review of Economic Policy 23 (4), 541–567.

Hall, B., Lerner, J. (2010). "Financing R&D and innovation". In: Hall, B., Rosenberg, N. (Eds.), Elsevier Handbook of the Economics of Innovation, vol. 2. Elsevier, Amsterdam.

Hall, B., Ziedonis, R. (2001a). "The patent paradox revisited: An empirical study of patenting in the U.S. semiconductor industry, 1979–1995". RAND Journal of Economics 32 (1), 101–128.

Hall, B., Ziedonis, R. (2001b). "The effects of strengthening patent rights on firms engaged in cumulative innovation: Insights from the semiconductor industry". In: Libecap, G. (Ed.), Entrepreneurial Inputs and Outcomes: New Studies of Entrepreneurship in the United States, vol. 13. Advances in the Study of Entrepreneurship, Innovation, and Economic Growth. Elsevier, Amsterdam.

Hall, B., Graham, S., Harhoff, D., Mowery, D. (2003). "Prospects for improving U.S. patent quality via postgrant opposition". In: Jaffe, J., Lerner, J., Stern, S. (Eds.), Innovation Policy and the Economy, vol. 4. MIT Press, Cambridge, MA, pp. 115–143.

Harhoff, D., Reitzig, M. (2004). "Determinants of oppositions against EPO patent grants: The case of biotechnology and pharmaceuticals". International Journal of Industrial Organization 22 (4), 443–480.

Harhoff, D., Wagner, S. (2005). "Modelling the duration of patent examination at the European Patent Office". Centre for Economic Policy Research Discussion Paper 5283.

Harhoff, D., Wagner, S. (2006). Business method patents in europe and their strategic use—Evidence from franking device manufacturers. Discussion Paper in Business Administration 1265 Munich School of Management, University of Munich, Munich, Germany.

Harhoff, D., Henkel, J., von Hippel, E. (2003). "Profiting from voluntary information spillovers: How users benefit by freely revealing their innovations". Research Policy 32 (10), 1753–1769.

Hedge, D., Mowery, D., Graham, S. (2007). "Pioneers, submariners, or thicket-builders: Which firms use continuations in patenting?" NBER Working Paper 13153 (June).

Heller, M., Eisenberg, R. (1998). "Can patents deter innovation? The anticommons in biomedical research". Science 280 (5364), 698–701.

Henkel, J. (2004). "The jukebox mode of innovation—A model of commercial open source development". Centre for Economic Policy Research Discussion Paper DP4507 (July).

Hopenhayn, H., Mitchell, M. (2001). "Innovation variety and patent breadth". RAND Journal of Economics 32 (1), 152–166.

Hopenhayn, H., Llobet, G., Mitchell, M. (2006). "Rewarding sequential innovators: Prizes, patents, and buyouts". Journal of Political Economy 114 (6), 1041–1068.

Horowitz, A., Lai, E. (1996). "Patent length and the rate of innovation". International Economic Review 37 (4), 785–801.

Horstmann, I., MacDonald, G., Slivinski, A. (1985). "Patents as information transfer mechanisms: To patent or (maybe) not to patent". The Journal of Political Economy 93 (5), 837–858.

Hunt, R. (2004). "Patentability, industry structure, and innovation". Journal of Industrial Economics 52 (3), 401–426.

Hylton, K. (2006). "Property rules and liability rules, once again". Review of Law and Economics 2 (2), 137–191.

Jaffe, A. (2000). "The U.S. patent system in transition: Policy innovation and the innovation process". Research Policy 29, 531–557.

Jaffe, A., Lerner, J. (2006). Innovation and Its Discontents: How Our Broken Patent System Is Endangering Innovation and Progress and What to Do About It. Princeton University Press, Princeton.
Janis, M. (2003). Experimental Use and the Shape of Patent Rights for Plant Innovation. University of Iowa School of Law, Iowa.
Kaplow, L. (1984). "The patent–antitrust intersection: A reappraisal". Harvard Law Review 97 (8), 1813–1892.
Katz, M., Shapiro, C. (1986). "Technology adoption in the presence of network externalities". Journal of Political Economy 94 (4), 822–841.
Katz, M., Shapiro, C. (1987). "R&D rivalry with licensing or imitation". American Economic Review 77 (3), 402–420.
Katz, M., Shelanski, H. (2006). "Merger policy and innovation: Must enforcement change to account for technological change?" In: Jaffe, A., Lerner, J., Stern, S. (Eds.), Innovation Policy and the Economy, vol. 1. MIT Press, Cambridge, MA, pp. 109–165.
Katz, M., Shelanski, H. (2007). "Mergers and innovation". Antitrust Law Journal 74, 1–86.
Kim, J., Marschke, G. (2005). "Labour mobility of scientists, technological diffusion, and the firm's patenting decision". RAND Journal of Economics 36 (2), 298–317.
Kitch, E. (1977). "The nature and function of the patent system". Journal of Law and Economics 20 (2), 265–290.
Klemperer, P. (1990). "How broad should the scope of patent protection be?" RAND Journal of Economics 21 (1), 113–130.
Kremer, M. (1998). "Patent buyouts: A mechanism for encouraging innovation". Quarterly Journal of Economics 113 (4), 1137–1167.
La Manna, M. (1994). "Research vs. development: Optimal patenting policy in a three-stage model". European Economic Review 38 (7), 1423–1440.
Lakhani, K., Wolf, R. (2005). "Why hackers do what they do? Understanding motivation and effort in free/open source software projects". In: Feller, J., Fitzgerald, B., Hissam, S., Lakhani, K. (Eds.), Perspectives on Free and Open Source Software. MIT Press, Cambridge, MA.
Langinier, C. (2005). "Using patents to mislead rivals". Canadian Journal of Economics 38 (2), 520–545.
Lanjouw, J., Lerner, J. (2001). "Tilting the table? The use of preliminary injunctions". Journal of Law and Economics 44 (2), 573–603.
Lanjouw, J., Schankerman, M. (2001). "Characteristics of patent litigation: A window on competition". RAND Journal of Economics 32 (1), 129–151.
Lanjouw, J., Schankerman, M. (2004). "Protecting intellectual property rights: Are small firms handicapped?" Journal of Law and Economics 47 (1), 45–74.
Layne-Farrar, A., Lerner, J. (2008). To Join or Not to Join: Examining Patent Pool Participation and Rent Sharing Rules. Harvard University, Cambridge, MA.
Lemley, M. (2001). "Rational ignorance at the patent office". Northwestern University Law Review 95 (4), 1497–1532.
Lemley, M., Shapiro, C. (2005). "Probabilistic patents". Journal of Economic Perspectives 19 (2), 75–98.
Lemley, M., Shapiro, C. (2007). "Patent holdup and royalty stacking". Texas Law Review 85, 1990–2049.
Lerner, J. (1995). "Patenting in the shadow of competitors". Journal of Law and Economics 38 (2), 463–495.
Lerner, J. (2005). "150 Years of patent office practice". American Law and Economics Review 7, 112–143.
Lerner, J., Tirole, J. (2002). "Some simple economics of open source". Journal of Industrial Economics 50 (2), 197–234.
Lerner, J., Tirole, J. (2004). "Efficient patent pools". American Economic Review 94 (3), 691–711.
Lerner, J., Tirole, J. (2005). "The economics of technology sharing: Open source and beyond". The Journal of Economic Perspectives 19 (2), 99–120.
Lichtman, D. (2006). Patent Holdouts in the Standard-Setting Process. Academic Advisory Council Bulletin 1.3. Progress and Freedom Foundation, Washington, DC.
Manelli, A., Regibeau, P., Rockett, K. (1994). "Trademark protection and post-patent competition". Cuadernos Economicos de ICE 57 (2), 115–139.
Matutes, C., Regibeau, P., Rockett, K. (1996). "Optimal patent design and the diffusion of innovations". RAND Journal of Economics 27 (1), 60–83.
Maurer, S., Scotchmer, S. (2002). "The independent invention defence in intellectual property". Economica 69 (276), 535–547.
Maurer, S., Scotchmer, S. (2004). "Procuring knowledge". In: Libecap, G. (Ed.), Intellectual Property and Entrepreneurship, vol. 15: Advances in the Study of Entrepreneurship, Innovation and Economic Growth. Elsevier, Amsterdam, pp. 1–31.
Maurer, S., Scotchmer, S. (2006a). "Profit neutrality in licensing: The boundary between antitrust law and patent law". American Law and Economics Review 8, 476–522.

Maurer, S., Scotchmer, S. (2006b). "Open source software: The new intellectual property paradigm". In: Hendershott, T. (Ed.), Handbook on Economics and Information Systems. Elsevier, Amsterdam Chapter 5.
Menell, P., Scotchmer, S. (2007). "Intellectual property". In: Polinsky, M., Shavell, S. (Eds.), Handbook of Law and Economics. Elsevier, Amsterdam.
Merges, R. (1999). Institutions for Intellectual Property Transactions: The Case of Patent Pools. School of Law, University of California at Berkeley, Berkeley.
Merges, R.P., Nelson, R.R. (1990). "On the complex economics of patent scope". Columbia Law Review 90, 839–916.
Merges, R., Menell, P., Lemley, M. (2006). Intellectual property in the new technological age (4). Aspen, New York.
Middendorf, W. (1981). What Every Engineer Needs to Know About Inventing, vol. 7. What Every Engineer Should Know. CRC Press, Boca Raton, FL.
Miller, A., Davis, M.H. (1983). Intellectual Property: Patents, Trademarks and Copyright in a Nutshell. West Publishing, St. Paul, MN.
Miyagiwa, K. (2007). "Collusion and research joint ventures". The Institute of Social and Economic Research, Osaka University, Discussion Paper 704.
Moser, P. (2005). "Do patent laws influence innovation? Evidence from nineteenth-century world's fairs". American Economic Review 95 (4), 1214–1236.
Nard, C., Morriss, A. (2006). "Constitutionalizing patents: From Venice to Philadelphia". Review of Law and Economics 2 (2), 223–321.
Noel, M., Schankerman, M. (2006). "Strategic patenting and software innovation". Centre for Economic Policy Research Discussion Paper 5701.
Nordhaus, W. (1969). Invention, Growth and Welfare: A Theoretical Treatment of Technological Change. MIT Press, Cambridge, MA.
O'Donoghue, T. (1998). "A patentability requirement for sequential innovation". RAND Journal of Economics 29 (4), 654–679.
O'Donoghue, T., Scotchmer, S., Thisse, J.-F. (1998). "Patent breadth, patent life, and the pace of technological progress". Journal of Economics & Management Strategy 7 (1), 1–32.
Pakes, A. (1986). "Patents as options: Some estimates of the value of holding European patent stocks". Econometrica 54 (4), 755–784.
Qian, Y. (2007). "Do additional national patent laws stimulate domestic innovation in a global patenting environment? A cross-country analysis of pharmaceutical patent protection, 1978–2002". Review of Economics and Statistics 89 (3), 436–453.
Regibeau, P., Rockett, K. (2007). "The relationship between intellectual property law and competition law: An economic approach". In: Anderman, S. (Ed.), The Interface Between Intellectual Property Rights and Competition Policy. Cambridge University Press, Cambridge, MA.
Sakakibara, M., Branstetter, L. (2001). "Do stronger patents induce more innovation? Evidence from the 1988 Japanese patent law reforms". RAND Journal of Economics 32 (1), 77–100.
Schankerman, M., Scotchmer, S. (2001). "Damages and injunctions in the protection of intellectual property". RAND Journal of Economics 32 (1), 199–220.
Schmidt, K. (2008). "Complementary patents and market structure". Centre for Economic Policy Research Discussion Paper 7005.
Scotchmer, S. (1991). "Standing on the shoulders of giants: Cumulative research and the patent law". The Journal of Economic Perspectives 5 (1), 29–41.
Scotchmer, S. (1996). "Protecting early innovators: Should second-generation products be patentable?" RAND Journal of Economics 27 (2), 322–331.
Scotchmer, S. (1998). "R&D joint ventures and other cooperative arrangements". In: Anderson, R., Gallini, N. (Eds.), Competition Policy and Intellectual Property Rights in the Knowledge-Based Economy. University of Calgary Press, Calgary.
Scotchmer, S. (1999). "On the optimality of the patent renewal system". RAND Journal of Economics 30 (2), 181–196.
Scotchmer, S. (2004). Innovation and incentives. MIT Press, Cambridge, MA.
Scotchmer, S., Green, J. (1990). "Novelty and disclosure in patent law". RAND Journal of Economics 21 (1), 131–146.
Segal, I., Whinston, M. (2007). "Antitrust in innovative industries". American Economic Review 97 (5), 1703–1730.
Sena, V. (2004). "The return of the Prince of Denmark: A survey on recent developments in the economics of innovation". The Economic Journal 114, F312–F332.
Severinov, S. (2001). "On information sharing and incentives in R&D". RAND Journal of Economics 32 (3), 542–564.

Shapiro, C. (2001). "Navigating the patent thicket: cross-licenses, patent pools and standard-setting". In: Jaffe, A., Lerner, S., Stern, S. (Eds.), Innovation Policy and the Economy, vol. 1. MIT Press, Cambridge, MA, pp. 119–150.
Shapiro, C. (2003). "Antitrust limits to patent settlements". RAND Journal of Economics 34 (2), 391–411.
Shapiro, C. (2004). "Patent system reform: Economic analysis and critique". Berkeley Technology Law Journal 19 (3), 1017–1047.
Shapiro, C. (2008). "Patent reform: Aligning reward and contribution". In: Jaffe, A., Lerner, J., Stern, S. (Eds.), Innovation and the Economy, vol. 8. MIT Press, Cambridge, MA, pp. 111–156.
Shavell, S., Ypersele, T. (2001). "Rewards versus intellectual property rights". Journal of Law and Economics 44, 525–547.
Siebert, R., von Graevenitz, G. (2005). "How licensing resolves hold-up: Evidence from a dynamic panel data model with unobserved heterogeneity". CEPR Discussion Paper No. 5436.
Sorenson, O., Fleming, L. (2004). "Science and the diffusion of knowledge". Research Policy 33 (10), 1615–1634.
Swanson, D., Baumol, W. (2005). "Reasonable and nondiscriminatory (RAND) royalties, standards selection, and control of market power". Antitrust Law Journal 73, 1–58.
Tandon, P. (1982). "Optimal patents with compulsory licensing". Journal of Political Economy 90 (3), 470–486.
Tao, Z., Wu, C. (1997). "On the organization of cooperative research and development: Theory and evidence". International Journal of Industrial Organization 15 (5), 573–596.
Trajtenberg, M. (2002). "Government support for commercial R&D: Lessons from the Israeli experience". In: Jaffe, A., Lerner, J., Stern, S. (Eds.), Innovation Policy and the Economy, vol. 2. MIT Press, Cambridge, MA, pp. 79–134.
Van Dijk, T. (2000). "Licence contracts, future exchange clauses, and technological competition". European Economic Review 44 (8), 1431–1448.
Van Zeebroeck, N., Van Pottelsberghe de la Potterie, B. (2008). "Filing strategies and patent value". Centre for Economic Policy Research Discussion Paper 6821.
Varian, H. (2005). "Copying and copyright". Journal of Economic Perspectives 19 (2), 121–138.
Vickers, J. (1986). "The evolution of market structure when there is a sequence of innovations". Journal of Industrial Economics 35, 1–12.
Von Graevenitz, G. (2007). Which reputations does a brand owner need? GESY Discussion Paper 215LMU, Munich, Germany.
Von Graevenitz, G., Wagner, S., Harhoff, D. (2008). Incidence and Growth of Patent Thickets—The Impact of Technological Opportunities and Complexity. Munich School of Management, Ludwig Maximilians University, Munich, Germany.
Walsh, J., Arora, A., Cohen, W. (2003). "Effects of research tool patents and licensing on biomedical innovation". In: Cohen, W., Merrill, S. (Eds.), Patents in the Knowledge-Based Economy. National Academies Press, Washington, DC, pp. 285–340.
Walsh, J., Cho, C., Cohen, W. (2005). "View from the bench: Patents and material transfers". Science 309 (5743), 2002–2003.
Waterson, M. (1990). "The economics of product patents". American Economic Review 80 (4), 860–869.
Weston, R. (1998). "A comparative analysis of the doctrine of equivalents: Can European approaches solve an American dilemma". IDEA: The Journal of Law and Technology 39, 35–91.
Whinston, M. (1990). "Tying, foreclosure, and exclusion". American Economic Review 80 (4), 837–859.
Whinston, M. (2001). "Exclusivity and tying in U.S. v. Microsoft: What we know and don't know". Journal of Economic Perspectives 15, 63–80.
Wright, B. (1983). "The economics of invention incentives: Patents, prizes, and research contracts". American Economic Review 73 (4), 691–707.

Chapter 8

STYLIZED FACTS IN THE GEOGRAPHY OF INNOVATION

MARYANN P. FELDMAN* *AND* DIETER F. KOGLER†

**Department of Public Policy*

University of North Carolina, Chapel Hill

North Carolina, USA

†Department of Geography & Planning, University of Toronto

Toronto, Ontario, Canada

Contents

Abstract

The geography of innovation describes the importance of proximity and location to innovative activity. As part of what has been termed the new economic geography, this area of research is less than 20 years old, and is now developed sufficiently so that the discussion can be organized around certain stylized and commonly accepted facts:

- Innovation is spatially concentrated.
- Geography provides a platform to organize economic activity.

Handbooks in Economics, Volume 01

DOI: 10.1016/S0169-7218(10)01008-7

- All places are not equal: urbanization, localization, and diversity.
- Knowledge spillovers are geographically localized.
- Knowledge spillovers are nuanced, subtle, pervasive, and not easily amenable to measurement.
- Local universities are necessary but not sufficient for innovation.
- Innovation benefits from local buzz and global pipelines.
- Places are defined over time by an evolutionary process.

The purpose of this chapter is to summarize recent work on innovation and location in light of these themes, and to consider how these stylized facts shed light on the broader process of technological change and economic growth.

While firms are one venue to organize economic activity, the resources required to generate innovation are typically not confined to a single firm, and geography provides another means to organize the factors of production. Geography is additionally a venue for complex multifaceted social relationships, and human community and creativity that are beyond the economic sphere. Economies are complex: highly integrated, globally interconnected, and highly agglomerated on centers of activity. There is always the temptation to analyze economic institutions and actors individually; however, the new economic geography literature considers the large context. Of course, once the analysis is open to consider geography there is a need to understand history, building a deep contextualized understanding of a place and the relationships that define it. The present review of the literature summarizes the advancements made in this stream of inquiry, but also indicates that many open avenues for research remain, thus encouraging others to contribute to the emerging field of economic geography.

Keywords

agglomeration economies, geography of innovation, knowledge spillovers, localization, new economic geography, urbanization

JEL classification: A33, O11, O18, O31, O33, R11

1. Introduction

Innovation has a decidedly geographic dimension that affects economic growth and technological change. The deliberate and unintended circulation of knowledge between economic actors and the role of physical proximity and colocation are pivotal in understanding the dynamics of the innovation process. The purpose of this chapter is to summarize recent work on innovation and location and consider how this may shed light on the broader process of economic growth and technological change.

Geography and place-specific interactions shape industries. Connoisseurs talk about *terroir*, a French term used to denote the special characteristics that geography bestows. The term can be translated literally as "dirt" but more poetically as a "sense of place." The term captures the total effect that the local environment has on the product, when the total effect is more than the sum of its parts and the effect is difficult to replicate (Feldman, 2009). For wine and coffee, it is the climate, angle of the sun, age of the stock, and growing and harvesting traditions that creates a unique product. Even the best vineyards experience different vintages, reflecting temporal variations. In addition, while quality winemaking is diffusing internationally, wines have become more complex and differentiated rather than homogeneous.

2. Stylized facts surrounding the geography of innovation: A road map

Historically, economic geographers examined the location of economic activity, considering the underlying determinants of the diffusion of innovation (Brown, 1981; Hägerstrand, 1967). Geographers were cognizant of the uneven spatial distribution of economic activity (Amin, 1994), concentrating on industrial restructuring which emphasized the loss of standardized manufacturing due to growing international competition (Harrison and Bluestone, 1988). However, the upside of this restructuring was the growth of technology-intensive industries and an increased emphasis on innovation and entrepreneurship.

This topic of inquiry has expanded to mainstream economics as a result of Krugman's (1991a,b) key observation linking geography and trade. He found that rather than converging, national economies became more divergent over time. This ran counter to the predictions of neoclassical growth theorists. Lucas (1988) and Romer (1990) challenge the assumption of constant or decreasing returns to scale by pointing out that knowledge is subject to increasing returns because of the externalities inherent in its production and use. Rather than diminishing, the value of knowledge actually increased, as a result of network effects (i.e., a larger number of participants increases the utility to any one user). In addition, nonexcludability (i.e., knowledge is accessible to those who invest in the search for it) and nonrivalry (i.e., knowledge can be exploited by many users simultaneously) are features of knowledge that further support the concept of increasing returns.

Porter's (1990) *Competitive Advantage of Nations* introduced geographic considerations to the field of management, exploring how firms benefit from localized competition. Porter extended his work on firm competitiveness to provide a model of a four-factor diamond, which explains "reinforcing supply and demand conditions, industry conditions, and related and supporting industries." Arguably, Porter provides a more formalized view of the national innovation systems literature (Lundvall, 1992; Nelson, 1993), which describes and analyzes the *gestalt* of innovation processes and technological change. Porter's diamond formalized the actors and relationships.

An extensive literature addresses the topic of geography of innovation and describes the importance of proximity and location to innovative activity. This has been termed the "new economic geography," an area of research that is less than 20 years old (Clark et al., 2000). This field is now developed sufficiently so that the discussion can be organized around certain stylized and commonly accepted facts:

- Innovation is spatially concentrated.
- Geography provides a platform to organize economic activity.
- All places are not equal: urbanization, localization, and diversity.
- Knowledge spillovers are geographically localized.
- Knowledge spillovers are nuanced, subtle, pervasive, and not easily amenable to measurement.
- Local universities are necessary but not sufficient for innovation.
- Innovation benefits from local buzz and global pipelines.
- Places are defined over time by an evolutionary process.

Each of these themes is developed in turn.

2.1. Innovation is spatially concentrated

Innovation exhibits a pronounced tendency to cluster both spatially and temporally. There is currently an active debate in the literature about whether the world is flat—that is to say whether opportunities are uniformly distributed, or if there are certain places at certain times that offer greater opportunity. Friedman's (2005) view of a flat world focuses on the impact of globalization, which is certainly significant and deserves attention. The argument of a flat world hails back to the neoclassical view that the economic activity takes place on a featureless plane, with the factors of production able to move frictionlessly between places. However, throughout human history we have observed that creative activity has been concentrated in certain places and at certain times; consider Florence under the Medici, Paris in the 1920s, England during the Industrial Revolution, Silicon Valley and even Wall Street in more recent times. For every generation, there is some location that captures the imagination as a locus of creative activity and energy. The activity may change, but the importance of place remains.

Global outsourcing allows firms to lower production costs; however, technologically sophisticated firms compete on the basis of differentiated performance and innovation. While firms are the entities that take ideas to the market and realize value from innovation, even the largest multinationals are embedded in ecosystems that support and sustain their activity (Gassler and Nones, 2008). These systems are globally connected, but the highest value activity is typically focused in certain locations. The literature alternatively defines collection of firms within one specialized industry or technology, concentrated within the same geographic area by a variety of names such as technopoles, clusters, etc. Marshall (1890) noted this tendency, citing three reasons behind it: an infrastructure of related and supporting industries; the presence of deep, specialized skilled labor pools; and the presence of nonpecuniary externalities due to proximity to a strong knowledge base that facilitates knowledge exchange. These factors are often analyzed relative to geographic concentrations.

Of all economic activity, innovation benefits most from location. Innovation is the ability to blend and weave different types of knowledge into something new, different and unprecedented that has economic value. Similar to art, innovation is a creative expression. However, unlike art, the measure of innovation is not in the eye of the beholder, but in acceptance within the marketplace that brings commercial rewards to the innovating entities and returns to society in terms of economic well being, prosperity, and growth.

Innovation is more geographically concentrated than invention, with invention defined as the first stage of the innovation process. Due to the creation of large patent databases, there are many studies that focus on invention, which should not be confused with innovation. The limitations of patents as an indicator are well known (Griliches, 1990; Scherer, 1984). Patents are geographically concentrated reflecting a concentration of research and development (R&D) activity. This does not necessarily translate into economic advantage for those locations. Feldman (1994) finds correlation of 0.8 in the location of new products introduced to the market and broad patent categories and a correlation of 0.7 between innovation and corporate R&D expenditures. Studies that draw inferences about innovation by focusing on invention should be interpreted with caution. Building on Jaffe (1989), Acs et al. (1994) find that new product introductions were more geographically concentrated than patents, with universities and industrial R&D as important inputs. Feldman (1994) demonstrates that the presence of other local factors such as related industry presence and specialized business services are also determinants of the ultimate realization of invention into product innovation.

Innovation is more geographically concentrated than production. Even after controlling for the geographic distribution of production, innovation exhibits a pronounced tendency to cluster spatially (Audretsch and Feldman, 1996a). The seminal dartboard approach of Ellison and Glaeser (1997) reiterates that geographic concentration is ubiquitous, but also demonstrates that most industries are only slightly concentrated. The aim was to capture the random agglomeration that a dart-throwing model would produce and to see if this was differentiated from industry-specific agglomerative forces resulting from spillovers and natural advantages. The findings indicate that some of the most extreme cases of industry agglomeration are mainly due to natural advantages, such as proximity to water, which would be especially the case if cost considerations concerning the shipment of heavy goods enter the equation of locational choice. However, a high degree of heterogeneity exists among the causes that lead to spatial clustering in most of the remaining industries.

Location matters most at the earliest stage of the industry life cycle. Once a good is at a mature stage of its life cycle costs of production become more important. The propensity for innovative activity to spatially cluster is subject to the industry life cycle, which indicates that there is a direct link between the localization of innovation and the maturity level of particular industries within a territory (Audretsch and Feldman, 1996b). Early stages of the industry life cycles are characterized by the importance of tacit knowledge. Once a product has become standardized and demand will support mass production, it is easier for an industry to disperse geographically.

While a distinction between novel and established products on the market is one way to differentiate between varying levels in the intensity of agglomeration forces, distinguishing between physical and service-oriented production units is certainly another useful approach. Traditional manufacturing, especially if it relies heavily on external production inputs is seemingly less footloose than service-oriented sectors where infrastructure demands and external capital investments are frequently low, and the most significant input factors leading to economic gain are highly skilled, mobile workers. A frequently used example of an innovative platform, which is geographically dispersed, is the open source community (von Hippel, 2001). The globally scattered network of members of this community is certainly impressive; however, it should be noted that most of the tools that are utilized to actually generate open source software products rely on hardware, but also on original source codices such as UNIX,[1] which at one point in time have been developed in particular agglomerated production centers. On the other end of the spectrum, even labor-intensive, low-technology sectors that have experienced a tremendous spatial shift away from highly industrialized places to developing countries, exhibit strong tendencies to cluster, once they are re-embedded into their new locational and institutional setting (Scott, 2006a).

New technologies and new industries, while offering the potential for economic growth, do not emerge fully developed, but begin rather humbly as scientific discoveries, suggestions by product users or suppliers or the novel idea from an entrepreneur. Initially, the commercial potential is unknown and only a few experts or lead users may appreciate its significance. Translating the discovery into commercial activity and realizing its economic potential entails a process that involves building an appreciation of what is possible among potential investors, customers, and employees, building a company and creating a value chain. Increasingly, there is recognition that what matters for place-specific industrial development is not necessarily resources or initial conditions, but the social dynamics that occur within a place and define a community of common interest around a nascent technology or emerging industry (Feldman and Romanelli, 2006). Certainly, this is the case for user-defined innovation (von Hippel, 1988, 2005). Community building, as opposed to insular scientific dialog, can be essential to regional industrial development by constructing a shared understanding and appreciation of the emerging technology (Lowe and Feldman, 2007).

When entrepreneurs confront new technological opportunities, they fashion solutions that adapt what they have on hand from what is easily accessible. The solutions they adopt are more likely to come from local sources—either through tapping networks of people working on similar things or through serendipitous encounters. Most importantly, entrepreneurs use local ingredients in creative and adaptive ways, thus entrepreneurship serves a conduit of knowledge spillovers (Audretsch and Keilbach, 2008, p. 1698). Solutions that appeared to work are repeated and fine-tuned, gradually evolving into accepted routines and operating procedures—the industrial recipes for the region. These recipes are adopted by institutions to define common practices and a common vision of the industry. This encourages further experimentation and adaptation. Knowledge of what does not work, what approaches have previously been tried, and led to dead ends, are part of this local knowledge (Eisenhardt and Martin, 2000; Feldman, 2000; Sitkin, 1992).

In contrast to the logic of specialization, there are benefits to cross-fertilization and collaboration (Jones, 2002). The costs of collaboration are simply lower due to geographic proximity. Geographic proximity promotes serendipity and chance encounters that suggest new uses, new solutions, and refinements. Diversity among industrial sectors within a jurisdiction is considered beneficial to innovative output (Feldman and Audretsch, 1999) and economic growth in general (Glaeser et al., 1992; Jacobs, 1969). However, a distinction between discrete and related variety might offer some further insight in this regard. Perhaps the most important influence on geographic proximity is underlying technical commonalities or related variety (Boschma and Iammarino, 2009). Related

[1] The UNIX system was initially developed in 1969 by a group of AT&T employees at Bell Labs (see http://www.unix.org for a detailed history and timeline). It would take until 1975 until it was widely available outside of Bell Labs.

variety is similar to the concept of Jacobs' externalities, or the relevant diversity between industrial activities that would create the transfer of ideas and spawn innovation. Certainly, much discussion of agglomeration economies focuses on industry localization, which may represent a production orientation or reflect the mature stage of an industrial life cycle, neither of which would be association with new ideas and novelty. Related variety on the other hand, which refers to sectors that have low cognitive distance in terms of their input mix (Frenken et al., 2007), is considered to stimulate innovation by means of spillovers between complementary sectors in the traditional sense of Jacobs-type externalities.

One decisive explanation in this context is offered by the increased importance of regionally embedded tacit knowledge (Maskell and Malmberg, 1999, p. 171). The importance of factors that are of implicit local and evolutionary nature, and therefore are not easily describable, such as the institutional setting, firm and market competencies, knowledge and available skills, and in particular the combination of all of these factors, which in a successful jurisdiction add up to more than the sum of the individual building blocks, is exaggerated in a global marketplace where widespread easy access to codified knowledge seems to be obtained universally. This is clearly a resource centric, or supply driven perspective on how an innovative economy is put in place, and suggests that given the right effort it may be possible to copy and emulate a successful example. However, localized learning processes, which are difficult to codify as they are substantially driven by tacit knowledge and thus regionally specific, need to be present in order to accomplish the continuous creation of improved and novel products and processes. It can be argued, *inter alia*, that knowledge is a fundamental resource, and therefore learning is an essential process that shapes the contemporary technological and innovation driven economy (Lundvall and Johnson, 1994, p. 23).

2.2. Geography provides a platform to organize innovative activity

Just as firms are one means to organize economic activity, geography also provides a platform to organize resources and relationships for economic activity. Beyond the natural advantages of resource endowments, proximity to markets, or climate, certain places have internal dynamics that increase the productivity of investments and result in higher innovation and creativity (Feldman and Romanelli, 2006; Rosenthal and Strange, 2003). These internal dynamics are socially constructed and involve a wide variety of actors. Most importantly, as it is difficult to predict future technological change and market evolution, the greater the number of individuals who are able to participate in creative endeavors, the higher the probability that a place, be it a city, region, or nation is able to capture the resulting benefits.

The resources required to produce innovation are typically not confined to the boundaries of a single firm. Firms frequently contract for external resources, and this can be done at great distances. The motivation is typically saving cost, but his strategy may result in a loss of dynamic efficiencies (Pisano, 1997). Contracting out elements of the value chain for innovative products means that any unexpected results that suggest future improvement or new opportunities are lost. When the value chain is more geographically concentrated there are opportunities for observation and interaction. At longer distance, the relationship is strictly contractual.

Three fundamental spatial scales can be considered to shape the geographic platform in which economic activity and innovation is organized: the global, the national, and the local scale. Some aspects that have a direct impact on innovation processes take place along the continuum of these scales whereas other factors are more specific to one particular scale. The shift from Fordism to post-Fordism, the advent of the information age resulting in globalization, the increasing flexibility of production processes and labor, and the change from a local or national to a global dominated marketplace all indicate that economic activity is now coordinated at the global scale (Dicken, 1998). This view encouraged some scholars to pronounced the "death of distance" (Cairncross, 1997), which they supported by the claim that in the future, economic activity will be dominated by global corporations that are footloose in the sense that they have no real home country affiliation and national identity (Ohmae, 1990). Multinational corporations (MNCs) are the focal entities in the investigation of global innovation activities at the firm level (Pavitt and Patel, 1999). It is recognized that both intra- and extra-organizational networks at the global scale significantly contribute to product and process innovations in MNCs; however, it is also noted that despite the apparent footloose character of MNCs, states or multinational organizations such as the European Union intervene and shape its choices regarding location and innovation strategies. While MNCs have the ability to relocate their sites of production and R&D at random, decisions are strongly guided by the availability of local resources, and the result is the actual innovation processes are still carried out predominantly in a few key regions (Rugman, 2000).

The increasing significance of the global scale seems to have eroded the sovereignty of national economies to the point where they are powerless players in a global game (Ohmae, 1995; Strange, 1997). In contrast some of the literature suggests that the role of nation states is now even more significant than in the past since "with fewer impediments to trade to shelter uncompetitive domestic firms and industries, the home nation takes on growing significance because it is the source of the skills and technology that underpin competitive advantage" (Porter, 1990, p. 19). The early development of the Internet offers an example of the potential instrumental role national public entities play in enabling and supporting innovation and technological change (Greenstein, 2007; Mowery and Simcoe, 2002; Rogers and Kingsley, 2004). Prioritized funding for the development of decentralized communication and transportation technologies by the US Department of Defense provided the initial idea, and substantial funding by the National Science Foundation in turn encouraged and expanded the involvement of a multitude of lead users (Kahn, 1994) transferring this potential technology from the public realm to the marketplace. This combined with favorable regulatory, antitrust, and intellectual property policies along with federal capital investment policies that allowed public entities to invest in venture capital are considered to have further added to the development of the WWW and spurred the development of commercial content and applications. The net result has been the creation of a general purpose technology associated with innovation and economic growth in the 1990s, which to a large extent can be attributed to policies at the national scale (Lipsey et al., 1998).

Recent work has established the importance of government investment in infrastructure to economic growth and competitiveness (Klein and Luu, 2003; La Porta et al., 1999). For example, the quality of transportation and communication infrastructures are frequently acknowledged in terms of trade in physical goods; however, they also directly shape the rate and timing of knowledge exchange that takes place between places (Parent and Riou, 2005). Advanced infrastructure potentially polarizes knowledge spillovers estimates to the extent where highly connected places engage and learn from each at a level that is much more intense than their relative spatial distance would suggest. On the other end of the spectrum, even if two places are closely situated to each other, but lack the support of an advanced transportation and communication infrastructures, knowledge spillovers will take place at a lower magnitude than estimated.

Out of all spatial scales that serve as the point of departure in the analysis of territorial innovation systems, it is probably the regional or local scale that has attracted most attention over the past decades, including participants form a wide range of disciplines. Two basic interpretations of the region as an innovation system have been offered. First, the region simply represents a subsystem of national or sector-based systems. Second, and perhaps more importantly, regions, even in the same national environment, primarily depend on local institutional capacity, which often leads to variances in the delivery of educational and regulatory services across nations. It is argued that as a result of specific advantages from locally rooted institutional capacities, in the form of tacit knowledge, the regional innovation system (Cooke, 1996; Maskell and Malmberg, 1999) is the most important factor for localized learning (Howells, 1996). Untraded interdependencies, which include tacit knowledge that is based on face-to-face exchange, routines, habits and norms, conventions of communication and interaction are considered important assets that shape the innovative potential of a region (Storper, 1997).

One of the primary reasons why regions, and in particular cities, have moved to the center of attention is based on the finding that inventors heavily rely on local information or knowledge as input factor for novel products or processes. Local variations in information available to decision makers exist, and in most instances information can be very costly to transfer from place to place. Such information "stickiness" can have a number of causes (von Hippel, 1994). First, this can be due to the attributes of the information itself, such as the way it is encoded (Nelson, 1982, 1990; Rosenberg, 1982). Second, information stickiness may be due to attributes of the information holders or seekers. The lack of "absorptive capacity" (Cohen and Levinthal, 1990) by a particular information seeker could limit their ability to acquire information due to the lack of certain tools or complementary information. Third, the availability of specialized organizational structures such as transfer groups (Katz and Allen, 1988) can significantly affect the information transfer costs between and within organizations.

In addition to the concept of "stickiness," the "communities of practice" literature (Brown and Duguid, 1991; Wenger, 1998) provides further insight into why innovators tend to use local information. This stream of thought recognizes the situated nature of knowledge as it is created by a community of individuals who have a shared practice or problem. One of the main arguments is that the ways people actually work usually differs fundamentally from the ways organizations describe that work in training programs or organizational charts (Granovetter, 1985). Knowledge is not only considered tacit, in the sense of it being not explicit (Nonaka, 1994), but also knowledge and knowing in general cannot be separated from an individual's engagement in the practicing of their practice (Cook and Brown, 1999). Communities of practice exist in a variety of settings and may develop improvements or innovations in products, services, and work practices in environments that have not much in common with the traditional geographic platform of economic organization, including newsgroups, cyber communities, or knowledge forums. This suggests that relational proximity might be a substitute for spatial proximity (Amin and Cohendet, 2004). However, there has to be a clear distinction between knowledge (i.e., technical expertise that leads to the development of a new product) and contents (i.e., random bytes of data or information that is interesting but of no particular economic value). While content is readily available, knowledge is best transmitted via face-to-face interaction. While it is possible for individuals to come together in temporary agglomerations, the more frequent and trusted interaction predominantly occurs in a collective place. Cities are key examples of such places of knowledge exchange, and in addition they are also primary places of creativity (Scott, 2006b) and dense locations of knowledge generation and spillovers (Feldman and Audretsch, 1999).

Innovation is inherently evolutionary, and in its fundamental nature consists of a multitude of socioeconomic interactions across different spaces and scales (Edquist et al., 1998). Geography provides a platform to organize these interactions, and focusing on only one spatial scale will perhaps not be sufficient to fully understand innovation processes (Bunnell and Coe, 2001). There are many distinctive parts that shape economic activity in a particular place (Feldman and Martin, 2005). This includes not-for-profit organizations, such as universities, research consortia, and standards setting organizations that play a significant role in affecting scientific opportunity and the diffusion of innovation, or other public entities, such as foundations that may fund research and help create markets. In addition there are also civil society organizations that create opportunities for discussion and engagement, and may formalize social networks, as well as intangible assets such as the reputation of a place in terms of business climate, or even the geopolitical setting that connects it to other political units. Institutional setting can be investigated at various spatial scales, local, regional, national, and global. Critical to understanding the dynamics of place is the interrelationships between the various spatial scales and institutions as national laws set the agenda for what lower levels of government may accomplish. Any attempts to replicate Silicon Valley are doomed to fail. What is needed is an understanding of the unique geographic platform on which innovation and economic growth is situated.

2.3. Places are not equal: Urbanization, localization, and diversity

The advantages that accrue to specific places are due to external economies of scope or agglomeration economies. When we refer to agglomeration economies, there are three different concepts to consider: urbanization, localization, and diversity economies.

Economic entities strive to obtain maximum output for a given set of inputs in order to gain comparative advantages in the market. Internal economies of scope, that is, improved efficiency due to product portfolio management, and scale, that is, increasing effectiveness in the utilization of the factors of production, to some extent explain variations in the performance of firms beyond a simple profitability framework (Bercovitz and Mitchell, 2007; Henderson and Cockburn, 1996), but what remains are a set of aspects benefiting the performance of such units that occur due to location, providing a comparative advantage that can be attributed to external economies of scope.

Urbanization economies refer to the component of agglomeration economies that focuses on the actual size of a place itself to explain varying levels of productivity, regardless of competition. Research in this stream has produced mixed results, but principally indicates that doubling the size of a city generally creates a productivity increase ranging from about 3% to 8% (Segal, 1976; Sveikauskas, 1975; Tabuchi, 1986). More recently, urbanization economies have been investigated in terms of inventive output rather than overall productivity levels. Bettencourt et al. (2007) find that large metropolitan areas have disproportionately more inventors than smaller ones, and they generate more patents, which indicates that increasing returns to patenting exist as a scaling function of city size. What remains to be validated is if larger metropolitan areas attract or generate more inventors, or both, than their smaller counterparts.

Localization economies on the other hand are attributed to the concentration of an industry at a particular place rather than agglomeration itself. The initial discussion of the advantages that derive from a densely spatial agglomerate of firms belonging to the same industry sector dates back to Marshall (1920). Three specific benefits are highlighted in this context: the spatial concentration of input–output linkages between buyer and supplier networks, the character of local labor pools with a high degree of specialization, and embodied knowledge spillovers that facilitate the diffusion of technical knowledge (Marshall, 1885). When localization and urbanization economies are investigated simultaneously the results point to a stronger impact of the former on productivity, but industry variations seem to exist (Henderson, 2003; Rosenthal and Strange, 2003). While the discussion regarding the relative importance of urbanization and localization economies to productivity and growth remains vibrant, contemporary research on the impact of external economies of scope frequently focuses on questions regarding specialization and diversity (Rosenthal and Strange, 2004).

Contrary to Marshall's findings, relating to urban specialization, Jacobs (1969) points to the significance of urban diversity as a source of external inputs that boost creativity and subsequently economic activity. The main argument in this context is that the diversity found in agglomerations fosters and enhances the cross-fertilization of ideas between industrial sectors. Although Marshall (1920, pp. 273–274) already recognized the inherent risk that a strictly localized industry produces in terms of vulnerability to external shocks in demand, or local labor uniformity that excludes certain segments of the population from participating, it was Jacobs' account that stressed the importance of diversity to economic development. In this context, development is considered growth through diversification, as the cross-fertilization of knowledge and technology between diverse sectors in the economy leads to the differentiation, diversification, and transformation of the underlying processes of production, which in turn directly influences total factor productivity (Ellerman, 2005). Much of the recent literature, which aims to investigate the effects of agglomeration on innovation and productivity, proceeds to situate external economies of scope, that is, localization and diversity, within a dichotomous framework, therefore generating a fundamental division between them (Baptista and Swann, 1998; Feldman and Audretsch, 1999; Glaeser et al., 1992). The results are mixed, and in some cases Marshall–Arrow–Romer (MAR) externalities[2] are considered more prevalent (Baptista and Swann, 1998), while in others, Jacobs' externalities (Feldman and Audretsch, 1999) are thought to dictate local knowledge spillover processes, and in some instances both types of externalities are found to be significant (Capello, 2002). The substantial variations in the findings are unsatisfactory and continue to fuel further research efforts in this context; however, they may also point to the necessity to review some of the underlying principles that guide research efforts on agglomeration economies, in particular we need to reconsider the potential inequality of places.

Marshallian externalities, which are concerned with intraindustry economies of localization, are different from Jacobs' externalities that refer to interindustry exchanges between different technologies and sectors within a particular metropolitan area, and therefore are not a mutually exclusive phenomenon (Ibrahim et al., 2009). Also, while the overall diversity of economic activity, and increasingly cultural activity (Scott, 2006b), within a metropolitan is pivotal in explaining performance resulting from the cross-fertilization of ideas and subsequent economic development, it does not rule out that

[2] Glaeser et al. (1992) formalized the MAR externality, which concerns knowledge spillovers between firms in an industry, based on the findings of Marshall (1890), Arrow (1962), and Romer (1990).

concurrently a progression of specialization, as indicated by a concentration of employment in a particular industry takes place within the very same city. Another aspect that should be considered is that diversity and specialization might play very different roles in terms of which kind of innovation they produce. The potential to generate radical, disruptive innovative output should be higher when very diverse sectoral knowledge bases are combined, while incremental innovation should demand specialized knowledge, which is necessary to improve existing technologies (Schumpeter, 1942). Seemingly the relationship between the relative importance of localized specialization and diversity to economic growth should not be characterized as a continuum from evolutionary to revolutionary innovation, but not as a dichotomous one (Christensen, 1997).

Finally, and most importantly, from a spatial perspective much work that attempts to investigate and differentiate between these two types of external economies of scope relies on aggregate data sources, and while presumably statistically significant, is not able to capture the substantial sectoral and spatial variety that exists at a particular place. Thus, much of the findings are not uncovering fundamental dimensions of innovation, but rather provide secondary results, which warrants caution to generalize and apply conclusions universally to all places (Scott, 2006b). This leads to yet another scope that should be considered in a comprehensive analysis of agglomeration economies, the temporal one (Rosenthal and Strange, 2003). Places are intrinsically evolutionary, which stresses the importance of historic events in shaping local economic activity. Research related to the static or dynamic nature of agglomeration economies provides evidence of the dynamic components of spillovers, but fails to fully address the underlying mechanism that facilitate them (Glaeser and Mare, 2001; Henderson, 1997).

Large metropolitan areas are among the most productive places (Ciccone, 2002; Ciccone and Hall, 1996; Feldman and Audretsch, 1999; Harris and Ioannides, 2000). Glaeser and Mare (2001) provide evidence that an urban wage premium, resulting from a combination of wage level and wage growth effect, exists, which potentially explains why workers in cities earn relatively more than their counterparts in nonurban areas. While it is unclear if a more efficient coordination of labor markets, or factors of accelerated learning, in cities cause higher levels of urban wage growth, this analysis also indicates that the workers who eventually leave cities continue to enjoy relative higher wage premiums. Recent evidence also suggests that cities may experience higher levels of productivity due to the positive relationship between hours worked and the density of professionals within an occupational group, in a particular metropolis. In addition to spatial agglomeration, the presence of rivals among professionals appears to further increase the number of hours worked, and subsequently productivity (Rosenthal and Strange, 2008). While the idea that cities are very vibrant and busy places is indeed not a new one, the actual notion of an "urban rat race," which conceptualizes the relationship between agglomeration and work output was first introduced by Akerlof (1976), not much attention has been given to concepts such as adverse selection processes or competition among professionals, as they relate to the work routines of professionals in urban agglomerations.

2.4. Knowledge spillovers are geographically localized

Knowledge is an ethereal concept that is perhaps best considered as embodied in human capital, which is individuals who are able to comprehend, integrate, and create new knowledge. Individual productivity is definitely influenced by location, that is, individuals with a given set of characteristics will have different levels of productivity depending on their location (Rigby and Essletzbichler, 2002).

The ground-breaking findings of Abramovitz (1956) and Solow (1957), which established that there was a large residual of aggregate productivity growth unexplained by capital accumulation, and Kuznets' (1962) pioneering research on the nature of inventions, have given rise to knowledge being named as the key economic asset that drives long-run regional and national economic performance. In particular, the concept of knowledge externalities or spillovers has become the focal interest in multiple disciplines concerned with research relating to the dynamics of location, and their impact on the processes of agglomeration of innovative activity. Contemporary research regarding knowledge spillovers in a spatial context takes into consideration and builds upon two commonly accepted facts. First, that innovative activity is concentrated in space (Feldman, 1994, 1999; Moreno et al., 2005), and second, that knowledge flows are geographically localized (Bottazzi and Peri, 2003; Branstetter, 2001; Jaffe et al., 1993; Maurseth and Verspagen, 2002; Sonn and Storper, 2008).

The significance of knowledge inputs in the generation of innovation and technological change, which in turn is followed by economic growth, combined with the fact that knowledge spillovers are considerably localized suggests that the performance of a jurisdiction heavily depends on what type and amount of knowledge will be produced internally. Strong connectivity between inventive agents within a regional community essentially benefits diffusion processes, but in more fundamental terms provides the exposure to new knowledge and perspectives that allows for increased creativity and innovativeness (Cowan and Jonard, 2004). In addition, regional network aggregation, that is, the connection of previous separate communities within a spatial cluster, creates opportunities for technological brokerage, which again will increase knowledge spillovers within a particular place (Burt, 2004).

The importance of geographic location as a factor for knowledge creation and innovative activity, in a world that is reliant on technology that provides instant communication may seem irrelevant and even paradoxical. After all, telecommunications technologies have triggered a virtual spatial revolution. Geographically dispersed activities may be linked electronically in real-time transactions. However, in this specific context, a sharp distinction between codified and tacit knowledge inputs has to be made. Codified knowledge is technical information that can be found in publications. It can be easily communicated through conventional media, and therefore has an extended spatial reach. Conversely, tacit knowledge constitutes the specific capabilities of individuals. To a large extent it describes the outcomes produced from social and institutional settings found within a particular place. This type of knowledge is best transferred through face-to-face interactions and, in general, is difficult to exchange over long distances (Gertler, 2003).

However, spatial proximity alone may not be sufficient enough for knowledge spillovers to occur. In addition, cognitive and social distance also has to be overcome by individuals and firms in order to engage in efficient knowledge exchanges that lead to learning processes and subsequent innovation. In other words, knowledge can only spill over if the involved parties exhibit an optimal cognitive distance (Nooteboom, 2000), because only then it will be possible to absorb and implement the external knowledge that in turn results in technological change and enables innovation (Cohen and Levinthal, 1990). Too much cognitive distance, which is the case when individuals or industries operate in very different knowledge bases and/or institutional settings, might prohibit communication and therefore entirely eliminate knowledge spillovers. On the other hand, too much cognitive proximity, which exists among firms that work in similar product portfolios and rely on related problem solving techniques, may result in spillovers that possess minimal value added, and in the worst case perhaps even diminish possible inventive advantages a firm may enjoy over its competitors. Related variety, not necessarily regional diversity or regional specialization, which refers to an optimal cognitive distance, is considered to be the most supportive factor for effective knowledge spillovers that actually lead to increased innovative output in a particular locality (Boschma and Iammarino, 2009; Frenken et al., 2007).

2.5. Knowledge spillovers are nuanced, subtle, pervasive, and not easily amenable to measurement

Significant considerations need to be taken into account in any investigation concerning knowledge externalities. First and foremost knowledge spillovers are nuanced in the sense that different variations exist. A broad distinction originates from Griliches (1979), who differentiates between two types of knowledge spillovers. On the one hand there are knowledge spillovers that are associated with the exchange of goods, they are labeled "rent spillovers," and refer to knowledge that is rival and excludable in nature. Alternatively there are those that arise purely from the process of R&D, which are identifies as "pure spillovers" or "idea-creating spillovers." This type of knowledge spillover refers to knowledge that is characterized by its nonrivalry and nonexcludability as it can be utilized by many users at the same time and is freely available. Due to these specific qualities, the nature of this type of knowledge is considered to resemble that of a public good (Arrow, 1962). There are some important implications resulting from these two particular properties of pure spillovers, which were first pointed out by Arrow (1962) in the context of public goods. First, nonexcludability implies that it is impossible to prevent someone from consuming it. In other words, if research results are disseminated through the regular channels of communication, for example, by means of publication in journals or books, or Web sites, knowledge enters the public realm, and therefore becomes available to anyone who searches for it. Second, nonrival knowledge or goods may not only be consumed by many individuals at the same time, but additional users of said knowledge, will not decrease the amount or quality available to others. In essence, readily available research results constitute a nonrival good, as their utility is not influenced by the size of the actual user group.

The intricate nature of knowledge and its associated spillovers pose a challenge when trying to draw such a strict dichotomous distinction. Upon closer examination pure knowledge externalities, frequently claimed to be the primary focus in this stream of research, turn out to be mediated by market mechanisms (Geroski, 1995), and therefore influence local firms' innovation opportunities indirectly through pecuniary rather than pure knowledge externalities (Breschi and Lissoni, 2001a). The complexities involved in describing true knowledge spillovers, that is, ideas that benefit research efforts in one industry or firm that originates from the results of previous undertaken research in another industry or firm, is reflected in the lack of a universal definition in the literature. In general, knowledge spillover can be considered, "intellectual gains by exchange of information for which no direct compensation to the producer of the knowledge is given or for which less compensation is given than the value of the knowledge" (Caniëls, 2000, p. 6).

Knowledge spillovers, while pervasive in shaping innovation, are also inherently subtle, making it a difficult phenomenon to identify and to measure. A general skepticism concerning the difficulty of measuring spillovers is reflected in Paul Krugman's statement "knowledge flows [...] are invisible; they leave no paper trail by which they may be measured and tracked, and there is nothing to prevent the theorist from assuming anything about them that she likes" (Krugman, 1991b, p. 53). Contrary to this view, Jaffe et al. (1993, p. 578) indicate that "knowledge flows do sometimes leave a paper trail, in the form of citations in patents," and, by this, suggest the research potential of this method as a means of studying the complex Webs of knowledge spillovers across locations, technologies, and time. Patent citation analysis, the study of citations made to previous patents, provides the opportunity to gain insights into the process of knowledge flows rather than just present a proxy measurement (Jaffe and Trajtenberg, 2002). Also, patent citations provide a way of exploring pure knowledge spillovers as they correspond to the nonrival property of knowledge that forms the foundation of endogenous growth, rather than pricing or pecuniary externalities that derive from the exchange of goods (Griliches, 1979).

The computerization of patent data by national patent offices, such as the United States Patent and Trademark Office (USPTO) or the European Patent Office (EPO) and further the provision of these data by commercial data vendors such as the Derwent World Patents Index offered by Thomson Reuters, have made patent data widely accessible to researchers. However, it was the first NBER US Patent Citations Data File (Hall et al., 2001),[3] and the original method presented in the seminal paper by Jaffe, Trajtenberg, and Henderson (1993, hereafter referred to as JTH), which especially enabled and inspired both geographers and economists to carry out a series of investigations that demonstrate the localization effect of knowledge spillovers, thus making patent citation analysis the most commonly used approach in this context. The goal set out in the original JTH experiment was to test whether knowledge spillovers are localized, by comparing the geographic location of patent citations with that of the cited patents. Further, the goal was also to measure the extent, it at all, that localization was present. To adjust for uneven patent output growth and varying levels of technology focus between spatial entities, a control sample was constructed. For each patent that cited a sample of original patents, a corresponding control patent was identified belonging to same technology class and as near as possible to the application date, to assure that the control patent closely resembled the citing patent in terms of technology and timing of the invention. The control patents, which are geographically matched with the original or cited patents, were then used as a baseline or reference value in the comparison of the frequency of geographic matches between the actual citing–cited patent pairs. The rational was to compare the localization of citations with that of similar patents that were not linked through citations to the original patents. The results confirmed that knowledge spillovers as indicated by patent citations are indeed localized. At the city[4] level citations were two to six times as likely to come from the same jurisdiction as control patents if self-citations were excluded from the analysis.[5]

The JTH control technique, which in its own class has somewhat become the standard methodology in the exploration of knowledge spillovers through patent data, has been employed in numerous similar studies since its inception (Almeida, 1996; Almeida and Kogut, 1997; Hicks et al., 2001), and it still applied today (Sonn and Storper, 2008), but has also been subject to substantial critique, in particular by Thompson and Fox-Kean (2005, henceforth TFK). The main concern put forward by TFK is that JTH's matched case-control methodology might not adequately control for existing patterns of industrial activity, which induces a systematic bias in the results, thus potentially portraying evidence that supports the localization of knowledge spillovers where really none exists. In essence, TFK outline two significant problems in this regard. First, matching of control patents to their citing counterparts is

[3] The National Bureau of Economic Research (NBER) is working on a major National Science Foundation (NSF) funded update and extension of these data; the new NBER patent data file is scheduled to be released in 2010.

[4] Cities refer to SMSAs, which are based on 1981 Standard Metropolitan Statistical Area definitions.

[5] Self-citations in Jaffe et al. (1993) refer to citing patents that are owned by the same organization as the originating patent; they do not represent an externality.

done by the broad three-digit USPTO technology classification, which suppresses "within-class" heterogeneity due to aggregation. Second, most patents contain several distinctive claims, each of which is assigned a different technology code in addition to the primary code used in the matching process. Again this makes matching a control patent a random task as it may not resemble the citing patent that is associated with the original patent. In summary, TFK question the level of precision by which the control patents eventually match their paired citing patents in terms of industrial similarity, something that could have a substantial effect on the final results as the derived "control frequency," which is used as a reference value in the evaluation of a geographic match between the citing and paired control patent to the original patent, might be erroneous. The results obtained by TFK, by applying subclasses rather than just the main three-digit classification in the selection process of control patents, show that there is no statistical support for intranational localization effects, but verifies JTH's earlier findings of localization at the country level. The reassessment concludes that in principal the JTH methodology is capable of indentifying the localization of knowledge spillovers, but only if controls are carefully selected based on the suggested detailed technology subclass classification (Thompson and Fox-Kean, 2005).

Henderson et al. (2005) provide comments on the reassessment of the JTH methodology carried out by Thompson and Fox-Kean (2005), in particular they point to the possibility that the lack of localization effects in the results, at the intranational scale, are due to a possible sample selection bias induced in the final step of the TFK test, where the sample is restricted to control patents whose primary subclass matches the primary subclass of the citing patent. The key problem, identified by Henderson et al. (2005), with the methodology applied in the TFK experiment relates to the missing justification as to why spillovers should only occur within the narrowly defined subclasses. An analysis that relies fundamentally on intratechnology flows would follow the argument of specialization, but at the same time would certainly exclude any possible evidence pertaining to knowledge spillovers from other technology sectors in the process of invention, thus rendering arguments for diversification inadequate. If knowledge spillovers are mainly intrasectoral, the impact of industrial specialization, and consequently the pattern of knowledge flows, is very different than in a system where technology spillovers flow easily between industries (Lucas, 1988). Controlling for the geographic and temporal distribution of "technology in order to identify knowledge spillovers is very tricky, and [...] the exercise in JTH can hardly be regarded as conclusive in that respect" (Henderson et al., 2005, p. 463). This suggests that further research is warranted.

Maurseth and Verspagen (2002) offer an alternative approach for capturing technological linkages, which has been an area of criticism in the JTH methodology, by constructing a regional compatibility index for all regions across Europe; however, they derived similar results, which also indicate that geography matters. Furthermore, studies that examine whether or not the strong proximity effect of knowledge spillovers found in macrolevel studies also holds true in a microcontext, generally confirm the proximity effect on spillovers, and find significant negative coefficients on the geographical and technological distance variables (Verspagen and Schoenmakers, 2004).

Patent citation analysis is not the only research framework that may be utilized to quantify knowledge externalities. An alternative stream of analysis focuses on the movement of people, and is based upon the idea that knowledge is embedded within an individual. In contrast to patent citation analysis, which focused on the mapping of codified knowledge, research undertaken in this context attempts to measure the flow of tacit knowledge (Polanyi, 1958), which is an equal pervasive but different part of knowledge spillovers. This type of knowledge is not only considered tacit, in the sense that it is not explicit (Nonaka, 1994), but there is also the understanding that knowledge is embedded in the individual, and cannot be separated from a person's functionality (Cook and Brown, 1999). For example, a study carried out by Zucker and Darby (1996) found that the agglomeration of star scientists (defined as highly productive individuals who have discovered a major scientific breakthrough) in the biotechnology field, directly results in a high concentration of new biotech ventures at the same location. In a similar study, Almeida and Kogut (1997) show that mobility patterns of star patent holders in the semiconductor industry match the transfer of knowledge, and therefore directly influence the geographic patterns of knowledge spillovers.

One of the problems that occurs when the mobility of individual (skilled) workers is used to demonstrate knowledge spillovers, is actually verifying how much pure knowledge spillovers, that is, appropriate knowledge that benefits the new firm to be more innovative, is generated. The seemingly tacit character of knowledge is questionable, because knowledge is actually embodied in human capital. In other words, what actually occurs when an individual moves from one firm to another is more of a knowledge transfer rather than a knowledge spillover. The literature surrounding *evolutionary economics* (Nelson and Winter, 1982) is particularly informative in this context.

This work discusses the forms of organizational knowledge that is embodied in firms' organizational routines, but not in individuals. Firm capabilities, in particular once they have demonstrated innovative success in the past, are eventually standardized, and therefore create an internal path dependency that is frequently rooted in local practices. However, at the same time that a firm develops greater path dependency, it also exhibits a lower receptivity to external knowledge sources. Thus, the effectiveness of knowledge sourcing in the form of highly skilled mobile workers relies on the degree of path dependency. Song et al. (2003) illustrate that mobile engineers who join a firm with stronger path dependence are less likely to build upon the knowledge of their previous firms, especially if the engineer's key area of expertise lies inside the core technology areas of the new firm. Furthermore, workers that embody relevant knowledge may tend to move locally for a number of reasons, such as risk aversion, localization sunk costs, and existing social ties, but because they are already embedded into local practices, regardless if they are concerned with technical or organizational routines, may not provide the desired knowledge transfer which is capable of inducing structural changes that lead to increasing levels of innovative output.

Of course, noncompete and nonsolicitation covenants, which are common practice in certain localities, potentially influence highly skilled labor mobility in that it becomes mainly of a cross-border nature, and therefore indirectly affect the quantity and direction of local knowledge spillovers (Stuart and Sorenson, 2003). Saxenian's (1994) work on Silicon Valley, which discusses some of these agreements, has shown that the absence of such covenants may result in a high rate of mobility between firms, something that is now considered a central contributing factor to the supportive entrepreneurship culture that has developed in Silicon Valley. Notably, the California noncompete provisions were not a strategic entrepreneurship policy enacted by California lawmakers. Gilson (1998, p. 5) notes, "Rather, the California prohibition dates to the 1870s, a serendipitous result of the historical coincidence between the codification movement in the United States and the problems confronting a new state in developing a coherent legal system out of its conflicting inheritance of Spanish, Mexican, and English law."

Of course, not all communities are defined by physical proximity. Recent studies suggest that members of scientific disciplines form epistemic communities, where they develop a great level of trust and communicate more frequently regardless of distance. An epistemic community is defined as a collective of individuals dedicated to the production of knowledge, with recognized expertise and competence in a particular domain, and who share a common understanding and language pertaining to problem solving within this specific area (Amin and Cohendet, 2004). The need to communicate about scientific findings that are relevant to a small spatially dispersed group indicated that there is greater interaction with this specific community than with other unrelated individuals in close physical proximity. Thus, social proximity potentially explains a significant share of knowledge externalities, and it is actually the borders of epistemic communities that define the scale of investigation rather than local or national borders when studying knowledge spillovers (Breschi and Lissoni, 2003; Singh, 2004). However, this does not explicitly imply that geography does not matter anymore in terms of knowledge diffusion and spillovers, as interpersonal networks are actually embedded in physical space (Singh, 2005). These may be considered the remnants of the prior geography–social relationships that were established by previous proximate contact (Bercovitz and Feldman, 2010).

In a more direct measure of knowledge spillovers, Feldman et al. (2009) employ indicators of knowledge sourcing provided by inventors obtained through survey data, rather than secondary sources such as patent citations. This type of analysis mimics closely the idea of Marshallian knowledge spillovers by investigating the underlying geographic distribution of knowledge generating agents. The findings show that even after controlling for the existing distribution of inventive activities, knowledge spillovers benefit from geographical proximity. This analysis provides further evidence for the breadth of knowledge spillovers at the microlevel, that is, individual inventors who are actually utilizing spillovers in the process of stimulating technological change.

Another approach used to explore knowledge spillovers is based on the finding that knowledge can be embodied in goods. As a result, knowledge externalities can be mapped using trade patterns (Feldman, 1999). However, trade pattern data is, in most cases, not available on a subnational level, which confines this stream of analysis to the international level (Coe and Helpman, 1995). In addition, trade patterns, even if they are considered in an international context, focus on what might be described as technology diffusion rather than knowledge diffusion (Jaffe and Trajtenberg, 1998), therefore this approach is interpreted as a reduced form of evidence of knowledge spillovers across international boundaries. It is acknowledged that bilateral trade flows are strongly correlated with various forms of communication and information transfer making it difficult to “distinguish the effect of pure knowledge flows from the effect of technology flows embodied in advanced capital goods sold from one country to another” (Jaffe and Trajtenberg, 2002, p. 200). In Chapter 19, Keller discusses in detail how import, export, and foreign direct investment (FDI) data can be employed in an analysis of international technology spillovers.

Certainly, we believe that receptivity to knowledge declines as physical space increases. Similar to the manner that Van Thunen’s urban rent gradient decreases from the city center, we expect that the transmission of knowledge declines over physical space. Face-to-face contact, social and cultural commonalities, a shared understanding and language in a specific technology field, are all attributes that hint at the localized character of knowledge spillovers. There is a broad consensus over the significance and the widespread use of the concept in the literature, but the inherently nuanced, subtle and pervasive character of the phenomenon, combined with the complexity to measure it, continues to generate disagreement, and therefore it seems that knowledge spillovers to a certain extent remain a black box, whose contents needs to be further investigated in order fully comprehend the localization of innovation processes (Breschi and Lissoni, 2001b).

2.6. Local universities are necessary but not sufficient for innovation

Universities are increasingly viewed as engines that are able to drive innovation and economic growth. In the knowledge economy universities are perceived as important input suppliers, both in terms of providing skilled labor and innovative ideas, but also as instrumental institutions that shape technological progress through various mechanisms as outlined by Salter and Martin (2001):

- Increasing the stock of knowledge
- Training skilled graduates
- Creating new instrumentation and methodologies
- Facilitating the formation of problem solving networks
- Increasing the capacity for problem solving
- Creating new firms

Universities have provided economic advantage to the regions they are located in since their onset, and reducing them to a simple factor of production ignores the fact that they have long been places of contemplation and exploration, unfettered inquiry, free expression, and public discourse, significantly shaping the sociocultural environment of regions and nations, thus building quality of place. Gertler and Vinodrai (2005) describe universities as anchors of creativity with the ability to attract highly skilled talent in the form or researches and students. These individuals potentially add to existing local knowledge assets in a region, which in turn strengthens local innovative competences.

The historical conceptualization of innovation, that portrays technological change as a linear process, places universities at the earliest stage of knowledge creation (Bush, 1945). However, since the shift from a closed innovation system to a more open one (Chesbrough, 2003), which is increasingly visible in the contemporary knowledge-based economy, universities are increasingly considered advanced production sites of applied research rather than just a provider of basic science findings. This has a significant impact on local economic growth as university spin-off firms are frequently acknowledged as one of the key drivers of technological change and subsequent economic growth, leading to the development of economic successful regions (Bercovitz and Feldman, 2006). Today, most advanced national economies strive to generate economic wealth by exploiting and diffusing public research by means of commercializing university research (Clarysse et al., 2005). In many cases, however, such endeavors have experienced limited success (Callan, 2001), and although comprehensive case studies of Silicon Valley, Route 128, and Research Triangle Park (RTP) highlight the supportive role of local universities, the literature points to the finding that research universities are a necessary, but not sufficient, condition for regional economic development.

Although universities are frequently considered one of the engines of growth (Feller, 1990; Miner et al., 2001), there are examples of prolific institutes that have either not been successful or do not actively participate in the pursuit of commercializing research findings, and yet their contribution to the advancement and dissemination of knowledge is profound and should not be underestimated (Feldman and Desrochers, 2004). After all, universities produce both the new ideas and skilled workers that are essential to innovation and economic growth, although the path is indirect. In contrast to commercial firms with a relatively simple profit motive, universities have complex objective functions that involve a variety of educational and societal objectives, as well as the interests of faculty members, students, politicians, and the larger scientific community. Also, the rate and direction of knowledge transfer and the actual strength and importance of linkages leading to university–industry partnerships vary significantly among industry sectors. In sectors where science plays a major role, as is the case in the biotechnology and information technology fields, the significance of university knowledge inputs, which preferably are readily accessible at a particular locality in order to gain a competitive advantage through localized knowledge spillovers, are certainly stronger than in other less knowledge-intensive sectors.

One particular important, and highly localized, transfer mechanism through which knowledge spillovers are realized is through knowledge exchanges that take place between people. Ideas embodied in individuals who possess particular skills, specific knowledge, and valuable know-how, have the potential to significantly influence the rate and direction of technological change at a particular locality. Universities play an important role in this context as they employ and train highly skilled scientific personnel. Of particular interest in this context are star scientists, which are defined by Zucker and Darby (1996) as highly productive individuals who discovered a major scientific breakthrough. Employing this criterion and based on the premise that such individuals embody the intellectual capital to enable the commercialization of advanced research findings, Zucker et al. (1998) investigate the impact of star scientists in the formation of New Biotech Entities (NBEs). The results show that the startup rate of NBEs is considerably higher in regions where the intellectual capital resides, that is, where outstanding scientists as measured by research output are located. Strong linkages between stars and NBEs are indicative of higher levels of productivity compared to regions where such linkages are missing. Overall, intellectual capital, as indicated by the number of stars and their collaborators in a given area, is considered the main determinant of where and when the US biotechnology industry developed (Zucker et al., 1998). In a similar attempt, Almeida and Kogut (1999) follow intellectual capital through interfirm mobility patterns of major patent holders in the semiconductor industry. One of the primary findings is that interfirm mobility of these highly skilled intellectuals significantly influences the transfer of knowledge, and that these transfer mechanisms are embedded in regional labor networks. However, Almeida and Kogut (1997, 1999) also show that the localization of knowledge diffusion displays considerable regional differences, thus variations in the spatial patterns of knowledge externalities exist. Particular strong localization effects are observed in Silicon Valley, one of the industry's most prolific regions. The investigation emphasizes the role small firms play, and in particular startups, as they display higher levels of research productivity than larger entities, which is more evident in the United States compared to other countries, and then again more so in specific industries (Pavitt et al., 1987; Scherer, 1984), as these entities are in particular sensitive to university research inputs that mostly take place in spatial proximity (Feldman, 1994). Following this argument, the impact of local university research may be a universal phenomenon, but is amplified in the context of firm formation, which in turn represents an integral segment of the economy driving technological change and subsequent economic growth.

The fundamental question regarding the role universities should play in an advanced economy is certainly accompanied by the concern that the current high level of emphasis on university–industry partnerships may be to the detriment of the significant role these institutions play for longer-term economic growth (Nelson, 2001). Chapter 6 provides an in-depth discussion of the significant role university research and public–private interaction play on innovation and economic growth.

2.7. Innovation benefits from local buzz and global pipelines

One of the main advantages to firms of locating in a cluster is that spatial proximity allows for a better exchange of tacit knowledge, an essential component in an innovative economy (Saxenian, 1994). However, recent empirical studies have begun to question the seemingly superior character of local versus global knowledge flows (Gertler, 2003; Malecki and Oinas, 1999), which indicates a certain dissatisfaction with the above line of reasoning. Based on this, Bathelt et al. (2004) have developed a concept that recognizes both the existence of a local buzz dynamic, which demonstrates the importance of just being there (Gertler, 1995), but also the significant role of extralocal sources of knowledge, that is, the pipeline structure (Owen-Smith and Powell, 2004). A significant benefit arising from urbanization economies is the inevitable exposure to a range of local knowledge bases of varying degrees of cognitive distance to a firm's core capabilities. Spatial proximity to a diverse set of activities initiates interactive learning processes along several dimensions (Malmberg and Maskell, 2006), namely through learning by interaction and by monitoring. Knowledge exchange in this context is frequently unintentional and serendipitous rather than mediated through market transactions. It is regular encounters and frequent face-to-face contact, facilitated through an organizationally and institutionally local embeddedness, which is further enhanced by shared socially constructed norms and conventions among the actors involved, that especially excels local learning processes and in turn creates what is labeled a buzz (Storper and Venables, 2004), or noise (Grabher, 2002), in a particular locality. In the context of localization economies, which are characterized by a high level of activities in a variety of rather closely related functions, as is the case in specialized clusters, the level of interaction is even further increased due to relational proximity of the actors involved. While local knowledge sourcing is certainly one of the determining factors affecting the performance of knowledge-intensive industries, competitive pressure and forces of globalization, among other factors, point to the necessity to harvest knowledge pools outside the local environment. Global knowledge search processes, however, are much more structured, formalized, and planned than in the case of localized learning (Bathelt et al., 2004). Determined by the spatial concentration of knowledge production in certain localities, the search for distant knowledge inputs by individual or firms is a conscious process that is only directed to specific places considered to possess particular competencies in a particular core activity (Bathelt, 2005a). To a certain extent this leads to a buzz-and-pipeline dichotomy where local knowledge flows are associated with tacit forms of knowledge, while global knowledge corresponds to more formalized or codified knowledge types. However, these arguments rest on a small base of empirical evidence (Gertler and Wolfe, 2006), and have been questioned more recently, especially in the context of particular industry sectors (Moodysson, 2008). In essence, while the search for new and useful knowledge inputs is a universal process, the way this is carried out at the local and global scales differs significantly.

Long-distance collaborations are certainly part of knowledge creation. As creative activity has become more complex teamwork has become more prominent. But does the fact that individuals collaborate over great distance indicate that distance is not important? Certainly new technologies have lowered the cost of long-distance collaborations. This begs the question of how such collaborations form. Bercovitz and Feldman (2010) examine teams of inventors to discern if collaboration is driven by prior employment relationships, prior social relationships or star attraction. They find that the majority of external members had some prior social relationship with internal members of the team—either as former colleagues or students or as long-time coauthors. These collaborations reflect the footprints of a former geography—that is to say that the collaboration reflects prior colocation. In only about 25% of the cases did the team come together without any prior working relationship such as previously being at the same institution or being a long-time coauthor. In the cases without a prior relationship what matters most was the star attraction of the external member. Highly cited individuals were more likely to be engaged in long-distance relationships. As found by Mansfield, industry collaboration was more likely to be local while academic collaboration took place at greater distance, confirming the relevance of epistemic communities. Moreover, having an external member was more likely to results in a producing economically valuable knowledge. This suggests that external members are brought in to address specific requirements. However, the most productive teams were internal to the organization and included novel combinations of inventors.

In the worst case scenario individuals and firms operate in a local vacuum, that is, they exhibit no, or very low levels of, local interaction with other actors working on similar problems in the same local setting, and, in addition, they also display a global void as they have limited capabilities in accessing geographic distant specialized knowledge pools that are relevant in their respective sector. The result is an inevitable decline of innovative output, resulting in a loss of productivity gains, and eventually economic stagnation. For example, Bathelt (2005b) illustrates that the lack of extralocal firm linkages and market relationships in the Leipzig media sector, combined with limited local networking and interactive learning activities within the cluster, have led to the actual decline of one of Germany's secondary media agglomerations, despite a favorable growth potential partially grounded in a historical context and recent national economic restructuring processes.

2.8. Places are defined over time by an evolutionary process

The remarkable growth of Silicon Valley, which is considered the archetype of high-tech industrial clusters, has made it the prototype for a wide range of policy initiatives aiming to replicate this success story in other regions. In a historical account, Moore and Davis (2004) emphasize that learning was the key process that led to the transformations that built Silicon Valley rather than specific institutions, single events, or even chance. Particular importance is attributed to the interplay of general and regional-specific growth factors as exemplified by the evolution of scientist-managers, which at the time was a nation-wide trend, but was especially amplified in the region due to the fact that many scientists learned about management at establishments such as Fairchild. These spatially concentrated learning processes caused a significant shift in business aptitude, and ultimately created an increased number of opportunities where the usually adverse task of risk-investment became an important regional quality resulting in a strong economic growth performance. Learning, in the perspective of industrial cluster formation, is considered a regionally embedded activity that is best facilitated in dense local social and institutional networks, where regional competition leads to severe selection processes, and in turn to efficiency by means of adaptation of enhanced production methods, as well as an accelerated rate of innovation. More importantly, learning is also an inherently evolutionary, cumulative, and most importantly, dynamic activity, and therefore any type of static cluster policy initiative is doomed to fail, especially if it does not incorporate local context, in which important processes such as the development of regional absorptive capacity are accentuated in a path-dependent fashion.

When considering the development of industrial clusters there are two diametrically opposing models. One model, practiced in China, relies on government dictating the growth of designated science cities (Hu, 2007). This is a very top-down approach to economic development that has been successful in Singapore and Taiwan: the central government dictates that a specific location will have a concentration of R&D and accomplishes this in a relatively short period of time. The verdict is still out as to whether these locations will be successful at creating a sustained competitive advantage given that innovation is more complex than simply conducting R&D. The other model occurs in the United States and other market economies and relies on self-organization and local initiative. In market economies the central government cannot dictate the actions of private companies but may only offer incentives to encourage firm location decisions and investments in R&D. The closest that we have to a government-induced clusters is RTP in North Carolina, which was the result of state and local government actions. RTP was a very long undertaking beginning in the 1920s and is now the largest research park in the world (Link, 1995). While there are many other examples of government trying to build clusters in market economies, the results typically look very different from what was originally intended (Leslie and Kargon, 1994).

Given that innovation is about the flows of people and ideas then institutional dynamics and political context certainly matters. In RTP, over time entrepreneurs left large firm employment or returned to the area to start firms, filling in a vibrant industrial landscape (Avnimelech and Feldman, 2009).

Causality is always difficult to discern: the attributes associated with fully functioning clusters are the result of their success, not the underlying cause. While it is always difficult to attribute causality and many policymakers search for the recipe for industrial cluster development and economic vitality there is evidence that cluster genesis is a social process (Braunerhjelm and Feldman, 2006). Indeed, many of the factors associated with success clusters such as venture capital or active university involvement lag rather than lead industrial viability (Feldman, 2001). What matter most is the entrepreneurial spark that takes hold and transforms a region. In the most successful places, entrepreneurs build institutions and shared resources that develop the cluster building the firm (Feldman and Francis—Building cluster while building a firm). Over time a social consensus develops about the potential of a new idea or a technology, new business models emerge and the place becomes about doing something unique and not easily replicated by other places (Lowe and Feldman, 2008). There are many attempts to model the stages of cluster formation (Avnimelech and Teubal, 2004; Maggioni et al., 2007).

Even Silicon Valley, the archetype of a technology-intensive cluster started from humble beginnings. Lécuyer (2005) examines the history of Silicon Valley to 1970 and documents how faculty and administrators at Stanford used proximity to local firms to build a major research program in solid-state electronics, which was the ultimate basis for the development of the computer industry in Silicon Valley. Levuyer offers a skill-based interpretation of the formation of Silicon Valley, noting that local entrepreneurs developed a unique know-how in the production of vacuum tubes and semiconductors. Important social innovations such as stock options provided a mechanism to attract and retain skilled workers.

While economic development officials and government planners want to define long term strategies, it is difficult, if not impossible to predict scientific discoveries, new technologies and new opportunities. IBM, an industry leader, underestimated the potential of the computer industry, creating an opportunity for new firms to create personal computers. Few people predicted the potential of the Internet and how it would change the way we access information and communicate. Moreover, successful entrepreneurs make their own luck, adjusting and adapting to survive. Instead of wisely considered, farsighted solutions, entrepreneurial activity is by necessity messy, adaptive and unpredictable. Economic development strategies need to be equally adaptive. The biggest problem is that it is impossible to predict which technologies are going to yield any payoff. By the time a new industry, for example, biotech or nanotechnology, has a defined name and is on its way to becoming a household name, it is probably too late for other places to decide that they will participate as major centers. Creating a cluster in a market economy is a messy social process. Designing an effective economic development strategy may be the ultimate local innovation.

3. Conclusions

Students in introductory classes are told that economics consists of three major questions: what to produce, how to produce, and for whom to produce. A fourth question that is increasingly important in the global economy is where to produce—where to locate the factors of production so that they are most efficient and productive. The study of the location of innovation is a subset of the question of where to locate; however, the character of place is not static and is constructed by the economic actors who locate there. While firms are one venue to organize economic activity, the resources required to generate innovation are typically not confined to a single firm, and geography is another means to organize the factors of production. But we should remember that geography is additionally a venue for complex multifaceted social relationships, and human community and creativity that are beyond the economic sphere.

Economies are complex: highly integrated, globally interconnected, and highly agglomerated on centers of activity. There is always the temptation to analyze economic institutions and actors individually; however, the new economic geography literature considers the large context. Of course, once the analysis is open to consider geography there is a need to understand history, building a deep contextualized understanding of a place and the relationships that define it.

This chapter has considered stylized facts related to the geography of innovation. While much is known there is still much to be done and many open avenues for research remain. We hope that this review will encourage others to contribute to the emerging field of economic geography.

Acknowledgments

The authors would like to thank Bronwyn Hall for comments and suggestions on this chapter. We have benefited from discussion with Stefano Breschi, Francesco Lissoni, and Franco Malerba.

References

Abramovitz, M. (1956). "Resource and output trends in the United States since 1870". American Economic Review 46, 5–23.

Acs, Z., Audretsch, J. D. B., Feldman, M.P. (1994). "R&D spillovers and innovative activity". Managerial and Decision Economics 15 (2), 131–138.

Akerlof, G. (1976). "The economics of caste and of rat race and other woeful tales". Quarterly Journal of Economics 90 (4), 599–617.

Almeida, P. (1996). "Knowledge sourcing by foreign multinationals: Patent citation analysis in the U.S. semiconductor industry". Strategic Management Journal 17, 421–431.

Almeida, P., Kogut, B. (1997). "The exploration of technological diversity and the geographic localization of innovation". Small Business Economics 9, 21–31.

Almeida, P., Kogut, B. (1999). "Localization of knowledge and the mobility of engineers in regional networks". Management Science 45 (7), 905–917.

Amin, A. (Ed.), (1994). Post-Fordism: A Reader. Blackwell, Oxford.

Amin, A., Cohendet, P. (2004). Architectures of Knowledge: Firms Capabilities and Communities. Oxford University Press, Oxford.

Arrow, K. (1962). "Economic welfare and the allocation of resources for inventions". In: Nelson, R.R. (Ed.), The Rate and Direction of Innovative Activity. Princeton University Press, Princeton, NJ, pp. 609–625.

Audretsch, D.B., Feldman, M.P. (1996a). "R&D spillovers and the geography of innovation and production". American Economic Review 86, 630–640.

Audretsch, D.B., Feldman, M.P. (1996b). "Innovative clusters and the industry life cycle". Review of Industrial Organization 11, 253–273.

Audretsch, D.B., Keilbach, M. (2008). "Resolving the knowledge paradox: Knowledge-spillover entrepreneurship and economic growth". Research Policy 37, 1697–1705.

Avnimelech, G., Feldman, M.P. (2009). "The industrial genesis of Research Triangle Park". Working Paper.

Avnimelech, G., Teubal, M. (2004). "Venture capital start-up co-evolution and the emergence and development of Israel's new high tech cluster". Economics of Innovation and New Technology 13 (1), 33–60.

Baptista, R., Swann, P. (1998). "Do firms in clusters innovate more?" Research Policy 27, 525–540.

Bathelt, H. (2005a). "Geographies of production: Growth regimes in spatial perspective (II)—Knowledge creation and growth in clusters". Progress in Human Geography 29 (2), 204–216.

Bathelt, H. (2005b). "Cluster relations in the media industry: Exploring the 'distanced neighbour' paradox". Regional Studies 39 (1), 105–127.

Bathelt, H., Malmberg, A., Maskell, P. (2004). "Clusters and knowledge: Local buzz, global pipelines and the process of knowledge creation". Progress in Human Geography 28, 31–56.

Bercovitz, J. B. L., Feldman, M.P. (2006). "Entrepreneurial universities and technology transfer: A conceptual framework for understanding knowledge-based economic development". Journal of Technology Transfer 31 (1), 175–188.

Bercovitz, J. B. L., Feldman, M.P. (2010). "The mechanisms of collaboration in inventive teams: Composition, social networks, and geography".

Bercovitz, J., Mitchell, W. (2007). "When is more better? The impact of business scale and scope of long-term business survival, while controlling for profitability". Strategic Management Journal 28, 61–79.

Bettencourt, L. M. A., Lobo, J., Helbing, D., Kühnert, C., West, G.B. (2007). "Growth, innovation, scaling, and the pace of life in cities". Proceedings of the National Academy of Sciences of the United States of America 104 (17), 7301–7306.

Boschma, R.A., Iammarino, S. (2009). "Related variety, trade linkages and regional growth in Italy". Economic Geography 85 (3), 289–311.

Bottazzi, L., Peri, G. (2003). "Innovation and spillovers in regions: Evidence from European patent data". European Economic Review 47, 687–710.

Branstetter, L.G. (2001). "Are knowledge spillovers international or intranational in scope? Microeconometric evidence from the U.S. and Japan". Journal of International Economics 53, 53–79.

Braunerhjelm, P., Feldman, M.P. (eds.) (2006/2007). Cluster Genesis: The origins and emergence of technology-based economic development. Oxford University Press, Oxford.

Breschi, S., Lissoni, F. (2001a). "Localised knowledge spillovers vs. innovative milieux: Knowledge "tacitness" reconsidered". Papers in Regional Science 80, 255–273.

Breschi, S., Lissoni, F. (2001b). "Knowledge spillovers and local innovation systems: A critical survey". Industrial and Corporate Change 10, 975–1005.

Breschi, S., Lissoni, F. (2003). "Mobility and social networks: Localised knowledge spillovers revisited". CESPRI Working Paper No. 142, pp. 1–27.

Brown, L.A. (1981). Innovation Diffusion. Routledge, London.

Brown, J.S., Duguid, P. (1991). "Organizational learning and communities-of-practice: Toward a unified view of working, learning, and innovation". Organization Science 2, 40–57.

Bunnell, T.G., Coe, N.M. (2001). "Spaces and scales of innovation". Progress in Human Geography 25, 569–589.

Burt, R.S. (2004). "Structural holes and good ideas". American Journal of Sociology 110 (2), 349–399.

Bush, V. (1945). Science: The Endless Frontier. U.S. Government Printing Office, Washington, DC.

Cairncross, F. (1997). The Death of Distance. Harvard Business School Press, Boston, MA.

Callan, B. (2001). "Generating spin-offs: Evidence from the OECD". Science Technology Industry Review 26, 13–56.

Caniëls, M. C. J. (2000). Knowledge Spillovers and Economic Growth—Regional Growth Differentials Across Europe. Edward Elgar, Cheltenham, UK.

Capello, R. (2002). "Entrepreneurship and spatial externalities: Theory and measurement". Annals of Regional Science 36, 387–402.

Chesbrough, H.W. (2003). Open Innovation: The New Imperative for Creating and Profiting from Technology. Harvard Business School Press, Boston, MA.

Christensen, C.M. (1997). The Innovator's Dilemma: When New Technologies Cause Great Firms to Fail. Harvard University Press, Cambridge, MA.

Ciccone, A. (2002). "Agglomeration effects in Europe". European Economic Review 46, 213–227.

Ciccone, A., Hall, R.E. (1996). "Productivity and the density of economic activity". American Economic Review 86 (1), 54–70.

Clark, G.L., Feldman, M.P., Gertler, M. (2000). Handbook of Economic Geography. Oxford University Press, Oxford.

Clarysse, B., Wright, M., Lockett, A., van de Velde, E., Vohora, A. (2005). "Spinning out new ventures: A typology of incubation strategies from European research institutions". Journal of Business Venturing 20 (2), 183–216.

Coe, D.T., Helpman, E. (1995). "International R&D spillovers". European Economic Review 39, 859–887.

Cohen, W.M., Levinthal, D.A. (1990). "Absorptive capacity: A new perspective on learning and innovation". Administrative Science Quarterly 35, 128–152.

Cook, D.N., Brown, J.S. (1999). "Bridging epistemologies: The generative dance between organizational knowledge and organizational knowing". Organization Science 10, 381–400.

Cooke, P. (1996). "Regional innovation systems: An evolutionary approach". In: Baraczyk, H., Cooke, P., Heidenreich, R. (Eds.), Regional Innovation Systems. UCL Press, London.

Cowan, R., Jonard, N. (2004). "Network structure and the diffusion of knowledge". Journal of Economic Dynamics and Control 28, 1557–1575.

Dicken, P. (1998). Global shift: Transforming the world economy. Paul Chapman, London.

Edquist, C., Hommen, L., Johnson, B., Lemola, T., Malerba, F., Smith, K. (1998). The ISE Policy Statement: The Innovation Policy Implications of the 'Innovation Systems and European Integration' (ISE) Research Project. Linköping University, Linköping, Sweden.

Eisenhardt, K.M., Martin, J.A. (2000). "Dynamic capabilities: What are they?" Strategic Management Journal 21, 1105–1121.

Ellerman, D. (2005). "Labour migration: A developmental path or a low-level trap?" Development in Practice 15 (5), 617–630.

Ellison, G., Glaeser, E.L. (1997). "Geographic concentration in U.S. manufacturing industries: A dartboard approach". Journal of Political Economy 105 (5), 889–927.

Feldman, M.P. (1994). The Geography of Innovation. Kluwer Academic, Boston, MA.

Feldman, M.P. (1999). "The New Economics of Innovation, Spillovers and Agglomeration: A Review of Empirical Studies". The Economics of Innovation and New Technology 8, 5–25.

Feldman, M.P. (2000). "Where science comes to life: University bioscience, commercial spin-offs, and regional economic development". Journal of Comparative Policy Analysis: Research and Practice 2, 345–361.

Feldman, M. (2001). "The Entrepreneurial Event Revisited: Firm Formation in a Regional Context". Industrial and Corporate Change 10 (4), 861–891.

Feldman, M.P. (2009). "The geography of emerging industries". Presentation to the Regional Studies Association, (March 22).

Feldman, M.P., Audretsch, D.B. (1999). "Innovation in cities: Science-based diversity, specialization, and localized competition". European Economic Review 43, 409–429.

Feldman, M.P., Desrochers, P. (2004). "Truth for its own sake: Academic culture and technology transfer at Johns Hopkins University". Minerva 42 (2), 105–126.

Feldman, M., Martin, R. (2005). "Constructing jurisdictional advantage". Research Policy 34 (8), 1235–1249.

Feldman, M.P., Romanelli, E. (2006). "Organizational legacy and the internal dynamics of clusters: The U.S. human bio-therapeutics industry, 1976–2002". Cinquièmes Journées de la Proximité: La proximité, entre interactions et institutions, Bordeaux, 28–30 June, p. 31.
Feldman, M.P., Gambardella, A., Mariani, M. (2009). Geographic knowledge spillovers as evidenced by patent inventors. Working Paper.
Feller, I. (1990). "Universities as engines of R&D-based economic growth: They think they can". Research Policy 19 (4), 335–348.
Frenken, K., Van Oort, F., Verburg, T. (2007). "Related variety, unrelated variety and regional economic growth". Regional Studies 41 (5), 685–697.
Friedman, T.L. (2005). The World Is Flat—A Brief History of the Twenty-First Century. Farrar, Straus and Giroux, New York, NY.
Gassler, H., Nones, B. (2008). "Internationalisation of R&D and embeddedness: The case of Austria". Journal of Technology Transfer 33 (4), 407–421.
Geroski, P. (1995). "Markets for technology: Knowledge, innovation and appropriability". In: Stoneman, P. (Ed.), Handbook of the Economics of Innovation and Technological Change. Blackwell, Oxford, pp. 90–131.
Gertler, M.S. (1995). "'Being there': Proximity, organization, and culture in the development and adoption of advanced manufacturing technologies". Economic Geography 71, 1–26.
Gertler, M.S. (2003). "Tacit knowledge and the economic geography of context, or the undefinable tacitness of being (there)". Journal of Economic Geography 3, 75–99.
Gertler, M.S., Vinodrai, T. (2005). "Anchors of creativity: How do public universities create competitive and cohesive communities?" In: Iacobucci, F., Tuohy, C. (Eds.), Taking Public Universities Seriously. University of Toronto Press, Toronto, ON, pp. 293–315.
Gertler, M.S., Wolfe, D. (2006). "Spaces of knowledge flows: Clusters in a global context". In: Asheim, B., Cooke, P., Martin, R. (Eds.), Clusters and Regional Development: Critical Reflections and Explorations. Routledge, London, pp. 218–235.
Gilson, R.J. (1998). "The legal infrastructure of high technology industrial districts: Silicon Valley, Route 128, and covenants not to compete". 10.2139/ssrn.124508 Stanford Law School, John M. Olin Program in Law and Economics, Working Paper No. 163. Available at http://ssrn.com/abstract=124508.
Glaeser, E.L., Mare, D.C. (2001). "Cities and skills". Journal of Labor Economics 19 (2), 316–342.
Glaeser, E.L., Kallal, H., Scheinkman, J., Shleifer, A. (1992). "Growth in cities". Journal of Political Economy 100, 1126–1152.
Grabher, G. (2002). "The project ecology of advertising: Tasks, talents and teams". Regional Studies 36 (3), 245–262.
Granovetter, M. (1985). "Economic action and social structure: The problem of embeddedness". American Journal of Sociology 91, 481–510.
Greenstein, S. (2007). "Innovation and the evolution of internet access in the United States". In: Aspray, W., Ceruzzi, P. (Eds.), The Commercialization of the Internet and Its Impact on American Business. MIT Press, Cambridge, MA.
Griliches, Z. (1979). "Issues in assessing the contribution of R&D to productivity growth". Bell Journal of Economics 10, 92–116.
Griliches, Z. (1990). "Patent statistics as economic indicator: A survey". Journal of Economic Literature 28, 1661–1707.
Hägerstrand, T. (1967). "The computer and the geographer". Transactions of the Institute of British Geographers 42, 1–19.
Hall, B.H., Jaffe, A.B., Trajtenberg, M. (2001). "The NBER patent citations data file: Lessons, insights and methodological tools". NBER Working Paper No. 8498, pp. 1–46.
Harris, T.F., Ioannides, Y.M. (2000). "Productivity and metropolitan density". Tufts University, Department of Economics Working Paper.
Harrison, B., Bluestone, B. (1988). The Great U-Turn: Corporate Restructuring and the Polarizing of America. Basic Books, New York, NY.
Henderson, J.V. (1997). "Externalities and industrial development". Journal of Urban Economics 42, 449–470.
Henderson, J.V. (2003). "Marshall's scale economies". Journal of Urban Economics 53, 1–28.
Henderson, R., Cockburn, I. (1996). "Scale, scope, and spillovers: The determinants of research productivity in drug discovery". RAND Journal of Economics 27 (1), 32–59.
Henderson, R., Jaffe, A., Trajtenberg, M. (2005). "Patent citations and the geography of knowledge spillovers: A reassessment: Comment". American Economic Review 95 (1), 461–464.
Hicks, D., Breitzman, T., Olivastro, D., Hamilton, K. (2001). "The changing composition of innovative activity in the US—A portrait based on patent analysis". Research Policy 30 (4), 681–703.

Howells, J. (1996). "Tacit knowledge, innovation and technology transfer". Technology Analysis & Strategic Management 8 (2), 91–106.

Hu, A.G. (2007). "Technology parks and regional economic growth in China". Research Policy 36, 76–87.

Ibrahim, S.E., Fallah, M.H., Reilly, R.R. (2009). "Localized sources of knowledge and the effect of knowledge spillovers: An empirical study of inventors in the telecommunications industry". Journal of Economic Geography 9 (3), 405–431.

Jacobs, J.M. (1969). The Economy of Cities. Wiley, New York, NY.

Jaffe, A.B. (1989). "Real effects of academic research". American Economic Review 79, 957–970.

Jaffe, A.B., Trajtenberg, M. (1998). "International knowledge flows: Evidence from patent citations". NBER Working Paper No. 6507, p. 42.

Jaffe, A.B., Trajtenberg, M. (2002). Patents, Citations, and Innovations: A Window on the Knowledge Economy. MIT Press, Cambridge, MA.

Jaffe, A.B., Trajtenberg, M., Henderson, R. (1993). "Geographic localization of knowledge spillovers as evidenced by patent citations". Quarterly Journal of Economics 108 (3), 577–598.

Jones, C.I. (2002). Introduction to Economic Growth. W. W, Norton, New York.

Kahn, R.E. (1994). "The role of government in the evolution of the internet". Communications of the ACM 37 (8), 15–19.

Katz, R., Allen, T.J. (1988). "Organizational issues in the introduction of new technologies". In: Katz, R. (Ed.), Managing Professionals in Innovative Organizations. Ballinger, Cambridge, MA.

Klein, P.G., Luu, H. (2003). "Politics and productivity". Economic Inquiry 41 (3), 433–447.

Krugman, P. (1991a). "Increasing returns and economic geography". Journal of Political Economy 99, 483–499.

Krugman, P. (1991b). Geography and Trade. MIT Press, Cambridge, MA.

Kuznets, S. (1962). "Inventive activity: Problems of definition and measurement". In: Nelson, R.R. (Ed.), The Rate and Direction of Inventive Activity: Economic and Social Factors. Princeton University Press, Princeton, NJ.

La Porta, R., Lopez-de-Silanes, F., Shleifer, A., Vishny, R. (1999). "The quality of government". Journal of Law, Economics, and Organization 15 (1), 222–279.

Lécuyer, C. (2005). Making Silicon Valley: Innovation and the Growth of High Tech, 1930–1970. MIT Press, Cambridge, MA.

Leslie, S., Kargon, R. (1994). "Electronics and the geography of innovation in post-war America". History and Technology 11, 217–231.

Link, A.N. (1995). Generosity of Spirit: The Early History of the Research Triangle Park. Research Triangle Foundation of North Carolina, Research Triangle Park, NC.

Lipsey, R.G., Bekar, C., Carlaw, K.I. (1998). "What requires explanation?" In: Helpman, E. (Ed.), General Purpose Technologies and Economic Growth. MIT Press, Cambridge, MA, pp. 15–54.

Lowe, N., Feldman, M.P. (2007). "Constructing entrepreneurial advantage: Consensus building, technological uncertainty and emerging industries". Cambridge Journal of Regions, Economy and Society 1 (2), 265–284.

Lowe, N., Feldman, M.P. (2008). "Constructing entrepreneurial advantage: consensus building, technological uncertainty and emerging industries". Cambridge Journal of Regions, Economy and Society 1 (2), 265–284.

Lucas, R. (1988). "On the mechanics of economic development". Journal of Monetary Economics 22, 3–39.

Lundvall, B. (1992). National systems of innovation. Towards a theory of innovation. Pinter, London and New York.

Lundvall, B.A., Johnson, B. (1994). The Learning Economy. Routledge, London.

Maggioni, M.A., Nosvelli, M., Uberti, T.E. (2007). "Space versus networks in the geography of innovation: A European analysis". Papers in Regional Science 86 (3), 471–493.

Malecki, E.J., Oinas, P. (Eds.), (1999). Making Connections—Technological Learning and Regional Economic Change. Ashgate, Aldershot.

Malmberg, A., Maskell, P. (2006). "Localized learning revisited". Growth and Change 37 (1), 1–18.

Marshall, A. (1885). The Present Position of Economics. Macmillan, London.

Marshall, A. (1890). Principles of Economics. Macmillan, London.

Marshall, A. (1920). The Principles of Economics. Macmillan, London.

Maskell, P., Malmberg, A. (1999). "Localised Learning and Industrial Competitiveness". Cambridge Journal for Economics 23, 167–186.

Maurseth, P.-B., Verspagen, B. (2002). "Knowledge spillovers in Europe: A patent citations analysis". Scandinavian Journal of Economics 104, 531–545.

Miner, A.S., Eesley, D.T., Devaughn, M., Rura-Polley, T. (2001). "The magic beanstalk vision: Commercializing university inventions and research". In: Bird Schoonhoven, C., Romanelli, E. (Eds.), The Entrepreneurship Dynamic: Origins of Entrepreneurship and the Evolution of Industries. Stanford University Press, Stanford, CA, pp. 109–146.

Moodysson, J. (2008). "Principles and practices of knowledge creation: On the organization of "buzz" and "pipelines" in life science communities". Economic Geography 84 (4), 449–469.

Moore, G.E., Davis, K. (2004). "Learning the silicon valley way". In: Bresnahan, T., Gambardella, A. (Eds.), Building High-Tech Clusters: Silicon Valley and Beyond. Cambridge University Press, Cambridge, UK, pp. 7–39.

Moreno, R., Paci, R., Usai, S. (2005). "Spatial spillovers and innovation activity in European regions". Environment and Planning A 37, 1793–1812.

Mowery, D.C., Simcoe, T. (2002). "Is the internet a US invention? An economic and technological history of computer networking". Research Policy 31 (8–9), 1369–1387.

Nelson, R.R. (1982). "The role of knowledge in R&D efficiency". Quarterly Journal of Economics 97, 453–470.

Nelson, R.R. (1990). What is public and what is private about technology? Consortium on Competitiveness and Cooperation Working Paper No. 90-9. Center for Research in Management, University of California at Berkeley, Berkeley, CA.

Nelson, R.R. (1993). National innovation systems: a comparative analysis. Oxford University Press, Inc., New York.

Nelson, R.R. (2001). "Observations on the post-Bayh–Dole rise of patenting at American universities". Journal of Technology Transfer 26 (1–2), 13–19.

Nelson, R.R., Winter, S.G. (1982). An Evolutionary Theory of Economic Change. Belknap Press of Harvard University Press, Cambridge, MA.

Nonaka, I. (1994). "A dynamic theory of organizational knowledge creation". Organizational Science 5, 14–37.

Nooteboom, B. (2000). Learning and Innovation in Organizations and Economies. Oxford University Press, Oxford.

Ohmae, K. (1990). The Borderless World: Power and Strategy in the Interlinked Economy. Free Press, New York, NY.

Ohmae, K. (1995). The End of the Nation State: The Rise of Regional Economies. Free Press, New York, NY.

Owen-Smith, J., Powell, W.W. (2004). "Knowledge networks as channels and conduits: The effects of spillovers in the Boston biotechnology community". Organization Science 15, 5–21.

Parent, O., Riou, S. (2005). "Bayesian analysis of knowledge spillovers in European regions". Journal of Regional Science 45 (4), 747–775.

Pavitt, K., Patel, P. (1999). "Global corporations and national systems of innovation: Who dominates whom?" In: Archibugi, D., Howells, J., Michie, J. (Eds.), Innovation Policy in a Global Economy. Cambridge University Press, Cambridge, MA, pp. 94–119.

Pavitt, K., Robson, M., Townsend, J. (1987). "The Size Distribution of Innovating Firms in the UK: 1945–1983". http://www.jstor.org/action/showPublication?journalCode=jinduecon. The Journal of Industrial Economics 35 (3), 297–316.

Pisano, G.P. (1997). The Development Factory: Unlocking the Potential of Process Innovation. Harvard Business School Press, Boston, MA.

Polanyi, M. (1958). Personal Knowledge: Towards a Post-Critical Philosophy. University of Chicago Press, Chicago, IL.

Porter, M.P. (1990). The Comparative Advantage of Nations. Free Press, New York, NY.

Rigby, D.L., Essletzbichler, J. (2002). "Agglomeration economies and productivity differences in US cities". Journal of Economic Geography 2, 407–432.

Rogers, J.D., Kingsley, G. (2004). "Denying public value: The role of the public sector in accounts of the development of the internet". Journal of Public Administration, Research & Theory 14 (3), 371–393.

Romer, P. (1990). "Endogenous technological change". Journal of Political Economy 64, 1002–1037.

Rosenberg, N. (1982). Inside the Black Box: Technology and Economics. Cambridge University Press, New York, NY.

Rosenthal, S.S., Strange, W.C. (2003). "Geography, industrial organization, and agglomeration". Review of Economics and Statistics 85 (2), 377–393.

Rosenthal, S.S., Strange, W.C. (2004). "Evidence on the nature and sources of agglomeration economies". In: Henderson, V., Thisse, J. (Eds.), The Handbook of Urban and Regional Economics, vol. 4. Elsevier, Amsterdam, pp. 2119–2172.

Rosenthal, S.S., Strange, W.C. (2008). "Agglomeration and hours worked". Review of Economics and Statistics 90 (1), 105–119.

Rugman, A. (2000). The End of Globalization. Random House Business Books, London.

Salter, A.J., Martin, B.R. (2001). "The economic benefits of publicly funded basic research: A critical review". Research Policy 30 (3), 509–532.

Saxenian, A. (1994). Regional Advantage Culture and Competition in Silicon Valley and Route 128. Harvard University Press, Cambridge, MA.
Scherer, F.M. (1984). Innovation and Growth: Schumpeterian Perspectives. MIT Press, Cambridge, MA.
Schumpeter, J. (1942). Capitalism, Socialism and Democracy. Harper, New York, NY.
Scott, A.J. (2006a). "The changing global geography of low-technology, labor-intensive industry: Clothing, footwear, and furniture". World Development 34 (9), 1517–1536.
Scott, A.J. (2006b). "Entrepreneurship, innovation and industrial development: Geography and the creative field revisited". Small Business Economics 26 (1), 1–24.
Segal, D. (1976). "Are there returns to scale in city size?" Review of Economics and Statistics 58, 339–350.
Singh, J. (2004). "Innovation and knowledge diffusion in the global economy". Ph.D. Thesis, Harvard School of Business Administration and Department of Economics.
Singh, J. (2005). "Collaborative networks as determinants of knowledge diffusion patterns". Management Science 51 (5), 756–770.
Sitkin, S. (1992). "Learning through failure: The strategy of small losses". In: Staw, B.M., Cummings, L.L. (Eds.), Research in Organizational Behavior, vol. 14. JAI Press, Greenwich, CT, pp. 232–266.
Solow, R.M. (1957). "Technical change and the aggregate production function". Review of Economics and Statistics 39, 312–320.
Song, J., Almeida, P., Wu, G. (2003). "Learning-by-hiring: When is mobility more likely to facilitate inter firm knowledge transfer?". Management Science 49 (4), 351–365.
Sonn, J.W., Storper, M. (2008). "The increasing importance of geographical proximity in knowledge production: An analysis of US patent citations, 1975–1997". Environment and Planning A 40, 1020–1039.
Storper, M. (1997). "Regional economies as relational assets". In: Lee, R., Willseds, J. (Eds.), Geographies of Economies. Arnold, London.
Storper, M., Venables, A.J. (2004). "Buzz: Face-to-face contact and the urban economy". Journal of Economic Geography 4 (4), 351–370.
Strange, S. (1997). The Retreat of the State. Cambridge University Press, Cambridge, UK.
Stuart, T.E., Sorenson, O. (2003). "Liquidity events and the geographic distribution of entrepreneurial activity". Administrative Science Quarterly 48, 175–201.
Sveikauskas, L. (1975). "The productivity of cities". Quarterly Journal of Economics 89, 393–413.
Tabuchi, T. (1986). "Urban agglomeration, capital augmenting technology, and labor market equilibrium". Journal of Urban Economics 20, 211–228.
Thompson, P., Fox-Kean, M. (2005). "Patent citations and the geography of knowledge spillovers: A reassessment". American Economic Review 95 (1), 450–460.
Verspagen, B., Schoenmakers, W. (2004). "The spatial dimension of patenting by multinational firms in Europe". Journal of Economic Geography 4, 23–42.
von Hippel, E. (1988). The Sources of Innovation. Oxford University Press, New York, NY.
von Hippel, E. (1994). "Sticky information and the locus of problem solving: Implications for innovation". Management Science 40, 429–439.
von Hippel, E. (2001). "Innovation by user communities: Learning from open source software". MIT Sloan Management Review 42 (4), 82–86.
von Hippel, E. (2005). Democratizing Innovation. MIT Press, Cambridge, MA.
Wenger, E. (1998). Communities of Practice. Cambridge University Press, New York, NY.
Zucker, L.G., Darby, M.R. (1996). "Star scientists and institutional transformation? Patterns of invention and innovation in the formation of the biotechnology industry". Proceedings of the National Academy of Sciences of the United States of America 93, 12709–12716.
Zucker, L.G., Darby, M.R., Brewer, M.B. (1998). "Intellectual human capital and the birth of U.S. biotechnology enterprises". American Economic Review 88, 290–306.

Chapter 9

OPEN USER INNOVATION

ERIC VON HIPPEL
MIT Sloan School of Management
Cambridge, Massachusetts, USA

Contents

Abstract

Almost 30 years ago, researchers began a systematic study of innovation by end users and user firms. At that time, the phenomenon was generally regarded as a minor oddity. Today, it is clear that innovation by users, generally openly shared, is a very powerful and general phenomenon. It is rapidly growing due to continuing advances in computing and communication technologies. It is becoming both an important rival to and an important feedstock for producer-centered innovation in many fields. In this chapter, I provide an overview of what the international research community now understands about this phenomenon.

Handbooks in Economics, Volume 01

DOI: 10.1016/S0169-7218(10)01009-9

Keywords

collaborative innovation, open innovation, user innovation

JEL classification: O31, O32, O34, O38

Ever since Schumpeter (1934) promulgated his theory of economic development, economists, policy-makers, and business managers have assumed that the dominant mode of innovation is a "producers' model." That is, it has been assumed that most important innovations would originate from producers and be supplied to consumers via goods that were for sale.

This view seemed reasonable on the face of it—producers generally serve many users and so can profit from multiple copies of a single innovative design. Individual users, in contrast, depend upon benefits from in-house use of an innovation to recoup their investments. Presumably, therefore, a producer who serves many customers can afford to invest more in innovation than any single user. From this, it follows logically that producer-developed designs should dominate user-developed designs in most parts of the economy.

However, the producers' model is only one mode of innovation. A second, increasingly important model is *open user innovation.* Under this second model, economically important innovations are developed by users and other agents who divide up the tasks and costs of innovation development and then *freely reveal* their results. Users obtain direct use benefits from the collaborative effort. Other participants obtain diverse benefits such as enjoyment, learning, reputation, or an increased demand for complementary goods and services.

User and open collaborative innovation is increasingly displacing producer innovation in many parts of modern economies (Baldwin and von Hippel, 2009). A growing body of empirical work clearly shows that users are the first to develop many and perhaps most new industrial and consumer products. In addition, the importance of product and service development by users is increasing over time. This shift is being driven by two related technical trends: (1) the steadily improving *design capabilities* (innovation toolkits) that advance in computer hardware and software make possible for users and (2) the steadily improving ability of individual users to *combine and coordinate* their innovation-related efforts via new communication media such as the Internet.

The ongoing shift of innovation to users has some very attractive qualities. It is becoming progressively easier for many users to get precisely what they want by designing it for themselves. Innovation by users also provides a very necessary complement to and feedstock for manufacturer innovation. And innovation by users appears to increase social welfare. At the same time, the ongoing shift of product-development activities from manufacturers to users is painful and difficult for many manufacturers. Open, distributed innovation is "attacking" a major structure of the social division of labor. Many firms and industries must make fundamental changes to long-held business models in order to adapt. Further, governmental policy and legislation sometimes preferentially supports innovation by manufacturers. Considerations of social welfare suggest that this must change. The workings of the intellectual property system are of special concern. Despite the difficulties, a user-centered system of innovation appears well worth striving for.

Today, a number of innovation process researchers are working to develop our understanding of open user innovation processes. In this chapter, I offer a review of some collective learning on this important topic to date.

1. Importance of innovation by users

Users, as I use the term, are firms or individual consumers that expect to benefit from *using* a product or a service. In contrast, manufacturers expect to benefit from *selling* a product or a service. A firm or an individual can have different relationships to different products or innovations. For example, Boeing is a manufacturer of airplanes, but it is also a user of machine tools. If one were examining innovations developed by Boeing for the airplanes it sells, Boeing would be a manufacturer-innovator in those cases. But if one were considering innovations in metal-forming machinery developed by Boeing for in-house use in building airplanes, those would be categorized as user-developed innovations and Boeing would be a user-innovator in those cases.

Innovation user and innovation manufacturer are the two general "functional" relationships between innovator and innovation. Users are unique in that they alone benefit *directly* from innovations. All others (here lumped under the term "manufacturers") must sell innovation-related products or services to users, indirectly or directly, in order to profit from innovations. Thus, in order to profit, inventors must sell or license knowledge related to innovations, and manufacturers must sell products or services incorporating innovations. Similarly, suppliers of innovation-related materials or services—unless they have direct use for the innovations—must sell the materials or services in order to profit from the innovations.

The user and manufacturer categorization of relationships between innovator and innovation can be extended to specific function, attributes, or features of products and services. When this is done, it may turn out that different parties are associated with different attributes of a particular product or service. For example, householders are the users of the switching attribute of a household electric light switch—they use it to turn lights on and off. However, switches also have other attributes, such as "easy wiring" qualities, that may be used only by the electricians who install them. Therefore, if an electrician were to develop an improvement to the installation attributes of a switch, it would be considered a user-developed innovation.

Both qualitative observations and quantitative research in a number of fields clearly document the important role users play as first developers of products and services later sold by manufacturing firms. Smith (1776) was an early observer of the phenomenon, pointing out the importance of "the invention of a great number of machines which facilitate and abridge labor, and enable one man to do the work of many." Smith went on to note that "a great part of the machines made use of in those manufactures in which labor is most subdivided, were originally the invention of common workmen, who, being each of them employed in some very simple operation, naturally turned their thoughts toward finding out easier and readier methods of performing it." Rosenberg (1976) explored the matter in terms of innovation by *user firms* rather than individual workers. He studied the history of the US machine tool industry, finding that important and basic machine types like lathes and milling machines were first developed and built by user firms having a strong need for them. Textile manufacturing firms, gun manufacturers, and sewing machine manufacturers were important early user-developers of machine tools.

Quantitative studies of user innovation document that many of the most important and novel products and processes in a range of fields have been developed by user firms and by individual users. Thus, Enos (1962) reported that nearly all the most important innovations in oil refining were developed by user firms. Freeman (1968) found that the most widely licensed chemical production processes were developed by user firms. von Hippel (1988) found that users were the developers of about 80% of the most important scientific instrument innovations and also the developers of most of the major innovations in semiconductor processing. Pavitt (1984) found that a considerable fraction of invention by British firms was for in-house use. Shah (2000) found that the most commercially important equipment innovations in four sporting fields tended to be developed by individual users.

Empirical studies also show that *many* users—from 10% to nearly 40%—engage in developing or modifying products. This has been documented in the case of specific types of industrial products and consumer products, and in large, multi-industry studies of process innovation in Canada and the Netherlands as well (Table 1). When taken together, the findings make it very clear that users are doing a *lot* of product development and product modification in many fields.

Table 1
Studies of user innovation frequency

Innovation area	Number and type of users sampled	% Developing and building product for own use
Industrial products		
1. Printed circuit CAD software[a]	136 user firm attendees at a PC-CAD conference	24.3
2. Pipe hanger hardware[b]	Employees in 74 pipe hanger installation firms	36
3. Library information systems[c]	Employees in 102 Australian libraries using computerized OPAC library information systems	26
4. Medical surgery equipment[d]	261 surgeons working in university clinics in Germany	22
5. Apache OS server software security features[e]	131 technically sophisticated Apache users (Webmasters)	19.1
Consumer products		
6. Outdoor consumer products[f]	153 recipients of mail order catalogs for outdoor activity products for consumers	9.8
7. "Extreme" sporting equipment[g]	197 members of four specialized sporting clubs in four "extreme" sports	37.8
8. Mountain biking equipment[h]	291 mountain bikers in a geographic region known to be an "innovation hot spot"	19.2
Multi-industry process innovation surveys		
26 "Advanced manufacturing technologies"[i]	Canadian manufacturing plants in nine manufacturing sectors (less food processing) in Canada, 1998 (population estimates based upon a sample of 4200)	28% developed; 26% modified
39 "Advanced manufacturing technologies"[j]	16,590 Canadian manufacturing establishments that met the criteria of having at least $250,000 in revenues and at least 20 employees	22% developed; 21% modified
Any type of process innovation or process modification[k]	Representative, cross-industry sample of 498 "high-tech" Netherlands SMEs	41% developed only; 34% modified only; 54% developed and/or modified

[a] Urban and von Hippel (1988).
[b] Herstatt and von Hippel (1992).
[c] Morrison et al. (2000).
[d] Lüthje (2003).
[e] Franke and von Hippel (2003b).
[f] Lüthje (2004).
[g] Franke and Shah (2003).
[h] Lüthje et al. (2002).
[i] Arundel and Sonntag (1999).
[j] Gault and von Hippel (2009).
[k] de Jong and von Hippel (2009).

Studies of innovating users (both individuals and firms) show them to have the characteristics of "lead users" (Herstatt and von Hippel, 1992; Lilien et al., 2002; Olson and Bakke, 2001; Urban and von Hippel, 1988). That is, they are ahead of the majority of users in their populations with respect to an important market trend, and they expect to gain relatively high benefits from a solution to the needs they have encountered there. The correlations found between innovation by users and lead-user status are highly significant, and the effects are very large (Franke and Shah, 2003; Lüthje et al., 2002; Morrison et al., 2000).

Since lead users are at the leading edge of the market with respect to important market trends, one can guess that many of the novel products they develop for their own use will appeal to other users too and so might provide the basis for products that manufacturers would wish to commercialize. This turns out to be the case. A number of studies have shown that many of the innovations reported by lead users are judged to be commercially attractive and/or have actually been commercialized by manufacturers.

Research provides a firm grounding for these empirical findings. The two defining characteristics of lead users and the likelihood that they will develop new or modified products have been found to be highly correlated (Morrison et al., 2004). In addition, it has been found that the higher the intensity of lead-user characteristics displayed by an innovator, the greater the commercial attractiveness of the innovation that lead user develops (Franke and von Hippel, 2003a). In Figure 1, the increased concentration of innovations toward the right indicates that the likelihood of innovating is higher for users having higher lead-user index values. The rise in average innovation attractiveness as one moves from left to right indicates that innovations developed by lead users tend to be more commercially attractive. (Innovation attractiveness is the sum of the novelty of the innovation and the expected future generality of market demand.)

2. Why many users want custom products

Why do so many users develop or modify products for their own use? Users may innovate if and as they want something that is not available on the market and are able and willing to pay for its development. It is likely that many users do not find what they want on the market. Meta-analysis of market-

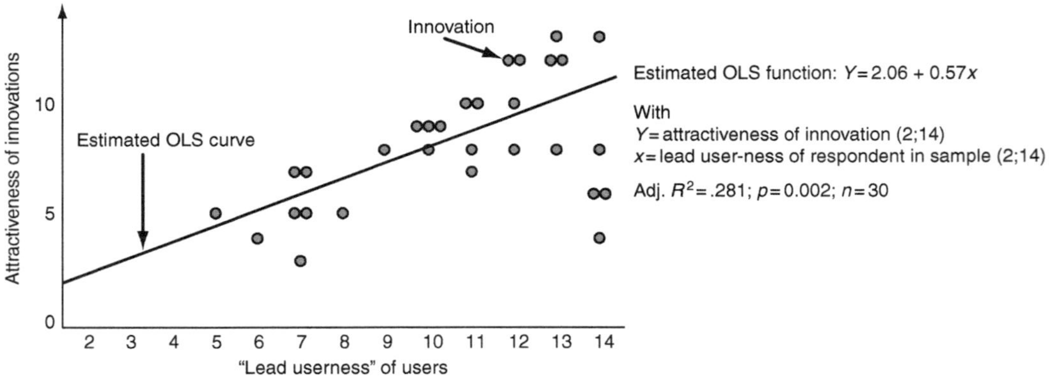

Figure 1. User-innovators with stronger "lead-user" characteristics develop innovations having higher appeal in the general marketplace (Source: Franke and von Hippel, 2003b).

segmentation studies suggests that users' needs for products are highly heterogeneous in many fields (Franke and Reisinger, 2003).

Mass producers tend to follow a strategy of developing products that are designed to meet the needs of a large market segment well enough to induce purchase from and capture significant profits from a large number of customers. When users' needs are heterogeneous, this strategy of "a few sizes fit all" will leave many users somewhat dissatisfied with the commercial products on offer and probably will leave some users seriously dissatisfied. In a study of a sample of users of the security features of Apache Web server software, Franke and von Hippel (2003b) found that users had a very high heterogeneity of need, and that many had a high willingness to pay to get precisely what they wanted. Nineteen percent of the users sampled actually innovated to tailor Apache more closely to their needs. Those who did were found to be significantly more satisfied.

3. Users' innovate-or-buy decisions

Even if many users want "exactly right products" and are willing and able to pay for their development, we must understand why users often do this for themselves rather than hire a custom producer to develop a special just-right product for them. After all, custom producers specialize in developing products for one or a few users. Since these firms are specialists, it is possible that they could design and build custom products for individual users or user firms faster, better, or cheaper than users could do this for themselves. Despite this possibility, several factors can drive users to innovate rather than buy. Both in the case of user firms and in the case of individual user-innovators, agency costs play a major role. In the case of individual user-innovators, enjoyment of the innovation process can also be important.

With respect to agency costs, consider that when a user develops its own custom product that user can be trusted to act in its own best interests. When a user hires a producer to develop a custom product, the situation is more complex. The user is then a principal that has hired the custom producer to act as its agent. If the interests of the principal and the agent are not the same, there will be agency costs. In general terms, agency costs are (1) costs incurred to monitor the agent to ensure that it (or he or she) follows the interests of the principal, (2) the cost incurred by the agent to commit itself not to act against the principal's interest (the "bonding cost"), and (3) costs associated with an outcome that does not fully serve the interests of the principal (Jensen and Meckling, 1976). In the specific instance of product and service development, a major divergence of interests between user and custom producer does exist: the user wants to get precisely what it needs, to the extent that it can afford to do so. In contrast, the custom producer wants to lower its development costs by incorporating solution elements it already has or that it predicts others will want in the future—even if by doing so it does not serve its present client's needs as well as it could.

A user wants to preserve its need specification because that specification is chosen to make *that user's* overall solution quality as high as possible at the desired price. For example, an individual user may specify a mountain-climbing boot that will precisely fit his unique climbing technique and allow him to climb Everest more easily. Any deviations in boot design will require compensating modifications in the climber's carefully practiced and deeply ingrained climbing technique—a much more costly solution from the user's point of view. A custom boot producer, in contrast, will have a strong incentive to incorporate the materials and processes it has in stock and expects to use in future even if this produces a boot that is not precisely right for the present customer. For example, the producer will not want to learn a new way to bond boot components together even if that would produce the best custom result for one client. The net result is that when one or a few users want something special they will often get the best result by innovating for themselves.

A model of the innovate-or-buy decision (von Hippel, 2005) shows in a quantitative way that user firms with unique needs (in other words, a market of one) will always be better off developing new products for themselves. It also shows that development by producers can be the most economical option when *n* or more user firms want the same thing. However, when the number of user firms wanting the same thing lies between 1 and *n*, producers may not find it profitable to develop a new product for just a few users. In that case, more than one user may invest in developing the same thing independently, owing to market failure. This results in a waste of resources from the point of view of social welfare. The problem can be addressed by new institutional forms, such as the user innovation communities that will be mentioned later.

It is important to note that an additional incentive can drive individual user-innovators to innovate rather than buy: they may value the *process* of innovating because of the enjoyment or learning that it brings them. It might seem strange that user-innovators can enjoy product development enough to want to do it themselves—after all, producers pay their product developers to do such work! On the other hand, it is also clear that enjoyment of problem solving is a motivator for many individual problem solvers in at least some fields. Consider, for example, the millions of crossword-puzzle aficionados. Clearly, for these individuals enjoyment of the problem-solving process rather than the solution is the goal. One can easily test this by attempting to offer a puzzle solver a completed puzzle—the very output he or she is working so hard to create. One will very likely be rejected with the rebuke that one should not spoil the fun. Pleasure as a motivator can apply to the development of commercially useful innovations as well. Studies of the motivations of volunteer contributors of code to widely used software products have shown that these individuals too are often strongly motivated to innovate by the joy and learning they find in this work (Hertel et al., 2003; Lakhani and Wolf, 2005).

4. Users' low-cost innovation niches

An exploration of the basic processes of product and service development shows that users and producers tend to develop different *types* of innovations. This is partially due to information asymmetries: users and producers tend to know different things. Product developers need two types of information to succeed at their work: need and context-of-use information (generated by users) and generic solution information (often initially generated by producers specializing in a particular type of solution). Bringing these two types of information together is not easy. Both need information and solution information are often very "sticky"—that is, costly to move from the site where the information was generated to other sites (von Hippel, 1994). It should be noted that the observation that information is often sticky contravenes a central tendency in economic theorizing. Much of the research on the special character of markets for information and the difficulty of appropriating benefit from invention and innovation has been based on the idea that information can be transferred at very low cost. (Thus, Arrow observes that "the cost of transmitting a given body of information is frequently very low.... In the absence of special legal protection, the owner cannot, however, simply sell information on the open market. Any one purchaser can destroy the monopoly, since he can reproduce the information at little or no cost." Arrow, 1962, pp. 614–615.)

When information is sticky, innovators tend to rely largely on information they already have in stock. One consequence of the resulting typical asymmetry between users and producers is that users tend to develop innovations that are functionally novel, requiring a great deal of user-need information and use-context information for their development. In contrast, producers tend to develop innovations that are improvements on well-known needs and that require a rich understanding of solution information for their development. Similarly, users tend to have better information regarding ways to improve use-related activities such as maintenance than do producers: they "learn by using" (Rosenberg, 1982).

This sticky information effect is quantitatively visible in studies of innovation. Riggs and von Hippel (1994) studied the types of innovations made by users and producers that improved the functioning of two major types of scientific instruments. They found that users are significantly more likely than producers to develop innovations that enabled the instruments to do qualitatively new types of things for the first time. In contrast, producers tended to develop innovations that enabled users to do the same things they had been doing, but to do them more conveniently or reliably (Table 2). For example, users were the first to modify the instruments to enable them to image and analyze magnetic domains at submicroscopic dimensions. In contrast, producers were the first to computerize instrument adjustments to improve ease of operation. Sensitivity, resolution, and accuracy improvements fall somewhere in the middle, as the data show. These types of improvements can be driven by users seeking to do specific new things or by producers applying their technical expertise to improve the products along known general dimensions of merit, such as accuracy.

The sticky information effect is independent of Stigler's (1951) argument that the division of labor is limited by the extent of the market. When profit expectations are controlled for, the impact of sticky information on the locus of innovation is still strongly evident (Ogawa, 1998).

If we extend the information-asymmetry argument one step further, we see that information stickiness implies that information on hand will also differ among *individual* users and producers. The information assets of some particular user (or some particular producer) will be closest to what is required to develop a particular innovation, and so the cost of developing that innovation will be relatively low for that user or producer. The net result is that user innovation activities will be *distributed* across many users according to their information endowments. With respect to innovation, one user is by no means a perfect substitute for another.

Table 2
Source of innovations by nature of improvement effected

	Innovation developed by			
Type of improvement provided by innovation	% User	User	Producer	Total
1. New functional capability	82	14	3	17
2. Sensitivity, resolution, or accuracy improvement	48	11	12	23
3. Convenience or reliability improvement	13	3	21	24
Total				64

Source: Riggs and von Hippel (1994).

5. Why users often freely reveal their innovations

The social efficiency of a system in which individual innovations are developed by individual users is increased if users somehow diffuse what they have developed to others. Producer-innovators *partially* achieve this when they sell a product or a service on the open market (partially because they diffuse the product incorporating the innovation, but often not all the information that others would need to fully understand and replicate it). If user-innovators do not somehow also diffuse what they have done, multiple users with very similar needs will have to independently develop very similar innovations—a poor use of resources from the viewpoint of social welfare. Empirical research shows that users often do achieve widespread diffusion by an unexpected means: they often "freely reveal" what they have developed. When we say that an innovator freely reveals information about a product or service it has developed, we mean that all intellectual property rights to that information are voluntarily given up by the innovator and all interested parties are given access to it—the information becomes a public good (Harhoff et al., 2003).

The empirical finding that users often freely reveal their innovations has been a major surprise to innovation researchers. On the face of it, if a user-innovator's proprietary information has value to others, one would think that the user would strive to prevent free diffusion rather than help others to free ride on what it has developed at private cost. Nonetheless, it is now very clear that individual users and user firms—and sometimes producers—often freely reveal detailed information about their innovations.

The practices visible in "open-source" software development were important in bringing this phenomenon to general awareness. In these projects, it was clear *policy* that project contributors would routinely and systematically freely reveal code they had developed at private expense (Raymond, 1999). However, free revealing of product innovations has a history that began long before the advent of open-source software. Allen (1983), in his study of the eighteenth-century iron industry, was probably the first to consider the phenomenon systematically. Later, Nuvolari (2004) discussed free revealing in the early history of mine pumping engines. Contemporary free revealing by users has been documented by von Hippel and Finkelstein (1979) for medical equipment, by Lim (2000) for semiconductor process equipment, by Morrison et al. (2000) for library information systems, and by Franke and Shah (2003) for sporting equipment. Henkel (2003) has documented free revealing among producers in the case of embedded Linux software.

Innovators often freely reveal because it is often the best or the only practical option available to them. Hiding an innovation as a trade secret is unlikely to be successful for long: too many generally know similar things, and some holders of the "secret" information stand to lose little or nothing by freely revealing what they know. Studies find that innovators in many fields view patents as having only limited value (Harhoff et al., 2003). Copyright protection and copyright licensing are applicable only to "writings," such as books, graphic images, and computer software.

Active efforts by innovators to freely reveal—as opposed to sullen acceptance—are explicable because free revealing can provide innovators with significant private benefits as well as losses or risks of loss. Users who freely reveal what they have done often find that others then improve or suggest improvements to the innovation, to mutual benefit (Raymond, 1999). Freely revealing users also may benefit from enhancement of reputation, from positive network effects due to increased diffusion of their innovation, and from other factors. Being the first to freely reveal a particular innovation can also enhance the benefits received, and so there can actually be a rush to reveal, much as scientists rush to publish in order to gain the benefits associated with being the first to have made a particular advancement.

6. Innovation communities

Innovation by users tends to be widely distributed rather than concentrated among just a very few very innovative users (Table 3). As a result, it is important for user-innovators to find ways to combine and leverage their efforts. Users achieve this by engaging in many forms of cooperation. Direct, informal user-to-user cooperation (assisting others to innovate, answering questions, and so on) is common. Organized cooperation is also common, with users joining together in networks and communities that provide useful structures and tools for their interactions and for the distribution of innovations. Innovation communities can increase the speed and effectiveness with which users and also producers can develop and test and diffuse their innovations. They also can greatly increase the ease with which innovators can build larger systems from interlinkable modules created by community participants.

Free and open-source software projects are a relatively well-developed and very successful form of Internet-based innovation community. However, innovation communities are by no means restricted to software or even to information products, and they can play a major role in the development of physical products. Franke and Shah (2003) have documented the value that user innovation communities can provide to user-innovators developing physical products in the field of sporting equipment. The analogy to open-source innovation communities is clear.

The collective or community effort to provide a public good—which is what freely revealed innovations are—has traditionally been explored in the literature on "collective action." However, behaviors seen in extant innovation communities fail to correspond to that literature at major points. In essence, innovation communities appear to be more robust with respect to recruiting and rewarding members than the literature would predict. The reason for this appears to be that innovation contributors obtain some private rewards that are not shared equally by free riders (those who take without contributing). For example, a product that a user-innovator develops and freely reveals might be perfectly suited to that user-innovator's requirements but less well suited to the requirements of free riders. Innovation communities thus illustrate a "private-collective" model of innovation incentive (von Hippel and von Krogh, 2003).

Table 3
User innovation is widely distributed: few users developed more than one major commercialized innovation

	Number of innovations each user developed					
User samples	1	2	3	6	NA	Sample (*n*)
Scientific instrument users[a]	28	0	1	0	1	32
Scientific instrument users[b]	20	1	0	1	0	28
Process equipment users[c]	19	1	0	0	8	29
Sports equipment users[d]	7	0	0	0	0	7

[a] von Hippel (1988, Appendix: GC, TEM, NMR Innovations).
[b] Riggs and von Hippel (1994, Esca and AES).
[c] von Hippel (1988, Appendix: semiconductor and pultrusion process equipment innovations).
[d] Shah (2000, Appendix A: skateboarding, snowboarding, and windsurfing innovations developed by users).
Source: von Hippel (2005, Table 7-1).

7. Adapting policy to user innovation

Is innovation by users a "good thing?" Welfare economists answer such a question by studying how a phenomenon or a change affects social welfare. Henkel and von Hippel (2005) explored the social welfare implications of user innovation. They found that, relative to a world in which only producers innovate, social welfare is very probably increased by the presence of innovations freely revealed by users. This finding implies that policymaking should support user innovation or at least should ensure that legislation and regulations do not favor producers at the expense of user-innovators.

The transitions required of policymaking to achieve neutrality with respect to user innovation versus producer innovation are significant. Consider the impact on open and distributed innovation of past and current policy decisions. Research done in the past 30 years has convinced many academics that intellectual property law is sometimes or often not having its intended effect. Intellectual property law was intended to increase the amount of innovation investment. Instead, it now appears that there are economies of scope in both patenting and copyright that allow firms to use these forms of intellectual property law in ways that are directly opposed to the intent of policymakers and to the public welfare (Foray, 2004). Major firms can invest to develop large portfolios of patents. They can then use these to create "patent thickets"—dense networks of patent claims that give them plausible grounds for threatening to sue across a wide range of intellectual property. They may do this to prevent others from introducing a superior innovation and/or to demand licenses from weaker competitors on favorable terms (Bessen, 2003; Shapiro, 2001). Movie, publishing, and software firms can use large collections of copyrighted work to a similar purpose (Benkler, 2002). In view of the distributed nature of innovation by users, with each tending to create a relatively small amount of intellectual property, users are likely to be disadvantaged by such strategies.

It is also important to note that users (and producers) tend to build prototypes of their innovations economically by modifying products already available on the market to serve a new purpose. Laws such as the (US) Digital Millennium Copyright Act, intended to prevent consumers from illegally copying protected works, also can have the unintended side effect of preventing users from modifying products that they purchase (Varian, 2002). Both fairness and social welfare considerations suggest that innovation-related policies should be made neutral with respect to the sources of innovation.

It may be that current impediments to user innovation will be solved by legislation or by policymaking. However, beneficiaries of existing law and policy will predictably resist change. Fortunately, a way to get around some of these problems is in the hands of innovators themselves. Suppose many innovators in a particular field decide to freely reveal what they have developed, as they often have reason to do. In that case, users can collectively create an information commons (a collection of information freely available to all) containing substitutes for some or a great deal of information now held as private intellectual property. Then user-innovators can work around the strictures of intellectual property law by simply using these freely revealed substitutes (Lessig, 2001).

This pattern is happening in the field of software—and very visibly so. For many problems, user-innovators in that field now have a choice between proprietary, closed software provided by Microsoft and other firms and open-source software that they can legally download from the Internet and legally modify as they wish to serve their own specific needs. It is also happening, although less visibly, in the case of process equipment developed by users for in-house use. Data from both Canada and the Netherlands show that about 25% of such user-developed innovations get voluntarily transferred to producers. A significant fraction—about half—being transferred both unprotected by intellectual property and without charge (de Jong and von Hippel, 2009; Gault and von Hippel, 2009).

Policymaking that levels the playing field between users and producers will force more rapid change onto producers but will by no means destroy them. Experience in fields where open and distributed innovation processes are far advanced show how producers can and do adapt. Some, for example, learn to supply proprietary platform products that offer user-innovators a framework upon which to develop and use their improvements (Jeppesen, 2004).

8. Diffusion of user-developed innovations

Products, services, and processes developed by users become more valuable to society if they are somehow diffused to others that can also benefit from them. If user innovations are not diffused, multiple users with very similar needs will have to invest to (re)develop very similar innovations which, as was noted earlier, would be a poor use of resources from the social welfare point of view. In the case of information products, users have the possibility of largely or completely doing without the services of producers. Open-source software projects are object lessons that teach us that users can create, produce, diffuse, provide user field support for, update, and use complex products by and for themselves in the context of user innovation communities. In physical product fields, the situation is different. Users can develop products. However, the economies of scale associated with manufacturing and distributing physical products give producers an advantage over "do-it-yourself" users in those activities.

How can or should user innovations of general interest be transferred to producers for large-scale diffusion? We propose that there are three general methods for accomplishing this. First, producers can actively seek innovations developed by lead users that can form the basis for a profitable commercial product. Second, producers can draw innovating users into joint design interactions by providing them with "toolkits for user innovation." Third, users can become producers in order to widely diffuse their innovations. We discuss each of these possibilities in turn.

To systematically find user-developed innovations, producers must redesign their product-development processes. Currently, almost all producers think that their job is to find a need and fill it rather than to sometimes find and commercialize an innovation that lead users have already developed. Accordingly, producers have set up market-research departments to explore the needs of users in the target market, product-development groups to think up suitable products to address those needs, and so forth. In this type of product-development system, the needs and prototype solutions of lead users—if encountered at all—are typically rejected as outliers of no interest. Indeed, when lead users' innovations do enter a firm's product line they typically arrive with a lag and by an unconventional and unsystematic route. For example, a producer may "discover" a lead-user innovation only when the innovating user firm contacts the producer with a proposal to produce its design in volume to supply its own in-house needs. Or sales or service people employed by a producer may spot a promising prototype during a visit to a customer's site.

Modification of firms' innovation processes to *systematically* search for and further develop innovations created by lead users can provide producers with a better interface to the innovation process as it actually works, and so provide better performance. A natural experiment conducted at 3M illustrates this possibility. Annual sales of lead-user product ideas generated by the average lead-user project at 3M were conservatively forecasted by management to be more than eight times the sales forecast for new products developed in the traditional manner—$146 million versus $18 million per year. In addition, lead-user projects were found to generate ideas for new product lines, while traditional market-research methods were found to produce ideas for incremental improvements to existing product lines. As a consequence, 3M divisions funding lead-user project ideas experienced their highest rate of major product line generation in the past 50 years (Lilien et al., 2002).

Toolkits for user innovation custom design involve partitioning product-development and service-development projects into solution-information-intensive subtasks and need-information-intensive subtasks. Need-intensive subtasks are then assigned to users along with a kit of tools that enables them to effectively execute the tasks assigned to them. In the case of physical products, the designs that users create using a toolkit are then transferred to producers for production (von Hippel and Katz, 2002). Toolkits make innovation cheaper for users and also lead to higher customer value. Thus, Franke and Piller (2004) in a study of a consumer wrist watches found the willingness to pay for a self-designed product was 200% of the willingness to pay for the best-selling commercial product of the same technical quality. This increased willingness to pay was due to both the increased value provided by the self-developed product and the value of the toolkit process for consumers engaging in it (Schreier and Franke, 2004).

Producers that offer toolkits to their customers can attract innovating users into a relationship with their firm and so get an advantage with respect to producing what the users develop. The custom semiconductor industry was an early adopter of toolkits. In 2003, more than $15 billion worth of semiconductors that had been designed using this approach were produced (Thomke and von Hippel, 2002).

Innovations developed by users sometimes achieve widespread diffusion when those users become producers—setting up a firm to produce their innovative product(s) for sale. Shah (2000) showed this pattern in sporting goods fields. In the medical field, Lettl and Gemnden (2005) have shown a pattern in which innovating users take on many of the entrepreneurial functions needed to commercialize the new medical products they have developed, but do not themselves abandon their user roles. New work in this field is exploring the conditions under which users will become entrepreneurs rather than transfer their innovations to established firms (Hienerth, 2004; Shah and Tripsas, 2004).

9. Summary

I summarize this overview chapter by again saying that users' ability to innovate is improving *radically* and *rapidly* as a result of the steadily improving quality of computer software and hardware, improved access to easy-to-use tools and components for innovation, and access to a steadily richer innovation commons. Today, user firms and even individual hobbyists have access to sophisticated programming tools for software and sophisticated CAD design tools for hardware and electronics. These information-based tools can be run on a personal computer, and they are rapidly coming down in price. As a consequence, innovation by users will continue to grow even if the degree of heterogeneity of need and willingness to invest in obtaining a precisely right product remains constant (Baldwin and von Hippel, 2009).

Equivalents of the innovation resources described above have long been available within corporations to a few. Senior designers at firms have long been supplied with engineers and designers under their direct control and with the resources needed to quickly construct and test prototype designs. The same is true in other fields, including automotive design and clothing design: just think of the staffs of engineers and model makers supplied so that top auto designers can quickly realize and test their designs.

But if, as we have seen, the information needed to innovate in important ways is widely distributed, the traditional pattern of concentrating innovation-support resources on a few individuals is hugely inefficient. High-cost resources for innovation support cannot efficiently be allocated to "the right people with the right information:" it is very difficult to know who these people may be before they develop an innovation that turns out to have general value. When the cost of high-quality resources for design and prototyping becomes very low (the trend we have described), these resources can be diffused very widely, and the allocation problem diminishes in significance. The net result is a pattern in which development of product and service innovations is increasingly shifting to users—a pattern that will involve significant changes for both users and producers.

References

Allen, R.C. (1983). "Collective invention". Journal of Economic Behavior & Organization 4 (1), 1–24.

Arrow, K.J. (1962). "Economic welfare and the allocation of resources of invention". In: Nelson, R.R. (Ed.), The Rate and Direction of Inventive Activity: Economic and Social Factors. A Report of the National Bureau of Economic Research, 609-625. Princeton University Press, Princeton, NJ.

Arundel, A., Sonntag, V. (1999). Patterns of Advanced Manufacturing Technology (AMT) Use in Canadian Manufacturing: 1998 AMT Survey Results. Research Paper No. 12. Science, Innovation and Electronic Information Division, Statistics Canada, Ottawa.

Baldwin, C.Y., von Hippel, E. (2009). "Modeling a Paradigm Shift: From Producer Innovation to User and Open Collaborative Innovation". MIT Sloan School of Management Working Paper 4764–09, http://papers.ssrn.com/sol3/papers.cfm?abstract_id=1502864.

Benkler, Y. (2002). "Intellectual property and the organization of information production". International Review of Law and Economics 22 (1), 81–107.

Bessen, J. (2003). "Patent thickets: Strategic patenting of complex technologies". Research on Innovation and Boston University School of Law, Working Paper.

de Jong, J.P.J., von Hippel, E. (2009). "Transfers of user process innovations to process equipment producers: A study of Dutch high-tech firms". Research Policy 38 (7), 1181–1191.

Enos, J.L. (1962). Petroleum Progress and Profits: A History of Process Innovation. MIT Press, Cambridge, MA.

Foray, D. (2004). Economics of Knowledge. MIT Press, Cambridge, MA.

Franke, N., Piller, F. (2004). "Value creation by toolkits for user innovation and design: The case of the watch market". Journal of Product Innovation Management 21 (6), 401–415.

Franke, N., Reisinger, H. (2003). "Remaining within cluster variance: A meta analysis of the "Dark" Side of cluster analysis". Vienna Business University, Working Paper.

Franke, N., Shah, S. (2003). "How communities support innovative activities: An exploration of assistance and sharing among end-users". Research Policy 32 (1), 157–178.

Franke, N., von Hippel, E. (2003a). "Finding commercially attractive user innovations". MIT Sloan School of Management, Working Paper No. 4402-03.

Franke, N., von Hippel, E. (2003b). "Satisfying heterogeneous user needs via innovation toolkits: The case of Apache security software". Research Policy 32 (7), 1199–1215.

Freeman, C. (1968). "Chemical Process Plant: Innovation and the World Market". National Institute Economic Review 45 (August), 29–57.

Gault, F., von Hippel, E. (2009). "The prevalence of user innovation and free innovation transfers: Implications for statistical indicators and innovation policy". MIT Sloan School of Management, Working Paper No. 4722-09, http://papers.ssrn.com/sol3/papers.cfm?abstract_id=1337232 (January).

Harhoff, D., Henkel, J., von Hippel, E. (2003). "Profiting from voluntary information spillovers: How users benefit by freely revealing their innovations". Research Policy 32 (10), 1753–1769.
Henkel, J. (2003). "Software development in embedded Linux—Informal collaboration of competing firms". In: Uhr, W., Esswein, E., Schoop, E. (Eds.), Proceedings der 6 Internationalen Tagung Wirtschaftsinformatik, vol. 2. Physica-Verlag, Heidelberg, pp. 81–99.
Henkel, J., von Hippel, E. (2005). "Welfare implications of user innovation". Journal of Technology Transfer 30 (1–2), 73–87.
Herstatt, C., von Hippel, E. (1992). "From experience: Developing new product concepts via the lead user method: A case study in a "low tech" field". Journal of Product Innovation Management 9 (3), 213–222.
Hertel, G., Niedner, S., Herrmann, S. (2003). "Motivation of software developers in open source projects: An Internet-based survey of contributors to the Linux kernel". Research Policy 32 (7), 1159–1177.
Hienerth, C. (2004). "The commercialization of user innovations: Sixteen cases in an extreme sporting industry". Vienna University of Economics and Business Administration, Working Paper.
Jensen, M.C., Meckling, W.H. (1976). "Theory of the firm: Managerial behavior, agency costs, and ownership structure". Journal of Financial Economics 3 (4), 305–360.
Jeppesen, L.B. (2004). "Profiting from innovative user communities: How firms organize the production of user modifications in the computer games industry". Department of Industrial Economics and Strategy, Copenhagen Business School, Copenhagen, Denmark, Working Paper No. WP-04.
Lakhani, K.R., Wolf, B. (2005). "Why hackers do what they do: Understanding motivation and effort in free/open source software projects". In: Feller, J., Fitzgerald, B., Hissam, S., Lakhani, K.R. (Eds.), Perspectives on Free and Open Source Software. MIT Press, Cambridge, MA.
Lessig, L. (2001). The Future of Ideas: The Fate of the Commons in a Connected World. Random House, New York, NY.
Lettl, C., Gemnden, H.G. (2005). "The entrepreneurial role of innovative users". Journal of Business and Industrial Marketing 20 (7), 339–346.
Lilien, G.L., Morrison, P.D., Searls, K., Sonnack, M., von Hippel, E. (2002). "Performance assessment of the lead user idea generation process". Management Science 48 (8), 1042–1059 (August).
Lim, K. (2000). "The many faces of absorptive capacity: Spillovers of copper interconnect technology for semiconductor chips". In: MIT Sloan School of Management, Working Paper No. 4110.
Lüthje, C. (2003). "Customers as co-inventors: An empirical analysis of the antecedents of customer-driven innovations in the field of medical equipment". Proceedings of the 32nd EMAC Conference, Glasgow.
Lüthje, C. (2004). "Characteristics of innovating users in a consumer goods field: An empirical study of sport-related product consumers". Technovation 24 (9), 683–695.
Lüthje, C., Herstatt, C., von Hippel, E. (2002). "The dominant role of local information in user innovation: The case of mountain biking". MIT Sloan School, Working Paper No. 4377-02.
Morrison, P.D., Roberts, J.H., von Hippel, E. (2000). "Determinants of user innovation and innovation sharing in a local market". Management Science 46 (12), 1513–1527.
Morrison, P.D., Roberts, J.H., Midgley, D.F. (2004). "The nature of lead users and measurement of leading edge status". Research Policy 33 (2), 351–362.
Nuvolari, A. (2004). "Collective invention during the British industrial revolution: The case of the Cornish pumping engine". Cambridge Journal of Economics 28 (3), 347–363.
Ogawa, S. (1998). "Does sticky information affect the locus of innovation? Evidence from the Japanese convenience-store industry". Research Policy 26 (7–8), 777–790.
Olson, E.L., Bakke, G. (2001). "Implementing the lead user method in a high technology firm: A longitudinal study of intentions versus actions". Journal of Product Innovation Management 18 (2), 388–395 (November).
Pavitt, K. (1984). "Sectoral patterns of technical change: Towards a taxonomy and a theory". Research Policy 13 (6), 343–373.
Raymond, E. (1999). The Cathedral and the Bazaar: Musings on Linux and Open Source by an Accidental Revolutionary. O'Reilly, Sebastopol, CA.
Riggs, W., von Hippel, E. (1994). "The impact of scientific and commercial values on the sources of scientific instrument innovation". Research Policy 23, 459–469 (July).
Rosenberg, N. (1976). Perspectives on Technology. Cambridge University Press, New York, NY.
Rosenberg, N. (1982). Inside the Black Box: Technology and Economics. Cambridge University Press, New York, NY.

Schreier, M., Franke, N. (2004). "Tom Sawyer's great law in action: Why users are willing to pay to design their own products via toolkits for user innovation and design". Vienna University of Economics and Business Administration, Working Paper.

Schumpeter, J.A. (1934). The Theory of Economic Development. Oxford University Press, New York, NY.

Shah, S. (2000). "Sources and patterns of innovation in a consumer products field: Innovations in sporting equipment". MIT Sloan School of Management, Working Paper No. 4105.

Shah, S., Tripsas, M. (2004). "When Do User-Innovators Start Firms? Towards A Theory of User Entrepreneurship". University of Illinois, Working Paper No. 04-0106.

Shapiro, C. (2001). "Navigating the patent thicket: Cross licenses, patent pools, and standard setting". In: Jaffe, A., Lerner, S., Stern, S. (Eds.), Innovation Policy and the Economy, vol. 1. MIT Press, Cambridge, MA, pp. 119–150.

Smith, A. (1776). "An Inquiry into the Nature and Causes of the Wealth of Nations". 1776; 5th ed., 1789. Modern Library Edition. Edited by Edwin Cannan. New York: Random House, 1937.

Stigler, G.J. (1951). "The division of labor is determined by the extent of the market". Journal of Political Economy 59 (3), 185–193 (June).

Thomke, S.H., von Hippel, E. (2002). "Customers as innovators: A new way to create value". Harvard Business Review 80 (4), 74–81.

Urban, G.L., von Hippel, E. (1988). "Lead user analyses for the development of new industrial products". Management Science 34 (5), 569–582.

Varian, H.R. (2002). "New chips can keep a tight rein on consumers". New York Times (July 4).

von Hippel, E. (1988). The Sources of Innovation. Oxford University Press, New York.

von Hippel, E. (1994). "Sticky information and the locus of problem solving: Implications for innovation". Management Science 40 (4), 429–439.

von Hippel, E. (2005). Democratizing Innovation. MIT Press, Cambridge, MA.

von Hippel, E., Finkelstein, S.N. (1979). "Analysis of innovation in automated clinical chemistry analyzers". Science and Public Policy 6 (1), 24–37.

von Hippel, E., Katz, R. (2002). "Shifting innovation to users via toolkits". Management Science 48 (7), 821–833.

von Hippel, E., von Krogh, G. (2003). "Open source software and the "private-collective" innovation model: Issues for organization science". Organization Science 14 (2), 209–223.

Chapter 10

LEARNING BY DOING

PETER THOMPSON

Department of Economics, Florida International University, Miami Florida, USA

Contents

Handbooks in Economics, Volume 01

DOI: 10.1016/S0169-7218(10)01010-5

Abstract

This chapter reviews the theoretical and empirical literature on learning by doing. Many of the distinctive theoretical implications of learning by doing have been derived under the assumption that the cost–quantity relationships observed in numerous empirical studies are largely the result of passive learning and some further require that passive learning is unbounded. The empirical literature raises doubts about both assumptions. When observed cost–quantity relationships indicate sustained productivity growth, factors other than passive learning are generally at work. When passive learning is the dominant factor, productivity growth is invariably bounded. Thus, empirically relevant theories incorporating learning by doing are hybrid models in which passive learning coexists with other sources of growth. But in such models, many of the distinctive implications of passive learning become unimportant. Moreover, passive learning is often an inessential component of long-run growth; to the contrary, too much learning can lead to stagnation.

Keywords

cost–quantity relationship, forgetting, knowledge spillovers, learning by doing, learning curves, passive learning, progress curves

JEL classification: D24, D92, F12, L11, L16, O3

1. Introduction

Learning by doing (LBD) is the colloquial name given by economists to the phenomenon of productivity growth associated with, but incidental to, the accumulation of production experience by a firm. The experience of a firm at any given age may be measured in a number of ways including, *inter alia*, the age of the firm, the cumulative prior output of the firm, the average tenure of its employees, or the average length of related work experience of its employees. The most popular implementation assumes that the current unit cost of a firm of age v, $c(v)$, is a decreasing function of its cumulative prior output, $y(v) = \int_0^v x(s)\,ds$; in much research, most especially in empirical and macroeconomic applications, a power rule of the form $c(v) = c(0)y(v)^{-\beta}$ is assumed.

The term LBD was in widespread use by the beginning of the twentieth century, largely motivated by its expanding popularity as a philosophy of educational method (cf. Dewey, 1897). Even in economics journals, for much of the century its context was limited to education. Not until Arrow (1962) was the term applied to firm learning, but thereafter its application to firms and even higher levels of aggregation quickly gained currency. Throughout the 1960s and 1970s, much of the focus of the literature was on documenting the importance and prevalence of LBD, especially in industrial settings. The literature in the late 1970s and through the next decade was dominated by theoretical work on the strategic implications of LBD; for a period much of this work was conducted in the context of industrial trade policy. Beginning around 1990, LBD factored prominently in macroeconomic models of endogenous growth. Most recently, the focus appears to have reverted to empirical work, which has mainly been concerned with identifying underlying sources of LBD.

One can point to several explanations for the prompt and sustained interest in LBD after Arrow's seminal paper. First, influential studies by Abramovitz (1956) and Solow (1957) had already established that technical change was a far more important source of long-run economic growth than had previously been realized. The consequent reduction of the theory of long-run growth to a time trend was intellectually unsatisfying and left economists with little to say about policy (Arrow, 1962, p. 155). LBD simultaneously appeared to offer a source of technical change that was intuitively plausible, that was susceptible to manipulation by appropriate policy intervention, and that did not increase the dimensionality of the optimization problems that economists needed to solve.

Second, LBD generated sufficiently distinctive implications for firm behavior and policy to sustain interest in models that incorporated it. For example, equating static marginal cost to marginal revenue is neither privately nor socially optimal; price-taking equilibria may not exist; and monopolies may be socially preferable to competitive markets. Competition policy is necessarily rather complicated in such circumstances, both in terms of philosophy (traditional antitrust policies may be unwise), and implementation (pricing below marginal cost need not signify predatory behavior). Moreover, LBD leads to hysteresis effects, where temporary shocks and policy interventions that alter output have permanent effects on productivity. Thus, not only the design of policy interventions but also their appropriate duration are more complicated in the presence of LBD.

Third, LBD appeared to be amply motivated by a large empirical literature, appearing predominantly in engineering and management fields, showing a robust relationship between cumulative output and unit costs. The nature of the relationship was often reported to be precisely or very nearly that indicated by the power rule. Wright's (1936) study of the cost–quantity relationship in aircraft manufacturing was the first to mention an organizational learning curve in the academic literature (see Figure 1), although

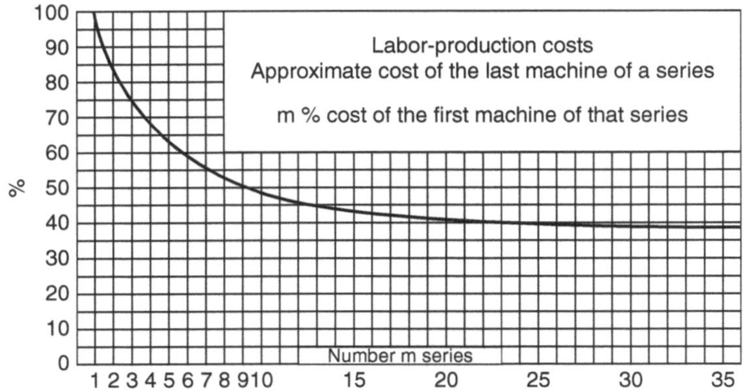

Figure 1. Wright's (1936) rendition of the learning curve. Wright provides no information about the data used to construct this figure, which may even have come from cross-sectional data obtained from different aircraft.

by this time the phenomenon appears to have already been well known in the aircraft industry. During World War II, the US Government incorporated expectations of strong organizational learning into the contracts it signed with aircraft manufacturers (Asher, 1956, p. 84) and shipbuilders (Lane, 1951), and studies released soon after the war showed that these expectations were well founded (Alchian, 1963 [1950]; Middleton, 1945; Montgomery, 1943; Searle, 1945). During the 1960s, many dozens of studies documented strong cost–quantity relationships in a broad range of industries. Some of these continued the practice of earlier studies, estimating changes in average costs over time (see, e.g., Baloff, 1966; Hirsch, 1952 on machine manufacturing, and Preston and Keachie, 1964 on radars); this activity attained industrial proportions when the Boston Consulting Group (1972) estimated hundreds of curves, and used them to promote a management strategy of maximizing market share. Around the same time other studies, beginning with Rapping (1965) and Sheshinski (1967a), began to estimate experience as an input in an otherwise conventional production function, also finding evidence of significant learning effects.

This chapter reviews theoretical research conducted over the last 40 years on the economic implications of LBD, as well as concurrent empirical research on its nature and importance. To summarize the plan of the chapter, it is useful to distinguish between different concepts of what Wright (1936) had rather generically called the cost–quantity relationship. I shall use the term *passive learning* to refer throughout this chapter to the conventional economic characterization of organizational LBD as an incidental and costless byproduct of a firm's production activities. A firm that increases productivity through passive learning will be said to move along an *experience curve*. I shall use the term *progress curve* to refer to the empirical relationship between current unit cost (or productivity) and a firm's cumulative experience. The term *cost–quantity relationship* will be used in the same way that Wright used it: to refer to the observed relationship between cumulative output and the average cost of producing that cumulative output. Finally, I reserve the rather special term *learning curve* for increases in productivity or, more generally, advances in knowledge, that individuals exhibit as they accumulate experience in a task.[1]

[1] Empirical work on the learning curve considerably predates that on progress curves. See Ebbinghaus, (1885) experiments on memory, Bryan and Harter's (1899) study of telegraph operators, and Book's (1908) study of typists.

The progress curve encompasses a broader range of sources of growth than does the experience curve. In addition to passive learning, it allows for research, innovation, product design changes, capital investment, and other costly activities that might, with the passage of time, enable a firm to become more productive. In turn, the cost–quantity relationship is a broader concept than the progress curve. Wright (1936, p. 124) offered three explanations for the cost–quantity relationship he had observed in his career as an aircraft engineer and executive. The first was the "improvement in proficiency of a workman with practice," characterized by the learning curve. Wright's other explanations were "the greater spread of machinery and fixture set up time in large quantity production," and "the ability to use less skilled labor as more and more tooling and standardization of procedure is introduced." These two are, of course, static scale economies, under which one would observe a cost–quantity relationship even in the absence of learning.

The distinctions between these concepts are not trivial. For example, if movement along the progress curve is driven solely by costly R&D, and not at all by passive learning, then equating static marginal cost to static marginal revenue is socially optimal. Similarly, if the cost–quantity relationship is purely a result of static scale economies, then it and the progress function have distinct economic implications. Consider, for example, an unanticipated transitory demand shock that raises the rate of production for a period of time. The progress function predicts a permanent decline in unit costs from this point forward. In contrast, static scale economies predict that transitory shocks have no effect on long-run costs. In the short-run, unit costs may bear a positive or negative relationship to output shocks, even when long-run average cost is declining, because the firm must respond to unanticipated shocks by moving along its short-run cost curve.[2] In summary, many of the distinctive (and intriguing) implications of LBD are lost when firm progress is not driven by passive learning.

I begin in Section 2 with a review of some theoretical implications of passive learning. The section considers, *inter alia*, its consequences for the pricing decision of a single firm, conditions for the existence of a competitive equilibrium, and its strategic implications. Many of the intriguing implications of learning turn out to depend upon auxiliary assumptions that may not hold. For example, it is a widely held belief that learning generates dynamic scale economies that are incompatible with price-taking equilibria; whether this is so depends on assumptions made about the static cost function, as well as assumptions about the form of the experience curve. Section 3 provides a selective review of empirical work. Two central questions emerge from the empirical literature. First, what fraction of the cost–quantity relationship is accounted for by passive learning? Second, is the contribution of passive learning unbounded as experience accumulates? The answers to these questions are clouded by considerable empirical difficulties caused in large part by the poor quality of data that have typically been available to researchers.[3] Early studies invariably indicated an important role for passive learning, and favored specifications consistent with unbounded productivity. But recent studies using highly detailed data have raised doubts about the conventional wisdom. The tenor of this newer literature is that relatively little of the cost–quantity relationship observed in industrial settings can be attributed to passive learning, and thus that much of the theoretical work on passive learning might be barking up the

[2] That is, with declining long-run average cost, small positive shocks to demand reduce average cost while sufficiently large shocks increase it.

[3] Perhaps it is more accurate to say that the answers are unusually demanding of the data, to an extent that strains even what high-quality datasets can offer.

wrong tree. In settings where LBD or passive learning is likely to be a major factor in the cost–quantity relationship, the likely conclusion is that it is bounded.

Section 4 reviews theories of learning, in two parts. The section first reviews theories that have attempted to generate a power rule for passive learning, before turning to a treatment of models with bounded learning. One lesson from these latter models is that alternative theories with potentially distinct policy implications may be exactly or nearly exactly observationally equivalent. Section 5 reviews macroeconomic models of economic growth, with a focus on models that incorporate bounded learning. These models are essentially hybrids involving the sequential introduction of generations of products or technologies, with passive learning within each generation. New generations are introduced either exogenously or as the result of some purposive activity distinct from passive learning, such as R&D, although there may also be learning spillovers across generations. In these hybrid models, passive learning is often an inessential component of long-run growth. To the contrary, too much passive learning can under certain circumstances lead to stagnation.

2. Microeconomic implications of passive learning

This section reviews some theoretical implications of passive learning under the conventional assumption that learning proceeds in lockstep with cumulative production volume. The review begins, in Section 2.1, with the pricing and output decisions of a single firm. Because passive learning generates dynamic increasing returns, much of the early theoretical literature confined itself to imperfect competition. However, perfect competition is compatible with passive learning when static marginal costs rise sufficiently rapidly; Section 2.2 reviews the conditions under which a price-taking equilibrium can exist. Sections 2.3 and 2.4 consider some implications of passive learning for industry concentration. Section 2.5 reviews some strategic implications of passive learning in imperfectly competitive markets. It begins by holding fixed the number of firms and exploring how passive learning influences pricing behavior when there are strategic considerations. The second part is concerned with the incentives that passive learning creates for incumbent firms to engage in predatory behavior designed to deter entry or promote exit. The section closes with a discussion of the robustness of results to alternative formulations of passive learning.

2.1. Pricing and output decisions

Let $x(t)$ denote the rate of output of a firm, $y(t) = \int_o^t x(s)\,\mathrm{d}s$ its cumulative output, $R(x(t))$ its revenues, and $c(x(t), y(t))$ its total costs. Assume $c_x \geq 0, c_y < 0$, and $c_{xy} < 0$; static marginal costs are nondecreasing at any level of experience, while experience lowers total and marginal costs at any output level. The firm has a planning horizon of T and faces an interest rate of r. Its objective is

$$V = \max_{\{x(t)\}_0^T} \int_0^T [R(x(t)) - c(x(t), y(t))]\,\mathrm{e}^{-rt}\mathrm{d}t, \tag{1}$$

subject to the constraint $\dot{y}(t) = x(t)$. Let $\lambda(t)$ denote the shadow price of experience. Equation (1) is a standard free-endpoint optimal control problem, so $\lambda(T) = 0$. The necessary condition for an interior maximum is

$$c_x(x(t)) = R'(x(t)) + \lambda(t), \tag{2}$$

and substituting the forward solution for the shadow price yields

$$c_x(x(t)) = R'(x(t)) - \int_t^T c_y[x(s), y(s)]\, \mathrm{e}^{-r(s-t)} \mathrm{d}s. \tag{3}$$

The optimal strategy sets marginal cost above marginal revenue by an amount equal to the discounted present value of the cost savings obtained from an increment to experience today. How large this wedge between static marginal cost and revenue is depends on the form of the cost function and the path that future output will take. Rosen (1972) was the first to study the problem. He was content to leave the form of $c(x(t), y(t))$ unspecified, and consequently he limited himself to deriving and discussing the condition (3). Spence (1981) considered the special case of a zero real interest rate and a constant static marginal cost, of the form $c(x(t), y(t)) = c_0\theta(y(t))x(t)$. Noting that $\mathrm{d}\theta/\mathrm{d}t = \theta'(y)\dot{y} = \theta'(y)x$, Spence's special case reduces Equation (3) to

$$c_0\theta(y(T)) = R'(x(t)). \tag{4}$$

This is Spence's well-known terminal marginal cost rule. The firm sets marginal revenue equal to the marginal cost that it will attain at the end of the planning horizon. As a result, price and output remain constant over the life of the firm, even though current marginal cost is falling. Current marginal cost consistently exceeds marginal revenue, although whether it also exceeds price at any point in time depends upon the elasticity of demand and the rate of learning.

The terminal marginal cost rule does not depend upon the precise form of the experience curve, but it is not robust to changes in the auxiliary assumptions. For example, if $r > 0$ and the planning horizon is infinite, Equation (3) becomes

$$rc_0 \int_t^\infty \theta(y(s))\, \mathrm{e}^{-r(s-t)} \mathrm{d}s = R'(x(t)), \tag{5}$$

so that marginal revenue is set equal to the annuitized discounted present value of all the marginal costs that will prevail in the future. In this case, as $\theta(y)$ is declining over time, marginal revenue declines monotonically along with current marginal cost.

When static marginal cost is not constant, the optimality condition cannot generally be written in a way more informative than has already been given in Equation (3). While it remains true that marginal revenue will be less than current marginal cost it turns out that marginal revenue is not necessarily nonincreasing over time: Cost-functions of the form $c(x(t), y(t)) = c_0 + h(x(t))\theta(y(t))$, with h an increasing convex function, induce monotonically declining paths for marginal revenue; functions of the form $c(x(t), y(t)) = c_0 + h(x(t)) + \theta(y(t))$ yield monotonically increasing paths (Clarke et al., 1982; example 1 in Petrakis et al., 1997); other functional forms can yield nonmonotonic paths.

In general, the firm's strategy is not socially optimal. But the divergence between the socially and privately optimal output paths is a result only of the market power of the firm. To see this, let $p(x)$ denote the inverse demand function. Assuming the interest rate and discount rate coincide, the planner maximizes

$$W = \int_0^T \int_0^{x(t)} p(v)\,\mathrm{d}v - c(x(t), y(t))\,\mathrm{e}^{-rt}\mathrm{d}t, \tag{6}$$

which yields the necessary condition

$$c_x(x(t)) = p(x(t)) - \int_t^T c_y(x(s), y(s))\mathrm{e}^{-r(s-t)}\mathrm{d}s. \tag{7}$$

The solutions to Equations (3) and (7) coincide if and only if $R'(x(t)) = p(x(t))$ for all t, that is if the firm is a price-taker. If the firm has market power, its optimal strategy involves less output and slower learning than the social planner prefers. The static exercise of market power induces a monopolist to reduce output relative to the social optimum, thereby reducing the rate at which experience is accumulated. Deviations from the static optimum depend on the size of the gains from cost reductions. When the demand curve is downward sloping, part of the social gains accrue to consumers, and so the planner gains more from a cost reduction than does a monopolist. Thus, both static and dynamic considerations induce deviations of the same sign between privately and socially optimal behavior. Put another way, passive learning exacerbates the suboptimality of monopoly output, but it does not create inefficiency on its own.

2.2. *Cost functions and price-taking behavior*

The welfare consequences of passive learning clearly depend in large part on the question of whether price-taking behavior can be sustained in equilibrium. The answer to this question in turn depends on the structure of marginal cost. Fudenberg and Tirole (1983) prove that a price-taking equilibrium does not exist when static marginal cost is constant, and this induces them to study, *inter alia*, the suboptimality of monopoly output. In contrast, Petrakis et al. (1997) show that price-taking equilibria can exist when static marginal cost is increasing. As one should expect from the discussion following Equation (7), the equilibria they analyze are socially efficient.

The intuition behind these results is straightforward. When static marginal cost is constant, learning has much the same impact on price-taking equilibria as does static increasing returns: it forces average cost below marginal cost and generates losses. For example, in Spence's special case, Equation (4), the optimality condition for a price taker is $c_0\theta(y(T)) = p$, but average cost over the life of the firm is $T^{-1}c_0 \int_0^T \theta(y(s))\,\mathrm{d}s > c_0\theta(y(T))$. Another way to think about the issue is to consider an arbitrary firm's problem in a two-period setting where all firms are *ex ante* identical. A price-taking equilibrium in period 2 requires a mass of atomistic firms producing with the same average cost and earning zero profit, which requires in turn that each firm had produced identical output in period 1. But this cannot be an equilibrium when static marginal cost is independent of scale. Any firm can choose to raise its output marginally in the first period by selling below cost; second-period cost is then strictly less than its competitors and so it captures the entire market. In contrast, when static marginal cost rises sufficiently rapidly, average cost over the life of the firm is no longer above marginal cost. On the one hand, increasing output today lowers future marginal costs, but the price of doing so is to raise current marginal cost. When a firm is behaving optimally, its marginal cost is locally increasing. As a result, a firm with lower costs captures a greater share of, but not all, the market, and a price-taking equilibrium can be sustained.

2.3. Endogenous heterogeneity

Passive learning can endogenously generate heterogeneous behavior among firms that are *ex ante* identical. Petrakis et al. (1997) show this in a deterministic two-period model with free entry and exit, which has three possible equilibria. In the first, all firms enter in period 1 and remain for the life of the industry. In the second, all firms enter in period 1, but some of them depart at the end of the first period. In the third, there is no exit at the end of the first period, and some firms enter only in period 2. There is no equilibrium that combines early exit and late entry. To see why this is the case, let $c(x(t), y(t))$ denote the cost function, and assume that static average cost is U-shaped. Thus, all firms that enter in period 1 face costs $c(x_{1i}, 0)$ in period 1 and $c(x_{2i}, x_{1i})$ in period 2. Further, let $p_m = \min_x c(x, 0)/x = c(x_m, 0)/x_m$ denote the minimum average cost for a firm with zero experience. Free entry implies that price cannot exceed p_m in either period, but it may be strictly less than this.

Consider first the equilibrium in which some firms exit after period 1, so that $p_1 = p_m$. Firms that exit early produce x_m and earn zero profit in period 1. For this to be optimal, the second-period price can be no greater than $\min_{x_2} c(x_2, x_m)/x_2 < p_m$. As a result, an equilibrium with early exit requires a strictly falling price, which is incompatible with late entry. Firms that remain in the industry produce $x_{1c} > x_m$ in order to benefit from passive learning; they consequently earn negative profit in the first period but recover this by earning positive profit in the second.[4] Thus, in an equilibrium with early exit, some firms initially produce more than others and sell at a price below their current average cost; these firms survive while the smaller firms exit. If the second-period price exceeds $c(x_2, x_m)/x_2$, there is no early exit. In this case, either $c(x_2, x_m)/x_2 < p_2 < p_m$, in which case there is no late entry, or $p_2 = p_m$ and some firms enter in period 2. Whenever there is late entry (and sometimes when there is not), $p_2 > p_1$, so passive learning is compatible with rising prices.

2.4. Learning and industry concentration

Passive learning is generally associated with increasing industry concentration. This is immediately apparent in Petrakis et al.'s analysis of price-taking equilibrium. Absent learning, *ex ante* identical firms have equal market shares at every point in time. Learning can induce *ex post* heterogeneity and consequently may increase concentration. Increasing concentration under passive learning appears

[4] The second-period costs of continuing firms must decline sufficiently as a result of producing $x_{1c} > x_m$ so that they can recover the first-period losses at a price satisfying $p_2 < \min_{x_2} c(x_2, x_{\mathrm{m}})/x_2$.

also to be a phenomenon of imperfectly competitive markets. Dasgupta and Stiglitz (1988) consider a duopoly with linear industry demand. They show that, even without allowing for strategic considerations, passive learning can amplify a small initial cost advantage for one of the firms, perhaps even to the point that the disadvantaged firm chooses to exit. These effects are most likely when firms are approximately myopic and the rate of learning does not decline too rapidly as experience is accumulated. Cabral and Riordan (1994) explore the same question in a differentiated duopoly model in which firms sell to a sequence of buyers with uncertain demands. They find that a sufficient condition for initial differences in the probability of securing the next sale to widen with the passage of time is that the discount rate be either very large or very small.

To abstract from strategic considerations (which will be considered in Section 2.5), I show here how initial differences in costs influence the evolution of concentration in a monopolistically competitive industry. Time is continuous, there is a continuum of firms indexed by $i \in [0, 1]$, industry revenues are set to unity, and the elasticity of substitution is denoted by $\sigma > 1$. Static marginal cost is constant and, following the notation of Section 2.1, satisfies $c_i(t) = c_i\theta(y_i(t))$ with $\theta'(y_i(t)) \leq 0$. For simplicity, I explore the consequences of passive learning under the extreme cases of myopia and no discounting.

Consider first myopia. Using standard calculations, demands are given by

$$x_i(t) = \frac{p_i(t)^{-\sigma}}{\int_0^1 p_j(t)^{1-\sigma} \mathrm{d}j}, \tag{8}$$

so the myopically optimal price is a constant markup over current marginal cost:

$$p_i(t) = \frac{\sigma c_i \theta(y(t))}{(\theta - 1)}.$$

Then firm i's share of industry revenues is

$$s_i(t) = \frac{[c_i\theta(y_i(t))]^{1-\sigma}}{\int_0^1 [c_j\theta(y_j(t))]^{1-\sigma} \mathrm{d}j}, \tag{9}$$

the growth rate of which satisfies

$$\begin{aligned} \frac{\dot{s}_i(t)}{s_i(t)} &= \frac{(1-\sigma)\theta'(y_i(t))x_i(t)}{\theta_i(y_i(t))} - \frac{(1-\sigma)\int_0^1 c_j^{1-\sigma}\theta(y_j(t))^{-\sigma}\theta'(y_j(t))x_j(t)\,\mathrm{d}j}{\int_0^1 [c_j\theta(y_j(t))]^{1-\sigma}\mathrm{d}j} \\ &= -\frac{(\sigma-1)^2\theta'(y_i(t))c_i^{-\sigma}}{\theta_i(y_i(t))^{1+\sigma}\int_0^1 [c_j\theta(y_j(t))]^{1-\sigma}\mathrm{d}j} - \mu(t), \end{aligned} \tag{10}$$

where $\mu(t) > 0$ is the loss of market share suffered by any firm as a result of learning by its competitors. As $y_i(0) = 0$ for all i, Equations (9) and (10) show that both market share and the growth rate of market share are initially decreasing in c_i. As a result, concentration must initially increase, but whether it continues to do so forever depends upon the functional form of the learning curve. In particular, if learning stops after some finite accumulation of experience (i.e., $\theta'(y) = 0$ for all $y > y^*$), then an early period of increasing concentration is followed by a period of decreasing concentration as initially disadvantaged firms catch up with the leaders.

Consider now the other extreme, where the discount rate is very small. In this case, too, passive learning is associated with greater concentration. To see this, assume a zero discount rate and a planning horizon of length T, so that Spence's terminal marginal cost pricing rule, $c_i\theta(y_i(T)) = R'(x_i(t))$, applies. Firm i sets a constant price equal to $p_i = \sigma c_i\theta(y_i(T))/(\sigma - 1)$ and, noting that $y_i(T) = x_i T$ under a constant pricing rule, demands are

$$x_i = \frac{[c_i\theta(x_i(T))]^{-\sigma}}{\int_0^1 [c_j\theta(x_j(T))]^{1-\sigma}\mathrm{d}j}. \tag{11}$$

Differentiating Equation (11) with respect to x_i and c_i, and evaluating at the symmetric equilibrium, yields

$$\frac{\mathrm{d}x_i}{\mathrm{d}c_i} = -\frac{(\sigma - 1)c^{-(1+\sigma)}\theta^{-\sigma}}{\int_0^1 c_j^{1-\sigma}\theta_j^{1-\sigma}\mathrm{d}j(1 + (\sigma - 1)x_i T\theta'/\theta)}. \tag{12}$$

As long as $(\sigma - 1)x_i T\theta'/\theta > -1$, which condition is necessary for concavity of the Hamiltonian, Equation (12) is negative. If there were no learning, (i.e., $\theta' = 0$), then the direct impact on output is simply $\mathrm{d}x_i/\mathrm{d}c_i = -(\sigma - 1)c^{-(1+\sigma)}\theta^{-\sigma}/\int_0^1 c_j^{1-\sigma}\theta_j^{1-\sigma}\mathrm{d}j$, as is evident from treating θ as a constant in Equation (11). The term $(1 + (\sigma - 1)x_i T\theta'/\theta)^{-1} > 1$ is the learning multiplier, showing that the increase in firm i's output resulting from a decline in its initial cost is greater in the presence of learning. Moreover, the multiplier is larger with stronger learning effects and a longer planning horizon.

2.5. *Strategic implications of learning*

Pricing and output decisions under passive learning with small numbers of firms are complicated by the potential for strategic behavior. As in the monopoly and price-taking settings, each firm continues to face a trade-off between current profits and investment in the form of overproduction to increase the rate of learning. But this trade-off is complicated by the fact that a firm's current output level influences its competitors' current and future output levels, the latter by altering the future structure of costs in the industry. Passive learning may also create motivations to overproduce with the intention of deterring potential future entrants, and to induce exit through predatory pricing.

Dynamic oligopoly models quickly become intractable, so much of the analysis has been conducting in specialized settings. As a consequence, some of the findings reported in this subsection are unlikely to be especially robust to perturbations in the auxiliary assumptions. Nonetheless, some results have been found to hold in several settings. First, there are a set of results that apply to industries with a fixed number of firms: passive learning in such markets appears to be procompetitive, raising output above the level that would be attained absent learning; output may fall over time even in settings in which monopoly output would unambiguously rise; and learning can lower industry profits even though it reduces costs and raises economic welfare. A second set of results concerns strategic behavior designed to deter entry and to force exit. In particular, passive learning induces aggressive pricing by incumbents to deter future entry, and it also creates a rationale for predation.

2.5.1. Fixed numbers of firms

Consider the two-period linear duopoly model developed by Fudenberg and Tirole (1983).[5] Denote the two firms by a and b and denote their outputs in period $i = 1, 2$ by x_i^A and x_i^B. Demand is $p_i = 1 - (x_i^A + x_i^B)$, and each firm's first-period unit cost is $c \in (0, 1)$. Second-period unit cost is given by $c_2^j = c - \lambda x_1^j, j = A, B$. In quantity competition, the second period is a standard static Cournot, in which average cost is a decreasing function of first-period output. Hence second-period output is increasing in the speed of learning and in first-period output.

Let β denote the discount factor. The Nash equilibrium for first-period output is

$$x_1^j = \frac{(1-c)(9+4\beta\lambda)}{27-4\beta\lambda^2}, \tag{13}$$

which is strictly increasing in λ for all but myopic firms. Thus, passive learning is associated with increased first-period output. It then follows that second-period costs are lower, and that second-period output is higher. This is, of course, equally true in the absence of strategic considerations. A more useful exercise, therefore, compares Equation (13) with the output that would be attained in a *precommitment equilibrium*, which Fudenberg and Tirole define as an equilibrium in which firms ignore the consequences of dynamic changes in the cost structure on their competitor's future output. First-period output under precommitment is

$$\tilde{x}_1^j = \frac{(1-c)(3+\beta\lambda)}{9-\beta\lambda^2}. \tag{14}$$

Although each firm ignores the effect of learning on its competitor's future output in the precommitment equilibrium, it continues to behave strategically with respect to current output and it takes into account the effect of its own first-period output on its own future cost. The degree to which passive learning alters strategic behavior in duopoly can therefore be summarized by the ratio $x_1^j/\tilde{x}_1^j$, which equals one when $\lambda = 0$, and is strictly increasing in λ. Thus, strategic considerations in the presence of passive learning promote competition in the first period and, by extension, in the second period as well. In fact, when the rate of learning is high and firms do not discount the future much, market performance is surprisingly good in the first period: if one allows $\beta\lambda^2$ to approach its upper limit of 3/4,[6] and sets the discount factor to unity, then the duopoly attains 32/33 of the competitive output.[7] The second-period

[5] The qualitative results here hold for n-firm oligopolies with equivalent auxiliary assumptions.

[6] "Conventional" comparative statics and stability require that $\beta\lambda^2 < 3/4$.

[7] Recall that in a static duopoly with linear demand and constant marginal cost, output is two-thirds of that attained in competition. This output level is attained when either $\lambda = 0$ or $\beta = 0$.

duopoly output remains at two-thirds the output level that would be attained under marginal cost pricing, although learning has of course reduced cost.[8]

It is a standard result that duopoly profits are inversely related to production costs. It is therefore somewhat surprising that, under a wide range of values for λ and β, passive learning *reduces* discounted lifetime profits, $\nu^j(\lambda, \beta) = \pi_1^j + \beta\pi_2^j$. In particular, if β is sufficiently large, then $\nu^j(\lambda, \beta)$ is decreasing in λ for all admissible rates of learning.[9] Profits are always increasing in λ in the precommitment equilibrium, so this surprising finding is clearly the competitive consequence of raising first-period output to influence the competitor's future output. There is no reason to expect this result to be especially robust, but Spence (1981) reports that in his model rates of return are generally lower when learning is rapid.

With constant static marginal cost, output rises monotonically in monopoly. It does so also in the duopoly precommitment equilibrium, but not in the subgame perfect equilibrium characterized by Equation (14). The strategic incentive to raise first-period output may be sufficiently strong that first-period output is higher than second-period output, even though costs have declined in the second period.

2.5.2. *Predation and entry deterrence*

The preceding analysis admits unavoidable fixed costs, which are irrelevant to outcomes (although they affect whether the duopoly would have been created in the first place). When there are avoidable fixed costs, however, strategic interactions are further complicated by the possibility of exit. More specifically, avoidable fixed costs create a motive for predation in the presence of learning. Cabral and Riordan (1997) explore this with a simple extension of Fudenberg and Tirole's two-period duopoly model.

Returning to our model, assume that firm a is committed to production in the second period, but firm b must pay a fixed cost, k, if it wishes to remain active. To ensure smoothness of the first-order conditions, assume that the fixed cost is stochastic with distribution and density functions $\Phi(k)$ and $\phi(k)$; its realization is observed at the end of the first period. If the realized fixed cost is sufficiently low, b remains active and payoffs in the second period are given by the duopoly profits

$$\pi_D^j = \left(\frac{1 - c + 2\lambda x_1^j - \lambda x_1^{-j}}{3}\right)^2, \quad j = A, B. \tag{15}$$

These payoffs are realized with probability $\Phi(\pi_D^B)$. With probability $1 - \Phi(\pi_D^B)$, firm b exits, leaving a to earn monopoly profits

$$\pi_M^A = \left(\frac{1 - c + \lambda x_1^A}{2}\right)^2. \tag{16}$$

[8] Spence (1981) obtains similar results for market performance, measured by the fraction of the maximum surplus archived, in his computational examination of a nonlinear oligopoly model. He reports performance rates of between 84% and 94%, and also finds performance is better the more rapid the learning rate. Interestingly, performance is not monotonically increasing in the number of firms.

[9] For modest values of β, $\nu^j(\lambda, \beta)$ first increases and then decreases with λ.

Taking the possibility of exit into account, the first-period necessary condition for firm a is

$$(1 - c - 2x_1^A - x_1^B) + \beta\left[\Phi(\pi_D^B)\frac{\partial\pi_D^A}{\partial x_1^A} + (1 - \Phi(\pi_D^B))\frac{\partial\pi_M^A}{\partial x_1^A}\right] = \beta\phi(\pi_D^B)\frac{\partial\pi_D^B}{\partial x_1^A}(\pi_M^A - \pi_D^A). \tag{17}$$

The first term on the left-hand side, when set to zero, is the usual static first-order condition for Cournot duopoly. The second term reflects the influence of learning on firm a's decision when the probability of b's exit is taken as given. Under the restrictions on λ and β given in the previous subsection, the left-hand side is strictly decreasing in x_1^A. The term on the right-hand side captures the incentive learning creates for predation; from Equation (15), $\partial\pi_D^B/\partial x_1^A < 0$, so this term is negative. Firm a is induced to increase output because doing so reduces b's profits under duopoly, and this increases the probability that b chooses to exit. Cabral and Riordan define the degree of predation as the difference between a's output given by Equation (17) and its output obtained after replacing the right-hand side of Equation (17) with zero. Noting that $\partial\pi_D^B/\partial x_1^A = -2\lambda\sqrt{\pi_D^B}$, predation is by this definition greater when the learning effect is stronger. In the absence of passive learning the right-hand side of Equation (17) is identically zero, and there is no incentive to engage in predatory pricing.

The preceding discussion might lead one to suppose that a firm will set its first-period price lower when its competitor faces a risk of exit. But this is by no means certain, because the possibility of exit induces two responses, one of them countervailing, from firm b. The first-order condition for b is given by

$$(1 - c - 2x_1^B - x_1^A) + \beta\frac{\partial\pi_D^B}{\partial x_1^B}[\Phi(\pi_D^B) + \phi(\pi_D^B)\pi_D^B] = 0. \tag{18}$$

On the one hand, the increase in profit that would correspond to a lower second-period cost is obtained only with probability $\Phi < 1$, which effect reduces b's first-period output. On the other hand, reducing second-period costs raises the probability of remaining in business by an amount that depends on ϕ. The term in braces is equal to unity when exit is not a possibility, but may sum to more or less than one when exit is possible. Consequently, the possibility of exit (and, more precisely, of avoiding exit by aggressive first-period pricing) may in fact raise b's first-period output, which in turn would reduce a's first-period output. This ambiguity is exacerbated when both firms face avoidable fixed costs.

It has previously been noted that in the presence of passive learning pricing below marginal cost does not constitute evidence of predation, and this creates difficulties for the implementation of antitrust policy. But Cabral and Riordan have shown how passive learning creates an incentive for predation that would not otherwise exist. The lesson might be that predation is more likely with passive learning, but proving it in court will be more challenging. But even if a plaintiff is successful in court, it is not clear what the appropriate remedy should be, because the welfare consequences of predation are ambiguous. Cabral and Riordan analyze the welfare consequences of prohibiting predation in their model. They find that consumer surplus may rise or fall when predation is outlawed. The intuition is straightforward. Predation reduces price in the first period, favoring consumers. In the second period, successful predation leads to monopoly pricing, which hurts consumers, but unit costs are lower than they would have been absent predation.

The principles behind entry deterrence are analogous to those behind predation. An incumbent monopolist increases output with the aim of reducing future costs, thereby limiting entry (see, e.g., Saunders, 1985; Scherer, 1980, pp. 250–252). Successful entry deterrence is associated with the maintenance of monopoly pricing, but its implications for consumer welfare are again ambiguous because future costs are lower than they would be absent the aggressive first-period pricing. However, as with much of the analysis in this section, one can develop market structures in which straightforward, intuitive, results do not hold. Hollis (2002) considers a two-period model in which firms learn at different rates, either because some firms are intrinsically better than others at learning or because some firms are further down a common progress curve. He shows that an incumbent firm with relatively little left to learn may be ambivalent about entry. While the incumbent would prefer no entry at all, it may prefer a lot of entry to a little: when there are just a few entrants, each may be able to learn a sufficient amount to become an effective competitor in the second period; but when there are many entrants, none learns much and so none becomes an effective competitor.

2.6. Alternative specifications of learning

So far, it has been assumed that passive learning is a product of a firm's own experience; that experience is best measured by cumulative output, rather than by alternatives such as elapsed time or cumulative investment; that learning remains proprietary; and that the effects of past experience are persistent. This subsection briefly considers some consequences of, and evidence in favor of, changing elements in this list of assumptions.

2.6.1. Spillovers

Most of the work on the strategic implications of passive learning assumes that what is learned remains proprietary. Ghemawat and Spence (1985), Stokey (1986), and Lieberman (1987a) have shown that many implications of passive learning, including first-mover advantages, the raising of entry barriers, and excess concentration are muted at a rate that varies inversely with the degree of learning spillovers. Moreover, when spillovers are sufficiently strong to effectively eliminate the incentives to deviate from static optimum pricing and output levels, prices fall in lockstep with static marginal costs. It may seem somewhat paradoxical therefore that many models exploring the implications of passive learning for strategic trade policy in large economies assume purely external learning (e.g., Krugman, 1987; Redding, 1999). However, in these cases, the usual assumption is that there are effective barriers to international knowledge diffusion, thereby enabling national policymakers to engage in strategic behavior.

The evidence points to the presence of significant learning spillovers in a variety of industries. Using survey data, Mansfield (1985) found that information about new processes and products in 10 industries surveyed had widely diffused within a year. Spillovers have also been found in econometric studies: Irwin and Klenow (1994) find them in semiconductors; Thornton and Thompson (2001) in wartime shipbuilding; Lieberman (1989) in chemicals; Foster and Rosenzweig (1995) in the adoption of high-yielding seed varieties; and Conley and Udry (2007) in the adoption of best practices by Ghanaian pineapple farmers. However, the reliability of evidence for spillovers is especially sensitive to problems of measurement error at the firm level. It is likely in many applications that firm-level experience is mismeasured, because cumulative output is measured with error or because it is only a proxy for a more appropriate but unobserved index of experience. The industry-wide experience assumed to give rise to learning spillovers is typically measured by average or total industry cumulative output. By construction, this variable suffers less from measurement error than does firm-level experience.

At the same time, it is positively correlated with firm-experience, not least because firms share correlated market conditions that influence output decisions. The result is that the coefficient on own-experience is attenuated, while the contribution to a firm's productivity of industry-wide experience is overstated.[10]

2.6.2. *Learning as a function of cumulated investment*

The earliest macroeconomic models of passive learning—Arrow (1962), Levhari (1966), and Sheshinski (1967b)—associated learning with cumulative investment rather than cumulative output. Sheshinski (1967a) observed that this is a plausible assumption because new investment changes the production environment and provides a stimulus for renewed learning. A similar argument was made much later by Mishina (1999), whose detailed study of the wartime production of the B17 heavy bomber led him to conclude that learning arose out of the new experiences afforded by scaling up plant capacity. It also seems reasonable to suppose that many of the consequences of passive learning would be robust to switching the engine of growth from output to investment: the excess of output over static optimum levels induced by passive learning is in fact often interpreted as a form of investment.

Nonetheless, linking learning to investment has received scant attention from microeconomic theorists.[11] One can conjecture why. First, by the time industrial organization theorists were beginning to turn their attention to passive learning in the early 1980s, there already existed a sizeable literature on the use of physical capacity as a strategic device, notably to deter entry but also to preempt existing rivals (Salop, 1979; Spence, 1977; Wenders, 1971, and others; see Lieberman, 1987b for a concise review). Second, it quickly emerged that the implications of strategic investment were, like passive learning, sensitive to auxiliary assumptions. For example, in a linear model it is not in the interests of an incumbent to invest in excess capacity following entry (Dixit, 1980), so investment in excess capacity prior to entry does not constitute a credible threat to potential entrants. However, this result can be reversed with an appropriately nonlinear demand curve (Bulow et al., 1985). Third, the empirical challenges involved in separating the learning effects of investment from scale economies in capacity expansion, or from vintage capital effects, must have seemed quite daunting.[12]

2.6.3. *Learning as a function of time*

If learning is a function of the passage of time spent producing, most of the strategic consequences of passive learning discussed earlier vanish. The intuition is straightforward: deviations of output from static optimization do not increase the rate of learning, so while the cost structure in the industry evolves

[10] Tambe and Hitt (2007) have tackled a similar problem involved in the measurement of knowledge spillovers resulting from investments in information technology. They obtained two distinct measures of IT capital, and argued that the measurement error in each is likely to be uncorrelated; this allowed one measure to serve as an instrument for the other. This approach to the measurement error problem in passive learning spillovers has not yet been attempted, probably because of the difficulty in identifying plausible candidate instruments.

[11] One notable exception is Jovanovic and Lach (1989). Their paper also studies the effects of spillovers, but does so in a non-strategic setting.

[12] One identification strategy is to contrast the effects of capacity *contractions* on productivity: scale economies in capital would be associated with a decline in productivity, while learning would not. Assuming capacity reductions are accomplished by retiring the oldest machines, vintage capital effects would induce a rise in productivity. Unfortunately, capital specificity ensures that significant declines in plant capacity are infrequent in most datasets.

over time this is, from the perspective of firms, an intrinsically exogenous process. There is, however, one notable exception: learning as a function of elapsed time continues to create first-mover advantages, and motivates early entry in oligopolies.

The evidence does not favor elapsed time over cumulative output or investment (Argote, 1993, p. 41). Investigating this is in principle as simple a matter as assessing the coefficients in the regression $\ln x = a + b \ln y + c \ln t + \varepsilon$. Collinearity produces imprecise results for the samples typically available in early studies. Panel data can expand the effective sample size, although it does so at the cost of constraining key parameters to be equal across units. Rapping (1965) exploits the panel structure of 15 shipyards engaged in the wartime construction of Liberty ships to assess the relative contributions of cumulative output and elapsed time on the current rate of output. Rapping allows for yard-specific level effects, but assumes the slope coefficients are equal across yards. His best-fit regression produces a coefficient of 0.26 on cumulative output and -0.03 on elapsed time, clearly favoring the conventional formulation of the progress curve.[13]

However, a caution is again in order. When rates of progress differ across units, panel techniques can provide spurious evidence in favor of the conventional formulation.[14] For example, if firm progress depends on elapsed time, but the rate of progress is different for each firm, a panel estimator that imposes the same coefficient on time for each firm but also includes cumulative output invariably indicates a significant impact of cumulative output. The reason is, much as in the well-known problem of confounding unobserved heterogeneity and contagion effects, cumulative output contains information about a firm's type.[15]

2.6.4. Forgetting

A sequence of papers by Argote, Epple, and colleagues (Argote et al., 1990, 1997; Darr et al., 1995; Epple et al., 1991, 1995) drew attention to evidence that unit costs frequently appear to increase during periods in which a firm experiences a decline in its volume of output. These researchers have argued that such reversals in productivity can be explained by a knowledge production function that allows for organizational forgetting.[16]

A simple formulation of this idea replaces cumulative output with effective experience, $E(t)$, so that current unit cost is given by $c(v) = c(0)E(v)^{-\beta}$. Experience is then assumed to increase with current output but to depreciate at a constant rate δ with respect to time:

$$\dot{E}(t) = x(t) - \delta E(t). \tag{19}$$

Estimates of the rate of depreciation suggest that organizational forgetting can be economically significant, although it varies widely across settings. Among pizza franchises, for example, Darr et al. (1995)

[13] Although Rapping's findings are consistent with the majority of the literature, there are exceptions. Levin (2000), for example, concluded that time spent producing automobiles is a better predictor of reliability than is cumulative output.

[14] Thompson (2007, Table A.1) shows that rates of progress varied widely across the Liberty shipyards.

[15] Concern with the confounding problem has an especially long history in count data, beginning with Greenwood and Yule (1920). A recent and important application to learning can be found in Wilcox (2006).

[16] Earlier studies had suggested interruptions to production may induce declines in productivity (Anderlohr, 1969; Baloff, 1970; Hirsch, 1956), but the more recent studies argue that organizational forgetting occurs even under conditions of continuous production.

found that knowledge depreciates at the astonishing rate of 17% a week, implying that "roughly one half of the stock of knowledge at the beginning of the month would remain at the end of the month." In wartime construction of Liberty cargo vessels, Argote et al. (1990) report that knowledge depreciated at the rate of 25% a month. However, other studies have either found evidence of much more modest rates of forgetting (e.g., Benkard, 2000; Thompson, 2007), or none at all (Ingram and Simons, 2002; Ohashi, 2005; Watkins, 2001).

Despite the mixed evidence, Benkard (2000) has called for theoretical efforts to investigate the strategic implications of organizational forgetting. The challenge was first taken up by Besanko et al. (2007), who add forgetting to Cabral and Riordan's (1994) duopoly model of learning and explore its implications for industry dynamics using the Markov perfect equilibrium framework developed by Ericson and Pakes (1995). A little reflection might lead one to suppose that forgetting, by undoing the gains from learning, attenuates the impact of learning on concentration and strategic behavior; as a result, one might further suppose that an industry with forgetting looks something like an average of an industry with no forgetting and an industry with no learning. These suppositions appear to be far from the truth.

Besanko et al. take pains to point out that forgetting "does not simply negate learning-by-doing"; to the contrary, forgetting enables the changes in the state of the industry (fully characterized in the Markov framework by the current unit costs of the two firms) to move forwards and backwards through the state space. The main consequence of this is that there may be multiple sunspot equilibria—as many as nine for some parameter configurations—even though Cabral and Riordan (1994) had already established uniqueness in the absence of forgetting. When there is no forgetting, it is inevitable that firms that do not exit eventually attain their terminal productivity, and this defined endpoint pins down a unique equilibrium path. At the other extreme, with an extremely high rate of forgetting, there can be little departure from initial costs, yielding a unique stationary equilibrium similar to that obtained in a duopoly without learning. But for intermediate rates of forgetting—especially rates similar to the rate of learning, multiple equilibria can be sustained by rational beliefs that different points on the learning curve can be sustained in the long-run. For example, if both firms believe that the long-run equilibrium involves two producers with little decline in costs, they have little incentive to price aggressively; as a result little net learning takes place and the beliefs are fulfilled in equilibrium. On the other hand, if both firms believe that the long-run equilibrium involves a single firm with low cost, both firms will induce this outcome by pricing aggressively in an attempt to be the surviving firm. In the latter case, Besanko et al. note, firms will price more aggressively in the presence of forgetting than in its absence.

3. Empirical evidence

The empirical literature on firm progress curves is distressingly large, consisting of literally thousands of reported progress curves in widely different industrial settings.[17] Much of this literature consists of somewhat naïve studies consisting of simple least-squares regressions of output or productivity on cumulative output or time, most often assuming a log-linear functional form. The naïve studies, because of their ubiquity, have shown us that the progress curve is a widespread phenomenon. They also reveal

[17] Asher (1956) provides a detailed review of the earliest work on airframe production. Between them, Yelle (1979), Argote and Epple (1990), Dutton and Thomas (1994), and Dar-El (2000, chapter 8) provide extensive references on subsequent literature.

that rates of progress vary dramatically across industries and firms, across products within firms, and even across different production runs of the same product within a firm (see Figure 2).

The estimation of progress curves induces a number of statistical problems that are in practice difficult to overcome. Prominent among them is the fact that progress curves relate two nonstationary variables, so the explanatory power of ordinary least-squares (OLS) regressions are inevitably high. Even so, out-of-sample predictions are often wide of the mark (Alchian, 1963 [1950]; Conway and Schultz, 1959; Hirsch, 1952, 1956), so estimated progress ratios are unreliable as a management planning tool. High coefficients of determination (in conjunction with an absence of guiding theory) also encouraged many researchers to express satisfaction with the appropriateness of the power rule specification. There are of course studies in which alternative specifications were considered, but these alternatives are usually nonnested. The resulting horse race between models, especially for short samples in which a terminal productivity had not been attained by the end of the sample period, is consequently reduced to a comparison of coefficients of determination that differ by margins of no real economic or statistical significance.[18]

The persistence of the power rule is all the more surprising in view of repeated evidence that after sufficient passage of time the rate of progress declines markedly, often to zero. Figures 3 and 4, which

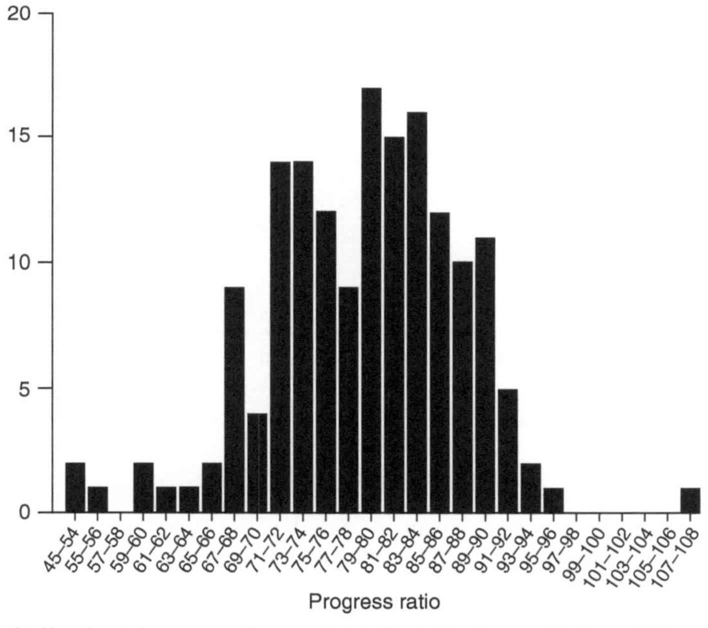

Figure 2. Distribution of 162 estimated progress ratios reported in 24 studies. Source: Dutton and Thomas (1994, Figure 1). Let y be cumulative output and $c(y)$ the current unit cost. The progress ratio is given by $c(2y)/c(y)$. In the power rule specification, $c = ay^{-b}$, the progress ratio is 2^{-b}.

[18] Feller (1940) pointed out long ago that it is difficult to discriminate between alternative growth functions.

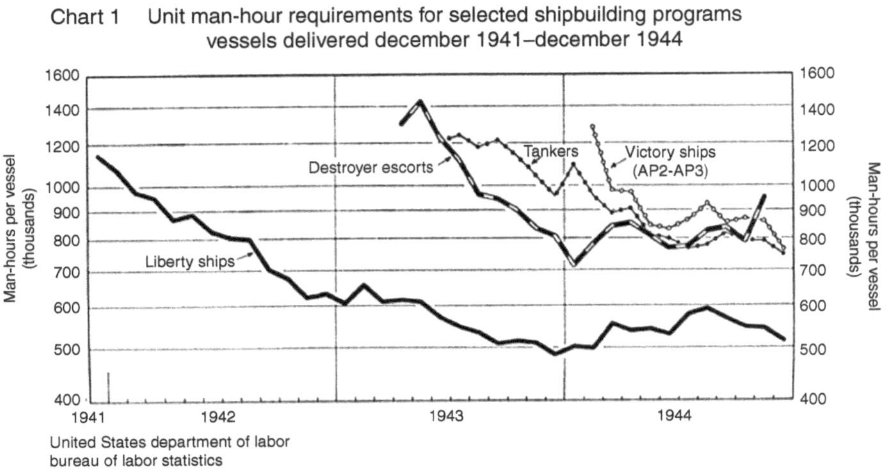

Figure 3. Progress curves for US wartime shipbuilding. Source: Searle (1945, Chart 1).

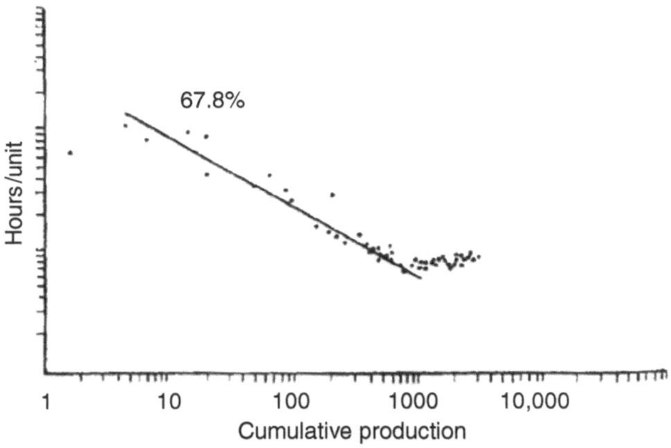

Figure 4. A progress curve for final assembly of a large electromechanical product. Source: Conway and Schultz (1959, Figure 9).

replicate plots from Searle (1945) and Conway and Schultz (1959), provide two neat early illustrations. In Conway and Schultz's paper, in fact, 6 out of 10 plots revealed compelling evidence that a terminal productivity had been attained and progress had stopped altogether. This study was one among several that led Baloff (1966) to assert that although the power rule curve may describe the startup phase in manufacturing, it does not describe a subsequent steady-state phase.

There is little reason to detain ourselves with further discussion of the early empirical literature, and the remainder of this section provides a selective review of the more recent empirical literature. Section 3.1 reviews attempts to measure learning in large, plant-level, datasets. Section 3.2 briefly discusses empirical studies of individual learning (by doing). Finally, Section 3.3 reviews small-sample evidence from detailed case studies that shed some further light on the role of passive learning, and on the difficulties involved in measuring the importance of passive learning in large samples.

3.1. Large sample evidence

Since confidential establishment data became available to researchers in the 1980s, a large body of evidence has accumulated showing that a firm's size increases with its age. New plants tend to be smaller than incumbent plants, but surviving plants grow most rapidly when young. In one of the best-known studies, Dunne et al. (1989) report that among 208,000 US manufacturing plants that survived any given 5-year period of observation, annual employment growth rates averaged 7.6% for plants under 5 years of age, 3.7% for plants aged 6–10 years, and 2.9% for plants 11–15 years of age. Comparable age effects have been observed in other multi-industry samples constructed from census data (Baldwin et al., 2000; Disney et al., 2000; Persson, 2002) in Dun and Bradstreet data (Evans, 1987a,b), Compustat data (Hall, 1987), and numerous specialized samples (e.g., Audretsch, 1991; Audretsch and Mahmood, 1995; Baldwin and Gorecki, 1991; Mata and Portugal, 1994; Wagner, 1994).

Although these findings have often been attributed to learning in young plants,[19] evidence for passive learning based on firm size is of limited value because the relationship between plant size and productivity is quite tenuous. For example, Baily et al. (1992) conclude that across 23 US manufacturing industries, productivity is in fact marginally *lower* in older plants than in younger plants. Bartelsman and Dhrymes (1998) restrict attention to the productivity rankings of plants in a large sample drawn from three US high-technology sectors. They also find that average productivity in young plants is marginally higher than in older plants. Similarly, Jensen et al. (2001) report that average labor productivity in their sample of manufacturing plants does not vary with age in any systematic fashion.

One candidate explanation for this disparity in the effects of age on size and productivity is that productivity data confound the effects of capital vintage and firm progress. On the one hand, new firms typically invest in technology of recent vintage, which raises their productivity relative to incumbents. Countervailing this vintage effect, older firms may have moved further down their progress curve. Jensen et al. (2001) conclude that these two effects have more or less the same magnitude in the Longitudinal Research Database (LRD). For example, the 1992 cohort of entering plants in US manufacturing was 51% more productive than the 1967 cohort had been when they entered; but the surviving plants in the 1967 cohort had by 1992 experienced an average productivity gain of 57%. Similar results hold for other entering years, so that in 1992 all cohorts of surviving firms had average productivity within 7% of the industry mean.[20]

Identifying age and vintage effects is not trivial. It is well known (cf. Hall, 1971) that productivity and output regressions with a full set of vintage and age effects cannot be identified along with a full set of time effects intended to capture industry-wide factors. Jensen et al. resolved this problem by assuming that time effects can be measured by industry-wide variables such as average labor productivity and

[19] Dunne et al., for example, motivate their empirical analysis by appeal to Jovanovic's (1982) model of learning and selection (reviewed in Section 4.2).

[20] This should not be a surprising equilibrium outcome. If vintage effects dominated learning effects, there would be few surviving firms from early cohorts; if learning effects dominated, there would be few late entrants.

total output; these are imperfectly correlated with time, which is then dropped from the regressions. Bahk and Gort (1993) also use the LRD to separate vintage and learning effects, but they adopt a different identification strategy. For each year of a plant's life they construct the current average vintage of capital out of its investment history. Doing so breaks the collinearity between vintage, time and age, especially among older plants, allowing Bahk and Gort to capture industry-wide effects with a time trend. Bahk and Gort found that plant age accounted for output growth among young plants equivalent to about 1% per year; this was somewhat less than half the estimated contribution of embodied technical change of physical capital.[21]

Jensen et al. are careful to note that their finding of significant age effects among surviving plants may be due to a number of reasons, including scale economies gained from expansion over time, equipment investment, selection effects, and of course passive learning. Bahk and Gort are rather more willing to identify age effects directly with passive learning, and they go further than others in attempting to decompose its sources. To do so, they estimate a production relation of the form

$$\ln y_{i\tau} = \beta_\tau + \beta_\tau^K \ln K_{i\tau} + \beta_\tau^L \ln L_{i\tau} + \beta_\tau^w \ln w_{i\tau} + \varepsilon_{i\tau} \tag{20}$$

on repeated cross sections of plants of the same age. In Equation (20), $y_{i\tau}$ is output of plant i at age τ, $K_{i\tau}$ is vintage-adjusted capital, $L_{i\tau}$ is labor, and $w_{i\tau}$ is the average wage, intended to measure general human capital. Bahk and Gort assert that passive learning can be inferred from increases over time in the estimated elasticities. They distinguish three potential sources of learning: manual or task learning accomplished by workers, learning how to use capital, and organizational learning that raises the productivity of employees by improving, *inter alia*, the match between worker skills and task requirements (cf. Prescott and Visscher, 1980). Plant level data does not allow manual learning to be distinguished from increases in human capital associated with changes in the composition of workers as a plant ages, so Bahk and Gort focus on learning how to use capital, measured by changes in β_τ^K, and organizational learning, measured by changes in β_τ^L and β_τ^w. Figure 5 summarizes the results of their decomposition, which despite the strong identifying assumptions met with only limited success. They found no evidence of organizational learning, both indicators of which first exhibited declines before rising modestly. In contrast, the elasticity of output with respect to physical capital rose markedly. Even for capital, though, "learning" appears to have been completed after 4 or 5 years. Moreover, as Bahk and Gort note, most of this apparent learning probably arises from the fact that capital goods are not initially fully installed and operational.

Studies using large samples have provided extensive evidence on the effects of plant and firm age on size and growth. But because of the tenuous link between age and productivity, these studies provide at best indirect evidence that passive learning may be taking place. Relatively few large sample studies directly measure productivity dynamics, and even fewer have attempted to measure the importance of passive learning. One challenge for large-sample studies is that researchers are really only able to measure movement along a firm's progress curve; they invariably lack the detailed data necessary to understand how much of this progress is driven by passive learning and how much is due to unmeasured

[21] Power (1998) develops this approach further by looking at productivity responses to spikes in investment. She finds a positive effect on productivity of plant age after controlling for investment spikes, but no effect of time that has elapsed since an investment spike.

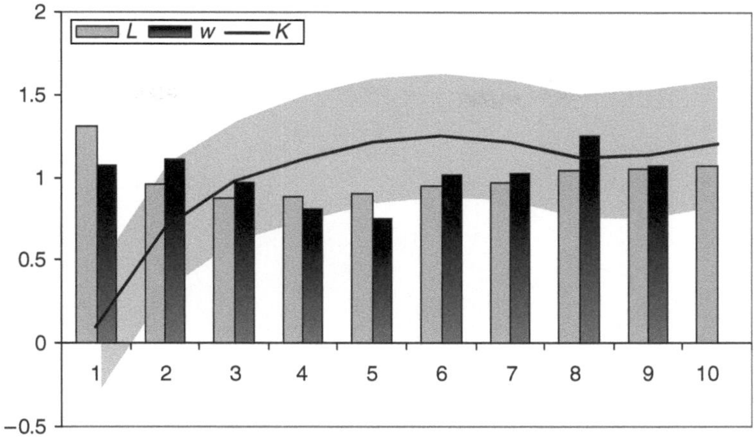

Figure 5. Decomposition of passive learning, Bahk and Gort (1993, Table 4). All coefficients normalized to indexes relative to their means over the 10 cross sections. Coefficients on L and w in column form. The line plots coefficients on K, with shaded 95% confidence interval.

factors. Unable to measure passive learning directly, these studies are also unable to shed much light on whether passive learning is short-lived and bounded.

3.2. Individual learning by doing

One possible way out of this impasse is to focus on special cases in which the context leads us to believe that progress is almost certainly dominated by passive learning. Unfortunately, these are almost invariably cases in which individual LBD is the focus of the study. Jovanovic and Nyarko (1995) collected together a number of datasets on LBD in commercial settings. Figure 6, which plots the productivity of new line-workers at a British munitions factory operating during World War I, illustrates the typical result that productivity quickly attains an upper bound (in this case within 4–5 weeks after initiating employment). Similar results have been obtained in studies of LBD among surgeons,[22] and in experimental settings (e.g., Mazur and Hastie, 1978).

For our purposes, however, studies of LBD have two limitations. The first is that the firm or plant can do better than the average performance of individuals because it can exploit variations in individual learning rates to reallocate workers to the most appropriate tasks (cf. Prescott and Visscher, 1980). This process may also be drawn out beyond the period in which individuals learn as the firm dismisses workers that have failed to learn and hires replacements who have yet to demonstrate their ability to learn. Second, in many of the settings examined learning is bounded by construction, so their findings about terminal productivities cannot readily generalize to other settings. This is especially true of medical applications, where the postsurgery complication rate or survival rate is the most common measure of performance.

[22] See Waldman et al. (2003) for citations to a small fraction of this extensive literature.

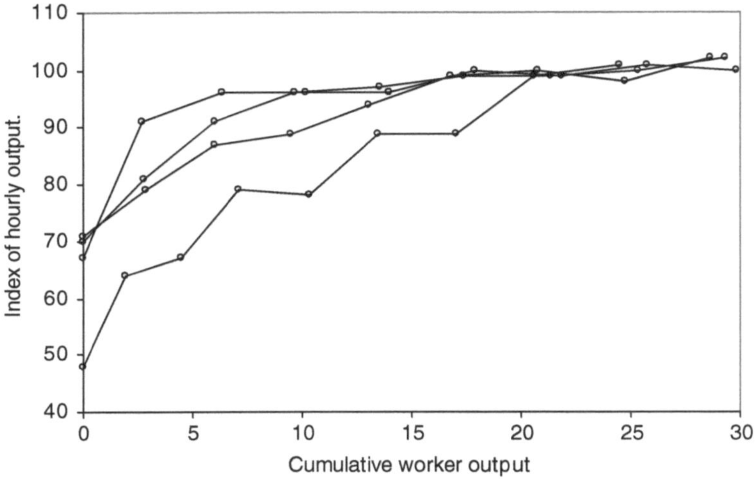

Figure 6. Average worker productivity (index) of new workers on four tasks at a munitions factory, plotted against average cumulative output per worker. Observations are recorded weekly, at the points indicated by circles. Source data: Jovanovic and Nyarko (1995, Table A-4).

3.3. *Case study evidence*

A second approach that avoids the limitations of studies based on large sample evidence involves case studies, typically of individual plants. Such case studies are potential sources of the detailed information that is missing from large-sample studies and that might provide rich insights into the sources of a firm's movement along the progress curve. Case studies become necessary here because the construction of data on these omitted sources of growth is extremely time-consuming. In this subsection, I describe two case studies in some detail; they are interesting in their own right, but they also illustrate two useful points. First, case studies frequently suggest that large sample studies are likely to mislead because much of what might be construed as passive learning is in fact the result of a variety of sometimes complex forces. Second, as will become equally evident, the very complexity of the forces identified in these case studies, while qualitatively revealing about the sources of growth, often make it difficult to measure the contribution of passive learning.

3.3.1. *Omitted variables*

The most obvious danger of large-sample studies, of course, is that measures of experience are correlated with variables known to be associated with rising labor productivity but that are simply not available. Their omission inevitably leads us to overstate the importance of passive learning (cf. Rosenberg, 1976).[23] For example, Thompson (2001) points out that earlier studies of the Liberty

[23] In much the same way as Abramovitz (1956, p. 10) urged caution in interpreting the Solow residual, strong measured passive learning effects may in fact be a measure of our ignorance.

shipbuilding program, which did not have access to data on the capital stock, constructed a crude proxy for capital that was essentially constant over time. The inaccuracy of this proxy is dramatically illustrated by the photographs in Figure 7. Thompson recovered capital stock data from the National Archives for 6 of the 13 Liberty shipyards and concluded that, for these yards, at least half of the

Figure 7. Top: The first Liberty ship keel being laid at Todd-Houston, May 1942. Because of the urgent need for rapid delivery, production of vessels began long before the yard was completed and all capital installed. Bottom: A fully operational shipyard, 2 years later. Source: Lane (1951). Originals in Records of the Historian's Office, Records of the US Maritime Commission, RG178, National Archives.

increase in output per worker was accounted for by capital deepening.[24] In a similar vein, Mishina (1999) undertook a closer look at Alchian's sample of aircraft factories, concluding that, *inter alia*, capital investments were a significant source of labor productivity growth in the production of the flying fortress bomber.

A particularly interesting case study by Sinclair et al. (2000) is revealing about the efforts sometimes needed to construct the necessary data. They investigated the sources of cost reductions for specialty chemicals manufactured by a Fortune 500 company. The company produced over one thousand different chemicals, but only a few batches of many of these were produced during their 30-month sampling period. Thus, their analysis focused on cost reductions for 99 chemicals that were each produced in at least 10 batches during the sample period. For these chemicals, Sinclair et al. had privileged access to a wealth of information, including batch-specific manufacturing costs and output, and (most remarkably) *chemical-specific* R&D expenditures. Equally important, they had access to personnel and to company records from which they were able to develop a sophisticated understanding of the firm's operational practices.

Sinclair et al. began by estimating a learning curve of the form $c_{ij} = \alpha y_{ij}^{-\gamma} t_{ij}^{-\beta_j} e^{u_{ij}}$, where c_{ij} is the unit manufacturing cost for the ith batch of chemical j, y_{ij} is the quantity produced in batch ij, and experience, t_{ij}, is measured by the time elapsed since the chemical was first produced. Column (1) of Table 1 reports the distribution of estimates, $\hat{\beta}_j$, for the 99 learning parameters. The average is 0.48 but the range is wide, with as many as one-third of the estimates indicating declining productivity. Sinclair et al. were able to identify four mutually exclusive groups of chemicals: 7 chemicals that were "campaigned,"[25] 13 that were affected by a project to reduce the frequency with which chemicals were sampled during the production process,[26] 25 that were the subject of formal R&D efforts, and a residual 59 chemicals that did not fall into any of these categories. Columns (2)–(5) summarize the distributions of estimated learning rates for each of these groups. The contrast between the first three groups and the residual group is quite remarkable: the learning rates for the seven campaigned chemicals are all in the upper tail of the distribution, with an average $\hat{\beta}_j$ of 1.4%; almost all the chemicals in the two groups that were affected by R&D returned positive values for $\hat{\beta}_j$, with averages exceeding 1.0. In contrast, the $\hat{\beta}_j$ in the residual group are centered on 0, with an average of -0.1.

Sinclair et al. made use of data not usually available to outside researchers to establish the dominant role of R&D in cost reduction. Chemicals that were not the subject of R&D effort experienced on average no cost reduction. It is of course possible that production experience revealed which chemicals had problems that might be addressed by R&D. But, Sinclair et al. noted, requests for process R&D on a particular chemical came from only two sources, neither of which were informed by production experience. The first was in response to an inability to meet industry specifications for the chemical.

[24] Bell and Scott-Kemmis (1990), Thompson (2001), and Thornton and Thompson (2001) catalog further omitted variables for which data are still unavailable.

[25] During the sample period, a large-volume product was launched that required the largest reactor. Seven chemicals were as a result displaced to smaller reactors. To minimize the effect on costs, each of these chemicals was produced in consecutive batches in the same reactor so that, *inter alia*, the small reactors would not need cleaning between batches. As a result, after controlling for the change in batch size, unit costs fell as a result of the displacement.

[26] A team was formed to study for each chemical which stages of the production process always seemed to run smoothly, and therefore did not need sampling. Thirteen products saw the number of samples reduced, and as a result registered sharp reductions in sampling costs.

Table 1
Distribution of OLS estimates of learning parameter for 99 specialty chemicals

	(1)	(2)	(3)	(4)	(5)
Range	All 99 products	7 campaigned products	13 affected by sampling reduction	25 products subject to R&D	Remaining 59 products
$\hat{\beta} < -1.4$	4	0	0	0	4
$-1.4 < \hat{\beta} < -1.0$	5	0	0	0	5
$-1.0 < \hat{\beta} < -0.6$	2	0	0	0	2
$-0.6 < \hat{\beta} < -0.2$	13	0	0	2	11
$-0.2 < \hat{\beta} < 0.2$	14	0	1	1	12
$0.2 < \hat{\beta} < 0.6$	21	0	4	5	12
$0.6 < \hat{\beta} < 1.0$	11	2	0	5	4
$1.0 < \hat{\beta} < 1.4$	10	2	4	3	1
$1.4 < \hat{\beta}$	19	3	4	9	3
Mean	0.48	1.41	1.04	1.20	−0.11

Source: Sinclair et al. (2000, Table 1).

The second, and more common, source was marketing and sales, which requested R&D after it identified a large potential demand if production could be scaled up and unit costs reduced. Finally, Sinclair et al. observed that if expected future demand conditions R&D expenditure and future demand is correlated with past demand, then past cumulative output will be negatively correlated with cost if data are not available to control adequately for costly R&D effort.

3.3.2. Institutional complexities: The upper weave room of Lawrence Company Mill No. 2, Lowell, MA

David (1973) documents an apparently clean example of passive learning in the Lawrence Manufacturing Company Mill No. 2, an integrated textile mills established in Lowell, MA, in 1834. David is careful to show that there was essentially no capital investment during 20 years that followed the founding of the mill. In particular, every loom that had been installed in 1834 was still in operation in 1856. Nonetheless, output per worker rose by an average 2% per year during this period. Recalling the Horndal Steel mill brought to economists' attention by Arrow (1962), David concluded that

> *"the evidence ... provides sufficient cause for American historians to insist that Horndal share with Lowell the honor ... in giving its name to the productivity effects of learning by doing in the context of a fixed industrial facility." (David, 1973, p. 142)*

The data available to David (from McGouldrick, 1968) consisted of plant level data on annual unit production costs. Using much more detailed records that survive in the Baker Library at Harvard, Lazonick and Brush (1985) also document a marked rise in output per worker. However, they reach more nuanced conclusions about its cause, in which passive learning plays only a modest role.

Lazonick and Brush's conclusions are driven by two significant changes in the composition of the mill's labor force between the late 1830s and the late 1850s. In the 1830s the labor force in the mill consisted primarily of "Yankee farm girls," who lived in boarding houses under paternalistic contractual arrangements with the mill. The farm girls were literate, but two characteristics limited their productivity. First, they tended to have little experience, it being the norm to abandon work in the mills upon marriage. Second, they frequently did not work in the mills during the summer, either returning to the farm to help during a busy time of year or taking summer teaching jobs.[27] Both characteristics limited the extent to which the farm girls could learn from experience. But they also limited the extent to which the mill's managers could extract effort from them. If work at the mill became too onerous, most employees had the option of returning to the farm.[28]

In the late 1830s the supply of farm girls began to fall behind demand. The number of mills in Lowell doubled between 1835 and 1847. At the same time the New England farming population was declining, both as a proportion of the labor force and in absolute numbers. Offsetting these changes, the population of Lowell rose markedly, primarily due to an influx of native-born families. The changing labor force, which remained predominantly female, created a more experienced labor force, thereby raising productivity.[29] However, Lazonick and Brush argue, managers remained constrained in their ability to extract greater effort, because native-born male heads of household were generally able to earn sufficient wages to support a family, and female workers continued to abandon the mills upon marriage.

In the mid-1840s, a second transformation of the labor force began, with an influx to New England of mostly Irish-born immigrants. The immigrants were mostly illiterate, and initially inexperienced in textiles. However, they had fewer outside options available to them. Irish heads of households could not earn a subsistence wage alone, and were unable to object to changes in work rules that intensified their work effort. Moreover, the existence of a "reserve army" of Irish workers made it increasingly difficult for native-born employees to resist intensification of effort, and rapidly forced them out of the mills. As a result the fraction of the labor force in Mill No. 2 that was not Irish declined from about 93% in 1845 to only 35% a decade later.

Thus, Lazonick and Brush argue, increases in output per worker were the result of two distinct processes. The first, until the mid 1840s, was primarily due to individual LBD. However, it required a compositional change of the labor force that raised the average experience level of the workforce for this individual learning to translate into rising productivity at the plant level. The second process was an increase in the intensification of effort made possible by the second demographic change in Lowell. Lazonick and Brush note two especially important pieces of evidence for their story. First, despite continuously rising labor productivity, real wages rose only until the mid-1840s. After that date, until the end of the sample period, they fell markedly. Second, a direct intensification of work effort is observed in what has been termed the "stretch-out" at Lowell. Between 1835 and 1842, most weavers were assigned two looms. In 1842, however, the number was raised to three, and in 1851 to four. The number of overseers in the mill, charged both with supervising the workers and intervening when

[27] In 1839–1840, 93% of summer teaching jobs in Massachusetts were held by women, compared with only 33% during the winter months.

[28] One could frame the language in terms of exploitation of labor or, more palatably for economists, in terms of the effect of the value of the outside option on equilibrium effort (e.g., Shapiro and Stiglitz, 1984).

[29] Lazonick and Brush do not report the trend in experience. However, evidence is available in Bessen (2003, Figure 1), which shows that the fraction of new hires at Lawrence Mill No. 2 who had previous experience in other mills rose from around 10% when the mill was opened, to around 50% in the mid 1840s.

there were problems with the machines, did not fall. As a result, effective monitoring of work effort increased.

Lazonick and Brush attempt to decompose the contributions to plant productivity of, *inter alia*, individual learning and effort intensification. The variables underlying these two contributions are not independent, so only a range could be provided. They concluded that between 4% and 14% of the variance of productivity can be explained by learning effects, while between 11% and 23% can be explained by effort intensification effects. Thus, they concluded,

> *"[the] results suggest that the production-relations hypothesis should be given at least as much attention as the learning by doing hypothesis in research into the 'Horndal effect'. There is more to the process of labor productivity growth than the technical development of inputs. Social influences on productivity growth must be considered as well." (Lazonick and Brush, 1985, p. 83)*

The story does not end quite here. Bessen (2003) revisited learning at Mill No. 2 yet again:

> *"Lazonick and Brush do not attempt to develop a complete picture of employers' motivations for these changes [in effort requirements] ... Employers could have hired allegedly docile Irish and 'low class' girls during the early decades, but did not .... More significantly ... the timing of this story is off." (Lazonick and Brush, 1985, p. 83)*

Bessen notes that the stretch-out from two to three looms per weaver occurred in 1842, *before* the influx of Irish immigrants had taken on significant proportions, but *after* the arrival in Lowell of significant numbers of native-born permanent residents. Bessen argues that the decision to stretch out the workers must have been driven by the greater work experience of the Yankee permanent residents. To support this claim, Bessen shows that workers assigned to just two looms learned more quickly than those assigned to three or four, although the latter eventually became more productive. Initial productivity for those working on two looms was about 25% of terminal productivity, and it took about 6 months to attain terminal productivity. For those working on three or four looms, initial productivity was less than 20% of terminal productivity, which took a year to attain. The profitability of the stretch-out therefore depended upon the labor turnover rate: workers must have been expected to remain in the job long enough to recoup the greater initial investment in human capital associated with assignment to more than two looms. Bessen calculates the profitability of the stretch-out directly as a function of the turnover rate: it was profitable in 1842, he concludes, but not in 1834.

Both studies agree that the transition from Yankee farm girls to Yankee permanent residents raised productivity because it increased average work experience in a setting in which LBD was important. Bessen continues this explanation into the second demographic transformation, while Lazonick and Brush turn to an explanation based on effort intensification. The timing of the first stretch-out seems to favor Bessen's story, but there was a second stretch-out in 1851 that, alongside the decline in real wages after the late 1840s, is consistent with Lazonick and Brush. Perhaps we should wait for yet another visit to the data from Mill No. 2 in order to decide between these stories. But regardless of the outcome of further research, one lesson from this case study is clear. Except perhaps in the earliest period after the founding of the Mill, even strong individual LBD cannot explain progress at the plant level without the broad process of social change that changed the composition of the labor force.

4. Models of passive learning

Although a variety of models that *contain* passive learning have been discussed, I have yet to describe any *theory* of learning. In this section, I discuss theories under two rubrics. Section 4.1 describes a model of learning that induces the familiar power rule. Section 4.2 describes models generating bounded learning.

4.1. Models with unbounded learning

Muth (1986) asserts that March and Simon (1958) were the first economists to develop a theory of organizational learning. They were followed by contributions from Crossman (1959), Levy (1965), Sahal (1979), and Roberts (1983). Muth succinctly describes each of them,[30] but then dismisses each of them in short order either because they fail to induce the power rule for passive learning, or because the way they do so "assumes the desired answer" (p. 952).

Muth (1986) constructs a theory of learning that leads to the power rule from somewhat deeper assumptions. He models a process of random sampling from a distribution of cost draws, where the current unit cost is the minimum of the draws. Let $F(c)$ denote the distribution of costs, let $F(c_0) = 0$, and let $\underline{c}_{(n)}$ denote the minimum of n draws from this sample. When sampling is random, the distribution of $\underline{c}_{(n)}$ is

$$G(\underline{c}_{(n)}) = 1 - [1 - F(c)]^n. \tag{21}$$

Let $u(c) = nF(c)$. For any draw, $\text{pr}\{c \leq x\} = \text{pr}\{u(c) \leq u(x)\}$, and so $\text{pr}\{\underline{c}_{(n)} \leq x\} = \text{pr}\{\min u(c) \leq u(x)\}$. It then follows that

$$G(\underline{c}_{(n)}) = 1 - \text{pr}\left\{ \min F(c) \geq \frac{u(\underline{c}_{(n)})}{n} \right\} = 1 - \left(1 - \frac{u(\underline{c}_{(n)})}{n}\right)^n. \tag{22}$$

The second line makes use of the fact that $F(c)$ is a distribution and therefore is uniformly distributed on [0, 1]. The large-sample approximation to Equation (22) is

$$G(\underline{c}_{(n)}) \approx 1 - \lim_{n\to\infty} \left(1 - \frac{u(\underline{c}_{(n)})}{n}\right)^n = 1 - \mathrm{e}^{-nF(\underline{c}_{(n)})}. \tag{23}$$

Muth then assumes that, at least near the left tail of the distribution, $F(c)$ can be approximated by the power function, $\alpha(c - c_0)^{1/\lambda}$. Then Equation (23) is a Weibull distribution:

[30] So I shall not describe them here.

$$G(\underline{c}_{(n)}) \approx 1 - \mathrm{e}^{-[(\alpha n)^{\lambda}(\underline{c}_{(n)} - c_0)]^{\lambda}}, \tag{24}$$

with expected value

$$E[\underline{c}_{(n)}] \approx c_0 + \Gamma\left(\frac{1+k}{k}\right)(\alpha n)^{\lambda}. \tag{25}$$

If one further assumes that the lower bound to cost is zero, and that the rate of sampling is proportional to the current rate of production, Equation (25) can be written in the form of the power rule,

$$E[\underline{c}_{(y)}] \approx Ay(t)^{-\lambda}. \tag{26}$$

The assumption that the lower tail of the distribution can be adequately approximated by a power function is less onerous than it may seem: in the context of applications to extreme value distributions, this assumption must hold as long as $F(c)$ has no mass point at zero.[31] Perhaps more onerous is the assumption that sampling is random out of $F(c)$: as costs decline, firms spend more and more time observing new ways of doing things that are far worse than the current state of the art. One might suppose instead that firms observe variations on how they are doing things today, in which case the distribution $F(c)$ should itself evolve over time. For example, firms may be able to eliminate from their search space any costs greater than, say, $\underline{c} + v$. Then, the sampling space has the distribution $F(\underline{c}+v)^{-1}F(c)$ for $c \in [0, \underline{c}+v]$, and zero otherwise. In this case, the auxiliary assumptions necessary to induce the power rule for passive learning become quite contrived.[32]

4.2. Models with bounded learning

4.2.1. Two simple models

One of the reasons that the power learning rule has remained popular has been the ease with which it can be empirically implemented. But mathematical psychologists long ago developed some equally straightforward models of bounded learning. Two specifications have been especially popular (Restle and Greeno, 1970, chapter 1). One is the replacement model, which models productivity, q_t, as

[31] In fact, the limiting distribution of the minimum [maximum] of a large set of independent draws is Weibull for any distribution that has a finite lower [upper] bound and does not having a mass point at that bound (e.g., Galambos, 1978).

[32] In Roberts (1983), agents are able to eliminate parts of the sample space. His model is an adaptation of the traveling salesman problem, in which it becomes progressively easier to eliminate from consideration whole subsets of the search space because they are known without inquiry to contain only routes that are longer than the fastest route currently known. Roberts applies his model to machine efficiency and devises a set of auxiliary assumptions that leads to the power rule for learning. Muth, however, objects that Roberts' assumptions might be reasonable for machine efficiency, but they cannot be justified in a model of manufacturing costs.

$$q_t = a - (a - b)(1 - \lambda)^{y_t - 1}, \tag{27}$$

where a, b, and λ are positive parameters and $y = 1, 2, \ldots,$ is cumulative output. The second is the accumulation model, which takes the form

$$q_t = \frac{b + a\lambda(y_t - 1)}{1 + \lambda(y_t - 1)}. \tag{28}$$

Both models predict an initial productivity of b, and a terminal productivity of a. The parameter λ governs the rate of learning. In the replacement model, $\lambda = (q_{i+1} - q_i)/(a - q_i)$, which is the change in productivity expressed as a fraction of the amount left to learn. The replacement model is a little more complicated, as $\lambda = (q_{i+1} - q_i)/(a - q_i - i(q_{i+1} - q_i))$ contains an extra term in the denominator.

The names for the two specifications are derived from two urn problems that generate these functions. There are two urns, A and B. Urn A contains a fixed number of marbles, of which a fraction b is red and a fraction $1 - b$ is white. On each trial, one marble is drawn from A. If it is red a 'correct' response is recorded. Urn B contains an infinite number of marbles, a fraction $a > b$ of which are red. In the replacement model, a fraction λ of the marbles currently in A are replaced after each draw by marbles drawn from B. Let q_t denote the probability of a correct response. Then $E[q_t|y_t]$ is given by Equation (27). In the accumulation model, a constant number of marbles equal to a fraction λ of the marbles initially in A are transferred from B to A. For this model, $E[q_t|y_t]$ is given by Equation (28).

4.2.2. *Bayesian models*

The urn problems are a rather abstract way of thinking about learning. Although they capture the idea that learning can be thought of as either the replacement of incorrect ways of doing tasks with correct ways of doing them, or as the accumulation of new skills on top of existing skills, the analogy has proved too loose for economists. Instead, economists have preferred to develop models based explicitly on Bayesian learning.

Bayesian models of learning take two main forms: learning about one's time-invariant ability to carry out a task, and learning how to accomplish the task. Both models may be applied to individual LBD, and to passive learning at the organizational level. Jovanovic (1979, 1982) pioneered the development of learning about ability at both individual and firm levels. Jovanovic (1979) studies the implications for job turnover of individuals learning about their ability to undertake a firm-specific task; individuals who discover they are not good at their current job leave to pursue other activities. This type of model is now commonly referred to as a model of learning about match quality. Jovanovic (1982) studies the implications for industry dynamics of firms learning about their production costs, which are stochastic but have time-invariant means; firms that learn they are low-cost producers expand, while firms that discover they are high-cost contract before exiting entirely.[33] Jovanovic, along with Yaw Nyarko, was

[33] Because firms are equally likely to receive bad news as good news, learning about firm ability need not be associated with rising productivity or output. However, selection removes the high-cost firms so that surviving firms are those that have grown in the past. Moreover, firms may still grow on average from learning about individual task ability, because they may be able to reallocate workers to tasks for which they are better suited (Prescott and Visscher, 1980) or to replace low-ability workers (Jovanovic, 1979).

also the first to apply a Bayesian approach to learning about how to accomplish a task (Jovanovic and Nyarko, 1995, 1996).

Let the current output of an agent with t periods of experience be given by

$$x_t = A_t h_t, \tag{29}$$

where A_t reflects the match quality, and h_t reflects task learning. The term A_t is given by

$$A_t = (\mu + u_t). \tag{30}$$

In each period, u_t is known to be a random draw from $N(0, \sigma_u^2)$, but μ is initially only known to be a random draw from the prior distribution $N(0, \sigma_\mu^2)$. One problem for the agent (and his employer) is to learn the value of μ. The task-learning term is given by

$$h_t = (1 - (\theta - z_t + \varepsilon_t)^2), \tag{31}$$

which consists of a target, θ, a decision in period t, z_t, and noise, ε_t. The target is fixed across periods, but initially is only known to be a draw from $N(0, \sigma_\theta^2)$. The noise in each period is known to be an i.i.d. random draw from $N(0, \sigma_\varepsilon^2)$. The second and third challenges for the agent are to learn θ while choosing in each period the optimal decision z_t.[34]

If $\sigma_\mu^2 = 0$, the match quality is known, and Equation (29) reduces to the pure task-learning model. Similarly, if $\sigma_\theta^2 = 0$, the target is known and Equation (29) is the pure match quality model. Each of these models is straightforward to analyze under the standard assumption that the agent learns by observing the sequence $\{x_\tau\}_{\tau=1}^{t}$. The combined model, however, is not easy to analyze under this same assumption. Nagypál (2007) greatly simplifies matters by assuming instead that in each period the agent observes A_t and h_t separately. This simplifying assumption will be maintained here.

Before proceeding, two features of the model that also contribute greatly to tractability merit comment. First, one signal is assumed to be observed in each period regardless of the level of output. There is, as a result, no incentive to deviate from the static optimum choice for z_t; this is clearly a departure from the usual assumption that the amount of information obtained depends on the rate of output. Second, all distributions are Normal with known variances. This assumption conveniently ensures that posterior variances depend on the number of signals received but not the realizations of those signals. Moreover, the variance is decreasing in the number of signals observed, so learning can be said to be monotonic.[35]

Consider first the task-learning function, h_t. The decision z_t is obviously known, so observing h_t is equivalent to observing a normally distributed signal, $s_t = \theta + \varepsilon_t$. Then, the agent's posterior variance of the target after t periods is (e.g., DeGroot, 1970, chapter 9)

$$E_{t-1}[\theta - E_{t-1}(\theta)]^2 = \frac{\sigma_\theta^2 \sigma_\varepsilon^2}{(t-1)\sigma_\theta^2 + \sigma_\varepsilon^2}. \tag{32}$$

[34] Distributions have been chosen to minimize notation. As a result, the model admits negative output; this could easily be corrected (without adding insight) by assuming output is an appropriate monotonic transformation of x_t.

[35] Pakes and Ericson (1998) point out that under alternative distributional assumptions agents may fail to learn the unknown parameters even after receiving an infinite number of signals. Thompson (2000) develops a model of growth with an alternative distributional assumption that does ensure learning.

The agent's decision in period t, is to set $z_t = E_{t-1}[\theta]$. It then follows that

$$E_{t-1}[h_t] = 1 - \frac{\sigma_\theta^2 \sigma_\varepsilon^2}{(t-1)\sigma_\theta^2 + \sigma_\varepsilon^2} - \sigma_\varepsilon^2. \tag{33}$$

Turning now to A_t, it is easy to see that

$$E_{t-1}[A_t] = \frac{\sigma_\mu^2 \sum_{\tau=1}^{t-1} A_\tau}{(t-1)\sigma_\mu^2 + \sigma_u^2}, \tag{34}$$

and hence that

$$E_{t-1}[x_t] = \left(\frac{\sigma_\mu^2 \sum_{\tau=1}^{t-1} A_\tau}{(t-1)\sigma_\mu^2 + \sigma_u^2}\right)\left(1 - \frac{\sigma_\theta^2 \sigma_\varepsilon^2}{(t-1)\sigma_\theta^2 + \sigma_\varepsilon^2} - \sigma_\varepsilon^2\right). \tag{35}$$

Equation (35) has endpoints of $E_0[x_1] = 0$ and $\lim_{t\to\infty} E_{t-1}[x_t] = \mu(1 - \sigma_\varepsilon^2)$, so the function is clearly bounded. Because the prior mean of μ is zero, Equation (35) yields (stochastically) positively sloped functions of t only when the realized value of μ exceeds zero. Taking averages over all individuals with a given strictly positive match quality, $E_{t-1}[x_t]$ is strictly convex, although for any individual both $E_{t-1}[x_t]$ and x_t are only stochastically increasing.

4.2.3. *Empirical discrimination*

It can be very difficult to distinguish between the two types of learning in field data. For example, assume there is no uncertainty about the target, so the task learning component is simply $(1 - \sigma_\varepsilon^2)$. Then, taking expectations of Equation (35) over all agents with match quality μ yields

$$E[E_{t-1}[x_t]|\mu] = \frac{\mu\sigma_\mu^2 (t-1)(1 - \sigma_\varepsilon^2)}{(t-1)\sigma_\mu^2 + \sigma_u^2}. \tag{36}$$

If instead there is no uncertainty about match quality (i.e., it is known to be μ), but there is task learning, then

$$E_{t-1}[x_t]|_\mu = \frac{\mu\sigma_\theta^2 (t-1)(1 - \sigma_\varepsilon^2)}{(t-1)\sigma_\theta^2 + \sigma_\varepsilon^2} + \frac{\mu\sigma_\varepsilon^2 (1 - \sigma_\theta^2 - \sigma_\varepsilon^2)}{(t-1)\sigma_\theta^2 + \sigma_\varepsilon^2} \tag{37}$$

Equation (36) has an initial value of zero. Imposing the same initial value on Equation (37) requires that $\sigma_\theta^2 + \sigma_\varepsilon^2 = 1$. Then Equation (37) simplifies to Equation (36), with σ_u^2 and σ_μ^2 replaced by σ_ε^2 and σ_θ^2, so both learning models yield much the same expression. Appropriate changes in $\sigma_\theta^2 + \sigma_\varepsilon^2$ allow for different prior means of μ. Thus, average behavior in the pure match quality model for individuals with any given μ can be replicated by the average behavior of a particular parameterization of the task-learning model.

To distinguish between the models, one must therefore move beyond their average behavior. Farber (1994) has done so by exploiting the fact that the two models have different implications for the hazard of job separation. When there is only learning about match quality, a poorly matched worker–firm pair waits to observe a number of signals before concluding there is sufficient evidence to warrant separation. After sufficient passage of time, only high-quality matches survive. Thus, the hazard of job separation rises before it falls. In contrast, when there is only (positive) task learning, the hazard falls monotonically with job tenure. In this case, rising expected productivity insulates the worker from exogenous shocks that make continued employment less attractive (either to the firm or to the worker). Farber uses a large sample from the National Longitudinal Survey of Youth to study the hazard of job separation as a function of job tenure. He reports that, consistent with the match quality model, the hazard rises before it falls, reaching a peak at about 3 months. Thus, Farber's analysis suggests that learning about match quality is dominant in the first few months of employment, although either type of learning may be more important thereafter.

Nagypál (2007) instead focuses on the effect of firm-specific price shocks on turnover. In the task learning model, negative price shocks primarily affect workers with limited tenure for reasons already explained. In contrast, negative price shocks adversely affect workers of all tenures in the match quality model. This is the result of two off-setting forces. On the one hand, selection implies that the match quality of workers with short tenure is on average lower and, as in the task learning model, this makes them susceptible to adverse shocks. On the other hand, workers with long tenure have a smaller option value of continuing the match because there is little for them to learn about its quality. This makes workers with long tenure susceptible to adverse shocks. Nagypál uses these insights to estimate the parameters of a structural model that embeds both types of learning. She finds, in contrast to Farber's results, that task learning is important in the first few months, while learning about match quality dominates at longer tenures. Indeed, task learning is all but complete after 6 months or so, but learning about match quality persists for up to 10 years. Nonetheless, Nagypál's point estimates suggest that the magnitude of task learning is much the greater of the two: about 80% of the estimated increase in average output is attributable to task learning.

These two studies appear to the only ones to date that attempt to discriminate between match quality and task learning. Their contrasting results, and limitations of both studies, should induce caution. Nagypál objects that estimates of the hazard during the first few months of tenure are especially susceptible to measurement error, so Farber's inferences are unreliable. At the same time, Nagypál's estimates of the rate of task learning are extremely imprecise (and cannot be distinguished statistically from zero). Thus, there appears room for more work on testing these theories of learning.

5. Passive learning and aggregate growth

This section addresses the consequences of passive learning for aggregate growth. Section 5.1 presents a simple one-sector model of growth driven by passive learning that contains some of the key features in the models developed by Arrow (1962) and Romer (1986a). These models predict that per capita income growth is positively related to the size of the population or to its growth rate, neither of which is consistent with evidence. Moreover, both models require that passive learning be unbounded, a feature inconsistent with empirical evidence. Section 5.2 therefore restricts attention to hybrid models in which learning within any given technology is bounded, but new technologies are introduced as a result of some mechanism distinct from passive learning.

5.1. A simple model

Let aggregate output be given by $Y(t) = A(t)K(t)^{\alpha}L(t)^{1-\alpha}$, where A is knowledge, K is aggregate physical capital and L is labor. Labor grows exogenously at the rate n, K evolves according to $\dot{K}(t) = sY(t)$, where s is the exogenous saving rate (there is no depreciation), and knowledge advances as a result of passive learning. In the conventional formulation, knowledge advances at a rate that depends upon cumulative output, but it is somewhat easier to follow Arrow (1962) and Romer (1986a) and link knowledge to cumulative investment, $A(t) = \beta K(t)^{\lambda}$. Dropping time arguments for compactness, constant returns to scale allows us to write the model in intensive form. Letting lower case letters denote per capita variables, per capita income is

$$y = Ak^{\alpha}, \tag{38}$$

and the equations of motion are

$$\frac{\dot{k}}{k} = sAk^{\alpha-1} - n \tag{39}$$

and

$$\frac{\dot{A}}{A} = \lambda n + \lambda\frac{\dot{k}}{k}. \tag{40}$$

Hence, per capita income grows at the rate

$$\frac{\dot{y}}{y} = \lambda n + (\lambda + \alpha)(sAk^{\alpha-1} - n), \tag{41}$$

so a steady state with constant per capita income growth requires that the growth rates of knowledge and capital are related by

$$\frac{\dot{A}}{A} = (1 - \alpha)\frac{\dot{k}}{k}. \tag{42}$$

This in turn implies that per capita income and the capital–labor ratio grow at the same rate. Equations (40) and (42) yield

$$\lambda n = (1 - \alpha - \lambda)\frac{\dot{k}}{k}. \tag{43}$$

What Equation (43) implies for long-run growth depends upon the auxiliary assumptions we choose to make. Consider first the choices made in Romer's (1986a) launch of the modern theory of endogenous growth. He sets $n = 0$, in which case steady state growth can only be sustained under the knife-edge assumption of *exactly* constant returns to scale in accumulable factors, $\alpha + \lambda = 1$.[36] If $\alpha + \lambda$ exceeds unity, even by a small margin, growth accelerates without limit and, moreover, infinite income is attained within a finite amount of time; if $\alpha + \lambda < 1$ the steady-state is one of stagnation. The knife-edge assumption demands that the learning parameter, λ, be exactly equal to the elasticity of aggregate output with respect to labor, $1 - \alpha$. As Solow (1994, p. 51) observed, "you would have to believe in the tooth fairy to expect that kind of luck."

Even with that kind of luck, Romer's model yields an unpalatable scale effect. When $n = 0$ and $\lambda = 1 - \alpha$, per capita income growth is

$$\frac{\dot{y}}{y} = s\beta K^{\lambda} K^{\alpha-1} L^{1-\alpha} = sL^{1-\alpha}. \tag{44}$$

The long-run growth rate is sensitive to policies that induce a permanent change in the saving rate (and in this sense the growth rate is said to be endogenous). Unfortunately, Equation (44) also implies that that "a country such as India should have an enormous growth advantage over a country such as Singapore" (Lucas, 1993, p. 263). There is, of course, no empirical support for this latter conclusion.

Arrow (1962) had assumed $\alpha + \lambda < 1$, which allows for positive population growth. Then, from Equation (43) we have

$$\frac{\dot{k}}{k} = \frac{\dot{y}}{y} = \frac{\lambda n}{1 - \alpha - \lambda}. \tag{45}$$

Per capita income growth does not depend on the saving rate (so Arrow's assumption produces a model without endogenous growth), but the model has the virtue that growth is no longer increasing in the scale of the economy. However, Arrow's formulation has the almost equally unpalatable implication that income growth is proportional to the rate of population growth. This prediction also finds no empirical support, at least in modern data (cf. Mankiw et al., 1992, Tables IV and V).[37]

This analysis is following a path that has been well-trodden by specialists in the "new growth theory." The scale effect inherent in Romer's specification of the passive learning model is also present in early models of R&D-driven endogenous growth (Aghion and Howitt, 1992; Grossman and Helpman, 1991; Romer, 1990) and was the subject of detailed criticism by Jones (1995a). Jones' (1995b) approach to eliminate the scale effect yielded a new class of R&D-driven growth models for which Jones coined the moniker "semiendogenous growth." These models turn out to be a translation into R&D of Arrow's passive learning model, and they too yield long-run per capita growth proportional to the rate of population growth. An alternative approach (Aghion and Howitt, 1998, chapter 12; Dinopoulos and Thompson, 1998; Peretto, 1998; Young, 1998) eliminates the scale effect while preserving the

[36] Romer's exposition is nominally more general than this. He notes that a steady state can exist in the presence of increasing returns as long as the rate at which knowledge can be accumulated has an upper bound (a formal proof is given in Romer, 1986b). Of course, the upper bound defines a point, below which there may be increasing returns to scale and above which the marginal product of experience is zero.

[37] Kremer (1993) offers some suggestive data linking population growth and income growth in very long-run data, but data preceding the demographic transition reflect in part Malthusian effects of income on population growth.

endogeneity of growth, but it does so at the price of introducing a second knife-edge assumption in addition to the one already in Romer's models.

The treatment of scale effects in endogenous growth models has consumed an inordinate amount of space over the last 15 years or so. However, models in which passive learning is the engine of growth figure nowhere in this literature. This is not because passive learning presents insurmountable technical obstacles (to the contrary, translating to passive learning the scale-free models of R&D-driven endogenous growth seems almost a trivial exercise), but rather because passive learning models took a different tack just about the time the first-generation R&D models were appearing.

5.2. *Hybrid models*

Empirical evidence has established that productivity gains from passive learning eventually dry up absent new sources of stimulation. As a result, a *plausible* model of long-run growth likely cannot be constructed out of a one-sector model with passive learning as the sole engine of growth. This realization led to the development of hybrid models that combine passive learning with the introduction of new, superior, vintages of technology. Within any vintage passive learning takes place, but at a rate that diminishes as experience is gained with that particular technology.

Hybrid models are of three kinds. First, there are models in which superior technologies are always available, but their adoption first requires the accumulation of experience in inferior technologies (e.g., Jovanovic and Nyarko, 1996; Parente, 1994; Stokey, 1988; Young, 1991). Second, there are models in which new technologies arrive at an exogenous rate (e.g., Chari and Hopenhayn, 1991). Third, there are models in which new technologies are developed through R&D (e.g., Stein, 1997; Young, 1993). Only in the first kind is passive learning the sole engine of growth; but such models fail to explain how superior technologies came to exist.[38] Models of the second kind more plausibly allow for sequential discovery of superior technologies, but they run the risk of leaving the engine of growth entirely unexplained.[39] Models of the third kind are most representative of what we may have in mind as a hybrid model.

How does the rate of passive learning influence the rate of aggregate growth in hybrid models? Almost any answer is correct, depending upon the auxiliary assumptions one chooses to make. An increase in the rate of learning may have no effect on long-run growth, it may increase it, or it may decrease it. Too much learning may lead to stagnation, and there may be stagnation that arises independently of the rate of learning. Passive learning may also induce clustering of innovations and, more generally, cyclical growth.

5.2.1. *Growth independent of the rate of learning*

I begin with a simple hybrid model of the second kind, developed by Lucas (1993). The model is one of a small open economy in which new goods are introduced continuously with respect to time at the constant rate γ. More recent vintages are superior in the sense that the world price of the newest goods

[38] But they are very useful for other questions. In particular, models of this kind have been used to assess the conditions under which firms will abandon a technology with which they have experience in favor of a superior, but unfamiliar, technology.

[39] Again, such models are very useful for other questions. For example, Chari and Hopenhayn (1991) explain which some firms choose to invest in inferior technologies while other (identical) firms adopt technologies at the frontier.

exceeds the price of the previous vintage by a constant proportion, e^{μ}, so the price of a good introduced at time v is $p(v) = e^{\mu\gamma v}$. The labor supply is normalized to unity, and output at time t of vintage v is given by the Ricardian technology:

$$x(v,t) = e^{\mu\gamma v} A(v,t)\phi(v,t), \tag{46}$$

where $\phi(v,t)$ is the fraction of the labor force employed in the production of vintage v. $A(v,t)$ advances as a result of within-vintage passive learning:

$$\dot{A}(v,t) = A(v,t)^{\lambda}\phi(v,t), \tag{47}$$

where $\lambda < 1$ is the learning parameter. Let $A(v,v) = \alpha > 1$ denote productivity at the time the good is introduced. From Equations (46) and (47), output of vintage v is given by

$$x(v,t) = e^{\mu\gamma v}\phi(v,t)\left[\alpha^{1-\lambda} + (1-\lambda)\int_v^t \phi(v,s)\,\mathrm{d}s\right]^{\lambda/(1-\lambda)} \tag{48}$$

Integrating over all vintages, aggregate output is given by

$$y(t) = \int_0^t e^{\mu\gamma v}\phi(v,t)\left[\alpha^{1-\lambda} + (1-\lambda)\int_v^t \phi(v,s)\,\mathrm{d}s\right]^{\lambda/(1-\lambda)} \mathrm{d}v. \tag{48a}$$

Assume that the distribution of labor over goods of different ages remains constant over time (as must be the case along a balanced growth path), and let $a = t - v$ denote the age of a good. Then the labor devoted to a good over its life, $\int_v^t \phi(v,s)\,\mathrm{d}s$ is the same as the labor devoted in the cross-section to all goods with age less than $t - v$. Let $\psi(a)$ denote employment on the good of age a, so that $\phi(v,t) = \psi(t-v)$, and let $\Psi(a)$ denote the corresponding distribution. Equation (48) can now be written as

$$y(t) = \int_0^t e^{\mu\gamma(t-a)}\phi(a)[\alpha^{1-\lambda} + (1-\lambda)\Psi(a)]^{\lambda/(1-\lambda)}\mathrm{d}a, \tag{49}$$

which yields the following aggregate rate of growth:

$$\frac{\dot{y}(t)}{y(t)} = \mu\gamma + \frac{\psi(t)[\alpha^{1-\lambda} + (1-\lambda)]}{\int_0^t e^{\mu\gamma(t-a)}\phi(a)[\alpha^{1-\lambda} + (1-\lambda)\Psi(a)]\,\mathrm{d}a}. \tag{50}$$

The integral term in Equation (50) is unbounded, so the asymptotic growth rate is

$$\lim_{t\to\infty}\frac{\dot{y}(t)}{y(t)} = \mu\gamma. \tag{51}$$

Growth is uniquely determined by the product of two exogenous parameters: the rate at which goods are introduced and the rate of increase across vintages in the value of goods. In particular the learning parameter, λ, has no bearing on the asymptotic growth rate. As Lucas points out, it has only a level effect in this simple model.

5.2.2. *Growth increasing in the rate of learning*

Lucas generalizes the simple model by allowing potential productivity on goods yet to be introduced to vary positively with experience gained in the production of older vintages. Assume that the initial productivity of a good introduced at time t, $A(t,t)$, depends positively on a weighted average of productivity on all goods previously introduced. The importance of prior productivity depends positively on a spillover parameter, $\theta < 1$, but older goods contribute less than recent goods at a rate determined by a decay parameter, $\delta > 0$:

$$A(t,t) = \theta \int_0^{\infty} \delta\gamma \, \mathrm{e}^{-\delta\gamma(t-v)} A(v,t) \, \mathrm{d}v. \tag{52}$$

Initial productivity is the weighted sum of current productivities on all goods produced since time zero, with weightings given by an exponential distribution over prior vintages with parameter $\delta\gamma$.

Substituting the solution for $A(v,t)$ from Equation (48) and setting $A(t,t) = \alpha$ yields

$$\alpha = \theta\delta\gamma \int_0^{\infty} \mathrm{e}^{-\delta\gamma a} [\alpha^{1-\lambda} + (1-\lambda)\Psi(a)]^{1/(1-\lambda)} \mathrm{d}a. \tag{53}$$

The implications of passive learning for long-run growth depend, yet again, on the auxiliary assumptions one makes. Equation (53) links initial productivity and the distribution of employment over vintages to the rate of product introduction, the spillover parameters, and the learning parameter. The model is clearly incompletely specified. Given the parameters of the model, initial productivity and the distribution of employment are presumably determined by equilibrium considerations, such as those explored in Stokey (1988), Young (1991), and Parente (1994). One might reasonably treat λ, θ, and δ as purely exogenous technological parameters, but it is less satisfactory to do the same for γ (and, probably, μ), which must depend upon innovative efforts undertaken *somewhere* in the world (e.g., Grossman and Helpman, 1991, chapters 11 and 12; Segerström et al., 1990; Stokey, 1991; Young, 1991).

Lucas leaves these elaborations to others, assuming that Equation (53) pins down γ as a function of the other, exogenous, parameters and functions. Noting that Equation (51) continues to define the asymptotic growth rate, if a solution to Equation (53) exists it satisfies

$$\frac{\dot{y}}{y} = \mu\gamma(\lambda, \delta, \theta, \Psi(a)), \tag{54}$$

where γ is increasing in λ and θ, and decreasing in δ. The rate of introduction of new goods is also greater if labor is concentrated in recent vintages, which confer greater spillover benefits to new goods than do older vintages. Thus, treating all parameters other than γ as given, learning spillovers enable the rate of passive learning to positively influence the long-run growth rate. It is easy to verify that the greater are the spillovers (i.e., the greater is θ or the smaller is δ), the greater is the influence of changes in the rate of learning on long-run growth.

Passive learning in the presence of spillovers raises long-run growth by inducing the economy to adopt new vintages more rapidly. This mechanism makes most sense if one assumes that new products are developed elsewhere, perhaps in advanced economies, and that the model applies only to developing economies some distance behind the technological frontier (and this is exactly the application Lucas focuses on). An alternative, also reasonable, assumption is that Equation (53) identifies α, with γ held fixed. But in this case, variations in the learning parameter have only level effects despite the presence of spillovers.

In advanced economies, the appropriate assumption likely lies between these two extremes, in the sense that both γ and α are endogenous. This is generally the case in models of the third kind, where the rate of innovation depends on the cost of R&D. In these models, however, a wide variety of outcomes are possible. Young (1993) studies the steady states of a model with domestic R&D and full learning spillovers across sectors. First, there is a steady state in which new products are invented and immediately enter into production. Second, when R&D costs are sufficiently low relative to the size of the market, a gap emerges between the time new products are invented and the time they enter into production; inventors wait until learning spillovers raise the productivity of the new good sufficiently to merit implementation. In both these equilibria, the growth rate is increasing in the rate of learning and decreasing in the cost of invention.

5.2.3. Stagnation independent of the rate of learning

Young's model has a third steady state, without innovation, which emerges when the cost of innovation is high relative to the size of the market. With the passage of time, learning on increasingly aged technologies ceases and so this steady state is one of stagnation. The learning rate does not figure in the conditions that determine the existence of the zero-growth steady state (which depends only on the size of the economy, the innovation cost, and the discount rate): if learning has been exhausted, its rate prior to exhaustion does not affect the present value of profits earned from sticking with the current technology. Young restricts his attention to steady-state analysis, so it is not known whether this zero-growth steady-state is attainable from arbitrary initial conditions, or from a set of initial conditions that is independent of the rate of learning.

5.2.4. Stagnation induced by learning

If there are insufficient learning spillovers across technologies, experience gained in learning can halt the adoption or development of new technologies altogether. A firm that has gained extensive experience on one good may find it more profitable to stick with the old technology than to adopt a new technology that would, with the passage of time, prove superior.

Lucas' model admits stagnation, but it is stagnation caused by the absence of productivity spillovers, not the presence of passive learning. To see this, note that Equation (53) yields a positive solution for γ if and only if

$$\theta > [1 + (1 - \lambda)\alpha^{\lambda-1}]^{1/\lambda-1}. \tag{55}$$

Intuitively, spillovers across vintages must be sufficiently large to maintain long-run growth. The right-hand side of Equation (55) has an upper bound at $\alpha/(\alpha + 1) < 1$, so under full spillovers (i.e., $\theta = 1$) growth is always positive. However, the right-hand side of Equation (55) is strictly *decreasing* in λ:

passive learning lowers the size of the spillovers necessary to sustain growth and makes stagnation less likely.[40]

Jovanovic and Nyarko (1996) have analyzed the economics of stagnation in some detail, using the single-agent task-learning model described in Section 4.2. They show that stagnation is more likely to occur if a firm has extensive experience with its current technology (as measured by the posterior variance of the current target), when spillovers across product generations are weak (as measured by the cross-product correlation in the targets) and when the difference between product generations in the terminal productivities is modest. The results are intuitive: the first characteristic raises the profitability of the current technology, while the second and third reduce the expected profitability of the new technology.

Jovanovic and Nyarko assume firms are myopic, comparing only the single-period payoff from sticking with the current technology and adopting the new one. For any given level of experience a high rate of learning raises the static payoff from sticking with the current technology more than it raises the one-period payoff from switching. In the limit when all experience is product-specific, the one-period payoff from switching is independent of the rate of learning. Thus, rapid learning unambiguously raises the probability of stagnation. However, myopia may not be an innocent assumption here, and this result may not hold when firms are forward-looking. While the value of sticking with the current technology is generally higher with rapid learning, so is the value of switching to a technology that is expected to be mastered quickly. Forward-looking firms will trade off these two consequences of rapid learning and the conditions under which one effect dominates remain unexplored.

5.2.5. Clusters and cycles induced by learning

Klenow (1998) considers the case of forward-looking firms, but focuses on the possibility of generating cycles. Myopic firms do not switch to a new technology until its initial productivity exceeds the current productivity of the old technology. Forward-looking firms note that switching at an earlier stage is a form of investment, enabling them to gain experience on a technology that will eventually be more productive. As a result, forward-looking firms switch to new technologies that initially yield lower productivity, generating cyclical productivity at the plant or firm level.[41]

Klenow's model is consistent with evidence that plants switching technologies initially have lower productivity (e.g., Cochran, 1960; Garg and Milliman, 1961; Yorukoglu, 1998), but for the mechanism to induce aggregate cycles, innovations must be coordinated in some way. In Shleifer (1986), innovations are coordinated because of aggregate demand externalities, but this mechanism for coordination induces countercyclical productivity. The development of a general purposes technology affecting multiple sectors may do the trick: Greenwood and Yorukoglu (1997), for example, have argued that the widespread adoption of information technology lay behind the productivity slowdown of the 1970s.

[40] Curiously, the possibility of stagnation does not depend upon the decay parameter, δ, or the distribution of employment across products, $\Psi(a)$, even though both affect the growth rate should it be positive.

[41] In Stein (1997), passive learning induces cycles through a rather different mechanism. Firm-specific learning makes it harder over time for potential entrants to invent and unseat the incumbent. Thus, potential entrants expend less effort on R&D when faced by a long-entrenched incumbent. If, eventually, the incumbent is replaced, the new incumbent is inexperienced and is more readily overturned by further innovations. As a result new potential entrants invest heavily in R&D, making rapid innovation more likely. In this way, innovations appear in clusters through a stochastic process characterized by contagion.

However, Basu et al. (2006) reject the GPT mechanism, concluding that sticky prices and not learning link recessions to technology improvements.

6. Concluding remarks

This chapter has reviewed the theoretical and empirical literature on LBD. Many of the distinctive theoretical implications of LBD have been derived under the assumption that the cost–quantity relationships observed in numerous empirical studies are largely the result of passive learning, and some further require that passive learning is unbounded. The empirical literature raises doubts about both assumptions. When observed cost–quantity relationships indicate sustained productivity growth, factors other than passive learning are generally at work. When passive learning is the dominant factor, productivity growth is invariably bounded. Thus, empirically relevant theories incorporating LBD are hybrid models in which passive learning coexists with other sources of growth. But in such models, many of the distinctive implications of passive learning become unimportant. Moreover, passive learning is often an inessential component of long-run growth; to the contrary, too much learning can lead to stagnation.

References

Abramovitz, M. (1956). Resource and Output Trends in the United States Since 1870. National Bureau of Economic Research, New York.

Aghion, P., Howitt, P. (1992). "A model of growth through creative destruction". Econometrica 60 (2), 323–351.

Aghion, P., Howitt, P. (1998). Endogenous Growth Theory. MIT Press, Cambridge, MA.

Alchian, A. (1963). "Reliability of progress curves in airframe production". Econometrica 31 (4), 679–693 (Originally Rand Corporation RM-260-1, 1950).

Anderlohr, G. (1969). "Determining the cost of production breaks". Management Review 58 (12), 16–19.

Argote, L. (1993). "Group and organizational learning curves: Individual, system, and environmental components". British Journal of Social Psychology 32, 31–51.

Argote, L., Epple, D. (1990). "Learning curves in manufacturing". Science 247, 920–924.

Argote, L., Beckman, S.L., Epple, D. (1990). "The persistence and transfer of learning in industrial settings". Management Science 36 (2), 140–154.

Argote, L., Epple, D., Rao, R.D., Murphy, K. (1997). The Acquisition and Depreciation of Knowledge in a Manufacturing Organization: Turnover and Plant Productivity. Carnegie Mellon University, Working Paper.

Arrow, K.J. (1962). "The economic implications of learning by doing". Review of Economic Studies 29 (3), 155–173.

Asher, H. (1956). Cost–Quantity Relationships in the Airframe Industry. RAND Corporation, Santa Monica, CA.

Audretsch, D.B. (1991). "New firm survival and the technological regime". Review of Economics and Statistics 60 (3), 441–450.

Audretsch, D.B., Mahmood, T. (1995). "New firm survival: New results using a hazard function". Review of Economics and Statistics 77 (1), 97–103.

Bahk, B.K., Gort, M. (1993). "Decomposing learning by doing in new plants". Journal of Political Economy 101 (4), 561–583.

Baily, M.N., Hulten, C., Campbell, D. (1992). "Productivity dynamics in manufacturing plants". Brookings Papers on Economic Activity: Microeconomics 1992, 187–267.

Baldwin, J.R., Gorecki, P. (1991). "Entry, exit and productivity growth". In: Geroski, P., Schwalbach, J. (Eds.), Entry and Market Contestability: An International Comparison. Basil Blackwell, Oxford, pp. 244–256.

Baldwin, J.R., Bian, L., Dupuy, R., Gellatly, G. (2000). Failure rates for new Canadian firms: New perspectives on entry and exit. Statistics Canada, Working Paper.

Baloff, N. (1966). "Startups in machine-intensive production systems". Journal of Industrial Engineering 31, 679–693.

Baloff, N. (1970). “Startup management”. IEEE Transactions on Engineering Management 17, 132–141.
Bartelsman, E.J., Dhrymes, P.J. (1998). “Productivity dynamics: U.S. manufacturing plants, 1972–1986”. Journal of Productivity Analysis 9 (1), 5–34.
Basu, S., Fernald, J.G., Kimball, M.S. (2006). “Are technology improvements contractionary?” American Economic Review 96 (5), 1418–1448.
Bell, R.M., Scott-Kemmis, D. (1990). The Mythology of Learning-by-doing in World War II Airframe and Ship Production. University of Sussex, Science Policy Research Unit, Working Paper.
Benkard, C.L. (2000). “Learning and forgetting: The dynamics of aircraft production”. American Economic Review 90 (4), 1034–1054.
Besanko, D., Doraszelski, U., Kryukov, Y., Satterthwaite, M. (2007). Learning-by-Doing, Organizational Forgetting, and Industry Dynamics. Northwestern University, Working Paper.
Bessen, J. (2003). “Technology and learning by factor workers: The stretch-out at Lowell, 1842”. Journal of Economic History 63 (1), 33–64.
Book, W.F. (1908). The Psychology of Skill. University of Montana Publications in Psychology Bulletin No. 53.
Boston Consulting Group. (1972). Perspectives on Experience. Boston Consulting Group, Boston, MA.
Bryan, W.N., Harter, N. (1899). “Studies in the telegraphic language: The acquisition of a hierarchy of habits”. Psychological Review 6, 345–375.
Bulow, J., Geanakoplos, J., Klemperer, P. (1985). “Holding idle capacity to deter entry”. The Economic Journal 85, 178–182.
Cabral, L. M. M., Riordan, M.H. (1994). “The learning curve, market dominance, and predatory pricing”. Econometrica 62 (5), 1115–1140.
Cabral, L. M. M., Riordan, M.H. (1997). “The learning curve, predation, antitrust, and welfare”. Journal of Industrial Economics 45 (2), 155–169.
Chari, V.V., Hopenhayn, H. (1991). “Vintage human capital, growth, and the diffusion of new technology”. Journal of Political Economy 99 (6), 1142–1165.
Clarke, F.H., Darrough, M.N., Heineke, J.M. (1982). “Optimal pricing policy in the presence of experience effects”. Journal of Business 55 (4), 517–530.
Cochran, E.B. (1960). “New concepts of the learning curve”. Journal of Industrial Engineering 11, 317–327.
Conley, T.G., Udry, C.R. (2007). Learning About a New Technology: Pineapple in Ghana. University of Chicago, Working Paper.
Conway, R.W., Schultz, A.Jr., (1959). “The manufacturing progress function”. Journal of Industrial Engineering 10 (1), 39–54.
Crossman, E. (1959). “A theory of the acquisition of speed skill”. Ergonomics 2, 153–166.
Dar-El, E.M. (2000). Human learning: From Learning Curves to Learning Organizations. Kluwer Academic Publishers, Dordrecht, The Netherlands.
Darr, E.D., Argote, L., Epple, D. (1995). “The acquisition, transfer, and depreciation of knowledge in service organizations: Productivity in franchises”. Management Science 41 (11), 1750–1762.
Dasgupta, P., Stiglitz, J.E. (1988). “Learning-by-doing, market structure, and industrial and trade policies”. Oxford Economic Papers 40 (2), 246–268.
David, P.A. (1973). “The ‘Horndal effect’ in Lowell, 1834–1856: A short-run learning curve for integrated cotton textile mills”. Explorations in Economic History 10 (1), 131–150.
DeGroot, M.H. (1970). Optimal Statistical Decisions. McGraw-Hill, New York.
Dewey, J. (1897). “My pedagogic creed”. School Journal 54, 77–80.
Dinopoulos, E., Thompson, P. (1998). “Schumpeterian growth without scale effects”. Journal of Economic Growth 3 (4), 313–337.
Disney, R., Haskel, J., Heden, Y. (2000). Entry, Exit and Establishment Survival in UK Manufacturing. Queen Mary and Westfield College, London Working Paper.
Dixit, A. (1980). “The role of investment in entry deterrence”. The Economic Journal 90, 95–106.
Dunne, T., Roberts, M.J., Samuelson, L. (1989). “The growth and failure of U.S. manufacturing plants”. Quarterly Journal of Economics 104 (4), 671–698.
Dutton, J.M., Thomas, A. (1994). “Treating progress functions as a managerial opportunity”. Academy of Management Review 9 (2), 235–247.
Ebbinghaus, H. (1885). Memory: A Contribution to Experimental Psychology. Columbia University, New York (Transl. by Henry A. Ruger and Clara E. Bussenius [1913]).

Epple, D., Argote, L., Devadas, R. (1991). "Organizational learning curves: A method of investigating intra-plant transfer of knowledge acquired through learning by doing". Organization Science 2 (1), 58–70.

Epple, D., Argote, L., Murphy, K. (1995). "An empirical investigation of the microstructure of knowledge acquisition and transfer through learning by doing". Operations Research 44 (1), 77–86.

Ericson, R., Pakes, A. (1995). "Markov-perfect industry dynamics: A framework for empirical work". Review of Economic Studies 62 (1), 53–82.

Evans, D.S. (1987a). "Tests of alternative theories of firm growth". Journal of Political Economy 95 (4), 657–674.

Evans, D.S. (1987b). "The relationship between firm growth, size, and age: Estimates for 100 manufacturing industries". Journal of Industrial Economics 35 (2), 567–581.

Farber, H.S. (1994). "The analysis of interfirm worker mobility". Journal of Labor Economics 12 (4), 554–593.

Feller, W. (1940). "On the logistic law of growth and its empirical verifications in biology". Acta Biotheoretica 5 (1), 51–66.

Foster, A.D., Rosenzweig, M.R. (1995). "Learning by doing and learning from others: Human capital and technical change in agriculture". Journal of Political Economy 103 (6), 1176–1209.

Fudenberg, D., Tirole, J. (1983). "Learning-by-doing and market performance". Bell Journal of Economics 14 (2), 522–530.

Galambos, J. (1978). The Asymptotic Theory of Extreme Order Statistics. Wiley, New York.

Garg, A., Milliman, P. (1961). "The aircraft progress curve—Modified for design changes". Journal of Industrial Engineering 12 (1), 23–27.

Ghemawat, P., Spence, A.M. (1985). "Learning curve spillovers and market performance". Quarterly Journal of Economics 100 (Suppl.), 839–852.

Greenwood, J., Yorukoglu, M. (1997). "1974". Carnegie-Rochester Conference Series on Public Policy 46, 49–95.

Greenwood, M., Yule, G.U. (1920). "An enquiry into the nature of frequency distributions representative of multiple happenings, with particular reference to the occurrence of multiple attacks of disease or repeated accidents". Journal of the Royal Statistical Society 83, 255–279.

Grossman, G.M., Helpman, E. (1991). Innovation and Growth in the Global Economy. MIT Press, Cambridge, MA.

Hall, R.E. (1971). "The measurement of quality changes from vintage price data". In: Griclihes, Zvi (Ed.), Price Indexes and Quality Change. Harvard University Press, Cambridge, MA, pp. 240–271.

Hall, B.H. (1987). "The relationship between firm size and firm growth in the U.S. manufacturing sector". Journal of Industrial Economics 35 (4), 583–606.

Hirsch, W.Z. (1952). "Manufacturing progress functions". Review of Economics and Statistics 34 (2), 143–155.

Hirsch, W.Z. (1956). "Firm progress ratios". Econometrica 24 (2), 136–143.

Hollis, A. (2002). "Strategic implications of learning by doing". International Journal of the Economics of Business 9 (2), 157–174.

Ingram, P., Simons, T. (2002). "The transfer of experience in groups of organizations: Implications for performance and competition". Management Science 48 (12), 1517–1533.

Irwin, D.A., Klenow, P.J. (1994). "Learning-by-doing spillovers in the semiconductor industry". Journal of Political Economy 102 (6), 1200–1227.

Jensen, J.B., McGuckin, R.H., Stiroh, K.J. (2001). "The impact of vintage and survival on productivity: Evidence from cohorts of U.S. manufacturing plants". Review of Economics and Statistics 83 (2), 323–332.

Jones, C.I. (1995a). "Time series tests of endogenous growth models". Quarterly Journal of Economics 110 (2), 495–525.

Jones, C.I. (1995b). "R&D-based models of economic growth". Journal of Political Economy 103 (4), 759–784.

Jovanovic, B. (1979). "Job matching and the theory of turnover". Journal of Political Economy 87 (5 part 1), 972–990.

Jovanovic, B. (1982). "Selection and the evolution of industry". Econometrica 50 (3), 649–670.

Jovanovic, B., Lach, S. (1989). "Entry, exit, and diffusion with learning by doing". American Economic Review 79 (4), 690–699.

Jovanovic, B., Nyarko, Y. (1995). "A Bayesian learning model fitted to a variety of empirical learning curves". Brookings Papers on Economic Activity. Microeconomics 1995, 247–305.

Jovanovic, B., Nyarko, Y. (1996). "Learning by doing and the choice of technology". Econometrica 64 (6), 1299–1310.

Klenow, P.J. (1998). "Learning curves and the cyclical behavior of manufacturing industries". Review of Economic Dynamics 1, 531–550.

Kremer, M. (1993). "Population growth and technological change: One million B.C. to 1990". Quarterly Journal of Economics 108 (4), 681–716.

Krugman, P.R. (1987). "The narrow moving band, the Dutch disease and the competitive consequences of Mrs. Thatcher: Notes on trade in the presence of dynamic scale economies". Journal of Development Economics 27 (1), 41–55.
Lane, F.C. (1951). Ships for Victory. Johns Hopkins University Press, Baltimore, MD.
Lazonick, W., Brush, T. (1985). "The 'Horndal Effect' in early U.S. manufacturing". Explorations in Economic History 22 (1), 53–96.
Levhari, D. (1966). "Extensions of Arrow's 'learning by doing'". Review of Economic Studies 33 (2), 117–131.
Levin, D.Z. (2000). "Organizational learning and the transfer of knowledge: An investigation of quality improvement". Organization Science 11 (6), 630–647.
Levy, F.K. (1965). "Adaptation in the production process". Management Science 11 (6), B136–B154.
Lieberman, M.B. (1987a). "The learning curve, diffusion, and competitive strategy". Strategic Management Journal 8 (5), 441–452.
Lieberman, M.B. (1987b). "Excess capacity as a barrier to entry: An empirical appraisal". Journal of Industrial Economics 35 (4), 607–627.
Lieberman, M.B. (1989). "The learning curve, technology barriers to entry, and competitive survival in the chemical processing industries". Strategic Management Journal 10 (5), 431–447.
Lucas, R.E. Jr. (1993). "Making a miracle". Econometrica 61 (2), 251–272.
Mankiw, N.G., Romer, D., Weil, D.N. (1992). "A contribution to the empirics of economic growth". Quarterly Journal of Economics 107 (2), 407–437.
Mansfield, E. (1985). "How rapidly does new industrial technology leak out?" Journal of Industrial Economics 35, 607–635.
March, J.G., Simon, H.A. (1958). Organizations. Wiley, New York.
Mata, J., Portugal, P. (1994). "Life duration of new firms". Journal of Industrial Economics 27, 227–246.
Mazur, J.E., Hastie, R. (1978). "Learning as accumulation: A reexamination of the learning curve". Psychological Bulletin 85 (6), 1256–1274.
McGouldrick, P.F. (1968). New England Textiles in the Nineteenth Century: Profits and Investment. Harvard University Press, Cambridge, MA.
Middleton, K.A. (1945). "Wartime productivity changes in the airframe industry". Monthly Labor Review 61 (2), 215–225.
Mishina, K. (1999). "Learning by new experiences: Revisiting the flying fortress learning curve". In: Lamoreaux, N., Raff, D.M., Temin, P. (Eds.), Learning by Doing in Markets, Firms, and Countries. University of Chicago Press, Chicago, IL (for NBER).
Montgomery, F.J. (1943). "Increased productivity in construction of Liberty vessels". Monthly Labor Review 57 (11), 861–864.
Muth, J.F. (1986). "Search theory and the manufacturing progress function". Management Science 32 (8), 948–962.
Nagypál, É. (2007). "Learning by doing vs. learning about match quality: Can we tell them apart?" Review of Economic Studies 74 (2), 537–566.
Ohashi, H. (2005). "Learning by doing, export subsidies, and industry growth: Japanese steel in the 1950s and 1960s". Journal of International Economics 66 (2), 297–323.
Pakes, A., Ericson, R. (1998). "Empirical implications of alternative models of firm dynamics". Journal of Economic Theory 79 (1), 1–45.
Parente, S.L. (1994). "Technology adoption, learning-by-doing, and economic growth". Journal of Economic Theory 63 (2), 346–369.
Peretto, P. (1998). "Technological change and population growth". Journal of Economic Growth 3 (4), 283–311.
Persson, H. (2002). The Survival and Growth of New Establishments in Sweden, 1987–1995. Stockholm University, Working Paper.
Petrakis, E., Rasmusen, E., Roy, S. (1997). "The learning curve in a competitive industry". RAND Journal of Economics 28 (2), 248–268.
Power, L. (1998). "The missing link: Technology, investment, and productivity". Review of Economics and Statistics 80 (2), 300–313.
Prescott, E.C., Visscher, M. (1980). "Organization capital". Journal of Political Economy 88 (3), 446–461.
Preston, L.E., Keachie, E.C. (1964). "Cost functions and progress functions: An integration". American Economic Review 54 (2), 100–107.
Rapping, L. (1965). "Learning and World War II production functions". Review of Economics and Statistics 47 (1), 81–86.
Redding, S. (1999). "Dynamic comparative advantage and the welfare effects of trade". Oxford Economic Papers 51 (1), 15–39.
Restle, F., Greeno, J.G. (1970). Introduction to Mathematical Psychology. Addison-Wesley, Reading, MA.

Roberts, P.C. (1983). "A theory of the learning process". Journal of the Operations Research Society 34 (1), 71–79.
Romer, P.M. (1986a). "Increasing returns and long-run growth". Journal of Political Economy 94 (5), 1002–1037.
Romer, P.M. (1986b). "Cake eating, chattering and jumps: Existence results for variational problems". Econometrica 54 (4), 897–908.
Romer, P.M. (1990). "Endogenous technological change". Journal of Political Economy 98 (5), S71–S102.
Rosen, S. (1972). "Learning by experience as joint production". Quarterly Journal of Economics 86 (3), 366–382.
Rosenberg, N. (1976). Perspectives on Technology. Cambridge University Press, Cambridge, UK.
Sahal, D.A. (1979). "A theory of progress functions". AIIE Transactions 11 (1), 29–39.
Salop, S.C. (1979). "Strategic entry deterrence". American Economic Review 69 (2), 335–338.
Saunders, R.S. (1985). "Learning by doing and dominant firm pricing strategy". Review of Industrial Organization 2 (1), 32–39.
Scherer, F.M. (1980). Industrial Market Structure and Economic Performance (second ed.). Rand McNally, Chicago, IL.
Searle, A.D. (1945). "Productivity changes in selected wartime shipbuilding programs". Monthly Labor Review 1132–1147 (December).
Segerström, P.S., Anant, T. C. A., Dinopoulos, E. (1990). "A Schumpeterian model of the product life cycle". American Economic Review 80 (5), 1077–1091.
Shapiro, C., Stiglitz, J.E. (1984). "Equilibrium unemployment as a worker discipline device". American Economic Review 74 (3), 433–444.
Sheshinski, E. (1967a). "Tests of the 'learning by doing' hypothesis". Review of Economics and Statistics 49 (4), 568–578.
Sheshinski, E. (1967b). "Optimal accumulation with learning by doing". In: Shell, Karl (Ed.), Essays on the Theory of Optimal Economic Growth. MIT Press, Cambridge, MA, pp. 31–52.
Shleifer, A. (1986). "Implementation cycles". Journal of Political Economy 94 (6), 1163–1190.
Sinclair, G., Klepper, S., Cohen, W. (2000). "What's experience got to do with it? Sources of cost reduction in a large specialty chemical's producer". Management Science 46 (1), 28–45.
Solow, R.M. (1957). "Technical change and the aggregate production function". Review of Economics and Statistics 39 (3), 312–320.
Solow, R.M. (1994). "Perspectives on growth theory". Journal of Economic Perspectives 8 (1), 45–54.
Spence, A.M. (1977). "Entry, capacity, investment and oligopolistic pricing". Bell Journal of Economics 10 (1), 1–19.
Spence, A.M. (1981). "The learning curve and competition". Bell Journal of Economics 12 (1), 49–70.
Stein, J.C. (1997). "Waves of creative destruction: Firm-specific learning-by-doing and the dynamics of innovation". Review of Economic Studies 64, 265–288.
Stokey, N.L. (1986). "The dynamics of industry-wide learning". In: Heller, W.P., Starr, R.M., Starrett, D.A. (Eds.), Essays in Honor of Kenneth Arrow Volume II. Cambridge University Press, Cambridge, UK.
Stokey, N.L. (1988). "Learning by doing and the introduction of new goods". Journal of Political Economy 96 (4), 701–717.
Stokey, N.L. (1991). "The volume and composition of trade between rich and poor countries". Review of Economic Studies 58 (1), 63–80.
Tambe, P., Hitt, L. (2007). The Effects of Measurement Error on Knowledge Spillover Estimates. Wharton School, University of Pennsylvania, Working Paper.
Thompson, P. (2000). "Learning from the experience of others: Parameter uncertainty and economic growth in a model of creative destruction". Journal of Economic Dynamics and Control 24 (9), 1285–1313.
Thompson, P. (2001). "How much did the Liberty shipbuilders learn? New evidence for an old case study". Journal of Political Economy 109 (1), 103–137.
Thompson, P. (2007). "How much did the Liberty shipbuilders forget?" Management Science 53 (6), 908–918.
Thornton, R.A., Thompson, P. (2001). "Learning from experience and learning from others: An exploration of learning and spillovers in wartime shipbuilding". American Economic Review 91 (5), 1350–1368.
Wagner, J. (1994). "The post-entry performance of new small firms in German manufacturing industries". Journal of Industrial Economics 42, 141–154.
Waldman, J.D., Yourstone, S.A., Smith, H.L. (2003). "Learning curves in health care". Health Care Management Review 28 (1), 41–54.
Watkins, P.N. (2001). "The persistence of learning and acquisition strategies". Acquisition Review Quarterly 15–30 (Winter).
Wenders, J.T. (1971). "Excess capacity as a barrier to entry". Journal of Industrial Economics 20 (1), 14–19.
Wilcox, N.T. (2006). "Theories of learning in games and heterogeneity bias". Econometrica 74 (5), 1271–1292.

Wright, T.P. (1936). "Factors affecting the cost of airplanes". Journal of the Aeronautical Sciences 3 (4), 122–128.
Yelle, L.E. (1979). "The learning curve: Historical review and comprehensive survey". Decision Sciences 10 (2), 302–328.
Yorukoglu, U.M. (1998). "The information technology productivity paradox". Review of Economic Dynamics 1 (2), 551–592.
Young, A. (1991). "Learning by doing and the dynamic effects of international trade". Quarterly Journal of Economics 106 (2), 369–405.
Young, A. (1993). "Invention and bounded learning by doing". Journal of Political Economy 101 (3), 443–472.
Young, A. (1998). "Growth without scale effects". Journal of Political Economy 106 (1), 41–63.

Chapter 11

INNOVATIVE CONDUCT IN COMPUTING AND INTERNET MARKETS

SHANE GREENSTEIN

Elinor and Wendell Hobbs Professor

Kellogg School of Management, Northwestern University, Evanston, Illinois, USA

Contents

Handbooks in Economics, Volume 01

DOI: 10.1016/S0169-7218(10)01011-7

Abstract

How has innovative and competitive behavior in computing and Internet markets evolved over the past half-century? In the first section of this review, I discuss these questions in light of six topics: the limited role for technology push; the diffusion of general-purpose technologies; the organization of proprietary platforms; the presence of asymmetric innovation incentives; the importance of market-oriented learning; and the localization of economic activity. Despite dramatic changes in outcomes, in the predominant product markets, and in the identities of leading sellers, the conditions of market structure shape innovative conduct in firms from one year to the next and, to a large extent, from one decade to the next, in many of the same economic terms.

In the second section, I closely examine the US commercial Internet experience in the 1990s. While the peculiar events that led to the invention of the commercial Internet explain some of the salient and unique features of the commercial experience, much innovative activity resembles conduct seen for many decades in computing markets. This analysis highlights three additional topics: the division of technical leadership; the rise of open organizational forms for coordinating platforms; and the extraordinary breadth of activity touched by the Internet. These three additional factors account for many of the novel aspects of innovative conduct in the Internet market.

Keywords

commercialization, computer hardware, computer software, diffusion, innovation, internet, invention, market conduct, technology

1. Introduction and overview

How has innovative and competitive behavior in computing and Internet markets evolved over the past half-century? This broad question does not and cannot have a simple answer for at least two reasons. First, the core determinants of behavior did not remain or stay constant over several decades. Second, commercial computing and Internet markets give rise to a variety of experiences that defy any single characterization.

Nonetheless, the question is worth asking because computing plays such a large role in the economy. Changes in computing now touch both the personal and professional lives of the vast majority of the work force. The basic experience of business computer users has undergone significant change over the last five decades. Starting from a small base of businesses in the 1950s, computing has diffused widely. In 1990, nominal investment in Information Technology (IT) goods totaled $131.5 billion, about 33% of private nonresidential equipment and software investment. By 2000, it was $406 billion and 44%.[1]

Similarly, the household experience with computing has also undergone significant change: It began from virtually nothing in the 1970s. Later, a 1995 survey found less than 20% of households had a personal computer (PC) (NTIA, 1995). In sharp contrast, an October 2003 survey found that 62% of respondents had a computer at home and an even larger percentage used a computer at work (Mankiw et al., 2005).

These events motivate a wide variety of microeconomic questions about innovative conduct in US commercial computing. In the first section of this review, I discuss these questions in light of six propositions commonly found in the themes of many studies:

- While technical frontiers in computing may stretch due to events reasonably described as "technology push," a more substantial amount of valuable innovation arises endogenously in response to market incentives and market-oriented events;
- The diffusion and development of computing resembles diffusion and development of a general-purpose technology (GPT), and as with such a technology, substantial costs arise from creating value by customizing the technology to the unique needs of users;
- The presence of computing platforms shapes incentives to innovate, and the unification or division of technical leadership shapes distribution of value within and between platforms;
- Leading incumbent firms and new entrants face differential incentives to innovate when innovation reinforces or alters market structure;
- Market-based learning activity plays an essential role in innovative conduct, especially in enabling exploration of multiple approaches for translating the frontier into innovative and valuable goods and services;
- The localization of economics activity leads to a concentration of some types of innovative conduct in a small set of locations.

These propositions highlight the continuity between different eras through the underlying economic links between market structure and producer conduct. Indeed, the central thesis of this review highlights

[1] See Doms (2004). After falling to lower levels in 2000 and 2001 and 2002, these levels came back up to almost the same levels in 2003 and for the next few years.

continuity, not change. Despite dramatic changes in outcomes, in the predominant product markets, and in the identities of leading sellers, the conditions of market structure shape innovative conduct in firms from one year to the next and, to a large extent, from one decade to the next, in many of the same economic terms.

In the second section, I closely examine the US commercial Internet experience in the 1990s—an analysis that illuminates the strengths and weaknesses of the established frameworks highlighted in the first section. From the outset of the commercial Internet many of its participants have maintained a strong sense about their exceptional nature, as if innovation within the existing value chain for the Internet defied established archetypes of innovation. This view raises a question about whether innovation within the Internet can be assessed with the same economic concepts used elsewhere in computing.

This essay will largely argue that it can be. While the peculiar events that led to the invention of the commercial Internet explain some of the salient and unique features of the commercial experience, much innovative activity resembles conduct seen for many decades in computing markets.

In demonstrating the continuity of economic links between market structure and producer conduct, however, I ultimately achieve almost the opposite—isolating a small set of unique economic factors from the recent era. I account for many of the novel aspects of innovative conduct in Internet markets with three interrelated propositions:

- Innovative conduct related to the commercial Internet did give rise to platforms, but it also gave rise to markets characterized by an extraordinarily high division of technical leadership. In turn, that resulted in an unprecedented dispersion of uncoordinated innovative conduct across a wide range of components affiliated with the Internet;
- Commercial Internet markets involve new organizational forms for coordinating firms with disparate commercial interests, such as open source platforms. Their presence and successful operation accounts for some salient unanticipated innovative conduct;
- The aspirations of entrepreneurs and incumbent firms in commercial Internet markets touched an extraordinarily large breadth of economic activity.

Throughout this review, the narrative will contain a slant toward events in the United States. This slant requires an explanation, since the computing industry today, and especially the commercial Internet, has reached a global scale in operation and in final service markets. This geographic bias partly reflects a pragmatic choice, choosing to compare changes over time, not across geographies. Though comparison between countries (e.g., Japan, United Kingdom, or Germany) can be done, such a comparison would widen the scope of the review and take it into too many topics. This review takes an approach that facilitates comparisons over time, concentrating on early computing firms and early commercial Internet participants who substantially, though not wholly, located in US regions. While this focus helps sharpen the contrast between distinct eras, it necessarily limits the scope of the review. It leaves open many comparative questions about the determinants of innovative behavior, that is, whether the propositions discussed here continue to usefully describe economic events outside of the US boundaries.

2. Innovation in commercial computing

From its military and research origins in the late 1940s, computing spread into the commercial realm and has since grown to include an extraordinary range of economic undertakings and a large fraction of US economic activity.[2] Many economists believe this expansion of applications for computing has been a driver of economic growth.[3] Many economic factors have shaped that movement. I begin with explanations that emphasize "technology push." Finding this approach inadequate in many respects, I inquire about other views that highlight the relationship between market structure and exploratory behavior.

2.1. Stretching the technological frontier and technology push

In popular discussions, advances in computing have become almost synonymous with advances in microprocessors. This is due to a 1965 observation by Gordon Moore, who cofounded and eventually became chairman at Intel: He foresaw a doubling of circuits per chip every 2 years. This prediction about the rate of technical advance later became known as "Moore's Law." In fact, microprocessors and DRAMS (dynamic random access memories), have been doubling in capability every 18 months over the last three decades.[4]

A similar pattern of improvement—though with variation in the rate—characterizes many other electronic components that go into producing a PC, server, or other equipment complementary with computers in many standard uses. This holds for disk drives, display screens, routing equipment, networking and communications equipment, operating systems, communications software, central switches, mainframes and microcomputers, storage devices, input devices, routers, modems, handheld devices, and Internet service provision, to name a few.[5]

Indeed, in virtually all applications in electronics, estimates have found extraordinarily rapid rates of improvement in the price per unit of quality of computing, no matter how it is measured. It is also common to measure increases in the ranges of new qualities—that is, to find increases in the number of new qualities provided. That is a robust finding, manifest across a range of computing equipment.[6] For example, in Trajtenberg's (1990) study of computer tomography, the cost of providing a basic scan declined dramatically. In addition, with each year the scanners increased their resolution and their ability to perform new services, achieving milestones that previously had not been possible at any price.

The constant improvement in performance supports the view that many changes in computing arise from "technology push." That is, the invention pushed out the technical or scientific frontier, leading

[2] Many authors have traced the long arc from research origins to commercial form. For overviews, see, for example, Flamm (1987), or Aspray and Campbell-Kelly (1996), among others.

[3] See the contrasting views of, for example, Gordon (2000), Jorgenson (2001), and Stiroh (2002).

[4] Moore's law has a long history, beginning with Moore (1965). See Flamm (2003) for a detailed analysis of the underlying components.

[5] See Jorgenson and Wessner (2005) for an extensive review of these many changes.

[6] There is a well-established literature on using hedonic price estimation to measure the rate of improvement in prices. See, for example, Triplett (1989) for a review of these estimates on large systems. See Berndt et al. (1995) for estimates for PCs. For more on similar trends in semiconductors, upstream to most computing equipment, see Aizcorbe (2006) or Aizcorbe et al. (2007).

other commercial actors to search for valuable uses. There is a grain of truth to this view, but it also requires proper qualification.

The supporting evidence is well known. Numerous prototypical technologies in computing found their way into products and services long after their invention. These inventions arose in university or commercial laboratories, sometimes as a by-product of basic scientific research goals and sometimes with no direct vision about their application to a valuable commercial activity. Then, these inventions spread through academic papers, by licensing of patents, or the movement of computer scientists and engineers into companies.[7]

Of these inventions, many arose from prototypes built with large subsidies from government funding. For example, the original investment by DARPA (Defense Advanced Research Agency) in the fundamental science of packet switching did not lead to any immediate practical commercial products. Years of sustained funding, however, led to a set of events that broadly subsidized the invention and operation of the basic building blocks of the Internet, such as the experiments that led to the definition of the Transmission Control Protocol/Internet Protocol (TCP/IP) stack and its practical implementation in a working communications and computing network. This funding occurred long before the commercial Internet was operational.

Indeed, for the many years the Internet was an engineering novelty, a fascinating invention used primarily by a small group of technically adept networking researchers.[8] Yet, sustained government subsidies helped push out the frontier over time. After the National Science Foundation (NSF) established the NSFNET, government subsidies funded new Internet connections around the country, helped increase its size, and rationalized operations so comparatively unsophisticated users could make use of the network. Only after widespread adoption did the impact on a substantial amount of research activity begin to become manifest.[9]

While these examples illustrate a role for technology push in conceptualizing innovation in computing, they give the misleading impression that Moore's Law and related phenomena are exogenous. It is as if breakthrough innovation in computing develops in a deliberate and sequential path, as if each starts as an invention, develops into a prototype, then finally morphs into a valuable product and service. On the surface this conceptualization must be false. After all, much of the behavior underlying the outcomes labeled as Moore's Law comes from firms with commercial motives, where the managers act in their own interest and in the interest of stock holders and others providing financing, who expect a return on their investments. These firms push out the frontier (in accordance with Moore's Law) because it serves their commercial interests, not because they have any desire or strategic interest in supporting industry-wide gains from widespread technological push.

[7] See accounts in, for example, Flamm (1987), Langlois and Mowery (1996), Waldrop (2001), or the overview in National Research Council (2003).

[8] For more on the state of worldwide networking prior to the commercialization of the Internet, see, for example, Quarterman (1989). For more on the origins of the Internet outside the United States, see Mowery and Simcoe (2002b).

[9] For example, by some measures, this communications technology had little impact on the conduct of science until the late 1980s. Agarwal and Goldfarb (2006) trace the impact of Bitnet, which was the predecessor to the Internet, on the coauthoring behavior of research engineers. While packet switching had been invented and implemented years earlier, Bitnet began diffusing to universities in the early 1980s, and had little impact on coauthoring behavior until the mid-1980s, and until later, most of the impact was incremental.

In practice, facets of computing technology and business coevolve as researchers and designers push forward the understanding about the costs and commercial value of achieving distinct technical outcomes, as embodied in products and service. Hence, it is more useful to conceive of much innovation taking place in the context of its anticipated effect on the use of computing in on-going economic activity, to analyze how market competition and feedback from user communities shape the direction and rate of commercial innovation. In short, technology push frameworks fall far short of yielding useful insights for economic analysts, managers, and policy makers.

There are numerous alternatives to technology push. We begin by conceptualizing computing as a GPT.

2.2. *Computing as a GPT*

Bresnahan and Trajtenberg (1995) define a GPT as a capability whose adaptation to a variety of circumstances raises the marginal returns to inventive activity in each of these circumstances. GPTs are associated with high fixed costs to inventing the technology and low marginal costs to use and reuse. This cost structure both (1) generates heavy early investment—which can occur before and during diffusion of the technology—and (2) leads to frequent repurposing of focal inventions. Rosenberg (1976) describes this as "the introduction of a relatively small number of broadly similar production processes to a large number of industries."[10]

The widespread use of computing can be interpreted as evidence of the first characteristic of a GPT—its extensive diffusion. Computing, or IT, ranges from aiding the automated tracking of transactions (a function necessary for automating billing, managing the pricing of inventories of airline seating, and restocking retail outlets in a geographically dispersed organization) to facilitating the coordination of information-intensive tasks, such as dispatching time-sensitive deliveries or emergency services. In addition, the second characteristic of a GPT is evidenced by computing's ability to improve the performance of, for example, advanced mathematical calculations, a function that is useful in activities so diverse as calculating the interest on loans and generating the estimates of underground geologic deposits. As a continued example of the repurposing of focal inventions, note that a closely related function to advanced mathematical computations is computer-aided precision, which improves the efficiency of processes ranging from manufacturing metal shapes to the automation of communication switches.[11]

The creation of intermediate goods like software or networking also shapes the valuation of adopting new IT applications and capital goods. Because it is used in business organizations, IT is deeply embedded in business processes. Accordingly, the business use of IT involves mutual adaptation between business processes and technology, an effort that can lead to large adjustment costs and slow learning about the most efficacious organization for linking inputs and outputs.[12] Part of the complexity

[10] The quote is from Ames and Rosenberg, the chapter on machine tools, in Rosenberg (1976). As noted in several essays in Rosenberg's book, these ideas have a long history in the studies of technology, and apply to many more industries than computing.

[11] See Cortada (2003) or McKinsey Global Institute (2001) for an analysis of a wide variety of applications.

[12] For an economic historians' perspective, see David (1990). Analysts from the information systems research community have highlighted determinants of these costs. See, for example, Attewell (1992), Fichman and Kemerer (1997), Bresnahan and Greenstein (1997), and Forman (2005). Forman and Goldfarb (2006) contain a summary of these factors as they applied to Internet-related technologies.

arises from the variety of applications for computing in modern economic activity. There may be sharing of noncapital investments across a wide array of processes, and, though the unit costs of sharing are lower for large organizations, the sharing usually does not occur instantaneously or without high coordination costs.[13]

The adoption and implementation of a GPT also can be costly because they lead to changes in other facets of an organization. For example, a GPT may motivate managers to reallocate decision-rights and discretion inside a large organization (Brynjolfsson and Hitt, 2000). This is especially so as local business units adapt IT to their local business processes—such as billing, account monitoring, and inventory management—or to the delivery of local services—such as retail sales, the delivery of financial data, and entertainment services. In this case, the boundary of the organization may change along with the adoption of IT, making a direct connection between organizational performance and IT difficult to trace.[14]

Nevertheless, researchers have tried to trace the links between investment in IT and productivity gains to an organization and the economy as a whole.[15] In general, appropriate microeconomic statistical evidence is difficult to find because such data must contain appropriate statistical variation at a sufficiently small unit of economic production.

In this sense, Atrostic and Nguyen (2005) have a rare and valuable study. They find evidence that use of networking technologies helped raise productivity in a wide array of manufacturing establishments in the late 1990s.[16] Bloom et al. (2007) also identify an interesting case for isolating the effects of IT use on establishment productivity, finding such evidence among US firms that acquire British firms. These establishments reinvest in their IT operations *after being acquired,* and that leads to measurable productivity advance.[17]

The economics of IT suggests that this evidence should be rare. Aside from simple lack of data with sufficient variance, even with its presence one might not expect a linear connection between changes in input and outputs. Computing frequently enables the invention of entirely new services and products that may or may not provide permanent or temporary competitive advantages.

Moreover, the changes may not be straightforward to measure in terms of value-added or total factor productivity. When new services are reasonably permanent, a private firm may see returns to the investment in the form of increases in final revenue or other strategic advantages. If a new product or service is quickly imitated by all firms, it quickly becomes a standard feature of doing business in a downstream market. The benefits from the new technology are rapidly passed on to consumers in the form of lower prices and better products. In this case, the benefits to a firm do not appear as an increase in revenues; but they exist nonetheless, in the form of losses avoided by the businesses in question.

[13] For an illustration in the adoption of machine tools, see, for example, Astebro (2002, 2004).

[14] This phenomenon results in the value of these investments manifesting in intangible investments. See, for example, Brynjolfsson et al. (2002).

[15] For a review of the micro- and macroeconomic evidence, see, for example, Jorgenson et al. (2005) or Draca et al. (2007).

[16] See also Nguyen and Atrostic (2006).

[17] For related evidence, see also Harrison et al. (2006).

2.3. Coinvention costs and creating value from GPTs

Coinvention costs are the various costs affiliated with customizing a technology to particular needs in specific locations at a point in time. A competitive supply of tools for coinvention activity can help lower but can never eliminate such costs. These costs shape the ultimate economy-wide cost from deploying and adopting a GPT.

The coinvention costs are frequently difficult to monetize, manifesting, as they do, as lost output, diversion of resources, or disruptions in routines and other "internal costs" inside an organization. Indeed, these should be higher with process improvements that lead to dramatic rearrangements of routine tasks, where firms must self-insure against unanticipated costs. As illustrated by David (1990) in his analysis of the electric dynamo inside manufacturing, computing was far from the first GPT to face such large adjustment costs from disruption of operations.

Bresnahan and Greenstein (1997) hypothesize that coinvention costs are at large-scale user installations driven up by complex or idiosyncratic organizational needs, which interfere with the use of generic solutions. They analyze the transition from mainframes to client–server architectures within establishments that already have mainframe computers. Their analysis provides a window on the factors that slow down adjustments inside an organization, because their sample includes many of the heaviest users of computing at the time and many of the establishments that initially adopted computing for business processes. Their findings emphasize the importance of the costs of inventing new uses for computing and adapting it to idiosyncratic and/or complex settings. Such costs slowed down the diffusion of a new technology, often to the users who could generate the highest benefits.

Coinvention costs are not borne solely by users during the deployment of a new GPT. Suppliers may incur them, as part of a strategic approach to developing a new service, limited by the idiosyncratic features of their own organizations and the market niches they serve. As GPTs diffuse, firms explore new ways to make viable businesses from providing services. In the case of dial-up Internet access, for example, the first generation of Internet service providers (ISPs) faced comparatively low incremental coinvention costs because they were complementary to the telephone system and they borrowed many practices from the bulletin board service market. Many firms quickly began generating revenue with just incremental action. Others pursued a variety of complementary businesses, trying to structure business to thrive or merely survive.[18]

An important open question concerns the presence (or absence) of "technological biases" in market outcomes as a result of firms supporting the adoption of innovating computing. In the simplest theory of an *unbiased* technology, its adoption alters the scale of production, without altering the proportion of inputs necessary to achieve that scale. In contrast, a biased technology may manifest itself in several ways, and, as a result, it is not straightforward to observe such biases.

For the first decades of computing, it was generally thought that computing favored substitution away from organizational forms that use less skilled labor, as in the canonical "labor-saving device." This perception was fostered by several popular images, such as the replacement of assembly line workers by robotic tools, or the replacement of a banking teller by an automatic teller machine.

[18] Greenstein (2000) explores such costs during the early growth of the Internet. The findings highlight that local conditions shaped the returns to these actions, with firms located in urban areas displaying a great variety of products and services.

The reality was more subtle than popular images suggest, and the diffusion of the PC made that apparent. More recent research tends to highlight skill-biased technical change.[19] That is, adoption can coincide with a disproportionately more intensive use of highly skilled labor, the returns to the higher value-added associated with more computing disproportionately goes to skilled labor instead of unskilled labor.[20]

Observing the effects of such substitution will be difficult if other aspects of the organization (such as the scale of operations, or valuable features of the end product) change at the same time. It is made further complicated if investors observe these changes and favor some types of investments over others, altering the implicit equity-based opportunity cost of making investment.[21]

Several microstudies of IT adoption illustrate the complexity of tracing the connection between adoption of IT coinventions and performance. In the trucking industry, for example, computing altered the allocation of transportation of goods. Hubbard (2000, 2003) examines the use of computing technologies to monitor the performance of trucks. Two distinct applications emerge as valuable, namely, tracking trucks in real time and auditing features of their performance after a completed task. If productivity increases come primarily from the truck being full more often—that is, facilitating matching of truck to prospective tasks, filling backhauls or mixing partial loads—then it is possible to identify the role of computing in bringing about this improvement.

The introduction of on-board computers also improves the ability to coordinate assets in different locations. On-board computing can improve monitoring of driver actions, leading to benefits such as reduced truck depreciation. Yet, fully realizing these improvements requires a rearrangement of the ownership of assets, so tracing all lines of causality in this industry remains challenging.[22]

Another example is how an IT improved the productivity of emergency response services. Athey and Stern (2002) examine the application of computing to emergency services, where value arises from giving timely information to dispatchers of ambulances (in their example). They focus on the consequences for heart attack patients. In particular, the enhanced 911 system (E911) allowed emergency dispatchers to pinpoint the location of a caller. Since timeliness is the key factor in emergency response to probable cardiac events, providing accurate information enabled a more rapid response. In this way, the introduction of E911 was shown to reduce the probability of a patient arriving with a high-risk pulse rate as well as the probability of mortality within 48 hours.

Research tends to focus on the factors that drive up coinvention costs, highlighting the myriad reasons why new technologies do not become employed rapidly. Yet, coinvention costs do not have to be high, and this can form an important part of the explanation for rapid diffusion of a GPT. For example, Forman (2005) examines the early adoption of Internet technologies at 20,000 commercial establishments from a few select industries. He concentrates on a few industries with a history of adopting frontier Internet technology and studies the microeconomic processes shaping adoption. He finds widespread use and adoption of basic technologies, such as e-mail and browsing, consistent with low coinvention costs for networks supporting these applications.

[19] For more on such biases, see, for example, Caroli and Van Reenen (2001), Bresnahan et al. (2002), Acemoglu et al. (2007), and Beaudry et al (2006) for PCs in particular.

[20] See, for example, Beaudry et al. (2006), Autor et al. (2003), Card and DiNardo (2002), Kreuger (1993).

[21] A rather provocative discussion of this last point about market valuation, see Shiller (2000).

[22] For example, see the analysis of Baker and Hubbard (2003, 2004).

Later, Forman et al. (2003a) extend Forman's work to all nonfarm private establishments in the US economy, surveying establishments with over 100 employees, where accurate and extensive data about the use of the Internet exist for roughly half the establishments of that size. Projecting from this survey to the economy as a whole, they estimate that by the end of 2000 close to 90% of all such establishments will have access to e-mail and browsing. Some industries had reached saturation, such a printing, parts supply, and many financial activities; other industries had high rates of adoption (over 80%), while others such as waste management, garden supply, and social assistance had lower rates. They conclude that coinvention costs were low for almost all industries because (1) PCs had already diffused, (2) supply of related services was widely available across the country, including low-density areas, and (3) the incremental costs of adopting these additional activities involved only a small set of steps.

Forman et al. (2005) examine the influence of factors on the marginal adopter. Holding all else constant, they show that business participation in the Internet is more likely in rural areas than in urban areas. This is particularly true for technologies that involve communication across establishments. Nevertheless, talk of the dissolution of cities is premature. Frontier Internet technologies for communication within an establishment appear more often at establishments in urban areas, even with industry controls. The difference between marginal and average rates is largely explained by differences in industry composition across major cities. More IT-intensive industries tend to cluster in urban areas. The effects of urban leadership and industry composition interact in a complementary way for advanced applications and that interaction exacerbates agglomeration in use.

While coinvention costs were low for basic browsing, they were much higher for any significant investment in enterprise computing. For example, building ERP systems consistent with IPs and integrating them into enterprise operations involved extensive customization to reflect each firm's production process, reporting norms, security and accounting procedures, and supplier relationships. Such investment could either provide managers with additional discretion or serve to centralize authority within firms.[23] Similar issues shaped a wide range of enterprise-level investments in Internet-enabled applications for procurement, distribution, inventory tracking, coordination of payroll, and so on.[24]

While coinvention costs help classify the costs of learning how to turn a GPT into something useful, analysis of innovative conduct in computing often requires something else. It needs to analyze the effects of specific institutional details on innovative conduct, such as the goals of the users, the procedural patterns of the industry-wide organizations, and the identity of the leading firms. We begin to develop that analysis next.

2.4. *Platform competition shapes economic incentives*

The direction of innovative opportunities is shaped by the relationships between firms, and those relationships are shaped by the presence of platforms. In any given era, computing markets are organized around platforms—a cluster of technically standardized components that buyers use together

[23] See, for example, Brynjolfsson and Hitt (2000) or Bloom and Van Reenen (2007).

[24] For further review of these and related studies, see Forman and Goldfarb (2006).

to perform the aforementioned wide range of applications. Platforms shape the incentives to pursue directions of innovative activity and, arguably, also its rate.[25]

Such platforms involve long-lived assets, namely, both components sold in markets (i.e., hardware and some software) and investments made by buyers (i.e., training and most software).[26] Important computing platforms historically include the UNIVAC, the IBM 360 and its descendents, the Wang minicomputers, IBM AS/400, DEC VAX, Sun SPARC, Intel/Windows PC, Linux, and, recently, TCP/IP-based client–server platforms linked together.

Vendors tend to sell groups of compatible products under umbrella strategies aimed at the users of particular platforms. In the earliest eras, the leading firms integrated all facets of computing and offered a supply of goods and services from a proprietary source. In later eras, the largest and most popular platforms historically included many different computing, communications, and peripheral equipment firms, software tool developers, application software writers, consultants, system integrators, distributors, user groups, news publications, and service providers. While some of these might take actions to serve proprietary interests, they all commit to the platform, and invest with the expectation that the platform will continue.

Platforms display a form of increasing returns that is sometimes given the labels "network effect" or "bandwagon effect."[27] That is, the value of participating in the platform grows as more participants commit to it. These benefits accrue through a variety of mechanisms: Users may benefit from participating in a large platform because large platforms display larger selection, lower prices, and greater opportunities to "mix and match" components from multiple suppliers. Vendors may benefit from participating in larger platforms because it provides them access to thicker demand for niche products and more accurate perceptions about the long-term viability of accumulated groups. Larger platforms also allow firms to specialize in innovating on a few areas while leaving other markets to specialists in other complements.

The emergence of platforms tends to coincide with the emergence of a standard bundle of components. That is, a standard bundle embodies a set of common arrangements of components for delivering services. Most users of a platform are similar in this respect.

This explanation also begins to hint at why some users and participants avoid platforms. The standard bundle may constrain the functionality. So, for example, technically adept users tend to favor different standard bundles than users that make up a mass market. Similarly, a standard bundle constrains a

[25] The word "platform," as used in engineering, is a different notion. In this context, the emphasis is on the formation of valuable interconnected economic relationships and how their presence alters incentives to undertake new innovative activity. See, for example, Bresnahan and Greenstein, 2000, or Gawer and Cusumano, 2002, among others.

[26] Within the computer industry, the user investments are often branded as "sweat equity" by sales forces who must be cognizant of their presence to make a sale. See the contrasting descriptions of Cortada (2003), and Shapiro and Varian (1998), where the former uses the industry vernacular and the latter uses the economic theory of "switching costs" to discuss much the same phenomenon.

[27] Jeffrey Rohlfs is credited with the earliest models of this phenomenon, which he motivated on his observations about the failed videophone at AT&T. Rohlfs (2001) contains numerous case studies, as well as an intellectual history of his thinking about bandwagon effects. Katz and Shapiro (1985, 1986) present a model of networks with endogenous pricing, calling this a "network externality." Their models consider a variety of settings where those externalities are internalized or not. Also see Farrell and Saloner (1985) for a related definition of network effects. For an overview of this approach, see Katz and Shapiro (2000).

vendor's ability to differentiate. For strategic reasons, occasionally vendors will try to break with standard bundles to achieve such differentiation.[28]

Until the early 1990s, platforms helped define the margins between most market segments, which were distinguished by the set of common functions in which a group of sellers/users shared an interest. These segments represented clusters of technical skills at firms and clusters of operations at users. Typically these shared interests corresponded to the size of tasks to be undertaken and the technical sophistication of the typical user. Mainframes, minicomputers, workstations, and PCs in decreasing order, constituted different size-based market segments.[29]

The most popular platform in the late 1980s and 1990s differed from the prominent platforms of earlier years. For example, the workstation appealed to technically sophisticated users, and typically employed advanced microprocessor power and some modified variant of a Unix operating system. Numerous companies competed with proprietary versions of hardware and software designs. Eventually the leading firm in this segment became SUN Microsystems, which employed a mix of proprietary and nonproprietary technologies that appealed to users and a large community of software application developers.[30]

The other popular small system was the PC. It began in the mid-1970s as an object of curiosity among technically skilled hobbyists. After a brief period of competition among designs within the segment, it became a common office tool after the entry of IBM's design. Unlike prior computing platforms, this one eventually has diffused into both home and business use. From the beginning, this platform involved thousands of large and small software developers, third-party peripheral equipment and card developers, and a few major players.[31]

More recently, control over standards has completely passed from IBM to Microsoft and Intel. Microsoft produces the Windows operating system and Intel produces the most commonly used microprocessor. For this reason the platform is often called *Wintel*.

The networking and Internet revolution in the late 1990s is responsible for blurring prior familiar distinctions. At first, these new technologies involved a combination of workstations and PCs hooked together with a local area network (LAN). These innovations made it feasible to build client–server systems within large enterprises and across ownership boundaries. Firms with dominant positions in the earliest client–server platforms included Novell, 3Com, Oracle, and Cisco.

Before client–server systems completely diffused to all enterprises, another innovation altered the path of development, the Internet. As a technical matter, the Internet involved a series of standard protocols that permitted a user to move data within and across networks as long as those networks employed the same protocol.

There were many new features to the commercial Internet, but two features especially stood out as a type of commercial computing network technology. First, the Internet was designed to have its

[28] Bresnahan and Yin (2007) provide an insightful analysis of the strategic imperative to achieve such differentiation in a standards battle and the factors that shape success and failure.

[29] See Bresnahan and Greenstein (1999) for a reinterpretation of computing history continuity and change in terms of its platforms.

[30] See, for example, the account by Baldwin and Clark (1997), and an update in Baldwin and Clark (2006).

[31] These events are well known and well documented. For example, see the accounts in Cringley (1992) or Frieberger and Swaine (1984), among many. For a statistical study of this competition, see Gandal et al. (1999). Bresnahan and Greenstein (1999) analyze these events in terms of platform economics.

intelligence at the end of the network. That is, users had to adopt applications in the PCs and workstations that were compatible with one another, but did not have to worry about any of the devices or protocols inside the network.[32]

Second, once the commercial Internet had diffused (by 1997 to all major cities in the United States), a remarkable set of new possibilities emerged: The Internet made it possible for users and vendors to move data across vast geographic distances without much cost, either in operational costs and/or in advanced set-up costs of making arrangements for transport of data. Together, those two features enabled enormous combinations of users and suppliers of data that previously would have required bilateral—and, therefore, prohibitively costly—agreements to arrange. In brief, it enabled a network effect where none had previously existed, involving participants who could not have previously considered it viable to participate in such a network.

Today such networking employs Internet-based computing systems connected across potentially vast geographic distances. This results in the emergence of a "network of networks,"[33] which employs a mix of nonproprietary designs, from such organizations as the IEEE (Institute of Electrical and Electronics Engineers), the IETF (Internet Engineering Task Force), the World Wide Web Consortium (W3C), and others that will be described in more detail in later sections. And these organizations coexist with providers of proprietary products and services. The latter are sponsored by firms such as Microsoft, SAP, Oracle, Google, Yahoo, who compete with each other, and use software built around their own designs, as well as around those from open sources, such as Linux, Apache, and MySQL, as well as others described in more detail in later sections.

2.5. Innovation within and between platforms

Platforms have existed in every era of computing. Platforms are significant because the direction of innovative activity takes place within the constraints of platform competition. Competition between platforms arises whether or not firms with proprietary interest (in a platform) either wholly or partly control the platform. Innovative activity contributes to altering the conditions of competition either between or within platforms. There have been few empirical studies of historical competition between platforms, primarily because such competition is infrequent. New proposals for platforms rarely develop past conceptualization into a commercial form that users will buy. In addition, once they are widely adopted, existing platforms tend to be hard to stop or slow down. As a result, markets do not give rise to many settings where several viable platforms last for long and compete.

Many of the studies of the mainframe era, for example, attempt to understand the rise of the IBM System 360 as the dominant computing platform. The interest is understandable, since this was the most lucrative commercial innovation in computing for several decades. The CEO (chief executive officer) at that time, Thomas Watson Jr., led a huge development effort, putting at risk virtually all the assets of the firm. This family of products succeeded in becoming the primary platform for the development of

[32] For an account of how this built up in e-mail design, see Partridge (2008). For an accessible discussion of the design principles behind the Internet and why recent events threaten their continuance, see Blumenthal and Clark (2001).

[33] See, for example, Noam (2001) for studies of "networks of networks." For a range of writing about the operations of this network and the numerous issues raised in trying to manage this at a large scale, see, for example, McKnight and Bailey (1997), Kahin and Keller (1995, 1997), Mansell and Steinmueller (2000), Compaine and Greenstein (2001), Cranor and Greenstein (2002), and Cranor and Wildman (2003).

numerous office-computing applications in banking, payroll, inventory accounting, and many other key innovative applications in the 1960s and 1970s.

Several studies have sought to understand either the circumstances that led to IBM's development or the consequences of its success. Katz and Phillips (1982), for example, focus on what led to IBM's success by highlighting the learning needed to change organizations and the product designs to accommodate the changing needs of customers. They find that large-scale computing had begun as a scientific and military pursuit, with most development funded by governments. The early technical leaders of computing for scientific and military applications had difficulty anticipating the operational requirements and redesigns valued by commercial users with different needs. The emergence of mass-market office users for computing changed the value of learning about commercial users, thereby altering the value of the firms that had been early pioneers.

Fisher et al. (1983a,b) focus on the consequences of IBM's success, by advancing a view that emphasizes the dynamic changes in the market place and the rewards to entrepreneurial commercial initiatives at leading firms, in this case, primarily IBM. They credit IBM management with creating and operating processes that translate the information gained from learning about new technologies and user experience into new product designs.[34]

There are two approaches to understanding how competition works within platforms. Consider first how the links inside a platform shape the conditions of innovation. Buyers and sellers become linked by their technical interdependence and their continuous economic relationships. As a result, it is common for these ecosystems of suppliers and users to evolve and revolve around a common bundle of components. Innovative opportunities and constraints thus depend on what new offering appears that departs from that standard bundle.

The ecosystems of vendors and users differ in their stability, turnover, and size, depending on the requirements of the services performed.[35] For example, consider the difference between the ecosystem for providing security software and that for providing printers, each of which are common components in a standard PC purchase today. The security market involves firms with a labor force that constantly works to predict the behavior of hackers and is potentially on-call and ready for immediate crisis when a virus spreads. In contrast, the printer markets contain participants more reminiscent of a group of typical manufacturing organizations, involving regular supply chains for parts and components, third-party distribution, and servicing of end products.

A second approach for understanding competition within platforms tries to incorporate recent circumstances into its method. In general, no firm controls all aspects of a single platform. Competitive analysts focus on understanding how ensembles of participants (within a platform) behave, either in competition with one another or in cooperation.[36] There are also several studies of format wars or

[34] See also Fisher et al. (1983a,b).

[35] See the extensive analysis in Messerschmitt and Szyperski (2003).

[36] For example, Bresnahan and Greenstein (1997) analyze such competition in the context of the flight between mainframe and client–server for large establishments. Bresnahan (1999) considers some of the barriers faced by "smart server–dumb client" platforms during an early period, offering a model of sequential entry from "one adjacent component market into another." As another example, Gandal et al. (1999) examine the competition between different platforms in the early PC market, highlighting the role of feedbacks between components markets.

standards wars, which may or may not involve large groups of firms providing a standard bundle, as in a platform.[37] The more recent literature also investigates why the era of single ownership over an entire platform ended. This brings us to a discussion of divided technical leadership.

2.6. *Divided technical leadership and innovation conduct*

In the 1950s, computing was a novel technology and only a handful of experts understood all its key features. During the past five decades, the dispersion of expertise has transformed dramatically to encompass a vast ensemble of participants from a variety of technical and commercial backgrounds with varying kinds of motives. This transformation coincides with a change in commercial conditions for computing. No longer does a small set of expert engineers understand *all* dimensions of computing. This does not imply that expertise has no value; rather, a small number of experts do not largely determine the rate and direction of technical change in all aspects of computing.

In other words, there has been a secular trend toward an increase in the number of firms that possess the necessary technical knowledge and commercial capabilities to bring to market some component or service of value to computing users. If a firm does not possess such capability, it can be easily acquired through market means, such as hiring a small team of qualified engineers. Bresnahan and Greenstein (1999) call this feature of market structure *divided technical leadership.*

To illustrate, consider some of the key innovations of distinct eras: In the 1960s, when technical expertise was not widely dispersed, the most lucrative innovation of commercial computing, the IBM System 360, arose from using designers and developers employed entirely by a single firm. It involved redesigning every major aspect of commercial computing. Although many of the inventions had arisen elsewhere, many also came from within IBM. More importantly, the key invention had not arisen anywhere—instead it involved putting all the other inventions together. Coordinating that large and diverse team pushed IBM to the boundaries of what any ambitious firm feasibly could manage. Its innovations thereafter employed various pieces of the System 360 for new applications, such as airline reservations systems or new account-tracking systems for financial users. Though it faced many imitators, IBM employed unique assets and products in producing, delivering, and servicing its own products, and that helped protect its innovation from imitation.

Consider, in contrast, divided technical leadership, which supported commercial initiatives by those specializing in supply of innovative components. Thus, broadly speaking, at the equipment layer one set of firms specialized in supply, while another distinct set of firms carried the data using frontier operations. Another set of firms brought new storage devices to market, yet another provided frontier software applications for users, and an even different set performed frontier services. This is sometimes called specialization in *horizontal* or *component* layers to distinguish it from competition between integrated systems.[38]

The specialization of supply frames one of the distinctive strategic issues of the modern era. Firms with quite different capabilities, specializing in one or a small set of components, cooperate with others

[37] On the VHS/Beta war, see Ohashi (2003), on the hardware/software interplay in DVD adoption, see Gandal et al. (2000), on the DiVX war, see Dranove and Gandal (2003), and on the 56K modem war, see Augereau et al. (2006).

[38] This was famously summarized by Andy Grove, CEO of Intel, who described the distinct layers of the PC industry. See Grove (1996).

at the boundary of their respective firms. In personal computing, for example, an array of distinct firms arose that specialized in supplying different parts of the PC (e.g., many firms provided the electronic components), while different firms provided the software. An entirely different set distributed the final product and became involved in servicing it. The benefits of allowing users to mix and match components and service outweighed most of the benefits of coordinating production entirely inside one firm.

Markets with dispersed technical leadership tend to contain a variety of firms. Here *variety* means the following: Firms use different commercial assets in different locations, different personnel with distinct sets of skills, different financial support structures with different milestones for measuring progress, and even different conceptual beliefs about the technical possibilities. In this sense, variety shapes the conditions of competition. One firm's assessment of the returns from innovating does not need to be the same as another's. Different assessments result in different methods for achieving the same commercial goals, which may lead to different costs, or different commercial goals altogether, such as targeting different customers.

Divided technical leadership also reduces barriers to entry in component markets, which, in turn, supports widespread availability of potential suppliers of a component or service. Thus, entry of new component firms into a market arises from one of three paths:

1. A component specialist in a one technical "area" may develop expertise on a "neighboring" area in the course of operations—for example, a hosting firm will learn plenty about security software. This firm may use existing personnel and assets in respective area to develop another specialty component that extends its existing business;
2. An entirely new entrant may arise, starting from scratch, assembling components made by others or hiring technical talent for a newly focused goal or newly reoptimized organizational form. Such firms often get financial backing from venture capitalists and view their goals in terms of a race to achieve a functional leap over all other firms in a specialized area;
3. A firm with a broad proprietary platform, such as Intel or Microsoft, may seek to enter by embedding similar functionality in one of its existing products, either through acquiring another specialty firm (who entered as 2) or developing their own proprietary version (as in 1).

Such competitive conditions can lead to a reduction in any specific supplier's ability to possess a unique set of assets or employ all the innovative experts in the world. Where once firms used technical prowess for commercial gain, now firms find their technical skills *commoditized,* whereby the firm becomes a supplier of a product (technical prowess) that is available from many vendors. These issues shaped the most lucrative computing innovations of the mid-1990s.

First consider Microsoft's Windows 95—an innovation that involved redesigning every major aspect of the operating system for PCs and pushed Microsoft to the boundaries of what its organization feasibly could manage.[39] Microsoft embedded many functions inside Windows 95 that had previously been provided by specialist software firms. Despite its internal capabilities, Microsoft's managers deliberately left an extraordinarily large amount of commercial computing untouched, especially in equipment and application markets where Microsoft had no products or strategic

[39] See, for example, Cusumano and Selby (1995).

interests. This was a rational decision that recognized both the limits of its own firm and the capabilities of others.[40]

Next, consider another lucrative innovation of the mid-1990s, upgrades to the microprocessor at Intel. Intel's managers repeatedly faced decisions about whether to initiate new projects or reinvest in existing projects, some of them in areas "on the motherboard of the PC," where this functionality complemented the microprocessor. When they reasoned that such investments would expand final demand for PCs, then they would invest, as they did, for example, in redesigning the bus for the PC.[41] Throughout the 1990s, managers chose to invest in interfaces that worked directly with their existing product line, while avoiding projects where plenty of other suppliers had the ability to reach the frontier at lower cost.[42]

One might summarize it thusly: While competition between platforms determines prices for customers deciding between platforms, divided technical leadership shapes the competition for, and division of, returns within a platform. These two margins differ, and rather distinct aspects of firm conduct determine each of them.

Divided technical leadership, along with the growth of the market to support specialization, also partly explains several other prominent features of the computing industry on a global scale, such as the increasing geographic concentration of some activities in some local areas. As many firms specialize in different components, supply chains for most computing hardware and software products no longer falls under control of one firm. The location of production becomes subject to competitive forces, and the identity of leading firms can change. This trend has been widely seen across many computing components, such as displays, mobile devices, and other hardware devices.[43] In addition, as in other manufacturing processes, the increasing use of sophisticated IT helps coordinate design and production involving firms from many countries and continents, which also contributes to spreading it to many locations. Divided technical leadership also affects software, a labor-intensive activity where coordination and monitoring costs have declined enough to support geographically dispersed production.[44]

Despite the emergence of divided *technical* leadership, computing markets do not display frequent turnover in the identity of *market* leaders. Why is that? I next examine persistence and racing as a means to answer this question.

2.7. Racing and persistence by incumbent and entrepreneurial entrants

Pricing at some level above unit cost requires something rare among specialized providers—for example, valued brand, frontier features, unique service, or better distribution. Firms take a variety of approaches to developing these assets in technology markets, even if they remain unique for only a short period. Said succinctly, firms face incentives to fund the search for and creation of innovations that use proprietary assets, such as existing effective distribution channels.

[40] These types of decisions are discussed at length in Bresnahan et al. (2009).

[41] In their own words, the managers would invest in trying to improve some aspects of the PC if it would "grow the pie for everyone." See the discussion in Gawer and Cusumano (2002).

[42] A review of these issues can be found in Henderson and Gawer (2007).

[43] For example, see Dedrick and Kraemer (2005) for a study of the sourcing for personal computers. See Hoetker (2006) on flat panel displays.

[44] See, for example, Mowery (1996) or Arora and Gambardella (2005).

Among these many actions, firms compete with one another to reach a unique technical accomplishment. This type of competition is often labeled *racing*. Racing arises in the midst of upgrades to existing products or during times when firms offer new product introductions. It has been a widely documented behavior, found in disk drives, printers, software, and other components, with considerable variance in the later performance of early winners of races.[45]

Despite frequent and sometimes dramatic technical improvements in the frontiers of technology, many features of the most common platforms in use tend to persist or change very slowly; this is labeled *persistence*. Persistence happens even when the frontier has far outstripped the most common technologies in use.

This observation raises the question: Why do incumbents or entrants *not* take advantage of every technological opportunity? What are the costs and benefits of racing and persistence? As it turns out, sometimes firms either cannot or will not do the same as those who race to the frontier. This is because many durable components make up platforms. Though old technology loses its status as a frontier application as it becomes obsolete in comparison to newer frontier products, it does not as quickly lose its ability to provide a flow of valuable services to users. In brief, a service may be valuable even when it is not on the frontier if it enhances and preserves the value of previous investments while simultaneously giving users access to some new functionality.

A "backward-compatible" upgrade or improvement is one that works with or remains compatible with existing equipment. The label is revealing. A "backward" technology is deliberately not one that moves "forward" to the frontier, but, instead, remains compatible with prior functionality. It also creates a demand for support and innovative service activities to reduce the costs of making the transition from old to new.

There is often a fundamental tension between aspirations toward the frontier and a backward-compatible upgrade. While this trade-off always exists, platform leaders face high opportunity costs for mismanaging the trade-off. There may be distinct adequate solutions to a customer's demand for backward-compatible components and for the frontier. Firms also might need different designs in order to satisfy both frontier and backward-compatible customers. The open question is whether most customers would be satisfied with the same design offered by a platform leader.

During the era of the large systems, for example, IBM faced a set of related issues after the success of the System 360. Instead of designing its next mainframe from scratch, it chose to introduce a family of systems, the System 370, which remained backward compatible with many of the investments made by users of the System 360. This design choice led to a lucrative business opportunity with many existing mainframe customers; however, it made it quite difficult for IBM to satisfy certain subsegments of users, particularly those focused on innovative high-speed computing.[46] Indeed, IBM faced essentially

[45] For example, Khanna's (1995) study of new product introductions in frontier computing documents such behavior. Stavins (1995) examines the incentives of PC firms to enter the frontier or the "middle" of the product space. Lerner (1997) examines whether hard-disk firms gain premiums in their pricing from developing new frontiers. Greenstein and Wade (1998) examine whether competitive setting shapes the incentives of firms to bring out new products. Cockburn and MacGarvie (2006) study how intellectual property shape the incentives of new software entrants; De Figueiredo and Kyle (2006) look at existing incumbents, principally Hewlett Packard, or at new entrants developing new frontiers of the product space for laser printers. Prusa and Schmitz (1994) present evidence that few of the early entrants into PC software survived thrusts into other segments.

[46] Accordingly, other platform providers, such as Control Data and Cray focused on them. See, for example, Fisher et al. (1983a,b).

the same trade-off for many years. In each case, IBM always made at least one option available that allowed its present customers, the vast majority of mainframe users, to upgrade without losing prior investments.

IBM's managers did try to extend the reach of its systems to appeal to new users, and they succeeded occasionally in these races and occasionally not. In these cases, IBM's competitors attempted to design systems that appealed to niche users, but the systems were not compatible with the platform. Notably, in races on the low end, where such upgrades were rarer and the demand for backward compatibility weaker, IBM experienced a variety of new entrants and competitors. IBM also tried a variety of different partially compatible systems to compete. None of them did especially well until the PC in 1981 and the AS400 some years later.[47]

A similar dilemma arose in the era of widely divided technical leadership in PCs. After the rise of 386 computing in PCs, IBM's leadership in PCs had diminished. Intel's management considered new designs for the microprocessor, particularly whether it should follow the lead of several other frontier firms and break with prior designs in order to reach the frontier. Although such a break would help Intel compete with others that demonstrated superior performance, it would sacrifice some backward-compatible functionality. Instead, after considerable debate, Intel chose to follow a path of backward-compatible improvement throughout the 1990s. In retrospect, this was a good decision for the firm's stockholders, even though it was far from obvious as the time.[48]

Aspects of this same dilemma arose during Microsoft's strategic thinking in the mid-1990s. For example, though Windows 95 was a new operating system for the PC, Microsoft eased the portability of software applications that had run on Windows 3.0 and 3.1, which sat on top of another operating system, DOS. Launched in August 1995, Windows 95 included a subwindow for porting files from the DOS environment, where vast majority of nonfrontier programs were still surviving. Microsoft went to such lengths for numerous reasons, but among them was the concern that—in the absence of such features—these existing users of DOS would not migrate to Windows 95 but would, instead, form a subgroup of users that another competitor in an operating system market might support and use as a base from which to grow.[49]

These last two examples allude to a subtle, but important feature of why the tension between racing and persistence is so central to the analysis of innovation and platform leadership: The incentives to meet demand for backward compatibility fall asymmetrically on existing and entrepreneurial entrants.

In one common scenario, an entrant seeks to imitate the incumbent with a backward-compatible offering that undercuts margins on one component. For example, RCA sought to imitate IBM's System 360 with a backward-compatible design, only to find that this was an expensive and ultimately unprofitable activity. As another example, DR-DOS sought to imitate Microsoft's DOS, only to find that getting access to distribution channels was expensive and difficult.[50]

[47] See the analysis in Bresnahan and Greenstein (1999) or Bresnahan et al. (2009).

[48] Grove (1996) describes the vigorous debate around this crucial decision, his initial desire to adopt a frontier design, RISC (reduced instruction set) instead of CISC (complex instruction set). A retrospective view is found in Tedlow (2006).

[49] For example, at that point in time Microsoft had recently experienced competition from OS2, an IBM operating system that closely resembled Windows 3.0 and exceeded its functionality, and from DR-DOS, an imitation of the DOS operating system.

[50] For an analysis of the issues Microsoft faced and their behavior, see Gilbert (1998).

The innovative outcome from this common scenario depends on the decisions made by incumbent firms. For many years, for example, IBM sought to compete with all entrants, experiencing considerable success in mainframe markets for corporate computing throughout the 1960s, 1970s, and, 1980s. In contrast, it had a more mixed experience in other segments when bringing innovations to market. For example, in the 1970s and 1980s, Digital Equipment Corporation had considerably more commercial success in factory floor computing and general-purpose minicomputers. Similarly, in the 1970s, Wang had more success in word processing—until the rise of the IBM PC in the mid-1980s.

The tolerance of early users for less-than-perfect products plays a role in the other common scenario. An entrant may seek to satisfy a segment of users that the incumbent firm's backward-compatible offering neglects after establishing its user base.[51] For example, many workstation firms in the mid- to late 1980s sought to satisfy the needs of users with demand for high-speed computation, a function that the common PC could not meet as efficiently and the mainframe could not provide as cheaply. It turned out that in the early 1990s, these same workstation firms were often the biggest supporters of client–server architectures, when these firms again tried to expand their functionality to fill a need that neither incumbent platform had met satisfactorily (though not for lack of trying).

In short, platform leaders have incentives to expand the scope of platforms from which they profit, and they have incentives to aspire to continuity in the use of that platform. Entrants, in contrast, have incentives to consider whether to commit to an existing platform, or join another that might compete with it. In turn, that translates into high incentives for incumbents to support design of new proprietary standards for an existing platform, but not nonproprietary standards that might lead to more competition between platforms. On the other hand, entrants of applications prefer to make them compatible with as many platforms as possible, which lead to incentives to work toward nonproprietary standards, or other technological tools to reduce the costs of supporting cross-platform applications.

The flow of events during more recent experience has also depended on the choice made by incumbent firms. For example, in the early 1990s, Microsoft devoted enormous organizational effort and energy to producing Windows 95, and it reaped enormous profits. At the same time, Microsoft's management misinterpreted events in the Internet, not recognizing how a series of innovations from different corners would lead to the viability of many businesses founded on browser-based computing. Accordingly, the company spent the better part of the mid- and late 1990s and beyond trying to make up for a comparatively late start, while also selling in large quantities Windows 95. Microsoft altered the direction of its investment, supporting some Internet technologies, defended the value of the investments it had already made.[52] Then, after the potential for the alternative platform built around Internet and Web standards diminished, it cut back on many of those same investments.[53]

In each platform, it is rare to observe more than a small number of firms acquiring leadership positions. It is unsurprising, then, that questions about how incumbent firms react to new entry and defend existing positions in valuable markets have attracted antitrust scrutiny. For example, IBM's

[51] For analysis of some of these issues, see, for example, Bresnahan and Greenstein (1997), Bresnahan and Yin (2007), Arora et al. (2006).

[52] This story is well documented in many places. See Cusumano and Yoffie (2000) for an account of Netscape's Founding.

[53] See Bank (2001) for primary account of this behavior. Bresnahan et al. (2009) provide a summary of Microsoft's actions.

behavior in peripheral markets attracted such attention in the 1950s and late 1960s. Intel and Microsoft's behavior in the mid-1990s also attracted such attention.[54] Did such attention alter innovative behavior? Almost certainly it did, though precisely how remains open as of this writing.[55]

2.8. *Economic experiments and market-based learning*

So far, I have highlighted that innovation in computing markets often does not begin with events in a laboratory nor does it follow a predictable sequential set of stages. Instead, activities outside of a laboratory often take primacy, such as innovations that arise from market experience. A series of studies of innovation in computing highlight how firms learn from their experience in markets, particularly when market experience alters knowledge about the value of a good or service or the costs for bringing a service to market. Following Rosenberg (1994) and Stern (2005), we label these events *economic experiments*.

Economic experiments involve more than just changing knowledge pertaining to technical invention; economic experiments may also change knowledge about business operations and organization that translate technology into economic value. By this broad definition, economic experiments encompass a wide range of market-based learning. The most common economic experiment is incremental in its technical scope and ambition. It aims at learning lessons with immediate consequences for a business. Though incremental, it can involve decisions of the utmost importance to the business, such as learning information about the pricing for a new service using a new technology.

From one perspective, this activity is mundane and almost routine. Managers would authorize the expenditure of resources, redirect personnel, alter a feature of an existing service, develop a new service, advertise a service or not, and then wait to find out whether these investments paid off in terms of additional revenue, market share, or pricing authority. Failure was not regarded automatically as a waste of resources if it led to valuable learning (e.g., a failed small-scale experiment could help managers avoid costly mistakes on a larger scale).

What do market participants learn from their experiments? They gain information that reduces uncertainty about the source of value in markets. From where does such uncertainty originate? Rosenberg's (1996) analysis of uncertainty in computing markets provides a suitable framework for understanding how experience shapes market-based learning in computing. In this structure, five factors prevent market participants from forecasting the future:

- primitive technology;
- unexpected complements;
- narrow search which yields applications with unexpected breadth;
- unanticipated systems;
- unpredictable user valuation.

[54] For an overview of differing viewpoints, see, for example, Schmalensee (2000), Fisher (2000), Henderson (2000), or Bresnahan (2004).

[55] See, for example, Gilbert (2006) and Baker (2007).

Uncertainty arises because technologies are primitive at the time of introduction. Market participants cannot anticipate the technology's use until it becomes more refined. For example, in 1975, it was difficult to imagine the use for an advanced PC in 1995, because the 1975 processors were incapable of doing more than add numbers. Only experience with faster processors contributed to understanding what the PC could achieve.

Market participants also cannot learn about key complements without experience. One invention motivates searches for complements, a search whose outcome may be difficult to predict. For example, improvements in microprocessors motivated inventions of complementary parts in the motherboard, software, printer devices, screens, and myriad input/output devices, which further motivated improvements in microprocessors. The result from such searches can be learned only after firms introduce new products.

A very subtle barrier to forecasting arises because a narrow search may generate a wider set of applications than were the original motivations for the search. For example, a short-range wireless networking technology—now popularly known as Wi-Fi—did not arise from a single firm's innovative experiment. Instead, there were many potential business applications for this standard; One of the earliest prototypes had been in wireless terminals[56] and another had been in a large-scale LAN for a university campus.[57]

After Wi-Fi was designed in the IEEE 802.11 committee, numerous businesses began directed experiments supporting what became known as *hot spots,* which was an innovative idea altogether for retail provision of wireless computing. A hot spot in a public space could be free—installed by a home-owner, maintained by a building association for all building residences, or supported by a café, restaurant, or library trying to support its local user base. Or, it could be subscription-based, with users signing contracts with providers, supported (usually with commercial arrangements) by a café', airport, restaurant, or other commercial entity. In any events, a hot spot was a use far outside the original motivation for the standard.

The Internet is an excellent example for another type of uncertainty, namely, unanticipated systems. Market experience is required to recognize the value of a system of complements. Several components may be comprised by a system that delivers a functionality whose value exceeds the value any component could deliver alone. In computing, for example, it was difficult to forecast how the PC, fiber optics, and appropriate software would provide functionality as the Internet. For many users it could not be appreciated until experienced.

Finally, user valuation of end products is difficult to predict and often cannot be understood until experienced. The history of the market is full of user valuations that a mere survey would not have revealed. For example, there were well-known examples of underestimates of mass-market enthusiasm for the back-office mainframe computer, general-purpose minicomputer, PC, browsing, mobile computing, and other facets of electronic commerce.

Economic experiments continue to shape recent events in markets within commercial computing, such as Internet access market. For example, at the very outset of the browser-based commercial Internet in 1995, many ISPs wrestled with fundamental decisions about how to commercialize the

[56] Vic Hayes, one of the earliest developers of wireless technologies and standards, and chair of the IEEE 802.11 committee during the 1990s, first developed wireless technologies for National Cash Register (NCR) (a subdivision of AT&T then, today a division of Agere Systems). In that capacity he first developed wireless terminals for stockbrokers. See Khariff (2003).

[57] See the description in Hills (2005) of the beginning development of the equivalent of a wifi network for the Carnegie Mellon campus in Pittsburgh, which starting in 1993.

Internet. Part of the confusion arose from uncertainty about whether to imitate the pricing norms in the bulletin board business, where users phoned into a single server acting as repository for content (hence, the server acted as a electronic equivalent to a bulletin board), or invent a new pricing model.[58] Ultimately, flat-rate pricing emerged in the United States. By 1997, ISPs offered service in every major US city, and many large firms had begun building national networks.[59]

This was but one of numerous experiments to resolve many open questions. For example, a crucial question at the outset concerned the design of the opening page—or, as it was subsequently labeled, *portal*—that users would see when they first clicked on their browser. Should it be a directory of Websites, as if it were the yellow pages for local Websites, or a search tool, permitting a wide variety of tastes, rented from another firm?[60]

The ISPs also varied in the range of services they offered. Different ISPs made distinct choices and learned different lessons about the trade-offs between these choices. No single choice dominated, and as firms learned more, perceptions about the costs and benefits of each changed over time.[61]

While economic experiments give rise to unexpected innovations in computing, most often the interplay of firms leads to outcomes that none of the firms individually intended. This observation motivates a close examination of learning externalities.

2.9. *Economic experiments with learning externalities*

Directed experiments are those undertaken by firms for their own purposes, while undirected experiments are those that arise from the interplay of many firms' actions. Learning externalities can arise from both types of experiments and in a variety of ways. Inter*firm* information externalities occur between firms. For example, one firm's directed experiment may teach another firm a lesson, or a set of actions may interact in an undirected experiment and teach every industry participant a lesson. Inter*temporal* externalities, however, occur over time. For example, the lessons of prior experiments may generate lessons on which further experiments are built. In practice, these two externalities are difficult to distinguish from one another.

The positive interfirm information externalities take one of two forms. In one case, what worked for one firm becomes known and imitated by others (e.g., success from an experiment at an ISP in one rural location in 1996 implies it might be profitable in another). Alternatively, what did not work for one firm becomes known and, therefore, avoided: For example, the difficulties with the first design for wireless computing became known from experiences in 1997, which caused equipment firms to delay building plans until a more suitable design emerged with institutional support for enforcing interoperability.[62]

In another form, one firms' failure can teach lessons that help another succeed. The history of Internet access is littered with examples of failures from which all other firms learned. For example, it is now

[58] This argument is fully developed in Greenstein (2007a,b).

[59] By the fall of 1996, there were over 12,000 local phone numbers in the United States to call for commercial Internet access, and more than 65,000 by fall 1998. See Downes and Greenstein (2002) for a description of the dial-up market, or Downes and Greenstein (2007) for an analysis for why some areas had more entry than others.

[60] See the discussion in Haigh (2007).

[61] This can be seen in Greenstein (2000). For a full summary, see Greenstein (2007a,b).

[62] Greenstein (2007a,b) discussed these examples at length.

accepted wisdom that users do not desire only a browser and phone numbers presented as if it were packaged software—as was first marketed by Spry networks in "Internet in a Box." Rather, users want ISPs that offer a different type of service with a different set of market features, combining local services with software tailored to their immediate demands (and tailored to some needs users do not know they even have). It is also accepted wisdom that mass-market users do not desire login names with acronyms that are difficult to recall or do not relate to natural language names, as was widely commercialized by CompuServe. Most users also value avoiding technically laborious set-up costs involving weeks of waiting, as was embedded in early data services, such as Integrated Services Digital Network (ISDN).[63] The list goes on.

Intertemporal externalities also lead to divergence between private costs and benefits and industry-wide costs and benefits. One party (in a directed economic experiment) or several parties (in an undirected economic experiment) assume the cost of generating lessons while many others gain the benefits later. That is, those who pay for lessons in an early market are not necessarily those who use them most profitably in a later market, but no contract between these firms governs the extent of direction of the early investment.

An important feature of intertemporal externalities is the asymmetries to the costs and benefits of generating lessons about commercial failure. Lessons about how to avoid commercial failure can be valuable, but the firm whose failure illustrates the lesson for others rarely, if ever, does so for that purpose, and almost never under contract with the others that (later) gain the benefit of the lessons learned from the failure. In an extreme case, a firm may learn a lesson, teach others from its failure, but go bankrupt before it is able to use that lesson. Even though the lesson was expensive to the stockholders of the firm that initiated the experiment, it was inexpensive to the survivors.

Intertemporal externalities also played a role in the early growth of the Internet. The browser gave many ISPs the confidence to open service for their areas.[64] The growing adoption of the Internet motivated many entrepreneurs to propose new businesses for venture capitalists to fund. The growing adoption of these services by business and households, in turn, motivated other software entrepreneurs to develop business that took advantage of developing electronic commerce, which also motivated further household adoption.[65] In brief, a series of largely uncoordinated, yet complementary, actions by buyers and suppliers reinforced the value of entry and adoption by each other in a positive direction. For a time, that motivated even more adoption and more entry (in the late 1990s) until the ceilings on that value creation became transparent to all, slowing the process. More specifically, after the spring of 2000 adoption by first users continued and increasing use by experienced users also increased, but the rate of entry of new firms declined.

In all these examples, no single firm initiated an economic experiment that altered the state of knowledge about how to best operate equipment or perform a service. Rather, many firms responded to localized user demand, demonstrations of new applications, tangible market experience, vendor reaction

[63] For example, in their estimates of demand for broadband, Savage and Waldman (2004) find that most users are willing to pay a considerable fee to avoid set-up hassles and achieve a reliable service.

[64] See the analysis in Downes and Greenstein (2002, 2007).

[65] Mowery and Simcoe (2002a), and Kenney (2000), contain analyses that place these events in the context of the national innovation system and the structure of Silicon Valley, respectively. See Goldfarb et al. (2005) for analysis of the dot-com entry wave in particular.

to new market situations, and other events that they could not forecast but which yielded useful insights about the most efficient business actions for generating value.

While directed experiments might have partially motivated the actions of any single firm, it would be an error to regard the lessons learned as singularly resulting from only one firm's actions. Instead, the interplay of firms, their actions, and their economic experiments yielded a form of serendipity in learning—learning that resulted from the unanticipated combination of lessons learned from several actions or sources.

2.10. Localization of innovative activity in computing

That innovators learn from one another is no secret to historians of technology. It arises frequently in descriptions of the history of the PC industry, for example.[66] Because learning externalities build over time, however, the accumulation tends to make the externalities geographically localized. In turn, this leads to the concentration of innovative activity in a small number of locations.

There are numerous reasons for this effect. For example, the tacit knowledge about the workings of a prototype cannot be transmitted easily without repeated face-to-face contact, a factor that tends to slow the spreading of information outside of a small geographic region. In addition, the early exploration phases of a new commercial market require giving enormous discretion to entrepreneurial actors or managers within divisions, without formal monitoring mechanisms. Venture capitalists supervising their investments (or managers supervising their employees) may prefer frequent "hands-on" contact, an activity that, once again, is easier in close proximity.

An additional factor shapes the character of a local network of firms, the labor market for technical talent. Firms that share a location (or reside in close proximity to one another) necessarily share a labor market. While the presence of many buyers of labor could bid up wages, the presence of a thick supply of labor also makes it easier to meet unique and potentially short-term demands for specialized skill.[67]

Indeed, the common name for the Santa Clara Valley, the Silicon Valley, recalls the era (1960s and 1970s) when many integrated circuit firms were founded there while sharing a labor market, input supply markets, and financial support structures. This enabled movement of new ideas between firms within close geographic proximity.[68]

[66] The early years were distinguished by entrepreneurial energy driving the segment's growth and by firms building innovations on top of other's innovations. See, for example, Cringley (1992) and Frieberger and Swaine (1984). Analysts also have emphasized the heavy reliance of third-party vendors on nonproprietary standards, which encouraged vertical disintegration in PC supply. Langlois and Robertson (1992, 1995) analyze the causes and consequences of such vertical integration in great depth for PCs and others facets of electronics.

[67] See, for example, Fallick et al. (2006) and Franco and Filson (2006).

[68] How such areas arise is distinct from questions about how they persist, and I focus on the latter. As for origins of the concentration of the integrated circuit industry, perhaps the best known historical cause was due to the founding of Shockley laboratories, founded by William Shockley in Santa Clara after he left Bell Laboratories, where he, John Bardeen, and Walter Brattain had invested the transistor. For a few years, this was an iconic commercial firm in electronics with a famous founder. With such status, it attracted considerable engineering and technical talent. However, Shockley's overbearing managerial style eventually drove many of the senior managers to leave the firm. Some of them went on to found their own companies, some that later became among the most famous in electronics, such as Fairchild and Intel.

Geographic localization has had several consequences for innovation in computing. For example, early events have had long-lasting consequences for the speed and direction of later innovation. A well-known illustration recounts events at Xerox's Laboratories in Palo Alto, as it was a site that developed key inventions for small-system office computing in the late 1970s. The (then dominant) photocopying company had established a laboratory with many leading researchers in computing. Prototypes for several key designs originated there, including the mouse, the graphical-user interface, and LAN. For a variety of internal reasons, the company was slow to commercialize on these, and, through personnel departures and information leakages, the ideas behind these inventions eventually moved to other nearby companies, such as 3Com and Apple Computer.[69]

Von Burg (2001) argues that sharing information with others was crucial for inviting many firms to make equipment. For example, Bob Metcalfe's design for the Ethernet was built on years of university research in data communications. Metcalfe became a part of Xerox research team. His dissatisfaction with some management decisions led him to initiate his own commercialization effort (3Com), where he found multiple ways to share the technological core of information with others. Subsequently, a community of firms and technologists grew up around the Ethernet standard, and as it became larger others became reassured about its continuation, attracting even more participants. As a result, this community became committed to Metcalfe's design and they collectively enjoyed the benefits of a network effect (pun intended). Eventually, other alternatives could not command much market share, which left the majority in favor of Metcalfe's Ethernet design.[70]

Localization has also shaped the spawning of new businesses. In North America these have tended to be concentrated in a small number of locations, such as the Boston area and Silicon Valley.[71] This does not mean all the significant young firms in the last 30 years start in the Valley—after all, for decades the largest large systems computer firm (IBM) was and continues to be headquartered in New York. Today the largest US PC hardware firm (Dell, founded in mid-1980s) is based in Texas, the largest PC software firm (Microsoft, founded in late 1970s) is based in Washington State, and for many years the largest dial-up national ISP (AOL, founded in mid-1980s) was based in the Virginia/Washington, DC area. It does mean, however, that many new firms are founded out of or near the Bay Area. In the last 30 years, this includes firms such as Oracle (late 1970s), SUN (early 1980s), 3Com (early 1980s), Cisco (mid-1980s), EBay (mid-1990s), Yahoo (mid-1990s), Google (mid- to late 1990s), and many others.

Localized learning displays self-reinforcing features. That is, one successful investment builds on another, with experienced workers and financial institutions continuing to create value. The same area became a nurturing nest for the boom in the PC markets of the late 1970s and early 1980s, the LAN boomlet of the late 1980s, and the dot-com boom of the 1990s. Some analysts have argued that the heavy concentration of invention in Silicon Valley arose from the inordinate extent of sharing of information and mobility among a talented workforce, giving the region greater potential to grow than the greater

[69] Apple Computer, for example, initiated programs in graphical-user interface and a mouse (famously) after Steve Jobs, then CEO of Apple, received a tour of the Xerox labs.

[70] In fact, a still-active standard-setting IEEE committee 802, which endorsed Metcalfe's design, continues to extend and improve it in directions far outside the scope of Metcalfe's original intent. The IEEE is a nonprofit consortium with representation from the industry. The IEEE also endorsed two competing technologies at the same time. Metcalfe's design, though, eventually became part of a suite of commercial products sold by many firms, including 3Com, which competed against alternative specifications developed by other firms, such as IBM.

[71] There is considerable writing on this topic. See, for example, Saxenian (1994), Kenney (2000), and Lee et al. (2000).

Boston area.[72] Small firms found the environment nurturing because they could take for granted the availability of many key inputs, a thick labor market for technical talent, financial help from venture capital, and up-to-date information about the latest technical trends.

There are countervailing forces pushing away from the concentration of supply, namely the geographic dispersion of users and the gains to suppliers from collocating next to them. As illustration, Arora and Forman (2006) examine the question of which services are tradable in the outsourcing of IT services. They analyze the outsourcing decisions of a large sample of 99,775 establishments in 2002 and 2004, for two types of IT services: programing and design, and hosting. Programing and design projects require communication of detailed user requirements in contrast to hosting, which requires less coordination between client and service provider. They show that the probability of outsourcing programing and design is increasing in the local supply of outsourcing, as should be expected if some nontradable or "local" component to programing and design services cannot be easily removed. In contrast, the decision to outsource hosting is insensitive to local supply except for users with security concerns.

Remarkably, over time, no single cluster alone has served the entire computer industry. Every facet of the supply chain for computing involves firms headquartered and operating in a much wider set of locations. Entry into facets of these markets has become an important phenomenon worldwide. The supply chain for many complementary components has also been associated with many firms in Western Europe and as well as in China, India, Ireland, Israel, Japan, South Korea, Singapore, and Taiwan. The software industry has spread to different areas of the world.[73] Even more widespread are computing service firms, which follow users dispersed across the globe.

Despite this geographic dispersion over the last five decades, US companies have retained leadership in generating new platforms and commercializing frontier technologies in forms that most users find valuable (Bresnahan and Malerba, 1999). Part of this results from the persistence of platform leadership for a time within a segment. In addition, US firms have historically been ascendant whenever platform leadership has changed. Nevertheless, this pattern seems likely to change in the twenty-first century, as non-US firms already have found leadership positions in producing components of many platforms and in related areas of electronics, such as consumer electronics, communication equipment, and specialized software.

The role of localization differs between creation and supply of new computing and its use and adoption. Adoption of frontier applications in computing is not necessarily localized, and on occasion has been much more geographically dispersed. A similar remark holds for the geographic supply for support services for frontier computing technology. As a result, there is little evidence of massively different adoption patterns for new computing across 50–100 major urban areas in the United States. While data do show slight biases in favor of a few areas, such as San Francisco and Boston, these are readily explained as a result of the composition of the work force or type of industry in the region.

More to the point, the differences between major urban areas are small in comparison to the more dramatic differences between major urban areas and slow growing small towns and/or poor rural areas. Those differences show up in PC use, as well as in Internet use and supply, at home and in business.[74]

[72] See, for example, Saxenian (1994) or Kenney and von Burg (1999).

[73] This movement has long antecedents. See, for example, Mowery (1996) or Arora and Gambardella (2005).

[74] For more on the geography of the Internet in the United States, see the reviews in Greenstein (2005) or Greenstein and Prince (2007). For recent evidence on PC use, see Beaudry et al. (2006).

3. The commercial Internet in the United States

The networking and Internet revolution in the late 1990s appears to be responsible for blurring once-familiar distinctions between segments of the computing market, geographic locations, and different technologies. For instance, the new networking technologies can build client–server systems within large enterprises and across ownership boundaries. Within these systems are Internet-based computing networks linking potentially vast geographic distances, and thereby supporting the emergence of a "network of networks" on an unprecedented scale.

What, if any, continuity is there in the economic determinants of innovative conduct? The growth of the Internet is a useful case study for illustrating that there is more continuity than meets the eye. Indeed, by identifying such continuity, this review ultimately highlights the very opposite—what is unique in the recent innovative experience.

To be sure, the answer is not obvious. The Internet is not quite like the mainframe or the PC. It is not just a single piece of equipment that embodies components from multiple suppliers. Though it helps move data between computers, it is also not quite like the LAN that attaches to existing computing. Its value chain is far more complex and involves many more firms. It is not quite like a new software application. It is not just a single program installed on a computer that generates a new set of functions.

The Internet is also not quite like any commercial communications network that came before it. It is partly a packet-switching network for moving data between computer clients. Yet, this does not fully describe the commercial form it took. A complex technology had to be embedded in a multilayered network, and many different participants operate its pieces. In addition, the Internet altered many different computer component and software markets simultaneously, so the boundaries of the computing market have changed. A hardware-based definition for the computing market was barely adequate in the 1960s and is no longer sufficient for economic analysis in the Internet era.

Moreover, its commercial diffusion looked quite unlike anything that had come before it. After years of development, a few applications were built that provided compelling value for tens of millions of decisions makers. In the popular imagination this happened overnight, with the creation of the browser. In fact, it had happened over more than two decades, starting from the first government funding at DARPA. The browser was but the last of many innovations, and, thankfully, a commercial marketplace for Internet services had been put in place just before it became available.

A brief review of the size of the Internet access economy gives a sense of how big demand for the Internet became, once it started to commercialize. The enormity of the Internet economy is discussed in Greenstein and McDevitt (2009), which analyzes but a small piece of it, the total revenues for Internet access since the late 1990s. The revenue affiliated with providing access is one of the largest categories of revenue out of the value chain for Internet services, and it is quite large. By 2006 total revenues have reached $39 billion. This is extraordinary for a technology that had few commercial service providers prior to 1989.

During this growth, the Internet began to accumulate more capabilities and functions, as a range of firms began to use pieces of the Internet to enhance services provided to paying customers. Over time, "the Internet" became a label for not only the Internet but also for all the applications that accumulated around the Internet, used pieces of the Internet, commercialized new functions for the Internet, and which together delivered an enormous array of services to a wide range of users.

Generally speaking, four types of rather different uses share the same capacity: browsing and e-mail, which tend to employ low bandwidth and tolerate delay; video downloading, which can employ high bandwidth and can tolerate some delay; voice-over IP and video-talk, which tend to employ high bandwidth and whose quality declines with delay; and peer-to-peer applications, which tend to use high bandwidth for sustained periods of time, and can tolerate delay, but, in some applications (such as Bit-Torrent) can impose delay on others.[75]

This range of uses and applications serves as cause for both celebration and consternation. The commercial Internet is not just an e-mail network for technically skilled users. It is an e-mail or instant messaging communications network for some, a gaming network for others, a source of news for others, and a distribution channel for video and musical entertainment for others. For many users it is also the principal media for engaging with geographically dispersed communities of friends.

Hence, it is challenging to begin at the basic starting point for empirical analysis—figuring out what to analyze and how to measure its change. This review will use the established frameworks, beginning with known historical facts about technology push prior to the commercialization of the Internet and then moving to a description of the Internet's diffusion and adoption patterns.

3.1. Stretching the frontier prior to commercialization

The Internet began to commercialize around 1992.[76] Within a few years, there was an explosion of commercial investment in Internet infrastructure in the United States. How did that occur? The transfer of the Internet from its research origins to a commercial form involved three somewhat interrelated events: The privatization of the Internet, the creation of the World Wide Web, and the commercialization of the browser. Together these set the stage for a surprising commercial explosion.

A fair reading of the history of each event would suggest that the inventors/initiators did not fully forecast the consequences of their own actions, and most industry insiders were surprised by the changes the commercialization of the Internet enabled.[77] This would suggest that a "technology push" interpretation of the early Internet is consistent with events. While that is partly so, it would be a mistake to go too far with such an interpretation, as if Internet technology was simply dropped on commercial markets like manna from heaven. The early events also cannot be understood apart from the institutional factors shaping commercial behavior at the time.

[75] This is explained in considerable detail in Ou (2008).

[76] While the commercialization of the Internet is sometimes dated to the development and implementation of NSF's privatization plan for the NSFNET in 1994–1995, that does not recognize the investments made by many early entrants prior to the final NSF plans. Attempts to privatize some assets affiliated with operating the Internet dates to the late 1980s. Those investments resulted in a confrontation between the commercial firms operating the NSFNET under contract with NSF, that is, IBM and MCI, and those firms, such as Sprint, PSINET, and UUNET, who also were building commercial services and sought to interconnect with other carriers without violating the NSF's Acceptable Use Policy (for traffic that had no direct relationship to research activity). A more proper dating for the beginning of commercialization was ending of this dispute. That occurred with the passage of the Scientific and Advanced Technology Act of 1992, Public Law Number 102-476, sponsored by Rick Boucher (D—9th District, VA), which amended the NSF acceptable use policy. See Kahin and McConnell (1997) or Hussain (2003).

[77] For a long detailed explanation for how unexpected factors shaped outcomes many different Internet markets, see, for example, David (2001) or Greenstein (2007b).

What became the Internet began in the late 1960s as a research project of the Advanced Research Projects Administration of the United States Department of Defense, the ARPANET. From these origins sprang the building blocks of a new technology for a communications network, one based on sending data where some amount of delay was tolerated. By the mid-1980s, the entire Internet used TCP/IP packet-switching technology to connect most universities and defense contractors.

Management for large parts of the Internet was transferred to the NSF in the mid-1980s. Through NSFNET, the NSF was able to provide connections to its supercomputer centers and a high-speed backbone. Since use of NSFNET was limited to academic and research locations, carriers who carried commercial traffic, such as UUNET, PSINET, and Sprint developed their own private backbones for corporations looking to connect their systems with TCP/IP (Kahn, 1995).

By the early 1990s, the NSF had developed a plan to transfer ownership of the Internet out of government hands and into the private sector. The plan for privatization was motivated by several factors. For example, it was forecast (correctly) that a privatized Internet would be more efficient than a government operated one, leading to lower costs for all users. There was also a concern that the NSF could not fund indefinitely the operations of the network, and it was thought that privatization would put the network on more stable financial footing. During the transition another issue arose: several of the private providers of data services were chafing under the NSF's "acceptable use" policy forbidding them to use government-owned assets for commercial purposes. Complete privatization also would remove this issue.

The privatization plan had three important elements. One key element was the operation of data-interchange points. There were precedents for such operations in federal operated data-interchange points. Among the earliest private arrangements for this was handled by the Commercial Internet eXchange (CIX), which permitted all parties to exchange data (to "peer" without charge) at locations supported through a group funding effort. NSF's privatization plan led to the opening of several more locations for data-interchange, called Network Access Points (NAPs). Altogether, at the outset these made it possible for multiple actors to enter into the networking business, with no dominant provider able to exclude any other.[78]

A second key element was the privatization of the domain name system, which until that point had been competently handled in a relatively informal operation. Subsequently, it became a large-scale thriving (and lucrative) commercial activity. Because it never lost some of its monopoly-like features in its commercial form, the public policies toward the domain system were controversial then and have remained so.[79]

The third and final key element was the shutdown of the NSFNET. When it shut down in 1995, only for-profit organizations were left running the commercial backbone, while a mix of commercial and nonprofit firms operated access points. With the Internet virtually completely privatized, its diffusion path within the United States was largely dependent on market forces and economic incentives.[80]

[78] See Hussain (2003) for a review of the changes in these over time.

[79] For different viewpoints, see, for example, Kesan and Shah (2001) or Mueller (2002).

[80] For a variety of perspectives about the path to privatization, see Abbate (1999), Kahin and McConnell (1997), Hussain (2003), Mowery and Simcoe (2002a,b), and Greenstein (2007b).

The NSF privatization plan put in place a scalable network. The operations for updating routing tables, exchanging data, obtaining domain names, and building applications would remain roughly the same even with many more users. Bottlenecks in the provision of capacity also did not materialize because private firms saw opportunities and acted on them, precluding any single firm from operating the network in its entirety and strategically blocking others.

One key early invention arose during privatization, the World Wide Web, and it is linked with a particularly important invention, specifically, the commercial browser. Tim Berners-Lee and Robert Cailliau built key parts of the World Wide Web and, in addition, Berners-Lee organized its pieces into the World Wide Web Consortium that standardized many protocols for a growing community of users. Those accomplishments took several years, and at the outset, even Berners-Lee did not forecast their large impact. Indeed, his goals at first were modest and focused on the needs of his employer and a research community he aspired to help.[81] As such, these circumstances fit one factor of Rosenberg's (1996) aforementioned frameworks—a narrow search with results of breadth outside the scope of the initial search.

Specifically, Cailliau and Berners-Lee were employed at CERN, a high-energy physics laboratory in Switzerland. They sought to make a program that aided the sharing of textual and nontext files among researchers with a program of sufficient generality to handle many different types of files. Though there had been years of discussion within computer science about how to design such a system, one of Berners-Lee's core insights was *not to* design a perfect system. Rather, he looked for a hypertext solution that met the needs of the local "constituency," that is, a solution that made it easier for technically oriented users (physicists who were not computer gurus) to send files easily to one another and make them available for downloading without knowing all the ins and outs of each computer system.

These inventions were useful—but not very useful—unless widely adopted. In 1991, Berners-Lee made two inventions available on shareware sites for free downloading: html (hyper-text markup language) and the URL (universal resource locator) a hypertext language and labeling system that made transfer of textual and nontextual files easier. Once installed in a host computer these were well suited to Berners-Lee's constituency in two specific senses: (1) It helped users organize transfers of known files and (2) it helped make files available to others without a tremendous amount of pretransfer searching. Trials and operations over the next several years helped further refine its operations in several ways described below.

As these inventions grew in use, Berners-Lee forecast the need for an organization to assemble and standardize pieces of codes into a broad system of norms for operating in the hypertext world. He founded the W3C for this purpose. In 1994, he left his employer and established the offices for the W3C in Cambridge, MA. This organization ultimately helped diffuse many of the software standards and tools that became important for operating on the commercial Web, such as html and the related tools for deploying it.

[81] See, for example, Berners-Lee and Fischetti (1999) and Gilles and Cailliau (2000).

3.2. Learning externalities and commercialization

Several researchers devised improved versions of browsers that worked on Unix operating systems. These were known among insiders and technically skilled programers in the research community. Improvements accumulated, and that set the stage for the invention of one additional complement, the commercial browser—an invention that in turn motivated subsequent further inventions.[82]

In 1992, a University of Illinois team situated at the National Center for Super Computing Applications (NCSA), an NSF funded research center that supported a super computer, sought to design an easy-to-use browser for nonresearchers. The NCSA supported a large social network of researchers, who regularly used shareware software and made it available to others. Because of this environment (and for other reasons), Mosaic, at first appeared to be a routine project. In this instance, the team of programers included Marc Andreeson, an undergraduate, and Eric Bina, an employee of NCSA and recent Masters graduate. They took increasing responsibility for the browser project over time, not only to program and design it, but to help it diffuse, to debug it, and to respond to requests for changes from users.

Mosaic had been built on the prior designs, borrowing many elements from them and the programers tailored the design to the new audience. The project had many features that made it easy to use. After success with their Unix-based project, they developed new features that were novel. They built a browser for a Windows-based systems, at that time the most widely used operating system worldwide for PCs. Up until that point it had not occurred to anyone in the technically adept community of Internet programers to write something of value for a nontechnical user.

The release of Mosaic browser began in early 1993, with the Windows-based browser coming later in the year. It became available on shareware sites aimed at sharing software among university users. Within a year, over a million downloads had occurred. As it grew in popularity, the managers at the University of Illinois, Urbana/Champaign realized this invention had great potential. They arranged for commercial licensing of the browser. They anticipated that the browser would diffuse into popular use through both shareware, which was free, and commercial licensing, which involved a more organized commercial support. Almost certainly those expectations would have been correct had a third channel not emerged. That third channel involved the student programers—but none of the faculty or other administrators—who had helped develop the Mosaic browser; the students decided to start a business around the same time as the university began its licensing program.

Specifically, Marc Andreessen, one of the lead programers on the Mosaic project, had graduated in December of 1993, moved to the west coast to take a software job, and subsequently grabbed the attention of Jim Clark, founder of Silicon Graphics from several years earlier. Clark had excellent established connections with the West Coast computing community. Clark's and Andreesen's relationship coalesced into a business plan in the spring of 1994. They called themselves the *Mosaic Communications Company* and sketched a plan to make money selling a browser. They received venture funding from the same venture capitalists that had backed Clark's earlier efforts, hired many of the same programers who had worked at NCSA in Champaign, and went on a crash course to become a large organization supporting worldwide use of their browser. Eventually this plan would expand far beyond

[82] A version called Viola would shape many of the decisions made by a team who explicitly aimed for a popular browser. See Gilles and Cailliau (2000).

the browser, blossoming into an extensive business plan to support a range of complementary activities around their own browser, server tools, and range of services. In effect, this program eventually aimed to make the licensing program of the university obsolete.[83]

The University's officially sanctioned channel was managed by a third party—a company known as Spyglass, located in Illinois with a history of commercializing NCSA inventions. Spyglass was given the right to license the trademarked name *Mosaic*. Spyglass eventually decided to defend its intellectual property, forcing Mosaic Communications Company not only to change their name to *Netscape*. The threat of further legal problems also made Netscape's programers take extra care not to overlap with the intellectual property owned by the University.[84]

Netscape's beta browser was released in the late fall of 1994, gaining publicity, followed by the first commercial release in the winter of 1995, and its IPO (initial public offering) in August of 1995. Netscape's business model and marketing efforts were wildly more successful than any other licensee's at catalyzing many other market participations. Among the results from its many effects, Netscape's activities were instrumental in motivating Bill Gates to reverse his previous position about staying out of the browser business.[85]

Indeed, the University of Illinois played a role there too. After failing to buy Netscape in the summer of 1995 or deter them from pursuing a strategy that conflicted with Microsoft's desire to lead all software application development, Microsoft set on a course to offer a browser. The fastest way to do that was through licensing Spyglass' browser, which had been licensed by farsighted Microsoft employees in January of 1995 (as part of an internal campaign to change the priorities of the organization). After doing so, Microsoft added a few features, rebranded the browser, and unveiled Internet Explorer in December 1995.[86]

The University of Illinois also served as the origin of the most widely used HTTP (hypertext transfer protocol) server, a collection of technologies that supported browsing, particularly SGI script, and use of web technologies. This eventually became known as the Apache server. The emergence of Apache was not the original intent of the university's administrators. The server was available for use as shareware, and many Web masters took advantage of it, adding improvements as needed.

By February 1995, however, the university was not keeping up with users, partially because key employees had left for Netscape. Many users of the NCSA server software sought to coordinate further improvements. NCSA tried to revive the software in April 1995, but, upon learning about the Apache effort, changed its plans to support a shareware version of the server software, and cooperated with Apache.[87] Out of these efforts arose the most widely used server on the commercial Internet and it became one of the most widely adopted open source projects after Linux.

[83] See, for example, Cusumano and Yoffie (2000).

[84] Netscape's management claimed it would have reprogrammed the browser from the ground up in any event, because they were developing software to support their long run goals, which required starting from scratch. However, the concerns about intellectual property made that goal a necessity rather than a luxury.

[85] See Bresnahan et al. (2009). For the original memo by Gate announcing this change in direction, see Gates (1995).

[86] See the account in Sink (2007). He argues that Spyglass' management was quite elated to get Microsoft as a licensee. After they licensed the browser Microsoft increasingly devoted more programmers to it over time, positioning itself as the firm to support other application development. All the other licensees eventually chose to either get support from Microsoft or Netscape; none continued with Spyglass.

[87] The name apache was a play on words, as the first February, 1995, effort involved bringing together a piece of software that involved many "patches." See http://httpd.apache.org/ABOUT_APACHE.html, accessed March 2007.

Both the browser and the server are good examples of the role of university-funded research breaking new technological ground that commercial firms overlooked. It is also good illustration of the value-added that commercial firms bring to products, refining them, branding them, and servicing them, contributing to the diffusion of a technology in a myriad number of ways.

It is also an example of localization in the commercialization of software production. Neither browser nor server technology moved to firms headquartered in Illinois, despite the seeding of a Spyglass, or the nearby presence of Chicago, the third largest city in the country and one that does not lack the appropriate infrastructure or labor market for technical talent. To be sure, production did stay inside the United States. Clark was already located in the Bay Area, and there was no debate that the new firm would stay there to take advantage of the existing cluster of software firms, venture capital finance, and labor market for technical talent. Their strategy depending on scaling their organization rapidly through hiring talent—arranging many business ventures and growing with the help of many complementary firms—and the location was a key part of that.[88]

Accordingly, while Netscape was the most important browser firm in the Internet, the core of development took place in California. After a few years, Microsoft's Internet Explorer became more important than Netscape, and for that period the key production for the browser was located just outside Seattle, where Microsoft located what became the Internet product and tools division.

As an interesting contrast, Apache was founded by several people and some of the key leaders were located in (and remain in) the San Francisco Bay Area. In some sense, however, it is not located anywhere in particular, remaining largely a virtual organization, with member and code coming from programers all over the world.

3.3. Localization and commercialization of Internet software production

As the browsers began to diffuse, entrants began exploring new businesses. Many of them were founded in the San Francisco Bay Area, a direct outgrowth of the entrepreneurial-orientation of the venture-funded community, who was looking to fund start-ups, and the preexisting labor market for technically skilled employees at other firms, who were becoming privy to the latest developments for the commercial web. These entrants included Yahoo!, Hotmail, Excite, EBay, Vermeer, and many others that anticipated building businesses on browser-based computing.[89]

The entrepreneurial movement at the time extended far beyond the San Francisco boundaries. For example, Bill Gates' memo of May 1995 entitled "The Internet Tidal Wave" advises other executives to examine Web pages by Lycos, Yahoo!, Oracle, Symantec, Borland, Adobe, Lotus, the *San Jose Mercury News*, Novell, Real Audio, Disney, Paramount, MCI, Sony, ESPN, and many other sites.[90]

[88] The reliance on external complementors is described in detail in Cusumano and Yoffie (2000). While they do not highlight the role of geographic localization in that strategy, it was plainly obvious to everyone that the users were geographically dispersed, many of the complementors would be geographically close, and most of the software vendors had an affinity for Netscape's competitive goals in opposition to Microsoft.

[89] With 20–20 hindsight it is too easy to treat this outcome as obvious. For insights into the uncertain experience of an entrepreneur in this early era, see Ferguson (1999).

[90] See Gates (1995). This memo announced Gate's intention to alter the strategy direction of the company and focus on Internet technologies. The list of sites were there to give other executives examples of sites Gates found exemplar in various respects.

Although the first eight of these companies were located in the Bay Area, the rest were not.[91] As another example, in May of 1995, *Boardwatch* magazine, the primary trade publication for the US bulletin board service market, listed over 700 price plans for getting commercial Internet access from local ISPs all over the United States.[92]

These types of events fueled expectations among industry insiders, futurists, and venture investors that substantial demand for the Internet at households and businesses would emerge quickly.[93] Many firms had began to initiate projects for converting part of their business to browser-based computing, especially among those technology firms whose livelihoods depending on racing into mass markets faster than others. The Netscape IPO took place in August of 1995 and it was wildly oversubscribed (or very badly underpriced), pushing the trading price multiple times higher on the initial day of trading.

With the Internet, the relationship between the investor community and entrepreneurial community took a different scale and pace than it had in prior technology-induced waves, such as with PCs, LANs, and client–server systems. In part, this was due to the breadth of perceived opportunities. Rather than being a brief race among several dozen firms to develop new components and related systems, the Internet invited a wide range of new thinking across many activities—in back-office computing, home computing, and information retrieval activities in numerous information-intensive industries, such as finance, warehousing logistics, news, entertainment, and more. Ultimately, the Internet motivated the entry of new entrepreneurial firms continuously until the spring of 2000, peaking in 1998 and 1999. The entrants ranged over the entire spectrum of businesses shaped by the emergence of IT in the prior decades.

A new data-carrier industry also arose. It involved a mix of existing firms and new ones, especially among ISPs. More than 5000 such firms arose in the United States by 1998, most of them small and specialized on serving a local area.[94] A few prominent firms emerged, principally AOL, Earthlink, Mindspring, Netzero, Level3, PSINET, and a few others. They competed with new divisions at established firms, such as AT&T and MCI.

Perhaps the most attention went to a category of entrant, generically known as the *dot-com*. The label came from a feature of the domain name system for the Internet, which initially designated five types of domain names: gov, net, org, edu, and com.[95] *Gov* was for government entities and *edu* for educational institutions. *Net, org*, and *com* were designated for nonprofit and private entities, organizations, and networks. *Com* became the most popular among commercial firms (by far), even for firms not based in the United States.

[91] As yet another illustration of the extent of geographic dispersion consider this example from a slightly later time, the analysis of commercial prospects in the Internet market—first performed in late 1995 by Morgan Stanley analysts Meeker and DuPuy (1996) after Morgan Stanley successfully handled Netscape's 1995 IPO. It highlights young Internet companies from California, such as Silicon Graphics, SUN, Cisco, Excite, Netcom, and Netscape, but also features plenty from all over the country, such as AOL, CompuServe, UUNet, Dell, Compaq, US Robotics, IBM, and others.

[92] This is a conservative estimate. Many more firms were list, but many did not provide pricing information. See Stranger and Greenstein (2007).

[93] For different perspectives on how US business reacted to the new opportunity, see, for example, Mowery and Simcoe (2002a,b), Kenney (2000), and Greenstein (2001).

[94] For detail, see Stranger and Greenstein (2007).

[95] Every country was assigned control over the allocation of domain names underneath its two-letter country code. The country code for the United States is US, but com, net, edu, gov, and org presume US sovereignty. See Mueller (2002) for a description of how this developed.

Dot-com entrants covered a wide range of new businesses. Initially a large number of entrants went into advertising-supported Websites or sites with no usage fees.[96] A number of sites covered directory and search activities, while others specialized in supporting conversations and information sharing for particular topics or groups.[97]

An entirely distinct group went into electronic commerce in various forms. Some sought to sell and distribute goods and services, such as books or travel services or apparel. Others sought to sell subscriptions to services, such as the *New York Times* crossword puzzle or regular updates of industry news. Still others tried to assemble groups of buyers and sellers, either in open auctions or for more tailored matching purposes.

At the time, there was a pervasive ideology about using the Internet to remake economic activities. Given the label *the new economy* in popular discussion, this vision argued for the Internet's exceptional nature. It argued that the Internet-enabled business processes which would remake the structure of production and redesign the locus of decision making within organizations. Its enthusiasts encouraged entry by new firms on the premise that many incumbent firms would be unable to adapt the new technologies to their existing businesses fast enough or effectively enough to compete with the new entrants.[98]

Despite such rhetoric, existing firms did not stand still. Many leading firms invested in new processes and new Websites to support their businesses. Still others hired consultants and experts to help with adjusting to the new potential and new commercial threats. Accordingly, a large consulting and advising industry grew in the late 1990s at companies such as Accenture and EDS and many others, as did a substantial set of firms for supply software infrastructure, such as hosting and Web design.

Although the entry and investment boom around the Internet resembled prior entrepreneurially led booms in new technical opportunities, such with PCs, LANS, and client–server networks, it also differed in size, scope, and aspiration. With the Internet some new entrants sought to create new businesses and remake value chains from production to distribution to final user. There was open discussion about changing the entire chain of actions supporting the delivery of any valuable information-intensive activity, such as music, publishing, news, financial activities, and entertainment.

For many investors a pervasive optimism in the late 1990s moved the otherwise enormous uncertainty about the source of value into the background. Nonetheless, there were open questions about how large the advertising-supported economy would become in its online format and how quickly online retail channels would become used by buyers and whether their growth would favor existing firms or incumbents. There were also questions about how to design for the most valuable features, computing

[96] See Goldfarb (2004).

[97] See Haigh (2007) for a discussion of this segment, and see Goldfarb et al. (2005) for discussion of the variety of entrants.

[98] Some of this view was founded on serious academic research, for example, Bower and Christensen (1997) or Christensen (1997), but over time it developed into an untested belief system that did not necessarily measure itself against results. While it justified numerous IT projects, ultimately these were judged against their productivity (e.g., see the summary in McKinsey, 2001). For a variety of perspectives, see, for example, Forman and Goldfarb (2006), Goldfarb et al. (2005), or Shiller (2000).

languages, and features to serve emerging usage patterns among the growing set of Internet users. In brief, this period was marked by economic experiments across a wide range of activities that overlapped with applications of computing, as well as any upstream or downstream activity related to it.

3.4. Platform development for the commercial Internet

As one might expect in a market with divided technical leadership, commercializing the Internet introduced a plethora of variants on developing platforms. There was considerable disagreement among participants about whether any of these were valuable, and about what to call each distinct strategy. Even with hindsight, there is disagreement about which strategic investments worked and why.

One approach to new platforms employed existing assets as much as possible, by trying to encourage existing users to employ a standard bundle of components that continued to use the proprietary assets of an existing firm. Firms with leading platform positions in the computing or data-carrier business before the Internet commercialized, such as IBM, MCI, Microsoft, Intel, Novel, 3Com and Cisco, initially sought ways to create value for their users through incremental innovation, while simultaneously attempting not to lose their leading positions. Beyond that generality, the details differed significantly.

I now explore some of the dramatic events arising from the actions that individual firms and types of firms undertook to develop successful platforms. I begin with IBM: At the outset of the 1990s, before the Internet commercialized, IBM's mainframe business had begun to decline significantly. This led the board to remove the CEO and break with precedent by hiring a CEO from outside the company, Louis Gerstner. Along with selling off several business units such as the networking carrier business, Gerstner chose to concentrate the firm's assets in advising companies how to implement IT in effective ways that created value. This approach used IBM's existing relationships with enterprises and grew the company into essentially a large consulting and systems integration practice, which grew even larger through mergers with smaller firms.[99]

Cisco behavior also drew attention in popular discussion for a few distinct reasons. It sought to become the leading provider of enterprise-level data equipment, through internal growth and designs out of its own considerable Research and Development department. Notably, though, Cisco also sought to increase its product line and set of personnel through the acquisition of small venture-funded start-ups. Most of these had one prototype product, if that, and fewer than 100 employees. Over the course of the late 1990s, Cisco bought over 80 such companies, forming an ecosystem with the venture capital community who started those firms.[100]

To be sure, venture-funded entry was not new for computing markets. The new phenomenon was the systematic strategy of larger firms to grow their portfolio of innovative projects through venture-funded firms. Firms, such as Cisco and JDS Uniphase, would monitor external innovative projects and "cherry pick" those that fit with their strategies, either through merger/acquisition or other forms of cooperation.[101]

[99] This account oversimplifies a enormous task. For more detail, see Gerstner (2002) or Carr (1999).

[100] There was no secret about what Cisco did, but the number of deals and their frequency was without precedent. See, for example, Paulsen (2001).

[101] For more on these type of strategies, see, for example, Paulsen (2001), Arora et al. (2001), Gans and Stern (2003), Gans et al. (2002), and Gawer and Cusumano (2002).

Among all incumbent computing firms who responded to the opportunities in the commercial Internet, Microsoft's actions perhaps gained the most attention in popular discussion because it generated a series of events known as the browser wars, which involved a dramatic confrontation between Netscape's browser business and Microsoft's. In addition, Microsoft's aggressive competitive tactics motivated a large federal antitrust case against it.

At a basic level, there was no mystery at all to Microsoft's goals. The company had gained an extraordinarily lucrative position in the PC operating system and applications markets, and a growing position in the lucrative networking operating systems market, and it had an incentive to protect that position by preventing any alternative platform from emerging, if it could.[102] One step to do that involved providing a browser, eventually named Internet Explorer. A second step involved converting most application developers to making programs compatible with that browser and no other. The second step, of course, reduced the value of Netscape's platform, but could not be achieved without first persuading users to employ Internet Explorer for most of their browsing, thereby giving Microsoft increasing ability to bargain with developers. Indeed, eventually Microsoft did achieve its first goal, making Internet Explorer the default browser on most PCs. Eventually they gained share of Internet Explorer that, for a time, exceeded 90% of all Internet users.[103]

If the goals, by themselves, were not controversial, the tactics were. Late in making crucial investments in Internet technologies, Microsoft was technically far behind its nearest rivals once it finally did begin to make investments to support its proprietary platform. Then, concerned that others would develop a persistent platform, Microsoft used its existing relationships with firms to insist on contractual obligations (such as first-screen restrictions, "*de facto*" exclusive deals, and controversial quid-pro-quos) that restricted innovative actions of OEMs. Those tactics came under close scrutiny, and once questioned, Microsoft reacted with an aggressive legal response instead of finding any point of compromise. That confrontation played a role in inducing the Department of Justice to bring a federal antitrust suit, which kept embarrassing revelations about the browser wars in the news long after Microsoft had achieved its commercial goals.[104]

Intel, like the previous examples, also altered its innovative priorities for computing in response to the new opportunities. Intel had crept into the motherboard business over the 1990s as it initiated a variety of improvements to the designs of computers using its microprocessors.[105] It also accelerated investments in manufacturing technology, resulting in what appeared to be the fastest rate of improvement in price/performance ever achieved.[106]

Another series of entrants sought to take advantage of the new opportunities to create value with Internet technologies: A large variety of different firms sought to match buyers with sellers, or match

[102] Gates (1995) provides a quite cogent analysis as to why the Internet could give rise to platforms that would undercut Microsoft's margins. In brief, a browser controlled by another firm could support a platform comprised of software available on the Internet that eliminated the unique position of Microsoft's operating system, reducing its value.

[103] See, for example, Cusumano and Yoffie (2000), and Bresnahan and Yin (2007).

[104] For different accounts and viewpoints, see, for example, Cusumano and Yoffie (2000), Henderson (2000), Fisher (2000), or Bresnahan (2004), Schmalensee (2000), or for a sardonic take with biting insight, see Chapman (2006), Chapter 10.

[105] See Gawer and Cusumano (2002) and Gawer and Henderson (2007).

[106] See, for example, Aizcorbe (2006) for analysis of the determinant of improvements, as well as Flamm (2003). See Aizcorbe et al. (2007) for why this helped computing equipment more than communications equipment.

users with similar communities of interest, or even match questioner and seeker with informative answers. Economists have labeled these different approaches either *intermediation* (Lucking-Reily and Spulber, 2001) or *two-sided markets* (Evans et al., 2006). These initiatives varied in their pricing structures, that is, in how they sought to generate monetary revenues from the new services offered. Some firms sought to make revenue from charging subscription fees (e.g., dating services), while others sought to make it through advertising next to the relevant information.

A related important innovative offering were *portals*. Portals became some of the most popular Websites among Internet users, accounting for the opening pages for most users, acting as a tool for organizing web surfing, and accounting for the highest share of an individual's time online. There were two approaches to providing portal services at first. One approach made it an extension of another service, and virtually all of these firms aimed at mainstream users. For example, Netscape and Microsoft made the opening Web page a default setting on their browser, directing considerable traffic to netscape.com and msn.com, respectively. Both were far down in use to AOL, which made its opening page an extension of its dial-up Internet service and organized its contents in an easy-to-use proprietary format—a strategy that became known as a *walled garden*.

Another approach was of stand-alone portals, and these differed in their appeal to technical and nontechnical users. Most early (and successful) portals in the mid-to-late 1990s provided directory services for the vast amount of content on the Web. Yahoo!, Lycos, Excite, and others took this approach. Yahoo! grew into a vast organizer of content and a popular destination. Excite sold itself to @home as part of a strategy to help cable Internet users. Late in the 1990s, a newly founded firm, Google, entered with a new approach to searching—ranking Web pages on the basis of the number of links to them, and supporting itself with advertising. This turned out to be popular with users; after the turn of the millennium, Google continued to grow this business and eventually surpassed the early leader, Yahoo!, in usage, revenue, and capitalized value.

This brief survey is the tip of the iceberg of how innovative computing enabled numerous new opportunities in electronic commerce and information-intensive activities, linking the market structure of computing and many other firms.[107] Retailing firms began developing strategies for differentiating their services from each other, and for customizing distinct versions of their activities to many customers. Firms in media markets—news organizations, music distributors, entertainment companies—began developing software solutions to restructure their services for the Internet. In turn, these opportunities generated a cornucopia of innovative efforts at software and hardware computing firms to support this new direction for value creation.[108]

Unlike the most Apocalyptic predictions after the episode in browsing, no dominant and proprietary Internet platform emerged after the first decade of the commercial Internet with control over the entire commercial value chain. That is, while many firms succeeded in making considerable revenue during this period of growth and in controlling niches or corners of the value chain, there was no single

[107] Other prominent platforms include those provided by Research in Motion (Blackberry), Apple (iPhone, iPod), Oracle (enterprise databases), EBay (auctions), Facebook (social relationships), as well as many others. Each one of these platforms deserves a longer description, and the reader should be clear that the absence of that here is due to space constraints, not their lack of importance.

[108] For more on the economic factors shaping competitive activity electronic commerce and technically intensive information industries, see Shapiro and Varian (1998), Varian (2000), Smith et al. (2000), Bakos and Brynjolfsson (2000), Bakos et al. (1999). For an overview of market structure, see Varian (2001).

proprietary firm largely shaping the direction of technical change in all aspects. And, yet, at the same time, something familiar was emerging: In each subcomponent or horizontal layer of the Internet, a few prominent firms made viable businesses selling components and services that the vast majority of users continued to use. This was a step toward becoming a standard bundle for what would someday be recognized as an Internet platform.

This observation can be restated as an aphorism. Since it first began to diffuse to the general public, the Internet has been called a "network of networks." Yet that phrase is misleading. It does not reflect how commercial behavior shaped the evolution of technology in the last decade and a half. Leading firms and their business partners view the commercial Internet through the same lens they view activities in the rest of computing. For them, the commercial Internet is a "network of platforms."

3.5. New forms of organizations to coordinate firm conduct

Some observers attributed the rapid accumulation of experimentation to the emergence of a new form of leadership for designing standards, one that involved collections of market participants. The standards committees that were responsible for designing key standards for the Internet were comprised of representatives from many firms and interested researchers from universities and other nonprofit organizations. Because undirected economic experiments are those undertaken by more than one firm working together, by definition, the committees participated in these types of experiments. This raised the profile of activities inside standards committees and it directed attention at different forms of consensus-oriented standards processes for designing standards accommodating a variety of complementary goods and services.

Ultimately, the accumulation of Internet industry knowledge depended on spreading the lessons learned from economic experiments. Further innovations then built on that knowledge, renewing a cycle of accumulated lessons from more experiments. This accumulation was a key driver of the market's evolution because it set the conditions for innovative behavior. Standards committees participated in this cycle and helped shape the Internet by affecting, for example, pricing, the quality of services, and the identity of leading firms.

Standards committees had always played some role in the computer market. Their role in the Internet was more notable for what it was *not:* These institutions were not beholden to the managerial auspices of AT&T or IBM. For that matter, these committees also did not simply ratify the design decisions of Intel, Microsoft, or Cisco, though all those firms sent representatives who had a voice in shaping outcomes.

The range of such important decisions shaped by standards committee was without precedent. The IEEE, for example, made designs that shaped the LAN market, modem, and wireless data communications markets, while the IETF made designs that shaped the operations of every piece of equipment using TCP/IP standards.[109] Many of these decisions went into use quickly, ensured that all complying components would interoperate, and had enormous consequences for the proprietary interests of firms. Never before had such a large industry had so much of its innovative activity shaped by collective firm decisions.

[109] Simcoe (2007) provides an overview of the operations of the standardization process at IETF and its changes as it grew.

Another notable feature of these committees was their governance structure. They were largely organized without much government directive or mandate, especially at the outset of the Internet. It is incorrect to say that the government was uninvolved: After all, the NSF and Department of Defense both had played a crucial role in starting and sponsoring organizations that managed and improved the operations of the Internet.[110] Often, government representatives were present and influential on specific features of design, and sometimes government provided crucial endorsements. It is just that at the outset of commercialization, governments did not put the force of law behind many of these standards, or insist that government employees have exclusive influence on design choices. Rather, the commercial Internet of the 1990s embodied the accumulation of multiple improvements suggested through a process of consensus in committees, and that consensus depended in large part on the functional ability of code to perform as claimed.[111]

Its importance can be understood in comparison to the next closest alternative. The next closest alternative design for global networking—the Open Systems Interconnection model, a.k.a., OSI seven-layer model—arose in the late 1980s in a process almost as different as could be any far-reaching standard. The OSI was a formal standard design for interconnecting networks that arose from an international standards body, reflecting the representation of multiple countries and participants. The network engineering community in the United States preferred their bottom-up approach to the OSI top-down approach, and, when given the opportunity, invested according to their preferences.[112] Indeed, pieces of the OSI model still live on today (how much is a point of some debate), but the locus of decision making over the direction of Internet standards resides firmly outside of what's left of the organization.

The lack of government involvement could also be seen in other aspects of the Internet in the United States. For example, although the Federal Communications Commission (FCC) had mandated a standard for digital television (as it had for color television), it refrained from mandating most Internet equipment design decisions. Just as the FCC had not mandated Ethernet design standards, so it let the spectrum become available for experiments by multiple groups who competed for *wireless* Ethernet standards, which eventually became Wi-Fi. Similarly, the FCC did not mandate a standard for modems other than to impose requirements that limited interference. It also did not mandate interconnection regulatory regime for Internet carriers in the 1990s, explicitly letting the firms innovate in the structure of their business dealings with one another, and evolve those dealings as they saw fit.[113]

Another notable innovative format for consensus decision making went by the name, *Open Source*. These organizations used variants on the General Public License (GPL), created by Richard Stallman. He called his creation *copy left* and he intended it as a contrast to the use of copyright. A GPL required all contributors to give up ownership rights to the collective group. Any contributor could use any

[110] This is especially true of the Internet Architecture Board and Internet Engineering Task Force, before it moved under the auspices of the Internet Society in 1992, where it remains today. See, for example, Abbate (1999) or Russell (2006).

[111] See Abbate (1999) for a history of the design of these protocols. See Partridge (2008) for a history of the processes that led to the development of e-mail, for example.

[112] See, for example, Russell (2006).

[113] The latter forbearance was deliberate. On the lack of interference in the design of the Ethernet, see von Burg (2001). On the design of 56K modems, see Augereau et al. (2006). On the lack of regulation for network interconnection, see the full discussions in, for example, Oxman (1999) or Kende (2000) or the account in Nuechterlein and Weiser (2005). More recent experience has departed from these trends, particularly in principles for regulating last-mile infrastructure. A summary of these departures is in Greenstein (2007a).

others' code and build on it, but only if no single individual and no firm claimed rights to exclude others from using their contribution.

The use of the GPL in the 1990s differed from Stallman's original goal. He had wanted to create a format that reduced the proprietary discretion restricting a user's ability to change computer code. By the mid-1990s and beyond, many Open Source licenses employed many variants that sought to accommodate commercial activity in one form or another, while still allowing users to contribute code and share with one another.

One well-known Open Source project was Linux, a basis for computer operating systems. It was begun by Linus Torvald in the early 1990s as a derivative, or "fix," to Unix. It was freely distributed, with alternating releases of a "beta" and "final" version. Starting around 1994–1995, about the same time as the commercialization of the Internet, Linux began to take off. What had started as a small project caught the attention of many Unix users both at commercial companies and elsewhere. Many users began converting from proprietary versions of Unix (often sold by the hardware manufacturers) and began basing their operating systems on Linux, which was not proprietary. Many of these same users began contributing back to the Linux project, strengthening the range of applications. This movement gained so much momentum that Linux-based systems became the most common server software other than Microsoft software. Many firms, such as Red Hat, began to turn a profit selling services and related components.[114]

As noted earlier, Apache was another early project founded to support and create "fixes" for the HTTP Web server originally written by programers at the NCSA. By 2006, more than 65% of Websites in the world were powered by the Apache HTTP Web Server.[115] Apache differed from many other Open Source organizations in that contributors "earned" the right to access the code. To be a contributor one had to be working on at least one of Apache's projects. By 2006, the organization had an active and large contributor base and seemed poised to continue indefinitely.

Another early Open Source project, MySQL, pursued a distinct model. From the outset, the organization behind MySQL aspired to make a profit. MySQL was a database with Website-powering, packaged software, and enterprise applications. Many small- to medium-sized companies utilized the free basic MySQL for their operations. This package was developed through an Open Source arrangement, in which no user paid for use of the software. Millions of copies have been distributed and companies pay only for the more advanced features of MySQL. The number of paying customers numbers in the tens of thousands; and the company only makes its revenue from these customers.

Perhaps the most well-known Open Source format was the least technical. It originated from something called a *wiki*. Developed in 1995 by Ward Cunningham, a software engineer from Portland, Oregon, wikis can either be used in a closed work group or used by everyone on the Internet. They originally were formed to replicate or make a variation on existing products or services, with the purpose of fixing bugs within the various systems. Accordingly, wikis were first developed and intended

[114] There is considerable writing about the growth of the production of Linux software and from a variety of perspectives. See, for example, Dalle et al. (2004), Lerner and Tirole (2002), Von Hippel (2005), West and Gallagher (2006), or Arora and Bokhari (2007). For an account and analysis of how many firms got on the Linux bandwagon, see, for example, Dedrick and West (2001) or Fosfuri et al. (2005).

[115] For more on the history and operation of Apache, see, for example, Mockus et al. (2005).

for software development, but had grown out of that first use and were now applied to a multitude of applications.

A particular popular application of wikis, Wikipedia, garnered worldwide attention. In the case of Wikipedia, the format was applied to the development of textual and nontextual content displayed on the Web. Wikipedia beat out Microsoft's Encarta for the honor of the Internet's top research site in 2005, a position that it has held thereafter, and it has been consistently ranked as a top 20 site for all Internet users. It is an online-only encyclopedia. The content is user-created and edited. As its homepage proudly states, it is "The Free Encyclopedia That Anyone Can Edit." The site has always been free of charge and never accepted advertising.[116]

These new organizational forms gave rise to a different leadership structure with regard to innovations for computing and the Internet—or what some observers regarded as the absence of a leadership structure. Unlike prior episodes of new uses for computing, after the first decade of the commercial Internet, no single firm or small group of firms emerged as the key drivers of technical change. Technical leadership did play some part in this behavior, but the additional and numerous Open Source initiatives across a range of applications were a driving force that led to this uncoordinated innovative behavior.

These experiments in new organizational forms gave rise to the belief, quite commonly expressed during the late 1990s, that the "new economy" would alter not only the use of computing, but also its production. It became apparent to savvy observers, however, that the extreme version of that prediction would not come to pass. Proprietary software continued to flourish, often coexisting with Open Source software while competing for some boundary uses. In fact, in many cases, such as Linux, the commercial firms actually prospered as part of the large coalition of firms supporting the Open Source software. In the case of MySQL, it was eventually purchases by Oracle.

Open questions festered, though: What types of innovative activity would remain proprietary and what applications were best organized around Open Source software? More broadly, if the fundamental economics of platforms still held in spite of divided leadership and the importance of nonproprietary standards development, then what form will new innovations take? In addition, how did focal developments arise in the absence of proprietary interest in sponsoring platforms?

These events illustrate several different mechanisms for the emergence of focal points for development other than sponsorship by a firm. For example, an individual with technical skill, a history of inventiveness, and, perhaps, celebrity status can gain authority over a range of standards, as Linus Torvald or Tim Berners-Lee did. Alternatively, a standardization institution with visionary leadership can take such a role, as several of the IEEE committees did for wireless Ethernet standards, and as the IETF did for a range of technologies in networking using the TCP/IP stack.

Moreover, both users and vendors have incentives to see focal points continue. So, as with any platform, these coalitions of supporting firms are difficult to stop once they gain a self-sustaining size. In brief, vendors continue to invest in these coalitions, as long as the hardware, software, and services embedded with these standards continue to provide functionality that participants find valuable.

[116] For more information, see Greenstein (2006).

3.6. Diffusion of the commercial Internet

From the previous sections, it is clear that like computing, the Internet was associated with high fixed invention costs and low marginal reuse costs. It also generated heavy early investment and led to frequent repurposing of focal inventions. In addition, the diffusion of the Internet followed the predictable regularities of a GPT. For example, it always takes time to move a frontier technology from a small cadre of enthusiastic first users to a larger majority of potential users. In this sense, the economic patterns found throughout the early diffusion of the Internet are general.

Nevertheless, the diffusion of the Internet also possessed some unique features. It thus far proceeded in two waves. There was a clear difference between low-speed/dial-up connection and high-speed/hard-wire connection. In the early 1990s, those with dial-up connection were considered at the frontier, but by the turn of the millennium, at households the new frontier consisted of high-speed connections, mainly through (digital subscriber line) DSL and cable modem supported access.

According to the National Telecommunications Information Administration (NTIA, 2004) study, as of 2003, approximately 61.8% of American homes owned a PC, with Internet participation rates at 54.6%. By 2006 Internet participation had reached 73% of households, with only a declining minority using dial-up connections.[117] These adoption rates suggest that the diffusion of each technology is moving into the late majority category of adopters, though there is considerable disagreement about how to portray the rate of adoption for the remaining households. Any entity (household, individual, or firm) is considered connected to the Internet if it has the capability of exchanging information with other entities via the physical structure of the Internet. Connections come at different speeds (56K dial-up vs. broadband) and from different types of suppliers (AOL vs. a telephone company). Figure 1 provides a visual representation of this diffusion to households, as well as the transition into broadband.

Data from 2001 show Internet usage to be positively correlated with household income, employment status, and educational attainment (NTIA, 2002). With regard to age, the highest participation rates were among teenagers, while Americans in their prime working ages (20–50 years of age) were also well connected (about 70%) (NTIA, 2002). Although there did not appear to be a gender gap in Internet usage, there did appear to be a significant gap in usage between two widely defined racial groups: (1) whites, Asian Americans, Pacific Islanders (approximately 70%); and (2) Blacks and Hispanics (less than 40%) (NTIA, 2002). Much of this disparity in Internet usage can be attributed to observable differences in education and income. For example, at the highest levels of income and education there were no significant differences in adoption and use across ethnicities.

Since the vast majority (87.6%) of PC owners had home Internet access, the marginal Internet adopter looked similar to the marginal PC adopter. For households, PC demand had two distinct populations: Those already owning a PC (repeat purchasers), and those that never owned a PC (first-time purchasers). Throughout the 1990s, two distinct Internet adoption patterns correlated with these types of PC demand. Either existing PC adopters converted to the Internet, or households bought PCs and converted to the Internet.

[117] See, for example, Greenstein and McDevitt (2009).

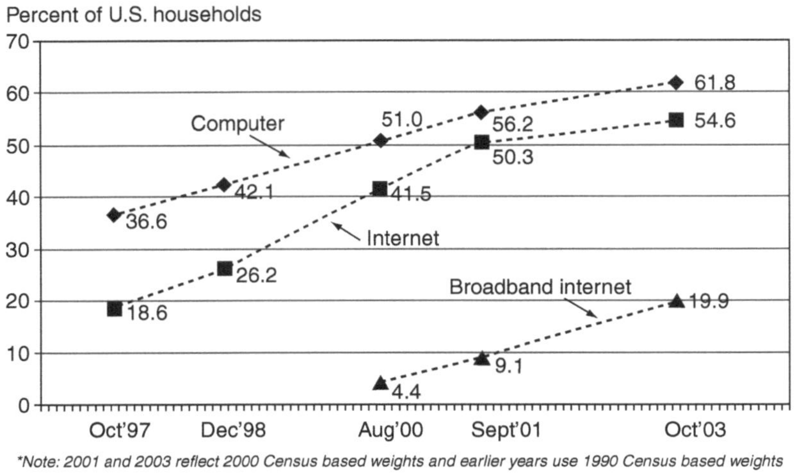

Figure 1. Percent of households with computers and Internet connections, selected years, 1997–2003*. Source: NTIA, 2004.

By 2001–2002, virtually all existing PC adopters had experience with the Internet at home. Accordingly, the diffusion process changed. There were large differences between existing users and new users in terms of the likelihood of buying a new PC (Prince, 2005). The demand for first-time purchasers was especially relevant since it represented the marginal adopters for PCs, and therefore, strongly resembled the marginal adopters of the Internet.[118]

As the diffusion of the PC moved deeper into mainstream use, the marginal PC adopter became a household with low marginal value for PC quality, high start-up costs, significant price sensitivity, and potential difficulty in determining *when* (not necessarily *if*) to buy. That is why the early Internet adoption experience provided little help for understanding user adoption in later periods. Quite a different set of factors shaped later adopter choices than did those for earlier adopters.[119] Similar reasoning partially explains the later appeal of cheaper devices for accessing the Internet, such as netbooks.

[118] In his paper, Prince describes three main determinants of the "divide" in PC ownership: heterogeneity (in the marginal utility of PC quality and PC holdings), start-up costs, and dynamics. His results indicate that the marginal utility of PC quality is strongly increasing in income and education, and strongly decreasing in age. Further, as prices fall and quality rises over time, the decision about whether to buy a new PC is complicated by the decision of when to buy a new PC. Finally, first-time purchasers are more price sensitive than repeat purchasers, and face large start-up costs. See also, Goolsbee and Klenow (2002), who emphasize the role of local network effects in motivating early adoption.

[119] Such an observation has led to distinct research approaches. For example, Sinai and Waldfogel (2004) investigate whether Internet adoption was motivated by desires to overcome isolation (particularly among minorities) or geographic distance (among those far from retail outlets). For more on urban/rural differences in connectivity, see Strover (2001) or Greenstein (2005). Some observers characterize the coming era as not one defined by access to computers, as in the past, but, instead, as one defined by differences in use of computers. Some users will display more sophistication than others, and this will shape differences in returns from investing in computing. See, for example, Hargettai (2003).

While dial-up connection has moved past the frontier stage and approached saturation point in the United States, broadband access approaches the frontier with some frictions preventing uptake. For a few years it was far from ubiquitous, though that is changing as of this writing.[120] As the volume and complexity of traffic on the Internet increases dramatically each year, the value of high-capacity and universal always-on broadband service is constantly increasing. Furthermore, broadband access enables providers to offer a wider range of bundled communications services (e.g., telephone, e-mail, Internet video, etc.) as well as promote more competition between physical infrastructure providers already in place.

In the earliest years of diffusion to households—that is, prior to 2002—the diffusion of broadband Internet access was very much supply-driven in the sense that supply-side issues were the main determinants of Internet availability and, hence, adoption. Cable and telephone firms needed to retrofit existing plants, which constrained availability in many places. In those years, the spread of broadband service was much slower and less evenly distributed than that of dial-up service. Highly populated areas were more profitable due to economies of scale and lower last-mile expenses. As building has removed these constraints, demand-related factors—such as price, bandwidth, and reliability—have played a more significant role in determining the margins between who adopts and who does not.[121]

As of October 2003, 37.2% of Internet users possessed a high-speed connection; the dominant types of broadband access were cable modems and DSL. In addition, broadband penetration has been uneven, as 41.2% of urban and 41.6% of central city households with Internet access used broadband compared to a rate of only 25.3% for rural households. Consistent with the supply-side issues, the FCC estimates that high-speed subscribers were present in 97% of the most densely populated zip codes at the end of 2000, whereas they were present in only 45% in the zip codes with the lowest population density (NTIA, 2002). By 2006 Internet participation had reached 73% of households, and the supply-side issues began to fade, with only the most low-density parts of the country lacking suppliers.

A similar (second) wave of investment occurred in many developed countries over the first decade of the new millennia. Figure 2 shows growth of subscribers per 100 inhabitants in Canada, United States, United Kingdom, Germany, France, Italy, and Japan, as well as the entire OECD. Though countries differ in the level of broadband use—partly due to household size and other factors, the similarities between them are more apparent. Adoption of broadband grew in all countries.

At this time firms are developing and deploying a wireless delivery channel for some Internet-related services. These options vary in speed, quality, and price. There have been data services from the major cellular carriers (e.g., Verizon, AT&T, and others) since the turn of the millennium, particularly for e-mail delivery to laptops. The most popular mechanism in the recent past was a simple device for delivery of e-mail (e.g., a Blackberry). More complex devices have gained popularity (e.g., iPhones and smart phones), and these have download speeds that begin to approach the low end of wire-line broadband speeds. Technological optimists forecast even faster download speeds from next generation wireless carriers (e.g., WiMax or LTE). There is still considerable uncertainty about how many of these services the market will support, about what price and sales levels will prevail, and, accordingly, what scale of deployment these prices and sales levels will support.

[120] Broadband is defined by the FCC as the capability of supporting at least 200 kilobytes per second in at least one direction (supplier and/or consumer), http://www.fcc.gov.

[121] Also see, for example, Savage and Waldman (2004).

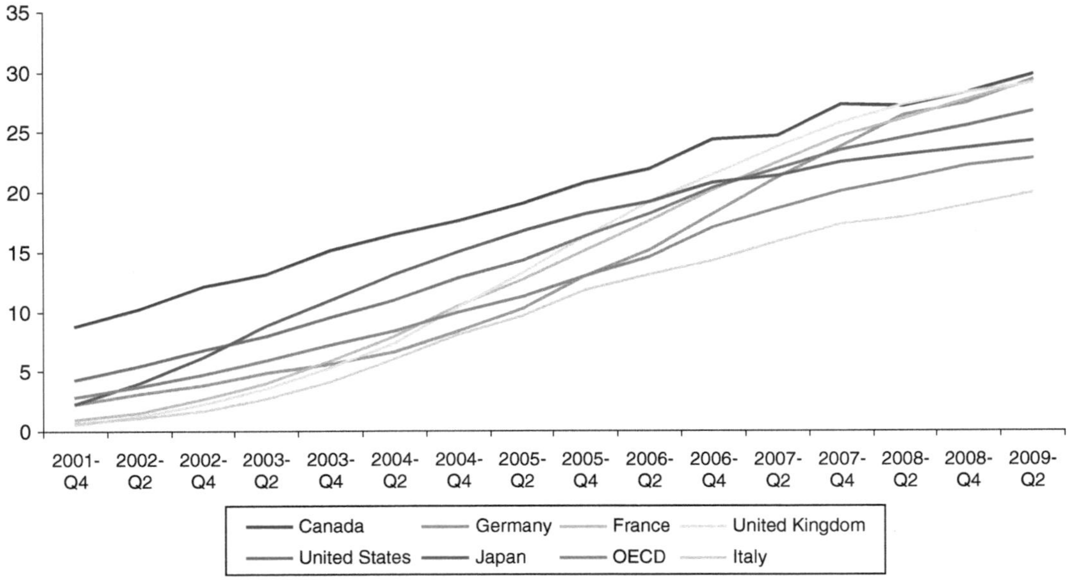

Figure 2. Broadband penetration, G7 countries through. Source: OECD Broadband Portal, http://www.oecd.org/sti/ict/broadband, **Table 1i.**

3.7. Coinvention and business processes for Internet technologies

Some industries are more information intensive than others and, thus, make a more intensive use of new IT developments, such as the Internet, in the production of final goods and services. Heavy computer technology–user industries have historically been banking and finance, utilities, electronic equipment, insurance, motor vehicles, petroleum refining, petroleum pipeline transport, printing and publishing, pulp and paper, railroads, steel, telephone communications and tires (Cortada, 1996).

The diffusion of the PC into business did not immediately alter those traditional rankings, but it did introduce computing into some industries that had previously been medium-intensive users, such as warehousing. The constant improvement in the quality of PCs combined with their falling prices along the entire range led to replacement and upgrades of existing systems, as well as to the addition of new uses for the PC.

The growth rates in real investment in computing equipment were extraordinary in the 1990s. Rates of investment in software reached 9.5% growth rates per year during 1990–1995, and 14.2% growth rates during 1995–2000. Computing equipment growth rates reached, respectively, 13.5% and 7.1% per year per period. Communications equipment growth rates reached 7.2% and 15.5% growth rates. All of these exceeded rates of growth in non-IT capital, which reached 6.8% and 4.9% per year over the same periods.

Business adoption of the Internet was partly responsible for some of the acceleration of investment in the late 1990s; and it came in a variety of forms. By the late 1990s, implementation for minimal applications, such as e-mail, was rather straightforward. It involved a PC, a modem, a contract with an ISP, and some appropriate software. In contrast, investment in the use of the Internet for an application module in a suite of Enterprise Resource Planning software, for example, was anything but routine during the latter half of the 1990s. Such an implementation included technical challenges beyond the Internet's core technologies, such as security, privacy, and dynamic communication between browsers and servers. Usually, organizational procedures also changed.[122]

A further motivating factor shaped business adoption: Competitive pressure. That is, there first may be a minimal level of investment necessary just to be in business. Second, there may be investments in the Internet that confer competitive advantage vis-à-vis rivals. Once again, these will vary by locations, industries, and even the strategic positioning of firms (e.g., price leader, high service provider) within those competitive communities.[123]

Forman et al. (2003a,b) measured national Internet adoption rates for medium and large establishments from all industries. They distinguish between two purposes for adopting, one simple and the other complex. The first purpose, labeled *participation*, relates to activities such as e-mail and Web browsing. This represents minimal use of the Internet for basic communications. The second purpose, labeled *enhancement*, relates to investment in frontier Internet technologies linked to computing facilities. These latter applications are often known as e-commerce, and involve complementary changes to internal business computing processes. The economic costs and benefits of these activities are also distinct; yet, casual analysis in the trade press tends to blur the lines between the two.

They show that adoption of the Internet for purposes of participation is near saturation in most industries. With only a few exceptional, laggard industries, the Internet is everywhere in medium to large business establishments. Their findings for enhancement contrast sharply. There is a strong urban bias toward the adoption of advanced Internet applications. The study concludes, however, that location, *per se*, does not handicap adoption decisions. Rather, the industries that "lead" in advanced use of the Internet tend to be disproportionately located in urban areas.[124] Related work suggests that small establishments in disadvantaged location may be unable to take advantage of the innovative opportunity due to lack of thick labor markets for technical talent.[125]

3.8. Unending economic experimentation

Innovation is experienced by forward looking participants, but understood only in retrospect, and usually only after considerable market experience. It is an exaggeration, but not much of one, to say that before events fully transpire there will be legitimate and passionate debate among participants about which model of value creations most accurately will predict near term events. Only economic experiments can resolve that uncertainty about value.

[122] See Forman and Goldfarb (2006), for a review of studies of Internet investment by business. See Doms (2004) for a review of the acceleration in business investment in PCs.

[123] As Porter (2001) argues, there are two types of competitive motives behind Internet adoption.

[124] See also Forman et al. (2005).

[125] See Forman et al. (2008).

As of this writing, the cyclic process of innovative activity continues in computing and the Internet. This can be seen in many places. For example, the dot-com boom has busted and entrepreneurially led entry has been reborn in a new wave called *web2.0* in sites such as You-Tube, Face Book, and MySpace, which take advantage of social networking among users. The Web is far from done experiencing new waves of entry for new applications, especially in the realm of software applications and other forms of electronic commerce.

In contrast, other layers of the industry have continued to undergo upheaval. Many of the early entrants into the directory business, for example, have lost market share and prominence. Google's innovative approach to searching the Internet and auctioning advertisements next to keywords has largely displaced many existing portals. Even the early leader, Yahoo, has lost some market share in relation to Google.[126]

Another change has begun at the layer of carriers. As users switch from dial-up to broadband Internet access, they also consider switching suppliers. This has led to a large decline in the prominence of AOL as a provider of access to households, and it has led to the ascendancy of the providers of broadband, largely local telephone companies (primarily Verizon, AT&T, and Qwest), and cable firms (primarily Comcast, Cox, Time-Warner, and a few others).[127]

The equipment market has stabilized, leaving Cisco in the dominant position in enterprise computing to serve data communications. Yet, many of Cisco's cousins, firms who grew spectacularly during the 1990s—such as JDS Uniphase, Corning, Lucent, Nortel, and 3Com, did not fare as well. They had to undergo large and painful adjustments in their operations because of the decline in demand (associated with the bursting of the dot-com bubble). Remarkably, the rate of entry of new equipment firms has significantly declined, leaving the incumbent firms more dependent on internal research and development activities than acquisitions (though that began to pick up again as the economy picked up in 2004 and beyond).

Intel also took a more aggressive role in designing PC. Increasingly over the 1990s Intel designed prototypes of these motherboards. By the early 2000 Intel was making some motherboards and encouraging many of its business partners to make similar designs.

In 2003 Intel announced *Centrino*, which marked a departure. It began embedding a Wi-Fi connection in all notebooks that used Intel Microprocessors. To be clear, this *did not* involve redesigning the Intel microprocessor, the component for which Intel is best known. It did, however, involve redesigning the motherboard for desktop PCs and notebooks by adding new parts. This redesign came with one obvious benefit: It eliminated the need for an external card for the notebook, usually supplied by a firm other than Intel, and installed by users (or OEMs—original equipment manufacturers) in an expansion slot. Intel also hoped that its endorsement would increase demand for wireless capabilities within notebooks by, among other things, reducing their weight and size, while offering users simplicity and technical assurances in a standardized function. Intel hoped for additional benefits for users, such as more reliability, less set-up difficulties, and less frequent incompatibility in new settings.

In brief, the Internet is undergoing a second wave of investment. It takes place in the presence of the near saturation of adoption by first-time users. That has been coincident with the presence of more capital deepening in business computing and in deepening operational to support it. These actions are

[126] These events are discussed more in Haigh (2007).

[127] These events are discussed more in Greenstein (2007a,b).

symptomatic that this second wave of investment, unlike the prior one, takes place in the presence of less uncertainty about the source of value.

The new organizational forms for designing new computing also face a series of new tests. Many firms found their business prospects too dependent on Linux to allow it to continue without some structured format, so together they established a firm-sponsored association to continue supporting changes to Linux, employing Linus Torvald and several others in a salaried position. Meanwhile, many new organizations continue to function and support developments, such as Apache, the World Wide Web, and the IETF. Yet, firms no longer naively leave these institutions alone. The standardization organizations find their committees filled with interested participants actively shaping future designs that might affect profitability. These institutions show signs of the stress by slowing down in their decision making, if they reach decisions at all. Perhaps that should also be cause for celebration, since it is a sign that the stakes are high.[128]

3.9. Continuity and change in innovative conduct

The diffusion of a new GPT, the Internet, and the World Wide Web, led to a wide variety of changes in computing markets. It was not the technology *per se*, however, that brought about the most dramatic change. If not technology, what distinguished the latter era from the earlier era?

There were many similarities. In many respects the economic opportunities and challenges resembled those found in prior episodes in computing markets. A new opportunity emerged from efforts at technology push in data networking, and the stretching of the frontier enabled opportunities for value creation as a GPT diffused. Different firms pursued the economic opportunities they perceived, limited by coinvention costs incurred by both buyers and sellers, and the boundaries of platform competition. Concerns about the emerging platform shaped difference in incumbent and entrepreneurial strategies. In addition, the inherent limits on learning through market experimentation shaped firm understanding about how to create value in this new unexpected commercial opportunity.

Three factors distinguish the Internet era from prior ones. First, the division of technical leadership cut across a wider array of activities than such prior innovative episodes as with the PC and LAN. As a result, a greater variety of market participants reacted to the opportunity. It also made for strange bedfellows. Firms that had little economic relationship to one another prior to the Internet, for example, such as a cable companies and an equipment firm like Cisco, or new firms like CNN and a portal like Yahoo, found themselves making deals and basing their growth projections on the outcomes of those deals. Business assumptions related to the structure of the value chain supporting valuable services and the appropriate innovative conduct for that structure had to be questioned and rethought.

The second distinguishing factor came from the new organizational forms for designing standards in advance of deploying functioning equipment, and, similarly, for altering designs already employed in functioning data networks. The Open Source movement was part of this change, and Linux and Apache, among others, received attention for good reason. Yet, that statement also understates the variety of different organizations for coordinating developments across firms in consensus forums, such as the

[128] For one interesting account of this slowdown at the IETF, see Simcoe (2007). For an account of the manipulation of hearings at the IEEE, see Mackie-Mason and Netz (2007).

IETF and IEEE and W3C. These actions changed the boundaries of platform competition, refocusing innovative competition on aspects of products and services other than proprietary features of design. In the short run, the effectiveness of these organizations actually reinforced preexisting tendencies to specialize in innovative activities as a source of differentiation and strategic advantage. It also raised questions, as yet unanswered, about the durable boundaries between developing proprietary and nonproprietary standards.

The third novel aspect of the commercial Internet involved its breadth of potential applications, leading to a greater breadth of aspiration among participants, and a corresponding change in uncertainty about the source of value from IT across a wide set of participants. The diffusion of the commercial Internet brought about the threat of lasting change in the structure of business, and not all of those changes occurred immediately. The symptoms of this breadth were everywhere in the late 1990s during the dot-com boom and the increasing globalization of production, and still remain in some forms for many firms afterwards. One symptom was that firms had to assign managers to follow developments (that they had previously ignored) in order to make thorough assessments about the direction of change in the source of value. For example, firms in the music distribution business had to follow developments at Internet start-ups, and firms at telephone companies had to understand the implications of the browser wars.

The combination of all three aspects perhaps led to the biggest surprise, widespread exploration by many players in a great many more applications than would have seemed possible or likely only a decade earlier. That is, the managers at firms who never had been big players in IT markets found themselves experimenting with nonproprietary and industry-wide standards-making institutions, facilitating negotiations between firms for key strategic issues. Such institutions coexisted alongside proprietary platforms, sometimes as competitors and sometimes as complements. Cisco, Intel, IBM and many Wi-Fi firms are active participants in these standards forums. Even (previously reluctant) firms such as Microsoft, AT&T, and Verizon have found it useful to participate and fund such activities. Completely new forums, such as the organization built to support Bluetooth, have hundreds of participants.

3.10. Economic conduct: Open questions

What are the economic determinants of firm conduct? What does the evolution of firm conduct over several decades tell us about those determinants?

Technology push continued to play a role over many decades. Has there been any diminution in the importance market incentives and market-oriented events in computing markets? It appears not. Market forces continue to shape the stretching of the frontier, the severity of change at any point in time, and the identities of leading firms. Coinvention continues to shape the deployment of general-purpose technologies, and market forces continue to shape the directions that users and firms pursue.

The experience of the last few decades highlights questions about the causes of variation in the costs of creating value. Why were these costs different for the PCs, LANs, and the Internet? Why did they differ over time? What economic mechanisms link such costs to the size of economic benefits from deploying new technology? These are some of the major unanswered questions highlighted by these events.

There also has been no decline in the economic role of computing platforms. Platforms shape economic behavior today as much as in any other time in the past. In part this happened because market participants have learned about the importance of platform strategies for creating value for firms, such as Intel and Microsoft. In part, changes in the unification or division of technical leadership within and between platforms has played a role in shifting value from one set of firms to another, as from Yahoo! to Google. How does the presence of platforms color the economic incentives of leading incumbent firms and new entrants to innovate? What mechanisms shape those incentives, and do these push firms to undertake innovation that reinforces or alters market structure? These are unanswered questions.

The role of market-based learning activity also has not diminished, but the economic behavior of the last few decades raises questions about the determinants of its prevalence and importance over time. How does behavior aimed at reducing uncertainty—such as the prevalence of platforms and localization of innovative conduct in a small set of locations—alter learning activities? Will other factors, such as the globalization of production and use, make substantial differences to related conduct? Similar questions apply to the unprecedented dispersion of uncoordinated innovative conduct that characterized the early diffusion of the Internet. Was that behavior merely an artifact of early diffusion or does it portends something like a permanent change in the norms for competitive behavior, resulting from coordinating participants with disparate commercial interests, such as found in Open Source platforms?

Another set of unanswered questions relate to the wide breadth of economic activity touched by computing. Many economic participants have built a network with a high degree of technical interrelatedness. Will the general gains to all parties from bringing routines into business activities overwhelm the discretion of innovative actors, limiting their innovative conduct? For the time being there appears to be no cessation in the never-ending nature of the investment in economic experiments, so that seems to imply that any breakthrough in widely used IT could have large economic effects on the entire economy.

Experimentation did not end with just one episode or the end of the dot-com bubble. Indeed, as of this writing, many questions remain open about the value of different aspects of IT in the long run, and firms continue to explore approaches to creating value. What is particularly notable is the lack of cessation of technically oriented entrepreneurialism. In the not so distant pass some participants had expected to undertake economic experiments all along, while other participants had previously thought they did not live in an entrepreneurial technology market. Instead, different expectations have melted away. Virtually all participants in these markets expect continual entrepreneurially led change, as well as its twin, the absence of economic tranquility.

Acknowledgments

I thank Tim Bresnahan, Chris Forman, Avi Goldfarb, Bronwyn Hall, Rebecca Henderson, Jeff Prince, Nate Rosenberg, Alicia Shems, Kristina Steffenson McElheran, and Scott Stern for comments. The Searle Foundation provided financial support. I am responsible for all errors.

References

Abbate, J. (1999). Inventing the Internet. MIT Press, Cambridge, MA.

Acemoglu, D., Aghion, P., Lelarge, C., Reenen, J.V., Zilibotti, F. (2007). "Technology, information and the decentralization of the firm". Quarterly Journal of Economics 122 (4), 1759–1799.

Agarwal, A., Goldfarb, A. (2006). "Restructuring research: Communications costs and the democratization of university innovation". Working paper, University of Toronto.

Aizcorbe, A. (2006). "Why did seminconduction price indexes fall so fast in the 1990s?" Economic Inquiry 44 (3), 485–496 (July).

Aizcorbe, A., Flamm, K., Khursid, A. (2007). "The role of semiconductor inputs in IT hardware price decline: Computers vs. communications". In: Berndt, E.R., Hulten, C.M. (Eds.), Hard-to-Measure Goods and Services: Essays in Honor of Zvi Griliches. University of Chicago Press, Chicago.

Arora, A., Bokhari, F. (2007). "Open versus closed firms and the dynamics of industry evolution". Journal of Industrial Economics 55 (3), 499–527.

Arora, A., Forman, C. (2006). "How local are IT outsourcing markets: Proximity and software programming". Available at http://ssrn.com/abstract=953254.

Arora, A., Gambardella, A. (2005). From Underdogs to Tigers: The Rise and Growth of the Software Industry in Brazil, China, India. Oxford University Press, Ireland and Israel.

Arora, A., Fosfuri, A., Gambardella, A. (2001). Markets for Technology: The Economics of Innovation and Corporate Strategy. MIT Press, Cambridge.

Arora, A., Caulkins, J., Telang, R. (2006). "Sell first, patch later: Impact of patching on software quality". Management Science 52 (3), 465–471.

Aspray, W., Campbell-Kelly, M. (1996). Computer: A History of the Information Machine. Basic Books, New York.

Astebro, T. (2002). "Noncapital investment costs and the adoption of CAD and CNC in U.S. metalworking industries". RAND Journal of Economics 33 (4), 672–688.

Astebro, T. (2004). "Sunk costs and the depth and probability of technology adoption". Journal of Industrial Economics LII (3), 381–399.

Athey, S., Stern, S. (2002). "The impact of information technology and job design on emergency health care outcomes". RAND Journal of Economics 33, 399–432.

Atrostic, B.K., Nguyen, S.V. (2005). "IT and productivity in US manufacturing: Do computer networks matter?" Economic Inquiry 43 (3), 493–506 (July).

Attewell, P. (1992). "Technology diffusion and organizational learning: The case of business computing". Organizational Science 3, 1–19 (February).

Augereau, A., Greenstein, S., Rysman, M. (2006). "Coordination versus differentiation in a standards war: 56K modems". Rand Journal of Economics 34 (4), 889–911.

Autor, D., Katz, L.F., Kearney, M.S. (2003). "The skill content of recent technological change: An empirical exploration". Quarterly Journal of Economics 118 (4), 1279–1334.

Baker, J. (2007). "Beyond Schumpeter vs. Arrow: How antitrust fosters innovation". Accessed at http://nber15.nber.org/~confer/2007/si2007/PRIPE/baker.pdf, Paper presented at NBER Summer Institute, Innovation Policy and the Economy.

Baker, G., Hubbard, T. (2003). "Make versus buy in trucking: Asset ownership, job design, and information". American Economic Research 551–572 (June).

Baker, G., Hubbard, T. (2004). "Contractibility and asset ownership: On-board computers and governance in US trucking". Quarterly Journal of Economics 1443–1480 (November).

Bakos, Y., Brynjolfsson, E. (2000). "Bundling and competition on the internet: Aggregation strategies for information goods". Marketing Science 19 (1), 63–82.

Bakos, Y., Brynjolfsson, E., Lichtman, D. (1999). "Shared information goods". Journal of Law and Economics 42, 117–155.

Baldwin, C.A., Clark, K.B. (1997). "Sun wars: Competition within a modular cluster". In: Yoffie, D.B. (Ed.), Competing in the Age of Digital Convergence. Harvard Business School Press, Boston, MA.

Baldwin, Carliss A., Clark, Kim B. (2006). Architectural Innovation and Dynamic Competition: The Smaller Footprint Strategy. Harvard Business School, Boston, MA, Working Paper.

Bank, D. (2001). Breaking Windows: How Bill Gates Fumbled the Future of Microsoft. The Free Press, London.

Beaudry, P., Doms, M., Lewis, E. (2006). "Endogenous skill bias in technology adoption: City-level evidence from the IT revolution". Federal Reserve Bank of San Francisco, Working paper 06-24.

Berndt, E.R., Griliches, Z., Rappaport, N. (1995). "Econometric estimates of price indexes for personal computers in the 1990s". Journal of Econometrics 68 (1), 243–268 (July).

Berners-Lee, T., Fischetti, M. (1999). Weaving the Web, The Original Design and Ultimate Destiny of the World Wide Web. Harper Collins, New York.

Bloom, N., Van Reenen, J. (2007). "Measuring and explaining management practices across firms and countries". Quarterly Journal of Economics 122 (4), 1351–1408.

Bloom, N., Sadun, R., Van Reenen, J. (2007). "Americans do I.T. better: U.S. multinationals and the productivity miracle". CEP Discussion Paper 788, April.

Blumenthal, M.S., Clark, D.D. (2001). "Rethinking the design of the internet: The end-to-end arguments vs. the brave new world". In: Compaine, B., Greenstein, S. (Eds.), Communications Policy in Transition: The Internet and Beyond. MIT Press, Cambridge, MA, pp. 91–139.

Bower, J., Christensen, C. (1997). "Disruptive technologies: Catching the wave". In: Brown, J.S. (Ed.), Seeing Differently: Insights on Innovation. Harvard Business Review Book, Boston, MA.

Bresnahan, T. (1999). "The changing structure of innovation in computing". In: Eisenach, J.A., Lenard, T.M. (Eds.), Competition, Convergence and the Microsoft Monopoly: Antitrust in the Digital Marketplace. Kluwer Academic Publishers, Boston.

Bresnahan, T. (2004). "The economics of the microsoft case". Working paper accessed at http://www.stanford.edu/~tbres/research.htm.

Bresnahan, T., Brynjolfsson, E., Hitt, L. (2002). "Information technology, work organization, and the demand for skilled labor: Firm-level evidence". Quarterly Journal of Economics 117, 339–376 (February).

Bresnahan, T., Greenstein, S. (1997). "Technical progress and co-invention in computing and in the use of computers". Brookings Papers on Economics Activity: Microeconomics 1–78.

Bresnahan, T., Greenstein, S. (1999). "Technological competition and the structure of the computer industry". Journal of Industrial Economics 1–40 (March).

Bresnahan, T., Greenstein, S., Henderson, R. (2009). "Schumpeterian competition and diseconomies of scope: illustrations from leading historical firms in computing". Working Paper accessed at http://www.kellogg.northwestern.edu/faculty/greenstein/images/research.html.

Brensnahan, T., Malerba, F. (1999). "Industrial dynamics and the evolution of firm's and nations' competitive capabilities in the world computer industry". In: Mowery, D., Nelson, R. (Eds.), Sources of Industrial Leadership. Cambridge University Press, Cambridge.

Bresnahan, T., Trajtenberg, M. (1995). "General purpose technologies: 'Engines of growth'?" Journal of Econometrics 65 (1), 83–108.

Bresnahan, T., Yin, P.-L. (2007). "Standard setting in markets: The browser wars". In: Greenstein, S., Stango, V. (Eds.), Standards and Public Policy. Cambridge University Press, Cambridge, UK, pp. 18–59.

Brynjolfsson, E., Hitt, L. (2000). "Beyond computation: Information technology, organizational transformation and business performance". Journal of Economic Perspectives 14 (4), 23–48.

Brynjolfsson, E., Hitt, L.M., Yang, S. (2002). "Intangible assets: Computers and organizational capital". Brookings Papers on Economic Activity: Macroeconomics 1, 137–199.

Card, D., DiNardo, J. (2002). "Skill-biased technical change and rising wage inequality: Some problems and puzzles". Journal of Labor Economics 20 (4), 733–783.

Caroli, E., Van Reenen, J. (2001). "Sill biased organizational change? Evidence from the British and French establishments". Quarterly Journal of Economics CXVI (4), 1449–1492.

Carr, D. (1999). IBM Redux: Lou Gerstner and the Business Turnaround of the Decade. Harper Collins, New York.

Chapman, M.R. (2006). In Search of Stupidity: Over Twenty Years of High Tech Marketing Disasters. Springer-Verlag, New York.

Christensen, C. (1997). Innovators Dillemma. Harvard Business School Press, Boston, MA.

Cockburn, I., MacGarvie, M. (2006). "Entry, exit and patenting in the software industry". NBER Working Paper No. W12563.

Compaine, B., Greenstein, S. (2001). Communications Policy in Transition: The Internet and Beyond. MIT Press, Cambridge, MA.

Cortada, J.W. (1996). Information Technology as Business History: Issues in the History and Management of Computers. Greenwood Press, Westport, CT.

Cortada, J.W. (2003). The Digital Hand: How Computers Changed the Work of American Manufacturing, Transportation and Retailing Industries. Oxford University Press, Cary, NC.

Cranor, L., Greenstein, S. (2002). Communications Policy and Information Technology: Promises, Problems, Prospects. MIT Press, Cambridge, MA.

Cranor, L., Wildman, S. (2003). Rethinking Rights and Regulations: Institutional Responses to New Communications Technologies. MIT Press, Cambridge.

Cringley, R.X. (1992). Accidental Empires. Harper Collins, New York.

Cusumano, M., Selby, R. (1995). Microsoft Secrets, How the World's Most Powerful Software Company Creates Technology, Shapes Markets and Manages People. Simon and Schuster, New York.

Cusumano, M., Yoffie, D. (2000). Competing on Internet Time: Lessons from Netscape and its Battle with Microsoft. Free Press, New York.

Dalle, Jean-Michel, David, Paul A., Ghosh, Rishab Aiyer, Wolak, Frank (2004). "Free & open source software developers and 'the economy of regard': Participation and code-signing in the modules of the Linux Kernel". Working paper, SIEPR, Stanford University, Open Source Software Project, Accessed on June, 2007, at http://siepr.stanford.edu/programs/OpenSoftware_David/NSFOSF_Publications.html.

David, P. (1990). "The dynamo and the computer: An historical perspective on the modern productivity paradox". American Economic Review 80, 2.

David, P. (2001). "The evolving accidental information super highway". Oxford Review of Economic Policy 17 (2), 159–187.

Dedrick, J., Kraemer, K.L. (2005). "The impacts of IT on firm and industry structure: The personal computer industry". California Management Review 47 (3), 122–142.

Dedrick, J., West, J. (2001). "Open source standardization: The rise of Linux in the network era". Knowledge, Technology and Policy 14 (2), 88–112.

De Figueiredo, J.M., Kyle, M.K. (2006). "Surviving the gales of creative destruction: The determinants of product turnover". Strategic Management Journal 27 (3), 241–264.

Doms, M. (2004). "The boom and bust in information technology investment". Federal Reserve Bank of San Francisco Economic Review 20–34.

Downes, T., Greenstein, S. (2002). "Universal access and local internet markets in the U.S". Research Policy 31, 1035–1052.

Downes, T., Greenstein, S. (2007). "Understanding why universal service obligations may be unnecessary: The private development of local internet access markets". Journal of Urban Economics 62, 2–26.

Draca, M., Sadun, R., Van Reenen, J. (2007). "ICT and productivity". In: Mansell, R., Avgerou, C., Quah, D., Silverstone, R. (Eds.), Oxford Handbook of ICTs. Oxford University Press, Oxford, UK.

Dranove, D., Gandal, N. (2003). "The DVD vs. DIVX standard war. Empirical evidence of network effects and preannouncement effects". Journal of Economics and Management Strategy 12 (3), 363–386.

Evans, D., Hagiu, A., Schmalensee, R. (2006). Invisible Engines. How Software Platforms Drive Innovation and Transform Industries. MIT Press, Cambridge, MA.

Fallick, B., Charles Fleischman, C., Rebitzer, J.A. (2006). "Job-hopping in silicon valley: Some evidence concerning the microfoundation of a high-technology cluster". Review of Economics and Statistics 88 (3), 472–481.

Farrell, J., Saloner, G. (1985). "Standardization, compatibility, and innovation". Rand Journal of Economics 16 (1), 70–83 (Spring).

Ferguson, C. (1999). High Stakes, No Prisoners: A winners Tale of Greed and Glory During the Internet Wars. Times Books, New York.

Fichman, R., Kemerer, C. (1997). "The assimilation of software process innovations: An organizational learning perspective". Management Science 43 (10), 1345–1363.

Fisher, F.M. (2000). "The IBM and microsoft cases: What's the difference?" American Economic Review 90, 180–183 (May 2000).

Fisher, F., McGowan, J.J., Greenwood, J.E. (1983a). Folded Spindled and Mutilated: Economic Analysis and U.S. vs. IBM. MIT Press, Cambridge, MA.

Fisher, F.M., McKie, J.W., Mancke, R.B. (1983b). IBM and the US Data Processing Industry, An Economic History. Praeger Publishers, New York.

Flamm, K. (1987). Creating the Computer: Government, Industry and High Technology. The Brookings Institution, Washington, DC pp. 203–234.

Flamm, K. (2003). "The new economy in historical perspective: Evolution of digital electronics technology". In: Jones, D.C. (Ed.), New Economy Handbook. Academic Press/Elsevier, San Diego, Chapter 2.

Forman, C. (2005). "The corporate digital divide: Determinants of internet adoption". Management Science 51 (4), 641–654.

Forman, C., Goldfarb, A. (2006). "Diffusion of information and communications technology to business". In: Hendershott, T. (Ed.), Economics and Information Systems, vol. 1. Elsevier, Amsterdam.

Forman, C., Goldfarb, A., Greenstein, S. (2003a). "Which industries use the internet?" In: Baye, M. (Ed.), Organizing the New Industrial Economy. Elsevier, Amsterdam, pp. 47–72.

Forman, C., Goldfarb, A., Greenstein, S. (2003b). "The geographic dispersion of commercial internet use". In: Cranor, L., Wildman, S. (Eds.), Rethinking Rights and Regulations: Institutional Responses to New Communications Technologies. MIT Press, Cambridge, pp. 113–145.

Forman, C., Goldfarb, A., Greenstein, S. (2005). "How did location affect adoption of the internet by commercial establishments? Urban density versus global village". Journal of Urban Economics 58 (3), 389–420.

Forman, C., Goldfarb, A., Greenstein, S. (2008). "Understanding inputs into innovation: Do cities substitute for internal firm resources?" Journal of Economics and Management Strategy 295–316.

Fosfuri, A., Giarratana, M., Luzzi, A. (2005). "Firm assets and investments in open source software products". Druid Working Paper No 05-10. Copenhagen Business School.

Franco, A., Filson, D. (2006). "Spin-outs: Knowledge diffusion through employee mobility". RAND Journal of Economics 37 (4), 841–860.

Frieberger, P., Swaine, M. (1984). Fire in the Valley: The Making of the Personal Computer. Osborne/McGraw Hill, Berkeley, CA.

Gandal, N., Greenstein, S., Salant, D. (1999). "Adoption and orphans in the early computer market: Chicken and eggs, hardware and software". Journal of Industrial Economics 87–105.

Gandal, N., Kende, M., Rob, R. (2000). "The dynamic of technological adoption in hardware/software systems: The case of compact disc players". RAND Journal of Economics 31, 43–61.

Gans, J., Stern, S. (2003). "The product market and the market for ideas: Commercialization strategies for technology entrepreneurs". Research Policy 32 (2), 333–350.

Gans, J., Hsu, D., Stern, S. (2002). "When does start-up innovation spur the gale of creative destruction?" RAND Journal of Economics 33 (4), 571–586.

Gates, B. (1995). "The internet tidal wave", http://www.usdoj.gov/atr/cases/ms_exhibits.htm, Government Exhibit 20, United States versus Microsoft accessed at.

Gawer, A., Cusumano, M. (2002). Platform Leadership: How Intel, Microsoft and Cisco Drive Innovation. Harvard Business School Press, Boston, MA.

Gawer, A., Henderson, R. (2007). "Platform owner entry and innovation in complementary markets: Evidence from Intel". Journal of Economics and Management Strategy 16 (1).

Gerstner, L.V. (2002). Who Says Elephants Can't Dance? Leading a Great Enterprise Through Dramatic Change. Harper Collins, New York.

Gilbert, R.J. (1998). "Networks, standards, and the use of market dominance: Microsoft (1995)". In: Kwoka, J., White, L. (Eds.), The Antitrust Revolution: The Role of Economics. (third ed.). Oxford University Press, Oxford.

Gilbert, R.J. (2006). "Looking for Mister Schumpeter: Where are we in the competition-innovation debate?" In: Jaffe, A., Lerner, S., Stern, S. (Eds.), Innovation Policy and the Economy, vol. 8. p. 159, MIT Press Cambridge, MA.

Gilles, J., Cailliau, R. (2000). How the Web was Born. Oxford University Press, Oxford.

Goldfarb, A. (2004). "Concentration in advertising-supported online markets: An empirical approach". Economics of Innovation and New Technology 13 (6), 581–594.

Goldfarb, B.D., Kirsch, D., Pfarrer, M.D. (2005). "Searching for ghosts: Business survival, unmeasured entrepreneurial activity and private equity investment in the Dot-com era". Robert H. Smith School Research Paper No. RHS 06-027, Available at SSRN, http://ssrn.com/abstract=825687.

Goolsbee, A., Klenow, P. (2002). "Evidence on learning and network externalities in the diffusion of home computers". Journal of Law and Economics XLV (2, part 1), 317–344 (October).

Gordon, R.J. (2000). "Does the new economy measure up to the great inventions of the past?" Journal of Economic Perspectives 14, 4.

Greenstein, S. (2000). "Building and delivering the virtual world". Journal of Industrial Economics 391–411 (December).

Greenstein, S. (2001). "Commercialization of the Internet: The interaction of public policy and private action". In: Jaffe, A., Lerner, S., Stern, S. (Eds.), Innovation Policy and the Economy.

Greenstein, S. (2005). "The economic geography of internet infrastructure in the United States". In: Cave, M., Majumdar, S., Vogelsang, I. (Eds.), Handbook of Telecommunication Economics, vol. II. Elsevier, Amsterdam.
Greenstein, S. (2006). "Wikipedia in the spotlight". Kellogg School of Management, Northwestern University, Teaching Case. Accessed at http://www.kellogg.northwestern.edu/faculty/greenstein/images/cases.htm.
Greenstein, S. (2007a). "Economic experiments and neutrality in Internet access markets". In: Jaffe, A., Lerner, J., Stern, S. (Eds.), Innovation, Policy and the Economy, *vol.* 8. MIT Press, Cambridge, MA.
Greenstein, S. (2007b). "The evolution of market structure for Internet access in the United States". In: Aspray, W., Ceruzzi, P. (Eds.), The Commercialization of the Internet and its Impact on American Business. MIT Press, Cambridge, MA.
Greenstein, S., McDevitt, R. (2009). "The broadband bonus: Accounting for broadband Internet's impact on U.S. GDP". NBER Working paper 14758, http://www.nber.org/papers/w14758.
Greenstein, S., Prince, J. (2007). "The diffusion of the Internet and the geography of the digital divide". In: Mansell, R., Quah, D., Silverstone, R. (Eds.), Oxford Handbook on ICTs. Oxford University Press, Oxford.
Greenstein, S.M., Wade, J.B. (1998). "The product life cycle in the commercial mainframe computer market, 1968–1983". Rand Journal of Economics 29, 772–789.
Grove, A. (1996). Only the Paranoid Survive: How to Exploit the Crisis Points That Challenge Every Company. Doubleday, New York.
Haigh, T. (2007). "Building the Web's missing links: Portals and search engines". In: Aspray, W., Ceruzzi, P. (Eds.), The Internet and American Business. MIT Press, Cambridge, MA.
Hargettai, E. (2003). "The digital divide and what to do about it". In: Jones, D.C. (Ed.), The New Economy. Academic Press, San Diego, CA, pp. 822–838.
Harrison, R., Griffith, R., Van Reenen, J. (2006). "How special is the relationship: Using the impact of the US R&D spillovers on British firms as a test of technology sourcing". American Economic Review 96 (5), 1859–1875 (December).
Henderson, R. (2000). "Declaration of Rebecca Henderson, United States v. Micros oft Remedies Papers", http://www.usdoj.gov/atr/cases/f4600/4644.htm.
Hills, A. (2005). Smart WiFi. Scientific American, October.
Hoetker, G. (2006). "Do modular products lead to modular organizations?" Strategic Management Journal 27, 501–518.
Hubbard, T. (2000). "The demand for monitoring technologies: The case of trucking". Quarterly Journal of Economics 533–560 (May).
Hubbard, T. (2003). "Information, decisions, and productivity: On-board computers and capacity utilization in trucking". American Economic Review 93, 1328–1353 (September).
Hussain, F. (2003). "Historic role of the commercial Internet eXchange router and its impact on the development of the Internet eXchange points [IXCs]". http://www.farooqhussain.org/projects/cixrouter19912001/index_html, accessed at July, 2007.
Jorgenson, D.W. (2001). "Information technology and the U.S. economy". American Economic Review 91 (1), 1–32 (March).
Jorgenson, D.W., Wessner, C.W. (Eds.), (2005). Deconstructing the Computer, Report of a Symposium. National Academies Press, Washington, DC.
Jorgenson, D.W., Ho, M.S., Stiroh, K. (2005). Productivity, Volume 3: Information Technology and the American Growth Resurgence. MIT Press, Cambridge, MA.
Kahin, B., Keller, J. (1995). Public Access to the Internet. MIT Press, Cambridge, MA.
Kahin, B., Keller, J. (1997). Coordinating the Internet. MIT Press, Cambridge, MA.
Kahin, B., McConnell, B. (1997). "Towards a public metanetwork; Interconnection, leveraging and privatization of government-funded networks in the United States". In: Noam, E., Nishuilleabhain, A. (Eds.), Private Networks Public Objectives. Elsevier, Amsterdam, The Netherlands, pp. 307–321.
Kahn, R. (1995). "The role of government in the evolution of the Internet". In: National Academy of Engineering, R. (Ed.), Revolution in the U.S. Information Infrastructure. National Academy Press, Washington, DC, pp. 13–24.
Katz, B., Phillips, A. (1982). "The computer industry". In: Nelson, R.R. (Ed.), Government and Technical Progress: A Cross-Industry Analysis. New York, Pergamon Press, pp. 162–232.
Katz, M., Shapiro, C. (1985). "Network externalities, competition and compatibility". American Economic Review 75 (3), 424–440.
Katz, M.L., Shapiro, C. (1986). "Technology adoption in the presence of network externalities". Journal of Political Economy 94 (4), 822–884.

Katz, Michael L., Shapiro, Carl (2000). "Systems competition and network effects". Journal of Economic Perspectives 8 (2), 93–115.
Kende, M. (2000). The Digital Handshake: Connecting Internet Backbones. Federal Communications Commission, Office of Planning and Policy, Washington, DC, Working Paper No. 32.
Kenney, M. (2000). Understanding Silicon Valley: The Anatomy of an Entrepreneurial Region. Stanford University Press, Stanford, CA.
Kenney, M., Von Burg, U. (1999). "Technology, entrepreneurship, and path-dependence: Industrial clustering in silicon valley and route 128". Industrial and Corporate Change 8, 67–103.
Kesan, J.P., Shah, R.C. (2001). "Fool us once, shame on you—Fool us twice, shame on us: What we can learn from the privatizations of the Internet backbone network and the domain name system". Washington University Law Quarterly 79, 89–220.
Khanna, T. (1995). "Racing behavior: Technological evolution in the high-end computer industry". Research Policy 24 (6), 933–958.
Khariff, O. (2003). Paving the Airwaves for Wi-Fi. Business Week, April 1.
Kreuger, A. (1993). "How computers have changed the wage structure: Evidence from microdata, 1984–1989". Quarterly Journal of Economics 108 (1), 33–60.
Langlois, R., Mowery, D. (1996). "Spinning off and spinning on (?): The Federal Government role in the development of the U.S. computer software industry". Research Policy 25, 947–966.
Langlois, R., Robertson, P.L. (1992). "Networks and innovation in a modular system: Lessons from the microcomputer and stereo component industries". Research Policy 21, 297–313.
Langlois, R., Robertson, P.L. (1995). Firms, Markets and Economic Change: A Dynamic Theory of Business Institutions. Routledge, London.
Lee, C.-M., Miller, W.F., Hancock, M.G., Rowen, H.S. (2000). The Silicon Valley Edge: A Habitat for Innovation and Entrepreneurship. Stanford University Press, Stanford, CA.
Lerner, J. (1997). "An empirical analysis of a technology race". RAND Journal of Economics 28, 228–247.
Lerner, J., Tirole, J. (2002). "Some simple economics of open source software". Journal of Industrial Economics 50 (2), 197–234.
Lucking-Reily, D., Spulber, D. (2001). "Business to business electronic commerce". Journal of Economic Perspectives 15 (1), 55–68.
Mackie-Mason, J., Netz, J. (2007). "Manipulating interface standards as an anticompetitive strategy". In: Greenstein, S., Stango, V. (Eds.), Standards and Public Policy. Cambridge Press, Cambridge, MA, pp. 231–259.
Mankiw, Gregory, Forbes, Kristin, Rosen, Harvey (2005). Chapter 6. Innovation and the information economy. Economic Report of the President. US Government Printing Office, Washington.
Mansell, R., Steinmueller, W.E. (2000). Mobilizing the Information Society Strategies for Growth and Opportunity. Oxford University Press, Oxford.
McKinsey Global Institute. (2001). U.S. Productivity Growth, 1995–2000: Understanding the Contribution of Information Technology Relative to Other Factors. McKinsey, Washington, DC.
McKnight, L., Bailey, J. (1997). Internet Economics. MIT Press, Cambridge, MA.
Meeker, M., DuPuy, C. (1996). The Internet Report. Harper Business, New York.
Messerschmitt, D.G., Szyperski, C. (2003). Software Ecosystems: Understanding an Indispensible Technology and Industry. MIT Press, Cambridge, MA.
Mockus, A., Fielding, R.T., Herbsleb, J.D. (2005). "Two case studies of open source software development: Apache and Mozilla". In: Feller, J. (Ed.), Perspectives on Free and Open Source Software. MIT Press, Cambridge, MA.
Moore, G. (1965). "Cramming more components onto integrated circuits". Electronics V38 (8), 114–117 (April).
Mowery, D. (1996). The International Computer Software Industry. A Comparative Study of Industry Evolution and Structure. Oxford University Press, Oxford.
Mowery, D.C., Simcoe, T.S. (2002a). "The origins and evolution of the Internet". In: Nelson, R., Steil, B., Victor, D. (Eds.), Technological Innovation and Economic Performance. Princeton University Press, Princeton, NJ, pp. 229–264.
Mowery, D.C., Simcoe, T.S. (2002b). "Is the Internet a U.S. invention? An economic and technological history of computer networking". Research Policy 37, 1369–1387.
Mueller, M.L. (2002). Ruling the Root: Internet Governance and the Taming of Cyberspace. MIT Press, Cambridge, MA.
National Research Council. (2003). Innovation in Information Technology. National Academy Press, Washington, DC.

Nguyen, S., Atrostic, B.K. (2006). "How business use information technology: Insights for measuring technology and productivity". Working paper 06-15, Center for Economic Studies, U.S. Census Bureau.
Noam, E. (2001). Interconnecting the Network of Networks. MIT Press, Cambridge, MA.
NTIA (National Telecommunications and Information Administration). "Falling through the net: A survey of the 'Have Nots' in rural and urban America. (1995). http://www.ntia.doc.gov/reports.html, assessed 17 March 06.
NTIA (National Telecommuncations and Information Administration). "A Nation online: How Americans are expanding their use of the Internet. (2002). http://www.ntia.doc.gov/reports.html, assessed 17 March 06.
NTIA (National Telecommunications and Information Administration). "A Nation online: Entering the broadband age. (2004). http://www.ntia.doc.gov/reports.html.
Nuechterlein, J.E., Weiser, P.J. (2005). Digital Crossroads: American Telecommunications Policy in the Internet Age. MIT Press, Cambridge, MA.
Ohashi, H. (2003). "The role of network effects in the US VCR market, 1978–1986". Journal of Economics & Management Strategy 12 (4), 447–494.
Ou, G. (2008). "A policy maker's guide to network management". The Information Technology and Innovation Foundation, http://www.itif.org/index.php?id=205.
Oxman, J. (1999). The FCC and the Unregulation of the Internet. Federal Communications Commission, Office of Planning and Policy, Washington, DC, Working paper 31.
Partridge, C. (2008). "The technical development of Internet e-mail". IEEE Annals of the History of the Computing 3–29 (April).
Paulsen, Ed. (2001). Inside Cisco: The Real Story of Sustained M&A Growth. Wiley, New York.
Porter, M. (2001). "Strategy and the Internet". Harvard Business Review 79 (3), 62–78.
Prince, J. (2005). Measuring the Digital Divide: Structural Estimation of the Demand for Personal Computers. Cornell University, New York. Mimeo.
Prusa, T.J., Schmitz, J.A. Jr., (1994). "Can companies maintain their initial innovative thrust? A study of the PC software industry". The Review of Economics and Statistics 76 (3), 523–540.
Quarterman, J.S. (1989). Matrix Computer Networks and Conferences. Digital Press, Bedford, MA.
Rohlfs, J. (2001). Bandwagon Effects in High Technology Industries. MIT Press, Cambridge, MA.
Rosenberg, N. (1976). Perspectives on Technology. Cambridge University Press, Cambridge, UK.
Rosenberg, N. (1994). "Economic experiments". Exploring the Black Box. Cambridge University Press, Cambridge.
Rosenberg, N. (1996). "Uncertainty and technology change". In: Landau, R., Taylor, T., Wright, G. (Eds.), The Mosaic of Economic Growth. Stanford University Press, Stanford, CA, pp. 334–356.
Russell, A.L. (2006). "Rough consensus and running code and the Internet-OSI standards war". IEEE Annals of the History of Computing 48–61, July–September.
Savage, S., Waldman, D. (2004). "United States Demand for Internet Access". Review of Network Economics 3 (3), 228–247.
Saxenian, A.L. (1994). Regional Advantage: Culture and Competition in Silicon Valley and Route 128. Harvard University Press, Cambridge, MA.
Schmalensee, R. (2000). "Antitrust issues in Schumpeterian industries". American Economic Review 90, 192–196 (May).
Shapiro, C., Varian, H.R. (1998). Information Rules. Harvard Business School Press, Boston, MA.
Shiller, R.J. (2000). Irrational Exuberance. Princeton University Press, New York.
Simcoe, T. (2007). "Delay and De jure standardization: Exploring the slow down in Internet standards development". In: Greenstein, S., Stango, V. (Eds.), Standards and Public Policy. Cambridge Press, Cambridge, MA, pp. 260–297.
Sinai, T., Waldfogel, J. (2004). "Geography and the Internet: Is the Internet a substitute or a complement for cities?" Journal of Urban Economics 56 (1), 1–24.
Sink, E. (2007). "Memoirs from the browser wars", http://biztech.ericsink.com/Browser_Wars.html. Published at accessed March 2007.
Smith, M.D., Bailey, J., Brynjolfsson, E. (2000). "Understanding digital markets: Review and assessment". In: Brynjolfsson, E., Kahin, B. (Eds.), Understanding the Digital Economy. MIT Press, Cambridge, MA, pp. 276–288.
Stavins, J. (1995). "Model entry and exit in a differentiated-product industry: The personal computer market". The Review of Economics and Statistics 77 (4), 571–584.
Stern, S. (2005). "Economic experiments: The role of entrepreneurship in economic prosperity". Understanding entrepreneurship: A research and policy report, http://research.kauffman.org/cwp/ShowProperty/web/CacheRepository/Documents/Research__Policy_Singles.pdf. Kaufman Foundation.

Stiroh, K. (2002). "Information technology and the U.S. productivity revival: What do the industry data say?" American Economic Review 92 (5)1559–1576.

Stranger, G., Greenstein, S. (2007). "Pricing in the shadow of firm turnover: ISPs in the 1990s". International Journal of Industrial Organization 26, 625–642.

Strover, S. (2001). "Rural Internet connectivity". Telecommunications Policy 25 (5), 331–347.

Tedlow, R. (2006). Andy Grove: The Life and Times of an American. Penguin Group, New York.

Trajtenberg, M. (1990). Economic Analysis of Product Innovation, The Case of CT Scanners. Harvard University Press, Cambridge, MA.

Triplett, J.E. (1989). "Price and technological change in a capital good: A survey of research on computers". In: Jorgenson, D.W., Landau, R. (Eds.), Technology and Capital Formation. MIT Press, Cambridge, MA, pp. 127–213.

Varian, H. (2000). "Versioning information goods". In: Kahin, B., Varian, H.R. (Eds.), Internet Publishing and Beyond. MIT Press, Cambridge, MA.

Varian, H. (2001). "High technology industries and market structure", http://www.ischool.berkeley.edu/~hal/papers/structure.pdf September, Working paper.

Von Burg, U. (2001). The Triumph of Ethernet: Technological Communities and the Battle for the LAN Standard. Stanford University Press, Stanford, CA.

Von Hippel, E. (2005). "Open source software projects as user innovation networks". In: Feller, J., Fitzgerald, B., Hissam, S. A., Lakhani, K. (Eds.), Perspectives on Free and Open Software. MIT Press, Cambridge, MA, pp. 267–278.

Waldrop, M. (2001). The Dream Machine: J.C.R. Licklider and the Revolution that Made Computing Personal. Penguin Books, New York.

West, J., Gallagher, S. (2006). "Patterns of innovation in open source software". In: Chesbrough, H., Haverbeke, W., West, J. (Eds.), Open Innovation: Researching a New Paradigm. Oxford University Press, Oxford, pp. 82–108.

Chapter 12

PHARMACEUTICAL INNOVATION

F.M. SCHERER

John F. Kennedy School of Government

Harvard University, Cambridge

Massachusetts, USA

Contents

Abstract

This chapter surveys the costs, risks, and challenges encountered in the progressive discovery and development of new pharmaceuticals. The changing methods by which drugs are discovered, the links between companies and academic science, the changing character of public regulation, and the sharp rise in the cost per new approved drug are analyzed. Determining which new drugs are both efficacious and safe poses classic statistical decision theory problems. Why patents are so important to drug developers is explored. A rent-seeking theory of new drug development is proposed to rationalize the high gross margins but only slightly supranormal returns on investment realized by pharmaceutical companies, and the puzzling economic welfare implications are investigated.

Handbooks in Economics, Volume 01

DOI: 10.1016/S0169-7218(10)01012-9

Keywords

innovation, parallel paths, pharmaceuticals, product quality regulation, profitability, research and development, virtuous rent-seeking

1. Introduction

The discovery and development of new pharmaceutical substances are among the most interesting of innovation processes. Unusually large privately financed expenditures on research and development (R&D) outlays are required to achieve a successful new product, and the pharmaceutical industry's R&D/sales ratios are extraordinarily high. The links to academic science and basic research performed in government laboratories are rich. The expectation of patent protection plays a more important role than in most other high-technology industries. New products must meet not only the test of market acceptance, but also survive rigorous scrutiny from government regulatory agencies. And the medical services market into which pharmaceuticals sell is itself unusually complex, with a significant fraction of consumers' purchases, at least in the wealthier nations, covered by insurance and hence subject to diverse moral hazard and adverse selection imperfections. Despite these problems, there is compelling evidence that the introduction of many new pharmaceutical products has yielded substantial net benefits in extending human lives and reducing the burden of disease (Lichtenberg, 2001, 2004, 2007; Long et al., 2006; Murphy and Topel, 2006).

Among these various characteristics, we focus preliminarily on one: the high ratio of pharmaceutical companies' R&D spending in relation to their sales. The clearest indicator of this trait comes from data systematically collected over the years by the US industry trade association, previously called the Pharmaceutical Manufacturers Association (PMA) and more recently the Pharmaceutical Research and Manufacturers of America (PhRMA). The solid line in Figure 1 shows aggregate PhRMA member company-financed R&D expenditures within the United States as a percentage of US sales for the years 1970–2007. From 12.4% in 1970 declining to 11.8% in recession year 1974, the spending ratio rose to a peak of 21.6% in 1996 before declining to 18% in 2001. Included for comparison is a less complete series (dotted line) of data on company-financed R&D outlays as a percent of sales for all manufacturing

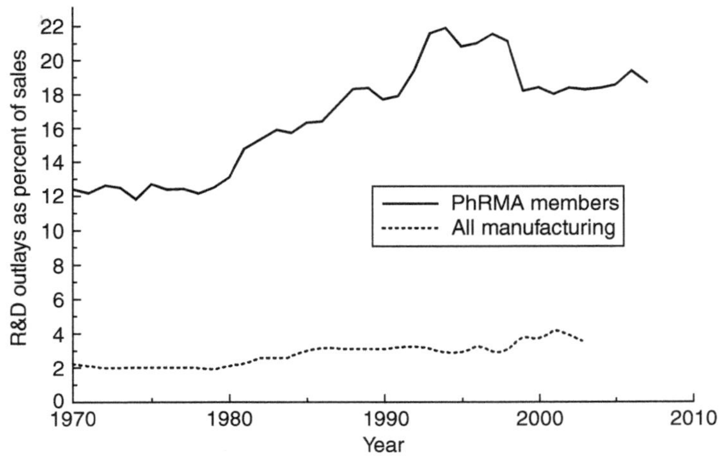

Figure 1. R&D as percentage of sales, pharmaceuticals, and all manufacturing.

corporations reporting R&D expenditures in periodic US National Science Foundation (NSF) surveys. Between 1999 and 2003, pharmaceutical R&D/sales ratios were nearly five times those of their all-manufacturing counterparts.

NSF survey data are often cited as an indicator of research intensity in pharmaceuticals. For the years 1999–2003, the average R&D/sales ratio for the NSF industry category "pharmaceuticals and medicines" was 9.2%, compared to 18.3% with the PhRMA series in Figure 1. However, the NSF figures are biased downward because they are assembled using what is called the whole company method, under which all the R&D outlays and sales for a company are assigned to the industry in which the company has its largest volume of sales. Thus, they are aggregations of pharmaceutical companies' R&D activity in their home industry along with data from the much less research-intensive toiletries, cosmetics, first-aid supplies, and insurance payment processing industries, among others. That the PhRMA data present a more accurate picture of what transpires in modern pharmaceutical manufacturing is revealed through data collected by the US Federal Trade Commission (1985, p. 31) for the 1970s in narrowly defined "lines of business." For "ethical drugs" in 1977, the reported company-financed R&D/sales ratio was 10.2%, compared to 11.3% in the contemporaneous PMA report (which excluded many older, i.e., generic, drug sales). Among the 220 lines of business for which the FTC reported 1977 data (and also for 1974–1976), ethical drugs had the *highest* company-financed R&D/sales ratio of any industry. For proprietary (e.g., over-the-counter) drugs, the comparable ratio in the 1977 FTC report was 2.9%, for toiletries and cosmetics, 2.5%, and for surgical and medical supplies, 3.8%.

From the R&D efforts of pharmaceutical companies has flowed a stream of new products, which is traced in Figure 2 through a count of the "new chemical entities" approved for marketing in the United States by the designated regulatory agency, the Food and Drug Administration. Excluded from the count are new formulations of existing products, combinations of previously approved entities, new uses of

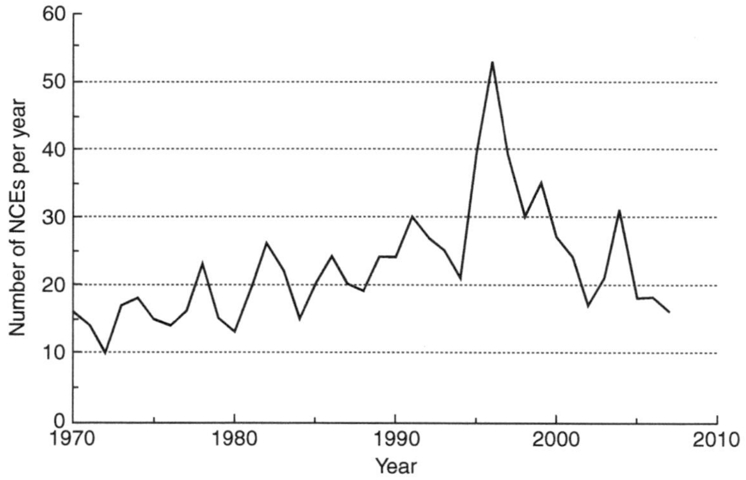

Figure 2. New chemical entities approved for marketing in the United States, 1970–2007.

approved products, and most new biological entities and vaccines (in principle, the output of a somewhat different industry covered elsewhere in this volume).[1] The unusually robust upward fluctuations for 1995 and 1996 came from backlog reductions on drugs awaiting approval by the Food and Drug Administration. When those 2 years are excluded, the average number of new chemical entities approved per year between 1970 and 2007 was 21.2, with a statistically significant upward time trend.

2. Time phases

It is customary to characterize new drug discovery and development in terms of time phases. The principal dichotomy is between the preclinical and clinical phases.

In the preclinical phase, research efforts are oriented at first toward isolating chemical or biological molecules that might have interesting therapeutic action *in vitro* and then testing for such action in diverse animals. The "animal model" tests often involve several different species, progressing over time from worms, mice, guinea pigs, and the like up the evolutionary scale to dogs and monkeys.

If those tests suggest that the drug might be efficacious and not cause toxic effects, the drug developer works to formulate the candidate in a form (e.g., pill) suitable for medical use and seeks permission to begin tests in human beings—that is, in the clinical trial phase. In the United States, formal approval for clinical testing occurs when the Food and Drug Administration issues a so-called IND—that is, investigation of new drug—permit. The clinical phase is, in turn, conventionally divided into three main phases. In Phase I, the drug is administered to a small number of subjects, sometimes with the target disease and sometimes not, to test for the safety of various dosages and (when diseased subjects are included) for preliminary indications of therapeutic efficacy. More careful and extensive tests for therapeutic efficacy are conducted in Phase II. If those tests are promising, the effort moves into Phase III, in which the drug is administered to at least two panels of patients who might be expected to benefit from the therapy. The number of subjects runs on average in the mid-thousands, and, especially for drugs targeted toward already curable diseases or that will be used for long-term therapy, can exceed 10,000. With the principal exception of drugs combatting diseases (such as AIDS) that could be lethal if untreated, the tests are double-blind, with half the subjects receiving the drug being tested and the other half an inert placebo or, less frequently, an alternative drug known to be effective against the disease. Under the double-blind approach, neither the recipients of the drug nor (to minimize subtle psychological influences) the administrators know whether a specific subject is receiving the drug being tested or a placebo. If the Phase III trials yield favorable results, the firm sponsoring the trials applies in the United States to the FDA (or in Europe to the European Medicines Agency[2]) for a new drug approval (NDA), submitting voluminous trial data to support its application. Testing often continues during and

[1] During the later years, the exclusion of new biological entities—an exclusion required to preserve longer term continuity—imparts a bias. From modest beginnings, the average number of new biologics approved between 2004 and 2007 was 9.5 per year—a 46% augmentation to the number of new chemical entities.

In assembling R&D statistics, the US Census Bureau and the National Science Foundation record biologically oriented companies as chemicals manufacturers when they sell substantial amounts of product but as research and development service providers when they have not yet marketed products.

[2] See Healy and Kaitin (1999). Before 2003, the organization was called the European Agency for the Evaluation of Medicinal Products.

after the approval interval through Phase IV trials, sometimes to answer unresolved questions posed by the regulatory agency and sometimes to provide additional evidence for the sponsoring company's planned marketing campaign.

Figure 3 presents a stylized characterization of the annual rate at which funds are spent in pharmaceutical discovery and development leading to a specific useful molecule and how the spending cycle is divided among the various phases. No overlap among the phases is assumed, although some overlap can in practice occur. The yearly spending rate is relatively modest in the early preclinical phases, rises sharply as clinical tests begin, peaks during Phase III human trials, and then tends to decline abruptly with application for regulatory approval and movement into Phase IV tests, if any.

Detailed analyses of selected drug development histories by DiMasi et al. (1991, 2003) provide among other things estimates of the attrition rates marking transition into successive clinical testing phases and ultimate marketing approval. For self-originated drugs brought into clinical testing by multinational pharmaceutical companies between 1970 and 1982, some 23% of the molecules entering Phase I testing ultimately gained marketing approval from government regulators following the completion of Phase III. For a later cohort initially tested in humans between 1983 and 1994, the estimated success rate was 21.5%.[3] In the later study, the attrition rate between Phase I and Phase II was 29%. The period of maximum hazard is Phase II, with an attrition rate of 56%. For the molecules surviving in Phase III, incremental attrition fell to 31.5%. (The over-all survival rate is found by multiplying 1 minus the attrition rates, e.g., $0.71 \times 0.44 \times 0.685 = 0.214$.) In pharmaceuticals, it would appear that substantial risks of total failure persist later in the R&D cycle than those in most

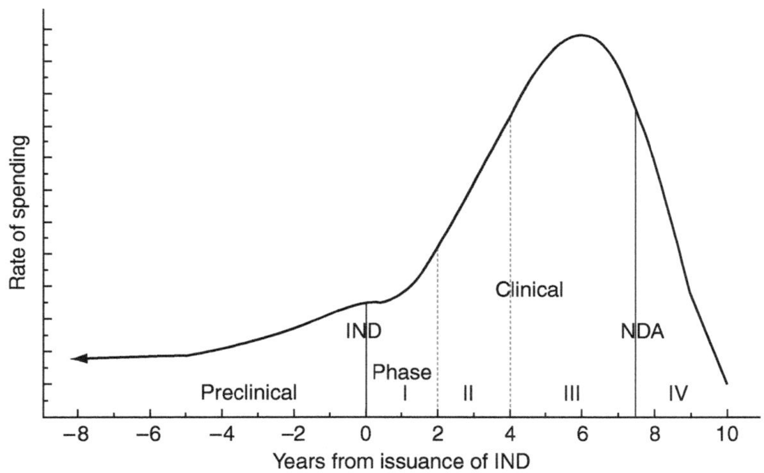

Figure 3. Rate of pharmaceutical R&D spending by phase.

[3] For more recent success rate estimates using a different sampling approach, see Tufts Center for the Study of Drug Development (2006).

other industries. More typically, as the rate of R&D spending rises, uncertainties abate, although uncertainties concerning market acceptance persist well past completion of the principal R&D tasks (Branscomb and Auerswald, 2001). Indeed, unless key technical uncertainties are resolved at relatively low spending levels, firms conducting the R&D in nonpharmaceutical fields will typically choose not to carry their efforts into higher levels of annual expenditure.

On average in recent US experience, Phases I–III span 6–7 years, with the regulatory approval process consuming an additional 1–2 years. On this more later. There is considerable variation from case to case; drugs for rare and incurable diseases are often tested and approved on expedited timetables. Even more variation is found in the length of preclinical research, and indeed, as we shall see, relevant events in the discovery process can sometimes be traced out to decades before testing in human beings commences.

3. Changing discovery methods

Over time immemorial, humans found through trial and error that certain naturally occurring substances had medicinal effects. Quinine, present in chinchona tree bark and first extracted by chemical methods in 1810, helped alleviate the symptoms of malaria. Vaccination, first with live smallpox toxin and then, thanks to Edward Jenner's experimentation in the 1790s, with a less dangerous vaccine based upon cowpox virus, helped eradicate the scourge of smallpox.

The bark of the white willow tree was known to provide relief against fevers and headaches. The active substance, salicylic acid, was extracted and identified in the 1830s. Salicylic acid, however, had severe side effects—ulcers and other gastric distress. Bayer A.G. of Germany had become one of the world's leading producers of synthetic dyestuffs, created through the manipulation and synthesis of organic chemical molecules. In 1896, Bayer established a laboratory to synthesize and test dyestuff formulations for medicinal effects in humans. One of its candidates, acetylsalicylic acid, proved to be as effective against fever and headaches as its parent molecule, salicylic acid, with far milder side effects. The new formulation was named "aspirin," which was patented, trademarked, licensed, and sold profitably by Bayer throughout the world (Mann and Plummer, 1991). The beginnings of the modern pharmaceutical industry can be traced to Bayer's work on aspirin.[4]

German dye-makers (amalgamated under the umbrella of I.G. Farben) continued to test coal-tar-based dyestuff variants for therapeutic effects, and in 1935, they discovered a wholly new class of so-called sulfa drugs, the first of which was sulfathiazole. The sulfa drugs proved to be remarkably effective against a range of bacterial diseases such as spinal meningitis, various forms of pneumonia, and gonorrhea. Variants were subsequently discovered to act as relatively safe diuretics, that is, to reduce tendencies toward high blood pressure.

There was also progress on the theoretical front. During the nineteenth century much was learned about the nature of cells, vaccines, and disease processes in the human body. In 1899, Paul Ehrlich became the director of the Institute for Experimental Therapy in Frankfurt/Main, Germany. He conceived an "affinity" theory of small organic molecules' target-specific binding to particular sites

[4] For various views on early pharmaceutical discovery approaches, see Schwartzman (1976, Chapter 2), Gambardella (1995), Werth (1994), and Pisano (2006, Chapter 2).

in living organisms such as cells in the human body and postulated that if one could find the right "magic bullet," diseased and disease-causing organisms could be destroyed without otherwise harming their human carrier. Salvarsan, one of the more than 600 molecules Ehrlich and colleagues synthesized and tested for therapeutic effects against syphilis, a widespread and essentially incurable disease, proved in 1908 to be effective.

The sulfa drugs provided a first line of defense against bacterial infections. A new line began to open up in 1928 when Alexander Fleming of London observed, but did not follow through on, the antibacterial action of a mold that had drifted onto and killed bacteria he was culturing in a Petri dish. His work was revitalized during the late 1930s by Howard Florey and Ernest Chain at Oxford University. Recognizing the antibacterial properties of Fleming's discovery, *Penicillium notatum*, and with financial support from the Rockefeller Foundation, they struggled to develop methods of producing the substance in quantities sufficient first to conduct tests in live subjects and then to use in general medical practice. The results were considered important to treating casualties from looming World War II, and so the British government approved their transmission to the US government. American defense authorities set in motion a major effort to produce penicillin in large quantities, eventually, by deep-vat fermentation in corn steep liquor—a process developed initially by a US Department of Agriculture laboratory in Peoria, Illinois. Contracts to produce penicillin were let to 20 chemical companies, which expanded production for military hospitals and simultaneously gained expertise in the technology of antibiotics.[5] This massive effort provided a major impetus to the emergence of a vibrant American pharmaceutical industry.

While Alexander Fleming's penicillin discovery lay dormant, Selman Waksman of Rutgers University began investigating whether naturally occurring spores might have antibiotic properties. With financial support from the Merck Company, Waksman and his students collected and tested against bacterial cultures approximately 10,000 soil samples. They made two discoveries—a new antibiotic, streptomycin effective among other things, against tuberculosis, and more importantly, a method for discovering even more new pharmaceuticals—the systematic screening of molds, fungi, and other substances occurring in nature.

With Waksman's success, the rapidly growing pharmaceutical industry had two main methods of identifying potential medicines—the screening of naturally occurring substances, plus the organic molecule synthesis approach pioneered half a century earlier by Bayer and Ehrlich. With each method, pharmaceutical action was ascertained empirically, that is, by testing the effects of a molecule on manifestations of disease. For antibiotic action, tests were initially conducted on cultures of bacteria growing in a Petri dish or a test tube—that is, *in vitro*. For anesthetic or tranquilizing action, initial tests might be conducted in earthworms. For blood pressure action and the like, the first target would be laboratory mice.

For the pharmaceutical industry, the antibiotic revolution also had unexpected negative consequences. Penicillin technology was widely diffused to facilitate rapid expansion of wartime production. Waksman obtained patent protection on streptomycin, but licenses were made available widely. Price competition soon emerged and then intensified, driving the prices of penicillin and streptomycin down

[5] On the economic history of antibiotics production, see US Federal Trade Commission (1958) and Kingston (2000).

sharply, in some cases, below average production costs. Producing the new "wonder drugs" was found to be unprofitable.[6]

Salvation came with new discoveries. Building upon what had been learned with penicillin and streptomycin, the pharmaceutical companies began synthesizing or modifying naturally occurring molecules to offer a new, more powerful line of antibiotics—the so-called broad-spectrum antibiotics, starting with aureomycin in 1948 and then encompassing several molecular variants. These could be patented, and they were sold for many years at prices several multiples of their production costs. Developing and patenting new pharmaceutical entities was found to be a profitable endeavor. Efforts to discover new and different kinds of pharmaceuticals proliferated, precipitating a rapid rise in R&D outlays.

The search for new and effective drugs in the 1950s and 1960s was preponderantly empirical and intuitive, through the screening of plausible alternative molecules. Where similar molecules had already exhibited therapeutic activity, the screens were typically narrow. Investigators tried to identify comparable effects from "me too" molecular variants. When new molecules had no clear therapeutic antecedents, the screens were broad, that is, covering a panoply of possible diseases. Schwartzman (1976, p. 60) reports that in 1970, PMA members prepared, extracted, or isolated for medical research 126,060 substances and tested for pharmacological action some 703,900 substances (many, presumably, duplicative), among which only about a thousand showed enough promise to be advanced through higher animal tests into human trials. One company interviewed for an interagency government inquiry put 20,000 compounds through a narrow screen for antibacterial activity in 1966 and carried roughly 4000 into further animal tests because of preliminary activity indications. The cost per individual screening test at that time was on the order of $50, while tests on animals such as guinea pigs and monkeys had a reported average cost of $10,000 (Harbridge House, Inc., 1967, pp. III-9 and IV-4). From such screening, sometimes called "random screening," companies developed extensive libraries of molecules with annotations on their effects, including serendipitous pharmacological activity. These were available for guidance when a search was begun for drugs that combatted a particular new medical problem of interest.

Gradually, as medical knowledge accumulated from research in hospitals, academic institutions, and industry, the search process narrowed to focus on molecules predicted on theoretical grounds to have desired therapeutic effects. Although earlier antecedents can be identified, this "rational drug design" approach is said to have flowered in the 1970s and early 1980s. Compare Gambardella (1995, Chapters 2 and 4) and Schwartzman (1996). A leading example was Tagamet, introduced by SmithKlineFrench in the late 1970s. Scientific research had shown that ulcers resulted from excess production of gastric acid in the stomach. Secretion of the acid was in turn instigated by histamine, an amine naturally present in the human body. Search for a therapy against ulcers more effective than traditional antacids such as sodium bicarbonate focused on finding agents that would block the acid-generating action of histamine. This narrowed the research agenda considerably, although trial-and-error research was still needed. SmithKlineFrench scientists synthesized and tested roughly 700 compounds over a period of 10 years before seizing upon the highly successful H2-antagonist Tagamet (chemical name, cimetidine). Its success in turn spurred others to explore molecular variants on Tagamet, which was soon surpassed by Glaxo's Zantac and Merck's Pepcid. These in turn were later overtaken by a different proton pump

[6] See US Federal Trade Commission (1958, Chapters 6 and 7).

inhibitor approach embodied in Astra's Prilosec. Schwartzman (1996) argues that the "rational drug design" approach was not as revolutionary a break as some claimed it to be, because scientific knowledge provided imperfect guidance and much screening of alternative molecules, to be sure, targeted screening, was necessary before therapeutically successful molecules were obtained. His criticism is valid, but it is also true that scientific knowledge at least narrowed the searches and limited the use of "try every bottle on the shelf" approaches. Gambardella (1995, p. 20) reports that some 5000 drugs had to be synthesized for early screens to achieve one marketable product and observes (p. 40) that as of the early 1990s, "attrition rates ... do not seem to have diminished."

A further step forward toward rational drug design came with the perfection of methods such as X-ray crystallography and nuclear magnetic resonance imaging for ascertaining the exact structure of proteins present in the human body that might serve as targets for hostile agents, and also the structure of the invading organisms.[7] One can then try to design therapeutic molecules whose three-dimensional profile meshes exactly with receptor sites in the target molecule, often a protein, as a key meshes with a lock, and which, having bound to the target, block molecular functions or changes adverse to individuals' health. Here the search narrows to a particular structure for the therapeutic agent. Again, however, empiricism is not eliminated. Many alternative molecules, identified among other ways through computer modeling, might bind to a particular target but have no desirable therapeutic effect or be toxic. Thus, search must continue for a molecule with the right configuration and also the desired therapeutic interaction with its target—a search much less precisely guided by received theory. For example, in its quest for a molecule that combatted the body's rejection of "foreign" kidney, liver, heart, and other tissue transplants with less severe side effects than the established inhibitor, cyclosporin, Vertex, a startup company pioneering the computer-aided design of therapeutic molecules, explored 367 different variants before finding one that bound to the receptor site appropriately and showed the hoped-for pharmacological effects.[8] As a skeptic of the rational design school observed:[9]

> *"A compound may be brilliantly designed—everything absolutely rational, but until that compound has been shown not only to do clinically what you want it to do, but to be safe, to be active orally, to stay around in the body, and not to give you nightmares, it's not a drug."*

An even newer approach is to use modern genetic methods to identify and synthesize therapeutic molecules. Clinical studies are sometimes able to determine where and how the lack of a particular protein (such as insulin, human growth hormone, or erythropoietin) in the human body leads to ill health. Even more recently, using high-speed DNA-sequencing techniques, researchers can identify gene sequences which, when present in the human body, increase the probability of serious diseases, or alternatively, to isolate the sequences associated with individuals who might be expected from heredity or life style to acquire a disease but do not. Those DNA sequences usually express specific proteins which underlie the disease mechanism. When the absence of a protein is likely to render a person disease-prone, the protein can be synthesized by recombinant methods—for example, splicing the

[7] X-ray crystallography images produced by Rosalind Franklin during the 1950s were crucial in the research by Francis Crick and James Watson ascertaining the structure of DNA.

[8] See Werth (1994, p. 251). Ultimately, no therapeutically successful molecule emerged from the effort.

[9] Werth (1994, pp. 215–216), quoting an unnamed Merck vice president.

relevant strand of DNA into *E. coli* bacteria and growing the modified organisms in fermentation cultures.[10] It can then be introduced, not always without difficulty, into the relevant organs of the human body. When a specific protein is found to increase the likelihood of disease, proteins or much smaller traditional organic molecules can be sought, as under rational drug design, to combat the action of those proteins or, as in the methods developed to combat HIV/AIDS, to interfere with the replication of harmful agents within them. These techniques probably provide paths to disease remediation with fewer blind alleys and detours than traditional screening approaches. However, the science is difficult, one cannot be sure whether a particular molecular modification will work safely, and the manufacturing techniques tend to be more difficult and error-prone than tried-and-true "small molecule" processes. Uncertainty is by no means eliminated.

The advance of science and technology has also facilitated new, higher powered methods of molecule screening.[11] Three main methods are of interest. First, through "combinatorial chemistry," fragments of molecules can be treated chemically and combined in a host of different ways on a single multiwell microtiter plate, each well yielding a distinct isomer of some larger organic molecule. The process of generating new and possibly interesting molecules is thereby accelerated. Second, interesting proteins and other organic molecule targets can be arrayed on similar plates with numerous wells, to which are added diverse molecular variants of possible therapeutic interest. From these tests and the methods devised to interpret their results, one can quickly screen to see which of many therapeutically interesting formulations bind to the targets. What would otherwise be a tedious process of testing for interactions one test tube or Petri dish at a time is accelerated in a kind of mass production.[12] Third, as suggested in the previous paragraph, the process of DNA-sequencing has been accelerated greatly, bringing the costs of sequencing an individual's DNA down to such modest levels that sequencing the DNA or fragments thereof for large numbers of individuals has become feasible. The data gained in this way can be mined for disease proclivities and other characteristics of therapeutic interest.

This third technique has brought still another possibility onto the horizon. The typical drug does not work in all individuals with a given ailment, and the administration of a drug can have adverse side effects in some recipients but not in others. These variations in drug receptivity are undoubtedly related to differences in the subjects' genetic endowments. The new science of "pharmagenomics" or "translational medicine" seeks to ascertain through large-scale DNA sequencing which individuals fit into which category.[13] In this way, individual drugs might be prescribed only for the individuals who will benefit from them and not be harmed by them, permitting better targeted drug therapy. It is proposed too that pharmagenomic screening might eliminate from clinical tests of new drugs the individuals likely to experience no effects or adverse side effects and therefore render drug testing processes more precise and less costly. The methods of pharmagenomics are so new that few if any successful applications have been reported. But there are great hopes for them.

[10] The key discovery was made by Stanley Cohen of Stanford University and Herbert Boyer of the University of California, San Francisco, during the 1970s. Three patents on the Cohen–Boyer gene splicing techniques were then licensed to hundreds of other organizations.

[11] For an excellent lay exposition of the principal advances, see Carr (1998). See also (2006) "New chips on the block." *The Economist* 24–26 (December 2); and (2005) "Mining DNA for biomarkers." *Business Week* 82–85 (September 5).

[12] For a more skeptical analysis of actual results, see Pisano (2006, p. 94).

[13] See Vernon and Hughen (NBER, 2005). According to the research director of a biotechnology firm, the human genome is believed to contain roughly 5000 pharmaceutically relevant genes, of which only 47 are targeted by the 200 best selling drugs of 2003. (2004) "Fixing the drugs pipeline." *The Economist* 37 (March 13).

More generally, the enormous advances that have been made in computer-aided structurally based drug design, low-cost molecular manipulation and screening, DNA screening, and recombinant genetics have inspired optimism about the possibility of a new "golden era" of pharmaceutical discovery. But the pot of gold has proved to be elusive. There have indeed been important breakthroughs, but they have been relatively scarce. As Figure 2 shows, after a burst of activity during the mid-1990s, the number of new pharmaceutical chemical entities introduced into US medical practice during the late 1990s and early years of the twenty-first century has been stagnant or perhaps even declined (Cockburn, 2006). Three influences were evidently at work. On one hand, the development and introduction of new drugs deplete the inventory of long-established chemical possibilities and raise the hurdles a new drug must clear in order to displace already efficacious existing drugs. On the other hand, advances in medical knowledge, laboratory methods, and instrumentation open up new possibilities that should in time lead to new and superior drugs. But third, the latter dynamics work with substantial lags and a good deal of uncertainty, engendering substantial, more or less random, fluctuations in the rate of new drug introduction. Optimists believe a golden era of pharmaceutical discovery is coming. When it will materialize is one of the remaining uncertainties.

4. Industry–academic science links

The evolution of pharmaceutical discovery away from unguided or at best intuitive random screening toward rational drug design and biological methods has led to increasingly rich linkages between the work of pharmaceutical companies on the one hand and academic science carried out in universities and governmentally supported research institutes on the other, both in the nations where the companies operate and across national boundaries. This has always been the case to some extent. The early work on sulfa drugs was conducted in German industrial laboratories by scientists trained at prominent German universities, which were at the time world leaders in chemical research and teaching. Penicillin moved quickly from the laboratories of Oxford University to numerous companies producing in quantity for the war effort. The first oral contraceptive was introduced by the G.D. Searle Company in 1960, a decade before the earliest date at which the trend toward rational drug design was said to begin. But a study by the IIT Research Institute (1968, pp. 58–72) for the US NSF revealed an intricate "tree" of scientific discoveries extending back to 1849 that laid a foundation for the Searle contraceptive and later improvements. The more recent changes lie mainly in the richness and closeness of the science–industry linkages and the magnitude of the science base on which the industry could draw. In 2006, for example, against the $34 billion of industry R&D expenditures within the United States reported by PhRMA members, the federally financed National Institutes of Health allocated a nearly comparable $28 billion to research intramurally and by outside grant recipients, much of it basic. A considerable but unmeasurable portion of NIH spending was for studies of direct or indirect interest to the discovery of new drugs. Among the knowledge "spillovers" traced by Adams and Clemmons (2008) through scientific journal article citations, drugs and biotechnology firms had five times the weighted citation volume from firm to university authors as the next most citation-intensive industry and nearly three times the volume of company scientist citations from firms to other firms.

Cockburn and Henderson (2000) studied the histories of the 21 new drugs introduced between 1965 and 1992 with the highest over-all therapeutic impact, as judged by industry experts. Among the 21, only five, or 24%, were developed with essentially no input from public sector research.[14] They contrast their results with an earlier analysis by Maxwell and Eckhardt (1990) concluding that 38% of an older sample of drugs was developed without public support. Cockburn and Henderson divided their sample of 21 into three categories—drugs discovered through essentially random screening, drugs that might be said to fall under the rational drug design rubric, and drugs discovered through fundamental science. All but one of the drugs in the latter two groups built upon key enabling discoveries from public science, while four of the seven random screening drugs did not have clear public science antecedents. For 18 drugs in the Cockburn–Henderson sample on which requisite timing data were available, the mean lag from the key enabling discovery to synthesis of an effective drug was 17.3 years, with a median value of 12.5 years, a minimum of 2 years, and a maximum lag of 54 years. Although public sector research played a seminal role in facilitating high-impact drugs, 14 of the 18 drugs on which information was available were first synthesized or, for naturally occurring substances, dug up, by private-sector firms. Plainly, a division of labor exists. Academic groups have comparative advantage in advancing the science underlying drug discovery and pharmaceutical companies excel at manipulating molecules into a form suitable for therapeutic use.[15]

Mansfield (1998) pursued a more aggregate approach toward identifying the importance of academic research to private-sector companies' innovations. He obtained from R&D laboratory heads in some 77 companies, operating in seven broad fields of technology, estimates of the percentage of their new products introduced during two time periods, 1975–1985 and 1986–1994, that "could not have been developed (without substantial delay) in the absence of recent academic research." For all fields, the average research-dependence fraction was 10% for the earlier innovations and 11% for the later group. "Drugs and medical products" had by far the highest average research-dependence ratio—27% for the earlier period and 31% for the later period. Supplementing the "could not have been developed" cohort, 13–17% of the drug and medical product innovations received "very substantial aid" from recent academic research—a virtual tie with "information processing" for first place among the seven groups. With his "recent" framing question, Mansfield found that the average lag from key academic research results to commercialization in drugs and medicines was 8.5–8.8 years.

Academic science is transformed into pharmaceutical innovations through richly interconnected networks.[16] Open science, to be sure, is available to pharmaceutical companies through journal articles and presentations at professional meetings. But in addition, there are tighter links. Pharmaceutical companies provide financial support for academic researchers, and their staffs sometimes perform joint research with academic researchers and coauthor articles with them. They also enter into cooperative research and development agreements (CRADAs) with government laboratories such as the US National Institutes of Health, permitting joint research, joint publication, and (under the Stevenson-Wydler Act of 1980, extended through a 1986 amendment), assignment of resulting patents to the companies. In recent years, many pharma companies have opened new laboratories in the vicinity of top

[14] Cockburn and Henderson leave the role of public science undecided for cyclosporine. But it is clear from Werth (1994, pp. 48–50), that academic and hospital investigators played key roles in determining that, with the proper dosage, the substance could be used as an immunosuppressant.

[15] See also Reichert and Milne (2002) and Ward and Dranove (1995).

[16] See, for example, Gambardella (1995), Cockburn and Henderson (1998), Cockburn et al. (1999), Zucker et al. (1998), Powell and Grodal (2005), and Pisano (2006, pp. 100–108).

academic research institutions in order to facilitate cooperation. Quick absorption of the newest scientific discoveries is facilitated when traditional pharmaceutical companies support their own active programs of basic research. In 1993, for example, drug companies reported that 13.6% of their total company-financed R&D budgets was devoted to basic research, as defined by the NSF—18% of the basic research spending of *all* industries covered by the NSF survey for that year. For the average research-performing company across all industries except pharmaceuticals, basic research was 6.3% of total company-financed R&D spending.[17]

Even closer links between academia and industry are seen in the emergence of hundreds of small new biotech firms, which tend to locate near academic centers, have academic scientists as their founding entrepreneurs, and count distinguished academic researchers as members of their boards of directors and/or scientific advisory councils. Traditional "Big Pharma" companies in turn license molecules discovered in biotech startups for later-stage commercial development or, with increasing frequency, acquire the biotech companies outright, securing full ownership rights in their development "pipeline" molecules and adding staff associated with them to their own R&D staffs (Kettler, 2000). In this way, they augment their inventories of interesting drug development candidates, among other things filling voids created when more traditional drug discovery approaches have yielded disappointing results.

An indication of the extent to which firms introducing new drugs to the US market depended upon others for early-stage discoveries is provided through a study undertaken by the author. For the five years 2001–2005, the Food and Drug Administration's Web site listing new medical entities approved for marketing during those years was searched. From information provided in the approval lists, the patents claimed by the drug developers as impediments to generic competition could be traced by searching the FDA's so-called "Orange Book." On the 85 new medical entities for which patent information was disclosed,[18] 251 applicable patents were found, or an average of 2.95 patents per molecule. Altogether, 47% of the patents were assigned at the time of their issue to companies with names (abstracting from obvious name changes due to large-company mergers) different from the company authorized by the FDA to begin commercial marketing of the sample drugs.[19] Patents issued in the earlier stages of development, that is, prior to January 2000, were more likely (54%) to be assigned originally to firms other than FDA approval recipients than patents issued in later years (38.4%). The difference is statistically significant. Evidently, the companies carrying out final-stage development and testing relied disproportionately upon outsiders for early-stage discovery. Among the 251 patents, 10.4% went to essentially academic institutions, that is, universities, hospitals, and independent research institutes. Seven percent went to universities, although a handful of the university assignments were joint with other institutions, including US government laboratories. Seven of the 251 patents had multiple organizational assignees and 10 had only individual inventors as assignees. Many of the nonacademic patent assignees were biotech companies, although an exact breakdown was not possible because information on companies that have not yet "gone public" is scarce. It cannot be ruled out that at least some of the assignees with names different from that of the company receiving FDA approval had common stock partially or wholly controlled by larger corporate parents, notably, the companies receiving FDA approvals.

[17] US National Science Foundation, Research and Development in Industry: 1993 (NSF 96-304), p. 45.

[18] Some drug categories are exempt from reporting their patent backgrounds.

[19] See also Pisano (2006, p. 102).

In an analysis covering an earlier and longer period, DiMasi (2000, p. 1177) found that 38.2% of the 691 new chemical entities approved in the United States between 1963 and 1999 originated from sources other than the company seeking FDA approval. For his sample, outside sources accounted for 28.4% of 1963–1969 NCEs and 39.1% of NCEs approved in the 1990s. This plus the results described in the previous paragraph suggest a rising trend in reliance upon outside discovery.

That drug discovery has become more science-based and hence more efficient, as argued by Gambardella (1995), might be consistent with a remarkable result in the relevant literature that, at least at the time this was written, had no clear explanation. A research team at Tufts University has published two leading empirical analyses of new drug discovery and clinical testing costs. The first focused on 93 new chemical entities introduced into human testing between 1970 and 1982; the second on 68 NCEs first tested between 1983 and 1994 (DiMasi et al., 1991, 2003). Detailed data were obtained from 10 to 12 pharmaceutical companies, and a consistent methodology was applied to allocate the costs of clinical trial failures to drugs that eventually succeeded in gaining approval and also to make the more difficult allocations of preclinical research costs to successful molecules. All entities in the sample were "self-originated," which means that the responsible companies did not license or buy discovered molecules from other companies, and hence presumably incurred the costs of discovery internally. The fraction of total constant-dollar preclinical outlays as a percentage of total preclinical plus clinical testing outlays, without adjustment for the cost of capital invested in the research, was as follows:

1970–1982 cohort	57.7%
1983–1994 cohort	30.0%

The decrease is remarkable.[20] One possible explanation is that drug discovery became more sharply focused and hence consumed a much smaller fraction of total R&D outlays. Another possibility is that in the later period prototypes, even if not actual molecules, were licensed in from outside science-based laboratories, although presumably this should have been ruled out by the sample design. A third possibility entails sampling variation or measurement error, although the differences seem too large to be explained in that manner alone. A fourth alternative is that for some reason clinical trial costs exploded during the later time period. We turn to that possibility now, although it must be admitted that an important mystery cannot be resolved here.

5. Clinical testing costs and regulation

Since 1938, when the Pure Food and Drug Act of 1906 was amended after approximately one hundred persons were killed by sulfanilamide adulterated with poisonous diethylene glycol (used for antifreeze), the interstate sale of new drugs was prohibited in the United States unless the would-be drug provider obtained a safety certification (an NDA) from the Food and Drug Administration. The FDA's powers were quite limited, and new drugs were often introduced into the market with claims of efficacy that were based on evidence that was more impressionistic than scientific. A demand for more stringent

[20] To the best of the author's knowledge, no explanation has appeared in published reports.

regulation emerged when thalidomide, a drug intended to combat morning sickness in pregnant women, caused severe birth defects in many infants born of women taking the drug. In Europe, where thalidomide had entered general use, some 8000 malformed babies were victims; in the United States, there were only nine known cases. The US Congress honored the FDA officer who had sidestepped regulations to keep thalidomide testing at low volumes in the United States, and in 1962, Congress passed the Kefauver–Harris Act to reform drug approval processes. It required the FDA to ensure that new drugs were not only *safe*, but also that they were *efficacious*, that is, that they actually had the therapeutic effects their makers claimed. An earlier loophole allowing full-scale marketing if the FDA did not act within 180 days of an application's filing was eliminated. The FDA in turn issued new regulations requiring that pharmaceutical producers seeking approval for their new drugs follow a strict regimen of clinical testing that adhered to scientifically grounded sample design, experimental control, and statistical inference norms. Among other things, the three-phase approach to clinical testing was introduced, and in Phase III, double-blind testing against a placebo became "the gold standard" for FDA oversight.

By the late 1960s, a hue and cry arose asserting that the new rules had drastically increased testing costs and that the number of new chemical entities receiving FDA approval had in turn been sharply reduced.[21] Four reasonably well contrived clinical trial cost estimates, all attempting to prorate the cost of testing molecules dropped at diverse clinical test phases, provide an overview of what happened.[22] All of the estimates are converted to year 2000 price levels using the gross domestic product implicit deflator:

Source	Test period	Average out-of-pocket cost per approved new chemical entity[a]
Mansfield	Late 1950s	\$5.4 million
Clymer	Late 1960s	\$40.2 million
DiMasi I	1970–early 1980s	\$65.7 million
DiMasi II	1983–late 1990s	\$282 million

[a] The Tufts group reports both out-of-pocket costs and costs capitalized at 9–11% interest rates to reflect the opportunity cost of capital tied up, sometimes for more than a decade, in testing. Capitalization roughly doubles the average NCE cost. It is the capitalized costs that are most frequently cited by pharmaceutical companies, although both alternatives have legitimacy. See US Office of Technology Assessment (1993).

There are two striking increases. The first is from the pre-1962 to the post-1962 period, as the Kefauver–Harris Act took hold, with estimates based in both cases on data from a single pharmaceutical firm, SmithKlineFrench.[23] The second, derived using a methodology consistent between periods, is for drugs entered into testing during the 1970s, as compared to those on which clinical testing began in 1983 or later.

[21] The latter result, which stands out in statistical time series longer than the one in Figure 2 (e.g, Scherer 1996, p. 351), was intended, for the FDA wished to discourage the proliferation of "me too" molecules that offered at best trivial incremental therapeutic benefits.

[22] The sources in order presented are Mansfield (1970), Clymer (1970), DiMasi et al. (1991), and DiMasi et al. (2003).

[23] Mansfield does not disclose his source, but his close relationship with SKF at the time makes it likely that the estimates are comparable within the same firm.

Grabowski et al. (1978) took advantage of a natural experiment to explore the reasons for the first apparent cost increase. In the United States, proof of efficacy was required after 1962; in Great Britain, proof of efficacy was added to the approval standard only in 1971. Between 1960–1961 and 1966–1970, inflation-adjusted drug development costs in Great Britain increased by a factor of 3 while they increased in the United States by six times. This result suggests that the change in regulatory regimes in the United States was responsible for a twofold cost increase, with the rest attributable to other factors such as fear of tort liability (following substantial thalidomide damages paid by Continental European firms) and companies' desire to differentiate new drugs from the large number of entities already on the market.

What happened to cause a fourfold increase in price-level-adjusted clinical test costs between the third and fourth samples remains, like the reason for the sharp relative fall in preclinical outlays, somewhat of a mystery. In part as a reaction to safety shortfalls, the FDA continued to strengthen its clinical evidence rules, leading to more complex trial protocols. As the number of molecules competing to be prescribed against the typical disease increased, companies sometimes found it advantageous on marketing grounds to conduct more Phase III and Phase IV tests than the FDA required. It is also possible that the gross domestic product deflators used to render outlays comparable in terms of purchasing power are inappropriate for measuring inflation of R&D costs, both generally and in pharmaceuticals. Some R&D costs, for example, for computer capabilities, rose much less than the rate of general inflation. The National Institutes of Health Biomedical Research and Development Price Index rose at an average annual rate of 6.53% between 1970 and 1990 while the GDP deflator was rising at 5.8%.[24] Moreover, clinical testing is a labor-intensive activity, and the wages of physicians, nurses, and related staff increased more rapidly than the GDP deflator. Much clinical testing is done in hospitals. The average cost per day of hospitalization rose at an average rate of 11% per year between 1970 and 1990. It is also possible, although no research on the matter is known, that hospitals using clinical testing activities as a "profit center" dumped some of their soaring overhead costs into the charges they levied on deep-pocketed pharmaceutical companies for clinical testing. And of course, sampling variation cannot be ruled out. The DiMasi et al. samples are weighted in favor of large pharmaceutical companies targeting their new drugs toward long-term therapy markets able to bear heavy testing costs and requiring extensive proof of long-term safety. Drugs aimed at acute but rare diseases are probably undersampled.[25]

[24] Derived from Table #304—NIH Research Grants for fiscal years 1950–2008, obtained as a spreadsheet from the NIH Office of External Research, http://report.nih.gov.

[25] For so-called "orphan" drugs and drugs targeting lethal diseases such as HIV/AIDS, the FDA typically authorized smaller clinical trial samples and shorter trial periods. On orphan drugs generally, see Shulman and Manocchia (1997). Love (2003) found average clinical trial costs of $34 million per approved entity between 1998 and 2000.

In recent years, the FDA has distinguished between "priority" and "standard" approval candidates—the former for molecules that offer significant improvements over existing approved therapies. Among the new medical entities approved between 2001 and 2005, 89% of the orphan drugs, that is, those treating diseases affecting fewer than 20,000 US residents, received priority designations, while only 38% of the nonorphan drugs were designated "priority."

For data on how testing costs vary over therapeutic categories within Tufts University samples, see DiMasi et al. (1995). The highest average costs were for nonsteroidal anti-inflammatories, which are often taken several times daily for extended periods. The lowest average costs were for anti-infectives, which are usually prescribed for only short periods.

Seeking to control their clinical testing costs or minimize internal bureaucracy, pharmaceutical companies have "outsourced" some of their clinical test tasks to independent contract research organizations (CROs) that specialize in test management, and more recently, they have sought to conduct tests in low-wage nations to the extent permitted by regulatory agency rules.[26] (The US FDA requires that some clinical trials be conducted within the United States.) Azoulay (2003) calculated that on average, pharmaceutical companies covered by a large survey outsourced to CROs 23% of their clinical test activity in 1999. He found too that outsourcing is most frequent for Phase IV trials, and that in earlier-phase trials, use of CROs is less common for tests and disease indications that require complex protocols and quick feedback from unexpected events than for relatively routine testing. In a possible reversion to practices abandoned in the 1970s, pharmaceutical companies have also reconsidered using prisoners in late-stage clinical trials.[27]

6. The decision-theoretic problem

Evaluating the results of clinical trials is a classic exercise in statistical decision theory. The Food and Drug Administration in the United States, and in the European Community since 1995, the European Medicines Agency, attempt to keep unsafe products off the market and to approve only products that can demonstrate their therapeutic efficacy. Because drugs have differing activity in diverse humans, clinical test results are statistically noisy. Decision makers for both the sponsoring company and the regulatory agency must try to sort out the true effects from the random variation. One must be wary of both Type I errors—concluding that the drug is safe and/or efficacious when it is not, and Type II errors—keeping a drug off the market when it actually will do good work. Tradeoffs are also required. Virtually every drug has some adverse side effects. Even century-old aspirin does; it can trigger ulcers, inhibit blood-clotting, and (more rarely) cause fatal Reye's syndrome. One must weigh the beneficial effects against the adverse effects, neither measured with certainty.

For the US Food and Drug Administration, the "gold standard" again has been the double-blind test of a new drug against placebos, for example, sugar pills. It is not always required, however. Denying an available potential remedy to patients with an otherwise fatal disease violates ethical canons, so the US FDA either allows new drugs to be tested on all clinical trial subjects (e.g., in the early years of the HIV/AIDS crisis) or uses as a comparative benchmark a drug whose therapeutic efficacy has already been established. It has been argued, for example, by Angell (2004), that, when they are available, established drugs should normally be used as the benchmark for comparison, because in a head-to-head competition, information may be generated that helps physicians choose among alternative therapies by comparing cost with benefits. This is not always a good idea, however. The largest trial of Merck's Cox-2 inhibitor pain reliever Vioxx was against an established over-the-counter drug, naproxen sodium (branded Naprosyn). The latter was known to have blood-thinning properties which reduce the severity or likelihood of strokes and heart attacks. When clinical trial subjects were found to have a higher propensity toward adverse cardiovascular events with Vioxx than with naproxen sodium, the inference

[26] (2005) "Drug companies cut costs with foreign clinical trials." *New York Times* (February 22); (2006) "Outsourcing." *Business Week* 58 (January 30); and Berndt et al. (2007).
[27] (2006) "Panel suggests using inmates in drug trials." *New York Times* 1 (August 13).

was drawn that naproxen sodium was having its well-known positive effect, and not that Vioxx was actually causing cardiac events. This was eventually found to be wrong, and when Vioxx was taken off the market in 2004, a torrent of tort litigation followed. An equally large trial against a placebo might have identified the adverse side effects from Vioxx more clearly, but left unclear Vioxx's superior record in avoiding ulcers as compared to naproxen sodium.

The decision-theoretic problems of clinical trial design are illustrated by the case of TPA (tissue plasminogen activator), a genetically engineered drug targeted against the blood clots that accompany heart attacks, versus an older, well-established drug, streptokinase (SKA).[28] The null hypothesis on efficacy would be that TPA was not more effective than a placebo or, under an alternative standard, than SKA. Carefully structured trials were conducted, and it was found that 6.3% of the subjects died after being injected with TPA plus heparin, while 7.4% died with SKA plus heparin. Assuming these values to reflect the true states of nature, Figure 4 shows how the statistical inference problem might evolve. The solid lines show the probability distribution of possible trial outcomes with samples of 1000 on each alternative—fairly typical of the Phase III sample sizes required during the early 1990s by the US FDA. If the death rate with a placebo exceeded 8%, trials with 1000 TPA injections would with high probability support a decision to allow marketing of TPA. The risk of a Type I error is quite small—the area under the right-hand tail of the TPA distribution above 8%. But the test would be insufficient to tell whether TPA was more effective than SKA. If both drugs were matched against each other, there would be roughly one chance in three that SKA would be found to be superior when in fact it is inferior. Recognizing this, TPA's developer, the biotech firm Genentech, chose to sponsor head-to-head clinical

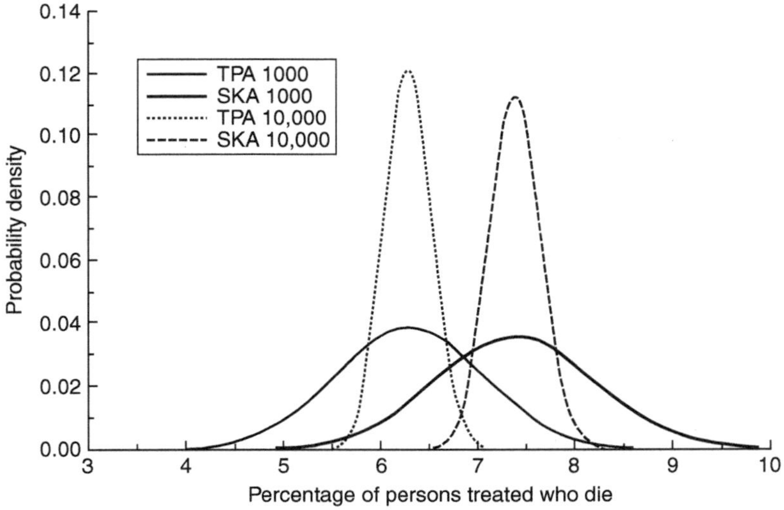

Figure 4. Testing TPA against streptokinase.

[28] This example is drawn from Scherer (1996, pp. 354–355).

trials with much larger samples than those required by the FDA. With samples of 10,000, we see in Figure 4, there is only a small probability, measured by the proportional area under the overlap between the two the dash–dash and dot–dot density functions, that one would err in concluding that TPA is not superior to SKA. With this evidence, TPA became the drug of choice for emergency treatment of heart attack and stroke patients, even with a TPA price 10 times the price at which SKA was marketed.

Similar problems pervade testing for adverse side effects. Many side effects are rare events, occurring with probabilities less than 0.01. A 12-week-long trial of Vioxx, for example, yielded only about 1.2 heart attack cases per thousand subjects, suggesting a probability (that undoubtedly would have risen with longer therapy) of 0.0012.[29] With the sample sizes typically required by the Food and Drug Administration, determining whether Vioxx actually caused heart attacks or whether the relatively few cases observed would have happened in any event was intrinsically difficult.

Given these difficulties, policy-makers and company officials may wish to commence full-scale marketing of a new drug, expecting that rare side effects will reveal themselves when the population using the drug numbers in the hundreds of thousands or millions. There are, however, two main problems with this approach. When adverse effects are rare, individual physicians administering the drug to their typically small number of patients will seldom be able to discern that observed complications were caused by the drug rather than something that would have happened in any event. And, busy as they are, they are unlikely to take the trouble to report the event to a central office processing data on the entire population. Given this, extraordinary diligence is required on the part of the agency responsible for overseeing drug safety, piecing together fragments of imperfect information from the field into a more coherent picture.

But here the wrong thing happened in the United States. The drug evaluation problem is sufficiently difficult that the US FDA took a long time making its decisions, once companies had deposited truckloads of clinical test information on its doorstep. During the 1980s, the time required for the FDA to make a definitive decision on companies' requests for an NDA averaged roughly 30 months. There were complaints of a "drug lag" relative to nations with less meticulous drug approval systems, and companies claimed that the regulation-induced delay of profitable sales impaired their incentives to sustain R&D efforts.

A solution to the problem was adopted by the US Congress in the Prescription Drug User Fee Act (PDUFA) of 1992. It allowed the FDA to levy user fees upon pharmaceutical companies, in part with a fee per application, partly through a fee proportional to the number of production sites licensed, and partly through fees rising with the number of new drug applications approved. The fees, eventually totaling $250 million per year, allowed the FDA to augment its new drug evaluation staff, in exchange for which it promised to reduce its decision-making lags to an average of 12 months.[30] The approval-linked component had potentially undesirable incentive effects, for the more drugs the FDA approved, the higher its revenues would be. Thus, it might be motivated to approve marginal drugs it would not favor if their approval did not yield additional revenue. The law was modified in 1997 to eliminate this incentive incompatibility by targeting an annual lump sum to be raised from drug approvals, with a

[29] (2005) "Evidence in Vioxx suits shows intervention by Merck officials." *New York Times* 1 (April 24).

[30] An irony of the legislation is that at the time, the FDA had an effort underway, called Project 007, to reduce decision-making lags without additional budgetary resources. Even though it was beginning to succeed, the FDA favored having substantially more budgetary resources under the PDUFA system. See US Food and Drug Administration (1993).

higher resultant fee per NDA, the fewer NDAs issued. Even so, another difficulty materialized. Congress made it clear that the FDA was not to divert the PDUFA revenues to activities other than new drug review, and when insufficient funds were separately appropriated to support all previously existing FDA functions, the FDA cut back on the resources it devoted to postmarketing surveillance. Its monitoring of postmarketing safety issues abated and responses to lethal side effects from already-approved drugs were delayed. Guided in part by a critical report from the Institute of Medicine (2006) of the National Academies of Science, corrective action was begun after the Vioxx crisis surfaced.

7. Uncertainty revisited

A recurrent theme in this essay has been the presence of uncertainty. As the research and testing process progresses, uncertainties are gradually mitigated. Several sources put the number of alternative molecules subjected to early screening at between 4000 and 10,000 in order to have a single approved drug at the end of the process.[31] According to PhRMA (2006, p. 4), the US industry association, a single approved drug emerges on average from five compounds entering clinical testing, 250 molecules subjected to animal and other laboratory tests, and 5000–10,000 molecules initially screened. As the number of drug candidates is winnowed, the costs of continued testing and hence the stakes in the game escalate.

Even when marketing approval is secured, risks do not vanish. As we have seen, severe safety hazards may become evident only when a drug has been accepted widely on the market. And approval is by no means synonymous with commercial success. Grabowski and Vernon (1990, 1994) have shown that the distribution of quasi-rents—that is, the surplus of revenues over variable production costs for individual drugs—is highly skew. Among any given 100 drugs introduced into the market, the top 10 by number realize from 48% to 55% of their cohort's discounted quasi-rents, while the least lucrative 80 out of 100 barely cover, or less than cover, their average capitalized research and testing costs. In its skewness, the profitability distribution for new drugs is similar to the distributions for most other new products, except that when one focuses only on approved drugs, one ignores the uncertainties that preceded approval and therefore obtains somewhat less skew outcome distributions than with samples that begin at earlier stages of the research and product innovation cycle (Harhoff and Scherer, 2000). Skewness of the distribution of rewards from innovation in turn makes it more difficult to hedge against risk by maintaining a portfolio of projects—a standard feature of high-technology investment strategies. Harhoff and Scherer (2000) demonstrate, for example, in a simulation analysis using the Grabowski–Vernon data, that even averaging over *all* the new pharmaceutical entities introduced into the US market for a total of 21 years, skewness and the random appearance of a few extreme values lead to fluctuations in *overall* annual industry gross profitability as high as plus or minus 25%. In other words, the overall industry portfolio is insufficiently diverse to eliminate significant profit variations.

To be sure, pharmaceutical companies have some bases for predicting before marketing begins whether their new drug will enter a market with blockbuster potential or occupy a niche in which quasi-rents will at best be modest. Among other things, first movers typically enjoy larger market shares than latecomers.[32] However, there are also surprises. The drug with the highest annual sales in

[31] See Gambardella (1995, pp. 20 and 40), Schwartzman (1996, p. 846), Pisano (2006, p. 56), and PhRMA (2006, p. 4).

[32] See, for example, Bond and Lean (1977) and Robinson and Fornell (1985).

pharmaceutical industry history, Lipitor (atorvastatin), was seen by its developers as at best a late entrant into the cholesterol-reducing statins drug market. With limited perceived prospects, the Lipitor project was on the verge of cancellation by its developer, Warner-Lambert, when a small clinical trial revealed, contrary to expectations, that it was more effective at given dosages than rival drugs.[33] Confirmation of this result plus a marketing decision to set Lipitor's price at half the price of the leading rival propelled Lipitor's sales to record levels, ahead of several competing molecules. Similarly, Abbott Laboratories' Hytrin (terazosin) was synthesized through a minor manipulation—the replacement of two pentane ring double bonds with single bonds in a quadruple-ring molecule—of an antihypertensive drug marketed by Pfizer. Its performance as an antihypertensive was unimpressive. But tests by academic researchers revealed serendipitously that Hytrin could ease the symptoms of benign prostate gland enlargement. It was retested for that use and approved, achieving annual sales in its category approaching a billion dollars per year—a result far beyond the expectations of the team that created it. On the other hand, drugs with the most optimistic prospects sometimes prove to have unacceptable side effects, crashing and burning after hundreds of millions of dollars have been spent for development and testing.[34]

8. The unique role of patents

The expectation of patent protection on new products plays a particularly important role in pharmaceutical R&D decision making. Levin et al. (1987) surveyed 650 corporate R&D managers, asking them inter alia to evaluate on a scale of 1 (not at all effective) to 7 (very effective) the effectiveness of patents as a means of protecting the competitive advantages from new products. From 17 pharmaceutical industry respondents, the average score was 6.53, compared to a response-weighted average of 4.33 for all 130 surveyed lines of business. Among the industries with more than one respondent, pharmaceuticals ranked second in its patent protection effectiveness score. This result is consistent with the findings of Mansfield (1986), who asked the top R&D executives of 100 US corporations what fraction of the inventions they commercialized between 1981 and 1983 would not have been developed in the absence of patent protection. For pharmaceuticals, the average was 60%; for all industries, 14%.

The importance of patents to pharmaceutical R&D decision makers stems not only from the large average investments in a typical new product and the many uncertainties lining the path to a new product approval. The differentiating factor is seen among other things through a comparison with another industry—aircraft—that taps a range of highly sophisticated technologies and spends billions of dollars developing the typical new product. For aircraft (both civilian and military), the average "effectiveness of product patents" score in the Levin et al. survey was 3.79—in the lowest third among 130 industry categories.

The key difference lies in the relative ease of imitation, that is, how difficult it would be with versus without patent protection, for new product imitators to launch their own competing products. Even without patents, the firm that would seek to imitate the Boeing 787 would have to build its own scale

[33] See Simons (2003).

[34] See (2006) "Failure of Pfizer cholesterol drug is major blow to promising approach." *Washington Post* (December 4); and Nocera, J. (2007). "The dangers of swinging for the fences." *New York Times* B1 (January 27).

models, perform wind tunnel tests, compile detailed engineering drawings and specifications for all structural parts, work out electronic system interfaces, construct full-scale test models, test them both on the ground and in flight for structural soundness and aerodynamic performance, and much else, spending very nearly as much as Boeing did to develop its 787. Presumably, it would have observed Boeing's design before undertaking the project, and by the time the imitator completed its developmental work, Boeing would be a decade ahead in sales and have progressed far down its learning curve, enjoying a substantial production cost advantage. But in pharmaceutical discovery and testing, much of the R&D is aimed at securing knowledge: knowledge of which molecules are therapeutically interesting, knowledge of which molecules work in animals, and most costly, knowledge as to whether a target drug is safe and efficacious in human beings. Once that knowledge is accumulated, absent patent protection, it is essentially there as a public good available to any interested party. Achieving it requires by recent US standards an investment measured in the hundreds of millions of dollars. But for most new drugs, and especially small-molecule drugs,[35] a would-be generic imitator could spend a few million dollars on process engineering and enter the market with an exact knock-off copy. Generic entry in turn could quickly erode the quasi-rents anticipated by a pharmaceutical innovator to repay its R&D investment. Hence the importance attributed to patents by drug companies.

This asymmetry between pharmaceutical innovators and imitators was not nearly as glaring during the early 1980s. Because of FDA and Supreme Court rulings, generic drug providers had to invest nearly as much per molecule in clinical testing to obtain marketing approval as the first-moving innovator.[36] Original developers also had problems. They typically sought patent protection just before beginning human tests, when probable "utility" could be documented, and at the completion of those tests, 30-month average decision-making lags at the FDA ate into the 17-year period over which their products were protected by patents. A grand compromise on these two points was achieved in the Hatch–Waxman Act of 1984.[37] It allowed patent holders to extend the lives of their patents, compensating for at least some of the period during which their new product introduction was delayed by regulatory oversight. It simultaneously reduced the clinical testing requirements for generic entrants once blocking patents had expired. Contrary to past precedents, the so-called Bolar Amendment also permitted generic drug developers to produce small quantities of the drug in question for their clinical trials before the drug's patents had expired so that they could complete their FDA paperwork and attempt entry as soon as patents expired. The generic entry provisions had a more dramatic impact, increasing the number of prescriptions filled generically in the United States from 19% in 1984 to 67% in the year 2007. The expectation of rapid generic entry following patent expiration in turn reinforced the incentive of pharmaceutical innovators to invigorate their R&D to compensate for impending profit losses, or, when that failed, to acquire other companies with better filled new drug development pipelines.

[35] That is, abstracting from biologicals, whose production tends to be more difficult and to entail more secret "black art."

[36] See Kitch (1973) and Bond and Lean (1977). A key Supreme Court ruling was U.S. v. Generix Drug Corp. et al. (1983). 460 U.S. 453.

[37] For a brief history, see Scherer (2009).

9. Profitability and research investments

With strong patent protection and well-differentiated products, pharmaceutical producers enjoy considerable discretion over the prices they set. Insurance coverage of drug outlays, expanded rapidly during the 1980s, reduced demand elasticities and conferred even more pricing power. One index for measuring pricing power is the price–cost margin (PCM), defined as:

$$\text{PCM} = \frac{\text{Sales} - \text{Material Costs} - \text{In-Plant Payroll Costs}}{\text{Sales}}.$$

For 459 four-digit manufacturing industries on which data were published for the year 1987, pharmaceuticals had the sixth-highest margin, at 61.4%. For all manufacturing industries, the average PCM was 30.5%. Similarly, for decades pharmaceutical producers appeared at or near the top of *Fortune* magazine's annual list of broad industry groups, ranked in order of after-tax profit returns as a percentage of stockholders' equity. In 27 of the years 1968–2006, the pharmaceuticals industry ranked either first or second among from 22 to 50 broad industry groups. Beginning already in the late 1950s, the drug makers were accused in public fora of profiteering at the expense of consumers. They argued in return that high profits were a reward for superior innovation and a necessary spur to investment in risky R&D.

Another more subtle defense led, after considerable repetition, to a large-scale analytic investigation by the US Office of Technology Assessment (1993). The basic argument was that, due to the R&D-intensity of pharmaceutical manufacturing and peculiarities in the way accepted accounting principles dealt with R&D outlays, reported profit returns on drug company assets and stockholders' equity were systematically overstated. Specifically, R&D outlays were recorded as a current year's expense when in fact they were investments yielding returns over decades following their incurrence. Ideally, they should be added to asset accounts and then depreciated only slowly. Ignoring their investment character understated drug company assets and hence, given the absolute R&D magnitudes and growth conditions experienced by drug companies, overstated profit ratios in which assets or stockholders' equity comprised the denominator.[38] A careful evaluation by the Office of Technology Assessment confirmed the validity of the underlying theory and concluded that, after appropriate accounting adjustments were made, pharmaceutical makers enjoyed returns on investment only 2–3 percentage points higher than the roughly 10% real cost of their financial capital.[39] And at least part of that differential could be attributed to the riskiness of drug companies' investments.[40] In other words, drug companies did not appear to be realizing extraordinary supranormal profit returns.

This conclusion left unsettled the specific behavioral dynamics that reconciled unusually high PCMs, atypically high R&D/sales ratios, and bottom-line returns on investment only moderately above

[38] For the relevant theory, see Stauffer (1971).

[39] See also Grabowski et al. (2002). Pisano (2007, pp. 112–118) presents evidence that the profit returns of publicly traded biopharmaceutical companies have in the net been negative. He attributes this result primarily to the lack of an appropriate business model. A complementary explanation might be that the venture capitalists who have been instrumental in financing most biotech startups are skewness lovers. See Scherer (2001b).

[40] Not explicitly recognized in the analysis was the difficulty of risk-hedging through portfolio maintenance with highly skew payoff distributions, as shown by Harhoff and Scherer (2000).

all-industry norms. Several studies of the links between profit potential and R&D investment have been published. I focus here on my own analysis, which has been brought as up-to-date as data availability permitted.[41] The profit potential is measured from US Census data for the "pharmaceutical preparations" industry as sales less materials purchases less in-plant payroll costs (including fringe benefits). Call this variable "gross margins." The data were adjusted for inflation to a 1992 = 100 price level base using the implicit GDP deflator. The coverage is for 1962 through 2004.[42] R&D data were spliced from various statistical reports of the Pharmaceutical Manufacturers of America and PhRMA. On the assumption that US members allocated resources internationally to the most favorable locations and that overseas R&D was influenced by profit prospects in the firms' largest single market, the United States, the R&D data include "R&D abroad" as well as domestic R&D.[43] Again, deflation was to 1992 price levels. For each deflated time series, an exponential growth trend was fitted using least-squares regression. For the gross margins variable, the average "real" rate of growth was 4.84% per year; for the R&D variable, 8.11%. Using the fitted trend, percentage deviations from the trend were computed. These are plotted in Figure 5, with margin deviations as a solid line and R&D deviations as a dotted line.

For at least the first three decades of the time series, the degree of coincidence is remarkable. When margins rise relative to trend, R&D rises in near tandem. The causation cannot plausibly run from R&D to margins, because R&D extends for a decade or so before marketing begins, and even then, it takes

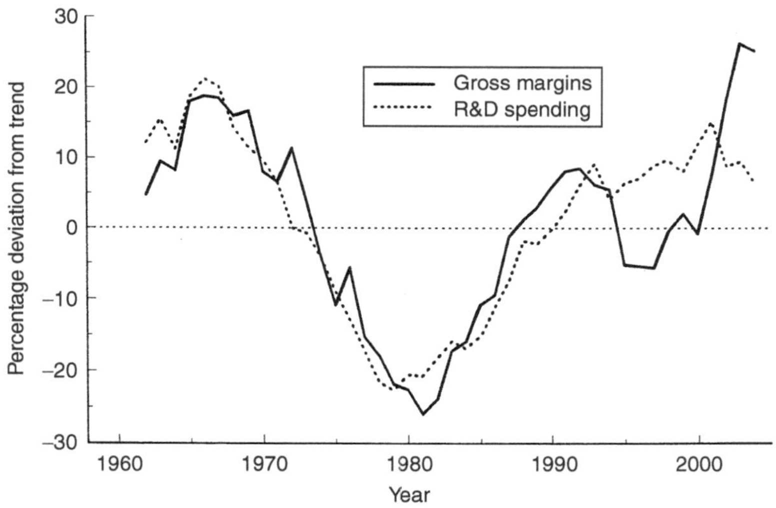

Figure 5. Trend-adjusted movements of pharmaceutical margins and R&D.

[41] Extended from Scherer (2001a). See also Lichtenberg (2004), Scherer (1996, p. 388 note), and Grabowski et al. (2002).

[42] A splice was necessary after 1996, when a new industrial classification was implemented and slight mismatches appeared between the new and old data.

[43] R&D performed abroad by the foreign divisions of foreign-owned member companies was excluded.

several years before sales and margins peak. The turning points are roughly coincident for the mid-1960s and the early 1990s, but R&D leads margins by 3 years for the early 1980s. Reconciliation might come from viewing margins as an imperfectly anticipated measure of profit expectations. Or companies might apply with some deviation a crude rule of thumb, raising R&D when margins increase and holding back its growth when their growth flags. The relationships appear to break down during the 1990s and early years of the new century. A possible explanation is that during the early 1990s, companies were under heavy pressure from the Clinton administration to curb their prices or face price controls. They argued against such policies, emphasizing the dependence of R&D investments on profits, and may have found it politic visibly to maintain R&D growth levels. Margin deviations rose sharply under a new and more conservative US president, while a decline in R&D growth may have been due to disappointing new product approvals. Compare Figure 2.

Whatever the exact causal dynamics, two things appear clear: (1) Correctly accounting for R&D as a long-lived investment tends to reduce substantially, if not to eliminate altogether, the inference that pharmaceutical companies are on average achieving supranormal profit returns. And (2), there are distinct links, both short-run and long-run, between gross margins and R&D investments. One possible theoretical explanation is that pharmaceutical companies are adhering to the Dorfman-Steiner (1954) theorem, which states that profit-maximizing investments in R&D (and also in drug promotion activity, generating nearly as much expenditure as R&D) are higher, the wider PCMs are. But under Dorfman–Steiner, one would not expect the nearly complete dissipation of profit margins for R&D and promotion.[44] An explanation in better accord with the evidence and consistent with received theory[45] is that pharmaceutical companies engage in competitive rent-seeking behavior—to be sure, of a virtuous character distinguishable from what some early theories emphasized. That is, when rents (PCMs) are high, the companies compete vigorously to capture them by increasing their R&D (and promotional) outlays, and indeed, the companies compete so vigorously, there is little left over in the end for supranormal profit. When rents decline, R&D outlays are also squeezed so that a competitive rate of return persists.

10. Implications for economic welfare

To the extent that these insights are anywhere near the mark, implications for economic welfare follow.

A simple version of the rent-seeking phenomenon is illustrated in Figure 6. Through R&D, a new product is created. Its existence gives rise to a new demand curve D* that did not exist before, and through process R&D (which we have touched only lightly in previous sections) a (constant) marginal production cost function M-MC* appears. With a patent monopoly limited in time, the responsible firm maximizes its profits by setting marginal revenue MR equal to marginal cost, quoting price OP* and offering output OQ*. During the on-patent period the firm realizes quasi-rents, also called producer's surplus, measured by the rectangular area P*WXM. This producer's surplus represents a welfare gain, and its expectation motivates investment in forward-looking R&D. In addition, the availability of the

[44] See also Stigler (1968). Compare Scherer (2004), which in a model with assumptions analogous to those of Dorfman–Steiner found equilibrium R&D to be approximately 29% of quasi-rents or 17% of sales revenue.

[45] See especially Tullock (1967), Barzel (1968), and Krueger (1974).

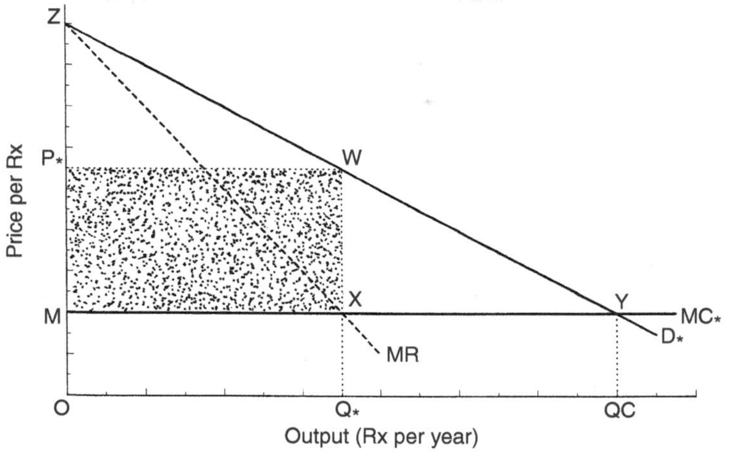

Figure 6. New product equilibrium with rent-seeking.

new product yields consumers' surplus measured by the triangular area P*WZ. When the patent expires, generic competition begins (before 1984, slowly), prices are driven to marginal cost OM, and output expands to OQ_C, giving rise to an additional surplus measured by triangle WYX and converting producer's surplus P*WXM into consumers' surplus. Public policy limits patent lives in part because, if producer's surplus is sufficient to induce R&D, policy makers do not want to delay indefinitely the realization of the consumers' surplus increment WYX (Nordhaus, 1969, pp. 70–90).

Now if competitive rent-seeking raises R&D costs to dissipate producers' surpluses totally, P*WXM can no longer be counted as a welfare gain during the period of patent monopoly. Rather, the gain is offset by R&D cost (dotted area). After patents expire, rectangular surplus P*WXM is not captured by product innovators, so it does not stimulate additional R&D. It becomes a pure surplus not offset by costs. Thus, the welfare gain pre-patent-expiry is only triangle ZWP* rather than the larger trapezoid ZWXM, and after patent expiry, the incremental welfare gain (beyond ZWP*) is trapezoid P*WYM. The net welfare gain in a competitive rent-seeking context is smaller while the patent is in force and larger incrementally after the patent expires. This, as simple models of cost-saving (process) invention have shown, leads to substantially shorter optimal patent lives—for example, in those models, as short as 1 year, compared to the much longer lives found in models without competitive rent dissipation.[46]

The cost-saving patented invention model, however, abstracts from important alternatives. In a market-oriented economy, private sector investments in R&D are driven (i.e., induced) by changes in demand conditions and by the fecundity of the science base. The simplest plausible model of these "demand-pull" and "science-push" influences is illustrated in Figure 7.[47] The R&D cost of achieving a

[46] See especially McFetridge and Rafiquzzaman (1986) for an analysis that focuses on cost-reducing (process) innovations rather than product innovations. For a more complex variant covering the product innovation case, see Scherer (2004).

[47] It is drawn from Scherer (2007a), summarizing research dating back to Scherer (1967). The competitive break-even notion was introduced by Barzel (1968).

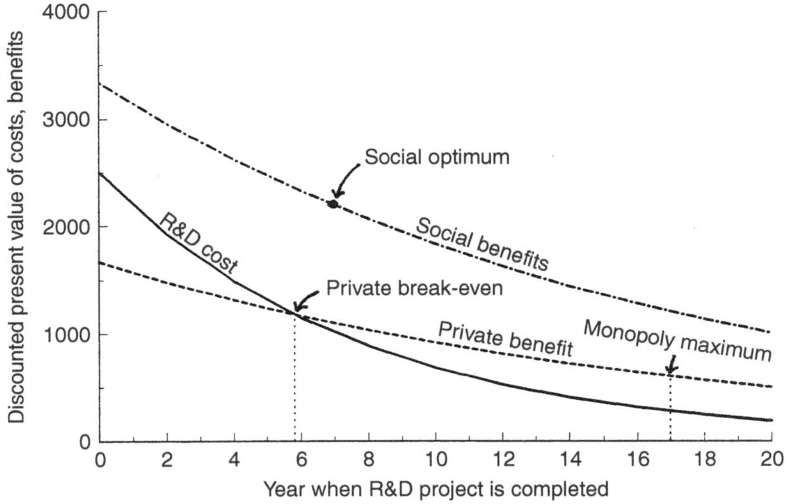

Figure 7. Optimal timing of innovation with science-push and demand-pull.

specific innovation is assumed to be 2500 (with appropriate trailing zeroes) and to decline with the advance of science at a steady rate, for example, as shown, at 3% per year. When discounted to year zero present value at a 10% interest rate, this is shown by a declining *R&D* cost curve (solid line). (A sudden scientific breakthrough would impart a one-time downward shift in the curve.) The demand side of the equation is modeled by a Private benefit function, which is the discounted present value of the quasi-rents (analogous to gross margins) appropriable by the innovator. The initial depth of the stream is 100 per year, growing with demand at 4% per year.[48] (A sudden spurt of demand, e.g., with the emergence of a new disease, would cause an upward jump in the curve.) Up to year 5.8 in Figure 7, discounted R&D costs exceed discounted private benefits, and a drug company with correct foresight would not undertake the development and testing effort. The first profitable instant—the "breakeven" point—occurs with the parameters assumed at year 5.8. However, a monopoly securely in control of the relevant therapeutic market would delay its R&D project (assumed for simplicity to consume at most 1 year) to year 17, when the discounted surplus of quasi-rents over R&D costs is maximized. But competition could force firms to accelerate their efforts or risk being preempted by either identified or inchoate rivals. Conceivably, the pace may be forced all the way forward to private break-even year 5.8. If such a competitive process operated, one would observe substantial R&D investments but on average no supranormal returns.

It remains to be asked, is the competitive acceleration of R&D desirable or undesirable from a broader economic perspective? The answer lies in the likelihood that even with patent monopolies, innovators

[48] For mathematical simplicity, the rents are assumed to continue in perpetuity, for example, under a perpetual patent. A finite patent life would shift the Private benefit function downward at all points and the *Social Benefits* function (to be explained shortly) upward.

are unlikely to appropriate as private gain all of the benefits from their innovations. There are also consumers' surpluses (e.g., triangle AWP* in Figure 6). The dot-dash line in Figure 7 assumes total social benefits—that is, producer's surplus plus consumers' surplus)—to be twice private benefits.[49] Given that assumed divergence and the other assumed parameters, the innovation date that maximizes the surplus of social benefits minus R&D costs is 7 years—close to the competitive break-even date. The larger the wedge between social and private benefits, ceteris paribus, the earlier the social welfare-maximizing date falls relative to the private break-even date.[50]

This lean model abstracts among other things from uncertainty. In fact, as we have seen, substantial uncertainties pervade all phases of the drug discovery and development process. When a safe and efficacious molecule cannot be identified in advance, pursuing parallel paths, that is, synthesizing and testing numerous alternative molecules, is often desirable. Figure 8 demonstrates a particularly simple version of the parallel paths strategy.[51] It assumes that all R&D projects can be carried out within a single year and are pursued simultaneously in year 1. Once a successful molecule is found, the innovator realizes diverse quasi-rents, constant over time and continuing through year 25.[52] Each R&D project is

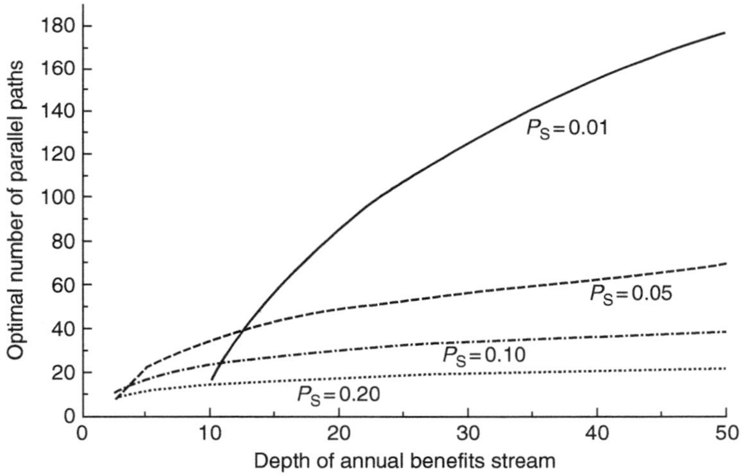

Figure 8. Optimal choice of parallel paths strategies.

[49] In the leading effort to measure this relationship, Mansfield (1977) found the median value of social benefits to be 2.25 times private benefits for a diverse sample of innovations (including no pharmaceuticals).

[50] When a is the rate of decline in R&D costs, g is the growth rate of annual-quasi rents, r is the discount rate, and social benefits are k times private benefits, the social optimum and private break-even coincide when $k = (r+a)/(r-g)$.

[51] It is adapted from Scherer (1966). More complex strategies involving a combination of parallel and series project scheduling are explored there and in Scherer (2007b).

[52] Multiple successes undoubtedly have incremental value, but the analysis assumes that one success suffices.

assumed to cost \$1 (with as many appended zeroes as reality requires), and the discount rate is 6%.[53] Uniform expected success probabilities per path (i.e., molecule) are assumed to range from 0.01 (approximating conditions in preclinical animal testing) to 0.2 (approximating the success rate of full-scale human trials). Plotted in Figure 8 are the profit-maximizing numbers of parallel paths as a function of the quasi-rent stream's depth.[54] One sees that the deeper the stream of anticipated benefits is, the larger is the optimal number of parallel paths. And the lower is the probability of success for any single path, the more sensitive the optimal number of paths is to differences in the depth of the quasi-rent stream.[55] These relationships appear to accord at least roughly with the reality of actual pharmaceutical industry behavior. For example, some 50–60 candidates for therapy against Alzheimer's disease—a scourge expected to consume \$100 billion in health care resources per year within the United States—were undergoing clinical trials in 2006.[56] Approximately 2000 cancer drugs were in development.[57] A successful drug in either category could yield substantial payoffs to both its innovator and society at large.

Parallel research paths may be pursued by an individual firm seeking a successful molecule among multiple candidates, or the parallelism may occur across competing firms at the market level. One possibility is that Firm A sees Firm B mounting an R&D project to develop new drug X, whereupon B initiates its own countervailing project to offer a variant of X and perhaps even to preempt B's innovation date. Cockburn and Henderson (1994) call such competition "racing" and, through interviews and an analysis of research focus data from pharmaceutical companies, find little support for it. Rather, they perceive investment decisions to be driven by the appearance of new technological opportunities and allocated among R&D laboratories on the basis of the firms' heterogeneous human capital capabilities. In this case, the more plausible chain of causation is that science-based or market demand changes create profit potentials, and, recognizing them sooner or later, companies compete vigorously to exploit them, in the process dissipating most or all of the attainable rents.

Supporting this view of the world, DiMasi and Paquette (2004) found that 72 first-in-class drugs approved in the United States between 1960 and 1998 were followed by at least 235 new drugs in the same narrow therapeutic categories by the year 2003.[58] And especially for later drug cohorts, the evidence pointed strongly toward parallel development and testing of drugs. Thus, in the 1990s, the average lag between the pioneer (i.e., first in class) and the first follower was 2.25 years—a period much too short for the follower to have initiated its R&D project only after observing the first-mover's success. The third mover followed the second during the 1990s by 2.5 years on average, the fourth

[53] Assuming at least one success in year 1, sales revenues and quasi-rents are assumed to begin in year 2.

[54] For annual quasi-rents of less than 10, there are no profitable single or parallel paths strategies with the assumed parameters. The average cost of R&D across all cases is 42, which is roughly 19% of the average discounted value of quasi-rents. A shorter assumed payoff period or higher discount rate would reduce the optimal number of parallel paths.

[55] The elasticities of optimal path numbers relative to quasi-rents are 1.32 for $P_S = 0.01$, 0.64 for $P_S = 0.05$, 0.41 for $P_S = 0.1$, and 0.32 for $P_S = 0.2$.

[56] See (2007) "Decoding Alzheimer's." *Business Week* 54 (January 8); (2007) "Closing in on Alzheimer's." *AARP Bulletin* 10 (June); and (2007) "Taking on Alzheimer's." *New York Times* 3-1, 4 (June 10). Note in Figure 8 that the optimal number of parallel paths for the deepest quasi-rent stream shown is 39 with a single-path success probability of 0.10, approximating the uncertainties in clinically testing drugs for a disease against which there are no effective therapies.

[57] (2007) "Cancer drugs take off." *Business Week* 71 (June 18).

[58] Recognizing that only one molecule in four or five survives clinical trials and receives FDA marketing approval, this means that the number of candidate molecules in clinical testing must have been on the order of four to five times 307 (= 72 +235), or approximately 19 parallel candidates per therapeutic category.

mover followed the third by 1.4 years. It seems virtually certain that parallel clinical testing paths were being pursued in response to perceived market opportunities.

Does the pursuit of parallel paths exhaust discounted quasi-rents, leading to only normal profits? A further analysis (Scherer, 2007b) revealed that over the range of probabilities spanned by Figure 8 and for all but the lowest annual quasi-rent potentials, there was a positive surplus of discounted quasi-rents less parallel R&D costs—over many parameter values, as high as 50–90% of the discounted quasi-rents. Thus, the observed tendency toward only small supranormal profits must result from interfirm competition, and not merely from an optimal response to uncertainty by any single company.

We conclude that the competitive rent-seeking observed in the pharmaceutical industry can help correct what otherwise might be market failures attributable to uncertainty and the disparity between social and privately appropriable benefits. Whether the "correct" amount of R&D, associated in part with the pursuit of parallel paths, is induced, cannot be determined from the lean theoretical assumptions embraced here. This is a problem on which additional research, both theoretical and factual, is much to be desired.

11. Developing new drugs and vaccines for third world diseases

Reverting to the simpler and less controversial assumption that pharmaceutical innovation is motivated by the lure of profits, a further dilemma presents itself. Rich consumers are able and willing to pay, either directly or through taxes and transfers, for an ample array of drugs to combat the diseases and debility afflicting them.[59] For the consumers in nations with very low *per capita* incomes, who tend to be concentrated in tropical areas harboring diseases such as malaria, sleeping sickness, and leishmaniasis seldom prevalent in the industrialized world, demand may be insufficient to yield quasi-rents inducing substantial investments in disease-alleviating R&D. A study by Medicins Sans Frontieres (2001) found that among 1393 new drug chemical entities introduced into world markets between 1975 and 1999, only 13 (or 15 counting tuberculosis drugs) were indicated for so-called "tropical" diseases. Clearly, the invisible hand falters in guiding research toward the needs of low-income populations.

There are several possible solutions. Prior to the Uruguay Round of international trade negotiations, concluded in 1994, many third-world nations (and some rich nations) did not offer patent protection on new pharmaceutical products. The resulting treaty required inter alia the provision of such patent rights in all World Trade Organization member nations by the year 2005 (later extended for the least-developed nations to 2016). One rationale was that this policy change would stimulate the development of medicines for tropical diseases, either by multinational pharmaceutical companies or enterprises based in low-income nations. (India, e.g., was home to several of the world's leading generic drug suppliers.) Whether this strategy will cause significant changes remains to be seen (Lanjouw, 1999, 2002). If under the logic of Figure 6, demand curves for drugs in low-income nations lie too close to marginal production cost functions, the pool of attainable quasi-rents will be too small to stimulate much development of tropical disease drugs by profit-seeking firms.

If private markets fail, a humanitarian case for governmental or philanthropic intervention exists. Governments and philanthropic agencies might intervene on either the supply side or the demand side.

[59] Antibiotics combatting rare but highly resistant bacteria may also face markets too small to induce much research and development. See Groopman (2008).

On the supply side, research and testing on tropical drugs might be conducted in government or government-supported laboratories, or grants could be issued to private corporations to subsidize the development of tropical disease therapies. The US Army's Walter Reed Hospital was once a leader in developing drugs to combat malaria and other tropical diseases. But as the desire to station American troops in tropical nations faded after the Vietnam War, so also did interest in developing such medicines. Thus, contracts and grants for altruistic motives remained the main supply-side recourse. Splendid work by the Gates Foundation, among others, has been done, but those activities, oriented thus far mainly toward basic research and therapeutic molecule discovery, are of too recent vintage to assess success. The alternative, especially when high-cost drug development and clinical testing stages are reached, is for governments to issue contracts to private enterprises—presumably, to the various pharmaceutical companies. Here the well-known agency-theoretic problems associated with national defense R&D contracting are encountered. Government agencies are not always adept at picking winning technological approaches, and indeed, given the uncertainties of drug discovery, one must be tolerant—although legislatures seldom are—of frequent failure. The choice problem is aggravated by the tendency of contract-seekers to exaggerate their chances of success at the early proposal stage and to underestimate the costs. Special contractual arrangements, such as cost-plus-fixed-fee contracts, may be necessary to transfer what would otherwise be unacceptable technological risks from private firms to the government sponsor. These often provide inadequate incentives for efficient operation and open the way to other moral hazards (Peck and Scherer, 1962).

An alternative to intervention on the technology-push side is for government agencies to create special demand-side incentives for R&D. An interesting and attractive approach was the advance purchase approach to inducing new vaccine development endorsed by the G-8 nations in 2005 and 2006 (Levine et al., 2005; Berndt and Hurvitz, 2005). Emphasizing the development of vaccines rather than traditional therapeutic pharmaceuticals was attractive because vaccines can prevent disease through one or very few inoculations, whereas treatment once a disease has taken hold often requires repeated and perhaps even life-long medical interventions that overstrain the healthcare delivery capabilities of low-income nations. One disadvantage of the vaccine approach is the particularly extensive and lengthy clinical testing required, since one cannot ethically tell in advance who would otherwise incur the target disease. The advantage of vaccines from the perspective of administration in low-income nations is a disadvantage for pharmaceutical companies, since each patient requires only one or a very few doses, which leaves much less demand than for medicines that will be administered once a day for many days or even years. Recognizing these problems, the G-8 proposal identified three target diseases—HIV/AIDS, malaria, and tuberculosis. For each, a generalized agreement to purchase 200 million doses at a prespecified subsidy of $15 per dose would be announced, that is, embodying a total commitment of $3 billion per disease, paid only if the goal of successful new vaccine development were achieved. The purchase would be conditional upon the development of effective vaccines, with efficacy judged against standards articulated by a coordinating committee and by the national health authorities of nations administering the vaccine (which would add their own more modest subsidies to the purchase price). Quantities above 200 million would be procured at prices to be negotiated through a process that remained unclear at the time the proposal was approved. At the time this essay was written, the G-8 governments had not made available the required financial commitments to induce development of *new* vaccines. However, funds were advanced for procurement of a pneumococcus vaccine already in the late stages of testing.[60] Thus, a judgment on the success of the advance purchase commitment

[60] (2007) "Wealthy Nations announce plan to develop and pay for vaccines." *New York Times* 3 (February 10).

approach would be premature. What is clear is that an important market failure persists with respect to incentives for the discovery and development of therapies effective against diseases uniquely threatening one to two billion inhabitants of low-income nations.

12. Conclusion

The pharmaceutical industry provides a fascinating laboratory for studying what we know and what we do not know about the economics of innovation. The industry has an extraordinary innovation record; it faces major risks and uncertainties in its efforts to solve new therapeutic problems; its links to academic science bases are unusually rich and deep; and the industry's responsiveness or lack thereof to economic stimuli is of considerable interest. That said, it must be admitted that there is much we still do not understand about the pharmaceutical innovation process. As always, more work remains to be done.

Acknowledgments

The author is indebted to members of the National Bureau of Economic Research productivity workshop, Christina Scherer, Joseph di Masi, and Bronwyn Hall for helpful comments.

References

Adams, J.D., Clemmons, J.R. (2008). "Industrial scientific discovery". National Bureau of Economic Research Working Paper No. 13823.

Angell, M. (2004). The Truth About the Drug Companies. New York, Random House.

Azoulay, P. (2003). "Acquiring knowledge within and across firm boundaries: Evidence from clinical development". National Bureau of Economic Research Working Paper 10083.

Barzel, Y. (1968). "Optimal timing of innovations". Review of Economics and Statistics 50, 348–355.

Berndt, E., Hurvitz, J. (2005). "Vaccine advance-purchase agreements for low-income countries". Health Affairs 24, 653–665.

Berndt, E., Cockburn, I., Thiers, F. (2007). "The globalization of clinical trials for new medicines into emerging economies". Paper presented at a UNU-MERIT conference, Maastricht, Netherlands.

Bond, R.S., Lean, D. (1977). Sales, Promotion, and Product Differentiation in Two Prescription Drug Markets. Federal Trade Commission staff report, Washington.

Branscomb, L., Auerswald, P. (2001). Taking Technical Risks. MIT Press, Cambridge, MA.

Carr, G. (1998). "The pharmaceutical industry". The Economist 21, 1–18, special survey (February).

Clymer, H.M. (1970). "The changing costs of pharmaceutical innovation". In: Cooper, J. (Ed.), The Economics of Drug Innovation. American University Center for the Study of Private Enterprise, Washington, pp. 109–124.

Cockburn, I.M. (2006). "Is the pharmaceutical industry in a productivity crisis?" Innovation Policy and the Economy 7, 1–32.

Cockburn, I.M., Henderson, R. (1994). "Racing to invest? The dynamics of competition in ethical drug discovery". Journal of Economics and Management Strategy 3, 481–519.

Cockburn, I.M., Henderson, R. (1998). "Absorptive capacity, coauthoring behavior, and the organization of research in drug discovery". Journal of Industrial Economics 46, 157–182.

Cockburn, I.M., Henderson, R. (2000). "Publicly funded science and the productivity of the pharmaceutical industry". Innovation Policy and the Economy 1, 1–34.

Cockburn, I.M., Henderson, R., Stern, S. (1999). "The diffusion of science driven drug discovery". National Bureau of Economic Research Working Paper 7359.

DiMasi, J.A. (2000). "New drug innovation and pharmaceutical industry structure: Trends in the output of pharmaceutical firms". Drug Information Journal 34, 1169–1194.
DiMasi, J.A., Paquette, C. (2004). "The economics of follow-on drug research and development". PharmacoEconomics 22, 1–14.
DiMasi, J.A., Hansen, R., Grabowski, H., Lasagna, L. (1991). "Cost of innovation in the pharmaceutical industry". Journal of Health Economics 10, 107–142.
DiMasi, J.A., Hansen, R., Grabowski, H., Lasagna, L. (1995). "Research and development costs for new drugs by therapeutic category". PharmacoEconomics 7, 152–169.
DiMasi, J.A., Hansen, R., Grabowski, H. (2003). "The price of innovation: New estimates of drug development costs". Journal of Health Economics 22, 151–185.
Dorfman, R., Steiner, P.O. (1954). "Optimal advertising and optimal quality". American Economic Review 44, 826–836.
Gambardella, A. (1995). Science and Innovation: The US Pharmaceutical Industry During the 1980s. Cambridge University Press, Cambridge, UK.
Grabowski, H., Vernon, J.M. (1990). "A new look at the returns and risks to pharmaceutical R&D". Management Science 36, 804–821.
Grabowski, H., Vernon, J.M. (1994). "Returns on new drug introductions in the 1980s". Journal of Health Economics 13, 383–406.
Grabowski, H., Vernon, J.M., Thomas, L.G. (1978). "Estimating the effects of regulation on innovation: An international comparative analysis". Journal of Law and Economics 21, 133–163.
Grabowski, H., Vernon, J.M., DiMasi, J. (2002). "Returns on research and development for 1990s new drug introductions". PharmacoEconomics 20 (Suppl. 3), 16–27.
Groopman, J. (2008). "Superbugs". The New Yorker 46–55 (August 11, 2008).
Harbridge House, Inc.. (1967). Patent Policy Study: Task IV, Drug Study. Harbridge House, Inc., Boston
Harhoff, D., Scherer, F.M. (2000). "Technology policy for a world of skew-distributed outcomes". Research Policy 29, 559–566.
Healy, E.M., Kaitin, K. (1999). "The European Agency for the Evaluation of Medicinal Products' centralized procedure for product approval: Current status". Drug Information Journal 33, 969–978.
IIT Research Institute. (1968). Technology in Retrospect and Critical Events in Science. National Science Foundation, Washington, DC.
Institute of Medicine, National Academies of Science. (2006). The Future of Drug Safety. National Academies of Science, Washington, DC.
Kettler, H.E. (2000). "Pharmaceutical and biotechnology industries coalescing". Presentation at the American Enterprise Institute, Washington, DC.
Kingston, W. (2000). "Antibiotics, invention, and innovation". Research Policy 29, 679–710.
Kitch, E.W. (1973). "The patent system and the new drug application". In: Landau, R.L. (Ed.), Regulating New Drugs. University of Chicago Center for Policy Study, Chicago, pp. 81–108.
Krueger, A.O. (1974). "The political economy of the rent-seeking society". American Economic Review 64, 291–303.
Lanjouw, J.O. (1999). "The introduction of pharmaceutical product patents in India". National Bureau of Economic Research working paper no. 6366.
Lanjouw, J.O. (2002). "Intellectual property and the availability of pharmaceuticals in poor countries". Center for Global Development working paper no. 5, Washington, DC.
Levin, R.C., Klevorick, A.K., Nelson, R.R., Winter, S. (1987). "Appropriating the returns from industrial research and development". Brookings Papers on Economic Activity 1987 (3), 783–820.
Levine, R., Kremer, M., Albright, A. (2005). Making Markets for Vaccines: Ideas to Action. Center for Global Development, Washington, DC.
Lichtenberg, F.R. (2001). "The benefits and costs of newer drugs: Evidence from the 1996 medical expenditure panel survey". National Bureau of Economic Research working paper 8147.
Lichtenberg, F.R. (2004). "Public policy and innovation in the U.S. pharmaceutical industry". In: Holtz-Eakin, D., Rosen, H. (Eds.), Entrepreneurship and Public Policy. MIT Press, Cambridge, MA.
Lichtenberg, F.R. (2007). "The impact of new drug launches on U.S. longevity and medical expenditure". American Economic Review 97 (Suppl.), 438–441.
Long, G., et al. (2006). "The impact of hypertension drugs on the number and risk of deaths, strokes, and myocardial infarctions". National Bureau of Economic Research working paper 12096.

Love, J. (2003). Evidence regarding research and development investment in innovative and non-innovative medicines. Consumer Project on Technology (CPTECH), Washington, DC.
Mann, C.C., Plummer, M.L. (1991). The Aspirin Wars. Harvard Business School Press, Boston.
Mansfield, E. (1970). "Discussion". In: Cooper, J. (Ed.), The Economics of Drug Innovation. American University, Washington, DC, pp. 149–154.
Mansfield, E. (1977). "Social and private rates of return from industrial innovations". Quarterly Journal of Economics 91, 221–240.
Mansfield, E. (1986). "Patents and innovation: An empirical study". Management Science 32, 173–181.
Mansfield, E. (1998). "Academic research and industrial innovation: An update of empirical findings". Research Policy 26, 773–776.
Maxwell, R.A., Eckhardt, S.B. (1990). Drug discovery: A case book and analysis. Humana Press, Clifton NJ.
McFetridge, D.G., Rafiquzzaman, M. (1986). "The scope and duration of the patent right and the nature of research rivalry". Research in Law and Economics 8, 91–129.
Medicins Sans Frontieres Access to Essential Medicines Campaign. (2001). Fatal Imbalance: The Crisis in Research and Development for Neglected Diseases. Medicins Sans Frontieres, Geneva, Switzerland.
Murphy, K.M., Topel, R.H. (2006). "The value of health and longevity". Journal of Political Economy 114, 871–904.
Nordhaus, W.D. (1969). Innovation, Growth, and Welfare. MIT Press, Cambridge, MA.
Peck, M.J., Scherer, F.M. (1962). The Weapons Acquisition Process: An Economic Analysis. Harvard Business School Division of Research, Boston, MA.
Pharmaceutical Research and Manufacturers of America. (2006). Pharmaceutical Industry Profile: 2006. Pharmaceutical Research and Manufacturers of America, Washington, DC.
Pisano, G.P. (2006). Science Business: The Promise, the Reality, and the Future of Biotech. Harvard Business School Press, Boston, MA.
Powell, W.N., Grodal, S. (2005). "Networks and innovators". In: Fagerberg, J., Mowery, D.C., Nelson, R. (Eds.), Oxford Handbook of Innovation. Oxford University Press, Oxford, pp. 56–85.
Reichert, J.M., Milne, C.P. (2002). Public and private sector contributions to the discovery and development of 'impact' drugs. Tufts University Center for the Study of Drug Development White Paper, Boston, MA.
Robinson, W.T., Fornell, C. (1985). "Sources of market pioneer advantages in consumer goods industries". Journal of Marketing Research 22, 305–317.
Scherer, F.M. (1966). "Time-cost tradeoffs in uncertain empirical research projects". Naval Research Logistics Quarterly 12, 71–82.
Scherer, F.M. (1967). "Research and development resource allocation under rivalry". Quarterly Journal of Economics 81, 359–394.
Scherer, F.M. (1996). Industry Structure, Strategy, and Public Policy. HarperCollins, New York.
Scherer, F.M. (2001a). "The link between gross profitability and pharmaceutical R&D spending". Health Affairs 20, 216–220.
Scherer, F.M. (2001b). "The innovation lottery". In: Dreyfuss, , R.et al. (Ed.), Expanding the Boundaries of Intellectual Property. Oxford University Press, Oxford, pp. 3–21.
Scherer, F.M. (2004). "A note on global welfare in pharmaceutical patenting". The World Economy 27, 1127–1142.
Scherer, F.M. (2007a). "Schumpeter and the micro-foundations of endogenous growth". In: Hanusch, H., Pyka, A. (Eds.), The Elgar Companion to Neo-Schumpeterian Economics. Edward Elgar, Cheltenham, UK, pp. 671–687.
Scherer, F.M. (2007b). Parallel Paths Revisited. John F. Kennedy School of Government, Harvard University, Cambridge, MA. Working Paper RWP07-040.
Scherer, F.M. (2009). "The political economy of patent policy reform in the United States". Journal on Telecommunications and High Technology Law 7, 167–216.
Schwartzman, D. (1976). Innovation and the Pharmaceutical Industry. Johns Hopkins University Press, Baltimore.
Schwartzman, D. (1996). "Has pharmaceutical research become more scientific?" Review of Industrial Organization 11, 841–851.
Shulman, S., Mannocchia, M. (1997). "The US orphan drug programme: 1983–1995". PharmacoEconomics 9, 312–326.
Simons, J. (2003). "The $10 billion pill". Fortune 20, 58–68 (January).
Stauffer, T.R. (1971). "The measurement of corporate rates of return: A generalized formulation". Bell Journal of Economics and Management Science 2, 434–469.
Stigler, G.J. (1968). "Price and non-price competition". Journal of Political Economy 72, 149–154.

Tufts Center for the Study of Drug Development. (2006). Impact Report. Tufts Center for the Study of Drug Development, Boston, MA, vol. 8, no. 3, pp. 1–4.

Tullock, G. (1967). "The welfare costs of tariffs, monopolies and theft". Western Economic Journal 5, 224–232.

U.S. Federal Trade Commission. (1958). Economic Report on Antibiotics Manufacture. U.S. Federal Trade Commission, Washington, DC.

U.S. Federal Trade Commission. (1985). Statistical Report: Annual Line of Business Report: 1977. U.S. Federal Trade Commission, Washington, DC.

U.S. Food and Drug Administration, Pilot Drug Evaluation Staff. (1993). "The 007 experiment—An approach to change". unpublished internal assessment, Rockville, MD.

U.S. Office of Technology Assessment. (1993). Pharmaceutical R&D: Costs, Risks, and Rewards. U.S. Office of Technology Assessment, Washington, DC.

Vernon, J.A., Hughen, W.K. (2005). "The future of drug development: The economics of pharmagenomics". National Bureau of Economic Research Working Paper 11875.

Ward, M.R., Dranove, D. (1995). "The vertical chain of research and development in the pharmaceutical industry". Economic Inquiry 33, 70–87.

Werth, B. (1994). The Billion Dollar Molecule. Simon & Schuster, New York.

Zucker, L., Darby, M., Armstrong, J. (1998). "Geographically localized knowledge: Spillovers or markets?" Economic Inquiry 36, 65–86.

Chapter 13

COLLECTIVE INVENTION AND INVENTOR NETWORKS

WALTER W. POWELL* *AND* ERIC GIANNELLA[†]

**Stanford University, Stanford*

California, USA

[†]*Xerox PARC, Palo Alto*

California, USA

Contents

Abstract

Collective invention occurs when competing organizations share knowledge about the design and development of new technologies. Such exchange and circulation of ideas and practices among communities of inventors was relatively common in the nineteenth century, most notably in geographically localized industrial districts. This collective system of innovation was eclipsed in the early and mid-twentieth century by the rise to prominence of the large corporate R&D lab. Recent decades, however, have seen the decline of stand-alone, internal corporate labs and the resurgence of collective efforts by networks of inventors, distributed across organizations and spanning distant locations. We draw on literatures in economics, innovation studies, management, and sociology to posit explanations for this

Handbooks in Economics, Volume 01

DOI: 10.1016/S0169-7218(10)01013-0

recent rise. Suggestive additional evidence is provided from comparative analyses of patent data from the 1970s and the present decade.

Keywords

collective invention, governance, networks, technological change

1. Introduction

Historians, sociologists, and economists who study innovation often differ in their emphases on the features of settings in which technical change occurs. For many business historians and economists, the "organizational synthesis" is the central story, as the large firm developed through linking investments in technology and corporate strategy (Chandler, 1977; Galambos, 1983). The corporate research laboratory, established after 1900 at General Electric, DuPont, Kodak, AT&T, RCA, and others, was created to bring the innovation process inside the large corporation, and provide a continuing basis for both control over and renewal of technological change (Carlson, 1991; Hounshell and Smith, 1988; Mowery, 1984; Reich, 1985). As Graham (2008) points out, even critics of the corporation viewed the large firm as the central force in technological change, although arguing that it also monopolized invention, repressed craft knowledge, and stifled the creativity of engineers (Noble, 1984).

Instead of focusing on the centrality of the firm, historians and sociologists of technology who have studied the evolution of industries have emphasized a different current of innovation. Systems technologies—electricity, the telephone, and its successors—have developed not because of a particular corporate champion or active commercial pursuit, but due to a collective "momentum," or the accumulation of investment and interest in a system's progress from a variety of participants (Bijker, 1987; Hughes, 1983, 1987, 1989; MacKenzie, 1990). These systems technologies were the combined product of research carried out by individual inventors, government and university researchers, and corporate labs. Many technological systems reflect a confluence of uncoordinated research efforts driven by intense and widespread interests that intersect around the development of a novel technology. As a new technology evolves in a growing web of social, economic, and artifactual parts, the primary control that individual firms have is to configure their own activities in light of the needs of these systems.

Alongside these two powerful currents, alternate modes for organizing the innovation process have persisted. In the nineteenth and early twentieth century, such alternatives typically involved craft-based models, based in local communities. In research on the blast furnace, Allen (1983) identified how a group of firms could produce "collective invention" by sharing information about the design and effectiveness of new technologies. From his studies of the disclosure of improvements in manufacturing processes within the iron industry, Allen suggested that the distinctive feature of collective invention is the exchange and circulation of ideas and practices among distributed networks of individuals located in diverse settings, rather than the housing of such efforts within the confines of particular firms. Building upon Allen's (1983) work, Nuvolari (2004: p. 348), in his study of Cornish steam pumping engines, defines collective invention as a *setting* in which: "competing firms release information freely to one another on the design and the performance of the technologies they have just introduced."

In Allen and Nuvolari's analyses, there are four contributors to technical change: R&D labs of private firms, nonprofit institutions, individual inventors, and collective invention. Allen (1983) and Nuvolari (2004) suggest that three propositions typify the setting of collective invention. First, technical change must be driven by primarily incremental improvements. Second, firms and other organizations must disclose any improvements they make. And third, firms must use the disclosed improvements to enhance the technology they have in common. We build on these insights, and connect them to recent work in the economics and sociology of technical change.

We begin by taking stock of the theoretical interests at stake in research on collective invention. In particular, collective invention has attracted much attention because it defies conventional wisdom about appropriability concerns; therefore, we suggest the link to intellectual property should be made explicit in defining collective invention. Next, it is important to highlight the tension implicit in the dual role of participants in collective invention as employees of competing organizations and as technologists who have personal or professional interests at stake in the overall advance of some technology. Thus, we distinguish between competing firms (or, more broadly, organizations including government and university labs), the loose network of inventors that cuts across these organizations, the growth of knowledge, and the actual improvement of technologies. We offer a substantively similar, but distinctly social definition of collective invention:

> *Collective invention is technological advance driven by knowledge sharing among a community of inventors who are often employed by organizations with competing intellectual property interests.*

This definition broadens the scope of collective invention to instances of university–university and university–industry interactions, and encompasses voluntary and informal associations that are often critical to economic activity (Granovetter, 2009). In addition, the role of patent pools and other collective agreements that further technical change are more amenable to analysis within this framework.

We should note at the outset that collective invention is merely the tip of the iceberg of increased knowledge sharing over the past several decades. Such disclosure of valuable information to competitors is much more pervasive than "pure" collective invention. Yet, because it represents one end of a continuum of knowledge-sharing regimes, collective invention offers fertile ground for empirical research and novel theorizing about the determinants of technological change.

Having defined and situated collective invention, we turn to its origins. We argue that the increasingly specialized division of labor makes it difficult to predict where complementary knowledge will arise—leading to greater knowledge sharing in order for participants to remain abreast of developments in the field. Additionally, we suggest that high expectations for a technology (i.e., technological opportunity) can lead individuals across firms and nonprofit organizations to contribute their efforts to a community endeavor that drives collective invention despite the lack of apparent economic gain to any particular organization.

Historical examples bear out the importance of collective invention in improving a number of notable technologies (Lamoreaux and Sokoloff, 2000; McGaw, 1987; Meyer, 2003; Scranton, 1997). A general lesson from numerous historical studies is that collective invention was an attempt to overcome the limitations of information access that accompanied extant economic and organizational structures. For some organizations, the inability to appropriate many types of technical improvements resulted in a lack of motivation to pursue internal research programs. Why invest in expensive exploratory efforts when the odds of capturing the fruits of research were low? Participation in collective efforts offered one solution. Many instances of collective invention today represent joint efforts at solving problems whose value cannot be appropriated by a single party, but which represent a bottleneck for the interdependent economic activities of participants. On the other side of the fence, some companies that are actively engaged in R&D may want their researchers to be involved in a larger technical community. Collective invention affords the chance for access to more diverse sources of knowledge, even if gaining control over these divergent ideas proves difficult.

With time, many knowledge-sharing practices associated with collective invention can become institutionalized as a set of norms or agreements (David, 2008; Merton, 1979; Sabel and Zeitlin, 1985). In the case of the diffusion of the Bessemer steel process, a patent license that nearly all manufacturers signed had a clause that required any subsequent operational improvements to be disclosed. This mandated sharing of knowledge led to the establishment of a small community of practice among engineers from different firms and launched a productivity race between participants from different firms (Allen, 1983: p. 11). A variety of practices—such as mutually respected prices, collective training programs, and technological standards, that spread risks and dampened competition were commonplace across industrial districts. Nuvolari's (2004) analysis of Cornish steam engines in the nineteenth century finds that the publication of advances in several trade outlets let to dramatic gains in the efficiency of the engines, due to the accumulation of myriad incremental improvements. Despite the variety of vibrant nineteenth century examples of collective invention, these efforts were largely displaced by the rise of the large corporate research and development (R&D) lab in the early twentieth century. For a time it seemed that these community efforts would be relegated to the annals of history.

Over the past 30 years, however, the large corporate R&D lab has fallen in prominence. Many of the most notable corporate labs have been shuttered and dismantled. A second wave of collective invention is now shaping the rate and direction of technological change in numerous technologically advanced industries (Freeman and Soete, 2009). These processes of distributed innovation characterize a wide array of contemporary industries, from the early origins of the computer to the development of software to the genesis and evolution of biotechnology. This transformation has been sparked by strategic, technical, and economic factors that influence the organization of innovative labor. Inventors with multiple contacts across organizations are more likely to be exposed to diverse ideas and benefit from them. Consequently, organizations attempt to position themselves in partnerships and alliances that foster connections across organizational boundaries, in hopes that novel ideas in one setting spark fresh approaches in another (Burt, 2004; Granovetter, 1973; Powell et al., 1996). Shared awareness of a technological frontier creates the circumstances for inventors to act in concert, regardless of the perceived tangible benefits for their organizations. The central technical drivers are shifts in technological opportunity, dictating the potential rate and direction of technological change (Malerba, 2007). The economic factors are demand (on economic *demand* vs. *need,* see Mowery and Rosenberg, 1979) and appropriability (Teece, 1986; Winter, 2006), which together represent necessary conditions for firms to invest in R&D.

Yet history and social structure also loom large, as many authors have noted (David, 2008; Scranton, 1993). The particularities of industry evolution and the historical organization of technical communities are deeply intertwined with economic and technical calculations. Whether nineteenth century glass making or blast furnaces, or the contemporary life sciences and open-source software, relationships within a community of inventors and researchers are influenced by a confluence of social, political, and economic forces. We summarize these disparate factors as follows:

1. The need to spread the costs of invention across multiple organizations.
   a. By implication, few participants possess a sufficient theoretical understanding to pursue new ideas without incurring the high costs of unguided trial and error.

2. The inability to appropriate innovations creates a discrepancy between the private value and social value of invention.
   a. The private value of invention is too low for some firms to pursue a technology individually, but individuals within these firms are able to recognize its potential benefits.
   b. Despite a lack of knowledge about demand and strong intellectual property rights, collective invention allows for continued improvement of technical performance.
3. The emergence of norms and identification of governance structures that encourage knowledge sharing among legally distinct parties.
4. Uncertainty about the direction a technology will evolve and the kinds of applications that may unfold encourage greater discussion within and across communities and provide an impetus for organizing.

In this chapter, we examine these and other reasons for the recent rise in collective invention. We look at the changing nature of technological opportunity, as well as factors shaping the organization and governance of innovative labor. One understudied aspect of collective invention is the growing fragmentation of the knowledge required for many promising technological opportunities, leaving relevant know-how spread across diverse organizations.

The knowledge boundaries of firms develop due to many social and economic processes that are unassociated with changes in technological opportunity. As Schumpeter (1942) argued, it would be naïve to expect firms to immediately and optimally adjust to changes in technology (Rosenberg, 2000). Indeed, it would be difficult to maintain that new technological knowledge is ever brought about under ideal circumstances for its evaluation, elaboration, and diffusion. By its very nature, new knowledge is, to varying degrees, at odds with the social structures in which it is discovered (Mokyr, 2005). Put differently, the inability to reconcile newly perceived goals with the internal and external distribution of knowledge for invention may, under certain circumstances, render collective invention a more viable option than internally funded R&D.

Unpredictable technical change also makes it more difficult for firms to house all the innovative labor required to pursue many technological opportunities. Such shortfalls in capability and opportunity can prompt some to make use of collective invention. Thus, to the extent that data for decision making overwhelms the machinery of hierarchical organization (Knudsen and Levinthal, 2007; Powell, 1990), collective invention becomes more prevalent. At the same time, for those companies with strong internal research capabilities that operate in domains in which technological futures are uncertain, collective invention provides an option to become involved in a broader effort of exploration and learning.

We organize our chapter around four arguments that account for the persistence of, and greater reliance on, collective invention:

1. As the stock of knowledge grows, the need to access specialized expertise outside the boundaries of individual organizations increases.
2. When the sources of potentially complementary knowledge become more diverse, engagement with external communities increases.
3. The emergence of new forms of governance makes collective invention less costly and still compatible with the goals of private enterprise.

4. Persistent interindustry variations in technological opportunities and social institutions result in marked differences across fields in the reliance on, and form of, collective invention.

Potentially complementary innovative labor spreads across a wider array of organizations as the stock of knowledge grows, making it more difficult for a single organization to possess requisite depth and breadth of expertise (Section 2). Intuitively, then, we would expect collective invention to expand as a result of the increasing complexity of products and processes and the narrower specialization of innovative labor. Put simply, one reason we see a resurgence of collective invention now is that there are more pieces to each puzzle and each player has fewer pieces.

The difficulty of identifying and absorbing complementary knowledge makes investments in access to diverse sources of knowledge more desirable. Because it is challenging to predict the spreading of organizational, technical, and geographic locations of relevant expertise and ideas, firms engage in collective invention to keep pace with recent developments (Section 3).

Collective invention is also fueled by the creation of governance structures that enable individuals from different organizations to share knowledge at lower cost and with reduced risks of misappropriation or malfeasance. Additionally, new technological and physical forms of organizing for collective invention help mitigate many of the challenges associated with asynchronous or remote coordination and collaboration (Section 4).

Finally, there are unique and persistent interindustry differences in the qualitative nature and magnitude of collective invention. These differences arise in part due to the distinctive social structures that characterize different industries and their divergent stages of technological evolution. These two factors alter the potential benefits that firms might hope to accrue, shifting the choice and mix of internal versus collective invention. Thus, interindustry differences in the use of collective invention stem from variation in technological opportunities, the uncertainty of technological trajectories, and the means of appropriating innovations that arise from collective knowledge. We discuss these differences, attending to the divergent norms found in various scientific and technical communities, which condition the creation and sharing of ideas (Section 5).

To add support for these arguments, we provide illustrative evidence from a number of technology-intensive industries. We also use patent data from key technology classes to add weight to our review of the literature, and gauge the extent of the changes over the course of recent decades.

2. The stock of knowledge has grown

Numerous arguments have been offered in recent decades that describe a transition from industrial society to a knowledge-based economy (Bell, 1973; Gibbons et al., 1994; Hicks and Katz, 1996; Powell and Snellman, 2004; Ziman, 1994 provide entry into these discussions). The relevance of these arguments for our purposes is their characterization of a marked change in the modern research enterprise. Collaboration—both domestic and international—has increased; and a more diverse set of organizations and nations are contributing to the stock of knowledge. In addition, the proportion of research that is interdisciplinary has grown, and key research funding agencies are now strongly behind efforts at translating basic research into application to solve pressing environmental and medical problems. The implications of these shifts toward greater collaboration and interdisciplinarity for collective invention are far-reaching.

Hicks and Katz (1996) were among the first scholars to use bibliometric evidence to examine the changing terrain of science. In an analysis of 376,226 publications between 1981 and 1991, they show notable growth in the average number of authors per paper, from 2.63 to 3.34, and a smaller uptick in the number of institutions and countries represented on each article. Their findings complemented earlier analyses of de Solla Price (1963), who chronicled the increasing importance of multiple authors in the chemical and physical sciences, areas he dubbed "Big Science." More recently, Wuchty et al. (2007), in a comprehensive analysis of 19.9 million articles and 2.1 million patents covering the late 1950s–2000, found that the increasing prevalence of multiple authors, or "team science," had extended from the physical sciences to chemistry, biology, engineering, the social sciences and even mathematics and the humanities.

The fields of medicine, biology, and physics have each shown at least a doubling in mean team size over the 45-year period from 1955 to 2000 (Wuchty et al., 2007: p. 1037). This growth in teamwork may well be triggered by an increase in knowledge specialization and the growing costs of doing research, but the number of authors on papers is also growing in fields where the overall number of researchers is growing less rapidly and costs are less a factor. Perhaps most consequential, Wuchty et al. (2007) find that, even after numerous relevant controls, papers by teams are cited more frequently and are much more likely to have high impact. In subsequent work, Jones et al. (2008) looked at a sample of 4.2 million papers published at US universities between 1975 and 2005, and observed that teams increasingly involve authors from multiple universities.

We add more empirical support for the argument that the stock of knowledge has grown in recent decades through a comparison of the number of inventors on patents from five US patent classifications across two time periods—1975–1979 and 2001–2005. We chose the technologies as useful indicators of older industries with a history of innovation (aerospace, pharmaceuticals), as well as sectors that came into prominence in the last quarter of the century (optical communications, semiconductors, and biotechnology). We obtained patent data from Delphion, a commercial patent search service owned by Thomson Reuters. We searched for all patents containing at least one US patent classification corresponding to our technology domains of interest, which we use for illustrative purposes. Table 1 is based on all patents filed over these time periods for each patent class. The *inventors* column contains the mean number of inventors. For example, there were an average of 1.5 inventors across 1118 Aerospace patents in the late 1970s, and 2.2 inventors on average on 1619 patents in the early years of this decade. With the exception of the new domain of biotechnology, which had a high rate of collective invention at its outset and continues to be highly collaborative, the organization of innovative labor appears to have shifted, with considerably more inventors per patent. This transition to multiple authors suggests a greater need to integrate a wider stock of knowledge. Biotechnology had its origins in the 1970s in university labs and continues today to be a science-driven field. Inventor teams in biotech are, not surprisingly, the largest of any technical area shown, suggesting that the functional diversity of these teams is also greatest.

While the number of inventors increased across the board, there are key differences that merit attention. Apart from biotechnology, semiconductor manufacturing processes and pharmaceutical compounds represent the greatest contrast. Semiconductor inventions are highly modularized by steps in the manufacturing process, which often correspond to a particular disciplinary foundation or the juncture between two disciplines. For example, much of modern semiconductor manufacturing is enabled by chemical engineering, optics, materials science, mechanical engineering, and optimization

Table 1
Number of inventors per patent in selected patent classes[a]

Patent class	1975–1979 Inventors	1975–1979 # Patents	2001–2005 Inventors	2001–2005 # Patents
Aerospace	1.5	1118	2.2	1619
Biotechnology	6.4	6533	6.5	22,881
Optical Comm.	1.6	511	2.4	6217
Pharm. Chem.	2.5	2467	4.3	7212
Semi. Mfg.	2.0	5630	2.7	79,069

[a]The technology areas correspond to the following patent classification titles and numbers:
Aerospace—Aeronautics and Astronautics; 244
Biotechnology—Chemistry: Molecular biology and microbiology; 435
Optical Comm.—Optical Communications; 398
Pharm. Chem.—Drug, bioaffecting and body treating compositions; 424
Semi. Mfg.—Semiconductor Device Manufacturing: Process; 438

and planning software. Each step of the process, such as the manufacture of masks for laser etching onto wafers, the design of robotic machinery, and the chemical baths used to remove support structures, represents a fairly distinct body of knowledge (Orton, 2004). Collaboration across these areas of expertise primarily occurs in order to coordinate across steps in the manufacturing process. Given that manufacturing is fairly decomposable into parts, the size of teams can stay relatively small, reflecting the reduced need to simultaneously solve a complex of problems.

In contrast, the field of pharmaceuticals often presents nondecomposable problems, which cannot be broken apart and addressed separately without significantly affecting the quality of the final result (Simon, 1962). Whereas the average number of inventors in semiconductor manufacturing processes increased from 2 to 2.7, pharmaceutical drug patents saw a larger jump in authors from 2.5 to 4.3. Economists often describe drugs as "discrete" technologies since they are not modular, whereas semiconductors and telecommunications equipment are called "complex" due to their many parts that need to be integrated (see Arora et al., 2001). Thus, the invention of pharmaceutical drugs cannot, for the most part, be cleanly divided across areas of expertise. Invention often requires intensive collaboration by organic chemists, microbiologists, and biochemists, as well as immunologists and pathologists in order to discover drug targets and potential drugs. Thus, the size of inventive teams depends on both the sheer amount of knowledge that needs to be integrated and the ways in which scientific and engineering training and expertise map onto technological problems.

Summary: We have presented a survey of some of the reasons that collaboration has increased in recent decades and relate these to an evolutionary logic of participation in collective invention. First, the knowledge required for involvement in any scientific and engineering domain has deepened, often leading to the involvement of a greater number of specialist researchers. Second, industries vary in their presentation of nondecomposable problems, but the tendency is for the interdependence of problem-solving activities to increase. Both of these trends help account for the shift in teams toward larger, more functionally diverse groups. The high costs of changing the knowledge boundaries of the organization to address the latest technological challenges make collective invention an attractive alternative. Thus, collective invention offers a medium for organizations to learn about and participate in technological advances that hold uncertain economic promise.

3. The sources of knowledge have become more diverse

Increasing specialization is the double-edged sword of technological change. On one side, it reflects the deepening of knowledge that can lead to a greater rate of technological advance. On the other, increasing specialization also suggests that the directions of technological advance have become path dependent due to extensive learning and organizational investments (Antonelli, 2007; Arthur, 1989; David, 1975, 1985). Not only do firms become less likely to change course in their R&D investments over time (Patel and Pavitt, 1997), but also they are less likely to recognize important new knowledge due to the blinders imposed by their past work (Cohen and Levinthal, 1989).

The tendency toward local search has long been noted as a problem for any research and development organization (March, 1991; March and Simon, 1958). A common issue raised by economists and management scholars is the extent to which learning in R&D is path dependent (David, 1985; Zollo and Winter, 2002), sowing the seeds for technological lock-out (Cohen and Levinthal, 1989; Henderson and Clark, 1990; Schilling, 1998). Rather than merely serving as a guide in research, the increasing depth and breadth of potentially relevant knowledge has exacerbated the challenge and complexity of commercial R&D (Nelson, 1982).

This challenge stems in part from identifying which sources of technological opportunity are relevant and deserve ongoing cultivation via the involvement of technical personnel. A *source of technological opportunity* provides information used in making new products or processes (Cohen et al., 2002; Klevorick et al., 1995; Malerba and Orsenigo, 1997). Not only have the sources of technological opportunity increased in contemporary times, these sources of knowledge are qualitatively different in form and content as well:

- Firms draw upon knowledge from more distant geographic locations (e.g., Gittelman, 2007; Johnson, 2006).
- Firms make more use of interindustry knowledge flows (e.g., Fung and Chow, 2002; Mansfield, 1982).
- Firms draw upon a broader array of scientific and technical domains (e.g., Cohen et al., 2002; Giuri et al., 2007; Levin et al., 1987).
- Firms make greater use of knowledge from universities and government labs (e.g., Branstetter and Ogura, 2005; Powell et al., 1996; see Foray and Lissoni, this volume).

As Antonelli (2001) suggests, collective knowledge is often the result of discovering latent complementarities among different sources. Given the widespread nature of technological opportunities, but the limited and costly means for appropriating returns from innovation, how do managers select where they will search? We posit that collective invention is a means for organizations to hedge their bets on technological futures. In addition to having the capacity to pursue a novel direction, collective invention enables contributing firms to be "in on the news" (Powell et al., 2005).

Collective invention may also serve as a form of knowledge "insurance" for organizations involved in overlapping technical domains. By sharing knowledge, organizations trade appropriability for access to unexpected technological opportunities. When previously intractable problems become decomposable through theoretical or technical advance, broader access to knowledge enables flexibility in factoring complementary advances into R&D (Brusoni et al., 2001; Rosenberg, 1982: pp. 104–119). In other words, collective invention is both a means for access to information and a coordinated way of developing relevant skills that aid in adapting to technical change (Cohen and Levinthal, 1994).

3.1. Costs of establishing knowledge access

The high costs of establishing access to a body of knowledge suggest that many organizations may prefer to merely pay the "maintenance costs" of ongoing sharing of knowledge in collective endeavors. As knowledge accumulates, the need for a specialized vocabulary, software and hardware tools, and unique theoretical models lead to the creation and branching of distinct epistemic communities (Knorr Cetina, 1999). Mokyr (2005) suggests that the larger the epistemic distance between technical communities, the greater the difficulties in communication and collaboration. Therefore, the tendency toward localized learning suggests that *potential*, more distant collaborators will find it both more time-consuming and difficult to simply establish a productive dialogue.

One of the principal challenges in forging new inventive collaborations is the acquisition of context- and technology-specific knowledge, rather than the general learning of new scientific facts or theories (Vincenti, 1990). Nelson and Winter (1982) argue that much of the knowledge of firms is embedded within routines. Because routines are the idiosyncratic result of many historical circumstances, articulating them systematically for transfer within and across organizations can be challenging (Arora et al., 2001; Von Hippel, 1994). Kogut and Zander (1993) provide evidence of this phenomenon in their analysis of 81 cases of technology transfer among firms in Sweden. They ask respondents to describe the technology being transferred across the dimensions of codifiability, teachability, and complexity, and to describe whether the technology was being transferred to outside firms or wholly owned subsidiaries. They found that transfers that occur to independent firms typically represented relatively codifiable and teachable knowledge, rather than tacit or novel ideas. Even in the case of joint-ventures in which companies may try to collaborate intensively to transfer knowledge, the costs were much greater than with intrafirm knowledge transfer.

Organizations attempt to articulate knowledge via standardized processes and documentation in order to make it more broadly useful internally, but this process itself can require learning and invention depending on the tacitness of the knowledge (Nonaka, 1994; Nonaka and von Krogh, 2009). Von Hippel (1994) refers to the context-dependent value of knowledge as *information stickiness*. He uses the term to describe the high costs that can be associated with extracting knowledge from organizational settings and routines in order to transfer it to a new context.

In a parallel vein, companies report that one reason for abandoning work on university-licensed technology is the challenge associated with knowledge transfer from the faculty inventors (Thursby and Thursby, 2003). Similarly, Jensen and Thursby (2001) find the most successful transfers of university technologies to a company were either more fully developed (e.g., in prototype stage vs. concept stage) or well-understood by the licensing firm, thus avoiding surprises in terms of incompatibilities between the firm's knowledge base and the university technology. Ongoing faculty participation was also found to be vital in the commercialization efforts. Thus, mature, formalized knowledge and a common "epistemic base" accelerate knowledge sharing. To the extent that organizations aim to transfer knowledge from particular sources, it would be reasonable to expect some level of participation by technical staff in associations that foster collective invention, such as standards bodies or communities of practice (Rosenkopf et al., 2001).

Given the many potential sources of technological opportunity and the relative invariance of appropriability mechanisms, merely knowing how and where to allocate research time is itself a dilemma. Thus, the new tightrope walk for managers is to simultaneously address appropriable short- and medium-term commercial opportunities while attending to the accumulation of internal expertise via participation in "open" activities such as collective invention.

3.2. Geographic dispersion of knowledge and collective invention

The need to access geographically localized knowledge suggests that firms will also engage in collective invention with distant parties to discover and gain access to complementary knowledge. Nevertheless, co-location is crucial to firm formation and innovation (Audretsch and Feldman, 1996; Whittington et al., 2009), hence the distance between individuals possessing complementary knowledge may delay the formation of projects aimed at creating near-term technology products. When research efforts are not aimed at commercializing a technology, individuals will disclose to the public domain, leading to a geographically dispersed accumulation of knowledge (see Breschi and Lissoni, 2009). To the extent that the stock of knowledge is diversifying, but complementarity is difficult to identify, we should see more geographically dispersed accumulation of technological opportunities.

Research on geographically distributed collaboration has found an increasing average distance of co-inventors over the past three decades. This development suggests that both the need to access distant knowledge and the lower costs of access via communications technologies are at play. In a study of US inventors, Johnson et al. (2006) find that the average distance of collaborators rose from 117 miles in 1975 to roughly 200 miles in 1999. Johnson and his colleagues found that rapidly advancing areas such as computers and biotechnology tended to exhibit more clustering than older industries such as textiles and mechanical devices, but even these new industries have begun to geographically spread in recent years (Johnson, 2006; Johnson et al., 2006). In Table 2, we return to the five patent classes for which we have collected data and look at the average geographic distance among co-inventors. We used the addresses of US-based co-inventors from patents to identify their respective cities and states of residence. We matched the city and state information to the US Geological Survey and computed the average distance by considering the distance that inventor *a* would need to travel to get to inventor *b*, inventor *b* would need to travel to get to inventor *c*, and so on. Thus, there is slight underweighting that occurs due to inventors who live in the same city—who have an average distance of zero. Nevertheless, all of the technology classes show evidence of greater geographic range, even when including only US inventors.

Greater distance among inventors does pose new challenges, however. Herbsleb et al. (2000) report that in commercial software engineering projects, greater distance is associated with significant delays

Table 2
Geographic dispersion of co-inventors in selected US patent classes[a]

	Avg. co-inventor distance (miles)		
Patent class	1975–1979	2003–2005	Percent (%) change
Aerospace	134	236	76
Biotechnology	147	285	94
Optical Comm.	161	215	34
Pharm. Chem.	101	252	150
Semi. Mfg.	153	222	45

[a] Gittelman (2007) finds that the average distance of biotechnology collaborators on scientific papers that contain corporate authors is 1500 miles when both international and United States are included. Gittelman's findings differ from ours and those of Johnson et al. (2006) for three reasons: the use of scientific papers as opposed to patents, the international focus (which accounted for 30% of co-authors in her data), and the use of organization rather than individual addresses.

and coordination problems. In a study of multidisciplinary, multisite National Science Foundation projects, Cummings and Kiesler (2005) found that increasing the number of disciplinary affiliations had no effect on coordination or research outcomes; instead, increases in the number of affiliated institutions posed larger collaboration obstacles. Thus, rather than epistemic distance posing the major difficulty for knowledge-based collaborations, much of the challenge of distance remains in the coordination difficulties that arise between organizations. Given that Asian countries, most notably China, Singapore, South Korea, and Taiwan, have increased their production of scientific papers (NSF S&EI, 2006), the challenges of distance may require new patterns of collaboration and competition among United States, European, and Asian scientists.

At the international level, research has examined the causes of increasing distributed collaboration (see chapter by von Hippel for more discussion of this literature). We touch on that aspect that relates to the uptick in collective invention. Saxenian and Sabel (2008) posit that the establishment of institutions such as venture capital, which support inventive activity by returning immigrants creates business and technical ties to their host nation. Saxenian (2006) suggests that these ties are mediated by first-generation immigrants who have maintained relationships in their home country, understand its culture, and can navigate local institutions. Kerr (2008) makes use of changes in US immigration quotas and a classification scheme for names of different ethnicities to study flows of knowledge back to immigrants' home countries. Even after controlling for the composition of inventor populations within detailed patent classifications, he finds that there are strong community effects in citations, with foreign researchers being 30–50% more likely to cite US-based inventors of their own ethnicity. This pattern is most pronounced in case of Chinese immigration. Shrum et al. (2007) demonstrate that multi-organizational collaborations in the field of high-energy physics (in which papers routinely contain hundreds of authors) are often facilitated by the standardization of laboratory procedures and well-established conventions about experimentation that enable far-flung teamwork despite individuals not being closely acquainted with one another.

Even in the United States, however, the growth in collaborative and interdisciplinary research does not proceed equally. Jones et al.'s (2008) research on the rapid expansion of cross-university teams also revealed increasing stratification. While the incidence of between-university collaboration has grown rapidly, the highest impact research had an elite university as one of the participants. And while policy pronouncements, such as the National Academy of Sciences (2004: p. 25) contention that interdisciplinary collaboration is needed to "address the great questions of science"... and the "societal challenges of our time," are increasingly common, it is the wealthiest universities that have been in the forefront of building interdisciplinary centers. Elite universities are most able to attract gifts for interdisciplinary centers from donors who are keen to build them. Consequently, while research activities spread, social distance still looms large. Even as research diffuses across organizational and disciplinary boundaries, elite universities in the United States are becoming "more intensely interdependent" (Jones et al., 2008: p. 1261). Consequently, the research efforts of top universities have become increasingly collaborative, and in many fields involve the joint participation of industry partners. Thus, universities often serve as a foundation upon which collective invention can arise.

Gittelman (2007) uncovered the interesting tendency for papers by geographically dispersed biotechnology collaborators to be cited less on patents by the firms affiliated with the papers, but cited more often on other scientific papers. In contrast, more geographically concentrated authors did not receive as many citations for their academic work, but garnered more references to their patents. Her interpretation of these competing results is that the geographic dispersion of knowledge varies markedly for public science and private science. The findings of Gittelman and others on the costs and benefits of accessing distant knowledge may suggest that geographically dispersed teams are better suited to more scientifically oriented work in which results are more foundational and relevant to a broader array of work. Furthermore, research at the scientific level is often more easily codified through formal language whereas work at the engineering level is often tacit, requiring co-location in order to be transmitted from one individual to another. These findings have important implications for collective invention, as its geographic range is a function of the tacit versus explicit nature of knowledge. In the case of high-energy physics, that range may be quite great, whereas in a craft-based setting, individuals may need to be co-located.

3.3. Collective knowledge versus competing artifacts?

One understudied theme in the literature on collective invention is the shifting focus on innovation and appropriability toward the level of the technological regime rather than the firm. Rather than focusing their efforts on similar technological competitors, organizations may have a greater incentive to first ensure the entrenchment of their technological regime in order to benefit from increasing returns to learning. A focus on appropriability at the level of the technical domain leads to greater specialization and to an organizational partitioning of commercial technologies. In some regimes, organizations compete for overlapping intellectual property, but create products that complement one another in the marketplace.

In such settings, competition occurs for scientific prestige and intellectual property, but in many instances of collective invention firms do not plan to address the same markets. Because these organizations compete for scope of intellectual property rights rather than market share, the stakes of knowledge sharing are much lower. These firms are jointly interested in the advance of a technical domain while they pursue different outlets for further elaboration of collective knowledge. Particularly during the establishment of a technology's commercial viability, survival of the technological regime itself may become a superordinate goal for the organizations invested in its research and commercialization.

To explore this idea, we collected data on the mobility of corporate researchers across industries. By mobility, we are not referring to job mobility, as is typical, but the movement of knowledge. We generated a sample of prolific inventors with over 10 patents in a "home" industry. We did this by matching the name of the patent assignee, or corporate owner, to an SIC code. We think of this as their industry of origin, and then we search the patent records to find patents by these inventors that were assigned to a firm in a different industry. We linked assignees to SIC codes using the NBER compustat—patent assignee matching file (Hall et al., 2001). When we limited our analysis to the 15 most heavily patenting industries, we were left with 572,000 patents. (Incidentally, these 15 industries accounted for 60% of the matched patents out of some 380+ industries.) We identified "unique" inventors based on a combination of matches from last, first, and middle names and their addresses. We found ~371,400 unique inventors through this method. Of these, we looked for inventors who had patented more than 10 times within one industry and at a single organization, resulting in 26,025 unique inventors. Our goal in deciding on these parameters was to set a high enough bar to ensure that inventors were full-time in engineering or research and that there were no name ambiguities that caused overestimation of movement across industries. Next, we looked at what industries inventors moved to after establishing expertise in their industry of origin (Table 3).

The exercise clearly shows marked differences across industries, a theme we will discuss in Section 5. For current purposes, note how widely inventors may travel starting from electronics, communications equipment, semiconductors, photography, and computers. In these information technology and computing fields, research is advancing on a very broad frontier, with a high likelihood of spillovers across industries. Few firms can have a hand in all these activities, instead technological progress is made collectively by an array of firms and public research organizations, while individual firms carve out narrower niches for themselves to hone in on. Not surprisingly, there is both intellectual and occupational mobility from radio and TV equipment to semiconductors and from chemicals to pharmaceuticals. The exercise is one illustration of how inventors and their research move across fields.

In many domains, public research is taking on a more active rather than supporting role in collective invention. The fruits of government and university research do not typically have an immediate bearing on private R&D, with the notable exception of the life sciences (Branstetter and Ogura, 2005; Powell et al., 1996; Rhoten and Powell, 2007). In a survey of industry managers, Cohen et al. (2002) found that university and government lab outputs were generally not seen as directly contributing to new project ideas. Instead, many managers emphasize the importance of intangible flows of knowledge, particularly contacts at conferences, faculty consulting, and hiring students. Branstetter and Ogura (2005) observe a strong increase in industry citations to university patents, even after controlling for changes in the propensity to cite and the available stock of knowledge to cite, but observe that the growth in industry–academy interaction is dominated by research related to the life sciences.

Much focus in recent years has been given to university–industry licensing, in part because many universities strive to find alternative sources of funding as federal research dollars have not kept pace with costs and industry support of basic science is still modest (Mowery et al., 2004; Powell et al., 2007). To be sure, there have been a number of notable successes where university licenses have generated significant income. Yet, as Zucker and Darby (1996) find, the distribution of commercial activity by academics is highly skewed. They suggested that star scientists, accounting for less than 1% of the population in biomedicine, produced over 20% of the publications. Nonetheless, we think such commercial involvement *per se* by universities plays only a limited role in collective invention, as the

Table 3
Industry researcher[a] co-patenting and movement across major industries

Industry of origin	SIC[a]	2800	2834	3571	3577	3663	3674	3711	3861	7370	7373
Chemicals & allied products	2800		59			1	2	1		2	
Pharmaceutical preparations	2834	15		2	1		1	1	7		2
Electronic computers	3571		4		5	5	36	8	15	18	4
Computer peripheral equipment	3577		1	2		4	14	1	20	29	1
Radio & TV broadcasting & communications equipment	3663		1	7			134	1	2	16	26
Semiconductors & related devices	3674	3		43	11	87		1	12	58	52
Motor vehicles & passenger car bodies	3711	2	1	12			3		15	1	
Photographic equipment & supplies	3861		10	60	27	10	9	31		21	1
Services-computer programming, data process.	7370		1	23	13	49	159	4	8		14
Services-computer integrated systems design	7373			3	2	28	88	2		6	

[a] Each inventor had to have more than 10 patents at one company in an industry of origin, which is displayed in the left column.

scale of such successes is rather modest. Moreover, successful licenses often represent an exclusive dyadic exchange between a university and a firm, rather than a collective or general-purpose license used by many.

Nonetheless, as Rosenberg (2000) points out, university research and training is broadly responsive to the needs of industry. And there are instances in which industry advances can trigger a series of complementary inventions by universities that absorb the new technology as a research tool or as an engineering system meriting its own study (Lenoir and Giannella, 2006; Rosenberg, 1982). The role of university science in private sector R&D is multifaceted. Thursby et al. (2009) consider the extent to which university faculty assign patents to nonuniversity entities. They find that roughly one-quarter of patents filed by university faculty are assigned to firms. They attribute this largely to faculty consulting. Murray (2002), in an analysis of the tissue-engineering field, reports that knowledge spills out of universities in myriad ways. In addition to consulting, scientific advisory board memberships, the exchange of research tools, and personnel movement in and out of laboratories are commonplace in this field. Fleming et al.'s (2007) analysis of inventor networks in Silicon Valley and Boston emphasized the critical bridging role of Stanford PhD graduates and a postdoctoral fellowship program at IBM's Almaden Labs in the larger Valley network, and the salience of MIT graduates in the Boston community. Whittington (2007), in a detailed study of inventor networks in the life sciences among Boston-area universities, research hospitals, and companies, found that a few key university laboratories and a small number of individual scientists who moved from universities to firms, or nonprofit institutes to firms, and vice versa, were the central nodes that tied a large ecosystem together and gave it vitality.

More direct participation in collective invention by universities has also increased. For example, the Biobricks project at MIT provides a repository for organizations to contribute knowledge about reusable genetic and proteomic structures. The license on the site enables firms to pursue private commercial interests using knowledge they obtain from the repository. In Gittelman's (2007) analysis of biotech firm co-authorships, she found that over 90% of the companies' research partners were universities or research institutions.

Many observers have noted the dramatic growth in university patenting, although there is debate over whether this represents an increase in valuable applied knowledge or herd-like behavior on the part of universities trying to signal their relevance to the private economy (Henderson et al., 1998; Owen-Smith and Powell, 2003; Ziedonis and Mowery, 2004). Sorting out the competing influences on universities is difficult, but there clearly is an upsurge in the quantity of university patents. We return to the five technology classes we have examined in previous tables and gauge the growth in the number of university and government patent assignees between 1975–1979 and 2001–2005. Table 4 shows how many patents were assigned to the government and universities during these two time periods, one three decades ago, one more recent. The two columns labeled *# patents* reflect the total number of patents filed by all individuals and organizations in each time period in each patent classification. We see an absolute increase in university involvement in every technical domain, but most notably in biotechnology, semiconductors, and to a lesser extent, pharmaceutical compounds. In contrast, absolute government patenting has only increased in biotechnology, in all other areas, government patenting decreased. Yet, the combined relative increase of government and university patenting tells a very different story. Compared with other patenting entities, universities and government labs only increased their activity in the fields of pharmaceuticals and biotechnology, whereas their activity declined as a fraction of overall activity in aerospace, optical communications, and semiconductors.

In addition to patents assigned to universities, we also looked at patenting by inventors who had been affiliated with a university on previous patents (Table 5). In particular, we looked at the industry of origin for patents in which an inventor had been matched to at least three university patents on the basis of first and last name, city, and state. This is a rather new line of inquiry, and we offer it as exploratory data. We capture one indicator of the cross-traffic between university scientists and private firms.

Table 4
University and government patenting in selected patent classes[a]

	1975–1979				2001–2005				
Patent class	Gov.	Univ.	Total	Gov. & Univ. Share (%)	Gov.	Univ.	Total	Gov. & Univ. Share (%)	Ratio T2/T1
Aerospace	145	3	1118	13.24	72	24	1619	5.93	0.45
Biotechnology	150	328	6533	7.32	373	3267	22,881	15.91	2.17
Optical Comm.	70	5	511	14.68	25	251	6217	4.44	0.30
Pharm. Chem.	36	62	2467	3.97	83	524	7212	8.42	2.12
Semi. Mfg.	210	93	5630	5.38	173	1297	79,069	1.86	0.35

[a] University patents were identified using a text query that matched terms such as university, college, (technology and institute), "regents of," "board of trustees," and others to standard USPTO assignee names. Government patents were identified using a text query that matched terms such as "government," "united states," "secretary of," "administration," "department of energy," "national science foundation," "national institutes," "national lab."

We have long known there are all manner of informal linkages between university science and industry (Colyvas, 2007; Murray, 2002; Rosenberg and Nelson, 1994), but this exercise helps show how these contacts translate into intellectual property. Whether these patents are the consequences of consulting agreements, faculty startup companies, postdoctoral fellows who move to industry, or technology "going out the back door," we cannot say. But the volume is not trivial, most notably in several key technical fields. While the overall number of patents is relatively small compared to the total for the industries of origin, we find that they follow a similar pattern regarding the division of innovative labor.

Table 5
Patents by university inventors assigned to publicly traded companies, 1975–2001[a]

SIC	SIC description	1975–1977	1978–1980	1981–1983	1984–1986	1987–1989	1990–1992	1993–1995	1996–1998	1999–2001
2834	Pharmaceutical preparations	35	23	20	53	68	117	473	367	523
2836	Biological products	0	1	2	26	38	41	309	208	335
3674	Semiconductors & related	0	2	2	6	42	75	129	158	349
3841	Surgical & medical instruments	0	9	12	17	11	32	112	135	125
1311	Crude petroleum & natural gas	3	5	14	20	17	19	28	16	239
3845	Electromedical & electrotherapeutic	2	3	12	13	27	40	62	95	162
7370	Services-computer programming	10	19	15	7	13	48	90	94	79
7373	Computer peripheral equipment	6	0	5	18	25	59	71	68	90
3577	Services-computer integrated system	3	6	8	5	5	18	70	143	87
2835	*In vitro & in vivo* diagnostics	0	2	1	1	9	10	145	118	55
2911	Petroleum refining	7	6	23	48	34	52	95	36	15
2821	Cleaning supplies, perfumes, cosmetic	7	10	3	9	9	7	60	64	80
2840	Plastic materials, synthetic resins	0	8	10	11	16	25	51	75	64
3570	Computer & office equipment	6	5	1	9	25	34	36	45	44
3861	Photographic equipment	11	5	21	18	17	30	9	31	52
7372	Chemicals & Allied Products	3	5	7	9	5	25	33	34	35
2800	Services-prepackaged software	0	0	0	2	0	0	4	71	77

[a] Each inventor had at least three patents assigned to a single university.

Technical opportunities and challenges in pharmaceuticals, biotechnology, semiconductors, and medical devices generate the most frequent interaction between industry and university science.

Summary: In this section, we suggested that the increasing diversity of sources of knowledge has important implications for collective invention. First, the risk of technological lock-in is greater for organizations in fields where the streams of knowledge required for invention are all rapidly advancing. Given the high costs of transferring knowledge across organizational and epistemic contexts, firms may use collective invention to maintain a dialogue with a broad community to hone their ability to transfer knowledge from potential sources of technological opportunity. Second, because knowledge can quickly become geographically localized, firms invest in collaborations to expand their reach—nevertheless, appropriable knowledge often requires co-location for its tacit transfer; whereas scientific collaborations can span greater distances. Geographic distance may result in greater deepening of formal knowledge (as opposed to tacit), which can in some circumstances create a larger stock of basic science that can be built upon. Finally, we point out that in many instances of collective invention intellectual property is at stake rather than product or service revenues. Firms interested in approaching different markets may share IP, thus limiting the scope of their claims, but they may make few concessions in the target markets they protect. Participants in collective invention may often see such engagements as complementary rather than mutually exclusive.

4. New forms of governance facilitate collective invention

Collective invention efforts depend on a social and organizational infrastructure for coordination. The complexity of most modern technologies requires the participation of many individuals from a practical standpoint, but the shared ethos of building something that people will use also encourages collaboration. Wray (2002) suggests that the increasing dependence of technical personnel on common equipment socializes scientists and engineers into norms of collective work. More generally, the development of communications and information technologies have greatly facilitated contact across geographic boundaries, leading in turn to the greater refinement of practices and norms of knowledge sharing (Cummings and Kiesler, 2007; Olson and Olson, 2004; Olson et al., 2008). We review how the governance of collective invention is shaped by the usage of new collaboration tools, social norms within a technical community, and the organizational form of collective invention efforts.

The basis for a technological community arises out of a set of common understandings. In his discussion on the stages of development of the electric grid in several countries, Hughes (1983) presents the idea that each stage is associated with a particular "culture of technology," that is, a set of values and ideas that orient inventors toward a common goal. These cultures of technology provide life within and among organizations toward the elaboration of a technical endeavor, what he termed "technological momentum." Mackenzie (1990) referred to technological momentum as an institutionalized form of technological change, created as participants mobilize to align political, social, economic, and technical structures around the survival of a technology. People not only build institutions to address technical uncertainties and obtain resources, but also invest their careers and credibility in the rapid alignment and pursuit of multifaceted goals.

Cultures of technology are important because they help explain the continuity of an underlying technical community despite temporal shifts in organizing for collective invention versus private R&D.

Allen's (1983) historical case of collective invention can be cast as a sustaining community at the intersection of private interests, or as a locus of accumulation for valuable knowledge. After knowledge advances for some time, internal or external participants can exploit the knowledge through network refunctionality. Research on the development of biotechnology in Boston, Massachusetts in the 1980s and 1990s showed that the initial anchors of the community were research universities, most notably MIT and later joined by Harvard, as well as such medical centers as Dana Farber Cancer Center and Massachusetts General Hospital (Owen-Smith and Powell, 2004). These public research organizations were connected to fledgling biotech companies through research partnerships and clinical trials. Over time, venture capital firms moved in, collaborations were forged with participants from around the globe, and an open community catalyzed private innovation. The imprint of public science remained, but the cluster of companies increasingly pursued more product-driven, dyadic alliances rather than exploratory research efforts. Leaky ties that previously served as the irrigation system for open collaboration were transformed into channels of private innovation.

Collective invention thus involves the combination of both open innovation and private interests. Participants move in and out of technical communities, and can use their connections for public or private gain. The important point, as Lakhani and Panetta (2007: pp. 104–105) observe in their work on open source, is that: "these systems are not "managed" in the traditional sense of the word, that is, "smart" managers are not recruiting staff, offering incentives for hard work, dividing tasks, integrating activities, and developing career paths. Rather, the locus of control and management lies with the individual participants who decide themselves the terms of interaction with each other." (See chapter by von Hippel for further discussion).

Hughes (1989) describes how the aerospace, computing, and communication industries acquired technological momentum with the injection of cash and the alignment of political and industrial interests behind the systems they produced. For example, in the case of communications, common goals were eventually institutionalized via the ITU's (International Telecommunications Union) implementation of standards that enabled regional telephone monopolies to interoperate. Systems engineers played the critical role in coordinating the development of various technological systems among dispersed organizations.

In general, participation in collective invention is typically voluntary and often the inventors themselves are highly substitutable. There are countless studies and surveys of why developers contribute to open-source software projects. As but one illustration, Lakhani and Wolf (2005) draw on an Internet survey of 684 developers across 287 different open-source projects to understand community participation, finding that enjoyment of the creative work is the most common and compelling motivation (this finding is even more striking given that 40% of their survey participants were paid to participate in open source). They find that addressing existing user needs, the intellectual challenges associated with programming, and learning are secondary drivers. With their intrinsic interest in the work itself and their common goals, open-source developers have been creative in developing effective governance structures.

At the group rather than individual level, another dynamic is at play that reinforces the drive to enlist and govern collective invention. Kling and Iacono (1988) argue that computerization (i.e., the deployment of information technology infrastructure) is not merely the result of a desire for efficiency. Instead, they suggest that an understudied aspect of computerization of the workplace is the mobilization of participants (early adopters) who advocate for the introduction of information systems. They do so by making appeals to ideologies that resonate within the organization, but which are often imported and translated in from the wider environment (Fligstein, 2001).

Different technologies call for different modes of governance. Collective invention can precede the rise of an industry that harnesses the accumulated technical knowledge of contributors or it can emerge as the by-product of existing inventive efforts. Effective governance structures typically address several problems: compatibility with the knowledge-sharing norms of distinct technical communities, responsiveness to both interesting and mundane technical challenges, and some means of coordination.

Meyer (2003) notes that IP barriers to collaboration can be confronted up-front via the use of licenses. Similarly, Gambardella and Hall (2006) find that some level of legal coordination is often needed for collective invention to be effective. In the case of software, the establishment of the General Public License (GPL) provided guidance to future inventors on how to contribute. As the lead developers emphasized that their contributions were collective goods, other followed suit using the GPL to advance the efforts of the community. In hardware, the use of patent pools and cross-licensing often presents a workaround to challenges in a narrow technical space, but the same type of practice can raise antitrust concerns if the patents are used to deter new competitors.

O'Mahoney and Ferraro (2007) find that individuals engaged in collective invention seek to establish formal mechanisms for exercising authority, but "cap" its power with democratic tools that allow for technical and organizational experimentation. They suggest that when members settle on a shared conception of authority, the result is often much more comprehensive than their original design. The governance systems of open-source communities have co-evolved with changing technical objectives and shared conceptions of authority.

Coordination can occur without a legal foundation, however. Ever since Marshall's (1920) evocative phrase, "the secrets of industry are in the air," researchers have focused on the productive relations that have typified some craft- and technology-based communities (Sabel and Zeitlin, 1997; Scranton, 1997). Foray and Perez (2006) emphasized the political factors that sustained an open technology in the eighteenth century silk industry in Lyon, France. Local elites were most concerned with the economic vitality of the region and the municipal government gave grants to inventors to support the sharing of new knowledge with the entire community of silk makers. They argue that although collective invention increased the risk of conflict, such disputes were dampened by common competitive pressures and the development of an ethos that encompassed contribution. Lamoreaux, Raff, and Temin (2003: p. 417) observed that in the era before the vertically integrated firm, "business people. . .industrial communities interacted socially as well as economically, and the resulting multidimensional relationships facilitated cooperation for purposes besides production."

Similarly, studies of contemporary high-tech clusters uncover various modes of private governance that create collective benefits. Perhaps most notably, interfirm job mobility, high rates of firm formation, and an ample supply of skilled technical labor are common to most thriving clusters (Bresnahan and Gambardella, 2004; Saxenian, 1994). Focusing more on the emergence of clusters as opposed to their persistence, Powell et al. (2009) analyze the three regions in the United States where biotech took off, along with eight locales with considerable endowments and resources where companies were created but clusters have not developed. They argue that participants have to take steps to pursue new technological trajectories well in advance of full knowledge of their potential. Such exploration, in the biotech case, was assisted by anchor tenants—public research organizations in Boston; venture capitalists, first-generation companies that encouraged scientists to publish, and university tech transfer offices committed to relationship building rather than revenue maximization in the San Francisco Bay Area; and nonprofit research institutes and a young university in San Diego—that emphasized open science, transparency in relationships, and a willingness to transpose practices and recombine them across the public, private, and nonprofit sectors. In regions where biotech did not grow, the dominant local anchors reinforced existing practices, acting as 800 lb. gorillas rather than catalysts.

Summary: The governance of collective efforts commonly begins with a shared ethos or compatible goals among participants. Often, the establishment of effective governance is often aided by widespread use of information technology, and organizational innovations that enable geographically distributed collaboration and address the intellectual property interests of participants. Once knowledge accumulates to the stage that tangible outcomes are possible, private interests may take hold and commercialize particular streams of technology that emerge from collective invention. The growing involvement of universities and associations in recent episodes of collective invention may alter this trajectory, however, keeping technology collective for a longer period of time. Finally, studies of regional science and technology clusters provide insights into how collective efforts can harness the energies of a diverse community of participants. For instance, a key ingredient appears to be the use of governance systems and relational contracting early in the development of a technology that can provide an interactional template that serves to promote collective invention.

5. Interindustry heterogeneity

Despite the increase in collective invention in recent decades, there remain many reasons for persistent interindustry differences in its form and prevalence (Breschi et al., 2000; Klevorick et al., 1995; Levin et al., 1987). The longevity of collective invention in many industries also suggests that the need to both explore and exploit rapidly expanding technological opportunities has reshaped intellectual property choices. In some instances, collective invention has led to appropriability strategies that serve the technological regime as well as the individual firm. In other words, rather than being singularly focused on immediate sources of revenue, firms may strive for the success of their interdependent R&D activities, at least during times of technological ferment.

5.1. Nature and relevance of collective knowledge

Nelson (1982: p. 468) suggested that in "industries marked by rapid sustained technological progress a good deal of the *logy* has been created within the firms themselves, yet made public." When knowledge is created by firms, as opposed to universities or individuals, it is at a point of "maturity" that makes it more likely to be relevant to other firms. In addition, the fact that firms in an industry openly share knowledge suggests that they are confident they have both organizational and legal means to pursue and protect these ideas. Nonetheless, industries differ markedly in the extent to which knowledge from one firm is complementary to the knowledge of another. Individuals require a common technical foundation in order for inventions to be easily learned and improved upon. Technical knowledge must be compatible across firms in order for collective invention to quickly take hold.

The variety of scientific sources underlying a technology can have a large impact on the speed of search (Klevorick et al., 1995). Collective invention provides a means for coping with this diversity of sources. For example, Cohen et al. (2002) report that the chemicals industry draws upon university research in chemistry and chemical engineering, whereas semiconductor firms draw upon a wider range of academic disciplines, including chemistry, physics, computer science, materials science, chemical engineering, electrical engineering, mechanical engineering, and math. The variety of scientific sources underlying a technology has a large effect on the number of useful directions that can be explored and the speed at which they can be pursued.

The development of biotechnology, with its extensive reliance on interorganizational collaboration in the R&D process, has been widely studied by scholars interested in innovation. Scientists from competing firms and public research organizations often share the "logy" (i.e., the theoretical understanding). Learning about a theoretical principle, or the basic idea of how a technique works, is often enough to stimulate search by a variety of firms. Diffusion of theory and appropriation of technique allows firms to search a richer opportunity landscape, while still profiting from their investments in research and development.

In order to provide an empirical basis for our discussion, we return to our analyses of five technological classes over two time periods. Table 6 offers insight into how concentrated the innovation process is across fields and over time. The first and third columns are counts of the number of organizations that account for 60% of the total patents in a specific technology classification. Columns two and four are the total number of patents filed in that classification during each time period. Only aerospace has seen a decline and become more concentrated. In the four other technology classes, the number of organizations has grown, burgeoning dramatically in biotech and pharmaceuticals.

5.2. The interorganizational decomposability of problems

The technical characteristics of problems faced by organizations also have important implications for organizational coordination and design. Technical problems sometimes "suggest" a search strategy (Hughes, 1983; Rosenberg, 1976; Vincenti, 1990). The nature of the knowledge underlying a technical problem may provide some clues about how to divide the process of search within or across organizations. The organization of innovative labor is partly driven by the complexity and decomposability of a problem.

Table 6
Number of organizations responsible for 60% of patents in selected patent classes

	1975–1979		2001–2005	
Patent class	Orgs. at 60th percentile	Total # of patents	Orgs. at 60th percentile	Total # of patents
Aerospace	81	1118	69	1619
Biotechnology	103	6533	261	22,881
Optical Comm.	24	511	40	6217
Pharm. Chem.	130	2467	655	7212
Semi. Mfg.	16	5630	24	79,069

Decomposability is defined as the ability to break apart a problem into subproblems that can be worked on independently (Simon, 1962). Most problems in organizations are not truly decomposable, but they are nearly so, suggesting that they can be divided up and coordinated without adversely affecting the final outcome. For example, the installation of a video card or a faster processor can cause overheating in a laptop. To prevent this, engineers must design in additional heat sinks and fans, making the laptop larger and heavier. Thus, the choice of one component constrains the selection of other components and the final design of the laptop, but this does not preclude the division of tasks in the organization. These diverse tasks can be performed by different organizations, with only modest need for common knowledge. Computer makers can buy standardized components from the same set of firms. Thus, the extent and complexity of technical interdependencies that need to be addressed throughout the search process determine the kinds of organizational arrangements suited to a technical problem (Nickerson and Zenger, 2004; Rivkin and Siggelkow, 2003).

Another aspect of search coordination that differs across technological regimes is the predictability of outcomes from search activities. Brusoni et al. (2001) study how and why system design problems are divided among groups based on the predictability and level of interaction among aircraft engine components. They find that predictable product interdependencies and an even rate of component change lead to independent entities that interact via market mechanisms. Predictable interdependencies and an uneven rate of change give rise to an interdependent sector of search and coordination via systems integration (e.g., the hard disk industry, in which manufacturers design the architecture and purchase components with standardized interfaces). When the product interdependencies are unpredictable but the rate of change is even, Brusoni et al. (2001) suggest that relatively independent organizations with coordination through systems integration will arise (e.g., the automotive industry, in which the architecture dictates some parts of the component design). When interdependencies are unpredictable and the rate of change is uneven, firms are more likely to vertically integrate (e.g., many mobile handset makers also manufacture infrastructure products such as base stations in order to exploit the highest end capabilities possible).

5.3. Feasibility of individual versus collective appropriability

Collective invention depends upon specific appropriability structures. The use of patents, trade secrets, complementary assets, and copyrights all have implications for both the degree of knowledge spillovers (improving technological opportunities for other firms) and the difficulty of circumventing barriers to using particular knowledge (Nelson, 2006). Because of their sequential influence on one another, opportunity and appropriability are inseparable in understanding the creation of fertile ground for collective invention.

The most pervasive influence of legal appropriability strategies on firm activities may lie in changing the costs and likelihood of pursuing particular technologies. Thus, many efforts at collective invention may in part seek to reduce the appropriability costs incurred by disjointed IP rights. Firms create paths of intellectual property reflecting their past and current research, which they can then either use to defend their products or as a tool to force competitors to cross-license. In some industries, there is no room for collective appropriability; hence firms may consciously avoid using the protected knowledge of their competitors in industries in which patents provide a key appropriability mechanism (Graham and

Sichelman, 2008; Lemley and Sampat, 2008). For instance, Lerner (1995) found that small biotechnology firms with high litigation costs actively avoid inventive activity in technology spaces that are associated with a large number of patents. Because these firms actively avoid the search paths of one another, it is unlikely they will engage in collective invention beyond peripheral or legally standardized aspects of their technologies.

In contrast, the paths of accumulated, legally claimed knowledge heavily overlap in the case of complex technology industries (Breschi et al., 2000; Hall and Zeidonis, 2007). Firms in industries with complex products cannot monopolize the intellectual property required for product development, nor can they realistically avoid infringing some of their competitors' patents (Cohen et al., 2000: pp. 13–14). Thus these firms are forced to cross-license due to "mutually assured destruction," which encourages negotiations and deters lawsuits (Allison et al., 2004). Because large patent holders in complex industries engage in and often encourage collective invention, it appears that patent pools and cross-licensing may clear the road for participation in collective invention.

Summary: We have posited some of the factors that account for the variation in the intensity of collective invention across industries. First, the number and quality of sources of technology opportunity vary by industry, so we should expect that turnover of sources, motivation to engage new sources, and ease of accessing new knowledge have implications for the possibility of collective invention. Second, some problems cannot be broken apart because they must be solved simultaneously by functionally diverse teams. These challenges are the least amenable to collective invention. Third, and very much related to the previous section on innovations in governance and organization, the long-run feasibility of appropriating returns from invention must be clear to motivate commitments at the organizational level to collective efforts.

6. Conclusion

The importance of collective invention has varied markedly across eras, locales, and technologies. We have emphasized the sharing of information across a network of participants as the central feature of collective invention. One notable point of departure between late nineteenth and early twentieth century examples and current ones is that in the earlier cases the participants were geographically concentrated, whereas in the present era this requirement for information exchange has relaxed, due to advances in both information technology and modes of governance.

Uncertainty surrounding the technical feasibility and economic viability of a technology create pressures upon firms, leading them to choose to carry out R&D activities internally, in tandem with parties facing similar constraints, or by sustained engagement with a wider community of practice. When universities and research institutes have a large hand in development of a technology, firms attempt to join in collective efforts. When firms create spillovers and incentives for outside organizations—such as universities and technical or research institutes—to pursue, the existing division of innovative labor serves to mold the set of future technological opportunities. In these circumstances, firms use collective invention as a means for obtaining data about the evolution of technology, which they employ to structure their research during times of regime emergence and stabilization. While this might appear to lead toward conservative technological ambitions, the story is not so straightforward. General awareness of uncertain requirements or trajectories can create the necessary space for scientists and engineers to collectively build the foundations for more radical technologies. It is in these times of change in technological regimes that a community ethos and broadly distributed participation characterizes collective invention.

The general lesson from our review of the diverse literatures we have drawn on is that collective invention is sparked when a new opportunity opens—either by an invention, the expiration of a patent, or general optimism about a technology but is accompanied by a lack of clarity about its possible trajectory. In this period of ferment, various participants emerge and often develop collective institutions—publications, workshops, standards, associations—that foster integration into a community of practice (Rosenkopf and Schilling, 2007; Rosenkopf and Tushman, 1998). As a flow of new inventions emerge, new firms appear, sensing opportunities. These new entrants' efforts to connect to the community act primarily as an "admission ticket" to access information that others possess (Powell et al., 1996). Over time, as technological uncertainty recedes, firms develop private R&D and focus on their own specific applications. Reliance on collective invention accordingly wanes.

This evolutionary view suggests a specific phase of technological evolution that marks the scope condition for collective invention. We concur that technological ferment is an enabling condition, but we are hesitant to accept that such a determinist position is the full story. In recent decades, we have seen collective invention efforts in both emerging fields—computers, software, biotech, and in sectors such as electronics where arms-length contracting among specialists employing standard technical interfaces was considered routine. Sturgeon (2002) describes the development of "modular production networks" among vertically specialized firms in the electronics field, and speculates that much closer collaboration in product design among independent firms is growing as products become more complex and less specialized. These contemporary examples suggest a wide range of efforts at collective invention that have as much to do with technical factors as the institutional arrangements in which they are addressed.

Recent developments highlight the dual elements in collective invention—such practices are promoted by both technological uncertainty *and* "situational particularities." Rather than to see collective invention as dictated solely by technological requirements, we also stress that it can emerge or be selectively adapted in different locales or branches of the same industry and across sectors. To the extent that social and political conditions facilitate connections to groups of inventors beyond their own, there will often be advantages that accrue to those who have early access to ideas and interpretations.

Yet, we have had little to say about the genesis of collective invention: the question of motivation. We have only speculated that motivation to improve a technology often leads individuals to find ingenious organizational means for carrying forth their ideas and sharing them with others. While the literature has much to say about the varieties of collective invention, its emergence remains understudied. Such efforts, while requiring an historian's skills and an economist's or sociologist's toolkit, are challenging, but nonetheless would offer critical insights into the individual and collective dynamics that spawn novelty.

Our charge has been to illuminate how diverse types of organizations engage in common problem solving on a technological frontier; the next step is to identify how these networks are composed of individual inventors. Deeper understanding of how networks of inventors form and how these individuals decide what to disclose to one another would offer insight into the viability of collective invention in the myriad circumstances of technological progress.

References

Allen, R.C. (1983). "Collective invention". Journal of Economic Behavior and Organization 4, 1–24.

Allison, J., Lemley, M., Moore, K., Trunkey, D. (2004). "Valuable patents". Georgetown Law Journal 92, 435–479.

Antonelli, C. (2001). The Microeconomics of Technological Systems. Oxford University Press, New York.

Antonelli, C. (2007). "Path dependence, localised technological change and the quest for dynamic efficiency". In: Antonelli, C., Foray, B., Hall, B., Edward Steinmuller, W. (Eds.), New Frontiers in the Economics of Innovation and New Technology: Essays in Honour of Paul A. David. Edward Elgar, Northampton, MA, pp. 51–69.

Arora, A., Fosfuri, A., Gambardella, A. (2001). Markets for Technology. MIT Press, Cambridge, MA.

Arthur, W.B. (1989). "Competing technologies, increasing returns, and lock-in by historical events". Economic Journal 99 (394), 116–131.

Audretsch, D.B., Feldman, M.P. (1996). "R&D spillovers and the geography of innovation and production". American Economic Review 86, 630–640.

Bell, D. (1973). The Coming of Post-Industrial Society. Basic Books, New York.

Bijker, W.E. (1987). "The social construction of bakelite: Toward a theory of invention". In: Bijker, W.E., Hughes, T.P., Pinch, T. (Eds.), The Social Construction of Technological Systems. MIT Press, Cambridge, MA, pp. 159–187.

Branstetter, L., Ogura, Y. (2005). "Is academic science driving a surge in industrial innovation? Evidence from patent citations". National Bureau of Economic Research Working Paper Series No. 11561, http://www.nber.org/papers/w11561 (Accessed January 19, 2010).

Breschi, S., Lissoni, F. (2009). "Mobility of skilled workers and co-invention networks: An anatomy of localized knowledge flows". Journal of Economic Geography 9, 439–468.

Breschi, S., Malerba, F., Orsenigo, L. (2000). "Technological regimes and schumpeterian patterns of innovation". The Economic Journal 110, 388–410.

Bresnahan, T., Gambardella, A. (Eds.), (2004). Building High-Tech Clusters. Cambridge University Press, Cambridge, UK.

Brusoni, S., Prencipe, A., Pavitt, K. (2001). "Knowledge specialization, organizational coupling, and the boundaries of the firm: Why do firms know more than they make?" Administrative Science Quarterly 46, 597–621.

Burt, R.S. (2004). "Structural holes and good ideas". American Journal of Sociology 110 (2), 349–399.

Carlson, W.B. (1991). Innovation as a Social Process: Elihu Thomson and the Rise of General Electric, 1879–1900. Cambridge University Press, New York.

Chandler, A. (1977). The Visible Hand. Harvard University Press, Cambridge, MA.

Cohen, W.M., Levinthal, D.A. (1989). "Innovation and learning: The two faces of R&D". Economic Journal 99, 569–596.

Cohen, W.M., Levinthal, D.A. (1994). "Fortune favors the prepared firm". Management Science 40, 227–251.

Cohen, W.M., Nelson, R.R., Walsh, J.P. (2000). "Protecting their intellectual assets: Appropriability conditions and why U.S. manufacturing firms patent (or not)". NBER Paper 7552.

Cohen, W.M., Nelson, R.R., Walsh, J.P. (2002). "Links and impacts: The influence of public research on industrial R&D". Management Science 48, 1–23.

Colyvas, J.A. (2007). "From divergent meanings to common practices: The early institutionalization of technology transfer in the life sciences at Stanford University". Research Policy 36, 456–476.

Cummings, J.N., Kiesler, S. (2005). "Collaborative research across disciplinary and organizational boundaries". Social Studies of Science 35, 703–722.

Cummings, J.N., Kiesler, S. (2007). "Coordination costs and project outcomes in multi-university collaborations". Research Policy 36 (10), 1620–1634.

David, P.A. (1975). Technical Choice, Innovative and Economic Growth: Essays on American and British Experience in the Nineteenth Century. Cambridge University Press, New York.

David, P.A. (1985). "Clio and the economics of QWERTY". American Economic Review 75, 332–337.

David, P.A. (2008). "The historical origins of 'open science': An essay on patronage, reputation and common agency contracting in the scientific revolution". Capitalism and Society 3 (2) Article 5.

de Solla Price, D. (1963). Little Science, Big Science. Columbia University Press, New York.

Fleming, L., King, C., Juda, A.I. (2007). "Small worlds and regional innovation". Organization Science 18, 938–954.

Fligstein, N. (2001). "Social skill and the theory of fields". Sociological Theory 19, 105–125.

Foray, D., Perez, L.H. (2006). "The economics of open technology: Collective organisation and individual claims in the 'fabrique lyonnaise' during the old regime". In: Antonelli, C., Foray, D., Hall, B., Edward Steinmuller, W. (Eds.), New Frontiers in the Economics of Innovation and New Technology: Essays in Honour of Paul A. David. Edward Elgar, Northampton, MA, pp. 239–254.

Freeman, C., Soete, L. (2009). "Developing science, technology and innovation indicators: What we can learn from the past". Research Policy 38, 583–589.

Fung, M.K., Chow, W.W. (2002). "Measuring the intensity of knowledge flow with patent statistics". Economics Letters 74, 353–358.

Galambos, L. (1983). "Technology, political economy, and professionalization: Central themes of the organizational synthesis". Business History Review 57 (4), 471–493.

Gambardella, A., Hall, B.H. (2006). "Proprietary versus public domain licensing of software and research products". Research Policy 35, 875–892.

Gibbons, M., Limoges, C., Schwartzman, S., Nowotny, H., Trow, M., Scott, P. (1994). The New Production of Knowledge: The Dynamics of Science and Research in Contemporary Societies. Sage, London.

Gittelman, M. (2007). "Does geography matter for science-based firms? Epistemic communities and the geography of research and patenting in biotechnology". Organization Science 18, 724–741.

Giuri, P., Mariani, M., et al. (2007). "Inventors and invention processes in Europe: Results from the PatVal-EU survey". Research Policy 36, 1107–1127.

Graham, M. B. W. (2008). "Technology and innovation". In: Jones, G., Zeitlin, J. (Eds.), The Oxford Handbook of Business History. Oxford University Press, Oxford, UK, pp. 347–373.

Graham, S. J. H., Sichelman, T.M. (2008). "Why do start-ups patent?" Berkeley Technology Law Journal 23, 1–34.

Granovetter, M. (1973). "The strength of weak ties". American Journal of Sociology 78, 1360–1380.

Granovetter, M. (2009). "Comment on "Capitalist entrepreneurship: Making profit through the unmaking of economic orders" (by Thorbjørn Knudsen and Richard Swedberg)". Capitalism and Society 4 (2) Article 8.

Hall, B.H., Ziedonis, R.H. (2007). "An empirical analysis of patent litigation in the semiconductor industry". University of California at Berkeley Working Paper.

Hall, B.H., Jaffe, A.B., Tratjenberg, M. (2001). "The NBER patent citation data file: Lessons, insights and methodological tools". NBER Working Paper 8498.

Henderson, R.M., Clark, K.B. (1990). "Architectural innovation: The reconfiguration of existing product technologies and the failure of established firms". Administrative Science Quarterly 35, 9–30.

Henderson, R., Jaffe, A.B., Trajtenberg, M. (1998). "Universities as a source of commercial technology: A detailed analysis of university patenting 1965–1988". Review of Economics and Statistics 80 (1), 119–127.

Herbsleb, J.D., Mockus, A., Finholt, T.A., Grinter, R.E. (2000). "Distance, dependencies, and delay in a global collaboration". In: Proceedings of the 2000 ACM Conference on Computer Supported Cooperative Work. ACM, Philadelphia, PA., pp. 319–328 http://portal.acm.org/citation.cfm?id=358916.359003 (Accessed September 22, 2009).

Hicks, D.M., Katz, J.S. (1996). "Where is science going?" Science, Technology Human Values 21, 379–406.

Hounshell, D.A., Smith, J.K. (1988). Science and Corporate Strategy—DuPont R&D: 1902–1980. Cambridge University Press, Cambridge, UK.

Hughes, T.P. (1983). Networks of Power: Electrification in Western Society, 1880–1930. Johns Hopkins University Press, Baltimore, MD.

Hughes, T.P. (1987). "The evolution of large technological systems". In: Bijker, W.E., Hughes, T.P., Pinch, T. (Eds.), The Social Construction of Technological Systems. MIT Press, Cambridge, MA, pp. 51–82.

Hughes, T.P. (1989). American Genesis: A Century of Invention and Technological Enthusiasm, 1870–1970. Viking, New York.

Jensen, R., Thursby, M. (2001). "Proofs and prototypes for sale: The licensing of university inventions". American Economic Review 91, 240–259.

Johnson, D.N. (2006). "It's a small(er) world: The role of geography in biotechnology innovation". SSRN eLibrary, http://papers.ssrn.com/sol3/papers.cfm?abstract_id=296813 (Accessed September 22, 2009).

Johnson, D. K. N., Siripong, A., Brown, A.S. (2006). "The demise of distance? The declining role of physical proximity for knowledge transmission". Growth and Change 37, 19–33.

Jones, B.F., Wuchty, S., Uzzi, B. (2008). "Multi-university research teams: Shifting impact, geography, and stratification in science". Science 322, 1259–1262 (November 21).

Kerr, W.R. (2008). "Ethnic scientific communities and international technology diffusion". Review of Economics and Statistics 90 (3), 518–537.
Klevorick, A.K., Levin, R.C., Nelson, R.R., Winter, S.G. (1995). "On the sources and significance of interindustry differences in technological opportunities". Research Policy 24, 185–205.
Kling, R., Iacono, S. (1988). "The mobilization of support for computerization: The role of computerization movements". Social Problems 35, 226–243.
Knorr Cetina, K. (1999). Epistemic Cultures: How the Sciences Make Knowledge. Harvard University Press, Cambridge, MA.
Knudsen, T., Levinthal, D.A. (2007). "Two faces of search: Alternative generation and alternative evaluation". Organization Science 18, 39–54.
Kogut, B., Zander, U. (1993). "Knowledge of the firm and the evolutionary theory of the multinational corporation". Journal of International Business Studies 24, 625–645.
Lakhani, K.R., Panetta, J.A. (2007). "The principles of distributed innovation". Innovations: Technology, Governance, Globalization 2, 97–112.
Lakhani, Karim R. and Robert Wolf. (2005). "Why Hackers Do What They Do: Understanding Motivation and Effort in Free/Open Source Software Projects". In Perspectives on Free and Open Source Software, edited by Joe Feller, Brian Fitzgerald, Scott Hissam and Karim Lakhani. Cambridge, Mass: MIT Press.
Lamoreaux, N.R., Sokoloff, K.L. (2000). "The geography of invention in the American glass industry, 1870–1925". Journal of Economic History 60, 700–729.
Lamoreaux, N.R., Raff, D. M. G., Temin, P. (2003). "Beyond markets and hierarchies: Toward a new synthesis of american business history". American Historical Review 108, 404–433 (April).
Lemley, M.A., Sampat, B.N. (2008). "Is the patent office a rubber stamp?" Emory Law Journal 58, 181–206.
Lenoir, T., Giannella, E. (2006). "The emergence and diffusion of DNA microarray technology". Journal of Biomedical Discovery and Collaboration 1, 11.
Lerner, J. (1995). "Patenting in the shadow of competitors". Journal of Law and Economics 38 (2), 463–495.
Levin, R.C., Klevorick, A.K., Nelson, R.R., Winter, S.G. (1987). "Appropriating the returns from industrial research and development". Brookings Papers on Economic Activity 1987, 783–831.
MacKenzie, D. (1990). Inventing Accuracy: A Historical Sociology of Nuclear Missile Guidance. MIT Press, Cambridge, MA.
Malerba, F. (2007). "Innovation and the dynamics and evolution of industries: Progress and challenges". International Journal of Industrial Organization 25, 675–699.
Malerba, F., Orsenigo, L. (1997). "Technological regimes and sectoral patterns of innovative activities". Industrial and Corporate Change 6, 83–118.
Mansfield, E. (1982). Technology Transfer, Productivity, and Economic Policy. W. W. Norton, New York.
March, J.G. (1991). "Exploration and exploitation in organizational learning". Organization Science 2, 71–87.
March, J.G., Simon, H.A. (1958). Organizations. Wiley, New York.
Marshall, A. (1920). Industry and Trade. Macmillan, London.
McGaw, J.A. (1987). Most Wonderful Machine: Mechanization and Social Change in Berkshire Paper Making. Princeton University Press, Princeton, NJ.
Merton, R.K. (1979). The Sociology of Science: Theoretical and Empirical Investigations. University Of Chicago Press, Chicago.
Meyer, P.B. (2003). "Episodes of collective invention". US Bureau of Labor Statistics Working Paper, http://papers.ssrn.com/sol3/papers.cfm?abstract_id=466880 (Accessed October 26, 2009).
Mokyr, J. (2005). "Long-term economic growth and the history of technology". In: Aghion, P., Durlauf, S.N. (Eds.), Handbook of Economic Growth. Elsevier, Amsterdam, pp. 1113–1180.
Mowery, D.C. (1984). "Firm structure, government policy, and the organization of industrial research: Great Britain and the United States, 1900–1950.". Business History Review 58 (4), 504–530.
Mowery, D., Rosenberg, N. (1979). "The influence of market demand upon innovation: A critical review of some recent empirical studies". Research Policy 8, 102–153.
Mowery, D., Nelson, R., Sampat, B., Ziedonis, A. (2004). Ivory Tower and Industrial Innovation. Stanford University Press, Stanford, CA.
Murray, F. (2002). "Innovation as co-evolution of scientific and technological networks: Exploring tissue engineering". Research Policy 31, 1389–1403.
National Academy of Sciences. (2004). Facilitating Interdisciplinary Research. National Academy Press, Washington, DC.

National Science Board. (2006). Science and Engineering Indicator. National Science Foundation, Arlington, VA.
Nelson, R.R. (1982). "The role of knowledge in R&D efficiency". Quarterly Journal of Economics 97, 453–470.
Nelson, R.R. (2006). "Reflections on "The Simple Economics of Basic Scientific Research": Looking back and looking forward". Industrial and Corporate Change 15, 903–917.
Nelson, R.R., Winter, S.G. (1982). An Evolutionary Theory of Economic Change. Harvard University Press, Cambridge, MA.
Nickerson, J.A., Zenger, T.R. (2004). "A knowledge-based theory of the firm–The problem-solving perspective". Organization Science 15, 617–632.
Noble, D. (1984). Forces of Production: A Social History of Automation. Knopf, New York.
Nonaka, I. (1994). "A dynamic theory of organizational knowledge creation". Organization Science 5, 14–37.
Nonaka, I., von Krogh, G. (2009). "Perspective–tacit knowledge and knowledge conversion: Controversy and advancement in organizational knowledge creation theory". Organization Science 20, 635–652.
Nuvolari, A. (2004). "Collective invention during the British Industrial Revolution: The case of the Cornish pumping engine". Cambridge Journal of Economics 28, 347–363.
O'Mahoney, S., Ferraro, F. (2007). "Governance in production communities". Academy of Management Journal 50, 1079–1106.
Olson, G.M., Zimmerman, A., Bos, N. (Eds.), (2008). Science on the Internet. MIT Press, Cambridge, MA.
Olson, J.S., Olson, G.M. (2004). "Culture Surprises in Remote Software Development Teams". Queue 1, 52–59.
Orton, J.W. (2004). The Story of Semiconductors. Oxford University Press, Oxford, UK.
Owen-Smith, J., Powell, W.W. (2003). "The expanding role of university patenting in the life sciences: Assessing the importance of experience and connectivity". Research Policy 32, 1695–1711.
Owen-Smith, J., Powell, W.W. (2004). "Knowledge networks as channels and conduits: The effects of spillovers in the Boston biotechnology community". Organization Science 15, 5–21.
Patel, P., Pavitt, K. (1997). "The technological competencies of the world's largest firms: Complex and path-dependent, but not much variety". Research Policy 26, 141–156.
Powell, W.W. (1990). "Neither market nor hierarchy: Network forms of organization". Research in Organizational Behavior 12, 295–336 (JAI Press).
Powell, W.W., Snellman, K. (2004). "The knowledge economy". Annual Review of Sociology 30, 199–220.
Powell, W.W., Koput, K.W., Smith-Doerr, L. (1996). "Interorganizational collaboration and the locus of innovation: Networks of learning in biotechnology". Administrative Science Quarterly 41, 116–145.
Powell, W.W., White, D.R., Koput, K.W., Owen-Smith, J. (2005). "Network dynamics and field evolution: The growth of interorganizational collaboration in the life sciences". American Journal of Sociology 110, 1132–1205.
Powell, W.W., Owen-Smith, J., Colyvas, J.A. (2007). "Innovation and emulation: Lessons from American universities in selling private rights to public knowledge". Minerva 45, 121–142.
Powell, W.W., Packalen, K., Whittington, K. (2009). "Organizational and institutional genesis: The emergence of high-tech clusters in the life sciences". In: Padgett, J., Powell, W. (Eds.), The Emergence of Organization and Markets. PrincetonUniversity Press, Chapter 13.
Reich, L. (1985). The Making of American Industrial Research: Science and Business at GE and Bell, 1876–1926. Cambridge University Press, New York.
Rhoten, D., Powell, W.W. (2007). "The frontiers of intellectual property: Expanded protection versus new models of open science". Annual Review of Law and Social Science 3, 345–373.
Rivkin, J.W., Siggelkow, N. (2003). "Balancing search and stability: Interdependencies among elements organizational design". Management Science 49, 290–311.
Rosenberg, N. (1976). Perspectives on Technology. Cambridge University Press, Cambridge, UK.
Rosenberg, N. (1982). Inside the Black Box: Technology and Economics. Cambridge University Press, Cambridge, UK.
Rosenberg, N. (2000). Schumpeter and the Endogeneity of Technology. Routledge, London.
Rosenberg, N., Nelson, R.R. (1994). "American universities and technical advance". Research Policy 23 (3), 323–348.
Rosenkopf, L., Schilling, M. (2007). "Comparing alliance network structure across industries". Strategic Entrepreneurship Journal 1, 191–209.
Rosenkopf, L., Tushman, M. (1998). "The coevolution of community networks and technology: Lessons from the flight simulation industry". Industrial and Corporate Change 7, 311–346.
Rosenkopf, L., Metiu, A., George, V.P. (2001). "From the bottom up? Technical committee activity and alliance formation". Administrative Science Quarterly 46, 748–772.

Sabel, C.F., Zeitlin, J. (1985). "Historical alternatives to mass production: Politics, markets, and technology in nineteenth-century industrialization". Past and Present 108, 133–176 (August).

Sabel, C.F., Zeitlin, J. (Eds.), (1997). World of Possibilities: Flexibility and Mass Production in Western Industrialization. Cambridge University Press, New York.

Saxenian, A.L. (1994). Regional Advantage. Harvard University Press, Cambridge, MA.

Saxenian, A.L. (2006). The New Argonauts: Regional Advantage in a Global Economy. Harvard University Press, Cambridge, MA.

Saxenian, A.L., Sabel, C. (2008). "Venture capital in the periphery: The new argonauts, global search, and local institution building". Economic Geography 84, 379–394 (October).

Schilling, M.A. (1998). "Technological lockout: An integrative model of the economic and strategic factors driving technology success and failure". Academy of Management Review 23, 267–284.

Schumpeter, J.A. (1942). Capitalism, Socialism, and Democracy. Harper & Brothers, New York.

Scranton, P. (1993). Proprietary Capitalism: The Textile Manufacturers of Philadelphia, 1880–1885. Cambridge University Press, New York.

Scranton, P. (1997). Endless Novelty: Specialty Production and American Industrialization, 1865–1925. Princeton University Press, Princeton, NJ.

Shrum, W., Genuth, J., Chompalov, I. (2007). Structures of Scientific Collaboration. MIT Press, Cambridge, MA.

Simon, H.A. (1962). "The architecture of complexity". Proceedings of the American Philosophical Society 106, 467–482.

Sturgeon, T.J. (2002). "Modular production networks: A new American model of industrial organization". Industrial and Corporate Change 11, 451–496 (June).

Teece, D.J. (1986). "Profiting from technological innovation: Implications for integration, collaboration, licensing and public policy". Research Policy 15, 285–305.

Thursby, J.G., Thursby, M.C. (2003). "Are faculty critical? Their role in university–industry licensing". NBER Working Paper 9991.

Thursby, J., Fuller, A.W., Thursby, M. (2009). "US faculty patenting: Inside and outside the university". Research Policy 38, 14–25.

Vincenti, W.G. (1990). What Engineers Know and How They Know It: Analytical Studies from Aeronautical History. Johns Hopkins University Press, Baltimore, MD.

von Hippel, E. (1994). ""Sticky Information" and the locus of problem solving: Implications for innovation". Management Science 40, 429–439.

Whittington, K.B. (2007). "Employment sectors as opportunity structures: Effects of location on male and female scientific dissemination". PhD Dissertation, Sociology, Stanford University.

Whittington, K.B., Owen-Smith, J., Powell, W.W. (2009). "Networks, propinquity, and innovation in knowledge-intensive industries". Administrative Science Quarterly 54, 90–122.

Winter, S.G. (2006). "The logic of appropriability: From Schumpeter to Arrow to Teece". Research Policy 35, 1100–1106.

Wray, K.B. (2002). "The epistemic significance of collaborative research". Philosophy of Science 69, 150–168.

Wuchty, S., Jones, B.F., Uzzi, B. (2007). "The increasing dominance of teams in production of knowledge". Science 316, 1036–1039 (May 18).

Ziedonis, A.A., Mowery, D.C. (2004). "The geographic reach of market and nonmarket channels of technology transfer: Comparing citations and licenses of university patents". In: Cantwell, J. (Ed.), Globalization and the Location of Firms. Edward Elgar, Northampton, MA.

Ziman, J.M. (1994). Prometheus Bound: Science in a Dynamic 'Steady State'. Cambridge University Press, New York.

Zollo, M., Winter, S.G. (2002). "Deliberate learning and the evolution of dynamic capabilities". Organization Science 13, 339–351.

Zucker, L.G., Darby, M.R. (1996). "Star scientists and institutional transformation: Patterns of invention and innovation in the formation of the biotechnology industry". Proceedings of the National Academy of Sciences 93 (23), 12709–12716.

PART III

COMMERCIALIZATION OF INNOVATION

Chapter 14

THE FINANCING OF R&D AND INNOVATION

BRONWYN H. HALL*,‡ *AND* JOSH LERNER†
**University of California*
Berkeley, California, USA
†Rock Center for Entrepreneurship
Harvard Business School, Boston
Massachusetts, USA
‡University of Maastricht,
Maastricht, The Netherlands

Contents

Handbooks in Economics, Volume 01

DOI: 10.1016/S0169-7218(10)01014-2

Abstract

Evidence on the "funding gap" for investment innovation is surveyed. The focus is on financial market reasons for underinvestment that exist even when externality-induced underinvestment is absent. We conclude that while small and new innovative firms experience high costs of capital that are only partly mitigated by the presence of venture capital, the evidence for high costs of R&D capital for large firms is mixed. Nevertheless, large established firms do appear to prefer internal funds for financing such investments and they manage their cash flow to ensure this. Evidence shows that there are limits to venture capital as a solution to the funding gap, especially in countries where public equity markets for VC exit are not highly developed. We conclude by suggesting areas for further research.

Keywords

cash flow, financing, innovation, liquidity constraints, R&D, venture capital

JEL classification: G24, G32, O32, O38

1. Introduction

It is a widely held view that research and development (R&D) and innovative activities are difficult to finance in a freely competitive market place. Support for this view in the form of economic-theoretic modeling is not difficult to find and probably begins with the classic articles of Nelson (1959) and Arrow (1962), although the idea itself was alluded to by Schumpeter (1942).[1] The main argument goes as follows: the primary output of resources devoted to invention is the knowledge of how to make new goods and services, and this knowledge is nonrival: use by one firm does not preclude its use by another. To the extent that knowledge cannot be kept secret, the returns to the investment in knowledge cannot be appropriated by the firm undertaking the investment, and therefore such firms will be reluctant to invest, leading to the underprovision of R&D investment in the economy.

Since the time when this argument was fully articulated by Arrow, it has of course been developed, tested, modified, and extended in many ways. For example, Levin et al. (1987) and Mansfield et al. (1981), using survey evidence, found that imitating a new invention in a manufacturing firm was not free, but could cost as much as 50–75% of the cost of the original invention. This fact will mitigate but not eliminate the underinvestment problem. Empirical support for the basic point made by Arrow concerning the positive externalities created by research is widespread, mostly in the form of studies that document a social return to R&D that is higher than the private level (Griliches, 1992; Hall, 1996; Chapter 24, this volume). Recently, a large number of authors led by Romer (1986) have produced models of endogenous macroeconomic growth that are built on the increasing returns principle implied by Arrow's argument that one person's use of knowledge does not diminish its utility to another (Aghion and Howitt, 1997).

This line of reasoning is already widely used by policymakers to justify such interventions as the intellectual property system, government support of R&D, R&D tax incentives, and the encouragement of research partnerships of various kinds. In general, these incentive programs can be warranted even when the firm or individual undertaking the research is the same as the entity that finances it. However, Arrow's influential paper also contains another reason for underinvestment in R&D, again one which was foreshadowed by Schumpeter and which has been addressed by subsequent researchers in economics and finance: the argument that an additional gap exists between the private rate of return and the cost of capital when the innovation investor and financier are different entities.

This chapter concerns itself with this second aspect of the market failure for R&D and other investments in innovation: even if problems associated with incomplete appropriability of the returns to R&D are solved using intellectual property protection, subsidies, or tax incentives, it may still be difficult or costly to finance such investments using capital from sources external to the firm or entrepreneur. That is, there is often a wedge, sometimes large, between the rate of return required by an entrepreneur investing his own funds and that required by external investors. By this argument, unless an inventor is already wealthy, or firms already profitable, some innovations will fail to be provided purely because the cost of external capital is too high, even when they would pass the private returns hurdle if funds were available at a "normal" interest rate.

In the following, we begin by describing some of the unique features of R&D investment. Then we discuss the various theoretical arguments why external finance for R&D might be more expensive that internal finance, going on to review the empirical evidence on the validity of this hypothesis and the

[1] See, for example, footnote 1, Chapter VIII of *Capitalism, Socialism, and Democracy.*

solutions that have been developed and adopted by the market and some governments, in particular the venture capital solution. Although we focus our attention on R&D in the first three sections of the chapter, much of what we discuss will apply to innovation investment more broadly defined. However, for reasons of data availability and measurement, the empirical literature has largely focused on R&D spending, at least up until now. The chapter concludes with a discussion of policy options.

2. R&D as investment

From the perspective of investment theory, R&D has a number of characteristics that make it different from ordinary investment. First and most importantly, in practice 50% or more of R&D spending is the wages and salaries of highly educated scientists and engineers. Their efforts create an intangible asset, the firm's knowledge base, from which profits in future years will be generated. To the extent that this knowledge is "tacit" rather than codified, it is embedded in the human capital of the firm's employees, and is therefore lost if they leave or are fired.

This fact has an important implication for the conduct of R&D investment. Because part of the resource base of the firm itself disappears when such workers leave or are fired and because projects often take a long time between conception and commercialization, firms tend to smooth their R&D spending over time, in order to avoid having to lay off knowledge workers. This implies that R&D spending at the firm level usually behaves as though it has high adjustment costs (Hall et al., 1986; Lach and Schankerman, 1988), with two consequences, one substantive and one that affects empirical work in this area. First, the equilibrium required rate of return to R&D may be quite high simply to cover the adjustment costs. Second, and related to the first, is that it will be difficult to measure the impact of changes in the costs of capital, because such effects can be weak in the short run due to the sluggish response of R&D to any changes in its cost. Brown and Petersen (2009b) offer direct evidence that US firms relied heavily on cash reserves to smooth R&D spending during the 1998–2002 boom and bust in stock market returns.

A second important feature of R&D investment is the degree of uncertainty associated with its output. This uncertainty tends to be greatest at the beginning of a research program or project, which implies that an optimal R&D strategy has an options-like character and should not really be analyzed in a static framework. R&D projects with small probabilities of great success in the future may be worth continuing even if they do not pass an expected rate of return test. The uncertainty here can be extreme and not a simple matter of a well-specified distribution with a mean and variance. There is evidence, such as that in Scherer (1998), that the distribution of profits from innovation sometimes has a Paretian character where the variance does not exist. When this is the case, standard risk-adjustment methods will not work well.

In spite of the problems suggested by the nature of uncertainty in this area, the starting point for the analysis of R&D investment financing has been the "neoclassical" marginal profit condition, suitably modified to take the special features of R&D into account. Following the formulation in Hall and Van Reenen (2000), we define the user cost of R&D investment ρ as the pretax real rate of return on a marginal investment that is required to earn a return r after (corporate) tax. The firm invests to the point where the marginal product of R&D capital equals ρ:

$$\text{MPK} = \rho = \frac{1 - A^{\text{d}} - A^{\text{c}}}{1 - \tau}(r + \delta - \Delta p_{\text{R}}/p_{\text{R}} + \text{MAC}) \tag{1}$$

τ is the corporate tax rate, δ is the (economic) depreciation rate, the term in p_{R} is the relative appreciation or depreciation of R&D capital, and MAC is the marginal adjustment cost.

In this equation, A^{d} and A^{c} are the present discounted value of depreciation allowances and tax credits, respectively. In most financial accounting systems, including those used by major OECD economies, R&D is expensed as it is incurred rather than capitalized and depreciated, which means that the lifetime of the investment for accounting purposes is much shorter than the economic life of the asset created and that A^{d} is simply equal to τ for tax-paying firms. Many countries have a form of tax credit for R&D, either incremental or otherwise, and this will be reflected in a positive value for A^{c}.[2] Note that when A^{c} is zero, the corporate tax rate does not enter into the marginal R&D decision, because of the full deductability of R&D.

The user cost formulation above directs attention to the following determinants of R&D financing:

1. Tax treatment such as tax credits, which are clearly amenable to intervention by policy makers
2. Economic depreciation δ, which in the case of R&D is more properly termed obsolescence. This quantity is sensitive to the realized rate of technical change in the industry, which is in turn determined by such factors as competition, market structure, and the rate of imitation. Thus, it is inappropriate to treat δ as an invariant parameter in this setting (Hall, 2005)
3. The marginal costs of adjusting the level of the R&D program
4. The investor's required rate of return r

The last item has been the subject of considerable theoretical and empirical interest, on the part of both industrial organization and corporate finance economists. Two broad strands of investigation can be observed: one focuses on the role of asymmetric information and moral hazard in raising the required rate of return above that normally used for conventional investment, and the latter on the requirements of different sources of financing and their differing tax treatments for the rate of return. Section 3 discusses these factors.

3. Theoretical background

This section of the chapter reviews the reasons that the impact of financial considerations on the investment decision may vary with the type of investment and with the source of funds in more detail. To do this, we distinguish between those factors that arise from various kinds of market failures in this setting and the purely financial (or tax-oriented) considerations that affect the cost of different sources of funds.

One of the implications of the well-known Modigliani–Miller (1958, 1961) is that a firm choosing the optimal levels of investment should be indifferent to its capital structure, and should face the same price

[2] See Hall and Van Reenen (2000) for details. For example, during the past three decades the United States has had an incremental R&D tax credit with a value for A^{c} of about 0.13 at the time of writing.

for investment and R&D investment on the margin. The last dollar spent on each type of investment should yield the same expected rate of return (after adjustment for nondiversifiable risk). A large literature, both theoretical and empirical, has questioned the bases for this theorem, but it remains a useful starting point.

Reasons why the theorem might fail in practice are several: (1) uncertainty coupled with incomplete markets may make a real options approach to the R&D investment decision more appropriate; (2) the cost of capital may differ by source of funds for nontax reasons; (3) the cost of capital may differ by source of funds for tax reasons; and (4) the cost of capital may also differ across types of investments (tangible and intangible) for both tax and other reasons.

With respect to R&D investment, economic theory advances a plethora of reasons why there might be a gap between the external and internal costs of capital; these can be divided into three main types:

1. Asymmetric information between inventor/entrepreneur and investor
2. Moral hazard on the part of the inventor/entrepreneur arising from the separation of ownership and management
3. Tax considerations that drive a wedge between external finance and finance by retained earnings

We discuss each of these reasons in separate sections below.

3.1. Asymmetric information problems

In the innovation setting, the asymmetric information problem refers to the fact that an inventor frequently has better information about the likelihood of success and the nature of the contemplated innovation project than potential investors. Therefore, the marketplace for financing the development of innovative ideas looks like the "lemons" market modeled by Akerlof (1970). The lemons' premium for R&D will be higher than that for ordinary investment because investors have more difficulty distinguishing good projects from bad when the projects are long-term R&D investments than when they are more short-term or low-risk projects (Leland and Pyle, 1977). When the level of R&D expenditure is a highly observable signal, as it is under current US and UK rules, we might expect that the lemons' problem is somewhat mitigated, but certainly not eliminated.[3]

In the most extreme version of the lemons model, the market for R&D projects may disappear entirely if the asymmetric information problem is too great. Informal evidence suggests that some potential innovators believe this to be the case in fact. And as will be discussed below, venture capital systems are viewed by some as a solution to this "missing markets" problem.

Reducing information asymmetry via fuller disclosure is of limited effectiveness in this arena, due to the ease of imitation of inventive ideas. Firms are reluctant to reveal their innovative ideas to the marketplace and the fact that there could be a substantial cost to revealing information to their competitors reduces the quality of the signal they can make about a potential project (Anton and Yao, 1998; Bhattacharya and Ritter, 1983). Thus, the implication of asymmetric information coupled with the

[3] Since 1974, publicly traded firms in the United States have been required to report their total R&D expenditures in their annual reports and 10-K filings with the SEC, under FASB rule No. 2, issued October 1974. In 1989, a new accounting standard, SSAP 13, obligated similar disclosures in the United Kingdom. Most continental European countries have not had such a requirement in the past, but this is changing as harmonized international standards come into force.

costliness of mitigating the problem is that firms and inventors will face a higher cost of external than internal capital for R&D due to the lemons' premium.

Some empirical support for this proposition exists, mostly in the form of event studies that measure the market response to announcements of new debt or share issues.[4] Both Alam and Walton (1995) and Zantout (1997) find higher abnormal returns to firm shares following new debt issues when the firm is more R&D-intensive. The argument is that the acquisition of new sources of financing is good news when the firm has an asymmetric information problem because of its R&D strategy. Similarly, Szewczyk et al. (1996) find that investment opportunities (as proxied by Tobin's q) explain R&D-associated abnormal returns, and that these returns are higher when the firm is highly leveraged, implying a higher required rate of return for debt finance in equilibrium.

3.2. Moral hazard problems

Moral hazard in R&D investing arises in the usual way: modern industrial firms normally have separation of ownership and management. This leads to a principal–agent problem when the goals of the two conflict, which can result in investment strategies that do not maximize the share value. Two possible scenarios may coexist: one is the usual tendency of managers to spend on activities that benefit them (growing the firm beyond efficient scale, nicer offices, etc.) and the second is a reluctance of risk averse managers to invest in uncertain R&D projects. Agency costs of the first type may be avoided by reducing the amount of free cash flow available to the managers by leveraging the firm, but this in turn forces them to use the higher cost external funds to finance R&D (Jensen and Meckling, 1976). Empirically, there seem to be limits to the use of the leveraging strategy in R&D-intensive sectors. See Hall (1990, 1994) for evidence that the LBO/restructuring wave of the 1980s was almost entirely confined to industries and firms where R&D was of no consequence. As we discuss Section 3.3, it is still true that R&D-intensive firms tend to have lower leverage than other firms on average.

According to the second type of principal–agent conflict, managers are more risk averse than shareholders and avoid R&D projects that will increase the riskiness of the firm. If bankruptcy is a possibility, both managers whose opportunity cost is lower than their present earnings and potential bondholders may wish to avoid variance-increasing projects which shareholders would like to undertake. The argument of the theory is that long-term investments can suffer in this case. The optimal solution to this type of agency cost would be to increase the long-term incentives faced by the manager rather than reducing free cash flow.

Evidence on the importance of agency costs as they relate to R&D takes several forms. Several researchers have studied the impact of antitakeover amendments (which arguably increase managerial security and willingness to take on risk while reducing managerial discipline) on R&D investment and firm value. Johnson and Rao (1997) find that such amendments are not followed by cuts in R&D, while Pugh et al. (1999) find that adoption of an Employee Stock Ownership Plan (ESOP), which is a form of antitakeover protection, is followed by R&D increases. Cho (1992) finds that R&D intensity increases

[4] See Campbell et al. (1997) for a description of this methodology, which infers the value of a firm's action when it is publicly announced by examining the market returns to a share of the firm's stock in the period surrounding the announcement.

with the share that managerial shareholdings represent of the manager's wealth and interprets this as incentive pay mitigating agency costs and inducing long-term investment.

Some have argued that institutional ownership of the managerial firm can reduce the agency costs due to free-riding by owners that is a feature of the governance of firms with diffuse ownership structure, while others have held that such ownership pays too much attention to short-term earnings and therefore discourages long-term investments. Institutions such as mutual and pension funds often control somewhat larger blocks of shares than individuals, making monitoring firm and manager behavior a more effective and more rewarding activity for these organizations.

There is some limited evidence that this may indeed be the case. Eng and Shackell (2001) find that firms adopting long-term performance plans for their managers do not increase their R&D spending but that institutional ownership is associated with higher R&D; R&D firms tend not to be held by banks and insurance companies. Majumdar and Nagarajan (1997) find that high institutional investor ownership does not lead to short-term behavior on the part of the firm; in particular, it does not lead to cuts in R&D spending. Francis and Smith (1995) find that diffusely held firms are less innovative, implying that monitoring alleviates agency costs and enables investment in innovation.

Although the evidence summarized above is fairly clear and indicates that long-term incentives for managers can encourage R&D and that institutional ownership does not necessarily discourage R&D investment, it is fairly silent on the magnitude of these effects, and whether these governance features truly close the agency cost-induced gap between the cost of capital and the return to R&D.

3.3. Capital structure and R&D

In the view of some observers, the leveraged buyout (LBO) wave of the 1980s in the United States and the United Kingdom arose partly because high real interest rates meant that there were strong pressures to eliminate free cash flow within firms (Blair and Litan, 1990). For firms in industries where R&D is an important form of investment, such pressure should have been reduced by the need for internal funds to undertake such investment and indeed Hall (1993, 1994) and Opler and Titman (1993) find that firms with high R&D intensity were much less likely to experience an LBO. Opler and Titman (1994) find that R&D firms that were leveraged suffered more than other firms when facing economic distress, presumably because leverage meant that they were unable to sustain R&D programs in the fact of reduced cash flow.

A more recent look at the consequences of these transactions is by Lerner et al. (2008). The authors investigate 495 buyout transactions where there was a patent application in the 9 years around the buyout. They find no evidence that LBOs are associated with a decrease in patenting. Relying on standard measures of patent quality, they find that patents granted to firms involved in private equity transactions are more cited (a proxy for economic importance), show no significant shifts in the fundamental nature of the research, and are more concentrated in the most important and prominent areas of companies' innovative portfolios, suggesting a refocusing on the core business, but not a reduction in innovative activity.

In related work using data on Israeli firms, Blass and Yosha (2001) report that R&D-intensive firms listed on the United States stock exchanges use highly equity-based sources of financing, whereas those listed only in Israel rely more on bank financing and government funding. The former are more profitable and faster growing, which suggests that the choice of where to list the shares and whether to finance with new equity is indeed sensitive to the expected rate of return to the R&D being undertaken. That is, investors supplying arms-length finance require higher returns to compensate them for the risk of a "lemon."

Although leverage may be a useful tool for reducing agency costs in the firm, it is of limited value for R&D-intensive firms. Because the knowledge asset created by R&D investment is intangible, partly embedded in human capital, and ordinarily very specialized to the particular firm in which it resides, the capital structure of R&D-intensive firms customarily exhibits considerably less leverage than that of other firms. Banks and other debtholders prefer to use physical assets to secure loans and are reluctant to lend when the project involves substantial R&D investment rather than investment in plant and equipment. In the words of Williamson (1988), "redeployable" assets (i.e., assets whose value in an alternative use is almost as high as in their current use) are more suited to the governance structures associated with debt. Empirical support for this idea is provided by Alderson and Betker (1996), who find that liquidation costs and R&D are positively related across firms. The implication is that the sunk costs associated with R&D investment are higher than that for ordinary investment.

In addition, servicing debt usually requires a stable source of cash flow, which makes it more difficult to find the funds for an R&D investment program that must be sustained at a certain level in order to be productive. For both these reasons, firms are either unable or reluctant to use debt finance for R&D investment, which may raise the cost of capital, depending on the precise tax treatment of debt versus equity.[5] Confirming empirical evidence for the idea that limiting free cash flow in R&D firms is a less desirable method of reducing agency costs is provided by Chung and Wright (1998), who find that financial slack and R&D spending are correlated with the value of growth firms positively, but not correlated with that of other firms.

3.4. Taxes and the source of funds

Tax considerations that yield variations in the cost of capital across source of finance have been well articulated by Auerbach (1984) among others. He argued that under the US tax system during most of its history the cost of financing new investment by debt has been less than that of financing it by retained earnings, which is in turn less than that of issuing new shares. More explicitly, if r is the risk-adjusted required return to capital, τ is the corporate tax rate, θ is the personal tax rate, and c is the capital gains tax rate, we have the following required rates of return for different financing sources:

Debt: $r(1-\tau)$ interest deductible at the corporate level
Retained earnings: $r(1-\theta)/(1-c)$ avoids personal tax on dividends, but capital gains tax
New shares: $r/(1\text{-}c)$ eventual capital gains tax

If dividends are taxed, clearly financing with new shares is more expensive than financing with retained earnings. And unless the personal income tax rate is much higher than the sum of the corporate and capital gains rates, the following inequalities will both hold:

[5] There is also considerable cross-sectional evidence for the United States that R&D intensity and leverage are negatively correlated across firms. See Friend and Lang (1988), Hall (1992), and Bhagat and Welch (1995).

$$(1-\tau) < \frac{1-\theta}{1-c} < \frac{1}{1-c} \tag{2}$$

These inequalities express the facts that interest expense is deductible at the corporate level, while dividend payments are not, and that shareholders normally pay tax at a higher rate on retained earnings that are paid out than on those retained by the firm and invested.[6] It implicitly assumes that the returns from the investment made will be retained by the firm and eventually taxed at the capital gains rate rather than the rate on ordinary income.

It is also true that the tax treatment of R&D in most OECD economies is very different from that of other kinds of investment: because R&D is expensed as it is incurred, the effective tax rate on R&D assets is lower than that on either plant or equipment, with or without an R&D tax credit in place. This effectively means that the economic depreciation of R&D assets is considerably less than the depreciation allowed for tax purposes—which is 100%—so that the required rate of return for such investment would be lower. In addition, some countries offer a tax credit or subsidy to R&D spending, which can reduce the after tax cost of capital even further.[7]

The conclusion from this section of the chapter is that the presence of either asymmetric information or a principal–agent conflict implies that new debt or equity finance will be relatively more expensive for R&D than for ordinary investment, and that considerations such as lack of collateral further reduce the possibility of debt finance. Together, these arguments suggest an important role for retained earnings in the R&D investment decision, independent of their value as a signal of future profitability. In fact, as has been argued by both Hall (1992) and Himmelberg and Petersen (1994), there is good reason to think that positive cash flow may be more important for R&D than for ordinary investment. Section 4 summarizes the results from empirical tests for this proposition.

4. Testing for financial constraints

The usual way to examine the empirical relevance of the arguments that R&D investment in established firms can be disadvantaged when internal funds are not available and recourse to external capital markets required is to estimate R&D investment equations and test for the presence of "liquidity" constraints, or excess sensitivity to cash flow shocks. This approach builds on the extensive literature developed for testing ordinary investment equations for liquidity constraints (Arellano and Bond, 1991; Fazzari et al., 1988). It suffers from many of the same difficulties as the estimates in the investment literature, plus one additional problem that arises from the tendency of firms to smooth R&D spending over time.

[6] A detailed discussion of tax regimes in different countries is beyond the scope of this survey, but it is quite a common in several countries for long-term capital gains on funds that remain with a firm for more than 1 year to be taxed at a lower rate than ordinary income. Of course, even if the tax rates on the two kinds of income are equal, the inequalities will hold. Only in the case where dividends are not taxed at the corporate level (which was formerly the case in the United Kingdom) will the ranking given above not hold.

[7] See Hall and Van Reenen (2000) for details.

The ideal experiment for identifying the effects of liquidity constraints on investment is to give firms additional cash exogenously, and observe whether they pass it on to shareholders or use it for investment and/or R&D. If they choose the first alternative, either the cost of capital to the firm has not fallen, or it has fallen but they still have no good investment opportunities. If they choose the second, then the firm must have had some unexploited investment opportunities that were not profitable using more costly external finance. A finding that investment is sensitive to cash flow shocks that are not signals of future demand increases would reject the hypothesis that the cost of external funds is the same as the cost of internal funds. However, lack of true experiments of this kind forces researchers to use econometric techniques such as instrumental variables to attempt to control for demand shocks when estimating the investment demand equation, with varying degrees of success.

The methodology for the identification of R&D investment equations is based on a simple supply and demand heuristic, as shown in Figure 1. The curve sloping downward to the right represents the demand for R&D investment funds and the curves sloping upward the supply of funds. Internal funds are available at a constant cost of capital until they are exhausted, at which point it becomes necessary to issue debt or equity in order to finance more investment. When the demand curve cuts the supply curve in the horizontal portion, a shock that increases cash flow (and shifts supply outward) has no effect on the level of investment. However, if the demand curve cuts the supply curve where it is upward sloping, it is possible for a shock to cash flow to shift the supply curve out in such a way as to induce a substantial

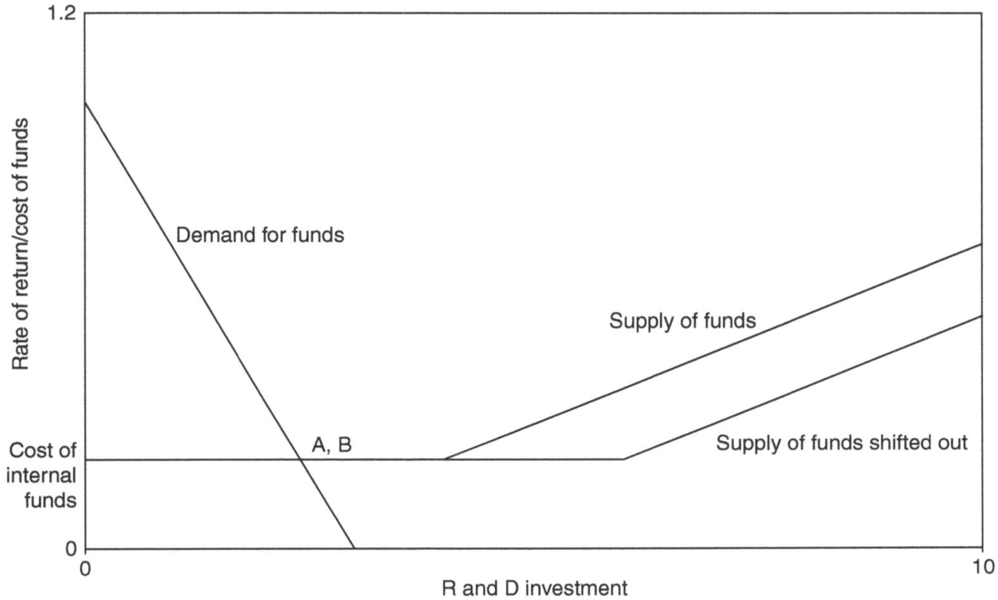

Figure 1. Unconstrained firm.

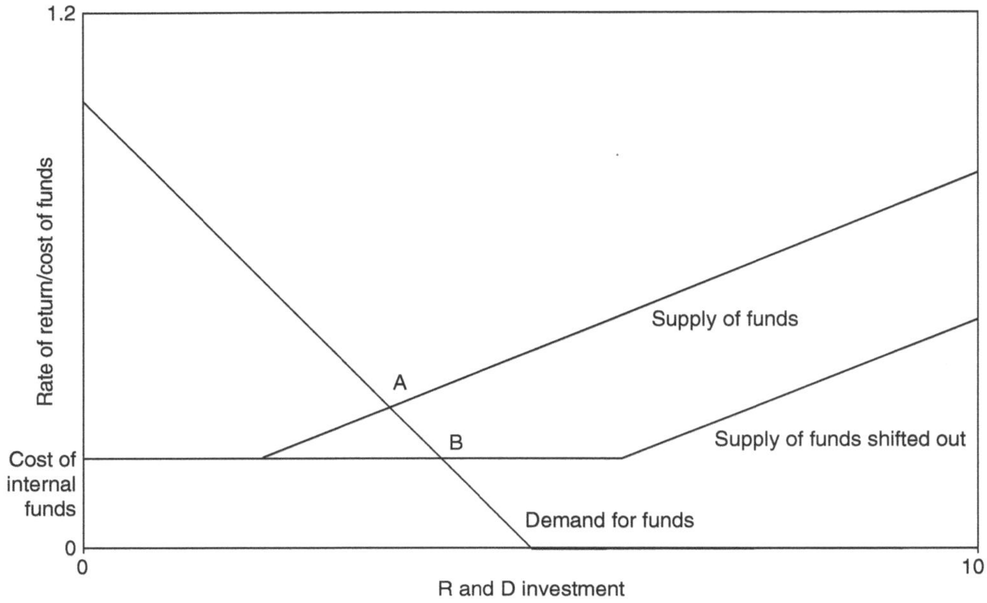

Figure 2. Constrained firm.

increase in R&D investment. Figure 2 illustrates such a case, where the firm shifts from point A to point B in response to a cash flow shock that does not shift the demand curve.

Econometric work that tests the hypothesis that financing constraints matter for R&D investment has largely been done using standard investment equation methodology. Two main approaches can be identified: one uses a neoclassical accelerator model with *ad hoc* dynamics to allow for the presence of adjustment costs, and the other an Euler equation derived from the forward-looking dynamic program of a profit-maximizing firm that faces adjustment costs for capital.[8]

The accelerator model begins with the marginal product equal to cost condition for capital:

$$\mathrm{MPK} = C \tag{3}$$

Assuming that the production function for the ith firm at time t is Cobb-Douglas, solving out the variable factors, and taking logarithms of this relationship yields

$$k_{it} = s_{it} + a_i - c_{it} \tag{4}$$

[8] A detailed consideration of the econometric estimation of these models can be found in Mairesse et al. (1999). See also Hall (1991).

where k = log(R&D capital), s = log(output or sales), and c = log(cost of R&D). a_i captures any permanent differences across firms, including differences in the production function.

Lagged adjustment of R&D capital to changes in its cost or expected future demand is allowed for by specifying an autoregressive distributed lag (ADL) for the relationship between capital and sales. For example, specifying an ADL(2,2) and approximating the growth of the capital stock Δk by $R/K-\delta$ yields an estimating equation of the following form:

$$\frac{R}{K}=f\left(\frac{R(-1)}{K(-1)}, \Delta s, \Delta s(-1), k(-2)-s(-2), \text{time dummies}, \text{firm dummies}\right) \tag{5}$$

The time dummies capture the conventional cost of capital, assumed to be the same for all firms. Note that any variations in R&D capital depreciation common to all firms will be in the time dummies, and any variations specific to a firm or sector but constant over time will be in the firm dummies. Firm-specific costs related to financing constraints are included by adding current and lagged values of the cash flow/capital ratio to this equation. Because of the presence of firm dummies, estimation is done using first differences of this equation, instrumented by lagged values of the right hand side variables to correct for the potential endogeneity of the contemporaneous values. In principle, this will also control for the potential simultaneity between current investment and the disturbance. However, if the firm's planning horizon for its R&D programs is long enough, as we might expect in the biotechnology area, for example, we might be concerned about the validity of lagged instruments.

The Euler equation approach begins with the following first-order condition for investment in two adjacent periods:

$$E_{t-1}\left[\text{MPK}_t + (1-\delta)(p_t + \text{MAC}_t) - (1+r)\left(\frac{\alpha_{t-1}}{\alpha_t}\right)(p_{t-1} + \text{MAC}_{t-1})\right] = 0 \tag{6}$$

where MAC denotes the marginal adjustment costs for R&D capital and α_t is the shadow value of investment funds in period t, which will be unity if there are no financing constraints. After specifying a Cobb-Douglas production function and quadratic adjustment costs, we obtain the following estimating equation:

$$E\left[\frac{R}{K} - \beta_1\frac{R(-1)}{K(-1)} - \gamma_1\frac{S}{K} - \beta_2\left(\frac{R}{K}\right)^2 - \text{time dummies} - \text{firm dummies}\middle| Z\right] = 0 \tag{7}$$

where Z is a set of appropriate instrumental variables. As in the case of the accelerator model, this equation is usually estimated in differenced form to remove the firm dummies, with lagged values of the right-hand side variables as instruments.

When financial constraints are present, the coefficient of lagged R&D investment in the Euler equation differs from $(1 + r)$ by the term (α_{t-1}/α_t). The implication is that when the firm changes its financial position (i.e., the shadow value of additional funds for investment changes) between one period and the next, it will invest as though it is facing a cost of capital greater than r (when the shadow value falls between periods) or less than r (when the shadow value rises between periods). Clearly this is a very difficult test to perform because (α_{t-1}/α_t) is not constant across firms or across time periods, so it cannot be treated as a parameter.

Three solutions are possible: the first is to model (α_{t-1}/α_t) as a function of proxies for changes in financial position, such as dividend behavior, new share issues, or new debt issues. The second is more *ad hoc*: recall that this term also multiplies the price p_t of R&D capital to create a firm-specific cost of capital. Most researchers simply include the cash flow to capital ratio in the model to proxy for the firm-specific cost of capital and test whether it enters in the presence of time dummies that are the same for all firms. This method assumes that all firms face the same R&D price (cost of capital), except for the cash flow effect.

The third possibility is to stratify firms in some way that is related to the level of cash constraints that they face (e.g., dividend-paying and nondividend paying firms) estimate separate investment equations for each group, and test whether the coefficients are equal. This last was the method used by Fazzari et al. (1988) in the paper that originated this literature. Note that these authors did not rely on the full Euler equation derivation, but used a version of the neoclassical accelerator model (the first model given above). See also Kaplan and Zingales (1997) for a critique of their approach, and Fazzari et al. (2000) for a response to the critique.

During the past several years, various versions of the methodologies described above have been applied to data on the R&D investment of US, UK, French, German, Irish, and Japanese firms. The firms examined are typically the largest and most important manufacturing firms in their economy. For example, Hall (1992) found a large positive elasticity between R&D and cash flow, using an accelerator-type model and a very large sample of US manufacturing firms. The estimation methodology here controlled for both firm effects and simultaneity. Similarly and using some of the same data, Himmelberg and Petersen (1994) looked at a panel of 179 US small firms in high-tech industries and find an economically large and statistically significant relationship between R&D investment and internal finance.

More recently, Brown et al. (2009) have shown that both cash flow and the issuance of public equity are very important for younger US firms during the 1990–2004 period, while they have little impact on mature firm R&D investment. They focus on the high-technology sector (drugs, office and computing equipment, communications equipment, electronic components, scientific instruments, medical instruments, and software), which accounts for almost all of the increase in R&D during this period, and use Euler equation methods with fixed firm effects and industry-level year dummies to remove most of the variation due to unobserved differences in firm characteristics and demand shocks across industry. A novel finding in this chapter and a companion paper by Brown and Petersen (2009a) is the increased importance of public equity issuance in financing R&D in the United States, which doubtless reflects a shift in expectations on the part of investors during this period.

Harhoff (1998) found weak but significant cash flow effects on R&D for both small and large German firms, although Euler equation estimates for R&D investment were uninformative due to the smoothness of R&D and the small sample size. Combining limited survey evidence with his regression results, he concludes that R&D investment in small German firms may be constrained by the availability of finance. Bond et al. (1999) find significant differences between the cash flow impacts on R&D and investment for large manufacturing firms in the United Kingdom and Germany. German firms in their sample are insensitive to cash flow shocks, whereas the investment of non-R&D-doing UK firms does respond. Cash flow helps to predict whether a UK firm does R&D, but not the level of that R&D. They interpret their findings to mean that financial constraints are important for British firms, but that those which do R&D are a self-selected group that face fewer constraints. This is consistent with the view that the desire of firms to smooth R&D over time combines with the relatively high cost of financing it to reduce R&D well below the level that would obtain in a frictionless world.

Mulkay et al. (2001) perform a similar exercise using large French and US manufacturing firms, finding that cash flow impacts are much larger in the United States than in France, both for R&D and for ordinary investment. Except for the well-known fact that R&D exhibits higher serial correlation than investment (presumably because of higher adjustment costs), differences in behavior are between countries, not between investment types, suggesting that they are due to differences in the structure of financial markets rather than the type of investment, tangible or intangible. This result is consistent with evidence reported in Hall et al. (1999) for the United States, France, and Japan during an earlier time period, which basically finds that R&D and investment on the one hand, and sales and cash flow on the other, are simultaneously determined in the United States (neither one "Granger-causes" the other), whereas in the other countries, there is little feedback from sales and cash flows to the two investments. Using a nonstructural R&D investment equation together with data for the United States, United Kingdom, Canada, Europe, and Japan, Bhagat and Welch (1995) found similar results for the 1985–1990 period, with stock returns predicting changes in R&D more strongly for the US and UK firms.

Bougheas et al. (2001) examined the effects of liquidity constraints on R&D investment using firm-level data for manufacturing firms in Ireland and also found evidence that R&D investment in these firms is financially constrained, in line with the previous studies of US and UK firms.

Brown (1997) argues that existing tests of the impact of capital market imperfections on innovative firms cannot distinguish between two possibilities: (1) capital markets are perfect and different factors drive the firm's different types of expenditure or (2) capital markets are imperfect and different types of expenditure react differently to a common factor (shocks to the supply of internal finance). He then compares the sensitivity of investment to cash flow for innovative and noninnovative firms in the United Kingdom. The results support the hypothesis that capital markets are imperfect, finding that the investment of innovative firms is more sensitive to cash flow.

The conclusions from this body of empirical work are several: first, there is solid evidence that debt is a disfavored source of finance for R&D investment; second, the "Anglo-Saxon" economies, with their thick and highly developed stock markets and relatively transparent ownership structures, typically exhibit more sensitivity and responsiveness of R&D to cash flow than continental economies; third, and much more speculatively, this greater responsiveness may arise because they are financially constrained, in the sense that they view external sources of finance as much more costly than internal, and therefore require a considerably higher rate of return to investments done on the margin when they are tapping these sources. However, it is perhaps equally likely that this responsiveness occurs because firms are more sensitive to demand signals in thick financial equity markets; a definitive explanation of the "excess sensitivity" result awaits further research.[9] In addition to these results, the evidence from Germany and some other countries suggests that small firms are more likely to face this difficulty than large established firms (not surprisingly, if the source of the problem is a "lemons" premium).

[9] It is also true that much of the literature here has tended to downplay the role of measurement error in drawing conclusions from the results. Measurement error in Tobin's q, cash flow, or output is likely to be sizable and will ensure that all variables will enter any specification of the R&D investment equation significantly, regardless of whether they truly belong or not. Instrumental variables estimation is a partial solution, but only if all the errors are serially uncorrelated, which is unlikely.

From a policy perspective, these results point to another reason why it may be socially beneficial to offer tax incentives to companies, especially to small and new firms, in order to reduce the cost of capital they face for R&D investment. Many governments, including not only those in the developed world (e.g., the United States and the United Kingdom), but also in the developing world (e. g., Chile, Brazil, and Argentina) currently have such programs. Such a policy approach simply observes that the cost of capital is relatively high for R&D and tries to close the gap via a tax subsidy. However, there is an alternative approach relying on the private sector that attempts to close the financing gap by reducing the degree of asymmetric information and moral hazard rather than simply subsidizing the investment. We turn to this topic in Section 5.

5. Small firms, startup finance, and venture capital

As should be apparent from much of the preceding discussion, any problems associated with financing investments in new technology will be most apparent for new entrants and startup firms. For this reason, many governments already provide some of form of assistance for such firms, and in many countries, especially the United States but also others such as Israel and Canada, there exists a private sector "venture capital" industry that is focused on solving the problem of financing innovation for new and young firms. This section of the chapter reviews what we know about these alternative funding mechanisms, beginning with then discussing the venture capital solution and then discussing public policy efforts. The discussion focuses on the United States for the most part, since the sector there is often the model for other countries, and most of the empirical evidence is based on US data.

Venture capital can be defined as independently managed, dedicated capital focusing on equity or equity-linked investments in privately held, high-growth companies. Typically, these funds are raised from institutional and wealthy individual investors, through partnerships with a decade-long duration. These funds are invested in young firms, usually in exchange for preferred stock with various special privileges. Ultimately, the venture capitalists sell these firms to corporate acquirers or else liquidate their holdings after taking the firms public.

The first venture firm, American Research and Development, was formed in 1946 and invested in companies commercializing technology developed during the Second World War. Because institutions were reluctant to invest, it was structured as a publicly traded closed-end fund and marketed mostly to individuals, a structure emulated by its successors.

By 1978 limited partnerships had become the dominant investment structure. Limited partnerships have an important advantage in the United States: capital gains taxes are not paid by the limited partnership. Instead, only the taxable investors in the fund pay taxes. Venture partnerships have predetermined, finite lifetimes. To maintain limited liability, investors must not become involved in the management of the fund.

Activity in the venture industry increased dramatically in early 1980s. Much of the growth stemmed from the US Department of Labor's clarification of Employee Retirement Income Security Act's "prudent man" rule in 1979, which had prohibited pension funds from investing substantial amounts of money into venture capital or high-risk asset classes. The rule clarification explicitly allowed pension managers to invest in high-risk assets, including venture capital.

The subsequent years saw both very good and trying times for venture capitalists. Venture capitalists backed many successful companies, including Apple Computer, Cisco, Genentech, Google, Netscape, Starbucks, and Yahoo! But commitments to the venture capital industry were very uneven, creating a great deal of instability. The annual flow of money into venture funds increased by a factor of 10 during the early 1980s. From 1987 through 1991, however, fund raising steadily declined as returns fell. Between 1996 and 2003, this pattern was repeated. Later in this chapter, we discuss the reasons behind this cyclicality.

Venture capital investing can be viewed as a cycle. In this section, we follow the cycle of venture capital activity. We begin with the formation of venture funds. We then consider the process by which such capital is invested in portfolio firms, and the exiting of such investments. We end with a discussion of open research questions, including those relating to internationalization and the real effects of venture activity.

5.1. Venture investing

The heart of the venture capital process is the connection between venture capitalists and the firms in which they invest. As discussed earlier, the economic and management literature emphasizes the informational asymmetries that characterize young firms, particularly in high-technology industries. These problems make it difficult for investors to assess firms, and permit opportunistic behavior by entrepreneurs after finance is received. Specialized financial intermediaries, such as venture capitalists, address these problems by intensively scrutinizing firms before providing capital and monitoring them afterwards.

Economic theory examines the role that venture capitalists play in mitigating agency conflicts between entrepreneurs and investors. The improvement in efficiency might be due to the active monitoring and advice that is provided (Cornelli and Yosha, 2003; Hellmann, 1998; Marx, 1994), the screening mechanisms employed (Chan, 1983), the incentives to exit (Berglöf, 1994), the proper syndication of the investment (Admati and Pfleiderer, 1994), or investment staging (Bergemann and Hege, 1998; Sahlman, 1990).

Staged capital infusion is the most potent control mechanism a venture capitalist can employ. The shorter the duration of an individual round of financing, the more frequently the venture capitalist monitors the entrepreneur's progress. The duration of funding should decline and the frequency of reevaluation increases when the venture capitalist believes that conflicts with the entrepreneur are likely.

If monitoring and information gathering are important—as models such as those of Amit et al. (1990) and Chan (1983) suggest—venture capitalists should invest in firms where asymmetric problems are likely, such as early-stage and high-technology firms with intangible assets. The capital constraints faced by these companies will be large and these investors will address them.

Gompers (1995) shows that venture capitalists concentrate investments in early-stage companies and high-technology industries where informational asymmetries are significant and monitoring is valuable. He finds that early-stage firms receive significantly less money per round. Increases in asset tangibility are associated with longer financing duration and reduce monitoring intensity, presumably because such assets increase the salvage value of the firm if the enterprise fails.

In a related chapter, Kaplan and Strömberg (2003) document how venture capitalists allocate control and ownership rights contingent on financial and nonfinancial performance. If a portfolio company performs poorly, venture capitalists obtain full control. As performance improves, the entrepreneur obtains more control. If the firm does well, the venture capitalists relinquish most of their control rights but retain their equity stake.

Related evidence comes from Hsu (2004), who studies the price entrepreneurs pay to be associated with reputable venture capitalists. He analyzes firms which received financing offers from multiple venture capitalists. Hsu shows that high investor experience is associated with a substantial discount in firm valuation.

Venture capitalists usually make investments with peers. The lead venture firm involves other venture firms. One critical rationale for syndication in the venture industry is that peers provide a second opinion on the investment opportunity and limit the danger of funding bad deals.

Lerner (1994a) finds that in the early investment rounds experienced venture capitalists tend to syndicate only with venture firms that have similar experience. He argues that, if a venture capitalist were looking for a second opinion, then he would want to get one from someone of similar or greater ability, certainly not from someone of lesser ability.

The advice and support provided by venture capitalists is often embodied in their role on the firm's board of directors. Lerner (1995) examines whether venture capitalists' representation on the boards of the private firms in their portfolios is greater when the need for oversight is larger, looking at changes in board membership around the replacement of CEOs. He finds that an average of 1.75 venture capitalists are added to the board between financing rounds when a firm's CEO is replaced in the interval; between other rounds 0.24 venture directors are added. No differences are found in the addition of other outside directors.

Hochberg (2005) studies the influence of venture capitalists on the governance of a firm following its initial public offering (IPO). Venture-backed firms manage earnings less in the IPO year, as measured by discretionary accounting accruals. Venture-backed firms also experience a stronger wealth effect when they adopt a poison pill, which implies that investors are less worried that the poison pill will entrench management at the expense of shareholders. Finally, venture-backed firms more frequently have independent boards and audit and compensation committees, as well as separate CEOs and chairmen.

So far, this section has highlighted the ways in which venture capitalists can successfully address agency problems in portfolio firms. During periods when the amount of money flowing into the industry grows dramatically, however, competition between venture groups can introduce distortions. This is shown in Figure 3, which shows relationship between venture returns and the amount invested in these funds. The returns are measured by the Sand Hill Index, which is a value-weighted and continuously invested index of the value of venture funded companies from their first round of institutional funding to their exit.[10] The money invested series (also from Sand Hill Econometrics) is the total dollars invested in the companies in the Sand Hill Index each month.

Gompers and Lerner (2000) examine the relation between the valuation of venture deals and inflows into venture funds. Doubling inflows leads to a 7–21% increase in valuation levels. But success rates do not differ significantly between investments made during periods of low inflows and valuations on the

[10] See the Sand Hill Econometrics Web site for details on the construction of this index. http://www.sandhillecon.com

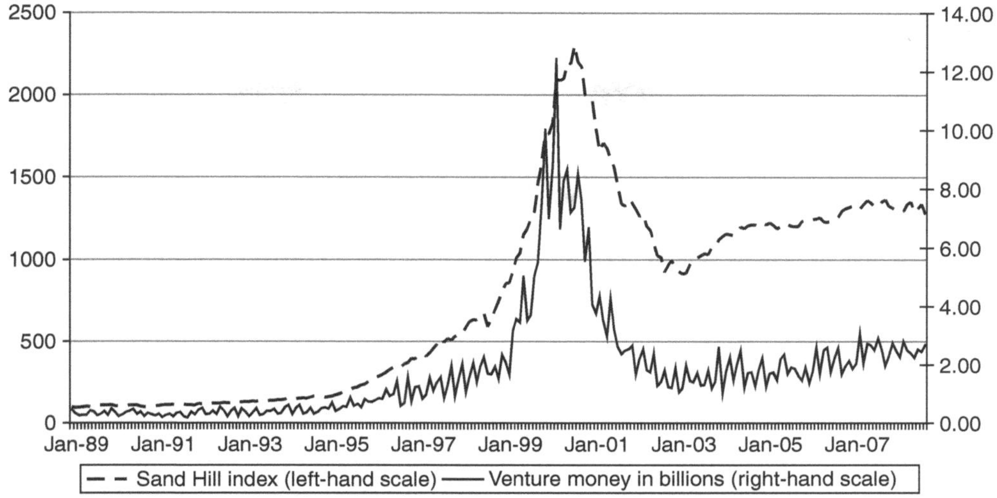

Figure 3. Relationship between venture returns and the amount invested. Source: Sand Hill Econometrics Website (2009).

one hand and those made in booms on the other. The results indicate that the price increases reflect increasing competition for investment, rather than changes in the expected returns.

5.2. Exiting

A third major area of research has been the process whereby venture funds exit investments. This topic is important because, in order to make money on their investments, venture capitalists must sell their equity stakes.

Initial research into the exiting of venture investments focused on IPOs, reflecting the fact that the most profitable exit opportunity is usually an IPO. Barry et al. (1990) and Megginson and Weiss (1991) document that venture capitalists hold significant equity stakes and board positions in the firms they take public, which they continue to hold a year after the IPO. They argue that this pattern reflects the certification they provide to investors that the firms they bring to market are not overvalued. Moreover, they show that venture-backed IPOs have less of a positive return on their first trading day, a finding that has been subsequently challenged (Kraus, 2002; Lee and Wahal, 2004). The authors suggest that investors need a smaller discount because the venture capitalist has certified the offering's quality.

Subsequent research has examined the timing of the exit decision. Several potential factors affect when venture capitalists choose to bring firms public. Lerner (1994b) examines how the valuation of public securities affects whether and when venture capitalists choose to finance companies in another private round in preference to taking the firm public. He shows that investors tend to take the firm public when the market value is high, relying on private financings when valuations are lower. Seasoned venture capitalists appear more proficient at timing IPOs. This finding is consistent with the work by Brown, Fazzari, and Petersen on the importance of public equity financing of R&D during the 1990s stock market boom.

Another consideration may be the venture capitalist's reputation. Gompers (1996) argues that young venture firms have incentives to "grandstand," or take actions that signal their ability to potential investors. Specifically, young venture firms bring companies public earlier than older one to establish a reputation and successfully raise new funds. Gompers shows that the effect of recent IPOs on the amount of capital raised is stronger for young venture firms, providing them with greater incentives to bring companies public earlier.

Lee and Wahal (2004) propose a variant of the "grandstanding" hypothesis: they posit that venture firms have an incentive to underprice IPOs. The publicity surrounding a successful offering will enable the venture group to raise more capital than it could otherwise. Lee and Wahal confirm this hypothesis by showing a positive relationship between first-day returns and subsequent fund raising by venture firms.

The typical venture firm, however, does not sell its equity at the time of the IPO. After some time, venture capitalists usually return money to their limited partners by transferring the shares to their investors, who are free either to hold or sell them. Gompers and Lerner (1998a) examine these distributions. After significant increases in stock prices prior to distribution, abnormal returns around the time of the distribution are negative. Cumulative excess returns for the 12 months following the distribution also appear to be negative. While the overall level of venture capital returns does not exhibit abnormal returns relative to the market (Brav and Gompers, 1997), there is a distinct rise and fall around the time of the stock distribution. The results are consistent with venture capitalists possessing inside information and with the (partial) adjustment of the market to that information.

A related research area is venture-fund performance. Kaplan and Schoar (2005) show substantial performance persistence across consecutive venture funds with the same general partners. General partners that outperform the industry in one fund are likely to outperform in the next fund, while those who underperform in one fund are likely to underperform with the next fund. These results contrast with those of mutual funds, where persistence is difficult to identify.

Cochrane (2005) estimates the returns of venture capital investments. He notes that many analyses of returns focus only on investments that go public, get acquired, or go out of business. Such calculations may produce biased returns by concentrating only on the portfolio's "winners" and outright failures, ignoring those firms that remain within the fund for longer periods. Cochrane develops a maximum likelihood estimate that uses existing data, but adjusts for these selection biases. While these papers—as well as Gompers and Lerner (1997) and Jones and Rhodes-Kropf (2003)—represent a first step toward understanding these issues, much more work remains to be done in this area.

5.3. Venture fund raising

Finally, research into the formation of venture funds has focused on two topics. First, the commitments to the venture capital industry have been highly variable since the mid-1970s. Understanding the determinants of this variability has been a topic of continuing interest to researchers. Second, the structure of venture partnerships has attracted increasing attention.

First, Poterba (1987, 1989) notes that the fluctuations could arise from changes in either the supply of or the demand for venture capital. It is very likely, he argues, that decreases in capital gains tax rates increase commitments to venture funds, even though the bulk of the funds are from tax-exempt investors. The drop in the tax rate may spur corporate employees to become entrepreneurs, thereby increasing the need for venture capital. The increase in demand due to greater entrepreneurial activity leads to more venture fund raising.

Gompers and Lerner (1998b) find empirical support for Poterba's claim: lower capital gains taxes have particularly strong effects on venture capital supplied by tax-exempt investors. This suggests that the primary mechanism by which capital gains tax cuts affect venture fund raising is the higher demand of entrepreneurs for capital. The authors also find that a number of other factors influence venture fund raising, such as regulatory changes and the returns of venture funds.

A second line of research has examined the contracts that govern the relationship between investors (limited partners) and the venture capitalist (general partner). Gompers and Lerner (1999) find that compensation for older and larger venture capital organizations is more sensitive to performance than that of other venture groups. Also, the cross-sectional variation in compensation terms for younger, smaller venture organizations is considerably lower. The fixed component of compensation is higher for smaller, younger funds and funds focusing on high-technology or early-stage investments. Finally, Gompers and Lerner do not find any relationship between the incentive compensation and performance.

The authors argue that these results are consistent with a learning model in which neither the venture capitalist nor the investor knows the venture capitalist's ability. With his early funds, the venture capitalist will work hard even without explicit pay-for-performance incentives: if he can establish a good reputation, he can raise subsequent funds. These reputation concerns lead to lower pay for performance for smaller and younger venture organizations. Once a reputation has been established, explicit incentive compensation is needed to induce the proper effort.

Covenants also play an important role in limiting conflicts in venture partnerships. Their use may be explained by two hypotheses. First, because negotiating and monitoring covenants are costly, they will be employed when monitoring is easier and the potential for opportunistic behavior is greater. Second, in the short run the supply of venture capital services may be fixed, with a modest number of funds of carefully limited size raised each year. Increases in demand may lead to higher prices when contracts are written. Higher prices may include not only increases in monetary compensation, but also greater consumption of private benefits through fewer covenants.

Gompers and Lerner (1996) show that both supply and demand conditions and costly contracting are important in determining contractual provisions. Fewer restrictions are found in funds established during years with greater capital inflows and funds, when general partners enjoy higher compensation. The evidence illustrates the importance of general market conditions on the restrictiveness of venture partnerships. In periods when venture capitalists have relatively more bargaining power—for instance, when there is a big increase in the funds being invested in venture funds—the venture capitalists are able to raise money with fewer strings attached.

5.4. *The globalization of venture capital*

While financial economists know much more about venture capital than they did a decade ago, there are many unresolved issues. We highlight here three promising areas, beginning with the globalization of the industry.

The rapid growth in the United States venture capital market has led institutional investors to look increasingly at private equity alternatives abroad. To date, however, outside of the United Kingdom

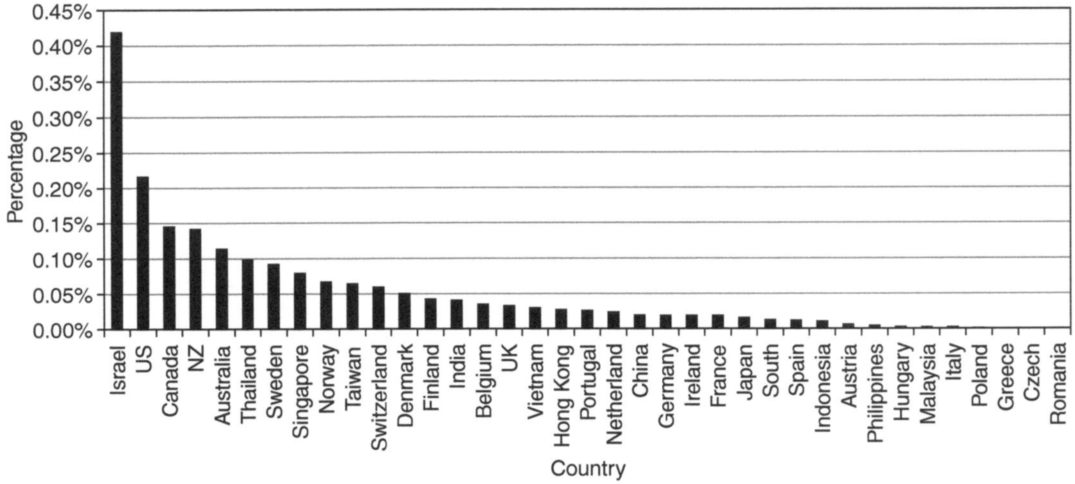

Figure 4. Percentage of venture investment over GDP various countries and regions.

(where performance of funds has been quite poor), Israel, Canada, and New Zealand, there has been little venture capital activity abroad. Figure 4 shows venture capital as a share of GDP in 2007 for a number of countries.[11] Black and Gilson (1998) argue that the key source of the US competitive advantage in venture capital is the existence of a robust IPO market. Venture capitalists can commit to transfer control back to the entrepreneur when a public equity market for new issues exists. This commitment device is unavailable in economies dominated by banks, such as Germany and Japan.

The rapid growth in the US venture capital market has led institutional investors to look abroad. In a pioneering study, Jeng and Wells (2000) examine the factors that influence venture fund raising internationally. They find that the strength of the IPO market is an important determinant of venture commitments, supporting Black and Gilson's hypothesis that the key to a successful venture industry is the existence of robust IPO markets. Jeng and Wells find, however, that the IPO market does not influence commitments to early-stage funds as much as those to later-stage ones. Much more remains to be explored regarding the internationalization of venture capital. Certainly, with a few exceptions such as Australia, China, India, and Japan, venture capital remains focused on the United States, as Figure 5 illustrates. Relative to the size of their GDP share, the European Union countries have almost no seed and startup funding when compared to the rest of the developed world.

A related question is why other financial intermediaries (such as banks) cannot duplicate these features of the venture capitalists, and undertake the same sort of monitoring. Economists have

[11] One potential source of confusion is that the term venture capital is used differently different in Europe and Asia. Abroad, venture capital often refers to all private equity, including buyout, late stage, and mezzanine financing (which represent the vast majority of the private equity pool in most overseas markets). In the US, these are separate classes. The data in Figures 3 and 4 are corrected for this fact and we confine our discussion of international trends to venture capital using the restrictive, US definition.

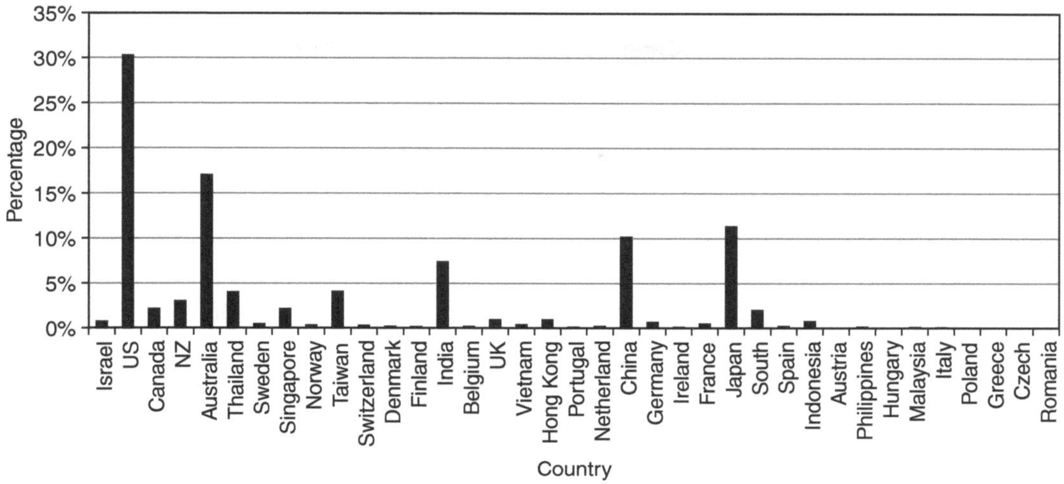

Figure 5. Country share of worldwide seed and startup venture capital funding in 2007.

suggested several explanations for the apparent superiority of venture funds in this regard. First, because regulations limit banks' ability to hold shares, at least in the United States, they cannot freely use equity. Second, banks may not have the necessary skills to evaluate projects with few collateralizable assets and significant uncertainty. Finally, venture funds' high-powered compensation schemes give venture capitalists incentives to monitor firms closely. Banks sponsoring venture funds without high-powered incentives have found it difficult to retain personnel.

5.5. The real effects of venture capital

A second area is even thornier: the impact of venture capital on the economy. While theorists have suggested a variety of mechanisms by which venture capital may affect innovation, the empirical record is more mixed. It might be thought that establishing a relationship between venture capital and innovation would be straightforward. For instance, one could look in regressions across industries and time whether, controlling for R&D spending, venture capital funding has an impact on various measures of innovation. But even a simple model of the relationship between venture capital, R&D, and innovation suggests that this approach is likely to give misleading estimates.

Both venture funding and innovation could be positively related to a third unobserved factor, the arrival of technological opportunities. Thus, there could be more innovation at times that there was more venture capital, not because the venture capital caused the innovation, but rather because the venture capitalists reacted to some fundamental technological shock which was sure to lead to more innovation. To date, only a handful of chapters have attempted to address these challenging issues.

The first of these papers, Hellmann and Puri (2000), examines a sample of 170 recently formed firms in Silicon Valley, including both venture-backed and nonventure firms. Using questionnaire responses, they find empirical evidence that venture capital financing is related to product market strategies and outcomes of startups. They find that firms that are pursuing what they term an innovator strategy (a classification based on the content analysis of survey responses) are significantly more likely to obtain venture capital and also obtain it more quickly. The presence of a venture capitalist is also associated with a significant reduction in the time taken to bring a product to market, especially for innovators. Furthermore, firms are more likely to list obtaining venture capital as a significant milestone in the lifecycle of the company as compared to other financing events.

The results suggest significant interrelations between investor type and product market dimensions, and a role of venture capital in encouraging innovative companies. Given the small size of the sample and the limited data, they can only modestly address concerns about causality. Unfortunately, the possibility remains that more innovative firms select venture capital for financing, rather than venture capital causing firms to be more innovative.

Kortum and Lerner (2000), by way of contrast, examine whether these patterns can be discerned on an aggregate industry level, rather than on the firm level. They address concerns about causality in two ways. First, they exploit the major discontinuity in the recent history of the venture capital industry: as discussed above, in the late 1970s, the US Department of Labor clarified the Employee Retirement Income Security Act, a policy shift that freed pensions to invest in venture capital. This shift led to a sharp increase in the funds committed to venture capital. This type of exogenous change should identify the role of venture capital, because it is unlikely to be related to the arrival of entrepreneurial opportunities. They exploit this shift in instrumental variable regressions. Second, they use R&D expenditures to control for the arrival of technological opportunities that are anticipated by economic actors at the time, but that are unobserved to econometricians. In the framework of a simple model, they show that the causality problem disappears if they estimate the impact of venture capital on the patent-R&D ratio, rather than on patenting itself.

Even after addressing these causality concerns, the results suggest that venture funding does have a strong positive impact on innovation. The estimated coefficients vary according to the techniques employed, but on average a dollar of venture capital appears to be three to four times more potent in stimulating patenting than a dollar of traditional corporate R&D. The estimates therefore suggest that venture capital, even though it averaged less than 3% of corporate R&D from 1983 to 1992, is responsible for a much greater share—perhaps 10%—of US industrial innovations in this decade. These findings have been supported by recent working paper by Mollica and Zingales (2007), who also use an instrumental variable approach based on state pension fund resources to look at the relationship of venture capital and innovation and find a strong relationship.

Some of the most interesting theoretical work in recent years has focused not on the question of whether venture capitalists spur innovation, but rather on the societal consequences of the relationship between venture-backed entrepreneurship and innovation. Landier (2006) presents a model in which entrepreneurial venture either succeed or fail on the basis of ability and luck.[12] He argues that as the venture progresses, the entrepreneur is likely to learn about the likely eventual success of the venture, but that the decision to continue or abandon the venture will not be the same in all environments. In particular, the decision depends critically on how expensive it would be to raise capital for a new venture from investors after a failure. In this setting, Landier shows, multiple equilibria can arise. If the

[12] See also Scharfstein and Gromb (2002) for a thoughtful theoretical analysis that touches on many of these issues.

cost of capital for a new venture after a failure is not very high, entrepreneurs will be willing to readily abandon ventures, and failure is commonplace but not very costly. Alternatively, if the cost of capital for failed entrepreneurs is high, only extremely poor projects will be abandoned. Thus, societies may differ dramatically in the prevalence of experimentation in high-risk, innovative ventures. But certainly, given the fact that even the question of whether venture capitalists make private returns which compensate them for the risk that they take on is controversial (Kaplan and Schoar, 2005), it is premature to conclude what the social returns are.

5.6. Government funding for startup firms

One provocative finding from Jeng and Wells's analysis is that government policy can dramatically affect the health of the venture sector. Researchers have only begun to examine the ways in which policymakers can catalyse the growth of venture capital and the companies in which they invest (Avnimelech and Teubal, 2004; Gilson, 2003; Irwin and Klenow, 1996; Lerner, 1999; Wallsten, 2000). Clearly, much more needs to be done in this arena.

Examples of such programs are the US Small Business Investment Company (SBIC) and Small Business Innovation Research (SBIR) programs. Together, these programs disbursed $2.4 billion in 1995, more than 60% of the amount from venture capital in that year (Lerner, 1998). In Germany, more than 800 federal and state government financing programs have been established for new firms in the recent past (OECD, 1995). In 1980, the Swedish established the first of a series of investment companies (along with instituting a series of measures such as reduced capital gains taxes to encourage private investments in startups), partly on the US model. By 1987, the government share of venture capital funding was 43% (Karaomerliolu and Jacobsson, 1999). Recently, the United Kingdom has instituted a series of government programs under the Enterprise Fund umbrella which allocate funds to small- and medium-sized firms in high technology and certain regions, as well as guaranteeing some loans to small businesses (Bank of England, 2001). There are also programs at the European level.

A limited amount of evidence, most of it US based, exists as to the effectiveness and "additionality" of these programs (see Lerner, 2009 for a review of the key programs and their evaluations). In most cases, evaluating the success of the programs is difficult due to the lack of a "control" group of similar firms that do not receive funding.[13] Therefore, most of the available studies are based on retrospective survey data provided by the recipients; few attempt to address the question of performance under the counterfactual seriously. A notable exception is the study by Lerner (1999), who looks at 1435 SBIR awardees and a matched sample of firms that did not receive awards, over a 10-year postaward period. Because most of the firms are privately held, he is unable to analyze the resulting valuation or profitability of the firms, but he does find that firms receiving SBIR grants grow significantly faster than the others after receipt of the grant. He attributes some of this effect to "quality certification" by the government that enables the firm to raise funds from private sources as well.[14]

[13] See Jaffe (2002) for a review of methodologies for evaluation such government programs. For a complete review of the SBIR program, including some case studies, see the National Research Council (2002).

[14] Also see Spivack (2001) for further studies of such programs, including European studies, and David et al. (2000) and Klette et al. (2000) for surveys of the evaluation of government R&D programs in general.

A series of papers by Czarnitzki and coauthors (Aerts and Czarnitzki, 2006; Almus and Czarnitzki, 2003; Czarnitzki and Hussinger, 2004) have looked at the performance of firms that receive public R&D subsidies in several European countries such as Belgium and Germany, using treatment effect analysis. They generally find that such subsidies do not completely displace private expenditure on R&D (i.e., they are *additional*) and that they are productive in the sense that they result in patenting by the firm. Hall and Maffioli (2008) survey a similar set of results for large Latin American economies and reach a more nuanced conclusion.

6. Conclusions

Based on the literature surveyed here, what do we know about the costs of financing R&D investments and the possibility that some kind of market failure exists in this area? Several main points emerge:

First, there is fairly clear evidence, based on theory, surveys, and empirical estimation, that small and startup firms in R&D-intensive industries face a higher cost of capital than their larger competitors and firms in other industries. In addition to compelling theoretical arguments and empirical evidence, the mere existence of the VC industry and the fact that it is concentrated precisely where these startups are most active suggests that this is so. The fact that ex post venture returns may lag the market, however, remains a puzzle and makes a clear-cut conclusion more complex.

Second, the evidence for a financing gap for large and established R&D firms is harder to establish. It is certainly the case that these firms prefer to use internally generated funds for financing investment, but less clear that there is an argument for intervention, beyond the favorable tax treatment that currently exists in many countries.[15]

Third, the VC solution to the problem of financing innovation has its limits: First, it does tend to focus only on a few sectors at a time, and to make investments with a minimum size that is too large for startups in some fields. Second, good performance of the VC sector requires a thick market in small and new firm stocks (such as NASDAQ) in order to provide an exit strategy for early stage investors. Introducing a VC sector into an economy where it is not already present is nontrivial as it requires the presence of at least three interacting institutions: investors, experienced venture fund managers, and a market for IPOs.

Fourth, the effectiveness of government incubators, seed funding, loan guarantees, and other such policies for funding R&D deserves further study, ideally in an experimental or quasiexperimental setting. In particular, studying the cross-country variation in the performance of such programs would be desirable, because the outcomes may depend to a great extent on institutional factors that are difficult to control for using data from within a single country.

Based on the survey of the literature presented here, other areas of interest for future research appear to be worthwhile. A longstanding debate in the literature is over the interaction between corporate governance and corporate finance and its impact on long-term investment, including investment in intangibles such as R&D. Although in principle one might have thought that financial markets focused

[15] It is important to remind the reader of the premise of this chapter: we are focusing *only* on the financing gap arguments for favorable treatment of R&D and ignoring (for the present) the arguments based on R&D spillovers and externalities. There is good reason to believe that the latter is a much more important consideration for large established firms, especially if we wish those firms to undertake basic research that is close to industry but with unknown applications (the Bell Labs model).

on quarterly performance, such as those in the Anglo-Saxon economies, would discourage such investment, this appears not to be the case, at least in the United States. However, for several large European countries, we have limited evidence that the required rate of return to R&D investment is perhaps somewhat lower than in the United States and the United Kingdom, especially when the firm has a large majority shareholder (see Chapter 24, this volume). This fact suggests that for these firms at least, the stability provided by concentrated ownership may encourage R&D. At the same time, the more fluid financial markets with active markets for corporate control seem to be better at financing new entrants, startups, and more overall investment in innovation. The future challenge is to understand more completely the interaction of financial market discipline with various forms of corporate governance and how this influences the organization and performance of innovation.

Acknowledgments

We are grateful to Bruce Petersen for a careful reading of the manuscript. Josh Lerner thanks the Harvard Business School's Division of Research for financial support. This chapter is based on material in Hall (2002) and Gompers and Lerner (2004).

References

Admati, A., Pfleiderer, P. (1994). "Robust financial contracting and the role for venture capitalists". Journal of Finance 49, 371–402.
Aerts, K., Czarnitzki, D. (2006). "The impact of public R&D—Funding in Flanders". Brussels, Belgium: IWT Study No. 54.
Aghion, P., Howitt, P. (1997). Endogenous Growth Theory. The MIT Press, Cambridge.
Akerlof, G.A. (1970). "The market for 'Lemons': Quality, uncertainty, and the market mechanism". Quarterly Journal of Economics 84, 488–500.
Alam, P., Walton, K.S. (1995). "Information asymmetry and valuation effects of debt financing". Financial Review 30 (2), 289–311.
Alderson, M.J., Betker, B.L. (1996). "Liquidation costs and accounting data". Financial Management 25 (2), 25–36.
Almus, M., Czarnitzki, D. (2003). "The effects of public R&D subsidies on firms' innovation activities: The case of East Germany". Journal of Business and Economic Statistics 21 (2), 226–236.
Amit, R., Glosten, L., Muller, E. (1990). "Entrepreneurial ability, venture investments, and risk sharing". Management Science 36, 1232–1245.
Anton, J.J., Yao, D.A. (1998). "The sale of intellectual property: Strategic disclosure, property rights, and incomplete contracts". Working paper, The Wharton School, University of Pennsylvania.
Arellano, M., Bond, S. (1991). "Some tests of specification for panel data: Monte Carlo evidence and an application to employment equations". Review of Economic Studies 58, 277–297.
Arrow, K.J. (1962). "Economic welfare and the allocation of resources for invention". In: Nelson, R. (Ed.), The Rate and Direction of Inventive Activity. Princeton, New Jersey.
Auerbach, A.J. (1984). "Taxes, firm financial policy, and the cost of capital: An empirical analysis". Journal of Public Economics 23, 27–57.
Avnimelech, G., Teubal, M. (2004). "Emergence and development of venture capital in Israel and the role of policy: A macro/microeconomic perspective". In: Bartzokas, A., Mani, S. (Eds.), Financial Systems, Corporate Investment in Innovation and Venture Capital. Edward Elgar, London.
Bank of England. (2001). Finance for Small Firms—An Eighth Report. Domestic Finance Division, Bank of England, London.

Barry, C., Muscarella, C., Peavy, J., III, Vetsuypens, M. (1990). "The role of venture capital in the creation of public companies: Evidence from the going public process". Journal of Financial Economics 27, 447–471.

Bergemann, D., Hege, U. (1998). "Venture capital financing, moral hazard, and learning". Journal of Banking and Finance 22, 703–735.

Berglöf, E. (1994). "A control theory of venture capital finance". Journal of Law, Economics, and Organizations 10, 247–267.

Bhagat, S., Welch, I. (1995). "Corporate research and development investments: International comparisons". Journal of Accounting and Economics 19, March–May.

Bhattacharya, S., Ritter, J.R. (1983). "Innovation and communication: Signaling with partial disclosure". Review of Economic Studies 50, 331–346.

Black, B., Gilson, R. (1998). "Venture capital and the structure of capital markets: Banks versus stock markets". Journal of Financial Economics 47, 243–277.

Blair, M.M., Litan, R.E. (1990). Corporate Leverage and Leveraged Buyouts in the Eighties. Brookings Institution, Washington, DC.

Blass, A.A., Yosha, O. (2001). "Financing R&D in mature companies: An empirical analysis". Working Paper, Bank of Israel, Tel Aviv University, and CEPR.

Bond, S., Harhoff, D., Van Reenen, J. (1999). "Investment, R&D, and financial constraints in Britain and Germany". Institute of Fiscal Studies Working Paper No. 99/5, London.

Bougheas, S., Goerg, H., Strobl, E. (2001). Is R&D Financially Constrained? Theory and Evidence from Irish Manufacturing. University of Nottingham, Nottingham.

Brav, A., Gompers, P. (1997). "Myth or reality? Long-run underperformance of initial public offerings; evidence from venture capital and nonventure capital-backed IPOs". Journal of Finance 52, 1791–1821.

Brown, W. (1997). R&D Intensity and Finance: Are Innovative Firms Financially Constrained? London School of Economics Financial Market Group, London.

Brown, J.R., Petersen, B.C. (2009a). "Why has the investment-cash flow sensitivity declined so sharply? Rising R&D and equity market developments". Journal of Banking and Finance 33, 971–984.

Brown, J.R., Petersen, B.C. (2009b). Cash Holdings and R&D Smoothing. University of Montana and Washington University, St. Louis.

Brown, J.R., Fazzari, S.M., Petersen, B.C. (2009). "Financing innovation and growth: Cash flow, external equity, and the 1990s R&D boom". Journal of Finance LXIV (1), 151–185.

Campbell, J.Y., Lo, A.W., MacKinlay, A.C. (1997). The Econometrics of Financial Markets. Princeton University Press, Princeton, New Jersey, USA.

Chan, Y. (1983). "On the positive role of financial intermediation in allocation of venture capital in a market with imperfect information". Journal of Finance 38, 1543–1568.

Cho, S. (1992). "Agency costs, management stockholding, and research and development expenditures". Seoul Journal of Economics 5 (2), 127–152.

Chung, K.H., Wright, P. (1998). "Corporate policy and market value: A *q* theory approach". Review of Quantitative Finance and Accounting 11 (3), 293–310.

Cochrane, J. (2005). "The risk and return of venture capital". Journal of Financial Economics 75, 3–52.

Cornelli, F., Yosha, O. (2003). "Stage financing and the role of convertible debt". Review of Economic Studies 70, 1–32.

Czarnitzki, D., Hussinger, K. (2004). "The link between R&D subsidies, R&D spending, and technological performance". Mannheim, Germany: ZEW Discussion Paper No. 04–56.

David, P.A., Hall, B.H., Toole, A.A. (2000). "Is public R&D a complement or substitute for private R&D? A Review of the Econometric Evidence". Research Policy 29, 497–529.

Eng, L.L., Shackell, M. (2001). "The implications of long term performance plans and institutional ownership for firms' research and development investments". Journal of Accounting, Auditing and Finance 16 (2), 117–139.

Fazzari, S.M., Hubbard, R.G., Petersen, B.C. (1988). "Financing constraints and corporate investment". Brookings Papers on Economic Activity 1988 (1), 141–205.

Fazzari, S.M., Hubbard, R.G., Petersen, B.C. (2000). "Investment-cash flow sensitivities are useful: A comment on Kaplan and Zingales". Quarterly Journal of Economics 115, 695–705.

Francis, J., Smith, A. (1995). "Agency costs and innovation: Some empirical evidence". Journal of Accounting and Economics 19 (2/3), 383–409.

Friend, I., Lang, L.H.P. (1988). "An empirical test of the impact of management self-interest on corporate capital structure". Journal of Finance 43, 271–283.

Gilson, R.J. (2003). "Engineering a venture capital market: Lessons from the American experience". Stanford Law Review 55, 1067.

Gompers, P. (1995). "Optimal investment, monitoring, and the staging of venture capital". Journal of Finance 50, 1461–1489.

Gompers, P. (1996). "Grandstanding in the venture capital industry". Journal of Financial Economics 42, 133–156.

Gompers, P., Lerner, J. (1996). "The use of covenants: An empirical analysis of venture partnership agreements". Journal of Law and Economics 39, 463–498.

Gompers, P., Lerner, J. (1997). "Risk and reward in private equity investments: The challenge of performance assessment". Journal of Private Equity 1, 5–12.

Gompers, P., Lerner, J. (1998a). "What drives venture fundraising?" Brookings Papers on Economic Activity-Microeconomics 149–192.

Gompers, P., Lerner, J. (1998b). "Venture capital distributions: Short- and long-run reactions". Journal of Finance 53, 2161–2183.

Gompers, P., Lerner, J. (1999). "An analysis of compensation in the U.S. venture capital partnership". Journal of Financial Economics 51, 3–44.

Gompers, P., Lerner, J. (2000). "Money chasing deals? The impact of fund inflows on private equity valuations". Journal of Financial Economics 55, 281–325.

Gompers, P., Lerner, J. (2004). The Venture Capital Cycle. MIT Press, Cambridge, Massachusetts.

Griliches, Z. (1992). "The search for R&D spillovers". Scandinavian Journal of Economics 94, S29–S47.

Hall, B.H. (1990). "The impact of corporate restructuring on industrial research and development". Brookings Papers on Economic Activity 1990 (1), 85–136.

Hall, B.H. (1991). "Firm-Level Investment with Liquidity Constraints: What Can the Euler Equations Tell Us?" NBER and the University of California at Berkeley. http://elsa.berkeley.edu/~bhhall/papers/BHH95%20euler.pdf.

Hall, B.H. (1992). "Research and development at the firm level: Does the source of financing matter?" NBER Working Paper No. 4096, June.

Hall, B.H. (1993). "R&D tax policy during the eighties: Success or failure?" Tax Policy and the Economy 7, 1–36.

Hall, B.H. (1994). "Corporate capital structure and investment horizons in the United States, 1976–1987". Business History Review 68, 110–143.

Hall, B.H. (1996). "The private and social returns to research and development". In: Smith, B. L. R., Barfield, C.E. (Eds.), Technology, R&D, and the Economy. Brookings Institution and the American Enterprise Institute, Washington, DC, pp. 140–183.

Hall, B.H. (2002). "The financing of research and development". Oxford Review of Economic Policy 18 (1), 35–51.

Hall, B. H. (2005). "Measuring the Returns to R&D: The Depreciation Problem". Annales d'Economie et de Statistique N° 79/80.

Hall, B.H., Maffioli, A. (2008). "Evaluating the impact of technology development funds in emerging economies: Evidence from Latin America". European Journal of Development Research 20 (2), 172–198.

Hall, B.H., Van Reenen, J. (2000). "How effective are fiscal incentives for R&D? A new review of the evidence". Research Policy 29, 449–469.

Hall, B.H., Griliches, Z., Hausman, J.A. (1986). "Patents and R&D: Is there a lag?" International Economic Review 27, 265–283.

Hall, B.H., Mairesse, J., Branstetter, L., Crepon, B. (1999). "Does cash flow cause investment and R&D: An exploration using panel data for French, Japanese, and United States firms in the scientific sector". In: Audretsch, D., Thurik, A.R. (Eds.), Innovation, Industry Evolution and Employment. Cambridge University Press, Cambridge, UK.

Harhoff, D. (1998). "Are there financing constraints for R&D and investment in German manufacturing firms?" Annales d'Economie et de Statistique 49 (50), 421–456.

Hellmann, T. (1998). "The allocation of control rights in venture capital contracts". RAND Journal of Economics 29, 57–76.

Hellmann, T., Puri, M. (2000). "The interaction between product market and financing strategy: The role of venture capital". Review of Financial Studies 13, 959–984.

Himmelberg, C.P., Petersen, B.C. (1994). "R&D and internal finance: A panel study of small firms in high-tech industries". Review of Economics and Statistics 76, 38–51.

Hochberg, Y. (2005). "Venture capital and corporate governance in the newly public firm". Unpublished working paper, Northwestern University.

Hsu, D. (2004). "What do entrepreneurs pay for venture capital affiliation?" Journal of Finance 59, 1805–1844.

Irwin, D., Klenow, P. (1996). "High tech R&D subsidies: Estimating the effects of Sematech". Journal of International Economics 40, 323–344.
Jaffe, A. (2002). "Building programme evaluation into the design of public research-support programmes". Oxford Review of Economic Policy 18 (1), 22–34.
Jeng, L., Wells, P. (2000). "The determinants of venture funding: Evidence across countries". Journal of Corporate Finance 6, 241–289.
Jensen, M.C., Meckling, W. (1976). "Theory of the firm: Managerial behavior, agency costs, and ownership structure". Journal of Financial Economics 3, 305–360.
Johnson, M.S., Rao, R.P. (1997). "The impact of antitakeover amendments on corporate financial performance". Financial Review 32 (4), 659–689.
Jones, C., Rhodes-Kropf, M. (2003). "The price of diversifiable risk in VC and private equity". Working paper, Columbia University.
Kaplan, S., Schoar, A. (2005). "Private equity performance: Returns, persistence, and capital". Journal of Finance 60, 1791–1823.
Kaplan, S., Strömberg, P. (2003). "Financial contract theory meets the real world: An empirical analysis of venture capital contracts". Review of Economic Studies 70, 281–315.
Kaplan, S., Zingales, L. (1997). "Do investment-cash flow sensitivities provide useful measures of financing constraints?'". Quarterly Journal of Economics 112, 169–215.
Karaomerliolu, D.C., Jacobsson, S. (1999). The Swedish Venture Capital Industry—An Infant, Adolescent, or Grown-up? Chalmers Institute of Technology, Goteborg, Sweden.
Klette, T.J., Moen, J., Griliches, Z. (2000). "Evaluating the effects of government-sponsored, commercial R&D: microeconomic studies". Research Policy 29, 471–495.
Kortum, S., Lerner, J. (2000). "Assessing the contribution of venture capital to innovation". RAND Journal of Economics 31, 674–692.
Kraus, T. (2002). "Underpricing of IPOs and the certification role of venture capitalists: evidence from Germany's Neuer Markt". Working Paper, University of Munich.
Lach, S., Schankerman, M. (1988). "Dynamics of R&D and investment in the scientific sector". Journal of Political Economy 97 (4), 880–904.
Landier, A. (2006). "Entrepreneurship and the stigma of failure". Unpublished working paper, New York University.
Lee, P., Wahal, S. (2004). "Grandstanding, certification and the underpricing of venture capital backed IPOs". Journal of Financial Economics 73, 375–407.
Leland, H.E., Pyle, D.H. (1977). "Informational asymmetries, financial structure, and financial intermediation". Journal of Finance 32, 371–387.
Lerner, J. (1994a). "Venture capitalists and the decision to go public". Journal of Financial Economics 35, 293–316.
Lerner, J. (1994b). "The syndication of venture capital investments". Financial Management 23, 16–27.
Lerner, J. (1995). "Venture capitalists and the oversight of private firms". Journal of Finance 50, 301–318.
Lerner, J. (1998). "'Angel' financing and public policy: An overview". Journal of Banking and Finance 22, 773–783.
Lerner, J. (1999). "The government as venture capitalist: The long-run effects of the SBIR program". Journal of Business 72, 285–318.
Lerner, J. (2009). The Boulevard of Broken Dreams: Why Public Efforts to Boost Entrepreneurship and Venture Capital Have Failed—And What to Do About It. Princeton University Press, Princeton.
Lerner, J., Sorensen, M., Stromberg, P. (2008). "Private equity and long-run investment: The case of innovation". Working Paper No. 14623, National Bureau of Economic Research.
Levin, R.C., Klevorick, A.K., Nelson, R.R., Winter, S.G. (1987). "Appropriating the returns from industrial research and development". Brookings Papers on Economic Activity 1987 (3), 783–832.
Mairesse, J., Mulkay, B., Hall, B.H. (1999). "Firm-Level Investment in France and the United States: An Exploration of What We Have Learned in Twenty Years". Annales d'Economie et de Statistique No. 55–56, 27–69.
Majumdar, S.K., Nagarajan, A. (1997). "The impact of changing stock ownership patterns in the United States: Theoretical implications and some evidence". Revue d'Economie Industrielle 82, 39–54.
Mansfield, E., Schwartz, M., Wagner, S. (1981). "Imitation costs and patents: An empirical study". Economic Journal 91, 907–918.
Marx, L. (1994). "Negotiation and renegotiation of venture capital contracts". Working paper, University of Rochester.
Megginson, W., Weiss, K. (1991). "Venture capital certification in initial public offerings". Journal of Finance 46, 879–893.

Miller, M.H., Modigliani, F. (1961). "Dividend policy, growth, and the valuation of shares". Journal of Business 34, 411–433.

Modigliani, F., Miller, M.H. (1958). "The cost of capital, corporation finance and the theory of investment". American Economic Review 48, 261–297.

Mollica, M., Zingales, L. (2007). "The impact of venture capital on innovation and the creation of new business". Unpublished working paper, University of Chicago.

Mulkay, B., Hall, B.H., Mairesse, J. (2001). "Investment and R&D in France and in the United States". In: Deutsche Bundesbank, J. (Ed.), Investing Today for the World of Tomorrow. Springer Verlag, Berlin and Heidelberg.

National Research Council. (2002). SBIR Program Diversity and Assessment Challenges. The National Academies Press, Washington, DC.

Nelson, R.R. (1959). "The simple economics of basic scientific research". Journal of Political Economy 49, 297–306.

OECD. (1995). Venture Capital in OECD Countries. Organization for Economic Cooperation and Devlopement, Paris.

Opler, T.C., Titman, S. (1993). "The determinants of leveraged buyout activity: Free cash flow vs. financial distress costs". Journal of Finance 48 (5), 1985–1999.

Opler, T.C., Titman, S. (1994). "Financial distress and corporate performance". Journal of Finance 49 (3), 1015–1040.

Poterba, J. (1987). "How burdensome are capital gains taxes? Evidence from the United States". Journal of Public Economics 33, 157–172.

Poterba, J. (1989). "Venture capital and capital gains taxation". In: Summer, L. (Ed.), Tax Policy and the Economy. MIT Press, Cambridge, MA.

Pugh, W.N., Jahera, J.S. Jr., Oswald, S. (1999). "ESOPs, takeover protection, and corporate decision making". Journal of Economics and Finance 23 (2), 170–183.

Romer, P.M. (1986). "Increasing returns and long run growth". Journal of Political Economy 94 (5), 1002–1037.

Sahlman, W. (1990). "The structure and governance of venture capital organizations". Journal of Financial Economics 27, 473–524.

Sand Hill Econometrics. "Benchmarking the returns to venture". (2009). Available at http://www.sandhillecon.com/pdf/SandHillWhitePaper.pdf.

Scharfstein, D. S., Gromb, D. (2002). "Entrepreneurship in Equilibrium". NBER Working Paper No. w9001.

Scherer, F.M. (1998). "The size distribution of profits from innovation". Annales d'Economie et de Statistique 49 (50), 495–516.

Schumpeter, J. (1942). Capitalism, Socialism, and Democracy. Harper and Row, New York (reprinted 1960).

Spivack, R.N. (2001). The Economic Evaluation of Technological Change. Conference Proceedings of the Advanced Technology Program. National Institute of Standards and Technology, Washington, DC.

Szewczyk, S.H., Tsetsekos, G.P., Zantout, Z.Z. (1996). "The valuation of corporate R&D expenditures: Evidence from investment opportunities and free cash flow". Financial Management 25 (1), 105–110.

Wallsten, S. (2000). "The effects of government-industry R&D programs on private R&D: The case of the Small Business Innovation Research program". RAND Journal of Economics 31, 82–100.

Williamson, O.E. (1988). "Corporate finance and corporate governance". Journal of Finance 43, 567–591.

Zantout, Z.Z. (1997). "A test of the debt monitoring hypothesis: The case of corporate R&D expenditures". Financial Review 32 (1), 21–48.

Chapter 15

THE MARKET FOR TECHNOLOGY

ASHISH ARORA* *AND* ALFONSO GAMBARDELLA†

**Fuqua Business School*

Duke University, Durham

North Carolina, USA

†*Department of Management and KITeS*

Bocconi University

Milan, Italy

Contents

Handbooks in Economics, Volume 01

DOI: 10.1016/S0169-7218(10)01015-4

Abstract

This chapter reviews the growing literature on the "market for technology," a broad term that denotes trade in technology disembodied from physical goods. The market for technology flourished during the nineteenth century in the United States. After several decades of relative decline, the market for technology has once again grown considerably in recent years, although the growth is uneven across sectors and across countries. Thus far, the literature has paid most attention to the supply of technology, and on the efficiency of market transactions in technology. A key contribution has been that the decision of firms to license depends on whether the revenues from licensing are higher than the rent-dissipation effect produced by increased competition in the licensor's product markets. The literature has featured several factors that condition the tradeoff between licensing revenue and rent dissipation. For instance, general-purpose technologies enable the potential licensors to sell technology in product markets distant from the product operations of the licensors, and thus are more likely to be licensed. Another stream of research has focused on the factors, such as intellectual property protection, that condition the efficiency of licensing contracts. The study of the demand for external technology is less developed, and is an open area for future research. Another exciting area for future research is the relationship between the product market and the market for technology, of which a special but important case is the division of labor between technology specialists such as biotech firms, and their customers downstream, in this instance, pharmaceutical firms. The area in the most urgent need of attention is research on the consequences of the market of technology, on the rate and direction of inventive activity, and on productivity growth. This will also require a deeper understanding of the microfoundations of the market for technology.

Keywords

division of labor, high-tech industries, markets for technology, patents, R&D

JEL classification: O3, L24, L26, M2

1. Introduction

A market for technology can yield important benefits. Trade in general expands the division of labor; trade in technology facilitates a division of labor in innovation. A division of labor yields the economies of learning and larger scale emphasized by Adam Smith, as well as a superior allocation of resources based on comparative advantage. An inventor need not acquire all the assets required to commercialize the invention and can instead license it to another firm better positioned to bring the innovation to market.[1] As well, a market for technology can lower entry barriers and increase competition in downstream product markets. Finally, in a world where commercialization is costly and slow, a market for technology diffuses technology more rapidly and increases productivity.

In this paper, we use the term "market for technology" in a broad sense. Strictly speaking, market transactions are arm's length, anonymous, and typically involve the exchange of a good for money. Most transactions for technology probably lack at least one of these criteria. For example, they may involve detailed contracts and be embedded within interfirm alliances, thus not be strictly anonymous, nor arm's length. A different perspective on markets analogizes them to centralized exchanges, including exchanges for trading contracts. Roth (2008) argues that well-functioning markets must be thick (many buyers and sellers), uncongested (each party can deal with many others on the opposite side), and safe (transacting outside or engaging in strategic behavior should not be profitable). The market for technology, at least as we know it, also fails the Roth test (Gans and Stern, 2010).

Imperfect though it might be, the market for technology has grown in recent years. Specific empirical estimates are discussed in greater detail below, but two empirical regularities are noteworthy. First, the market for technology has grown steadily in size since the mid-1980s. This is shown by the increase in annual licensing and royalty payments, the rise in the percentage of startups intending to license as a way to derive profit from some or all of their inventions, and the growing number of firms and organizations that specialize as intermediaries in the market for technology.[2]

Although the growth in the market for technology over the past quarter of a century marks a change over the relatively quiescent period that preceded it, this is not a secular trend. A series of papers, by Naomi Lamoreaux, Kenneth Sokoloff, and colleagues has demonstrated the existence of a vibrant market for patents and patent licensing in America during the mid- and late nineteenth century. However, by the early part of the twentieth century, patent licensing began to diminish. Winder (1995) describes the widespread use of licensing of inventions in harvesting machinery in North America in the late nineteenth century, but also notes that licensing diminished after the 1880s. International licensing was also found to be more important in the nineteenth century. Important inventions, such as the ammonia soda process patented by Ernst Solvay in 1861, were licensed extensively internationally. However, foreign direct investment by multinational corporations appears

[1] Lamoreaux and Sokoloff (1996) show that growth in patent assignments before grant, their measure of trade in patents, coincided with growth in specialization in invention.

[2] These intermediaries include firms such as yet2.com, which runs an online site where technologies can be traded, Oceantomo, which runs online patent auctions, Intellectual Ventures, which acquires patent portfolios and contracts with inventors to develop inventions and technologies, and IP Bewertung, which provides several similar services in Europe. A yet different type of intermediary includes financial firms, such as Royalty Pharma, that acquire interests in future royalty streams.

to have been the dominant mode of international technology flows in the twentieth century (see also Chapter 3, Vol. 2).

Second, the market for technology is much more extensive in North America, and some limited evidence produced by Khan and Sokoloff (2004) suggest that this exists even during the nineteenth century. Figure 1, taken from Khan and Sokoloff (2004) indicates that over three quarters of patents granted in America in the 1870s and 1880s were assigned (indicating that the patent was traded), whereas fewer than one-third of patents in the United Kingdom were assigned or licensed. Since many more patents were granted in America, this gap is even more noteworthy. More recent data discussed below suggest that a gap, although perhaps smaller, remains.

These trends raise a number of related questions. Why has trade in technology been so limited in the twentieth century and what has caused the apparent growth since the 1980s? Why did a flourishing market for technology in nineteenth century America more or less vanish, only to rise again more than three quarters of a century later? Why is the market for technology more extensively developed in America than elsewhere? And finally, when do technology markets matter for the rate and direction of technical activity, for the evolution of industries, or for the rate of productivity growth? Any proposed answers must address the fundamental questions about the nature and functioning of the market for technology, namely who participates in them, under what conditions, and with what consequences.

We begin by clarifying what we mean by *markets for technology* in the next section. Section 3 reviews the microfoundations of the market for technology—why companies license technology and the factors that condition their demand for external technology. Section 4 provides some estimates of the size of the market for technology. Section 5 reviews the literature on the factors that condition the efficiency—and

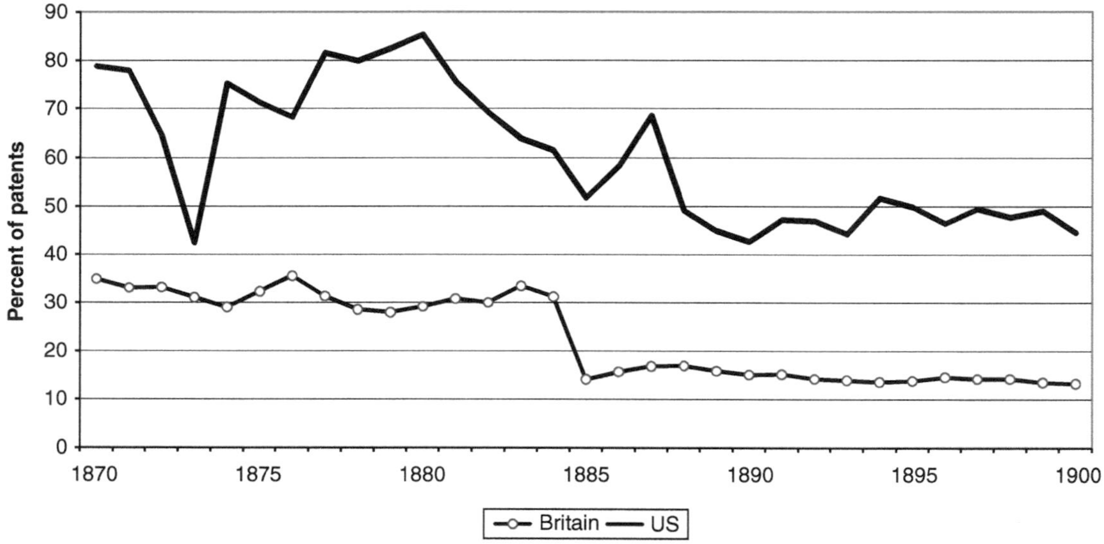

Figure 1. The market for inventions. The figure shows the percentage of patents assigned for the US, and patents assigned or licensed for the UK. Source: Khan and Sokoloff (2004).

hence the extent—of markets for technology, with particular focus on the role of intellectual property protection. Section 6 discusses the division of innovative labor that a market for technology can make possible. Section 7 concludes by highlighting unresolved questions and topics for further research.

It is also important to delineate some topics that we shall not discuss in this chapter. We shall not analyze university licensing. Though scholars have used it to examine issues related to licensing more broadly (e.g., Jensen and Thursby 2004; Mowery et al., 2001; Thursby and Thursby, 2002), the literature on university licensing is more closely related to how licensing does and does not comport with the objectives of the university. (See also Foray and Lissoni's (2010), this volume). Space constraints also preclude coverage of the extensive literature on R&D joint ventures and technology alliances. Finally, we shall only touch upon the literature on international technology licensing, mainly because it has been extensively covered in a number of places (see, for instance, Arora et al., 2008; Hoekman et al., 2005). Cross-licensing and other antitrust aspects of technology licensing are not covered for the same reason (see, for instance, Gilbert and Shapiro, 1997).

2. The market for technology: Definition and scope of our analysis

Technology comes in very different forms, and no general definition will fit. We will not define technology, treating it instead as an imprecise term for useful knowledge, rooted in engineering and science, which usually also draws on practical experience from production. Technology can take the form of "intellectual property" (e.g., patents), or intangibles (e.g., a software program, a design), or it can be embodied in a product (e.g., a prototype, a device like a chip designed to perform certain operations), or it can be a technical service.

The way technology is traded reflects the peculiar nature of technology as an economic asset. While pure forms of licenses (e.g., patent licensing or licensing of chip designs) are common, technology transfer is also frequently accompanied by the transfer of associated artifacts and know-how. In other cases, the supplier–buyer relationship is an R&D or codevelopment contract. The buyer may have to invest effort and resources to shape the technology to its needs (i.e., codevelopment), or fund the research of a liquidity constrained technology supplier.

Technology can also be exchanged through joint ventures and through the acquisition of firms. We exclude here these modes of interfirm technology flow.[3] Acquisitions, and to a lesser extent joint ventures, involve issues specific to the market for *firms*. Thus, though we shall contrast market transactions with processes within the firm, it is not to dispute the existence of hybrid forms but to sharpen the exposition. We also distinguish between ex-ante contracts (i.e., contracts for R&D) and ex-post contracts (i.e., contracts for existing technology). The distinction is especially important from a transaction cost perspective, since ex-ante contracting potentially creates greater contracting problems.[4]

[3] Interfirm movement of technology can also occur through labor mobility, which we also ignore.

[4] Barring Mowery's study of contract R&D firms and their decline (Mowery, 1984), the empirical literature on contract R&D is limited. Mowery emphasizes the need for potential buyers of R&D services to have considerable in-house capability. He also notes that if contracts are incomplete, the buyer becomes increasingly vulnerable to opportunistic behavior as the R&D supplier progressively acquires more buyer-specific knowledge. Arora and Merges (2004) emphasize the reverse; as the buyer learns the supplier's know-how, it renders the supplier vulnerable to holdup.

Table 1
A simple typology of markets for technology

	Existing technology	Future technology or component for future
Horizontal market/ transactions with actual or potential rivals	Union carbide licensing unipol polyethylene technology to huntsman chemicals	Sun licensing Java to IBM; R&D partership between rivals (e.g., see Hagedoorn, 2002)
Vertical market/licensing to nonrivals	Licensing of IP Core in semiconductors	R&D agreements or other technological alliances; Affymax licensing combinatorial drug discovery technology to pharmaceutical companies

In sum, a market for technology refers to transactions for the use or creation of technology. It includes transactions ranging from full technology packages (patents and other intellectual property, along with know-how and services) to bare-bones patent licensing. It also includes transactions involving knowledge that is not patented but embodied in artifacts such as designs, software, or technical services. It can involve parties in the same product markets or vertically related suppliers and buyers, and the contracts involved can vary in simplicity and design. It can involve the transfer of existing knowledge or contracts for the creation of new knowledge. Most of the literature reviewed below, both theoretical and empirical, focuses on some subset of the market for technology.

Table 1 summarizes our definition of the markets for technology in the form of a simple two-by-two typology, along with canonical examples for each case. Technologies can be sold to firms in the same product-market (horizontal transactions) or to firms operating downstream (vertical markets). The market for technology can involve existing technologies that are licensed, or it can be the market for contract R&D and associated alliances, more properly thought of as the market for "future" technologies, sometimes called the "market for innovation."

3. The microfoundations: Why do companies license?

3.1. Gains from trade

The literature has tended to separate analysis of why firms choose to license out and license-in technology. We follow this division here. However, the conceptual starting point is with the gains from trade. Gains from trade in technology have three sources. First and foremost, technology is "infinitely expansible," to use the term coined by Dasgupta and David (1994). Simply put, it is a good thing if one does not have to reinvent the wheel. Thus, expanding the use of technology will create gains which have to be balanced against the potential loss due to the decreased exclusivity of access. This aspect is particularly salient (and well understood as such) in international technology licensing, and in the discussion of general-purpose technologies (GPT).[5]

[5] In passing, we note that this point is more commonly discussed using the related concept of nonrivalry. However, in most cases of interest, technology is in fact a rival good because exclusive access to it is more valuable than access shared with others. Even when it is a rival good, however, technology can be infinitely expansible in the sense that a wheel does not have to be reinvented.

The second source of gains from trade is comparative advantage. As discussed in the context of a division of labor, sometimes the inventor of a technology is not best equipped to develop or commercialize it. Engaging in commercialization may even retard innovation, by diverting attention and changing the nature of the organization.[6] Licensing to another firm with a comparative advantage in manufacturing and marketing will yield gains to both parties.

The third source of gains is more obvious. For instance, a firm may develop a technology that it does not wish to use but which is applicable elsewhere, and can gainfully license it (or sell it). Some licensing is undoubtedly of this nature, but it does not require much explanation. There are few studies that explicitly take a "gains-from-trade" approach to analyzing the market for technology. Instead, most studies analyze either why a firm licenses its technology to others, or, less frequently, when a firm uses external technology (in-licensing).

3.2. Supply: Determinants of technology licensing

The literature has analyzed a variety of reasons for firms to license their technology. The early literature on licensing focused on the optimal licensing behavior of the monopolist inventor once it has developed and patented a new technology or production process (see Gallini and Wright, 1990; Kamien and Tauman, 1986). Katz and Shapiro (1986) analyze the optimal number of licensees for a single technology holder who does not compete in the product market. Rockett (1990) develops a model where the technology holder also produces the product but faces entry after its patents expire. He also shows that a technology holder will optimally license an inefficient potential entrant to foreclose entry by a more efficient firm. Gallini (1984) also provides a model where licensing is strategically used to deter entry.

In addition, firms license as parts of standard-setting bodies or to promote their technology as a dominant standard (see, e.g., Shapiro, 2000). Firms may choose to license some technology to provide incentives to potential adopters. For instance, Corts (2000) provides a model where a firm may optimally commit to innovate by licensing the production of the ancillary product to another firm, even when licensees are inefficient. The intuition is that innovation may require substantial redesign of the ancillary product, entailing costs that an integrated firm will internalize. When potential adopters have to coinvest for an innovation to be successful, an integrated firm may be tempted to free-ride on their investment. Knowing this, potential adopters are reluctant to coinvest. A firm can credibly commit to innovate, therefore, by licensing to other producers of the ancillary products. Similarly, Shepard (1987) shows that firms may license to enhance demand, in essence protecting potential buyers against having to deal with a monopolist supplier.

[6] Lamoreaux and Sokoloff (2005, p. 17) relate the story of Elmer Perry, who started The Sperry Electric Light, Motor, and Car Brake Company in 1883, to commercialize his dynamo. "Although the company launched Sperry's career as an inventor, it left him little time and energy for creative pursuits. Indeed, the 19 patents he applied for during his 5 years with the company amounted collectively to half his *annual* average over a career as an inventor that stretched from 1880 to 1930."

3.2.1. Licensing revenue versus rent-dissipation effects

The foregoing papers have usually assumed a single technology holder and that the technology holder is also the monopoly producer of the good. They ignore competition among technology holders and also typically ignore the very likely situation that the technology holder competes with other producers in the product market. These simplifying assumptions imply that licensing is typically not profitable, but instead can only be attractive to serve some other strategic purpose. However, the example of firms such as Texas Instruments, IBM, and Union Carbide, which earned millions of dollars from licensing technology, points to the possibility that even large, well-established, firms may directly profit from their technology by licensing it, rather than merely embody it in their own output.

Arora and Fosfuri (2003) develop a framework to understand the decision of firms to sell technology, and how product market and technology market competition condition this decision. In their model, multiple technology holders compete, both in the technology market and in the product market. Technologies are not perfect substitutes for each other, and neither are the goods produced from the technology. In deciding whether to license or not, the technology holder has to balance the revenue from licensing and the rent-dissipation effect produced because licensing will increase product-market competition. As a result, factors that enhance licensing revenue or that reduce rent dissipation will encourage licensing.

This tradeoff depends upon competition in the product market. If the licensee operates in a "distant" market, rent dissipation is small compared to when the licensee is "nearby." For example, the licensee may operate in a geographical market in which the licensor finds it costly to operate, for example, because the licensor does not have the complementary downstream assets. Similarly, the technology could be used for a different type of product that the licensor may not produce. Arora and Fosfuri note that product-market competition enhances licensing because rent dissipation falls faster than licensing revenues as product market competition increases. Indeed, as is well known, a monopolist will not license. Consistent with this, Lieberman (1989) finds that licensing was less common in concentrated chemical products, and the limited licensing that did take place was by outsiders (nonproducers and foreign firms).

Arora and Fosfuri also point out that licensing is more likely when products are homogeneous rather than differentiated. If products are differentiated, a licensee is closer in the product space to the licensor than to other producers, so that the rent dissipation felt by the licensor is greater than if the product is homogenous. Put differently, by licensing, a technology holder imposes a greater negative (pecuniary) externality on other producers when the product is homogenous. Consistent with this, Fosfuri (2006) finds that licensing is lower in markets where technology-specific product differentiation is high.

The Arora–Fosfuri framework also implies that smaller firms are more likely to license, because they suffer less from the rent-dissipation of additional competitors. The logic is apparent in the extreme case in which the licensor has no stakes in the downstream markets, and thus has no product-market rents to worry about. This is also consistent with the observation that technology suppliers often do not produce in the product markets for which they supply technology, as is the case in biotechnology (Arora and Gambardella, 1990), semiconductors (Hall and Ziedonis, 2001), software security (Giarratana, 2004), and chemical engineering (Arora and Gambardella, 1998). This implication is also consistent with Teece (1986,1988) in that control of downstream assets makes licensing less likely. The point is confirmed by McGahan and Silverman (2006), Ford and Ryan (1981), and more recently by Kollmer and Dowling (2004), who show that licensing is less likely if firms have downstream assets. Similarly, Fosfuri (2006) finds a negative effect of downstream assets on licensing in chemicals.[7]

This is exemplified by the different ways in which BP Chemicals approached acetic acid and polyethylene licensing in the 1980s. In acetic acid, BP Chemicals had strong proprietary technology, but licensed very selectively, typically only in markets it would otherwise be unable to enter. By contrast, in polyethylene, BP had less than 2% of the market. Although BP had good proprietary technology as well, there were several other sources of polyethylene technology. Accordingly, BP licensed its polyethylene technology very aggressively, competing with Union Carbide which was the market leader in licensing polyethylene technology.

By relaxing the assumption of a single technology holder, Arora and Fosfuri (2003) point to the importance of competition among technology suppliers. For instance, BP initially tried not to license even polyethylene technology in Western Europe, where it had a substantial share of polyethylene capacity. However, other licensors continued to supply polyethylene technology to Western Europe, resulting in BP losing potential licensing revenue without any benefits in the form of restraining entry. BP's response was to also offer its technology for license. The direct implication is that the market for technology feeds on itself: competition from one technology holder promotes licensing by others.

3.2.2. Licensing decisions in the long-run

Gambardella and Giarratana (2009) generalize the Arora and Fosfuri framework by emphasizing the interplay between the generality of the technology and the fragmentation of the product markets. Generality of the technology makes it attractive to "distant" user firms, which implies that revenues from licensing can be earned from firms in product markets different from that of the technology holder. Because the markets are distant in product space, the rent dissipation is small, which raises the incentives to license.

Gambardella and Giarratana (2009) jointly consider both the licensing decision and the decision on the range of product markets that the technology holder will enter. The key assumption is that technology can be deployed in more product markets than is profitable for the technology holder to serve directly. The contrast between the generality of technology and the narrowness of product market assets is significant. Several scholars have observed that firms frequently "know more than they make" (Brusoni et al., 2001; Gambardella and Torrisi, 1998), suggesting that technology has broader economies of scope than marketing and manufacturing assets, which creates opportunities for licensing. This logic applies a fortiori to GPT, which are so broadly applicable that few firms are likely to exploit all applications.

In the longer run, the decision to supply technologies depends upon the market for downstream assets involved in the commercial application of technology. The interactions between the two can lead to complex patterns, as illustrated by the history of licensing in farm machinery in the United States between 1850 and 1910 (Winder, 1995). Winder (1995) shows that in the 1850s there was considerable technology licensing in this industry even though a typical harvester had many different components,

[7] However, firm size also comes with broad scope of activities, and thus the relationship between size of the firm and probability to license out is U-shaped: small firms and large firms are more likely to license out their patented inventions, a finding also reported by Zuniga and Guellec (2008) and Motohashi (2008). Larger firms may be more likely to develop technologies in which they have limited interest, or operate in markets where they face competition from other licensors.

each individually protected by patents. Fragmentation of the product market, due to high transport costs, meant that innovators would typically license technology to producers in geographically distinct markets. The result was that there were many producers in a market, and a typical harvester embodied innovations from multiple innovators.

Over time, this modular system changed to one that was like many vertical silos, with competition between silos, but with licensing still prevalent within silos (e.g., type 1 harvester had many components, with different firms producing this type of harvester licensing designs and technologies to each other, but not to producers of type 2 harvesters). By the 1890s, this licensing regime disappeared and product markets consolidated, with a couple of dominant producers controlling both the technology and production.

Winder (1995) links the disappearance of the licensing regime to changes in technology (steel instead of iron, which mean that small foundries could no longer produce parts and larger scale, steel-using, factories were required), which in turn meant that small machinery producers had higher costs. However, Winder's explanation ignores the reductions in transport costs and greater integration of hitherto geographically distinct markets, which were likely very important. As markets integrate, reducing the "distance" between markets, the incentives for larger scale production are enhanced and the incentives to license are reduced. Put differently, the asymmetry between the scope implied by technology and that implied by the production and marketing capabilities of the firm diminished, reducing the gains from trade from licensing. Additional support can be found in Lamoreaux and Sokoloff (2005), who note that as US market integrated in the latter part of the nineteenth century, independent inventors that had hitherto sold multiple licenses for their invention, while also manufacturing for their local market, were forced to either license to a single firm or contemplate manufacturing for the entire national market.

Modeling the interaction between the product market and the technology market, plus the possible coevolution of the two, is an area ripe for additional research. Given the daunting complexity of theoretical models, simulation-based models may provide useful insights (see, for instance, Malerba et al., 2008). Focusing on the long-run, decisions regarding entry into product markets and technology markets naturally leads to the literature on specialization and division of labor, which we cover in Section 6.

3.3. Demand

The demand for technology licenses has received less attention in the literature compared to the willingness or desires of firms to license. We ignore factors that condition the demand for technology in general, and focus on the factors that condition the demand for external technology.

One situation in which firms license external technology is when their internal efforts do not bear fruit (or the firm did not invest in research in the first instance). For instance, Higgins and Rodriguez (2006) show that pharmaceutical firms with thinner product pipelines were more likely to acquire external technology. This perspective, though undoubtedly correct, is also limited. Technology differs from conventional goods in an important but underappreciated respect: Knowledgeable buyers of technology are at a marked advantage compared to buyers that lack such knowledge. This means that buyers have to be technically sophisticated themselves, so that the demand for technology may be confined to small subset of firms, at least until the technology itself becomes highly standardized.

3.3.1. Absorptive capacity

It is now standard in the literature to refer to the notion of "absorptive capacity" to mean that the ability of a firm to use technology depends on its internal technical competence. Cohen and Levinthal (1989) develop a model in which this internal competence is related to whether (and how much) the firm conducts R&D internally. There is no licensing in Cohen and Levinthal's model, and the external technology is absorbed through spillovers from the research of other firms. However, the idea of absorptive capacity can be applied quite directly in developing a firm-level demand for technology. In a similar spirit, Rosenberg (1990) asks "Why firms do basic research (with their own money)?" He notes that an important reason for making these investments, despite the low levels of private appropriability of basic research, is that by performing basic research firms are better equipped to understand knowledge produced by others.[8]

Arora and Gambardella (1994b) develop these ideas further. They distinguish between "ability to utilize" and "ability to evaluate." The ability to utilize denotes the ability of a firm to extract value from the technology, and requires technical competence as well as downstream assets such as manufacturing and marketing. The ability to evaluate denotes the ability of the firm to judge the value of the technology. This is a second dimension of absorptive capacity, which is more closely related to the technical and scientific capability of the firm. While both these dimensions of absorptive capacity increase the value that the firm can extract from external technology, they have different implications for the demand for external technology. Arora and Gambardella (1994b) show that firms with greater ability to utilize will demand more external technologies (i.e., more likely to license). However, firms with higher ability to evaluate will demand fewer external technologies, even though the expected value for the technologies that they demand will be higher. The intuition for this result is that technology acquisition is like purchasing a real option, in which the licensing fees paid to acquire a technology are substantially smaller than the investments in development, manufacturing, and marketing to use the technology. Firms that are better able to judge will optimally acquire fewer options.

3.3.2. Internal R&D and the demand for technology: Other considerations

Internal R&D has another, more obvious, impact on the demand for external technology. Consistent with Mowery's observations about the danger of buyers of contract R&D services becoming "locked in" to their technology suppliers, Gans and Stern (2000) develop a model where the potential buyer engages in R&D to increase bargaining power in licensing negotiations (see also Ulset, 1996). Insofar as internal efforts are successful, this will reduce the demand for external technology. Sometimes, learning how to use and maintain external technology may require as much effort as creating the technology itself, as is sometimes the case with software. In such cases, a firm may optimally choose to develop technology internally even

[8] Many studies use the idea of absorptive capacity, broadly defined. For instance, Forman et al. (2008) use data on almost 87,000 US establishments and look at their decision to adopt advanced Internet technologies. They find that establishments with a larger number of software programers are more likely to adopt the technology. However, when the establishment is located in large cities, the effect of internal programers on adoption is smaller. In other words, internal programers are complementary to external technology, but less so in bigger cities, perhaps because larger cities offer greater possibilities of using external software programers to adapt Internet technologies for the firm's needs. The point is that if firms want to buy the technology, they need to have internal competences in the broadly defined area of the technology.

when it is possible to license in external technology. When the internal R&D effort becomes significant enough, the firm may choose to develop the technology internally instead. Cohen and Kepler (1992) develop a model in which the benefits of investing in R&D are proportional to sales. Though they do not analyze licensing, the small firms in their model are better off licensing external technology while larger firms may prefer to develop technology internally. More broadly, Arora and Gambardella (1994a) note that because technology buyers are also likely to have internal R&D, the dynamics of technology markets are more complicated, which is a potentially fruitful area for future research.

Motivated by the "make-buy" perspective, in which internal R&D is a substitute for external technology (Pisano, 1990; Williamson, 1985), there are a number of studies that estimate the demand for licensing, usually as part of an effort to determine whether licensing is a substitute for internal R&D or not. For the most part, these studies find that internal R&D and licensing are complements rather than substitutes. For instance, Cassiman and Veugelers (2006) find that Belgian firms view R&D and external technology acquisition as complements. The complementarity is especially marked in firms that invest in basic research.

Firms may also choose not to license in technology for strategic reasons. Rotemberg and Saloner (1994) develop a model where a firm rationally chooses a Not-Invented-Here strategy of explicitly excluding external technology to provide incentives to its own employees to innovate. While there is much discussion of the Not-Invented-Here syndrome among practitioners and industry observers, to our knowledge there is little research in economics on the topic, the reasons why firms may be affected by it, and its consequences.[9]

4. The size of the market for technology

4.1. The world market for technology since the mid-1990s

Arora et al. (2001a) review studies that quantify the size of the market for technology in the 1990s. Despite different data sources and methods, the estimates provided by these studies are remarkably similar: In the mid-1990s, the annual value of transactions in the market for technology was $25–35 billion in the United States, and about $35–50 billion globally.

A survey by the British Technology Group, based on interviews of 133 R&D intensive firms and 20 universities in Europe, North America, and Japan, estimated that expenditures on technology licenses amounted to 12%, 5%, and 10% of the total R&D budgets, respectively, for each region. These percentages were also used to estimate the order of magnitude of the size of the markets for technology in each of the three regions. In 1996, OECD figures indicate that North America spent $027 billion on R&D, the European Union $132 billion, and Japan $83 billion. This implied that the size of the market for technology was approximately $25 billion in North America, $6.6 billion in Europe, and $8.3 billion in Japan, and would put the total world market for technology at about $40 billion in the mid-1990s.

[9] There is little doubt that here practice is far ahead of scholarship. For instance, a leading pharmaceutical firm, Glaxo, has explicitly declared that it will rely upon external technology for a significant fraction of its products in the future. The actual behavior of other pharmaceutical firms indicates that Glaxo is not an exception.

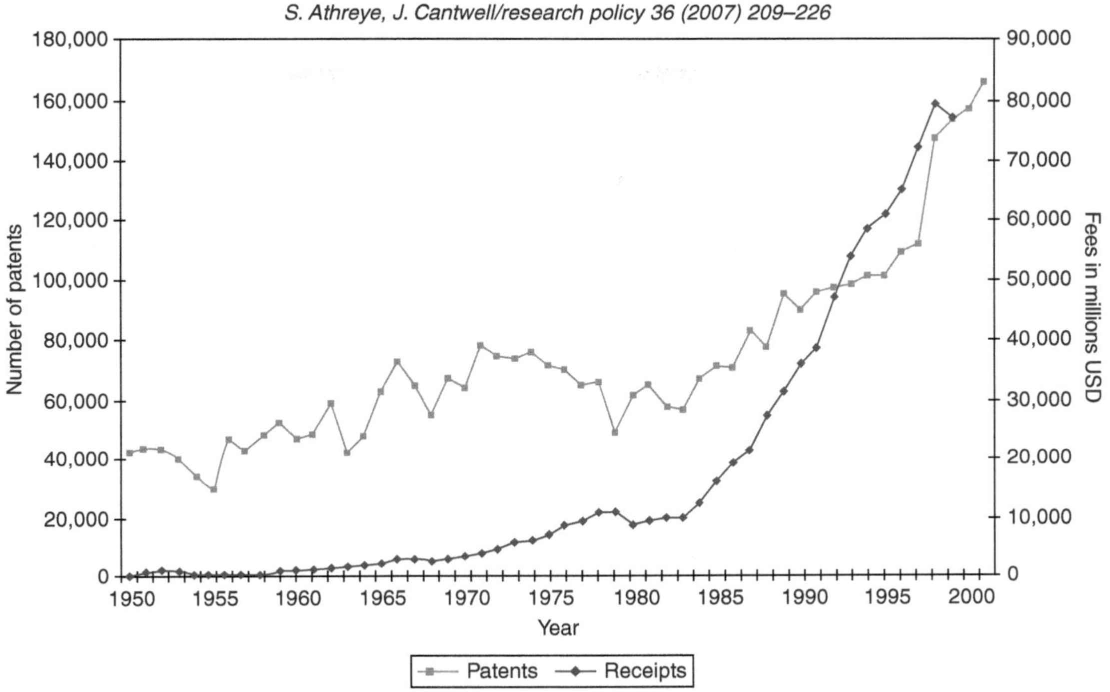

Figure 2. Growth in non-US held patents and worldwide royalty and license and revenues. Source: Athreye and Cantwell (2007).

Since Arora et al. (2001a), two additional estimates have been generated. Athreye and Cantwell (2007) analyzed trends over time in international royalty and licensing revenues worldwide between 1950 and 2003. For 1950–1970, they used the IMF Balance of Payments Yearbook and for 1970–2003 they used the World Development Indicators (WDI) database. Figure 2 reports their chart of the world licensing payments and receipts between 1950 and 2003. The estimates reported by Athreye and Cantwell tend to be on the higher end of spectrum. For example, they set the world market for technology at $55–60 billion in the mid-1990s. For 2000, they size the world market for technology at $90–100 billion.

The Athreye and Cantwell figures also indicate strong growth in the international flow of licensing fees and royalties. Adjusting for changes in coverage, we computed that royalty payments and receipts increased at 8.7% and 7.0% in 1980–1990 and 9.8% and 5.6% in 1990–2003, substantially higher than the growth rate of the world GDP, which was 3.3% on average for 1980–1990 and 2.8% for 1990–2003.[10]

The data on international royalty flows suggest that markets for technology have grown over the last two decades. However, there are two potentially offsetting effects. First, the bulk of these transactions

[10] See Table 4.1 of the World Development Indicators (2005).

may be among affiliated entities rather than market transactions. Data from the United States indicate that transactions among unaffiliated entities account for fewer than one-third of the licensing and royalty receipts of American firms. For instance, in 2007, the latest year for which data were available,[11] the total receipts of US firms from royalties and licensing fees for industrial processes and products amounted to $37.4 billion. Of this, $7.9 billion, or about 21%, came from unaffiliated entities. The share of unaffiliated transactions has fluctuated over the years, and no clear trend is discernable, which suggests that the cross-border market for technology is considerably smaller than the $100 billion reported by Athreye and Cantwell.

A second offsetting effect is that the figures for licensing fees and royalties used in Athreye and Cantwell (2007) include payments for packaged software, trademarks, and copyrights. Data from the United States suggest that although licensing and royalty receipts have grown strongly, at over 10% per annum on average, payments for industrial processes and products, which correspond mostly closely to the market for technology, have grown far more slowly. Correspondingly, the share of payments for industrial processes and products has steadily dropped, from around 70% in 1987 to 33% in 2007. However, Figure 3 shows that even accounting for these, cross-border flows of technology between unaffiliated parties has grown steadily.

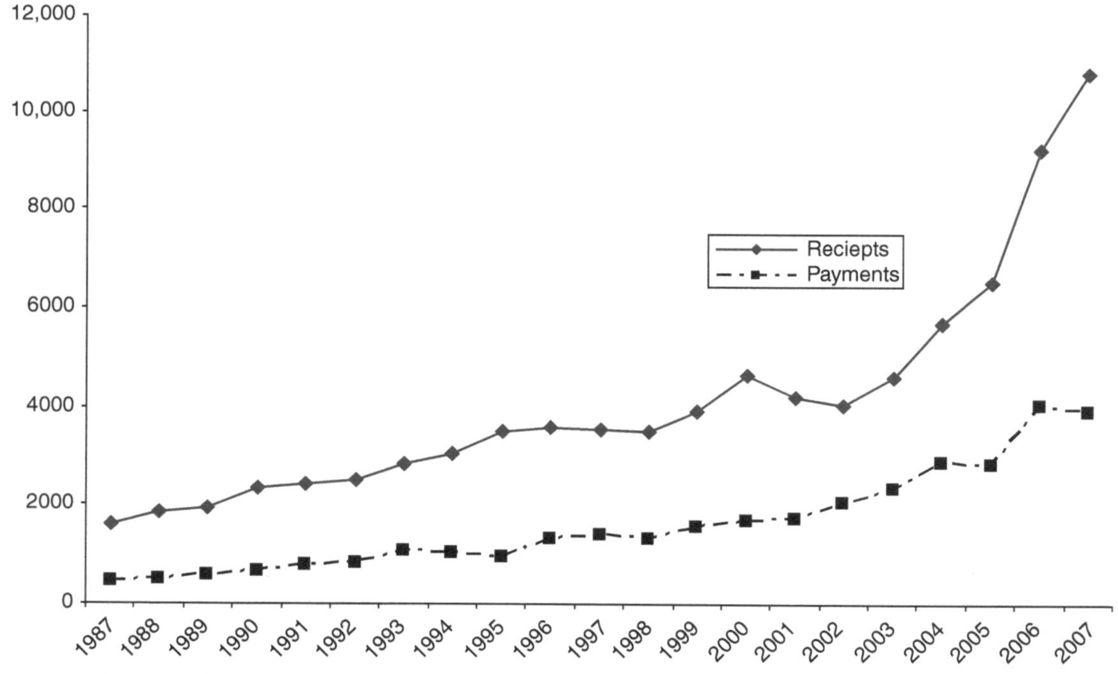

Figure 3. International licensing royalties for industrial processes, unaffiliated transactions only, United States, 1987–2007, $ millions. Source: Table 4.22. Royalties and License Fees, 2007, US http://www.bea.gov/international/intlserv.htm

[11] See Table 4.22. Royalties and License Fees (2007). US http://www.bea.gov/international/intlserv.htm

The most authoritative estimates of the size and growth of markets for technology, although only for the United States, are provided by Robbins (2006), based on confidential tax data. Robbins estimates that domestic income from licensing intellectual property was $92 billion in 2002. She follows Arora et al. (2001a) and assumes that the proportion of technology licensing, as opposed to licensing of trademarks, copyrights, and packaged software, is the same as that in cross-border transactions, which implies that licensing of industrial processes amounted to $66 billion.[12] Of this, about $50 billion was earned domestically, and the remaining was earned from overseas. If one assumes that the United States accounts for 60% of the global market for technology, this would imply that the global market for technology in 2002 was about $100 billion. Using the same method, Robbins produces estimates of $27.4 billion for 1995, $29.4 billion for 1996, and $31.8 billion for 1997 for US corporate supply of IP-licensing of industrial processes, which are very close to the estimates provided by Arora et al. (2001a) using transaction data. These imply a growth rate of about 13% per annum, somewhat faster than the growth rate estimated by Athreye and Cantwell.

A recent OECD survey confirms both that established firms have increased their propensity to license-in and to license-out new technologies, and that the market for technology is disproportionately larger in the United States (Sheehan et al., 2004). The survey, which was administered in 2003, covered 105 firms in Europe (68 firms), North America (20), and Asia-Pacific (17, mostly from Japan). Most firms were large—only 20% had fewer than 1000 employees. Almost 60% of the firms interviewed reported increased inward and outward licensing during the previous decade. Moreover, North-American and Japanese firms reported licensing more frequently than European firms, consistent with the findings of the British Technology Group survey discussed earlier.

In sum, the evidence suggests that markets for technology are of significant size and have grown over the last decade. They appear to be the most extensive in the United States, followed by Japan, with Europe lagging both. Undoubtedly, the robust economic growth over this period, particularly in information and communication technologies, and the huge growth in research and development expenditures in life sciences have contributed greatly to the growth of technology markets. Since 2002, ICT growth has slowed, as have investments in life sciences research and development. It is highly likely, therefore, that markets for technology have also grown more slowly since then, and perhaps even declined somewhat.

[12] Cockburn and Henderson (2003) asked 81 IP managers from a range of industries to estimate the value of IP assets. These estimates implied that patents, trade-secrets, and know-how account for about three quarter of the value of intellectual property, and trademarks and copyrights for 18% and 9%, respectively. If one believes that licensing of industrial processes involves licensing of patents, know-how, and trade-secrets, then Arora et al. (2001a) and Robbins (2006) effectively assume that the latter account for about 72% of all licensing royalties. In other words, the share of licensing of industrial processes in all licensing is remarkably close to the estimated share of patents, know-how, and trade-secrets in total intellectual property reported in Cockburn and Henderson.

4.2. Firm-level evidence

A 2003 OECD survey indicates that established companies worldwide are more likely to license-in and to license-out in the pharmaceutical, and the information and communications technology (ICT) industries (Sheehan et al., 2004). Licensing has been common in the chemical industry, at least since World War II (e.g., Anand and Khanna, 2000; Arora and Gambardella, 1998; Cesaroni, 2003). There is a large literature that has studied licensing between biotech firms and pharmaceutical companies (see, for instance, Gambardella, 1995). A more recent survey (Zuniga and Guellec, 2008) reports a U-shaped relationship, with both very small and very large firms indicating higher rates of out-licensing than firms in between. Zuniga and Guellec (2008) analyze a representative sample of patent-filing firms in 2007: 600 European and 1600 Japanese. The results show that patent licensing is widespread among patenting firms. Nearly a fifth of the European companies license patents to nonaffiliated partners, whereas more than a quarter do so in Japan.

Zuniga and Guellec (2008) find that among the European and Japanese firms that patent and license, a very large fraction of patents are licensed. For instance, nearly 50% of the European firms that did some licensing to unaffiliated parties report that they licensed more than 80% of their patent portfolio, while of Japanese firms that report some licensing to unaffiliated parties, around 40% claim to have licensed more than 80% of their portfolio. The survey further finds that although both cross-border licensing and cross-licensing are important, neither type of licensing accounts for all the licensing activity reported. Nearly, two-thirds of European and over 85% of the Japanese firms that license report that less than 20% of their licensing is cross-border. Nearly, 80% of European firms, and a slightly higher share of Japanese firms, report that less than 20% of their patents involved in licensing are cross-licensed. Thus, the licensing activity reported in this survey is more than simply cross-licensing and is further supported by the finding that over 40% of the European firms that license report that know-how transfer is involved in more than 20% of their licensing deals, and a third of the firms report that know-how transfer was involved in more than 40% of their licensing deals. Japanese firms appear to participate less intensively in patent licensing deals that also involve know-how transfer: only a quarter report that know-how transfer was involved in more than 20% of their licensing deals, and only one-sixth report that more than 40% of their licensing deals involved know-how transfer.

Furthermore, licensing activity appears to have increased between 2003 and 2006. Of the European firms reporting licensing in 2006, about 45% reported an increase in licensing revenues or the number of licensing deals, although only 8% reported a dramatic increase in either. Only 3% of the firms reported a decrease, with most (slightly more than 50%) indicating no change.

Overall, the data indicate that licensing transactions have increased since the mid-1990s, with some evidence that non-American firms are catching up with their American rivals. The data also indicate that, though substantial in absolute value, licensing as an activity is still not central to the innovation process, although with some notable exceptions such as biopharmaceuticals. Nor, once again with notable exceptions such as chemicals and petroleum refining, is licensing the dominant form of technology flows across firms. These findings, that technology markets have grown but are still limited in extent over industrial and geographical scope, necessitate the discussion of the factors that are responsible.

5. Factors that condition the market for technology

Following Arrow (1962), economists have emphasized asymmetry of information as the key barrier to trade in technology. Though asymmetric information may well be important, lack of information, or uncertainty, is surely a more important problem. Whereas asymmetric information creates problems when agents behave in self-interested ways, the nature of technology creates problems for a market for technology even absent such behavior. Uncertainty about technical success and commercial applicability, the difficulty in specifying a technology and valuing it, and the challenge of locating potential trading partners may be more serious problems than asymmetric information. In plain words, lack of information may be a much bigger problem than differences in access to information. For the most part, however, the literature has paid insufficient attention to the problem of insufficient information, with disproportionate attention to the issue of asymmetric information.

5.1. Cognitive limitations

Uncertainty poses a significant barrier to the market for technology. Unlike specific products or services, technology is hard to pin down. This is especially true when technology is not codified, and is embedded in people or machines. For example, improvements in a production process or in a service may be hard to define and codify with precision. In these cases, the object of the transaction is ill-defined to begin with, and this ambiguity makes it harder to trade in the improved process.

The difficulties are not only contractual. Discovering who has relevant technology and the price at which they may make it available (if at all) is also difficult. Understanding what they have and how to use it amplifies the problem. Conversely for a seller, identifying potential buyers can be problematic, and once a prospective partner has been identified, settling on the price can be no less challenging.

Problems of price discovery are not unique to markets for technology. For instance, a common approach used in the valuation of startup firms is the price paid for comparable firms. Although no two firms are identical, often they are similar enough for one to be used as a benchmark. However, using comparables begs the question inasmuch as it assumes a reasonably liquid market for acquiring startups. Lamoreaux and Sokoloff (1996, 2001) describe how the rise of a market for patents in the United States in the nineteenth century involved the growth of supporting institutions such as intermediaries that helped spread information about patents (e.g., patent agents, patent lawyers, and even publications such as the Scientific American, which reported available patents for sale). Their work underlines the important role that patents play in this market. Patents provide a document that clearly defines the object of exchange, and represents a focal point of the transaction. Second, patents clearly define the intellectual property rights of the two parties, thus avoiding potential ambiguities. Third, the patent offices themselves, along with patent agents and lawyers, can be a focal institution for organizing technology trade.

One problem in the market for technology is that the knowledge to be traded is often partially inarticulable (Winter, 1987) in part because the knowledge is largely based on empirical observation and experience, rather than understood through general principle. Arora and Gambardella (1994a) argue that the increase in the extent to which industrial technologies are based in science (including engineering sciences), and the use of advanced instruments and computers is reducing the fraction of "inarticulable" technology. Thanks to advances in computer technology, including software, many technical problems (e.g., in design, semiconductors, biotechnology, and many other industries) can be defined in logical terms (e.g., mathematical language) and captured in software. Interestingly, there are useful synergies with patents in facilitating technology transactions. Codified technology is easier to patent. Conversely, an increasing appreciation of intellectual property rights encourages codification of innovations.

New technologies are often surrounded by commercial uncertainty (Rosenberg, 1996). Simply put, it is difficult to know what applications the technology can have. This raises the search costs of both buyers and suppliers and leads to considerations of option values rather than actual values, and renders potential transactions subject to a variety of biases that human beings are prone to when faced with uncertainty. The net result is that technology transactions are more imperfect and harder to accomplish.

A special and important case in this context is GPT. Technology trade involving GPT has many of the features that we have just described. There is uncertainty about their applications. Often GPT emerges ex-post, as people realize that a technology created for certain purposes can also be used for other applications. Not only is there uncertainty about the applications but also that the potential users have to invest to learn if the technology is useful to them. For example, Maine and Garnsey (2006) tell the story of Hyperion Catalysis, which has developed special applications of fullerenes, a carbon allotrope discovered in 1985. The firm struggled to find uses for the new materials, and systematically explored applications in a number of industries, including automotive, aerospace, and power generation, through alliances with manufacturers. Today, it produces more than 40 products for these three distinct industries. Thoma (2009) describes a similar process in the case of Echelon, a company that has developed a universal electrical controller technology (LonWork) for diverse applications including a wide range of manufacturing, and heating and cooling systems for buildings.

A common thread running through these examples is that judging the technical merit of the technology or innovation often draws upon a very different set of expertise from that required to judge its applicability to a particular end use. Bresnahan and Greenstein (1996) note that creating new software technology requires expertise in computer science and software engineering. Understanding how the technology can be best used requires not just only the technical expertise, but also management skills and industry expertise. Both are separate, though not independent, sources of uncertainty, which make it significantly more difficult to contract for technology. Nonetheless, there are also significant advantages to specialization with a GPT. As Bresnahan and Trajtenberg (1995) point out, no individual user sector firms will have the cognitive breadth to see the common elements between what they are doing and what firms in other users sectors are doing. Therefore, each sector will not develop the "general" technology. Instead, it will be content to only develop the application of the technology specific to the sector. This is both more costly and also reduces the potential for learning across different applications.

5.2. Contractual limitations

Much of the economics literature has focused on the difficulties in writing contracts for technology trade, particularly in contracting for R&D, that is, for technology that is not yet developed (e.g., Mowery, 1983).

Teece (1988) notes three problems associated with R&D contracts. First, because the output of R&D is ill-defined, and hard to predict ex-ante, the parties have to write very detailed contracts specifying several contingencies, raising contracting costs. Second, these contracts call for exchange of information between the buyer and the supplier that they would prefer to keep secret. Third, because of the set-up costs of R&D contracting, or the tight linkages between buyers and suppliers (e.g., because of the need to exchange information), these contracts may be subject to lock-in. That is, once they are set-up, it is hard for both parties to exit the relationship, with implied potential for opportunistic renegotiations. The need to monitor the execution of the contract by the buyer may also require substantial administration costs. But this is like setting up an internal monitoring structure, making little difference between activities that are integrated within the firm, rather than acquired contractually from independent parties. All these factors make R&D contracting particularly costly, thereby encouraging integration of R&D in firms that also conduct the downstream manufacturing and commercial operations.

Zeckhauser (1996) provides a more recent restatement of the problems in contracting for technology in general. In particular, he alludes to problems of asymmetric information and contractual difficulties. He contends that "[c]ontracting to provide technological information (TI) is a significant challenge." Specifically he notes that (i) TI is difficult to count and value and is often sold at different prices to different parties. (ii) To value TI, it may be necessary to "give away the secret." (iii) TI is often bundled into products, such as a computer chip, which reduces efficiency. (iv) The sellers' superior knowledge about TI's value makes buyers wary of overpaying. Notice that most of these considerations apply to many types of modern goods and services, including art and music. Most of the attention in the literature, however, has been focused on the so-called lemons problem, namely that the seller has private information about value.

5.2.1. Asymmetric information and the market for lemons

Arrow (1962) articulated the problem faced by a potential buyer having to pay for information whose value he was unable to judge—the asymmetry in information would introduce inefficiency into the market for technology. Akerlof (1970) showed that this kind of asymmetry in information, plausibly present in the market for used cars, can prevent a market from functioning altogether, as "lemons" drive out good used cars.

The lemons problem in technology trade may not be as serious a problem as some economists believe. Not only are there contracting solutions that can mitigate the problem, in some cases, institutional arrangements may minimize information asymmetries. For instance, in pharmaceuticals, clinical trials reveal a great deal of information about the likely market value of the drug under development. Patents themselves disclose information about the innovation. The lemons problem is probably more serious in international technology transfer, especially between advanced and less advanced countries. In this case, there are barriers to the circulation of information, and a gap in expertise between the two parties. The problem is less severe when both parties operate in the same market or industry wherein technical information circulates, and the levels of technical expertise are similar.

Second, the key assumption of the lemons problem—namely, that the licensor holds useful private information—may not always be sensible. Sometimes the potential licensee may hold more significant private information about the potential applications of the technology. In addition, integrators, such as Boeing or the present day pharmaceutical firms, often embody the in-licensed technology in a larger system, whose characteristics they understand better than others. If so, the buyer may be better able, than the supplier, to evaluate the technology.

Empirical investigation of the lemons problem in licensing is difficult and the few extant studies are from the pharmaceutical sector.[13] Pisano (1997) finds that compounds developed internally are more likely to succeed than in-licensed compounds. Guedj (2005), though not explicitly testing for the lemons effect, finds that projects financed by pharmaceutical companies but developed by biotech firms are more likely to fail than projects developed by pharmaceutical firms. These findings are consistent with in-licensed compounds being drawn from an inferior distribution than those developed internally by the licensee, though other interpretations are also possible. On the other hand, Danzon et al. (2005) find that compounds developed in alliances (roughly equivalent to licensed compounds) have a lower probability of failure in clinical trials. Notice, moreover, that a lemons problem requires that in-licensed compounds be systematically inferior to those that the *licensor* kept for itself. Arora et al. (2009a) develop a structural model of drug development in pharmaceuticals, and find that licensed compounds are drawn from the same distribution as the internally generated compounds of the licensor. Although the empirical literature is both scant and inconclusive, our sense is Lamoreaux and Sokoloff (1999: p. 2) were right when they noted that "... scholars have overemphasized the information problems associated with contracting for new technological developments in the market."

5.3. Patents and the market for technology

Arrow's own solution to the problem of buying a pig in a poke was to appeal to intellectual property protection. If protected, the seller could disclose the details to potential buyers, mitigating the problem. This close relationship between patenting, the market for technology, and specialization in invention is reflected in trends in patenting and measures of the market for technology. Lamoreaux and Sokoloff note that patenting per capita in America rose during the nineteenth century, peaked in the early twentieth century, and then declined thereafter, closely mirroring trends in individual inventorship and in trade in patents. After the mid-1980s, patenting per unit of R&D investment in the United States changed course and began to rise, very close in time to the resurgence in markets for technology as well.

However, know-how and trade-secrets are important complements for patented technology. Robbins (2006) reports that in 2002, the sector NAIC 533 (lessors of nontangible property) earned $7.6 billion from patent licensing in the United States. The firms in this sector are likely pure patent holding companies, or specialized organizations set up by firms in other industries to license patents. Thus, of the $66 billion in technology licensing in the United States, about 12% was accounted for by pure patent licensing and the remainder by technology licensing, comprising patents, unpatented technology, know-how, and technical services.

Arora (1995) shows that patent protection can additionally improve the efficiency of licensing contracts that also require the provision of know-how and technical services, which has been shown to be an important component of licensing contracts (Contractor, 1981; Taylor and Silberston, 1973). He models the case where, along with the technology, the licensor also has to transfer know-how. Given the difficulty in objectively verifying that the know-how is provided, the licensor has an incentive to skimp,

[13] Evidence for the lemons' problem in financing development of the technology itself is surveyed in Chapter 14, this volume.

since providing such know-how services is costly. Conversely, insofar as some payments are conditional on the provision of the know-how, the licensee has an incentive to withhold payment, claiming inadequate know-how was provided.

The model shows that these problems can be solved by staggering the payment to the licensor over time, and by relying on the property rights on the technology. The buyer's value depends on the technology and the know-how. While the know-how that is transferred cannot be withdrawn, by withdrawing the rights to use the technology, the licensor does have a hostage because the know-how without a license to the patent is of diminished value. In some cases, the bundling with other complementary inputs, such as specialized machinery can provide a similar role (e.g., Arora, 1996). And, as Zeckhauser (1996) notes, technology is frequently sold by embodying in artifacts such as computer chips or software (provided without source code) to overcome the problem.

The empirical literature provides mixed evidence on the relationship between patent protection and technology-licensing contracts. Using a sample of 118 MIT inventions, Gans et al. (2002) find that the presence of patents increases the likelihood that an inventor will license to an incumbent rather than enter the product market by commercializing the invention. Dechenaux et al. (2009) link patent characteristics to outcomes in a sample of 805 MIT inventions licensed to private firms. They find that licenses based upon stronger patents are more likely to be commercialized. Anand and Khanna (2000) find that in the chemicals sector, where patents are believed to be more effective, there are more technology deals, a larger fraction of these are arm's length, involving exclusive licenses and a larger fraction of licensing is for future technologies rather than existing technologies. In contrast, Cassiman and Veugelers (2002) do not find that more effective patents encourage Belgian firms to enter into collaborative R&D arrangements.

Evidence from cross-national data is similarly mixed. Some studies find a positive association between patents and licensing. Yang and Maskus (2001) report a strong positive relationship between improved IPR regimes and licensing by US multinational corporations. Analyzing data on international technology-licensing contracts of Japanese firms, Nagaoka (2002) finds that weak patent regimes are associated with an increase in the fraction of transfers to an affiliate (such as a subsidiary), rather than to an unaffiliated firm. Smith (2001) finds that US firms are more likely to export or directly manufacture rather than license technology in countries with weak patent regimes. A study using French data finds that exports of technology services are greater to countries with more effective patent protection, although only for higher income countries (Bascavusoglu and Zuniga, 2002). Arora (1996) used a sample of 144 technology-licensing agreements signed by Indian firms where the provision of three technical services—training, quality control, and help with setting up an R&D unit—serve as empirical proxies for the transfer of know-how.[14] He found that the probability of technical services being provided was higher when the contract also included a patent license or a turnkey construction contract.

Other studies, however, cast doubts on the link between patent protection and the extent or form of technology licensing. Fink (1997) finds a very weak relationship using German data. Similarly, Fosfuri (2004) does not find that patent protection significantly affects the extent or channel of technology flow (through joint-venture, direct investment or licensing) in the chemical industry. These studies are plagued by the problem of measuring the effectiveness of patent protection, and typically rely upon a

[14] Mendi (2007) finds that technical assistance is bundled together with the transfer of know-how in Spanish technology import contracts.

widely used index, the Ginarte–Park index, which is based on legal provisions, rather than the actual enforcement of patents. A recent study by Branstetter et al. (2006) exploits changes in patent regimes in countries pressured by the United States. Using detailed data on the technology royalty payments received by US firms, and controlling for country, industry, and firm fixed effects, they find that stronger patent protection does not increase the transfer of technology by US multinationals to unaffiliated parties. However, it does increase the flow of technology to affiliates. Thus, despite much improved measures and a more careful design, this study too reflects the mixed nature of evidence on the topic.

Arora and Ceccagnoli (2006) provide a potential resolution of this mixed evidence. They argue that when licensing is attractive, then patent protection does indeed facilitate licensing. However, for firms with the ability to commercialize technology themselves, patent protection also increases the payoffs to commercialization. Analyzing data from a comprehensive survey of R&D performing firms in the United States, they find that patent protection increases licensing, but only for firms that lack complementary manufacturing capabilities. Hall and Ziedonis (2001) provide similar evidence from the semiconductor industry: all else being equal, small design specialists are more likely to patent, and case study evidence suggests that they do so to license their technologies. Gans et al. (2008) further note that patent licensing occurs predominately during a small time interval, near the date of the patent grant, because a patent resolves some transaction costs in the technology trade, such as uncertainty about the scope attributed to the patent and asymmetric information. Fosfuri et al. (2008) provide empirical evidence that firms that are better protected by software patents are more likely to exchange information in an open source software environment.

The OECD survey by Sheehan et al. (2004) also found that licensing influences patent strategies. They report that firms ranked "revenues from licensing" as the third most important reason for patenting. There are important differences across regions consistent with markets for technology being better developed in North America. First, the importance of licensing in patent strategies is higher for the North-American than European and Asian-Pacific firms. Second, revenue from licensing was mentioned to be very important by 39% of the ICT firms and 27% of biopharmaceuticals firms. A much lower fraction of firms in remaining sectors considered licensing to be a very important motivation for patenting.

In sum, patent protection increases the efficiency of technology-licensing contracts. However, stronger patent protection may also reduce incentives to license in some instances, thereby potentially offsetting the increase in transaction efficiency.

5.3.1. The problem with patents

Some authors have argued that excessively fragmented patent holdings can actually retard the rate at which new technologies are introduced into the market, by encouraging patent holders to hold up innovations in the hope of trying to extract more rents (e.g., Heller and Eisenberg, 1998; Lemley and Shapiro, 2007). They point out that many modern innovations are complex and build upon multiple elements, each capable of being patented separately and independently. When these patents are not held by a single entity, whoever wants to develop the technology needs to collect the rights from the different patent holder, potentially allowing a single patent holder to "hold up" the innovation. Foreseeing this problem, potential integrators may be reluctant to invest in the first instance. More generally, fragmented property rights can potentially lead to a what Heller and Eisenberg (1998) dub "the tragedy of the anticommons" where, instead of no one controlling the use of a common resource as in the well-known "tragedy of the commons," too many people hold a veto (see also Chapter 7, this volume for a more extensive discussion of patent pools and patent thickets).

In a recent study, Cockburn et al. (2008) find that IT firms facing more fragmented IP landscapes have higher licensing costs. In the life sciences, empirical evidence suggests that although patent proliferation has created challenges, it has not as yet become a serious problem, in part because it is possible to work around some of the problems.[15] Walsh et al. (2003) report on interviews with about 70 life sciences companies about the problem, and found that although fragmented patent rights were often encountered, the companies managed to resolve the problem by licensing, working around the patents, or simply by ignoring the problem altogether. Murray and Stern (2007) find that scientific papers see a decline in citations after the associated research is patented, which they interpret as evidence in favor of the anticommons retarding scientific growth. However, a detailed survey by Walsh et al. (2007) of academic researchers in life sciences reported that patents had limited impact on academic research. Only scientific projects with commercial objectives appear to be influenced by patenting by others, which is entirely understandable since existing patents would reduce the commercial, but not the scientific, value of such projects.

However, that the problem can be solved does not mean that it does not exist. Indeed, Merges and Nelson (1990) and Scotchmer (1991) have argued that the short-sighted use of even one patent can impede innovation where a technology is cumulative (i.e., where invention proceeds largely by building on prior invention). Merges and Nelson (1990) relate the case of radio technology where the Marconi Company, De Forest, and De Forest's main licensee, AT&T, arrived at an impasse that lasted about 10 years and was only resolved in 1919 when RCA was formed at the urging of the Navy. In aviation, Merges and Nelson (1990) argue that the refusal of the Wright brothers to license their patent was compounded as improvements were patented by others. Ultimately, World War I forced the Secretary of the Navy to intervene to work out an automatic cross-licensing arrangement. The theoretical literature on cumulative innovation and patent protection is discussed in Chapter 7, this volume.

5.3.2. Patents and nonmarket institutions for technology flows

Technology can also be traded outside the market. In a seminal paper, Allen (1983) describes what he called "collective invention" in the Cleveland district in Britain during the second half of the nineteenth century. During this period, Cleveland saw an active exchange of technical information about blast furnaces. Though many technologies were patented, the firms nonetheless transferred technology and information in meetings and conferences without contracts or royalty payments. Nuvolari (2004) documents a similar phenomenon in the mining industry in Cornwall, in the early nineteenth century. In a series of papers, von Hippel (1987) details instances of information sharing in the late twentieth century as well. He documents active know-how trading networks among engineers working in rival firms in the US steel minimill industry, which managers tolerated because they believed such sharing was broadly beneficial because it enabled their engineers to gain from the experience of others.

[15] Indeed in Japan, where there are many more patents per product across the entire manufacturing sector than in the United States, licensing and cross-licensing are commonplace (Cohen et al., 2002).

Allen showed that collective inventions depended on mobility of personnel and other channels through which know-how leaked out. Not only was it costly to plug these channels, it appeared that the firms realized that such know-how sharing was mutually beneficial, enabling them to compete against producers in other regions. As Allen notes, know-how sharing was more likely when the higher productivity produced by sharing benefited firms in the region but not firms outside the region. Thus, for example, Nuvolari (2004) notes that improvements in the average aggregate performance of Cornish engines also increased the value of the Cornish ore deposits and that similarly, improvements in the performance of the blast furnaces in Cleveland increased the value of Cleveland iron mines. Second, sharing was likely when problems were common. Indeed, von Hippel (1987) reports that specialty steel mills did not share know-how, because each mill tended to have processes specific to the products it produced. It appears that when the know-how related to proprietary products, it was less likely to be shared, reminiscent of the findings about licensing (rather the lack thereof) in differentiated product industries in Arora and Fosfuri (2003).

The absence of patenting, and of markets in the knowledge more generally, seems important for information sharing. Nuvolari (2004) notes that the collective sharing of technical know-how by steam engineers in Cornwall followed the lapse of the Watts–Bolton patents. Information sharing appears to rely upon barter: In von Hippel's (1987) case studies, managers tolerated and even encouraged the barter of know-how but any attempts to monetize the transactions would surely bring swift punishment. Nonmarket mechanisms for information sharing and diffusion rely upon collectively held norms that can rupture when the market intrudes. Dasgupta and David (1994) discuss the importance of norms of disclosure in governing what they call the Republic of Science. When academic research is also motivated by commercial considerations, the considerations of profit maximization and the academic norms of open disclosure (information sharing) can conflict. Indeed the finding reported by Murray and Stern (2007), that scientists are less likely to cite papers with an associated patent, in conjunction with that reported by Walsh et al. (2007) that academic scientists working to discover drugs (and who intend to file patents on their findings themselves) pay close attention to patents, suggest that commercial considerations can severely erode academic norms. Patents are not the source of commercial considerations but doubtless make them more salient.

Modern day incarnations of collective invention—open source communities—are typically vigilant about enforcing norms. Gambardella and Hall (2006) develop a theoretical model in which sharing norms are unstable when members can use the jointly developed invention to make money, even if members directly enjoy contributing to the joint project. In open source software projects, a mechanism such as a GPL license (which ensures that any software incorporating the jointly developed software must itself be made available under a GPL license) makes deviating from the norm less remunerative, making collective development more likely. In sum, although patents can facilitate trade in technology, they can also undermine the viability of some nonmarket institutions that facilitate the flow of knowledge.

5.4. *Contracting for technology without patents*

The literature suggests that patents can overcome the potential problem of asymmetric information. However, in a series of papers, James Anton and Dennis Yao show that competition among potential buyers can be leveraged to mitigate the problem as well. Anton and Yao (1994) develop a model in which an inventor cannot obtain a patent and neither can she commercialize it herself. Instead, she must sell the idea to buyers. The problem is that buyers are uncertain about whether the idea is valuable or not. Anton and Yao show that one solution is for the seller to disclose the idea to one buyer. If the buyer does not pay for a good idea, the inventor can credibly threaten to disclose it to the other buyer, thereby destroying some of the rents to the first buyer. What makes the model work is that a potential buyer of an invention values exclusive access to it, which makes eminent sense for ideas or inventions. There is another sense in which Anton and Yao's model is specifically about inputs that are "infinitely expansible" but not nonrival. The Anton and Yao model would not apply, for instance, to a truffle of unknown value. If the only way to determine its value is for the potential buyer to eat the truffle, then the seller cannot credibly threaten to sell it to another buyer if not paid. This paper provides a different but complementary explanation for the importance of a competitive product market for encouraging specialized technology suppliers (see also the discussion of GPT and the importance of product-market competition below).

In a subsequent paper, Anton and Yao (2002) analyze a situation where the invention can be disclosed in parts. Once again, the invention is not patented, and buyers value exclusivity. The value of the invention is conceived of as know-how, whose use increases the probability of successfully using the invention. Buyers do not know the value of the invention, that is, they do not know how much know-how the seller has. Although the "blackmail" strategy is still useful in preventing a buyer from expropriating the know-how, it is not enough. Rather, inventors must now signal the quality of their know-how by partially disclosing it. The better the know-how, the more is publicly disclosed (although more is also left undisclosed because "better" know-how is simply *more* know-how in this model). In order to signal the quality of the invention, sellers must also be willing to have some "skin the game," agreeing in essence to pay the buyer if the invention does not succeed and accepting a share of the payoff if the invention does succeed. Paying the buyer for an unsuccessful invention, or providing a warranty, requires capital, pointing to another link between the market for technology and capital markets. Instances of such warranties are rare, perhaps because successful inventions depend upon the efforts and investments of the buyer, not simply the quality of the idea provided by the seller. Thus, by conditioning the payments the seller receives on successful outcomes provides the right incentives to the seller but also weakens those of the buyer, thereby potentially jeopardizing success of the invention.[16] A warranty by the seller against failure will further attenuate the buyer's incentives to invest, and is probably why such warranties are rare.

5.5. *The structure of licensing contracts*

The suspected inefficiency of licensing contracts has attracted some theoretical and empirical research. Anton and Yao's work is an example of the application of mechanism design theory to the problem of the market for technology. There is a sizable literature that focuses on the structure of licensing contracts, such as whether licensing contracts are exclusive or not, and whether they have sales royalties or fixed fees, as well as other contractual provisions. A pioneering study by Caves et al. (1983)

[16] This is a special case of the Marshallian share-cropping problem—unless the inputs are contractible, contracting on output alone is suboptimal (see Cheung, 1968).

documented imperfections in the market for licenses. Gallini and Wright (1990) show that performance-based royalties may allow separation between high-value and low-value innovations, when it is commonly known that a higher value innovation will result in greater output than a lower value innovation (see also Macho-Stadler et al., 1996). Beggs (1992) obtains a similar result in a model in which it is the licensor who lacks information about the "type" of the licensee. Kamien (1992) provides a survey of the theoretical literature.

There are a number of empirical studies on the structure of licensing contracts, mostly based on data from Europe, Brazil, and Japan. This literature shows that the vast majority of licensing contracts involve performance-based royalties, often in combination with fixed fees. For example, Macho-Stadler et al. (1996) found royalty provisions in 72% of 241 Spanish technology transfer contracts while Bessy and Brousseau (1998) found such provisions in nearly 83% of French contracts. Empirical studies of licensing contracts are only weakly related to the theories about the structure of licensing contracts and sometimes yield contrasting findings.[17] Contractor (1981) finds that royalty rates tend to vary very little across licensing contracts for any given industry, and are typically established by "rule of thumb." Nagaoka (2005) analyses Japanese data from the period 1981–1998 across 32 sectors. He finds that high royalties are more likely to be observed when the licensing contract also includes patents. However, Villar (2004) finds that, in a sample of 925 licensing agreements in Spain, the parties are more likely to agree on fixed payments when the technology is patented. More recent attempts to test the insights from contract theory or transaction cost theory to understand the structure of licensing contracts are provided in Bessy et al. (2008) and Brousseau et al. (2007). These studies lack sources of exogenous variation that would identify how observed licensing contracts reflect underlying contract design issues.

6. Consequences of the existence of markets for technology

6.1. The division of innovative labor

One consequence of the existence of well-functioning markets for technology is that they create incentives for vertical specialization. This is just a straightforward application of the classical theory of division of labor. Indeed, as Table 2 shows, in the United States, the revenues of establishments that supply scientific R&D services (NAIC 5417) are sizable: around $75 billion in 2004 and $85 billion in 2005. These establishments are highly R&D intensive, and perform about 5% of the total industrial R&D.

This is consistent with other data reported by the NSF which indicate that contract R&D (the bulk of which was contracted to other companies) grew from 3.7% of total company funded R&D in 1993 to 5.6% in 2003, the latest year for which data are available. The pharmaceutical sector stands out in the extent to which R&D was outsourced, with 13.2% R&D outsourced in 2005.[18] These data clearly point to the substantial specialization in R&D, which is a rough indicator of the extent of what we call the

[17] In a more recent study, Dechenaux et al. (2009) relate the features of university licensing contracts, such as milestone payments to the special problems in licensing embryonic technologies. Embryonic technologies involve a combination of the need to share risk, discourage the licensee from shelving the technology, and the need to involve the inventor in subsequent development.

[18] See NSF, Science and Technology Indicators (2008). Appendix table 4–51.

Table 2
Estimated total revenue and R&D for US establishments classified in selected service industries in 2004 and 2005
(Billions of current dollars)

		Revenue		Total R&D	R&D as % of total industrial R&D	R&D/sales
Service industry	NAICS code	2004	2005	2005		
Professional, scientific, and technical services (except Notaries)	54	966	1058	32.0	14.2%	3%
Scientific R&D services	5417	74.8	81.5	12.3	5.4%	15%
R&D in physical, engineering, and life sciences	54171	70.0	76.4			

Source: Science and Engineering Indicators 2008, tables 2, 4–20, NSF 07-335.

division of innovative labor. It is also likely that the United States is in the vanguard of this trend. Comparable data, if available, would likely show a less extensive division of labor in Europe and Japan.

Consistent with the rise of technology specialists, large firms account for a steadily smaller fraction of R&D performed in the United States. Figure 4 shows that the share of nonfederal R&D accounted for by large firms, defined at those with more than 25,000 employees, has fallen steadily from around two-thirds in 1980 to slightly more than one-third in 2005. Over the same period, small firms, defined as those with fewer than 500 employees, have increased their share from 6% to around 18%. Firms in the next size category (500–999 employees) have seen a similar increase. Doubtless this reflects changes in the industrial structure in the United States, but it also points to the growing ability of small firms to appropriate rents from innovations, perhaps through the licensing to others.

This type of specialization reflects the tendency toward progressive specialization as markets expand. George Stigler had argued that when an industrial activity, such as the production of new technology, has large fixed costs, restricting the provision of that activity to a single specialist producer who can serve the entire market will yield the greatest economies of scale (Stigler, 1951). However, the various imperfections in the market for technology imply that the cost of acquiring external technology must be counted against the potential benefits of specialization. Intuitively, the benefits of specialization increase with the size of the market, but as Bresnahan and Gambardella (1998) point out, the size of the market for the technology specialist is different from the size of the market for the product. They show that the relevant size of the market for technology is the number of different applications or buyers (breadth) rather than the intensity of demand of the average application. Simply put, a large firm can produce technology more cheaply than acquiring it externally, once the cost of adapting external technology is included.

Bresnahan and Gambardella (1998) develop a model with several downstream firms which do not compete (and thus can be thought of as downstream applications) and one upstream supplier of technology. Downstream firms can either develop a dedicated technology or buy from the technology supplier. Technology development requires a fixed cost, and technology developed by a downstream firm can only be used internally. On the other hand, the technology of the upstream supplier is a GPT applicable to all downstream firms, but it needs to be adapted at a cost which increases with the intensity

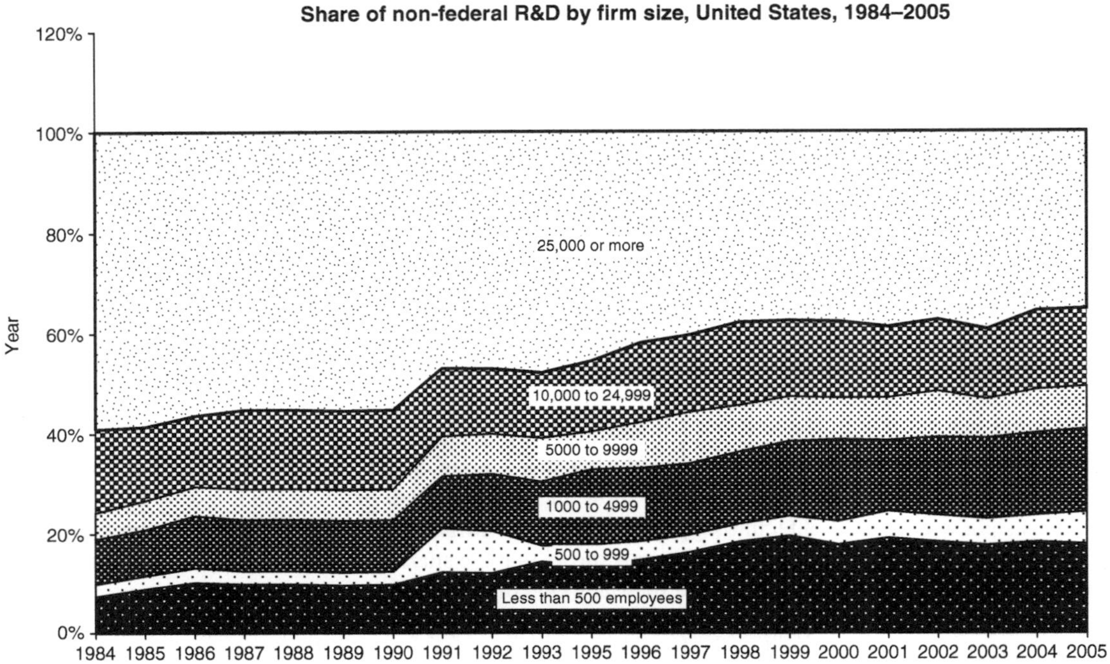

Figure 4. Share of Nonfederal R&D by Firm Size, United States, 1984–2005. Source: NSF Science and Technology Indicators, various years.

of use. Thus, downstream firms with greater intensity of demand (i.e., "large buyers") develop dedicated technology, whereas smaller firms buy the GPT. A crucial insight is that as an existing market is divided among a greater number of producers, the benefits of a division of labor grow. As a result, firms that were hitherto producing technology internally may switch to buying. By increasing demand for the technology supplier, this type of market broadening also encourages the supplier to invest in making the technology more general, reducing the cost of adaptation. Gambardella and Giarratana (2009) find that division of innovative labor and the generality of technology go hand in hand: Specializing as a technology supplier is more attractive when the technology is more general purpose.

Arora et al. (2009b) test the predictions of Bresnahan and Gambardella using data from the chemical plant engineering sector. In their model, large chemical firms (those investing in more than one plant) choose whether to design the plant internally or engage an external supplier of design and engineering services, labeled SEFs. Small firms either use an SEF or do not enter the market. They generalize the model by allowing the number of SEFs operating in a market to depend on the demand for their services, and therefore depend upon the decisions of potential buyers, that is, the chemical firms. Consistent with the theoretical predictions in Bresnahan and Gambardella (1998), they find that the number of SEFs increases when the market expands through an increase in the number of potential buyers but not when market expansion is due to an increase in the average size of buyers.

6.2. Entry and competition upstream and downstream

Markets for technology enhance entry and competition in both the upstream technology supplier industry and the downstream product industry.

Without markets for technology, a company that can develop a new technology will be unable to enter the market, unless it also able to invest in the far more costly and risky assets to develop and commercialize the innovation. Both Table 2 and Figure 4 show the increasing role of small firms and technology suppliers in the innovation system in America. Notice that the resurgence of the market for technology coincides with the increasing importance of R&D services suppliers and small firms. It also coincides with the boom in patenting in the United States, reversing a long period of decline. Figure 5 shows that after falling steadily from the 1960s, US patent applications per R&D dollar reversed trend in the mid-1980s. As discussed, patents enable the technology specialists to appropriate the rents from their innovations (see Hall and Ziedonis, 2001 for semiconductors, and Cockburn and MacGarvie, 2006, for software).

In the United States, specialized intermediaries, such as Royalty Pharma, buy future royalty streams from licensed inventions from small firms and universities, bolstering the ability of inventive firms to sustain themselves without having to participate in commercialization. Thus, while patents are often seen as an instrument for restraining competition, they have features that may also enhance it. Hall and Lerner discuss the role of patents in financing innovation in Chapter 14, this volume in greater detail.

In addition to facilitating the entry of technology specialists, technology markets also stimulate entry and competition in the downstream product markets. The availability of technology lowers entry costs into the product market, particularly for firms that lack internal R&D capability for innovation or even

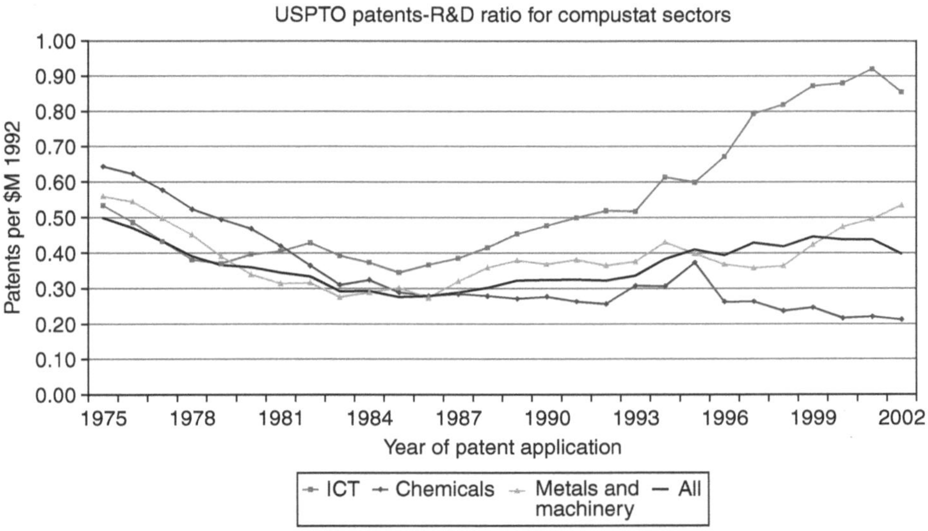

Figure 5. US patents per million dollars of R&D, 1975–2002. Source: Bronwyn Hall, private communication.

quick imitation. Moreover, specialized technology suppliers have an incentive to offer complementary services and know-how, and to reduce the cost of absorbing and using the technology.

The impact of licensing on entry is evident in the chemical industry, which has a long history of licensing of chemical processes (Arora and Gambardella, 1998). Lieberman (1989) finds that licensing was less common in concentrated chemical products, and that when licensing was restricted, there was less entry. In a related study of 24 chemical product markets, Lieberman (1987) reports that patenting by outsiders was associated with a faster decline of product price, once again suggesting that patenting by outsiders encouraged entry in the product market. Arora et al. (2001b) provide more direct evidence that specialized technology suppliers facilitate downstream entry. Using data on the chemical plants built during the 1980s in 38 less developed countries (LDC), they find that the number of specialized suppliers (SEFs) increases both the total number of plants in a market (a country sector pair), as well as the fraction that are based on externally supplied technology.[19] Simply put, a market for technology enhances competition downstream by making technology available more broadly and cheaply, enabling the entry of firms that would not enter otherwise.

By making technology less scarce, technology markets reduce the value of technology as a critical competitive asset. Competitive advantage must be sought in other assets, which are located downstream. Thus, firms try to differentiate products created with similar and relatively widely available technologies. The ability to create a specific product or market niche then becomes critical for success. Consistent with this, Arora and Nandkumar (2007) find that in the information security software industry, technology markets raise the value of marketing capabilities in ensuring the survival of firms, while diminishing the value of technical capabilities.

The discussion in this section also highlights that a division of innovative labor is a mechanism for creating spillovers that are transmitted to other parts of the system via the upstream sector of technology suppliers (see also Bresnahan and Trajtenberg, 1995). In brief, positive shocks to downstream industries (e.g., an increase in demand or the development of complementary technologies) induce positive shocks upstream (e.g., higher productivity or new technologies), which are then transmitted to the other downstream sectors served by the technology supplier industry. The link between two seemingly unrelated downstream sectors occurs because the shock to one sector raises the productivity of the upstream sector which then enhances the productivity of the other sector to which it is applied. For example, growth in the first world chemical market gives rise to specialized technology suppliers, the SEFs, which subsequently supplied LDC markets, contributing to the growth of the chemical industry. The link with the upstream SEF is key for transmitting the shocks from one product market to the other.

Importantly, these spillovers can also occur across sectors. In his study of the US machine tool sector in the nineteenth century, Rosenberg (1976) noted that the various downstream industries using machine tools arose at different times. For instance, firearm manufacturing emerged earlier than sewing machines, typewriters, or bicycles. The growth of the firearm industry spurred the development of

[19] Conversely, in Klepper's (1996) model of industry shake-outs, a key entry barrier is new firms' inability to enter by innovating. The returns to process innovation are proportional to size, and entrant size is eventually too small compared to incumbents. A market for technology would enable process specialists to enter with process innovations, although other features of the model would imply that downstream producers would still face rising barriers to entry over time.

metal cutting and shaping machines. Bicycle production required metal cutting operations that were very similar to those of the firearm industry (e.g., boring, drilling, milling, planning, grinding, polishing, etc.—see Rosenberg, 1976: 16), and thus the bicycle industry could rely upon the suppliers of metal cutting machines that were already serving the larger firearm industry. What the suppliers had learned in producing metal cutting machines for the firearm producers did not have to be learned again to supply bicycles producers. The commonality in the learning process across the industries, or what Rosenberg called "technological convergence," was critical for the transmission of growth, but required the intermediation of an upstream sector.

6.3. The rise and decline and the rise once again?

Recall that Lamoreaux and Sokoloff had documented an extensive market for technology in the United States in the nineteenth century, which declined by the end of the century. By World War II, innovation in the United States was dominated by the in-house laboratories of large corporations, a trend that continued well into the 1960s. Data discussed earlier indicate that the market for technology has revived, certainly by the beginning of the 1990s, and likely somewhat earlier. Mowery (1983) and Teece (1988) argue that increasing contracting problems, principally due to asymmetric information, undermined the market for technology in the nineteenth century.

Lamoreaux and Sokoloff (2005) take issue with this view. Instead, they argue that the market for technology in the United States in the nineteenth century was closely related to the existing division of innovative labor between independent inventor-entrepreneurs and manufacturers who relied upon them for inventions and improvements. Thus, the decline of the market for technology is, in their view, rooted in the decline of the individual inventor. Individual inventorship declined, in turn, because invention became increasingly rooted in science and engineering, rather than practical experience alone. In their sample of prolific patentees, their so-called "great inventors," they find a marked increase in the educational attainments of inventors born after 1865. They further argue that this increasing technical education requirement must have limited entry into independent invention, resulting in a situation where inventors either had to seek employment with large firms, or commercialize their inventions themselves, although on a much larger scale than before. Raising large amounts of capital was difficult, especially for inventors without an established track record. Thus, larger firms with superior access to national capital markets had a marked advantage in financing innovation. In other words, Lamoreaux and Sokoloff (2005) suggest that a combination of increasing cost of R&D and contracting problems in the capital market rather than in the market for technology were behind the decline of the market for technology in the nineteenth century.

Aghion and Tirole's (1994) model also rationalizes a capital-constraint story. In their model, both the buyer and seller (the R&D unit, in their exposition) provide inputs that contribute to a successful invention. They show that when the seller's inputs are noncontractible but the seller is cash constrained, the buyer may end up in control, even when it would be more efficient to give control to the R&D unit. Thus, financial constraints may limit the division of innovative labor. Lerner and Merges (1998) provide evidence from biotechnology licensing and R&D contracts to show that control rights tend to favor the buyer, who is also financing the R&D, when the financial position of the R&D performing firm is weak.

Our discussion suggests a complementary explanation, which appeals to the changes that were taking place on the demand side. The early twentieth century was also a time of significant market integration, leading to the rise of the great Chandlerian firms. At a minimum, this consolidation in production, even while accompanied by growth in population, would lead to deeper, rather than broader, markets for a potential technology supplier. Following Bresnahan and Gambardella (1998), this would imply lower gains from specialization in technology supply. Indeed, in their empirical study of the division of labor in the chemical engineering sector, Arora et al. (2009b) find that as the share of large firms in a market increases, fewer small firms enter, resulting in fewer specialized suppliers. Note that the Anton and Yao (1994, 2002) theory yields a similar prediction: a reduction in competition among potential buyers reduces the ability of the inventor to appropriate rents from her invention, thereby reducing the number of innovators.

The resurgence of markets for technology in the 1980s can be explained by the same set of factors. The tremendous growth in the scope and sophistication of capital markets, particularly for financing young, technology-based, ventures, surely helped mitigate the challenges that entrepreneurial inventors faced. Equally, the growing science and engineering basis of technical change, along with an accommodating public policy, improved the efficacy of patent protection. Arora and Gambardella (1994a) argue that improvements in instrumentation (particularly information technology) strongly complemented the use of scientific knowledge, contributing to a greater tradability of knowledge, and also increased the scope of new technologies.[20] Furthermore, changes in the composition of industrial activity have broadened the potential market for technology, complementing the greater generality of innovation, which would favor specialized suppliers of technology.

These considerations also suggest that the United States, with its long tradition of widely accessible patent protection, especially for small inventors, would provide more hospitable environment for a market for technology to thrive. However, other than Khan and Sokoloff's comparison of costs of patenting in the nineteenth century Britain and the United States (Khan and Sokoloff, 2004), we are not aware of any systematic studies on why markets for technology have not grown as vigorously outside the United States.

7. Conclusions and avenues for further research

Despite the many challenges it faces, trade in (disembodied) technology has grown steadily over the last two decades, and is now sizable. Its extent and spread has been uneven, both across regions, and across industries. With some exceptions, there is little known about what factors condition the extent of markets of technology and how these vary across industries and technologies, or across space and time. Explaining this variation is an important opportunity for further research.

[20] After examining a variety of political economy explanations, Kortum and Lerner (1999) conclude that the spurt in patenting in the United States after 1984 cannot be attributed to policy changes, such as the establishment of the Court of Appeals of the Federal Circuit. Instead, they suggest that a broad based increase in research productivity, as well as changes in the management of research, is a more likely explanation. However, Hall and Ziedonis (2001) show that the increase is partly due to patent portfolio races in the semiconductor sector whose cause was rooted in the increased strength of patents induced by the early 1980s policy changes.

At the risk of oversimplification, the focus in the literature has been on the transaction, and on the costs of the transaction relative to alternatives. There has been much less on the broader context of the transaction, conforming to the view in which transactions in technology are *ad hoc*, the exception rather than the norm. The steady growth in the volume of trade in technology makes it important to understand the market for technology, not simply the particularities of the transactions.

A particularly important aspect of the market for technology is the growth of firms that specialize in supplying technology. The determinants of a division of innovative labor (including the nature of technology), the conditions of intellectual property protection, and the industry structure in the product market, are all important topics of further research. The special role of GPT in the innovation process alerts us both to the potential importance of a division of labor and to the potential perils of studying an industry in isolation from where it draws its inputs, including technology.

Another potentially fruitful area for additional research is how the internal organization of firms interacts with markets for technology. Although there are some prescriptions offered in management books (e.g., Chesbrough, 2003), an analytical and empirical exploration of how the internal organization of firms conditions their participation in the market for technology, and conversely, how markets for technology are likely to affect how firms are organized, and in particular, how R&D is managed inside firms.

The most glaring lacuna is probably on the consequences of markets for technology, particularly for growth in productivity and for industry structure. Most economists would agree that trade is mutually beneficial, that it improves resource allocation and increases efficiency. Easing the conditions for trading industrial inputs, such as technology, should have important and measurable effects. The few studies reviewed here suggest that they lower entry barriers and increase competition. Scattered evidence from the literature on international technology diffusion (see Chapter 3, Vol. 2) also points to potential impact on productivity growth, although the evidence is mixed and the role of technology trade in that is even less clear. A systematic examination of how markets for technology affect the rate and direction of inventive activity is therefore urgently needed.

References

Aghion, P., Tirole, J. (1994). "The mangement of innovation". Quarterly Journal of Economics 109 (4), 1183–1209.

Akerlof, G.A. (1970). "The market for 'Lemons': Quality uncertainty and the market mechanism". Quarterly Journal of Economics 84 (3), 488–500.

Allen, R. (1983). "Collective invention". Journal of Economic Behavior and Organization 4, 1–24.

Anand, B.N., Khanna, T. (2000). "The structure of licensing contracts". Journal of Industrial Economics 48 (1), 103–135.

Anton, J.J., Yao, D.A. (1994). "Expropriation and inventions: Appropriable rents in the absence of property rights". American Economic Review 84 (1), 190–209.

Anton, J.J., Yao, D.A. (2002). "The sale of ideas: Strategic disclosure, property rights, and contracting". Review of Economic Studies 69 (240), 513–531.

Arora, A. (1995). "Licensing tacit knowledge: Intellectual property rights and the market for know-how". Economics of Innovation and New Technology 4, 41–59.

Arora, A. (1996). "Contracting for tacit knowledge: The provision of technical services in technology licensing contracts". Journal of Development Economics 50 (2), 233–256.

Arora, A., Ceccagnoli, M. (2006). "Patent protection, complementary assets, and firms' incentives for technology licensing". Management Science 52 (2), 293–308.

Arora, A., Fosfuri, A. (2003). "Licensing the market for technology". Journal of Economic Behavior and Organization 52 (2), 277–295.
Arora, A., Gambardella, A. (1990). "Complementarites and external linkages: The strategies of the large firms in biotechnology". The Journal of Industrial Economics 38 (4), 361–379.
Arora, A., Gambardella, A. (1994a). "The changing technology of technical change: General and abstract knowledge and the division of innovative labour". Research Policy 23, 523–532.
Arora, A., Gambardella, A. (1994b). "Evaluating technological information and utilizing it: Scientific knowledge, technological capability, and external linkages in biotechnology". Journal of Economic Behavior and Organization 24 (1), 91–114.
Arora, A., Gambardella, A. (1998). "Evolution of industry structure in the chemical industry". In: Ashish Arora, R.L., Rosenberg, Nathan (Eds.), Chemicals and Long-Term Economic Growth. Wiley, New York.
Arora, A., Merges, R.P. (2004). "Specialized supply firms, property rights and firm boundaries". Industrial and Corporate Change 13 (3), 451–475.
Arora, A., Nandkumar, A. (2007). Securing their Future? Entry and Survival in the Information Security Industry. NBER Working Paper No.13634.
Arora, A., Fosfuri, A., Gambardella, A. (2001a). Markets for Technology: The Economics of Innovation and Corporate Strategy. MIT Press, Cambridge, MA.
Arora, A., Fosfuri, A., Gambardella, A. (2001b). "Specialized technology suppliers, international spillovers and investments: Evidence from the chemical industry". Journal of Development Economics 65 (1), 31–54.
Arora, A., Fosfuri, A., Gambardella, A. (2008). "Patents and the market for technology". In: Maskus, K.E. (Ed.), Intellectual Property Rights, Growth and Trade. Amsterdam, Oxford, Elsevier Press, pp. 123–156.
Arora, A., Gambardella, A., Magazzini, L., Pammoli, F. (2009a). "A breadth of fresh air? Firm type, scale, scope and selection effects in drug development". Management Science 55 (10), 1638–1653.
Arora, A., Vogt, W., Yoon, J. (2009b). "Is the division of labor limited by the extent of the market?: Evidence from chemical industry". Industrial and Corporate Change 18 (5), 785–806.
Arrow, K. (1962). Economic Welfare and the Allocation of Resources for Invention, The Rate and Direction of Inventive Activity: Economic and Social Factors: 609–625. Princeton University Press, Princeton, NJ.
Athreye, S., Cantwell, J. (2007). "Creating competition? Globalisation and the emergence of new technology producers". Research Policy 36 (2), 209–226.
Bascavusoglu, E., Zuniga, M.P. (2002). Foreign Patent Rights, Technology and Disembodied Knowledge Transfer Cross Borders: An Empirical Application. University of Paris I, France.
Beggs, A.W. (1992). "The licensing of patents under asymmetric information". International Journal of Industrial Organization 10 (2), 171–191.
Bessy, C., Brousseau, E. (1998). "Technology licensing contracts: Features and diversity". International Review of Law and Economics 18 (4), 451–489.
Bessy, C., Brousseau, E., Saussier, S. (2008). "Payment schemes in technology licensing agreements: A transaction cost approach" http://economix.u-paris10.fr/pdf/workshops/2008_contracting/Bessy-Brousseau-Saussier-Apri_08.pdf. Unpublished working paper.
Branstetter, L.G., Fisman, R., Foley, C.F. (2006). "Do stronger intellectual property rights increase international technology transfer? Empirical evidence from U.S. firm-level panel data". Quarterly Journal of Economics 121 (1), 321–349.
Bresnahan, T.F., Gambardella, A. (1998). "The division of inventive labor and the extent of the market". In: Helpman, E. (Ed.), General Purpose Technologies and Economic Growth. MIT Press, Cambridge, MA, pp. 253–281.
Bresnahan, T.F., Greenstein, S. (1996). "Technical progess and co-invention in computing and in the use of computers' brookings papers on economic activity". Microeconomics 1–78.
Bresnahan, T.F., Trajtenberg, M. (1995). "General purpose technologies: 'Engines of growth'?" Journal of Econometrics 65 (1), 83–108.
Brousseau, E., Coeurderoy, R., Chaserant, C. (2007). "The governance of contracts: Empirical evidence on technology licensing agreements". Journal of Institutional and Theoretical Economics 163 (2), 205–235.
Brusoni, S., Prencipe, A., Pavitt, K. (2001). "Knowledge specialization, organizational coupling, and the boundaries of the firm: Why do firms know more than they make?" Administrative Science Quarterly 46 (4), 597–621.

Cassiman, B., Veugelers, R. (2002). "R&D cooperation and spillovers: Some empirical evidence from Belgium". American Economic Review 92 (4), 1169–1184.
Cassiman, B., Veugelers, R. (2006). "In search of complementarity in innovation strategy: Internal R&D and external knowledge acquisition". Management Science 52 (1), 68–82.
Caves, R.E., Crookell, H., Killing, J.P. (1983). "The imperfect market for technology licenses". Oxford Bulletin of Economics and Statistics 45 (3), 223–248.
Cesaroni, F. (2003). "Technology strategies in the knowledge economy: The licensing activity of Himont". International Journal of Innovation Management 7 (2), 1–23.
Chesbrough, H.W. (2003). Open Innovation: The New Imperative for Creating and Profiting From Technology. Harvard Business School Press, Boston, MA.
Cheung, S. N. S. (1968). "Private property rights and sharecropping". Journal of Political Economy 76 (6), 1107–1122.
Cockburn, I.M., Henderson, R. (2003). The IPO Survey on Strategic Management of Intellectual Property. Conference Report. Intellectual Property Owners Association, Washington, DC, November 2003.
Cockburn, I.M., MacGarvie, M. (2006). Entry, Exit and Patenting in the Software Industry. NBER Working Papers 12563.
Cockburn, I.M., MacGarvie, M., Muller, E. (2008). Patent Thickets, Licensing and Innovative Performance. ZEW Discussion Papers: 08-101.
Cohen, W.M., Kepler, S. (1992). "The anatomy of industry R&D intensity distributions". The American Economic Review 82 (4), 773–799.
Cohen, W.M., Levinthal, D.A. (1989). "Innovation and learning: The two faces of R&D". Economic Journal 99 (397), 569–596.
Cohen, W.M., Goto, A., Nagata, A., Nelson, R., Walsh, J.P. (2002). "R&D spillovers, patents and the incentives to innovate in Japan and the United States". Research Policy 31 (8–9), 1349–1367.
Contractor, F.J. (1981). International Technology Licensing. Lexington, MA, D C Heath and Company.
Corts, K.S. (2000). "Focused firms and the incentive to innovate". Journal of Economics and Management Strategy 9 (3), 339–362.
Danzon, P.M., Nicholson, S., Pereira, N. (2005). "Productivity in pharma-biotechnology R&D: The role of experience and alliances". Journal of Health Economics 24 (2), 317–339.
Dasgupta, P., David, P. (1994). "Towards a new economics of science". Research Policy 23, 487–521.
Dechenaux, E., Thursby, M., Thursby, J. (2009). "Shirking, sharing risk and shelving: The role of university license contracts". International Journal of Industrial Organization 27 (1), 80–91.
Fink, C. (1997). "Intellectual property rights and U.S. and German international transactions in manufacturing industries". Unpublished manuscript. On file with Duke Journal of Comparative and International Law.
Foray, D., Lissoni, F. (2010). "University research and public–private interaction". In: Hall, B., Rosenberg, N. (Eds.), Handbook of Economics of Innovation. Amsterdam, Elsevier.
Ford, D., Ryan, C. (1981). "Taking technology to market". Harvard Business Review 59 (2), 117–126.
Forman, C., Goldfarb, A., Greenstein, S. (2008). "Understanding the inputs into innovation: Do cities substitute for internal firm resources?" Journal of Economics and Management Strategy 17 (2), 295–316.
Fosfuri, A. (2004). "Determinants of international activity: Evidence from the chemical processing industry". Research Policy 33 (10), 1599–1614.
Fosfuri, A. (2006). "The licensing dilemma: Understanding the determinants of the rate of technology licensing". Strategic Management Journal 27 (12), 1141–1158.
Fosfuri, A., Giarratana, M.S., Luzzi, A. (2008). "The penguin has entered the building: The commercialization of open source software products". Organization Science 19 (2), 227–291.
Gallini, N. (1984). "Deterrence through market sharing: A strategic incentive for licensing". American Economic Review 74 (5), 931–941.
Gallini, N.T., Wright, B.D. (1990). "Technology transfer under asymmetric information". RAND Journal of Economics 21 (1), 147–160.
Gambardella, A. (1995). Science and Innovation: The US Pharmaceutical Industry During the 1980s. Cambridge University Press, Cambridge, MA.
Gambardella, A., Giarratana, M.S. (2009). General Technologies, Product-Market Fragmentation, and the Market for Technology: Evidence from the Software Security Industry. draft-Bocconi University, Milan.
Gambardella, A., Hall, B.H. (2006). "Proprietary vs public domain licensing in software and research products". Research Policy 35 (6), 875–892.

Gambardella, A., Torrisi, S. (1998). "Does technological convergence imply convergence in markets? Evidence from the electronics industry". Research Policy 27 (5), 445–463.
Gans, J.S., Stern, S. (2000). "Incumbency and R&D incentives: Licensing the gale of creative destruction". Journal of Economics and Management Strategy 9 (4), 485–511.
Gans, J.S., Stern, S. (2010). Is There a Market for Ideas? Industrial and Corporate Change. Forthcoming.
Gans, J.S., Hsu, D., Stern, S. (2002). "When does start-up innovation spur the gale of creative destruction?" RAND Journal of Economics 33 (4), 571–586.
Gans, J.S., Hsu, D., Stern, S. (2008). "The impact of uncertain intellectual property rights on the market for ideas: Evidence from patent grant delays". Management Science 52 (5), 982–997.
Giarratana, M.S. (2004). "The birth of a new industry: Entry by start-ups and the drivers of firm growth. The case of encryption software". Research Policy 33 (5), 787–806.
Gilbert, R., Shapiro, C. (1997). "Antitrust issues in the licensing of intellectual property: The nine no-no's meet the nineties". Microeconomics 1997, 283–349 Brookings Papers on Economic Activity.
Guedj, I. (2005). Ownership vs. Contract: How Vertical Integration Affects Investment Decisions in Pharmaceutical R&D. Unviveristy of Texas Austin Business School, Texas. Unpublished paper.
Hagedoorn, J. (2002). "Inter-firm R&D partnerships: An overview of major trends and patterns since 1960". Research Policy 31 (4), 477–492.
Hall, B.H., Ziedonis, R.H. (2001). "The patent paradox revisited: An empirical study of patenting in the US semiconductor industry 1979–1995". RAND Journal of Economics 32 (1), 101–128.
Heller, M., Eisenberg, R. (1998). "Can patents deter innovation? The anticommons in biomedical research". Science 698–701.
Higgins, M.J., Rodriguez, D. (2006). "The outsourcing of R&D through acquisitions in the pharmaceutical industry". Journal of Financial Economics 80 (2), 351–383.
Hoekman, B.M., Maskus, K.E., Saggi, K. (2005). "Transfer of technology to developing countries: Unilateral and multilateral policy options". World Development 33 (10), 1587–1602.
Jensen, R., Thursby, M. (2004). "Proofs and prototypes for sale: The licensing of university inventions". American Economic Review 91 (1), 240–259.
Kamien, M.I. (1992). "Patent licensing". In: Aumann, R.J., Hart, S. (Eds.), Handbook of Game Theory with Economic Applications, vol. 1. Elsevier Science, North Holland, pp. 331–354.
Kamien, M., Tauman, Y. (1986). "Fees versus royalties and the private value of a patent". Quarterly Journal of Economics 101 (3), 471–491.
Katz, M.L., Shapiro, C. (1986). "How to license intangible property". Quarterly Journal of Economics 101 (3), 567–589.
Khan, B.Z., Sokoloff, K.L. (2004). "Institutions and democratic invention in 19th-century America: Evidence from "Great Inventors," 1790–1930". American Economic Review 94 (2), 395–401.
Klepper, S. (1996). "Entry, exit, growth, and innovation over the product life cycle". The American Economic Review 86 (3), 562–583.
Kollmer, H., Dowling, M. (2004). "Licensing as a commercialisation strategy for new technology-based firms". Research Policy 33 (8), 1141–1151.
Kortum, S., Lerner, J. (1999). "What is behind the recent surge in patenting?" Research Policy 28 (1), 1–22.
Lamoreaux, N.R., Sokoloff, K.L. (1996). "Long-term change in the organization of inventive activity". Proceedings of the National Academy of Sciences of the United States of America 93 (23), 12686–12692.
Lamoreaux, N.R., Sokoloff, K.L. (1999). "Inventors, firms, and the market for technology in the late nineneteenth and early twentieth centuries". In: Lamoreaux, N.R., Raff, D. M. G., Temin, P. (Eds.), Learning by Doing in Markets, Firms, and Countries. University of Chicago Press, Chicago, IL.
Lamoreaux, N.R., Sokoloff, K.L. (2001). "Market trade in patents and the rise of a class of specialized inventors in the 19th-century United States". The American Economic Review 91 (2), 39–44.
Lamoreaux, N.R., Sokoloff, K.L. (2005). The Decline of the Independent Inventor: A Schumpterian Story? NBER Working Paper 11654.
Lemley, M.A., Shapiro, C. (2007). "Patent holdup and royalty stacking". Texas Law Review 85 (7), 1991–2049.
Lerner, J., Merges, R.P. (1998). "The control of technology alliances: An empirical analysis of the biotechnology industry". Journal of Industrial Economics 46 (2), 125–156.

Lieberman, M.B. (1987). "Patents, learning by doing, and market structure in the chemical processing industries". International Journal of Industrial Organization 5 (3), 257–276.
Lieberman, M.B. (1989). "The learning curve, technological barriers to entry, and competitive survival in the chemical processing industries". Strategic Management Journal 10 (5), 431–447.
Macho-Stadler, I., Martinez-Giralt, X., Perez-Castrillo, J.D. (1996). "The role of information in licensing contract design". Research Policy 25 (1), 43–57.
Maine, E., Garnsey, E. (2006). "Commercializing generic technology: The case of advanced materials ventures". Research Policy 35 (3), 375–393.
Malerba, F., Nelson, R., Orsenigo, L., Winter, S. (2008). "Vertical integration and disintegration of computer firms: A history-friendly model of the coevolution of the computer and semiconductor industries". Industrial and Corporate Change 17 (2), 197–231.
McGahan, A.M., Silverman, B. (2006). "Profiting from technological innovation: The effect of competitor patenting on firm value". Research Policy 35 (8), 1222–1242.
Mendi, P. (2007). "Trade in disembodied technology and total factor productivity in OECD countries". Research Policy 36 (1), 121–133.
Merges, R.P., Nelson, R.R. (1990). "On the complex economics of patent scope". Columbia Law Review 90 (4), 839–916.
Motohashi, K. (2008). "Licensing or not licensing? An empirical analysis of the strategic use of patents by Japanese firms". Research Policy 37 (9), 1548–1555.
Mowery, D. (1983). "The relationship between intrafirm and contractual forms of industrial research in american manufacturing, 1900–1940". Explorations in Economic History 20, 351–374.
Mowery, D.C. (1984). "Firm structure, government policy, and the organization of industrial research". Business History Review 58 (4), 504–531.
Mowery, D.C., Nelson, R.R., Sampat, B.N., Ziedonis, A.A. (2001). "The growth of patenting and licensing by U.S. universities: An assessment of the effects of the Bayh–Dole act of 1980". Research Policy 30 (1), 99.
Murray, F., Stern, S. (2007). "Do formal intellectual property rights hinder the free flow of scientific knowledge? An empirical test of the anti-commons hypothesis". Journal of Economic Behavior and Organization 63 (4), 648–687.
Nagaoka, S. (2002). "Impacts of intellectual property rights protection on international licensing: Evidence from the licensing contracts of Japanese industry", Working paper 02–04, Institute of Innovation research, Hitotsubashi University.
Nagaoka, S. (2005). "Determinants of high-royalty contracts and the impact of stronger protection of intellectual property rights in Japan". Journal of the Japanese and International Economies 19 (2), 233–254.
Nuvolari, A. (2004). "Collective invention during the British Industrial Revolution: The case of the Cornish pumping engine". Cambridge Journal of Economics 28 (3), 347–363.
Pisano, G. (1990). "The R&D boundaries of the firm: An empirical analysis". Administrative Science Quarterly 35, 153–176.
Pisano, G. (1997). R&D Performance, Collaborative Arrangements, and the Market for Know-How: A Test of the 'Lemons' Hypothesis in Biotechnology. Harvard Business School, Cambridge, MA. Mimeo.
Robbins, C. (2006). Measuring Payments for the Supply and Use of Intellectual Property. Bureau of Economic Analysis, U.S. Department of Commerce, Washington, DC.
Rockett, K.E. (1990). "Choosing the Competition and Patent Licensing". RAND Journal of Economics 21 (1), 161–171.
Rosenberg, N. (1976). Technological Change in the Machine Tool Industry, Perspectives on Technology. Cambridge University Press, Cambridge, New York, pp. 11–31.
Rosenberg, N. (1990). "Why do firms do basic research (with their own money)?" Research Policy 19 (2), 165.
Rosenberg, N. (1996). "Uncertainty and technology change". In: Landau, R., Taylor, T., Wright, G. (Eds.), The Mosaic of Economic Growth. Stanford University Press, Stanford, CA, pp. 334–356.
Rotemberg, J.J., Saloner, G. (1994). "Benefits of narrow business strategies". American Economic Review 84 (5), 1330–1349.
Roth, A.E. (2008). "What have we learned from market design? Hahn lecture". Economic Journal 118, 285–310.
Scotchmer, S. (1991). "Standing on the shoulders of giants: Cumulative research and the patent law". Journal of Economic Perspectives 5 (1), 29–41.
Shapiro, C. (2000). "Navigating the patent thicket: Cross licenses, patent pools, and standard setting". NBER Innovation Policy and the Economy 1 (1), 119–150.

Sheehan, J., Martinez, C., Guellec, D. (2004). Understanding Business Patenting and Licensing: Results of a Survey, Patents, Innovation and Economic Performance—Proceedings of an OECD Conference: Chapter 4. Paris, France, Organisation for Economic Co-operation and Development.

Shepard, A. (1987). "Licensing to enhance demand for new technology". RAND Journal of Economics 18 (3), 360–368.

Smith, P.J. (2001). "How do foreign patent rights affect U.S. exports, affiliate sales, and licenses?" Journal of International Economics 55 (2), 411–439.

Stigler, G.J. (1951). "The Division of Labor is Limited by the Extent of the Market". The Journal of Political Economy 59 (3), 185–193.

Taylor, C.A., Silberston, A.Z. (1973). D.A.E. Monograph 23. The Economic Impact of the Patent System: A Study of the British Experience. University of Cambridge, Cambridge.

Teece, D.J. (1986). "Profiting from technological innovation". Research Policy 15, 285–305.

Teece, D.J. (1988). "Technological change and the nature of the firm". In: Dosi, G., Freeman, C., Nelson, R., Soete, L. (Eds.), Technological Change and Economic Theory. London, Printer Publishers, pp. 257–281.

Thoma, G. (2009). "Striving for a large market: Evidence from a general-purpose technology in action". Industrial and Corporate Change 18 (1), 107–138.

Thursby, J.G., Thursby, M.C. (2002). "Who is selling the ivory tower? Sources of growth in university licensing". Management Science 48, 90–104 (1 Special Issue on University Entrepreneurship and Technology Transfer).

Ulset, S. (1996). "R&D outsourcing and contractual governance: An empirical study of commercial R&D projects". Journal of Economic Behavior and Organization 30 (1), 63–82.

Villar, M.C. (2004). The structure of payments of technology licensing agreements, The case of Spain. Unpublished Manuscript. European University Institute, Florence.

von Hippel, E. (1987). "Cooperation between rivals: Informal know-how trading". Research Policy 16 (6), 291–302.

Walsh, J.P., Arora, A., Cohen, W.M. (2003). "Working through the patent problem". Science 299 (5609), 1020.

Walsh, J.P., Cohen, W.M., Cho, C. (2007). "Where excludability matters: Material versus intellectual property in academic biomedical research". Research Policy 36 (8), 1184–1203.

Williamson, O. (1985). The Economic Institutions of Capitalism: Firms, Markets, Relational Contracting. Free Press, New York and London.

Winder, G.M. (1995). "Before the corporation and mass production: The licensing regime in the manufacture of North American harvesting machinery, 1830–1910". Annals of the Association of American Geographers 85 (3), 521–552.

Winter, S. (1987). "Knowledge and competence as strategic assets". In: Teece, D.J. (Ed.), The Competitive Challenge: Strategies for Industrial Innovation and Renewal. Harper and Rowe, New York, NY.

Yang, G.F., Maskus, K.E. (2001). "Intellectual property rights and licensing: An econometric investigation". Weltwirtschaftliches Archiv 137 (1), 58–79.

Zeckhauser, R. (1996). "The challenge of contracting for technological information". Proceedings of the National Academy of Sciences of the United States of America 93 (23), 12743–12748.

Zuniga, M.P., Guellec, D. (2008). Survey on Patent Licensing: Initial Results from Europe and Japan. OECD, Paris.

Chapter 16

TECHNOLOGICAL INNOVATION AND THE THEORY OF THE FIRM: THE ROLE OF ENTERPRISE-LEVEL KNOWLEDGE, COMPLEMENTARITIES, AND (DYNAMIC) CAPABILITIES

DAVID J. TEECE

Thomas W. Tusher Professor in Global Business

Haas School of Business, University of California

Berkeley, California, USA

Contents

Handbooks in Economics, Volume 01

DOI: 10.1016/S0169-7218(10)01016-6

Abstract

The firm is the central actor for the effectuation of innovation and technological change. The large industrial laboratories of the previous century have given way to more organizationally and geographically diffuse sources of technology, placing even greater emphasis on the coordination skills of managers. Dynamic capabilities are the skills, procedures, organizational structures, and decision rules that firms utilize to create and capture value. Managers must be able to sense opportunities, craft a business model to capitalize on them, and reconfigure their organizations, and sometimes their industries, as the business environment and technology shift. The key employees in this regard are experts (literati and numerati), whose management requires limited hierarchy, flexible teams, and performance-based incentives. To encompass these realities, the theory of the firm needs to be augmented to account for opportunity as well as opportunism, coordination beyond the boundaries of the firm as well as within it, variations in the level of capability across firms, and the frequent superiority of the firm over markets for the creation, transfer, and protection of intangible assets. Complementarities and cospecialization are advanced as two emerging concepts of particular relevance to a theory of the innovating enterprise earning above-normal returns.

Keywords

appropriability, business ecosystems, cospecialization, creativity, dynamic capabilities, entrepreneurs, hierarchy, innovation, intangible assets, literati, numerati, organizational structure, strategic management, transaction costs

1. Introduction

The advanced economies of Europe and the United States have gone through a significant transformation over the last half-century. The industrial age, characterized by a supply driven logic (build it and the customers will come) and relying on mass production has given way[1] to more customer-centric logic,[2] characterized by better customer information, rapid feedback cycles, and denser interfirm relationships. At the same time, the organization of innovation is being transformed by an increase in the geographic and organizational diversity of the sources of productive knowledge,[3] and by new ways of organizing.

In innovation-driven economies intangible assets, including relationship capital, are critical to the creation and production of new goods and services. In such economies, it is well recognized that the firm is a key, if not the most important, institution, through which technological change is effectuated. Despite the stark reality of this situation, the economic theory of the firm is still, as Rosenberg (1982) pointed out more than a quarter century ago, a "black box," when it comes to displaying (and understanding) the processes which result in the creation of new products and services and their profitable commercialization. Economics as a discipline may have had success with developing an understanding of the consequences of technological change, but the firm-level and market determinants are still enigmatic.

While the business enterprise plays a large role in determining the rate, direction, and nature of the commercially relevant technological change, the firm's ecosystem, including supporting institutions and legal structures, remains of great importance too, but is omitted in much theorizing about enterprise performance. Likewise, economic theories of the firm often blot out considerations of capability augmentation, technology transfer, and management, despite their great importance in today's industrial landscape. In particular, technology-driven firms have had to face the problem of how to manage and integrate the output of highly skilled experts (*literati* and *numerati*) across countries, time zones, and organizational boundaries. Management involves not only motivating talent and ensuring the job gets done; there is also a strategic component—what tasks to assign, what priorities to set, what resources to use, and where to get them from. To respond to these challenges, business enterprises need to develop capabilities and deploy them on a global basis. Economic theory has barely begun to recognize this.

This chapter does not attempt to survey the economics of technological change for which there is now a large literature. Instead, discussion is restricted to more neglected topics that relate primarily to innovative activities and their management inside the firm. These include:

[1] See Piore and Sabel (1984) who were among the first to trumpet the end of mass production.

[2] Today's era is not one in which big business has to be held in check by big labor, as Galbraith (1952) claimed was necessary; rather big business is held in check by customers who have choices, and by small businesses that compete with them, using domestic or offshore production bases. Global competition and dispersion in enterprise capabilities has enabled these trends. Because of these developments, large-scale established firms like Citibank, IBM, AT&T, Sears and General Motors have had their very existence threatened; and they are transforming in order to survive. Meanwhile, companies with barely a quarter century of history, like Apple, Cisco, Dell, Google, Microsoft, and Wal-Mart are now very significant firms on the industrial landscape. In this new world, entrants, not incumbents, tend to create most of the new jobs.

[3] See, for example, Teece (2000). There is no doubt that technological innovation is the primary driver of economic growth and was also critical to the outcomes in both world wars.

1. A historical sketch of R&D and innovative activities by firms over the past 100 years, describing at a general level how the organizational model associated with R&D activity in the private sector has changed over the past century.
2. The landscape in which firms innovate, including the institutional setting, market structure (which has attracted inordinate attention from economists), and the technology environment.
3. An introduction to the concept of the firm's capabilities and the dynamic capabilities framework.
4. Notes on the internal structures and incentives conducive to innovation, including the management of experts.
5. A theory of the innovating firm; how the firm not only solves contracting problems but also develops and deploys capabilities for the creation and management of know-how. The theory of the innovating firm advanced here does not displace transaction cost theory. Rather it builds a capabilities dimension into the Coase-Williamson theory of the firm.

2. The organization and environment of enterprise-level R&D

In the advanced economies of North America, Europe, and Asia, the business firm is at the core of the system of technological innovation.[4] The emergence and growth of industrial research and development (R&D) during the twentieth century, first in the United States and later in Europe, must rank as one of the most important economic developments in modern history.[5]

This enterprise-level R&D has steadily evolved in response to competitive opportunities and pressures. Its internal organization and relation to external actors have changed completely. Today, innovation takes place in a transformed global landscape from that of a century, or even 20 years, ago.

Industrial R&D is the activity in which the talents of scientists and/or engineers, the numerati, are harnessed to create new products, processes, and services. "R&D" encompasses several classes of activities that can occur in any order and across multiple organizations. There is basic research, which is aimed purely at the creation of new knowledge. Its purpose is to create new understandings of phenomena. There is applied research, which is work expected to have a practical, but not a commercial, payoff. Development is when the technologies behind a product or service are integrated and honed toward commercial application. Boundaries among these activities are quite fuzzy, and the manner in which they have been organized and linked has changed over time.

The roots of (American) industrial research can be found in the early nineteenth century when independent inventors like Eli Whitney (cotton gin) and Charles Goodyear (vulcanization of rubber) set out to commercialize their own inventions, often earning only meager profits. Later in the century, science began to be applied to industries such as dyestuffs, chemicals, electricity, and telecommunications.

Corporate research laboratories first appeared in the German chemical industry in the late nineteenth century, following the enactment of strong patent protection that put a stop to rampant imitation

[4] This is not to discount the critical role of supporting institutions as discussed in Nelson (1993) and elsewhere, but the emphasis in this chapter is on the internal operations of the firm.

[5] More detailed accounts of this history can be found in Mowery and Rosenberg (1989) and Hounshell (1996).

(Hounshell and Smith, 1988, p. 4). The first organized research laboratory in the United States was established by the inventor Thomas Edison in 1876. In 1886, an applied scientist by the name of Arthur D. Little started his firm, which became a major technical services/consulting firm to other enterprises.

Corporate laboratories on the German model began to appear in the United States soon after the Sherman Antitrust Act of 1890 steered companies to look for new ways to gain an advantage over rivals. Significant R&D labs were founded in the years before World War I at Eastman Kodak (1893), B.F. Goodrich (1895), General Electric (1900), Dow (1900), DuPont (1902), Goodyear (1909), and American Telephone and Telegraph (AT&T; 1907).

Independent research organizations like Arthur D. Little and the Mellon Institute continued to grow during the early twentieth century, but were surpassed by the rapid expansion of in-house research (Mowery, 1983). However, the many technology contracting problems and the efficiencies achievable from integration with manufacturing meant that external R&D could only serve as a complement, not a substitute, for in-house research (Armour and Teece, 1980).

The founding of formal R&D programs and laboratories stemmed in part from competitive threats. For instance, AT&T at first followed the telegraph industry's practice of relying on the market for—that is, it outsourced—technological innovation. However, the expiration of the major Bell patents and the growth of large numbers of independent telephone companies helped stimulate AT&T to organize Bell Labs to generate inventions and innovations internally. Competition likewise drove George Eastman to establish laboratories at Kodak Park in Rochester, New York, to counteract efforts by German dyestuff and chemical firms to enter into the manufacture of photographic chemicals and film.

During the early years of the twentieth century, the number of research labs grew dramatically. By World War I there were perhaps as many as one hundred industrial research laboratories in the United States. The number tripled during World War I, and industrial R&D even maintained its momentum during the Great Depression. The number of scientists and research engineers employed by these laboratories grew from 2775 in 1921 to almost 30,000 by 1940.

The interwar period saw some of the labs make significant advances in basic research. In 1927, Clinton Davisson began his work at Bell Labs on electron diffraction. His work led to a Nobel Prize in physics in 1937. At DuPont, Wallace Carothers developed and published the general theory of polymers, and went on in 1930 to create synthetic rubber, and then a strong, tough, water-resistant fiber called nylon. These technological breakthroughs were in and of themselves of great importance, but it took time and money to leverage them into marketable products. For instance, over a decade elapsed from the beginning of research in super polymers to the production of nylon on commercial terms.

Building on wartime success, including the Manhattan Project (to create the atomic bomb), the era of big science began after World War II, fueled by the optimism that well-funded scientists and engineers could produce technological breakthroughs that would benefit the economy and society. University scientists, working together with the engineers from corporate America, had indeed produced a string of breakthrough technologies including radar, antibiotics, the digital electronic computer, and atomic energy. The dominant intellectual belief of the immediate postwar period was that science-driven research programs would ensure the development of an endless frontier of new products and processes. The development of the transistor at Bell Labs gave strength to this view, and many firms augmented their commitments to industrial R&D, including a small portion of purely basic research. In 1956, IBM established a research division whose mandate included world-class basic research.

As international tensions increased during the Cold War, government funding grew considerably. In 1957, government funding of R&D performed by industry eclipsed the funding provided by the firms themselves. By 1967, it went back the other way, with private funding taking the lead. By 1975, industry funding of industry-conducted R&D was twice the federal level and the ratio was expanding.

Government procurement was perhaps even more important to the technological development of certain industries, as it facilitated early investment in product facilities, thus easing the cost of commercialization. The newly emergent electronics industry in particular was able to benefit from the Defense Department's demand for advanced products. By 1960, the US electronics industry had come to rely on the federal government for 70% of its R&D dollars (which may have cost US firms their leadership in consumer electronics as they became preoccupied with the more performance-oriented requirements of the US military).

By the early 1970s, management was beginning to lose faith in the science-driven approach to innovation, primarily because few blockbuster products had emerged from the research funded during the 1950–1970s. Competition became more global, leaving firms less certain of cash flow from their domestic market for funding R&D. New technology was not converted into new products and processes rapidly enough, confronting many companies with the paradox of being leaders in R&D and laggards in the introduction of innovative products and processes. The fruit of much R&D was appropriated by domestic and foreign competitors, and much technology languished in research laboratories. In telecommunications, Bell Labs' contribution to the economy at large far outstripped its contribution to AT&T. In the semiconductor industry, Fairchild's large research organization contributed more to the economy through the spin-off companies it spawned than to its parent. Xerox Corporation's Palo Alto Research Center made stunning contributions to the economy in the area of the personal computer, local area networks, and the graphical user interface that became the basis of Apple's Macintosh computer (and, later, of Microsoft's Windows). Xerox shareholders were well served too, but most of the benefits ended up in the hands of Xerox's competitors or of companies in adjacent industries.

Different modes of organization and different funding priorities were needed. Knowledge throughout the firm had to be embedded in new products promptly placed into the marketplace. A new way of conducting R&D and commercializing new products was needed.

By the 1980s and 1990s, a new model for organizing research became apparent. First, inside large corporations, R&D activity came to be decentralized, with the aim of bringing it closer to users and customers. By the mid-1990s, Intel, the world leader in microprocessors, was spending over $1 billion per year on R&D, yet did not have a separate R&D laboratory. Rather, development was conducted in the manufacturing facilities. Intel did not invest in fundamental research at all apart from its funding of university research and some research activities located on or near university campuses.

Second, many companies were looking to the universities for much of their basic or fundamental research, maintaining close associations with the science and engineering departments at the major research universities. Indeed, the percentage of academic research funded by industry, which had declined to 2.5% by 1966, rose steadily to 7.4% in 1999, declining since then to about 5%, its level in the early 1980s (National Science Board, 2008, Appendix Table 4-3). Strong links between university research and industrial research are present in electronics (especially semiconductors), chemical products, medicine, and agriculture. For the most part, university researchers are insufficiently versed in the particulars of specific product markets and customer needs to help configure products to the needs of the market.

Third, corporations have embraced horizontal, vertical, and lateral alliances involving R&D, manufacturing, and marketing in order to get products to market quicker and leverage off complementary assets and capabilities already in place elsewhere. A variant on this strategy is the new product-oriented corporate acquisition, employed as a vital complement to in-house R&D, perhaps most notably by Cisco, which has spent billions to acquire dozens of companies with products that had been recently placed into the market (Mayer and Kenney, 2004). It is important to note, however, that outsourced R&D is a complement, not a substitute, to in-house activities. Outsourcing and codevelopment arrangements had become common by the 1980s and 1990s (e.g., Pratt & Whitney's codevelopment programs for jet engines, or the IBM-Sony-Toshiba alliance for the development of the Cell processor) as the costs of product development increased, especially after the antitrust laws were modified to recognize the benefits of cooperation in R&D and related activities. Cooperation was also facilitated by the emergence of capable potential partners in Europe and Japan.

These developments meant that at the end of the twentieth century, R&D was being conducted in quite a different manner from how it was organized at the beginning. Many corporations had closed, or dramatically scaled back, their central research laboratories, including Westinghouse, RCA, AT&T, US Steel, and Unocal to name just a few. Alliances and cooperative efforts of all kinds were of much greater importance.[6] Many firms are now sourcing much of their innovation externally, following an "open" innovation model (Chesbrough, 2006).

Moreover, much of the momentum for commercializing innovations had shifted to venture capital-funded "start-ups." By the 1980s, private venture funds began to have a transformative effect on the US industrial landscape, particularly in biotech and information technology. They dramatically increased the funds that were available to, as well as the professionalism of, entrepreneurs.

In many ways these new, agile venture-funded enterprises still depended on the organized R&D labs for their birthright. Some start-ups were exploiting technological opportunities that incumbents had considered and rejected.

The long lead time needed to commercialize early stage research (and the potential for leakage to domestic and foreign rivals) was difficult for management to justify. Venture funds were also generally uninterested in funding exploratory research. This has left basic and applied research in some sectors (like communications) with a diminished funding base. Some observers fear that society is "eating its seed corn."[7]

The organization of R&D in the last half-century also becomes multinational in scope.[8] The result is that, by 2000, domestic as well as multinational firms employ numerati and literati in a globally distributed fashion to (1) develop localized products and services closer to offshore users, (2) take advantage of specialized sources of creativity and innovation, and (3) source development services from low-cost providers. It is particularly noteworthy that, since the late 1990s, United States and European companies have been establishing satellite R&D facilities in China and India at a high rate. However,

[6] Economists generally—and antitrust authorities in particular—have become more receptive to the notion that technology alliances among competing firms can bring societal as well as private benefits (Baumol, 2001; Jorde and Teece, 1990; Katz and Ordover, 1990; Teece, 1992).

[7] Interview with Dr. William Spencer, former head of R&D at Xerox Corporation.

[8] Although the offshore R&D trend has accelerated in recent years with the advent of improved talent pools in industrializing countries, enhanced telecommunications, and liberalized trade, the use of foreign R&D labs has deep roots. See Mansfield et al. (1979).

the trend toward greater globalization should not be exaggerated. Globalization of innovation (as distinct from manufacturing) has not yet been significant in all industries, and it often involves lower level activities based on technologies developed closer to company headquarters.

3. The innovating firm in context

The above short history makes it apparent that firms exist and innovate neither in isolation nor in some "flat" world of uniformly and globally distributed capabilities. Before analyzing the nature and organization of the dynamically innovating firm, one must understand the external factors that affect such firms. Firms operate with a balance—sometimes favorable, sometimes unfavorable—of help or hindrance from domestic and local institutions. Another element of the context for innovation is market structure. The technological environment in which a firm innovates is yet a third cluster of factors which shape (and are shaped by) innovation.

3.1. The ecosystem for innovation

Several important literatures address factors in the firm's external environment which impact firm-level innovative performance. These literatures are not themselves well integrated, and bear labels such as national systems of innovation, regional systems of innovation, clusters, and ecosystems. This section does not attempt to review this literature, but merely highlights some of the key elements.

The basic argument of the literature is that firm-level innovation depends on the supply of skilled workers (who are not entirely mobile internationally), universities (for access to both highly educated talent and faculty research), financial institutions (especially venture capital), the legal system (especially intellectual property law and employment law), the supply base (including complementors), the domestic market, and the presence of other firms in the same or related industries. Figure 1 displays factors and their interaction.

While institutional structures can have national identities, they may have regional identities, too. Work on national and regional systems argues for defining what might be thought of as national and regional business ecosystems supporting innovation (Nelson, 1993). The evidence supporting the concept is considerable, with Silicon Valley being a classic case (Saxenian, 1996).

Economic historians have always given considerable weight to the role of institutions and government in economic growth at the national level (e.g., Abramovitz, 1986; Nelson, 1982), but very few studies connect the performance of particular firms to key elements in the ecosystem. However, vignettes and anecdotes abound. In the US civilian aircraft industry, for example, foreign technology and government procurement were vital inputs to domestic innovation. Boeing and others in the United States accessed developments in jet engine technology that had occurred in the United Kingdom and Germany in the creation of their own aircraft (Phillips, 1971). Boeing leveraged its subsequent success with the KC-130 jet tanker built for the Air Force into a civilian version—the Boeing 707—and captured a lead in global market share that lasted until the emergence and growth of Airbus. The nascent semiconductor industry also benefited from the willingness of the US military to buy advanced products at premium prices.

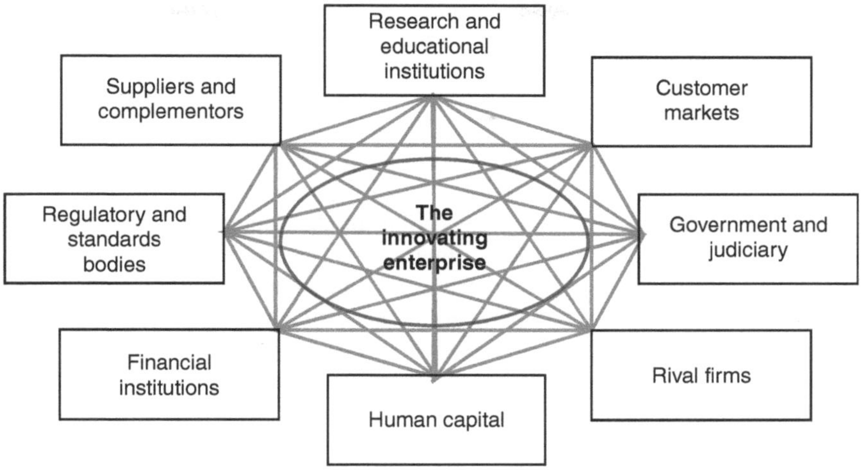

Figure 1. The innovation ecosystem.

Another prominent example of the ecosystem impacting innovation is the Internet. The basic technology and structure of the Internet has its origins in university research applied in the late 1960s by Bolt, Beranek, and Newman, a contractor to the US Department of Defense, to build a network connecting researchers with government contracts to government-sponsored computers in order to maximize resource utilization. ARPANET gradually extended its reach around the world and was merged in 1983 with similar networks to form the Internet.

3.2. Market structure as a determinant of innovation

A less important context for innovation, although one which has received an inordinate amount of attention by economists over the years, is market structure, particularly the degree of market concentration. Indeed, it is not uncommon to find debate about innovation policy among economists collapsing into a rather narrow discussion of the relative virtues of competition and monopoly. Clearly there is much more at work, including the various elements of the ecosystem noted above. Subsequent sections will identify various dimensions of internal firm structure and management that also impact the rate and direction of innovation.

There has been considerable debate and scholarly attention to the role of market structure on determining firm-level innovation. Schumpeter was among the first to declare that perfect competition was incompatible with innovation. The hypothesis often attributed to him (see Schumpeter, 1942, especially Chapter VIII), posits that profits accumulated through the exercise of monopoly power (assumed to be correlated with large firms) are a key source of funds to support risky and costly innovative activity. These predictions, even as a matter of theory, are not well grounded in the financial realities of the firm (Kamien and Schwartz, 1978).

Any theory of market power as a funding mechanism for innovation in specific markets is further unshackled if the multiproduct (multi-industry) firm is admitted onto the economic landscape. The multiproduct structure allows the allocation of cash generated anywhere to be directed to high-yield purposes everywhere inside the firm. The fungibility of cash inside the multiproduct firm thus unlocks any causal relationship between market power (which is a market-specific concept) and innovation.

The Schumpeterian notion that small entrepreneurial firms lack adequate financial resources for innovation seems at odds with his earlier views (1934) on entrepreneur-led innovation and seems archaic in today's circumstances where venture capital-funded enterprises play such a large role in innovation (Gompers and Lerner, 2001). From time to time, public equity markets have also funded relatively early stage biotech and Internet companies with minimal revenues and negative earnings.

Another setback for the various Schumpeterian market structure-innovation hypotheses is that the logic can run the other way: namely, that innovation shapes market structure. Success garnered from innovation can lead to market concentration, as it has with Intel and Microsoft, and as it once did with the Ford Motor Company and Xerox.

Various reviews of the extensive literature on innovation and market structure generally find that the relationship is weak or holds only when controlling for particular circumstances (Cohen and Levin, 1989; Gilbert, 2006; Sutton, 2001). The emerging consensus (Dasgupta and Stiglitz, 1980; Futia, 1980; Levin and Reiss, 1984, 1988; Levin et al., 1985; Nelson and Winter, 1978) is that market concentration and innovation activity most probably either coevolve (Metcaffe and Gibbons, 1988) or are simultaneously determined. Context (stage in the industry life cycle; technological environment) is likely to matter.

3.3. The technological environment

An important consideration shaping innovation is the technological environment that surrounds and shapes the firm's technological activity.

One prominent feature of the environment is the abundance (or scarcity) of technological opportunities. In an industry with lots of technological opportunities, innovation is expected to be relatively easy due to a lower expected development cost and/or a plentiful supply of relevant and available knowledge. For example, university and government (funded) research in science and technology help create vibrant technological environments with multiple sources of new technology, fueling venture-funded new businesses. Biotech is a case where US government funds distributed through the National Institutes of Health have helped to create technological opportunities which are then seized upon and developed further by new venture-funded startups. While most of these companies fail, enough survive to impact the structure of the pharmaceutical industry.

Technology opportunities may shed light on the market structure enigma from the previous section. A leading textbook (Scherer and Ross, 1990, p. 645) notes that "the structure-to-innovation linkage probably operated over a much shorter time span than the innovation-to-structure linkage." This second linkage is expected to be stronger in industries with rich technology opportunities. The idea is that concentration is more conducive to innovation in slow-moving fields, whereas technological opportunity, which can give rise to radical breakthroughs, favors newcomers, not incumbents. These refinements seem plausible; however, recent empirical work suggests that the relationship between technological opportunities and market leadership are far from straightforward (Fai, 2007).

A broader concept than "technological opportunities" is that of "technological regimes." Nelson and Winter (1982) claimed that knowledge and opportunity are determined by the underlying "technological regime" (p. 258). Studies have identified various types of technological regimes without settling on a common definition (Acs and Audretsch, 1990; Malerba and Orsenigo, 1993; Pavitt et al., 1987; Shane, 2001; Winter, 1984). Variables that have been considered in these studies include technological opportunity, the stage of the technology's life cycle, appropriability, cumulativeness, complexity, and capital requirements.

Across the range of opportunities encompassed by a given technological regime, innovation tends to occur along trajectories within a technological paradigm (Dosi, 1982; Teece, 2008).[9] The paradigm includes a definition of those problems currently identified as important and the patterns of solutions (principles) that should be applied. To this, firms bring their internal routines and other tacit knowledge for problem solving, but the paradigm serves as a (soft) constraint on the innovations likely to occur, until a disruptive innovation leads to a new paradigm.

Just how and why some firms tap into technological opportunities remains somewhat enigmatic; the microanalytics of these decisions are not well explained by economic theory, or by any other theory for that matter. The various economic theories of innovation pay very little attention to factors inside the firm. An effort to remedy the situation is commenced in the next section.

4. Resources, competences, and dynamic capabilities

To better understand the nature of the innovating firm, several concepts developed in the strategic management literature are relevant. However, the terminology of resources, competences, and capabilities has never been standardized. To reduce confusion, this section defines the key terms used in this chapter.

4.1. Resources/competences

Resources are firm-specific assets that are difficult, or impossible, to imitate. They are stocks, not flows. They could be tangible but are more likely to be intangible. Such assets are idiosyncratic in nature, and are difficult to trade because their property rights are likely to have fuzzy boundaries and their value is context dependent. As a result, there is unlikely to be a well-developed market for resources/competences; in fact, they are typically not traded at all. They are also generally difficult to transfer among firms. Examples include intellectual property, process know-how, customer relationships, and the knowledge possessed by groups of especially skilled employees.[10]

[9] This notion of a consensus-based technology trajectory lies in the background of a common innovation model in economics, the patent race (e.g., Reinganum, 1981).

[10] While the industrial workforce has always contained individuals with high education and/or exceptional talent, the economic significance of such literati and numerati has become more important as the traditional sources of firm profitability have been undermined (Albert and Bradley, 1997, p. 4). The nature and management of the firm's "expert talent" are discussed below in Section 6.

Competences are a particular kind of organizational resource. They result from activities that are performed repetitively, or quasi-repetitively. Organizational competences enable economic tasks to be performed that require collective effort. Organizational competences are usually underpinned by organizational processes/routines. Indeed, they represent distinct bundles of organizational routines and problem-solving skills.[11]

In short, ordinary competence defines sufficiency in performance of a delineated organizational task. It is about doing things well enough, or possibly very well, without attention to whether the economic activity is the right thing to do. Competences can be quantified because they can be measured against particular (unchanging) task requirements. The level of a competence can be benchmarked; the assessment of a competence does not require that the activity be aligned with the firm's environment and other assets/competences.

Some processes undergirding competence are formal, others informal. As employees address recurrent tasks, processes become defined. The nature of processes is that they are not meant to change (until they have to). Valuable differentiating processes may include those that define how decisions are made, how customer needs are assessed, and how quality is maintained.

As an organization grows, its capabilities are embedded in competences/resources and shaped by (organizational) values. Organizational values define the implicit norms and rules of the organization. They determine how it sets priorities with respect to how employees and affiliates work together.

While economics has often modeled firms as homogeneous, or asymmetric only in their access to information, the "resource-based view" of the firm recognizes the unique attributes of individual firms. The resources framework has developed in the management literature, building on Penrose (1959), Rubin (1973), and others. In the 1980s, a number of strategic management scholars, including Rumelt (1984), Teece (1980b, 1982, 1984), and Wernerfelt (1984) began theorizing that a firm earns rents from leveraging its unique resources, which are difficult to monetize directly via transactions in intermediate markets. This in turn gave rise to the analysis of learning and knowledge management as the means to develop and augment new, hard-to-imitate resources.

Since markets are a great leveler, firms can build long-term profitability in what we normally think of as competitive markets mainly from the ownership and orchestration of nontradable (intangible) assets. If an asset or its services are traded in a market, it can be accessed by all who can pay. The range of domains in which competitive advantage can be built narrows as more and more activities become outsourceable. The Internet and other recent innovations have vastly expanded the number and type of goods and services that are readily accessed externally.

The set of intangible assets that remains especially difficult—although not impossible—to trade consists of knowledge assets and, more generally, resources. Knowledge assets are tacit to varying degrees and costly to transfer (Teece, 1981). The market for know-how is also riddled with imperfections, which favors internalization to capture strategic value in that certain assets are more valuable to one firm than another. Assets that have such special value are referred to here to as "strategic assets."

A firm's resources, which can include knowledge and intellectual property, are significant potential sources of advantage. As already noted, "resources" are a stock, not a flow. However, resources must be constantly renewed (Teece, 2009). The need for renewal is amplified in fast-moving environments such

[11] Organizational competences have their roots in the work of Simon (1947), Nelson and Winter (1982), Winter (1988), Teece et al. (1994), and Dosi et al. (2000).

as those characteristic of high-tech sectors (e.g., computers). However, a need to renew resources can also occur in "low-tech" industries (e.g., life insurance).

4.2. Resources/competences, "strategic" assets, and price theory

Because the value of a resource/asset is context dependent, the market for such assets is generally thin. Two powerful economic implications follow: when one is able to secure strategic resources/assets through purchase, they may be bought for less than they are worth to the buyer because this may be considerably more than they are worth to the seller (the converse in also true). Put differently, from some perspectives, the market need not fully price strategic assets/resources. Accordingly, "abnormal" or "supernormal" profits can flow, at least for a while, from securing (either by purchase or through "building" internally) such assets.

This situation (i.e., abnormal returns) has its roots not in luck but in the possession of other complementary or cospecialized assets (which creates unique contexts), along with sensing (including search) and seizing (including good execution). The market may not lead to the necessary "coordination." It is well understood that the price system's normal asset allocation role is unlikely to occur properly when asset values depend on idiosyncratic combinations. Achieving such value-enhancing combinations is discussed below under the concept of dynamic capabilities. The entrepreneurial manager, not the Walrasian arbitrageur, achieves the microlevel coordination that economic theory (and the economy) requires.

An implication of this approach is that input or factor markets are not fully efficient. Mainstream price theory expounds the view that with (perfect) competition it is impossible to purchase something for less than it is worth or for less than the long-term costs of producing it. However, and without appealing to monopsony theory, it is often possible to secure (by buying or building) something for less than it is worth (to the buyer/owner) if one has superior information, or if one owns related specific assets for which there is no established market.

Note that thin markets are ubiquitous when one is referencing intangible assets/intellectual property and "resources." Hence, the context in which the phenomenon at hand is common includes situations where intangible assets are used intensively.

Economic theory has yet to recognize these lacunae, and explore implications for the theory of the firm. Hints about what is being articulated here can be found in Richardson (1972) and in the literature on entrepreneurship (e.g., Kirzner, 1997). Teece (1980b, 1981) made the case that the market for know-how—a critical "resource"—was riddled with inefficiencies which would blunt market exchange, and favor internal organization. It is well recognized that it is generally hard for a firm to earn better than a competitive return if factor markets are efficient (Barney, 1986). But, if one uses what Oliver Williamson refers to as "main case reasoning," then the (implicit) "main case" where intangible assets are created, transferred, combined, and used intensively, is that the market is not perfectly competitive and all public information need not be reflected in the prices of specific assets, tangible or intangible.

Even if prices did reflect all information, the thin market phenomenon referenced here would still result in wide bands for "competitive" prices if firms are heterogeneous and innovation and product differentiation are ubiquitous. This is the setup implicitly adopted in the strategic management literature (Denrell et al., 2003; Rumelt et al., 1991; Teece and Winter, 1984). Modern auction theory (e.g., Klemperer, 2002) likewise recognizes that assets will not achieve their full value in an auction if there is only one buyer. What is missing is an effort to tie these disparate threads in the literature to a theory of the firm. The concept of dynamic capabilities is a framework to move the theory of the firm in that direction.

4.3. Dynamic capabilities

Dynamic Capabilities are the firm's ability to integrate, build, and reconfigure internal and external resources/competences to address and shape rapidly changing business environments (Teece et al., 1997, 1990). The goal is to generate abnormal returns. Dynamic capabilities may sometimes be rooted in certain change routines (e.g., product development along a known trajectory) and analysis (e.g., of investment choices). However, they are more commonly rooted in creative managerial and entrepreneurial acts (e.g., pioneering new markets). They reflect the speed and degree to which the firm's idiosyncratic resources/competences can be aligned and realigned to match the opportunities and requirements of the business environment. An organization with strong dynamic capabilities can achieve abnormal returns because markets do not price them at their value to the buyer if the buyer possesses complementary and, in particular, cospecialized assets.

The essence of resources/competences as well as dynamic capabilities is that they cannot generally be bought; they must be built. As noted above, dynamic capabilities measure the capacity to align and realign, and resources/competences are integrated and reintegrated so that they are tuned to the business environment. Sensing, seizing, and transforming are particular attributes of firms that enable them to evolve and coevolve with the business environment. Such capabilities are critical to long-term profitability (Teece, 2007b).

Sensing and seizing are similar to two activities discussed in the management literature as potentially incompatible inside a single organization: exploration and exploitation (March, 1991). Exploration (e.g., research on a potentially disruptive technology) has a longer time horizon and greater uncertainty than exploitation (e.g., selling mature products). The two types of activities require different management styles; one solution is an "ambidextrous organization" where two separate subunits with different cultures are linked by shared company-wide values and senior managers with a broad view—and appropriate incentives (O'Reilly and Tushman, 2004).

As discussed above, a firm's basic competences, if well honed, enable it to perform efficiently the activities that it sets out to perform. However, whether the enterprise is currently making the right products and addressing the right market segment, or whether its future plans are appropriately matched to consumer needs and technological and competitive opportunities, is determined by its dynamic capabilities. Dynamic capabilities, in turn, require the organization (especially its top management) to develop conjectures, validate them, and realign assets and competences for new requirements. They enable the enterprise to profitably orchestrate its resources, competences, and other assets.

Dynamic capabilities are also used to assess when and how the enterprise is to ally with other enterprises. The expansion of trade has enabled and required greater global specialization. To make the global system of vertical specialization and cospecialization (bilateral dependence) work, there is a need (indeed an enhanced need) for firms to develop and align assets and to combine the various elements of the global value chain so as to develop and deliver a joint "solution" that customers value.[12]

[12] Cospecialization has strong implications for organization and strategy (Teece, 2007a).

Not infrequently, the innovating firm(s) will be required to create a market, such as when an entirely new product is offered to customers, or when new intermediate products must be traded. Dynamic capabilities, particularly the more entrepreneurial competences, are a critical input to the market creating (and cocreating) processes.[13]

To summarize, dynamic capabilities reflect the capacity a firm has to orchestrate activities and resources/assets within the system of global specialization and cospecialization. They also reflect the firm's efforts to create/shape the market in ways that enable value to be created and captured. This often requires extending, modifying, or, if necessary, completely revamping what the enterprise is doing so as to maintain a good fit with (and sometimes to transform) the ecosystem and markets that the enterprise occupies. Microfoundations and organizing principles have been laid out elsewhere (Teece, 2007a). A brief summary is provided below for the major (dynamic) capability categories.

5. A dynamic capabilities view of the firm

5.1. General

As discussed earlier, the proper functioning of an economy experiencing change (whether driven by innovation or anything else) requires resources (including intangible assets) and ordinary competences/capabilities to get the job done (i.e., to produce the products/services customers want and to do so efficiently and expeditiously). Such an economy will also need either a new set of firms to produce what customers want next (and what technology allows next), or it will need existing firms to morph in order to both shape and address new opportunities and threats.

For an existing enterprise, there is a requirement to first identify new opportunities and threats and then to shape and reshape the enterprise—and possibly elements of the market environment itself. The capacity to reengineer the enterprise and its product offerings, its internal activities, and its external relationships is what we mean by dynamic capabilities. Externally, they also involve managing/pacing the coevolution of suppliers, competitors, and complementors.

Dynamic capabilities require entrepreneurial activity; but dynamic capabilities use the current "platform" of the enterprise. Dynamic capabilities are not simply manifestations of "intrapreneurship," although this may be an element. Dynamic capabilities have both external and internal organizational dimensions.

5.2. Innovation and change

Innovation and change are sometimes conceived as a two-step procedure—invention and commercialization (Mansfield, 1974). In fact, Teece (2007a) suggests that it is more realistic to view continuous renewal as requiring an ongoing set of activities and adjustments that can be divided into three clusters: (1) identification and assessment of an opportunity (*sensing*), (2) mobilization of resources to address an opportunity and to capture value from doing so (*seizing*), and (3) continued renewal (*transforming*).

[13] The entrepreneurial creation and cocreation of markets is often required to ensure the appropriability of returns from innovation (Pitelis and Teece, 2009).

Table 1
Activities conducted to create and capture value (organized by clusters of dynamic capabilities)

	Sensing	Seizing	Transforming
Creating value	Spotting opportunities Identifying opportunities for research and development Conceptualizing new customer needs and new business models	Investment discipline Commitment to research and development Building competencies Achieving new combinations	Achieving recombinations
Capturing value	Positioning for first mover and other advantages Determining desirable entry timing	Intellectual property qualification and enforcement; Implementing business models Leveraging complementary assets Investment or coinvestment in "production" facilities	Managing threats Honing the business model Developing new complements

Collectively, these are a firm's *dynamic capabilities*. They are required if the firm is to sustain itself as markets and technologies change.

One could imagine that a market economy would allow individuals and organizations to specialize in one of the three capability clusters. However, the markets for opportunities, inventions, and know-how are riddled with inefficiencies and high transaction costs, and most entrepreneurs are forced to bundle these activities (i.e., do all three).[14]

The relative importance of the competences and adjustment mechanisms that constitute sensing, seizing, and transforming varies according to circumstance. To simplify the analysis of dynamic capabilities even further, they can be grouped into two essential classes of activities: creating value and capturing value (see Table 1).

Dynamic capabilities are most relevant in a regime of rapid change, a condition that prevails in a growing number of industries. The global economy has undergone drastic changes that have accelerated the rhythm at which firms innovate. The decreased cost of communication and data flow, the reduced barriers to trade, and the liberalization of labor and financial markets in many parts of the world are forcing firms to confront agile and/or low-cost competitors early in the product cycle. This in turn has caused firms to undertake a major revision of their innovation strategies, such as the closure or downsizing of large industrial research labs described earlier and a corresponding greater reliance on open innovation.[15]

The following sections present dynamic capabilities in greater detail organized within the subsets of activities related to creating and capturing value.

[14] The market for opportunities is imperfect due both to problems of conveying the merits of ideas and also because of opportunism, which can lead to the "lemons" problem identified by Akerlof (1970). In general, entrepreneurs will be reluctant to "sell" or simply license ideas they believe are undervalued. The outcome thus tends toward internalization. For an early statement of some of these issues, see Teece (1981).

[15] This is in some ways the mirror image of the "Second Industrial Revolution," when earlier improvements to communications (telegraph) and transportation (railroad) induced a period of vertical integration on a continental scale with an emphasis on in-house R&D (Chandler, 1990).

5.3. Creating value with innovation

Despite its obvious importance, a theory of how firms create value is largely missing from the standard economics literature. To the extent it is addressed, the industrial organization literature dwells almost entirely on the funding of R&D, figuring (implicitly) that the R&D expenditure is the main driver of innovation. However, R&D activity is only one of several factors likely to determine the generation of new ideas.[16] The concept of dynamic capabilities—the sensing, seizing, and transformation that ongoing innovation requires—provides a broader framework to help one understand how firms create value.

Sensing is an entrepreneurial activity—whether conducted by a new or an existing firm—that involves the identification and conceptualization of opportunities both within and beyond prevailing technological paradigms (Teece, 2008). It involves cognition. As markets evolve, changes in consumer needs, product technologies, and the competitive positioning of other companies can threaten a firm's existing position or open the possibility of a new or better one. In some cases, as stressed by Kirzner (1973), the entrepreneur/manager may have differential access to existing information relative to rivals. More often, sensing opportunities involves scanning, interpretation, and learning across technologies and markets, both "local" and "distant", that are also visible to rival firms (March and Simon, 1958; Nelson and Winter, 1982).

In reality, management teams often find it difficult to look beyond a narrow search horizon tied to established competences. Henderson (1994) cites General Motors, Digital Equipment, and IBM as companies that faced major problems from becoming trapped in their deeply ingrained assumptions, information filters, and problem solving strategies.

Seizing an opportunity requires investments in development via further creative and/or combinatorial activity that addresses the opportunity with new products, processes, or services. It may involve building a necessary new competence or identifying an appropriate external alliance that can secure access to one.

As the global sources of invention and innovation become dispersed, it is less likely that the enterprise can rely on internal R&D, even in very large firms. As a result, services and intangibles that formerly needed to be built internally are outsourced, at least partially. Declines in the cost of computing and communications have facilitated collaboration with suppliers and other elements of the innovation ecosystem (Teece, 1989). This means that markets open up for at least some items of know-how and intellectual property. The expansion of outsourcing has increased the viability of an "open innovation" approach (Chesbrough, 2006). With open innovation, a firm identifies and exploits new technologies and creative capacities developed both inside and outside the boundaries of the firm.[17]

An open innovation approach can even be used to leverage assets that were not previously organized for the purpose. The movie rental firm Netflix, for example, ran a 3-year contest for improving the accuracy of its movie recommendation algorithm by more than 10%, with the winner retaining ownership of the solution, apart from a compulsory free license to Netflix. Over 51,000 contestants

[16] The literature on cumulative innovation, with its emphasis on optimal patent policies (e.g., Scotchmer 1991), captures some of the larger context for innovation, as does that on learning from customers (e.g., von Hippel, 1998). Section 3, above, discusses the broader range of factors that shape innovation.

[17] Open innovation does not necessarily mean that property rights are not sought, protected, and respected.

from 186 countries stepped up to the challenge. The $1 million prize was awarded in September 2009 to a team that had come together for the purpose, and Netflix promptly launched a follow-on competition (Ortutay, 2009). Netflix is not unique. A start-up, Innocentive.com, is a clearing house for this "crowdsourcing" approach where companies can post a research problem along with the amount they are willing to pay for a solution. By mid-2009, it had awarded more than $4 million for over 500 solutions.[18]

Transformation of the firm itself is the third capability required for creating (and capturing) value. Past efforts at sensing and seizing delineate a path for the creation of value, but over time the firm still needs to periodically consider (and reconsider) its own "fit" to the opportunities it plans to exploit. Management must assess the coherence of the firm's business model, asset structure, and organizational routines with respect to its environment. Yet commitment to existing processes, assets, and problem definitions makes this extremely hard to do, especially in a firm that is currently performing satisfactorily.

Organizational innovation can allow the firm to escape unfavorable path dependencies. However, reconfiguring the firm is often costly in terms of both money and morale. When such innovation is incremental, routines and structures can probably be adapted gradually. Radical organizational innovation can potentially be accommodated by a "break out" unit where new capabilities are established before being introduced to the firm as a whole (Teece, 2000).

Organizational innovation has a long history. As Chandler (1962, 1977) and Williamson (1975, 1981) have chronicled, the large, multidivisional (M-form) organization has its roots in the development of line-management hierarchies by the nineteenth-century railroads, which needed a system to manage a continent-spanning organization[19]. In the twentieth century, large corporations such as DuPont and General Motors gradually shifted from a functionally organized (U-form) structure to a multidivisional structure (M-form) that relieved top management of responsibility for operational details. Related innovations such as the conglomerate and the multinational forms allowed organizations to span a wider array of activities and locations than ever before.

Organizational innovation and change continue, with the benefits of greater decentralization being "rediscovered" as the enterprise grows. John Chambers, the CEO of US network equipment company Cisco Systems, described how the management structure of Cisco changed some 15 years after its founding: "In 2001, we were like most high-tech companies—all decisions came to the top 10 people in the company, and we drove things back down from there" (McGirt, 2008). It seems that Cisco now has a more decentralized and collaborative management system, with a network of councils and boards entrusted and empowered to launch new businesses, and incentives to encourage executives to work together flexibly. Chambers claim that "these boards and councils have been able to innovate with tremendous speed. Fifteen minutes and one week to get a [business] plan that used to take six months" (ibid).

Organizational innovation is not only an important form of creating value but of capturing it as well. Armour and Teece (1978) showed that the petroleum industry firms that first adopted M-form structures retained a profit advantage until the innovation was eventually replicated throughout the industry by the early 1970s.

[18] Information from http://www.innocentive.com/crowd-sourcing-news/innocentive-at-a-glance/. Accessed October 8, 2009.

[19] For treatments of organizational innovation and its diffusion, see Armour and Teece (1978), Teece (1980), and the discussion below.

5.4. Capturing value (profiting) from innovation

Companies that are narrowly focused on creating value will not perform well commercially. Invention without a commercialization strategy and access on competitive terms to complementary assets is unlikely to lead to commercial success. Although it is possible to disseminate some innovations (e.g., software over the Internet) without using complementary assets, most industrial innovations are of no benefit to consumers unless considerable resources and complementary assets are mobilized for production, distribution, and promotion. Many engineering-driven companies' brilliant ideas have never found (or created) a market.

Value capture requires selecting the right timing for market entry. In some cases, it is beneficial to be a first mover while in others it may be more advantageous to exploit a gap left by a pioneer. "Seizing" is the core competence cluster for capturing value and is encompassed by the Profiting from Innovation (PFI) framework, which is discussed below. Successful strategies to capture value require choosing an appropriate mechanism for the protection of intellectual property (e.g., trade secrets vs. patents), deciding which activities must be performed by the firm or procured in the market—discussed in Section 7.3—and crafting a business model.

A business model (Chesbrough and Rosenbloom, 2002; Teece, forthcoming) defines a product's value proposition for customers and how the firm will convert that to profit.[20] A business model defines an organizational and financial architecture which embraces and integrates in a consistent fashion (1) the feature set of the product or service; (2) the benefit (value proposition) from consuming/using the product or service; (3) the market segments to be targeted; (4) the "design" of revenue streams and cost structure; (5) the way products/services are to be combined and offered to the customer; and (6) the mechanisms by which value is to be captured.[21]

Google, developer of the leading Internet search engine, founded in 1998, provides clear examples of these business model elements in action. Initially, the company's investments in proprietary search algorithms and computing resources made it the most popular search engine on the Internet, but these innovations did not translate directly to profits. In late 2000, the company began auctioning ads linked to specific keywords (a system similar to that already employed by a competing search site, GoTo.com). Google recognized that part of its appeal was the minimalist design of its web site, and it has limited ads on the site. It was Google's combination of innovation, awareness of how it provided value both to search users and to advertisers, and a system for turning the advertising into revenue—and then into profit—that provided the foundation for the company's ongoing success.

[20] An e-mail from Bill Gates that became public during the Oracle-PeopleSoft merger provides insight into Microsoft's business model for software. In a 2002 message to Microsoft managers, he wrote:
"A product with high share generates a common sense around it.
A common sense that Community Colleges train on that product.
A common sense that temporary workers know the product.
A common sense that certification in the product is a valuable thing.
A common sense that the industry can exchange data or aggregate data using schema specific to that product.
A common sense that someone doing something new should move to that product.
A common sense in terms of how the press covers the product and its development..."

[21] Economics has for the most part not investigated business models. Some specific cases have been analyzed, especially the "bundling" or "tying" of goods for joint sale, typically discussed in an antitrust context (e.g., Adams and Yellen, 1976), and the provision of public goods (e.g., Demsetz, 1970).

To seize the opportunities created by innovation, innovators must excel at understanding not only customer needs, but also the possible future evolution of technology, costs, and customer willingness to pay. Even a successful business model, however, is insufficient to assure sustained profitability when imitation is easy. When hard to imitate—or if used to pioneer a winner-take-all market—a business model can be a source of sustained profitability.[22]

The business model also encompasses a firm's strategy toward its rivals. Positioning within fast-moving industries often takes the form of a standards competition, either in the market (e.g., Windows vs. Mac) or through political maneuvering within a cooperative organization (e.g., the International Organization for Standardization).[23] Control of a successful standard has numerous potential benefits, including licensing revenue, privileged access to new technologies, and influence over technology trajectory.

Seizing and transforming capabilities allow firms to refine and expand their business models in order to exploit new opportunities or defend against new competitive threats. They are the means by which organizations remake parts of themselves, possibly redrawing the firm's boundaries to respond to changes in the business environment. A reformulation of the business model may require radical shifts in the supply chain, asset ownership, or sales channels to ensure continued/improved value capture.

In fast-moving market and technology environments, firms must be ready to continuously reinvent themselves. Netflix, introduced above, is a good example of a firm that has gone through multiple business models in a short period of time (Teece, forthcoming). The company launched an online rental service in 1998, when video rental stores were the standard outlet for home viewing. At its initial launch, the Netflix business model was based on a pay-per-rental service with customers selecting the rentals online and Netflix shipping the movies directly to their homes. By 1999, it was clear to management that Netflix was failing. Later that year, the company launched a monthly fee plan that was subsequently amended to enable subscribers to rent any number of DVDs per month subject to a limitation on the number of DVDs that could be out at any one time, which led to tremendous growth. However, the CEO, Reed Hastings, recognized the potential that streaming media over the Internet had to undermine the company. After several false starts, he began a streaming service over multiple devices already in the home including Xbox 360 game consoles and Tivo digital video recorders. The service is finding acceptance with consumers, but Netflix has yet to work out licensing deals with the movie studios to offer its full catalog online (Roth, 2009). In the meantime, Netflix's national network of dozens of distribution centers—key to rapid delivery of DVDs via the mail—is leased, not owned, permitting the company to remain flexible for the future.

5.5. The profiting from innovation framework

Over the past two decades, our understanding of value capture from innovation and the link to firm strategy has expanded dramatically. A stream of research has stressed the importance of the architecture of the enterprise (especially the boundaries of its ownership and its control of complementary assets)

[22] Markets that can produce winner-take-all outcomes include those in which network externalities (Katz and Shapiro, 1986), switching costs (Klemperer, 1987), or learning economies (Krugman, 1987) confer a substantial incumbent advantage.

[23] David and Greenstein (1990) provide a review of the extensive literature on the economics and competitive consequences of compatibility standards.

for improving the chances of sustainable success when new technologies are commercialized. The role of supporting institutions and public policy—especially appropriability regimes—has also been highlighted.

This body of work has come to be known as the PFI framework[24] and was the topic of a special issue of Research Policy in 2006 (vol. 35, no. 8). PFI addressed a puzzle that had not been well explained in the previous literature, namely: why do highly creative, pioneering firms often fail to capture the economic returns from innovation? The original framework (Teece, 1986) cites several examples (e.g., EMI in CAT scanners, Bowmar in calculators), and the phenomenon does indeed endure. The first-generation PC manufacturers all but disappeared from the scene (and even IBM, which pioneered the Microsoft-Intel PC architecture, exited the business in 2005 by selling its PC business to a Chinese company, Lenovo). Xerox (PARC) and Apple invented the graphical user interface, but Microsoft Windows dominates the PC market with its follow-on graphical user interface. Netscape invented the browser, but Microsoft captured more of the market. Apple's iPod was not the first MP3 player, but it has a commanding position in the category today. Merck was a pioneer in cholesterol-lowering drugs (Zocor), but Pfizer, a late entrant, secured a superior market position with Lipitor.

At first glance, it is tempting to say that these examples reflect the result of Schumpeterian gales of creative destruction where winners are constantly challenged and overturned by entrants.[25] Indeed, entrants with potentially disruptive innovations are almost always waiting in the wings, but many of the cited cases involved mostly incremental/imitative entrants rather than the radical breakthroughs typically invoked in accounts of Schumpeterian competition.

More importantly, there is ample variance in the outcomes from entry, with many cases where first or early movers captured and sustained significant competitive advantage over time. Genentech was a pioneer in using biotechnology to discover and develop drugs, and 30 years later was the second largest biotechnology firm (and, the most productive in its use of research and development dollars) right up to its acquisition by Hoffmann-La Roche in 2009. Intel invented the microprocessor and still has a leading market position more than 30 years later. Dell pioneered a new distribution system for personal computers and, despite recent challenges and many would-be imitators, remained the leader until it was bypassed by Hewlett-Packard in 2007. Toyota's much studied "Toyota Production System" has provided the auto maker a source of competitive advantage for decades despite numerous and sustained attempts at imitation, with the company finally becoming the world's biggest car manufacturer in 2008.

The PFI framework provides an explanation as to why some innovators profit from innovation while others lose out—often to rank imitators—and why it is not inevitable that the pioneers will lose.

The fundamental imperative for profiting from an innovation is that unless the inventor/innovator enjoys strong natural protection against imitation and/or strong intellectual property protection, then the

[24] The core paper in the Profiting from Innovation (PFI) framework is Teece (1986). The intellectual origins of the framework can be traced to Williamson (for his work on contracting), Abernathy and Utterback (for their work on the innovation life cycle), to economic historians like Nathan Rosenberg and Alfred Chandler (for their work on complementary technologies), to Nelson and Winter (for their work on the nature of knowledge), and to Schumpeter (for his focus on the need for value capture). See Winter (2006) for a review of PFI's intellectual origins.

[25] There is a long literature on the role of new entrants in dislodging established firms. See, for instance, Anderson and Tushman (1990), Clark (1985), Henderson and Clark (1990), and Christensen (1997).

potential future stream of income is at risk. The relevant appropriability regime is thus critical to shaping the possible outcomes.

Appropriability regimes can be "weak" (innovations are difficult to protect because they can be easily codified and legal protection of intellectual property is ineffective) and "strong" (innovations are easy to protect because knowledge about them is tacit and/or they are well protected legally). Regimes differ across fields of endeavor, not just across industries or countries.

The degree to which knowledge about an innovation is tacit or easily codified also affects the ease of imitation, and hence appropriability. The tacitness of knowledge varies to some extent over the product cycle. New products and processes are often highly nuanced. Thus in the preparadigmatic phase of technological innovation (Abernathy and Utterback, 1978; Teece, 1986), the tacit component is likely to be high. Once a dominant design emerges, the rate of change of product design slows, and there is then the opportunity, if not the need, to codify technology. However, more rapid rates of innovation mean that there may be no time to codify (make explicit) new knowledge even when it is technically feasible to do so.

Patents can in some cases be used to slow rivals and generate profits. However, patents rarely, if ever, confer strong appropriability, outside of special cases such as new drugs, chemical products, and rather simple mechanical inventions (Levin et al., 1987). Many patents can be "invented around" at modest costs (Mansfield, 1985; Mansfield et al., 1981).[26] They are especially ineffective at protecting process innovation. Often patents provide little protection because the legal and financial requirements for upholding their validity or for proving their infringement are high, or because, in many countries, law enforcement for intellectual property is weak or nonexistent.

The inventor of a core technology can also seek complementary patents on new features and/or manufacturing processes, and possibly on designs. The way the claims in the patent are written also matters. Of course, the more fundamental the invention, the better the chances that a broad patent will be granted, and granted in multiple jurisdictions around the world.

While a patent is presumed to be valid in many jurisdictions, validity is never firmly established until a patent has been upheld in court. A patent is merely a passport to another journey down the road to enforcement and possible licensing fees. The best patents are those that are broad in scope, have already been upheld in court, and cover a technology essential to the manufacture and scale of products in high demand.

In some industries, particularly where the innovation is embedded in processes, trade secrets are a viable alternative to patents. Trade secret protection is possible, however, only if a firm can put its product before the public and still keep the underlying technology secret. Many industrial processes, including semiconductor fabrication, are of this kind.

The conundrum that managers confront beyond protecting the innovation itself is at least twofold. Firstly, most innovations require complementary products, technologies, and services to produce value in consumption. Hardware requires software (and vice versa); operating systems require applications (and vice versa); digital music players require digital music and ways of distributing digital music (and

[26] Mansfield et al. (1981) found that about 60% of the patented innovations in their sample were imitated within 4 years. In a later study, Mansfield (1985) found that information concerning product and process development decisions was generally in the hands of at least several rivals within 12–18 months, on average, after that decision is made. Process development decisions tend to leak out more than product development decisions in practically all industries, but the difference on average was found to be less than 6 months.

vice versa); mobile phones need mobile phone networks (and vice versa); web browsers and web search engines require web content (and vice versa); airlines require airports (and vice versa). In short, technology must be embedded in a system to yield value to the user/consumer. Value capture becomes more difficult if other entities control required elements of the system.

Secondly, the delivery of product/process innovation requires the employment not just of complements but of many inputs/components up and down the vertical chain of production. Hence, when the inventor/innovator is not already in control of the necessary inputs/components, the profitability of the inventor/innovator will be considerably compromised by whatever economic muscle is possessed by owners of required inputs/components. The firm must be prepared to change its assessment over time as the identity of the bottleneck asset may change due to innovation elsewhere in the system. The implications of these complementary asset and value chain considerations for the boundaries of the firm are addressed below in Section 7.3.

An obvious implication of this framework is that the firm's endowment of expert talent (literati and numerati), however brilliant, does not by itself guarantee that the organization will capture much of the value from innovation. Absent quality entrepreneurial managers, good intellectual property protection, some control over complementary assets, an appealing value proposition to the customer, and a good business model, superb performances by literati, numerati, and other employees are likely to be in vain.

6. Innovation and internal structure/management

As discussed above, dynamic capabilities are underpinned by organizational competences, which in turn are underpinned by human resources and other assets. This section considers the nature and management of the firm's key personnel—especially its highly trained specialized talent—and their impact on performance.

6.1. General considerations

The question arises as to how the (strategic) management of human resources can support competences and dynamic capabilities, and thereby assist in building and maintaining a sustained profit advantage.[27] Becker and Huselid (2006) note that the most pressing theoretical challenge facing the strategic management of human resources is the unpacking of a "black box," specifically, that which describes the logic linking the firm's human resources architecture and its performance (p. 899).

The dynamic capabilities framework can help illuminate the causal links between human resources and economic performance. Before outlining this approach further, some general observations are in order.

The first observation is that the stock of human capital readily available to the firm (i.e., its employees and affiliates) cannot meaningfully be thought of as a dynamic capability itself. Dynamic capabilities are organizational. An organizational capability does not stem from the mere presence on the payroll of

[27] Raymond Miles (2007) notes that US scholars were among the early leaders in studying and describing effective managerial and organizational approaches to knowledge creation, sharing, and utilization. However, practice in the field has fallen short of the theory outlined in the textbooks. Miles goes on to give a remarkably good overview of basic management issues.

talented individuals; rather, it derives from ways in which competences are combined and employees interact in productive combinations.

A second observation is that the manner in which human resources need to be managed is task-specific. The three clusters of competences and adjustment mechanisms identified in the dynamic capabilities framework—sensing, seizing, and transforming—require somewhat different human resource management practices. Moreover, sensing, seizing, and transforming are not necessarily sequential; they are likely to be taking place simultaneously across the enterprise, especially if it is multidivisional/multiproduct. In an enterprise with dynamic capabilities, selecting the relevant human resource management practices and procedures is likely to itself be a demanding task.

The next section introduces the experts who help devise and execute the firm's strategy: the numerati and the literati. Entrepreneurs are involved, too. Subsequent sections consider the management of top talent and appropriate incentive systems. The literati and numerati are unlikely to be productive and satisfied in a traditional hierarchical organization, being compensated in traditional ways, and having compensation put at risk for events beyond their control.

6.2. Literati, numerati, and entrepreneurs

There are three categories of talent required for innovation: the literati, the numerati, and entrepreneurial managers. The first two are closely related. The literati and the numerati are the highly educated "classes" of specialists. The literati tend to have both undergraduate and, usually, graduate education in arts and sciences, economics, business, or law. The numerati are likewise highly educated, with capabilities in mathematics or statistics, information systems, computer science, engineering, or accounting and finance. Both groups synthesize and analyze, but the former tend to be more specialized at synthesis and the communication of ideas. The latter excel at analysis, especially of large data sets. Both groups of expert talent are important to today's knowledge economies.[28] Both groups earn top quartile salaries.

The third category is entrepreneurial managers. As Baumol and Strom (2007) note. . ."A close look at the extraordinary economic growth of the last two centuries, however, suggests that the market mechanism does not do its work without the input of individual actors—the entrepreneurs who bring cutting edge innovation to market" (p. 233). Indeed, in fast-paced, globally competitive environments, consumer needs, technological opportunities, and competitor activity are constantly in a state of flux. Opportunities open up for both newcomers and incumbents, putting the profit streams of incumbent enterprises at risk. As discussed in Teece et al. (1997), the path ahead for some emerging marketplace trajectories is easily recognized. In microelectronics this might include miniaturization, greater chip density, and compression and digitization in information and communication technology. However, most emerging trajectories are hard to discern. For instance, when will 3D flat screen technology emerge? Will it be first on small panels, or on large-panel public display monitors? Sensing (and shaping) new opportunities are very much a scanning, learning, creative, and interpretive activity at

[28] In many cases, firms need to tap the skills of numerati and literati externally, via strategic alliances and other knowledge networks, often with formal contracts to spell out specific details about the types of interaction and knowledge sharing that will take place (Mayer and Teece, 2008). However, this section restricts its attention to the expert talent over whom a firm's managers exercise direct authority.

which, by definition, entrepreneurs excel. Investment in research and the related activities that require expert talent is a necessary complement to this activity.

Kirzner (1979) and Shane (2003) analyze entrepreneurship as a process of discovering opportunities. While this is one component of entrepreneurship, as already noted, entrepreneurship is not just a search for opportunities. It is also about the proactive creation of them (through research and development), the accurate assessment of them, and the mobilization of resources to address them.

The work of the entrepreneur (or entrepreneurial manager) includes organizing resources to explore and develop those opportunities, and forming a team with the requisite complementary skills to develop and execute a business model. Understanding just how the various inputs in a creative exercise are likely to respond and coevolve together is decidedly complex. The economic function involves direction setting (strategy) and coordination. Performing this well is likely to involve deep understandings of market opportunities and the technical, physical, and human constraints of the resources at hand.

The most challenging human resources to be managed here are the numerati and literati, who have become an even more important resource to the business enterprise in recent decades (Reich, 2002). Firms must pay great attention to understanding how best to attract, retain, and motivate their most productive literati and numerati. Studies show that the most productive and eminent scientists are strongly motivated. Almost all have good stamina in the sense that they work hard in the pursuit of long-run goals (Fox, 1983, p. 287).[29] Creative activity involving such expert talent is necessary to design and develop new products, services, and business models. Creativity is a difficult process to manage, as it cannot be forced. Creative people may need some direction, but they cannot be micromanaged. As Gil and Spiller (2007) note, "high-level creative activity can only be fostered, it cannot be coerced" (p. 244). This is as true for research and development activity as it is for the arts.

However, because it is difficult to monitor and measure the output of creative individuals, there are also hazards for an enterprise, or any money source that is financing creative activity. Gil and Spiller refer to one class of these as dynamic hazards. The creative individual can potentially have good ideas/breakthroughs and leave the organization where these ideas were developed in order to commercialize them in a context where it may not be necessary to share the rewards with the previous capital provider. Gil and Spiller point out that these are "transaction hazards quite different from the standard transaction cost framework" (p. 245). The fundamental organizational "problem" associated with managing creative activity stems from the nature of creative work: high uncertainty and informational asymmetries (Caves, 2000). The problem is not relieved by internalization, as is the case with many high transaction cost situations (Coase, 1937; Tadelis, 2007; Williamson, 1975, 1985).

6.3. Teams

Although the internal structure of the organization appears to matter for innovation, it has been neglected in much economic analysis. In particular, there is little if any attention to organization design issues as they relate to promoting creativity and inventiveness.

[29] Empirical studies on scientists and engineers suggest that high performers are absorbed, involved, and strongly identified with their work. They also have a preoccupation with ideas, not people. Early in their lives, they show autonomy, independence, and self-sufficiency. They are self-motivated. To maintain their productivity, they do not generally require other people to approve their work.

If firms are to cut time-to-market for new products and processes, cross-functional interaction must take place concurrently, rather than sequentially. Cross-functional teams and cross-departmental networks must be instituted without causing information overload. If such activity becomes too unstructured, it augments rather than displaces bureaucracy. Cross-functional teams should have well-defined goals, subject to redefinition as needed, and draw on the requisite knowledge wherever it may be located.

Teams have become increasingly important to science and engineering tasks because of increased specialization and a corresponding need to integrate individual capabilities. While the numerati and literati value professional autonomy, they are nevertheless willing to collaborate when they perceive that collaboration will yield benefits. Even in the days of Thomas Edison, the use of multidisciplinary research teams was important to the solution of complex technological problems.[30]

Because it is very hard to measure both the inputs and outputs of team members, managers often seek to build a high commitment culture to help effectuate the necessary activity (Baron and Kreps, 1999). In fact, employee motivation appears to be more important than raw competence for outcomes (Katz, 2004).

With expert teams, the identity of the team leader/captain is likely to be of considerable importance. For all to succeed there must be mutual respect between and among experts and leaders.

The very notion of what constitutes a team may be different for creative tasks than for routine operations. When team requirements are too heavy, decision cycles lengthen, expenses mount, and the organization adopts an inward focus. Nelson (1962) notes that team structure in the development of the transistor was broadly inclusive: "several people outside the team also interacted in an important way... teamwork... did not mean a closely directed project" (p. 578).

Teams need not emphasize consensus and compromise, which tend to endorse the status quo. Innovation is often ill served by consensus-driven structures, as the new and the radical will almost always appear threatening to some constituents. Rather, the aim of expert teams should be to achieve excellence while giving some degree of liberty to individualism. Certain especially creative and exceptionally talented individuals can be given special recognition. Hence, team building with top talent is somewhat different from certain aspect of everyday team building. Table 2 summarizes some of the differences between traditional teams and such "virtuoso teams" (Fischer and Boynton, 2005).

A key feature of expert-led teams is that they are likely to be quite fluid. Indeed, not everything is appropriately organized in teams. Rather, groups need to form, get their work done, and disband or move onto other project teams. It is desirable to keep project teams small.

Put differently, one cannot simply assume that more is better when it comes to collaboration. Consensus and participatory leadership is not always a good thing, particularly when the issues are complex and there is considerable asymmetry in the distribution of talents on the team. The right voices need to be heard. Unproductive collaboration can sometimes be more dangerous than missed opportunities for collaboration.

[30] "Treated in many accounts of his life as an inspired, lone inventor, Edison was in fact a research and development manager... At its height, the Menlo Park laboratory had a total of some 40 employees, ranging from glassblowers and machinists to physicists and chemists" (Hounshell and Smith, 1988, pp. 2–3).

Table 2
Key differences between traditional teams and virtuoso teams

Team characteristics	Traditional teams	Virtuoso teams
Membership	Members chosen based on who has available time	Members chosen based on expertise
Culture	Collective	Collective and individual
Focus	Tight project management. "On time" and "on budget" more important than content	Ideas, understanding, and breakthrough thinking emphasized
Clients	Mundane	Sophisticated
Intensity	High/medium	High
Stakes	Low/medium	High

Source: Drawn from Fischer and Boynton (2005).

6.4. *Hierarchy*

The methods of (light touch) management appropriate to the literati and numerati involve a break with classical notions of the employment relation.

In Coase (1937), the employment relation was defined as one of authority, in which individual employees "agree to obey the directions of an entrepreneur within certain limits" (p. 391). If the relationship is less expensive than hiring the same skills via the price system, then this provides the rationale for internal organization. However, the Coasian conception does not extend to *how* the authority should be exerted.

Alchian and Demsetz's (1972) analysis of the employment relation was different and is in some ways more relevant to innovative organizations. Their claim is that the *raison d'etre* of the firm is team production. According to them, managers do not have any power of fiat or authority that the marketplace does not have. Managers monitoring team behavior detect shirking, and align reward to performance. There is no need in their model for the employee to surrender control, as was assumed in the Coase (and the Simon, 1951) model of the employment relationship. The existence of the firm flows from its ability to enable cooperative activity (i.e., the assessment and effectuation of combinations of employees to achieve goals) superior to that available in a market setting. But it does not follow that the manager has authority over employees beyond that which it exercises over external contractors.

Although Alchian and Demsetz identify team activity as the justification of the employment relationship, their development does not describe the nature of team activity well, particularly for the management of expert teams. The advantage to doing creative work in an internal setting is not just the ability to effectuate cooperation better than the market but also the ability of the firm to (1) organize financial resources thereby insulating top talent from the need to raise money themselves, (2) shape and maintain high commitment cultures to "regulate" teaming, and (3) build the team (identify the needed skills, choose suitable candidates, and provide parameters within which the team will function).

The numerati and literati value autonomy, so traditional command-and-control structures are unlikely to elicit their best performance. Their autonomy is also congruent with optimal resource allocation because of management's limited information processing bandwidth. In this regard, Nelson (1962) studied the development of the transistor at Bell Labs and noted:

> *"... the type of interaction we have noted in the transistor project requires that individuals be free to help each other as they see fit. If all allocation decisions were made by a centrally situated executive, the changing allocation of research effort called for as perceived alternatives and knowledge change would place an impossible information processing and decision making burden on top management. Clearly the research scientists must be given a great deal of freedom..." (p. 569).*

Strongly authoritarian management that suffocates initiative is anathema. In creative organizations, the evidence shows that management must have a "light" touch, that is, to provide "soft" rather than "hard" direction. Otherwise potentially fruitful combinations of expert talents may be suppressed and creativity will be compromised. Difficult and granular technical tradeoffs and judgments that are needed for problem solving must be made by "front line" professionals themselves and can rarely be sensibly ascertained and then imposed by management.

Accordingly, for innovation to occur, management usually needs to be decentralized/distributed and take the supporting role. Traditional notions of management relying heavily on authority and decisions driven from the center are unlikely to work well in organizations that are highly innovative. Reliance on hierarchy becomes more useful in execution phases of a project, as operations become more routinized or the environment more stable, than in creative phases of innovation and in technological regimes of rapid change (Burns and Stalker, 1961).

As described earlier, Cisco Systems adopted a decentralized structure that appears to have increased its ability to develop and deploy innovations. According to McGirt (2008), Cisco is now "a distributed idea engine where leadership emerges organically, unfettered by a central command" (p. 93). While most efforts are led from below (decentralized or distributed), some are still led from the center. Chambers puts the ratio at 70/30 (p. 135).

The point here is a simple one: in fast-paced complex environments where there is heterogeneity in customer needs, it is very difficult for the firm to be responsive if it has a highly centralized command-and-control structure. Moreover, with a highly talented workforce, excessive centralization can shut down local initiative and creativity. The organizational challenge is to connect individual initiatives to the overall corporate strategy/goals without building an expensive and initiative-sapping hierarchy inside the firm. Every member must act as a responsible decision maker within their professional domain, and there must also be strong leadership in the top management team.

Managing professionals, especially high-level expert professionals, requires rejection of traditional heavy-handed hierarchical structures that may work in more stable industries. Indeed, consistent with the analysis here, Quinn et al. (1996) go so far as to say that it is often necessary to invert the traditional hierarchy in order to create the organizational structures that successful professionals will accept. This is consistent with Teece (2003). With an inverted hierarchy, the job of the manager is to provide support. This proposition overturns some traditional notions of control, if not traditional notions of principal and agent.

In some purely creative environments, it is indeed the highly skilled experts that hire "bosses" rather than the other way around. The Hollywood agency model for creative talent was an early manifestation. As explained by Albert and Bradley (1997), the stars themselves, beginning with Newman, Streisand,

and Poitier, broke away from the studios to create their own production company, First Artists. A key element of First Artists' strategy was to create a climate in which leading actors can control their professional environment and lives. The artists put a professional manager in place, but the manager's mandate was clearly to effectuate the artist's view of how films should be produced. There have been many independent production companies founded since, with varying degrees of success.

University faculties have some similar attributes. The faculty arguably hires their Dean since the Dean generally serves at the sufferance of the faculty, at least in some of the major research universities on the west coast of the United States.

In short, creative and highly skilled knowledge workers, be they scientists, engineers, medical doctors, professors, or economists, desire high autonomy and can be self-motivated and self-directed because of their deep expertise. The university environment caters for this with the tenure system—requiring the discharge of teaching, research, and service obligations by faculty, but allowing the individual faculty member considerable discretion as to whether and when tasks (other than class meetings) are performed.

Expert talent is also likely to be functionally elitist, at least to some small degree. One corollary is that expert talent will be reluctant to accept authority from managers who are not, or have not been, respected professionals themselves. According to Quinn et al. (1996), this is "why most professional firms operate as partnerships and not as hierarchies" (p. 72). Any "power" that individual leaders have should stem from professional and personal respect gained through professional success and through creating and maintaining an open, honest, and transparent culture.

In short, when the modern organization employs many highly skilled individuals, it has to create an organization of colleagues and associates. The W.L. Gore Company, inventor of Goretex, is a well-known case of an innovative organization which has dropped all hierarchical designations. Everyone, including the *de facto* chief executive officer, is an "associate"; the nomenclature of hierarchy has been abandoned.

Implemented properly, the distributed leadership approach is not an abdication of managerial responsibility and good governance. It is just the opposite. The executive leadership team should be responsible to the Board of Directors and to shareholders, as well as to employees and other constituents.

In environments where stimulating creativity is important to enterprise success, management's role is to forge incentive alignment, to expedite resource availability, and to remove barriers standing in the way of professionals doing their work, so long as that work is consistent with the organization's goals. Of course, strong accountability is still required from the literati and the numerati but it can rarely be gained in the traditional manner; figuring out how to achieve this requires new forms of compensation rarely discussed in the literature. Compensation arrangements that recognize differences but reward cooperation must be designed and implemented.

6.5. Incentive systems

Reich (2002, p. 107) has observed that talented and ambitious people can earn more today, relative to the median wage, than could talented and ambitious people in the industrial era. Larger and more open or "contestable" markets are the reasons why dispersion in earnings has increased. The higher rewards that top talent can command stems from the value which now seems to flow from creative, analytical, and "rainmaking" abilities of leading professionals. In particular, the skills to help solve complex problems, to help make critical decisions or resolve complex disputes, and to identify and exploit opportunities command high value.

Intrinsic motivators (e.g., intellectual challenge) are sometimes found to be more important than direct inducements such as compensation for worker performance (Hayton, 2005; Sauermann and Cohen, 2008). Nevertheless, potentially complex compensation issues for expert talent need to be addressed in the context of innovation.

The human resource management literature tends to want to bring uniformity to human resource management practices across the organization. The rationale for this is that (1) employees will judge the system as unfair if disparities in compensation open up and (2) it is more complex to manage an organization if there is variety in human resource management systems and practices. The latter observation may be true, but variety may be unavoidable because building different capabilities requires different systems, and organizations/enterprises usually need to be ambidextrous to create and capture value. In the expert context, pay differentials—even among members of a team—ought not to be an issue so long as they are based on performance and not purely discretionary.

Where financial rewards are directed in ways that are highly subjective, competition takes place to move up the organizational hierarchy. Seniority in the hierarchy allows more personal freedom, control over discretionary resources, and is the confident path to higher compensation. The politics of pay become part of everyday life. People jostle to claim credit, even at the expense of colleagues. A good deal of time and effort is spent posturing in order to appear valuable to the organization, through the eyes of the boss. Eventually the need to do excellent work, much less take risks, gets lost sight of.

Innovation has particular challenges because one must create incentives that promote sensing, seizing, and transforming, and the incentive design likely to aid one might handicap the others. The underlying incentive design problem is even more complicated because the three tasks differ in their measurability and/or timelines. Consider, for instance, sensing and seizing. The first involves highly creative activity with medium- to long-term benefits for the enterprise; the latter, although it also involves creative elements, is more about delivering on the current strategy. The transformational dynamic capabilities, for example, honing the business model, are also highly creative. The metrics for the more creative tasks are necessarily looser than for more traditional executive roles, creating a tension between the need for creative autonomy and the need to offset poor measurement with tighter control.

The better the performance measures available, the less costly it is to provide strong incentives for the activity in isolation. Poor measurement, which might, for example, be caused by unforeseeable external events, means that employees (and employers) face uncontrollable risk which may interfere with respect to achieving rewards. This uncertainty is costly if employees are risk averse.

One approach to the incentive design problem is to simplify the design of the underlying organization by having separate subunits focusing on sensing, on seizing, or on transforming. However, many strategic and organizational contexts will not afford the opportunity for this, nor do all analysts of organization design consider it optimal.[31] More integrated approaches include relying on especially versatile managers with a wide range of capabilities, or on moving executives in and out of particular jobs as the nature of the work changes. However, organizational integration of the three tasks exacerbates the challenge of designing incentives to allow multiple dexterities to flourish in the business

[31] Raisch et al. (2009) provide an overview of the "ambidexterity" literature that debates the degree to which "exploration" and "exploitation" (March, 1991) must be organizationally integrated.

enterprise. Of course, stock options are one vehicle for rewarding multiple activities that jointly create value, so long as those incentive investments are long-term, that is, the options/warrants/restricted stock units have a long vesting period.

Innovating firms face the challenge of finding ways to effectuate cooperation and avoid hierarchical friction. Traditional organizations, when faced with a lack of clear performance metrics, increase rules, directives, and monitoring, which only hampers innovative activity. If firms can indeed provide satisfactory ways of objectively measuring relevant aspects of employee performance, they then can provide greater autonomy by using incentives to begin a virtuous circle of work freedom and high reward.

In this regard, "sensing" (an entrepreneurial activity) is particularly difficult to calibrate. Because of this, it is difficult to start up a new business inside an existing enterprise. It is often easier for those with good sensing skills to become independent entrepreneurs, or to be otherwise associated in the formation of a new business.

Alchian and Demsetz (1972) are rightly skeptical that high-end specialized services can be organized under traditional employment structures because of imperfect monitoring of individual performance. As they put it, "while it is relatively easy to manage or direct the loading of trucks by a team of dock workers when input activity is so highly related in an obvious way to output, it is more difficult to manage and direct a lawyer in the preparation and presentation of a case" (p. 786). Others have suggested the partnership form is the response to this problem, as partners can monitor each other, although this is available in only a limited number of innovation contexts (e.g., consultancies).

As a complement to good incentive design—which is innately difficult to effectuate in the innovation context—managers must inculcate culture/values to bring about greater alignment among the interests and employees. In the organizational behavior literature, such cultures are referred to as "high commitment" cultures. Most are associated with Total Quality Management or Japanese styles of organization (Baron and Kreps, 1999). Elements of high commitment cultures that are relevant to innovation include functional flexibility of employees, systemic approach to solving problems, employee/team empowerment in decisions, and responsiveness to (internal or external) customer needs. Establishing a high commitment culture is a valuable complement to strong incentives. It may also be the low-cost way to proceed.

Table 3 tabulates some of the ways in which traditional firms are likely to be different from dynamically competitive ones with respect to incentives and the management of (expert) human resources.

Table 3
Contrasting views of the business enterprise

Organizational characteristics	Industrial model	Knowledge model (for literati and numerati)
Financial incentives	Base + discretionary bonus salary	Metrics based compensation; limited management discretion
Hierarchy	Deep	Shallow
Leadership	Centralized	Distributed
Work	Segmented	Collaborative
People	Cost	Asset
Basis of control	Authority	Influence and example
Assumptions about individuals	Opportunistic	Honorable

7. Towards a theory of the innovating firm

7.1. Context

As explained above, fundamental changes in the global economy are changing the way firms innovate. More open and competitive trading regimes have increased the importance of know-how and other intangible assets. There are significant implications for the theory of the firm, if such a theory is to connect meaningfully with the contemporary economy.

This section begins by introducing some of the theories of the firm that have emerged outside mainstream economics. Subsequent sections use the dynamic capabilities framework to reconsider the "problems" for which firms are the solution, showing the complementarity of the contracting and capabilities perspectives. The final section argues that a more complete theory of the firm will recognize that firms exist in part to compensate for weak or nonexistent markets for know-how. For the economic system to work, entrepreneurs and managers are required to orchestrate the resources/competences needed for creating and capturing the value of an innovation. Absent managers and management, economic theory cannot explain the evolution and growth of the economy.

One would hope that the theory of the firm would provide some insight into firms as they exist today. Unfortunately, whether one uses the lens of transaction costs (e.g., Coase, 1937; Williamson, 1985), ownership perspectives (e.g., Hart and Moore, 1990), incentive perspectives (e.g., Holmstrom and Milgrom, 1994), or other "modern" theories of the firm, nicely summarized and illustrated by Roberts (2004), the many theories available today still seem to caricature firms, at least those engaged in innovation. Mainstream economics must reconceptualize how markets and market processes relate to the theory of the firm if economic theory is to be both relevant and rigorous.

Furthermore, as Gibbons (2005) has noted, many theories of the firm today can more properly be characterized as theories of the boundaries of the firm. Gibbons further points out, following Cyert and March (1963), that the term "theory of the firm" is more apt for descriptive and prescriptive models of firms' decision making processes. Gibbons provides an excellent survey of four theories of the firm that he calls (1) rent seeking, (2) property rights, (3) incentive systems, and (4) adaptations. He makes oblique reference to the resources/capabilities approaches which he indicates "have mouth watering potential implications" and he "expects them to play key roles in future formal theories of the firm." This section and those that follow are designed to turn some of Gibbons' perceived potential into actuality. The capabilities approach recognizes values in all four streams and incorporates some ideas from each.

To help overcome blatant deficiencies in the standard production–function theory of the firm, transaction cost economics arose. This is now being combined with knowledge-based theories of the firm. Williamson himself sees the "relation between competence and governance as both rival and complementary—more the latter than the former" (1999, p. 406). Knowledge-based theories indirectly respond to the issues raised by Winter (1988), Demsetz (1988), and others. Emanating from the field of strategic management (e.g., Teece, 1982, 1986; Wernerfelt, 1984), these theories show some capacity to inform the theory of the modern firm.

However, theories developed in strategic management do not explicitly endeavor to yield a theory of the (nature of the) firm. Rather, they theorize about how competitive advantage can be developed and maintained, and how supernormal profits can be earned. As discussed above, the resources perspective indicates how rents can flow, at least for a finite period, from the possession and protection of scarce and difficult-to-imitate assets, or "resources." Nevertheless, resources and capabilities theories can provide insights into the nature of firms, at least those firms that survive in regimes of rapid technological change. Accordingly, they are developed in more detail below.

The dynamic capabilities framework is now well known in the strategic management field. It transcends narrower perspectives and illuminates many issues, including the firm's desirable boundaries. The central concerns of the dynamic capabilities framework—sensing opportunities, seizing them, and transforming firms (and markets) to build and maintain competitive advantage—can provide insights to inform both boundary and decision making issues. Some of these insights are outlined below.

7.2. Dynamic capabilities, cospecialization, and transaction costs

Coase (1937) in his classic article on the nature of the firm described firms and markets as alternative modes of governance, with a profit seeking orientation leading to choices being made so as to minimize transaction costs. The Coasian firm has a simple decision making calculus that supposedly determines the firm's boundaries. The boundaries of the firm are set by bringing transactions into the firms so that the marginal costs of organizing inside the firm are equilibrated with the costs associated with transacting in the market.[32]

A substantial literature has emerged since Coase's landmark 1937 article on the relative efficiencies of firms and markets. This literature, greatly expanded by Nobel Laureate Oliver Williamson (1975, 1985) and others, has come to be known as transaction cost economics. It analyzes the relative efficiencies of governance modes: markets and internal organization, as well as intermediate forms or organization such as strategic alliances.

Contractual difficulties associated with asset specificity are at the heart of the relative efficiency calculations in transaction cost economics. When irreversible investments in specific assets are needed to support efficient production, then the preferred organizational mode is internal organization. Internal organization minimizes exposure to the hazards of opportunistic recontracting and allows more flexible adaptation (Williamson, 1975, 1985).

In some ways, but not in others, the dynamic capabilities approach is consistent with a Coasian perspective. It conceptualizes the firm and markets as alternative modes of governance. However, the selection of what to organize (manage) internally versus via alliances or versus the market depends on the availability and the nontradability of assets, capabilities, and to some extent on what Langlois (1992) has termed "dynamic transaction costs."[33]

The notion of "nontradability" advanced here does not precisely match Coasian or Williamson concepts of "transaction costs." There is nevertheless a strong relationship between specific assets and nontraded or thinly traded assets. However, there are reasons why assets are not traded (or are thinly traded) that do not relate to asset specificity and transaction costs as such. For example, there may simply be no viable business model for licensing certain types of know-how.

[32] Another key feature of the Coasian firm was his emphasis on authority and the employment relationship as the backbones of the enterprise. The discussion on teams and hierarchy in Sections 6.3 and 6.4 implicitly undermine this dimension of the Coasian firm—at least for the innovating firm.

[33] Langlois (1992) defines dynamic transaction costs as "the costs of persuading, negotiating, coordinating and teaching outside suppliers" (p. 113).

Indeed, many companies will simply not license "strategic" technological assets, especially not to direct competitors. The reason, at one level, is because a contract cannot be written that would compensate the licensor for the likely loss of customers if the licensee uses the licensor's technology to compete against the licensor. Theoretically, a licensor ought to be indifferent between own sales and the sales of a licensee if the royalty rate is set to enable royalties to equalize with lost profits. However, such arrangements are rarely, if ever, seen, in part because there is likely to be ambiguity with respect to which customers and what sales are actually lost to the licensee. Accordingly, it is uncommon in the actual world to see exclusive licenses (to direct competitors) when the licensor is able to sell in the same territory. At another level, it may simply be because there are differences in expectations with respect to the profit potential associated with the use of the technology. There are also likely concerns with respect to whether the licensor or the licensee will capture the "learning by using" know-how associated with exploiting the technology. Negotiating, contractually specifying, and monitoring the sharing arrangements are also likely, as Williamson's framework suggests, to be very difficult. [34]

In short, the business model that firms use to capture value from innovation is usually one that involves manufacturing and selling products that contain new knowledge. It is rare that firms will rely entirely on an unbundled business model in which patent/trade secret licensing is used as a mechanism to capture value from know-how. Rambus, Inc, and Dolby Labs are among the exceptions.

In capabilities-based theories of the firm, the concept of cospecialization is particularly important (Teece, 1986). Assets that are cospecialized to each other need to be employed in conjunction with each other, usually inside the firm (Teece, 1980b). Cospecialization and the organizational challenges associated with achieving scope economies and seizing new opportunities is not the emphasis in the pathbreaking scholarship of Ronald Coase, Armen Alchian, Harold Demsetz, or Oliver Williamson. However, it is a phenomenon that requires (theoretical) attention. Some is provided below.

Cospecialized assets are the building blocks of firms. Building and assembling cospecialized assets inside the firm (rather than accessing them through a skein of contracts) is not done primarily to guard against opportunism and recontracting hazards, although in some cases that may be important. Instead, because effective coordination and alignment of assets/resources/competences is important, but difficult to achieve through the price system, special value can accrue to achieving good alignment. This is more easily done inside the firm. Achieving such alignment through internalization goes beyond what Barnard (1938) has suggested as the functions of the executive—which he sees in achieving cooperative adaptation.

The imperative for internalization is not just a matter of minimizing Williamsonian transaction costs. Rather, at least in the dynamic capabilities framework, the distinctive role of the (entrepreneurial) manager is to "orchestrate" cospecialized assets. Performed astutely and proactively, such orchestration can: (1) keep cospecialized assets in value-creating alignment, (2) identify new cospecialized assets to be developed through the investment process, and (3) divest or run down cospecialized assets that no longer yield special value. These goals cannot be readily achieved through contracting mechanisms in

[34] Accordingly, Coca-Cola is unlikely to license its secret formula, and W.L. Gore is unlikely to license the technology behind Gore-Tex fabrics to anyone other than its wholly or partially owned subsidiaries. Intel and TSMC will likewise be reluctant to license their key semiconductor processes to competitors, except with severe restrictions and circumstances of high trust. Brands that signal particular values (e.g., Lexus, Tiffany's) are likewise rarely licensed, partly for contractual reasons, partly for other reasons.

part because of dynamic transaction costs (the costs of negotiating, etc.) but also because there may not be a competent entity to build or "supply" the assets that are needed. In short, capabilities must often be built, they cannot be bought, and there is limited utility in labeling this conundrum as a transactions cost problem.

Rather than stressing opportunism (although opportunism surely exists and must be guarded against), the emphasis in dynamic capabilities is on building specialized assets (that cannot be bought) and on change processes (to keep the enterprise aligned with its business environment). These processes include, research and development, remolding the business architecture, asset selection, and asset orchestration. In dynamic capabilities, "small numbers" bargaining is at the core, as in Williamson (1975). Importantly, the emphasis in dynamic capabilities is not just on protecting value from recontracting hazards; it is also on creating the assets that in transaction cost economics become the object of rent appropriation.

The basic unit of analysis for dynamic capabilities is not the transaction (as in transactions cost economies) but the innovating firm and the (largely intangible) specific assets it creates and controls. To the extent the emphasis in dynamic capabilities is on deals and contracts (explicit or implicit) it is less concerned with avoiding opportunism and more concerned with embracing opportunity. However, there is also considerable emphasis on "production," learning, and innovation. These considerations are largely absent from alternative theories of competitive advantage and from alternative theories of the firm.

7.3. The boundaries of the innovating firm

7.3.1. General

Where a firm draws its boundary is one of the fundamental parameters that a theory of the firm must address.[35] The firm's decision on how to delineate and implement a suitable business model for commercializing innovation and achieving economic rents is another important part of dynamic capabilities. Formulating and implementing a strategy is yet another.

The commoditization of certain services such as back office operations (e.g., testing, telemarketing, benefits management, record keeping, and IT management) has greatly expanded the menu of make-or-buy options facing a firm, and heightens the need to have a theory which can predict the boundaries of innovating firms. The growing range of potential suppliers itself reflects greater global distributions of capabilities. This both expands and complicates the managerial choices of where and by whom activities from managing R&D to after-sales service are to be performed. Moreover, as the dynamic capabilities framework makes clear, this choice must be periodically reevaluated.

Economic theory has so far failed to capture core considerations that are critical to where management decides to draw the boundaries for innovating firms. According to Coase, it is a simple calculus: internalize until the marginal cost of doing so equates with the marginal cost of not doing so. With

[35] A theoretical framework that endeavors to account for the horizontal boundaries of the overall corporation, based on learning, path dependencies, technological opportunities, the selection environment, and the firm's position in complementary assets, can be found in Teece et al. (1994). The firm's ongoing reassessments of its coherence in product space are part of its dynamic capabilities.

Williamson, it is a matter of making sure that internal governance costs are in equilibrium with (asset specificity-driven) transactions costs—other things equal. But other things are often not equal, appropriability issues are likely to be paramount, and internal production costs and other manifestations of capability may depend endogenously on the governance modes chosen.

7.3.2. Capabilities, complementary assets, and intellectual property

The PFI framework introduced in Section 5.4, which builds on the insights from the contracting approach of Coase and Williamson, considers a richer set of factors as relevant to choosing the firm's boundaries. These include intellectual property rights, complementary assets, and time to market considerations. (see also Jacobides et al., 2006). This section will go even further and discuss opportunity preservation, technology pacing, and capability building.

The PFI framework from the beginning considered some factors beyond contracting ones. Teece (2006) summarizes PFI's rules by saying that firms should rely on markets unless there are

> *"compelling reasons to internalize. Such reasons could be grounded in one of two major circumstances: (a) cospecialization, which would lead to transaction costs if heavy reliance was made externally [i.e., on externally provisioned assets/services]; (b) shoring up the appropriability situation by building or buying complementary assets which the innovation would likely drive up in value, or that were otherwise important to getting the job done" (p. 1140).*

The dynamic capabilities framework identifies yet additional factors, most notably whether the firm's competences/complementary assets are sufficiently advanced to enable it to competitively self-supply the required inputs or services. Chandler (1992) noted that during the Second Industrial Revolution the "initial move forward into distribution and marketing by entrepreneurs was that often suppliers and distributors had neither sufficient knowledge of the novel complex products nor the facilities required to handle them efficiently. This is why so many of the new companies met their needs by building almost immediately a national marketing and distribution network staffed by their managers and workers" (p. 87).

The distribution of capabilities is not uniform across firms in an industry.

Nor need suppliers and distributors have the capabilities in place to meet the needs of innovators. Hence, when industries are new, it is often necessary for the developer/manufacturer to integrate upstream/downstream not for transaction cost reasons, but for entrepreneurial and "capability" reasons. That is, there may simply not be an established enterprise with the requisite capabilities able to supply and distribute the innovator's products. Vertical integration upstream and downstream then becomes a necessity, not strictly for transaction cost reasons, but because there simply are not qualified parties available with whom one can contract. [36] As mentioned earlier, capabilities cannot always be bought; they sometimes must be built. Capability considerations help explain the differing interpretations of Oliver Williamson and Alfred Chandler over the backward integration of certain large US companies early in the twentieth century. Chandler (1992) puts it this way:

[36] Once the firm's architecture of supply and distribution had been crafted, its managers must provide the orchestration, or "system integration" function. The prevalence of outsourcing has made this integration function a strategic competence of the first order (Pisano and Teece, 2007; Prencipe et al., 2003).

Williamson (1985, p. 119) notes that:

> *"'Manufacturers appear sometimes to have operated on the mistaken premise that more integration is always preferable to less.' He considers backward integration at Pabst Brewing, Singer Sewing Machine, McCormack(sic) Harvester, and Ford 'from a transaction cost point of view would appear to be mistakes.' But when those companies actually made this investment, the supply network was unable to provide the steady flow of a wide variety of new highly specialized goods essential to assure the cost advantages of scale. As their industries grew and especially as the demand for replacement parts and accessories expanded, so too did the number of suppliers who had acquired the necessary capabilities" (p.89).*

Chandler goes on to note that:

> *"The point is that an understanding of the changing boundaries of the firm required an awareness of the specific capabilities of the firm and the characteristics of the industry and market in which it operates* at the time *the changes were made. Many of the first-movers in the new capital-intensive industries which integrated forward into distribution and marketing and backward into control of supplies did so on an international scale. Knowledge gained in the creation of a wholesaling or direct marketing organization at home led to building a comparable one in foreign markets" (ibid, emphasis in the original).*

Chandler's historical analysis is very consistent with a dynamic capabilities theory. Perhaps his accounts would be better couched in transaction costs terms, but it is not immediately apparent how that would be done.

7.3.3. Opportunity "management"

In economic theory today, the general outsourcing logic relies on Williamsonian considerations of asset availability/specificity and expected contingent moves in prices. For example, when a firm must make a relationship-specific investment in order to work with a supplier, it exposes those nonredeployable assets to subsequent recontracting hazards that can be eliminated by vertical integration (Williamson, 1985). These are important insights, but they need to be supplemented by an understanding of technological and market hazards of a different kind. Also, opportunity needs emphasis alongside opportunism.

In business transactions when new technology is at stake, a less understood set of hazards (and opportunities) may arise.[37] This class of contracting hazards (and opportunities) stems not so much from the extraction of quasi-rents, but from the need to guard future strategic opportunities from certain competitors.

Situations may arise when a vertically integrated firm has the ability to use its upstream technological prowess to deny a downstream rival access to a patent that one must practice in order to engage in certain future technological and commercial opportunities.[38] An innovator's ability to pace, direct, control, and guard the development of new products and technologies poses risks to competitors.

[37] This section is based in part on de Figueiredo and Teece (1996).

[38] This notion can also be viewed as a dynamic extension of the raising rival's cost literature (Salop and Scheffman, 1983). However, the predicament analyzed is unlikely to require antitrust intervention.

Even when firms leave research and development/new product development to nonintegrated suppliers, the downstream firm may then have no choice but to purchase critical components from a supplier who also emerges as a competitor.

A subtler form of hazard is the inability to pace or direct the evolution of new products[39] that depend on a supplier's proprietary technology. If a firm has no input into a supplier's development process, the supplier might be able to independently shape the trajectory of the technology. Transaction cost economics would posit that such hazards can best be understood as contracting issues. However, one can question whether transaction costs and recontracting hazards are the core issues; rather, it is that outsourcing may lead to the loss of opportunities to accumulate critical competences important to the firm's overall new product development strategy. Theoretically, contracts might be written that would require royalty-free grantbacks of any trade secrets accumulated. However, such arrangements are rare.[40] Opportunity management may require investment in own R&D, rather than relying on the efforts of suppliers.

7.3.4. Coordination of complementary assets and systems integration

Another reason that a firm faces hazards when relying on an external supplier for complementary innovation is the difficulty associated with accomplishing coordination of complementary assets and activities. This is related to what Richardson (1960) and Williamson (1975) have called "convergence of expectations." Investment (in research and development) must be coordinated between upstream and downstream entities, and this is difficult to effectuate using contractual mechanisms.

Coordination is of greatest concern when innovation is systemic (Teece, 1988). Systemic innovation requires harmonized action by all parties (e.g., the development of new cameras and film which instant photography required). When there is asymmetry in capabilities between firms, achieving harmonization is difficult. Boeing discovered this to its cost when it decided to rely on a global array of suppliers to develop parts for its new 787 Dreamliner as a cost-sharing measure; some suppliers lacked the capabilities to develop parts of the necessary quality, and Boeing had cut back its monitoring capability. Deficits in the capabilities of suppliers resulted in years of delay (Michaels and Sanders, 2009). It is not clear, from the perspective of theory, whether this is best viewed as a contracting problem or a capabilities issue. However, the latter appears to be more powerful.[41] The Boeing experience echoes Lockheed's experience three decades earlier when the L1011 wide bodied plane was delayed by the failure of Rolls Royce to develop and deliver on time the RB211 jet engine

[39] The software industry provides an illustration of how an integrated firm can pace technological development downstream of its operating system. Microsoft develops its operating systems in-house. It also develops applications while looking to others for additional applications. These independent application designers rely on Windows to run their applications. Thus, Windows acts as a constraint on some of the technological features of the downstream application (e.g., protocols for data exchange). Microsoft's ability to pace the upstream technology and its ability to use its operating system technology in its applications software has helped it to become one of the dominant players in applications.

[40] An exception is Pilkington, which for many years had such terms in its float glass license arrangements.

[41] As a Boeing executive, Jim Albaugh, noted in explaining delays with the Dreamliner: "while Boeing's commercial division had a fine record as a manufacturer, the defense unit had far more experience with the complex development required for an aircraft such as the 787, which has a much larger amount of lighter composite materials than normally used in commercial planes ... when you only do a development program every decade or so, I think you lose some of those capabilities and some of the knowledge" (Clark, 2009).

for the L1011, effectively putting Lockheed out of the civilian aircraft industry. This was not an exercise of opportunism by Rolls Royce; rather it reflected Rolls Royce lack of ability to achieve ambitious technological goals.

Teece (1996, 2000) and Chesbrough and Teece (1996) have analyzed the difficulties in coordinating the development of complementary technologies when pursued independently and coordinated by contract.[42] Delays are frequent and need not result from strategic manipulation; they may simply flow from uncertainty, limited capabilities, and divergent goals among the parties.[43]

Autonomous innovations, which do not require coordinated activities between parties, can occur within one organization's boundaries and then be "plugged in" to the bigger project. Autonomous innovations are pervasive when standards are present, such as the open architecture of the IBM personal computer.

Outsourcing components used in new products and new systems also raises hazards of technology leakage to competitors. Arrow (1962) first brought to light the disclosure problem in the market for know-how and others have since elaborated on this and related technology transfer problems (Goldberg, 1977; Teece, 1981, 1985, 1986). Appropriability hazards are of concern when property rights are difficult to establish. The leakage can occur vertically (upstream and downstream) as well as horizontally (Silverman, 1996).

Proprietary knowledge that leaks from buyer (supplier) to supplier (buyer) in the course of fulfilling a purchase contract is especially problematic when the supplier (buyer) is integrated downstream (upstream). The argument is of course symmetric. Although an independent supplier who obtains knowledge from the buyer may choose to integrate into the downstream product, the likelihood that this will occur is small. However, a firm which is already vertically integrated downstream and supplies a downstream competitor may be able to take the know-how that has leaked to its upstream division and incorporate it into the downstream products and processes relatively quickly.[44]

In the presence of these hazards, maintaining technological control of the innovation trajectory sometimes requires vertical integration (including heavy investment in R&D). When this is not possible, other strategies for (re)shaping the industry's architecture must be pursued, for example, through corporate venture investments in the supply base to build a competitive market for key complements (Pisano and Teece, 2007).

[42] These dynamic coordination issues are very different from the rent extraction of concern in the economics literature on innovation. In Farrell and Katz (2000), for example, a monopolist may extract so much rent from the firms selling a competitively supplied complement that their innovation is suboptimal even from the monopolist's perspective.

[43] MIPS encountered this with their failed attempt to promote their Advanced Computing Environment (ACE) to compete with Sun's Scalable Processor Architecture (SPARC). MIPS set up alliances with Compaq, DEC, Silicon Graphics, and other firms to pursue a RISC-based computing standard. However, soon after DEC and Compaq announced that they were going to reduce their commitment to ACE, the alliance fell apart because MIPS could not pick up the slack in some of the upstream activities. It failed both to develop competencies in key aspects of the technology and to create a common expectation for the alliance (Gomes-Casseres, 1994).

[44] The term "leakage" does not mean that intellectual property rights have necessarily been violated. This leakage is the quite legal imitation and emulation that take place in the normal course of business.

7.4. The fundamental economic "problems" to be solved by the (innovating) firm

As earlier sections made clear, the fundamental problems solved by the innovating firm are not just coordination to overcome high transaction costs (and other issues flowing from incomplete contracts) but also the design and implementation of opportunity and value capture strategies and mechanisms. These strategies and mechanisms can help solve the appropriability problems and help create the new organizational capabilities needed to address new opportunities as they arise. These theoretical challenges require the joining of transaction cost economics and capabilities theory. The problems associated with creating and capturing value are as important as coordination and incentive design in defining the nature of the (innovating) firm.

Likewise, the economic problem being addressed here has little to do with incentive design and principal-agent problems. Managing expert talent (literati and numerati) has less to do with metering and monitoring to detect and punish *opportunism* than it has to do with detecting, monitoring, and metering *opportunity*.

Alchian, Demsetz, and Williamson have all emphasized opportunistic free riding as one organizing principle. Clearly, it is an important issue. Williamson assumes, correctly so, that human actors are boundedly rational, self-interest seeking, and opportunistic. The dynamic capabilities framework emphasizes other (arguably less ubiquitous and unevenly distributed but nevertheless more salient) traits of human nature: (1) entrepreneurship and pursuit of high-risk/high-reward opportunities, and (2) foresight and acumen.

Williamson (1999a) appears to recognize that skills and foresight are not uniformly distributed. He quotes businessman Rudolf Spreckels—"Whenever I see something badly done, or not done at all, I see an opportunity to make a fortune." Williamson comments: "Those instincts, if widely operative, will influence the practice and ought to influence the theory of economic organization" (p. 1089). This statement invites a capabilities-based theory of the firm.

There are other differences between transaction cost and capabilities perspectives. Williamson makes the transaction the unit of analysis, with (the degree of) asset specificity a key explanatory variable in organizational design. In the dynamic capabilities framework, complementary assets and the degree of their cospecialization are important explanatory variables. The firm is the focus, if not the unit of analysis.

The utility of transaction cost economics and related frameworks for make-buy-ally and related governance decisions are not in dispute. But transaction cost economics leaves us without an understanding of the distinctive role of the manager. Executives must not only choose governance modes (between market arrangements, alliances, and internal organization); they must also understand how to design and implement different governance structures, to coordinate investment activities, to design and implement business models, and to choose appropriability strategies.

A dynamic capabilities/knowledge-based theory of the firm is not completely at odds with Coase, Williamson, Hart, Moore, and others. In the dynamic capabilities framework, opportunism is not held in abeyance, nor are principal-agent and incentive issues ignored. But the essence of the innovating firm lies in the generation, configuration, and leveraging of knowledge assets and organizational capabilities to allow the owners (shareholders) to create and capture value.

While the understanding of the existence and growth of the firm can be assisted by transaction cost theory, the advantages of organizing economic activity inside the firm go well beyond savings in transaction costs, however, these are manifested. Advantages also flow from the ability of entrepreneurial managers to combine idiosyncratic cospecialized assets not just to achieve "scope economies," but to create and capture value by offering distinctive services (solutions) to customers while solving the firm's appropriability problems. Over reliance on the transaction cost economics apparatus can add unnecessary baggage. For instance, if one wanted to understand issues surrounding creating value, not simply protecting value created, transaction costs can only go part of the way. The firm's routines for sensing, seizing, and transforming can provide a basis for profitability well beyond the avoidance of contracting costs and hazards.

There is empirical evidence that even outsourcing decisions do not depend on transaction cost (asset specificity) considerations alone. Studies show that "system effects" such as interdependencies and complementarities (Monteverde and Teece, 1982)[45] and capability advantages (Argyres, 1996) impact economic organization in a statistically significant manner.[46] These studies seem to indicate that boundary placement influences production learning and impacts R&D efficiency (Armour and Teece, 1980), resulting in lower costs and superior innovation potential.

What then is the role of managers in the theory of the innovating firm? They are not primarily micromanaging creative people so as to stamp out opportunistic behavior. Nor are they merely engaged in adaptive sequential decision making. Rather, they are helping the organization to create and implement the systems and structures that enable the firm to sense opportunities, execute on them, and transform as the environment changes, which inevitably it will.

Opportunism is controlled not just through metrics and monitoring, but also through high commitment cultures/values. Innovative firms typically need strong values because it is harder in the loosely structured internal environments that innovation requires to define and measure performance and implement rigid controls. Incentive issues are powerful as well; creative and entrepreneurial activity need to be encouraged and rewarded.

The transaction cost economics perspective clearly needs dynamic capabilities, and vice versa. The complementarity between capabilities-based views and contractual/transaction costs/property rights views is hopefully apparent. It has been remarked on by this author elsewhere, as well as by others (e.g., Foss, 1996).[47] Transaction cost economics implicitly assumes what might be referred to as capabilities neutrality. In transaction cost economics, so-called "production costs"—which might be thought of as a proxy for the firm's level of (operational) capability—are assumed to be the same across organizational types so that the choice between market and nonmarket arrangements swings entirely on transaction/governance costs. This assumption is a natural connection point to capabilities theory, which clearly indicates that the level of capabilities is itself a function of managerial activity/excellence

[45] This article is often cited as reflecting empirical support for transactions cost economies, which indeed it does. But the variable for systems effects has more explanatory power and is consistent with the capabilities perspective advanced here (see text below).

[46] Monteverde and Teece have been cited most extensively as providing the first empirical support for transaction cost economies. However, a little noticed feature of the econometrics is that systems effects and firm effects are more powerful explanatory variables.

[47] The Profiting from Innovation framework (Teece, 1986) illustrates how a contracting framework is useful as a tool for building a (dynamic) capabilities-based theory of the firm (see also Winter, 2006).

(or lack thereof). Differences in capabilities can lead to wide disparities in "production" costs within an industry. The field of strategic management is built on the recognition that firms are different—not just as to governance, but with respect to other features too (Rumelt et al., 1991)—and that this drives performance differences.

The (dynamic) capabilities framework, which posits that knowledge assets and their (dynamic) management have become central to profit maximization in an era of globalized commerce and information, suggests a new theory of the firm, one that is consistent with the observation of Marshall (1898, p. 213) that "capital consists in a great part in knowledge and organization: and of this some part is private property and the other part is not. Knowledge is our most powerful engine of production—organization aids knowledge." The proposed new capabilities-based theory opens up the black box of the firm and injects into economic theory new considerations which are generally not central to the theory of the firm as commonly presented.

7.5. Recapping complementarities, cospecialization, and the scope of the (innovating) firm

The theory of the innovating firm has benefited, and can benefit further, from a more rigorous exploration of the concepts of complementarities and cospecialization. The earliest use of the idea of complementarities in economics can be traced to Edgeworth (1881). Early applications in the economic development literature include Hirschman (1958) and in the innovation literature can be found in Rosenberg (1979, 1982) and Teece (1986). Work on complementarities in a strategic context includes Teece (1980b), Milgrom and Roberts (1990a,b), and Miller (1988).

Rosenberg (1979) notes: "Time and again in the history of American technology it has happened that the productivity of a given invention has turned on the availability of complementary technologies... these linkages are both numerous and of varying degrees of importance" (pp. 26–27). Furthermore, "the growing productivity of industrial economies is the complex outcome of large numbers of interlocking, mutually reinforcing technologies, the individual components of which are of very limited economic consequences by themselves. The smallest relevant unit of observation, therefore, is seldom a single innovation but, more typically, an interrelated clustering of innovations" (pp. 28–29).

Complementarities exist when various activities reinforce each other in such a manner that performing multiple activities together lowers/(raises) cost, increases economies/(diseconomies) of scope, or otherwise improves/(depresses) payoffs.[48] More technically, complementarities exist when the mixed partial derivatives of a cost function or a payoff function provide positive returns at the margin associated with one variable increasing as the levels of other variables increase too. Doing more of one activity increases the returns from doing more of another. The aggregate economic value achieved by combining two or more complementary factors therefore exceeds the value that would be achieved by applying these factors in isolation.

Of course, as pointed out by Teece (1980b), this in and of itself has no direct implication for the theory of the (boundaries of the) firm, although it has powerful implications for economic organization

[48] The notion of complements has gained mathematical tractability through the concept of supermodularity (Milgrom and Roberts, 1994; Topkis, 1978, 1987). For an excellent review of the literature, see Ennen and Richter (2009). This is discussed below.

more generally. The existence of positive complementarities indicates the advantage of having separate activities occur together. However, without more structure to the concept, one cannot predict where the individual firm boundaries should lie because contractual arrangements exist that, in theory, can enable joint activities to take place absent common ownership of the parts.

While the importance of complementarities is now being recognized, the approach still needs additional specificity (with respect to causal relationships among key constructs) to allow it to morph fully into a falsifiable theory. Put differently, a robust theory of complementarities that provides economic insight is yet to emerge. While there is little doubt that complementary relationships exist among heterogeneous factors inside the firm (and that these can impact firm performance), the contexts in which such interactions occur is yet to be adequately specified. However, some evidence has been assembled. Monteverde and Teece (1982), while testing for the importance of asset specificity in predicting outsourcing decisions for GM and Ford, also found that a "systems effect"—defined as "the degree to which any given component's design affects the performance or [system-level integration] of other components" (p. 210)—was statistically significant in explaining GM and Ford's outsourcing decisions. The longstanding notion of strategic "fit" is obviously consistent with notions of complementarity.

It should be noted that the notion of complementarity can be applied at a high level of aggregation, as with the Toyota System of production. It can also be applied at a high level of specificity, such as the complementarity between the (integrated) design and manufacture of automobile components. An example is the complementarity in design between an automobile's exterior grill and its headlamp assemblies (Monteverde and Teece, 1982). Parmigiani and Mitchell (2009) use the example of automobile dashboards, which they note typically consist of multiple, interrelated, complementary components. Both levels of aggregation seem to provide insights, suggesting the power and generality of insights from the concept of complementarity.

Complementarities expressed through their mathematical corollary (supermodularity) break from classical economics. Most classical economics models of production recognize only traditional "factors of production" like labor and capital and assume homogeneity with respect to the distribution of these factors among firms. The standard production function sees no benefit from the use of particular inputs—in the sense that, apart from diminishing returns related to fixed factors, there is no special significance to the identity of particular factors of production (Teece and Winter, 1984). Moreover, everything is infinitely divisible—indeed, twice differentiable—and firms maximize some objective function subject to constraints. Complementarity does not require divisibility; changes in one variable may require discrete (nonincremental) changes in another.

With production functions of the standard kind, decision makers need only equate marginal revenues to marginal cost and they will deliver global maxima in output. There are serious issues with this theory surrounding the search for, and the discovery of, a global maximum, if one exists. Complementarity modeled as supermodularity enables some departures from this extreme caricature by at least recognizing local maxima. It also accepts that payoff functions may be discontinuous. Design choices are recognized as being discrete and not necessarily continuous. These perspectives have received endorsement by organizational ecologists and strategic management scholars including Levinthal (1997), Porter and Siggelkow (2008), and Teece (2007a).

However, capabilities theory at present runs the risk of providing more *ex post* rationalization than *ex ante* guidance with respect to the particulars of the requirements—with Teece (1986, 2006) being possible exceptions since these papers are quite explicit about the contexts in which complementary assets are important for capturing value from innovation. These papers are also able to specify when complementary assets should be included inside the boundaries of the enterprise, as discussed in Section 7.3, above.

7.6. The "nature" of the innovating firm

Knowledge-based theories of the firm see business organizations as accumulating capabilities in path-dependent ways. Recognizing, creating, and exploiting complementarities is very much at the core of what firms do. Sustained "abnormal" or "supernormal" profitability occurs because factor markets for certain types of assets (particularly intangibles and idiosyncratic physical and human assets) are not fully efficient. To take full advantage and earn superior profits, firms need to sense, seize, and transform in ways that exploit inefficient factor markets. Indentifying and securing combinations and permutations of assets which enable the enterprise to address customer needs is key.

As firms build the microfoundations needed to sense, seize, and transform, all the while exploiting complementarities, they lay the foundations for sustained above-average profitability. There is nothing in Ronald Coase's or Oliver Williamson's work to explain how firms identify and exploit complementarities and develop competitive advantage. This raises the question of how the Coase/Williamson conceptualizations of the firm relate to dynamic capabilities.

As stated earlier, the knowledge and contracting perspectives are complementary theories/frameworks. No theory of the firm can ignore contractual issues. But neither Coase nor Williamson see a firm as a pure nexus of contracts. Nor do they see the firm as merely "social communities in which individual and social expertise is transformed into economically useful products and services by the application of a set of higher order organizing principles" (Kogut and Zander, 1992).

There is clearly a way for knowledge-based theories and transaction cost perspectives to be brought together. Arrow (1974) provided a commanding and potentially unifying insight. He observed that the reason firms exist is not simply due to high transaction costs; rather, markets in some situations simply do not work and there is market "failure."[49] One can do a thought experiment and conclude that if the transactions were forced into a market, transaction costs in such circumstances would be very high; but it is perhaps simpler to just recognize that there are many circumstances where internal organization is clearly a necessary and superior way to organize, and it is desirable for innovative activity to take place inside a firm orchestrated by entrepreneurial managers embedded in some kind of management structure.

For purposes of building a theory of the innovating firm, it is important to specify the contexts in which these market failures are prevalent. The most important (and also the most under-researched) domain within which organization inside the firm is likely to be necessary is the creation, transfer, protection (appropriability), and orchestration (so as to exploit complementarities) of know-how and other intangibles. As noted more than two decades ago (Teece, 1981): "unassisted markets are seriously

[49] Arrow (1969) acknowledged that in some cases markets might simply not exist. Williamson (1971), in his best known statement on market failure, which he still endorsed 28 years later Williamson (1999b), restricted his attention to those that were "failures only in the limited sense that they involve transaction costs that can be attenuated by substituting internal organization for market exchange" (p. 114).

faulted as institutional devices for facilitating trading in many kinds of technological and managerial know-how. The imperfections in the market for know-how for the most part can be traced to the nature of the commodity in question" (p. 84). The market is also imperfect as a tool to create know-how. One can "buy in" technology more easily than one can have it created through a contractual agreement and then transfer it in. "Creation" must frequently be done internally, even though external sourcing is usually a necessary complement to own development.

One must recognize that it is only after industrially relevant know-how is first created that it can be traded (via licensing arrangements). Even once its created, mutually beneficial trades frequently do not happen because the property rights covering knowhow may be poorly defined (fuzzy),[50] the asset difficult to transfer, or its use difficult to meter. Internal resource allocation within the firm (a managerially directed activity) is the only viable alternative.

Moreover, because of complementarities and cospecialization, many intangible assets may be more valuable when they can coevolve in a coordinated way with other assets. The ability to assemble unique configurations of cospecialized assets, as in the case of systemic innovation (Teece, 2000), can therefore enhance value. Rosenberg (1979) seems to go further and argues that such coordination and clustering is necessary for value to be created.

In a globalized, knowledge-based economy, firms can secure short-term advantage from the coordination of bundles of difficult-to-trade assets and competencies, at least when such assets are scarce and difficult to imitate. Advantage that is sustainable over a longer term, however, can only flow from unique abilities possessed by business enterprises to continuously shape, reshape, and orchestrate those assets to create new technology, to respond to competition, achieve critical market mass, exploit complementarities, and serve changing customer needs. The particular (nonimitable) orchestration capacity of a business enterprise—its dynamic capabilities—is the irreducible core of the innovating firm. It cannot be reproduced simply by assembling a constellation of contracts.

Fundamentally, business firms know how to do things. Most figure out how to adapt and possibly even shape their environment to some (small) degree. As noted earlier, even Harold Demsetz was willing to see the firm as a repository of knowledge.

However, it is not clear that many economists are willing as yet to recognize the implications of firms being repositories of knowledge and instruments for learning. One exception is Winter (1982) who correctly notes, "it is the firms, not the people who work for the firms, that know how to make gasoline, automobiles, and computers" (p. 76).

Organizational capabilities explain why an enterprise is more than the sum of its parts. They also help explain why the profits of the enterprise cannot be completely competed away in factor markets. Employees can come and go to a certain extent and the organization can continue without interruption.

Mainstream theory too often takes production functions and production sets as given, ignores complementarities and cospecialization, and fails to explain capabilities and heterogeneity among firms even in the same industry. Mainstream theory also completely sidesteps the problem of how firms actually perform the tasks of storing the knowledge that underlies productive competence, transferring it internally (or externally), augmenting it in value-enhancing ways, and identifying and exploiting complementarities.

[50] See Teece (2000) for discussion of the fuzzy boundaries associated with intellectual property rights.

In short, managers often create great value by assembling particular constellations of complementary and cospecialized assets, especially knowledge assets, inside the enterprise to produce highly differentiated and innovative goods and services that customers want. This process of identifying, assembling, and orchestrating constellations of complementary and cospecialized assets is a fundamental function of management—and points to the fundamental "nature" of the innovating firm. It is different from the Coasian firm.

8. Conclusion

This chapter has endeavored to motivate and shape a theory of the innovating firm consistent with descriptions of the firm that business historians like Alfred Chandler have provided. Historians remind us that innovation is central to the role of the enterprise in modern society. Accordingly, a theory of the firm that fails to reflect these dimensions is unlikely to have utility for business strategy and public policy analysis.

The good news is that the theory advanced here does not require one to displace all of mainstream theory. Mainstream approaches can be augmented with now well-established concepts from transaction cost economics, from the economic and organizational theories of complements, and from (dynamic) capabilities theory. Innovation scholars and industrial organization theorists will hopefully demonstrate over the next decade how organization theory, strategic management theory, and innovation theory can inform each other, while also benefiting from the study of business history and industrial organization.

Acknowledgments

I thank Greg Linden for help in content and in editing. Gary Pisano and John de Figueiredo, along with the editors Nate Rosenberg and Bronwyn Hall, also provided helpful insights and comments for which I am much indebted. Ken Arrow, Henry Chesbrough, Giovanni Dosi, Connie Helfat, Dan Lovallo, James March, the late Edwin Mansfield, David Mowery, Richard Nelson, Gary Pisano, Christos Pitelis, Oliver Williamson, Sidney Winter, and many others provided decades of inspiration and ideas.

References

Abernathy, W.J., Utterback, J.M. (1978). "Patterns of industrial innovation". Technology Review 80 (7), 40–47.

Abramovitz, M. (1986). "Catching up, forging ahead, and falling behind". Journal of Economic History 46 (2), 385–406.

Acs, Z., Audretsch, D. (1990). Innovation and Small Firms. MIT Press, Cambridge, MA.

Adams, W.J., Yellen, J.L. (1976). "Commodity bundling and the burden of monopoly". Quarterly Journal of Economics 90 (3), 475–498.

Akerlof, G.A. (1970). "The market for lemons". Quarterly Journal of Economics 84 (3), 488–500.

Albert, S., Bradley, K. (1997). Managing Knowledge Experts, Experts, Agencies and Organizations. Cambridge University Press, Cambridge, MA.

Alchian, A.A., Demsetz, H. (1972). "Production, information costs and economic organization". American Economic Review 62 (2), 777–795.

Anderson, P., Tushman, M.L. (1990). "Technological discontinuities and dominant design". Administrative Science Quarterly 35 (4), 604–633.

Argyres, N. (1996). "Evidence on the role of firm capabilities in vertical integration decisions". Strategic Management Journal 17 (2), 129–150.

Armour, H.O., Teece, D.J. (1978). "Organizational structure and economic performance: A test of the multidivisional hypothesis". Bell Journal of Economics 9 (18), 106–122.

Armour, H.O., Teece, D.J. (1980). "Vertical integration and technological innovation". Review of Economics and Statistics 62 (3), 470–474.

Arrow, K.J. (1962). "Economic welfare and the allocation of resources of invention". In: National Bureau of Economic Research, K.J. (Ed.), The Rate and Direction of Inventive Activity: Economic and Social Factors. Princeton University Press, Princeton, NJ, pp. 609–625.

Arrow, K.J. (1969). "The organization of economic activity: Issues pertinent to the choice of market versus nonmarket allocation". In: The Analysis and Evaluation of Public Expenditures: The PPB System. U.S. Government Printing Office, Washington, DC, pp. 47–64.

Arrow, K.J. (1974). The Limits of Organization. Norton, New York.

Barnard, C.I. (1938). The Functions of the Executive. Harvard University Press, Cambridge, MA.

Barney, J.B. (1986). "Strategic factor markets: Expectations, luck, and business strategy". Management Science 32 (10), 1231–1241.

Baron, J.N., Kreps, D.M. (1999). Strategic Human Resources: Frameworks for General Managers. Wiley, New York.

Baumol, W.J. (2001). "When is inter-firm coordination beneficial? The case of innovation". International Journal of Industrial Organization 19 (5), 727–737.

Baumol, W.J., Strom, R.J. (2007). "Entrepreneurship and economic growth". Strategic Entrepreneurship Journal 1 (3–4), 233–237.

Becker, B.E., Huselid, M.A. (2006). "Strategic human resources management: Where do we go from here?" Journal of Management 32 (6), 898–925.

Burns, T., Stalker, G.M. (1961). The Management of Innovation. Tavistock, London.

Caves, R.E. (2000). Creative Industries: Contracts Between Art and Commerce. Harvard University Press, Boston, MA.

Chandler, A. (1962). Strategy and Structure. MIT Press, Cambridge, MA.

Chandler, A. (1977). The Visible Hand: The Managerial Revolution in American Business. Belknap Press, Cambridge, MA.

Chandler, A. (1990). Scale and Scope: The Dynamics of Industrial Capitalism. Belknap/Harvard University Press, Cambridge, MA.

Chandler, A. (1992). "Organizational capabilities and the economic history of the industrial enterprise". Journal of Economic Perspectives 6 (3), 79–100.

Chesbrough, H. (2006). Open Business Models. Harvard Business School Press, Boston, MA.

Chesbrough, H., Rosenbloom, R.S. (2002). "The role of the business model in capturing value from innovation: Evidence from xerox corporation's technology spin-off companies". Industrial and Corporate Change 11 (3), 529–555.

Chesbrough, H., Teece, D.J. (1996). "When is virtual virtuous: Organizing for innovation". Harvard Business Review 74 (1), 65–73.

Christensen, C.M. (1997). The Innovator's Dilemma. Harvard Business School Press, Boston, MA.

Clark, K.B. (1985). "The interaction of design hierarchies and market concepts in technological evolution". Research Policy 14 (5), 235–251.

Clark, P. (2009). "Boeing to test fly Dreamliner by year's end". Financial Times, Nov.15. http://www.ft.com/cms/s/0/9e52eeec-d208-11de-a0f0-00144feabdc0.html.

Coase, R.A. (1937). "The nature of the firm". Economica 16 (4), 386–405.

Cohen, W.M., Levin, R.C. (1989). "Empirical studies of innovation and market structure". In: Schmalensee, R.L., Willig, R.D. (Eds.), Handbook of Industrial Organization, 2. North Holland, Amsterdam.

Cyert, R.M., March, J.G. (1963). A behavioral theory of the firm. Prentice-Hall, Englewood Cliffs, N.J.

Dasgupta, P., Stiglitz, J. (1980). "Industrial structure and the nature of innovative activity". Economic Journal 90 (358), 266–293.

David, P.A., Greenstein, S. (1990). "The economics of compatibility standards: An introduction to recent research". Economics of Innovation and New Technology 1 (1), 3–41.

de Figueiredo, J.M., Teece, D.J. (1996). "Mitigating procurement hazards in the context of innovation". Industrial and Corporate Change 5 (2), 537–559.

Demsetz, H. (1970). "The private production of public goods". Journal of Law and Economics 13 (2), 293–306.

Demsetz, H. (1988). "The theory of the firm revisited". Journal of Law, Economics, and Organization 4 (1), 141–161.

Denrell, J., Fang, C., Winter, S.G. (2003). "The economics of strategic opportunity". Strategic Management Journal 24 (10), 977–990.

Dosi, G. (1982). "Technological paradigms and technological trajectories: A suggested interpretation of the determinants and directions of technical change". Research Policy 11 (3), 147–162.
Dosi, G., Nelson, R.R., Winter, S.G. (Eds.), (2000). The Nature and Dynamics of Organizational Capabilities. Oxford University Press, New York.
Edgeworth, F.Y. (1881). Mathematical Physics: An Essay on the Application of Mathematics to the Moral Sciences. Kegan Paul, London.
Ennen, E., Richter, A. (2009). The Whole is More than the Sum of Its Parts—Or Is It? A Review of the Empirical Literature on Complementarities in Organizations. European Business School, Wiesbaden, Germany European Business School Research Paper No. 09-07.
Fai, F. (2007). "A structural decomposition analysis of technological opportunity, corporate survival, and leadership". Industrial and Corporate Change 16 (6), 1069–1103.
Farrell, J., Katz, M.L. (2000). "Innovation, rent extraction, and integration in systems markets". Journal of Industrial Economics 48 (4), 413–432.
Fischer, B., Boynton, A. (2005). "Virtuoso teams". Harvard Business Review 83 (7), 116–123.
Foss, N. (1996). "Knowledge based approaches to the theory of the firm: Some critical comments". Organizational Science 7 (5), 470–476.
Fox, M.F. (1983). "Publication productivity among scientists: A critical review". Social Studies of Science 13 (2), 285–305.
Futia, C.A. (1980). "Schumpeterian competition". Quarterly Journal of Economics 94 (4), 675–695.
Galbraith, J.K. (1952). American Capitalism: The Concept of Countervailing Power. Houghton Mifflin, Boston.
Gibbons, R. (2005). "Four formal(izable) theories of the firm?" Journal of Economic Behavior and Organization 58 (2), 200–245.
Gil, R., Spiller, P.T. (2007). "The organizational dimensions of creativity: Motion picture production". California Management Review 50 (1), 243–260.
Gilbert, R. (2006). "Looking for Mr. Schumpeter: Where are we in the competition-innovation debate?" In: Jaffe, A.B., Josh Lerner, S., Scott Stern, S. (Eds.), Innovation Policy and the Economy, vol. 6. MIT Press, Cambridge, MA.
Goldberg, V. (1977). "Competitive bidding and the production of precontract information". Bell Journal of Economics 8 (1), 250–261.
Gomes-Casseres, B. (1994). "Group versus group: How alliance networks compete". Harvard Business Review 4, 62–74 (July–August).
Gompers, P., Lerner, J. (2001). "The venture capital revolution". Journal of Economic Perspectives 15 (2), 145–168.
Hart, O., Moore, J. (1990). "Property rights and the nature of the firm". Journal of Political Economy 98 (6), 1119–1158.
Hayton, J.C. (2005). "Promoting corporate entrepreneurship through human resource management practices: A review of empirical research". Human Resource Management Review 15 (1), 21–41.
Henderson, R.M. (1994). "Managing innovation in the information age". Harvard Business Review 72 (1), 100–106.
Henderson, R.M., Clark, K.B. (1990). "Architectural innovation: The reconfiguration of existing product technologies and the failure of established firms". Administrative Science Quarterly 35 (1), 9–30.
Hirschman, A.O. (1958). The Strategy of Economic Development. Yale University Press, New Haven, CT.
Holmstrom, B., Paul Milgrom, P. (1994). "The firm as an incentive system". American Economic Review 84 (4), 972–991.
Hounshell, D.A. (1996). "The evolution of industrial research in the United States". In: Rosenbloom, R.S., Spencer, W.J. (Eds.), Engines of Innovation: U.S. Industrial Research at the End of an Era. Harvard Business School Press, Boston, MA.
Hounshell, D.A., Smith, J.K. (1988). Science and Corporate Strategy: Du Pont R&D, 1902–1980. Cambridge University Press, New York.
Jacobides, M.G., Knudsen, T., Augier, M. (2006). "Benefiting from innovation: Value creation, value appropriation and the role of industry architectures". Research Policy 35 (8), 1200–1221.
Jorde, T.M., Teece, D.J. (1990). "Innovation and cooperation: Implications for competition and antitrust". Journal of Economic Perspectives 4 (3), 75–96.
Kamien, M.I., Schwartz, N.L. (1978). "Self financing of an R&D project". American Economic Review 68 (3), 252–261.
Katz, R. (2004). "Motivating professionals in organizations". In: Katz, R. (Ed.), The Human Side of Managing Technological Innovation. Oxford University Press, New York.
Katz, M.L., Ordover, J.A. (1990). "R&D cooperation and competition". In: Brookings Papers on Economic Activity—Microeconomics. Brookings Institution, Washington, DC, pp. 137–203.

Katz, M.L., Shapiro, C. (1986). "Technology adoption in the presence of network externalities". Journal of Political Economy 94 (4), 822–841.
Kirzner, I.M. (1973). Competition and Entrepreneurship. University of Chicago Press, Chicago.
Kirzner, I.M. (1979). Perception, Opportunity and Profit: Studies in the Theory of Entrepreneurship. University of Chicago Press, Chicago.
Kirzner, I.M. (1997). "Entrepreneurial discovery and the competitive market process: An Austrian approach". Journal of Economic Literature 35 (1), 60–85.
Klemperer, P. (1987). "Markets with consumer switching costs". Quarterly Journal of Economics 102 (2), 375–394.
Klemperer, P. (2002). "What really matters in auction design". Journal of Economic Perspectives 16 (1), 169–189.
Kogut, B., Zander, U. (1992). "Knowledge of the firm, combinative capabilities, and the replication of technology". Organizational Science 3 (3), 383–397.
Krugman, P. (1987). "The narrow moving band, the dutch disease, and the consequences of Mrs. Thatcher: Notes on trade in the presence of scale economies". Journal of Development Economics 27 (1–2), 41–55.
Langlois, R.N. (1992). "Transaction-cost economics in real time". Industrial and Corporate Change 1 (1), 99–127.
Levin, R.C., Reiss, P.C. (1984). "Tests of a Schumpeterian model of R&D and market structure". In: Griliches, Z. (Ed.), R&D, Patents, and Productivity. University of Chicago Press, Chicago, pp. 175–204.
Levin, R.C., Reiss, P.C. (1988). "Cost reducing and demand creating R&D with spillovers". RAND Journal of Economics 19 (4), 538–556.
Levin, R.C., Cohen, W.M., Mowery, D.C. (1985). "R&D appropriability, opportunity, and market structure: New evidence on some Schumpeterian hypotheses". American Economic Review 75 (2), 20–24.
Levin, R.C., Klevorick, A.K., Nelson, R.R., Winter, S.G., Gilbert, R., Griliches, Z. (1987). "Appropriating the returns from industrial research and development". Brookings Papers on Economic Activity 1987 (3), 783–831.
Levinthal, D.A. (1997). "Adaptation on rugged landscapes". Management Science 43 (7), 934–950.
Lotka, A.J. (1926). "The frequency distribution of scientific productivity". Journal of the Washington Academy of Science 16 (12), 317–323.
Malerba, F., Orsenigo, L. (1993). "Technological regimes and firm behavior". Industrial and Corporate Change 2 (1), 45–71.
Mansfield, E. (1974). "Technology and technological change". In: Dunning, J. (Ed.), Economic Analysis and the Multinational Enterprise. George Allen and Unwin, London.
Mansfield, E. (1985). "How rapidly does new industrial technology leak out?" Journal of Industrial Economics 34 (2), 217–223.
Mansfield, E., Teece, D.J., Romeo, A. (1979). "Overseas research and development by US-based firms". Economica 46 (182), 187–196.
Mansfield, E., Schwartz, M., Wagner, S. (1981). "Imitation costs and patents: An empirical study". Economic Journal 91 (364), 907–918.
March, J.G. (1991). "Exploration and exploitation in organizational learning". Organization Science 2 (1), 71–87.
March, J.G., Simon, H.A. (1958). Organizations. Wiley, New York.
Marshall, A. (1898). Principles of Economics (fourth ed.). MacMillan, London.
Mayer, D., Kenney, M. (2004). "Economic action does not take place in a vacuum: Understanding Cisco's acquisition and development strategy". Industry and Innovation 11 (4), 299–325.
Mayer, K.J., Teece, D.J. (2008). "Unpacking strategic alliances: The structure and purpose of alliance versus supplier relationships". Journal of Economic Behavior and Organization 66 (1), 106–127.
McGirt, E. (2008). "Revolution in San Jose". Fast Company 131, 90–93.
Metcaffe, J.S., Gibbons, M. (1988). "Technology variety and organization: A systematic perspective on the competition process". In: Rosenbloom, R.S., Burgelman, R.A. (Eds.), Research on Technological Innovation, Management and Policy. JAI Press, Greenwich, CT, pp. 153–193.
Michaels, D., Sanders, P. (2009). "Dreamliner production gets closer monitoring". WSJ.com October 8.
Miles, R.E. (2007). "Innovation and leadership values". California Management Review 50 (1), 192–201.
Milgrom, P., Roberts, J. (1990a). "The economics of modern manufacturing: Technology, strategy, and organization". American Economic Review 80 (3), 511–528.
Milgrom, P., Roberts, J. (1990b). "Rationalizability, learning and equilibrium in games with strategic complementarities". Econometrics 58 (6), 1255–1277.

Milgrom, P., Roberts, J. (1994). "Complementarities and systems: Understanding Japanese economic organization". Estudios Economicos 9 (1), 3–42.

Miller, D. (1988). "Relating porter's business strategies to environment and structure: Analysis and performance implications". Academy of Management Journal 31 (2), 280–308.

Monteverde, K., Teece, D.J. (1982). "Supplier switching costs and vertical integration in the automobile industry". Bell Journal of Economics 13 (1), 206–213.

Mowery, D.C. (1983). "The relationship between intrafirm and contractual forms of industrial research in American manufacturing, 1900–1940". Explorations in Economic History 20 (4), 351–374.

Mowery, D.C., Rosenberg, N. (1989). Technology and the Pursuit of Economic Growth. Cambridge University Press, New York.

National Science Board. (2008). Science and Engineering Indicators 2008. National Science Foundation, Arlington, VA.

Nelson, R.R. (1962). "The link between science and invention: The case of the transistor". In: National Bureau of Economic Research, R.R. (Ed.), The Rate and Direction of Inventive Activity. Princeton University Press, Princeton, NJ.

Nelson, R.R. (Ed.), (1982). Government and Technical Progress: A Cross-Industry Analysis. Pergamon Press, New York.

Nelson, R.R. (Ed.), (1993). National Systems of Innovation. Oxford University Press, New York.

Nelson, R.R., Winter, S.G. (1978). "Forces generating and limiting concentration under Schumpeterian competition". Bell Journal of Economics 9 (2), 524–548.

Nelson, R.R., Winter, S.G. (1982). An Evolutionary Theory of Economic Change. Harvard University Press, Cambridge, MA.

O'Reilly, C.A., Tushman, M.L. (2004). "The ambidextrous organization". Harvard Business Review 82 (4), 74–81.

Ortutay, B. (2009). Netflix Awards $1M Prize to Improve Movie Picks. Associated Press, September 21. http://www.sfgate.com/cgi-bin/article.cgi?f=/n/a/2009/09/21/financial/f071427D69.DTL

Parmigiani, A., Mitchell, W. (2009). "Complementarity, capabilities, and the boundaries of the firm: The impact of within-firm and interfirm expertise on concurrent sourcing of complementary components". Strategic Management Journal 30 (10), 1065–1091.

Pavitt, K., Robson, M., Townsend, J. (1987). "The size distribution of innovating firms in the UK: 1945–1983". Journal of Industrial Economics 35 (3), 297–316.

Penrose, E.G. (1959). The Theory of the Growth of the Firm. Wiley, New York.

Phillips, A. (1971). Technology and Market Structure: A Study of the Aircraft Industry. Heath Lexington Books, Lexington, MA.

Piore, M.J., Sabel, C.F. (1984). The Second Industrial Divide. Basic Books, New York.

Pisano, G.P., Teece, D.J. (2007). "How to capture value from innovation: Shaping intellectual property and industry architecture". California Management Review 50 (1), 278–296.

Pitelis, C.N., Teece, D.J. (2009). "The (new) nature and essence of the firm". European Management Review 6 (1), 5–15.

Porter, M., Siggelkow, N. (2008). "Contextuality within activity systems and sustainability of competitive advantage". Academy of Management Perspectives 22 (2), 34–56.

Prencipe, A., Davies, A., Hobday, M. (Eds.), (2003). The Business of Systems Integration. Oxford University Press, Oxford.

Quinn, J.B., Anderson, P., Finkelstein, S. (1996). "Managing professional intellect: Making the most of the best". Harvard Business Review 74 (2), 71–80.

Raisch, S., Birkinshaw, J., Probst, G., Tushman, M.L. (2009). "Organizational ambidexterity: Balancing exploitation and exploration for sustained performance". Organization Science 20 (4), 685–695.

Reich, R. (2002). The Future of Success: Working and Living in the New Economy. Vintage Books, New York.

Reinganum, J.F. (1981). "Dynamic games of innovation". Journal of Economic Theory 25 (1), 21–41.

Richardson, G. B. H. (1960). Information and Investment. Oxford University Press, London.

Richardson, G. B. H. (1972). "The organisation of industry". Economic Journal 82 (327), 883–896.

Roberts, J. (2004). The Modern Firm: Organizational Design for Performance and Growth. Oxford University Press, New York.

Rosenberg, N. (1979). "Technological interdependence in the American economy". Technology and Culture 20 (1), 25–50.

Rosenberg, N. (1982). Inside the Black Box: Technology and Economics. Cambridge University Press, New York.

Roth, D. (2009). "NetFlix everywhere: Sorry cable, you're history". Wired 17 (10). http://www.wired.com/techbiz/it/magazine/17-10/ff_netflix.

Rubin, P.H. (1973). "The expansion of firms". Journal of Political Economy 81 (4), 936–949.

Rumelt, R.P. (1984). "Towards a strategic theory of the firm". In: Lamb, R.B. (Ed.), Competitive Strategic Management. Prentice-Hall, Englewood Cliffs, NJ.

Rumelt, R.P., Schendel, D., Teece, D.J. (1991). "Strategic management and economics". Strategic Management Journal 12 (Winter Special Issue), 5–29.

Salop, S.C., Scheffman, D.T. (1983). "Raising rival's costs". American Economic Review 73 (2), 267–271.
Sauermann, H., Cohen, W.M. (2008). What Makes Them Tick? Employee Motives and Firm Innovation. National Bureau of Economic Research, Cambridge, MA NBER Working Paper No. W14443.
Saxenian, A. (1996). Regional Advantage: Culture and Competition in Silicon Valley and Route 128. Harvard University Press, Cambridge, MA.
Scherer, F.M., Ross, D. (1990). Industrial Market Structure and Economic Performance. Houghton Mifflin, Boston, MA.
Schumpeter, J.A. (1934, 1980). The Theory of Economic Development. Oxford University Press, London.
Schumpeter, J.A. (1942). Capitalism, Socialism, and Democracy. Harper & Brothers, New York.
Scotchmer, S. (1991). "Standing on the shoulders of giants: Cumulative research and the patent law". Journal of Economic Perspectives 5 (1), 29–41.
Shane, S. (2001). "Technology regimes and new firm formation". Management Science 47 (9), 1173–1190.
Shane, S. (2003). A General Theory of Entrepreneurship: The Individual-Opportunity Nexus. E. Elgar, Northampton, MA.
Silverman, B.S. (1996). "Technical assets and the logic of corporate diversification". Ph.D. Dissertation, University of California, Berkeley.
Simon, H.A. (1947). Administrative Behavior. Macmillan, New York.
Simon, H.A. (1951). "A formal theory of the employment relationship". Econometrica 19 (3), 293–305.
Sutton, J. (2001). Technology and Market Structure: Theory and History. MIT Press, Cambridge, MA.
Tadelis, S. (2007). "The innovative organization: Creating value through outsourcing". California Management Review 50 (1), 261–277.
Teece, D.J. (1980a). "The diffusion of an administrative innovation". Management Science 26 (5), 464–470.
Teece, D.J. (1980b). "Economies of scope and the scope of the enterprise". Journal of Economic Behavior and Organization 1 (3), 223–247.
Teece, D.J. (1981). "The market for know-how and the efficient international transfer of technology". Annals of the Academy of Political and Social Science 458 (1), 81–96.
Teece, D.J. (1982). "Towards an economic theory of the multiproduct firm". Journal of Economic Behavior and Organization 3 (1), 39–63.
Teece, D.J. (1984). "Economic analysis and strategic management". California Management Review 26 (3), 87–110.
Teece, D.J. (1985). "Multinational enterprise, internal governance, and industrial organization". American Economic Review 75 (2), 233–238.
Teece, D.J. (1986). "Profiting from technological innovation". Research Policy 15 (6), 285–305.
Teece, D.J. (1988). "Technological change and the nature of the firm". In: Dosi, G., Freeman, C., Nelson, R., Silverberg, G., Soete, L. (Eds.), Technical Change and Economic Theory. Pinter, London.
Teece, D.J. (1989). "Inter-organizational requirements of the innovation process". Managerial and Decision Economics 10 (Spring Special Issue), 35–42.
Teece, D.J. (1992). "Competition, cooperation, and innovation: Organizational arrangements for regimes of rapid technological progress". Journal of Economic Behavior and Organization 18 (1), 1–25.
Teece, D.J. (1996). "Firm organization, industrial structure, and technological innovation". Journal of Economic Behavior and Organization 31 (2), 193–224.
Teece, D.J. (2000). Managing Intellectual Capital: Organizational, Strategic, and Policy Dimensions. Oxford University Press, Oxford.
Teece, D.J. (2003). "Expert talent and the design of (professional services) firms". Industrial and Corporate Change 12 (4), 895–916.
Teece, D.J. (2006). "Reflections on profiting from innovation". Research Policy 35 (8), 1131–1146.
Teece, D.J. (2007a). "Explicating dynamic capabilities: The nature and microfoundations of (sustainable) enterprise performance". Strategic Management Journal 28 (13), 1319–1350.
Teece, D.J. (2007b). "Managers, markets, and dynamic capabilities". In: Helfat, C., Finkelstein, S., Mitchell, W., Peteraf, M., Singh, D.J., Teece, D.J., Winter, S. (Eds.), Dynamic Capabilities: Understanding Strategic Change in Organizations. Blackwell, Oxford, UK.
Teece, D.J. (2008). "Dosi's technological paradigms and trajectories: Insights for economics and management". Industrial and Corporate Change 17 (3), 507–512.

Teece, D.J. (2009). Dynamic Capabilities and Strategic Management: Organizing for Innovation and Growth. Oxford University Press, New York.
Teece, D. J. (forthcoming). "Business Models, Business Strategy and Innovation." Long Range Planning. DOI: 10.1016/j.lrp.2009.07.003
Teece, D.J., Winter, S.G. (1984). "The limits of neoclassical theory in management education". American Economic Review 74 (2), 116–121.
Teece, D.J., Pisano, G., Shuen, A. (1990). "Firm capabilities, resources, and the concept of strategy". Center for Research in Management. University of California, Berkeley, CCC Working Paper 90–8.
Teece, D.J., Rumelt, R., Dosi, G., Winter, S. (1994). "Understanding corporate coherence: Theory and evidence". Journal of Economic Behavior and Organization 23 (1), 1–30.
Teece, D.J., Pisano, G., Shuen, A. (1997). "Dynamic capabilities and strategic management". Strategic Management Journal 18 (7), 509–533.
Topkis, D.L. (1978). "Minimizing a submodular function on a lattice". Operations Research 26 (2), 305–321.
Topkis, D.L. (1987). "Activity optimization games with complementarity". European Journal of Operations Research 28 (3), 358–368.
von Hippel, E. (1998). "Economics of product development by users: The impact of 'sticky' local information". Management Science 44 (5), 629–644.
Wernerfelt, B. (1984). "A resource-based view of the firm". Strategic Management Journal 5 (2), 171–180.
Williamson, O.E. (1971). "The vertical integration of production: Market failure considerations". American Economic Review 61 (2), 112–123.
Williamson, O.E. (1975). Markets and Hierarchies. The Free Press, New York.
Williamson, O.E. (1981). "The modern corporation: Origins, evolution, attributes". Journal of Economic Literature 19 (4), 1537–1568.
Williamson, O.E. (1985). The Economic Institutions of Capitalism. The Free Press, New York.
Williamson, O.E. (1999a). "Strategy research: Governance and competence perspectives". Strategic Management Journal 20 (12), 1087–1108.
Williamson, O.E. (1999b). "Some reflections". In: Carroll, G.R., Teece, D.J. (Eds.), Firms, Markets, and Hierarchies: The Transaction Cost Economics Perspective. Oxford University Press, New York.
Winter, S.G. (1982). "An essay on the theory of production". In: Hymans, S.J. (Ed.), Economics and the World Around It. University of Michigan Press, Ann Arbor.
Winter, S.G. (1984). "Schumpeterian competition in alternative technological regimes". Journal of Economic Behavior and Organization 5 (3–4), 287–320.
Winter, S.G. (1988). "On coase, competence, and the corporation". Journal of Law, Economics, and Organization 1 (4), 163–180.
Winter, S.G. (2006). "The logic of appropriability: From Schumpeter to Arrow to Teece". Research Policy 35 (8), 1100–1106.

AUTHOR INDEX OF VOLUMES 1 AND 2

n indicates citation in a footnote.

SUBJECT INDEX OF VOLUMES 1 AND 2

FSC
www.fsc.org
MIX
Papier | Fördert gute Waldnutzung
FSC® C083411

Kolibri 360 GmbH
Ferdinand-Jühlke-Straße 7
99095 Erfurt, Deutschland
produktsicherheit@kolibri360.de

Kolibri 360 GmbH
Ferdinand-Jühlke-Straße 7
99095 Erfurt, Deutschland
produktsicherheit@kolibri360.de